Université Joseph Fourier

Les Houches

Session LXXXIV

2005

Particle Physics Beyond the Standard Model

Lecturers who contributed to this volume

Pierre Binétruy
Emilian Dudas
Tony Gherghetta
Christophe Grojean
JoAnne Hewett
Antonio Masiero
Yosef Nir
Valery Rubakov
Yael Shadmi
Alexei Smirnov
Alessandro Strumia
Peter Tinyakov

ÉCOLE D'ÉTÉ DE PHYSIQUE DES HOUCHES

SESSION LXXXIV, 1–26 AUGUST 2005

ÉCOLE THÉMATIQUE DU CNRS

PARTICLE PHYSICS
BEYOND THE STANDARD MODEL

Edited by

Dmitri Kazakov, Stéphane Lavignac and Jean Dalibard

ELSEVIER

Amsterdam – Boston – Heidelberg – London – New York – Oxford
Paris – San Diego – San Francisco – Singapore – Sydney – Tokyo

Elsevier
Radarweg 29, PO Box 211, 1000 AE Amsterdam, The Netherlands
The Boulevard, Langford Lane, Kidlington, Oxford OX5 1GB, UK

First edition 2006

Permissions may be sought directly from Elsevier's Science & Technology Rights Department in
Oxford, UK: phone (+44) (0) 1865 843830; fax (+44) (0) 1865 853333; email: permissions@elsevier.
com. Alternatively you can submit your request online by visiting the Elsevier web site at http://
elsevier.com/locate/permissions, and selecting *Obtaining permission to use Elsevier material*

Notice
No responsibility is assumed by the publisher for any injury and/or damage to persons or property as
a matter of products liability, negligence or otherwise, or from any use or operation of any methods,
products, instructions or ideas contained in the material herein. Because of rapid advances in the
medical sciences, in particular, independent verification of diagnoses and drug dosages should be
made

Library of Congress Cataloging-in-Publication Data
A catalog record for this book is available from the Library of Congress

British Library Cataloguing in Publication Data
A catalogue record for this book is available from the British Library

ISBN-13: 978-0-444-52814-8

For information on all Elsevier publications
visit our website at books.elsevier.com

Working together to grow
libraries in developing countries

www.elsevier.com | www.bookaid.org | www.sabre.org

ELSEVIER   BOOK AID
            International   Sabre Foundation

Transferred to Digital Printing 2010

École de Physique des Houches

Service inter-universitaire commun
à l'Université Joseph Fourier de Grenoble
et à l'Institut National Polytechnique de Grenoble

Subventionné par le Ministère de l'Éducation Nationale,
de l'Enseignement Supérieur et de la Recherche,
le Centre National de la Recherche Scientifique,
le Commissariat à l'Énergie Atomique

Previous sessions

Publishers:
- Session VIII: Dunod, Wiley, Methuen
- Sessions IX and X: Herman, Wiley
- Session XI: Gordon and Breach, Presses Universitaires
- Sessions XII–XXV: Gordon and Breach
- Sessions XXVI–LXVIII: North Holland
- Sessions LXIX–LXXVIII: EDP Sciences, Springer
- Sessions LXXIX–LXXXIII: Elsevier

Organizers

KAZAKOV Dmitri, Bogoliubov Laboratory of Theoretical Physics, Joint Institute for Nuclear Research, 141 980 Dubna, Moscow region, Russia

LAVIGNAC Stéphane, SPhT, CEA/Saclay - Orme des Merisiers, 91191 Gif-sur-Yvette cedex, France

DALIBARD Jean, LKB/ENS, 24 rue Lhomond, 75231 Paris cedex 05, France

Lecturers

BINÉTRUY Pierre, APC, Université Paris VII, 2 place Jussieu, 75251 Paris cedex 05, France

CARENA Marcela, Fermi National Accelerator Laboratory, P.O. Box 500, Batavia IL 60510-0500,USA

DUBOVSKY Sergei, CERN, Department of Physics, Theory Division, 1211 Genève 23, Switzerland

DUDAS Emilian, Centre de Physique Théorique, Ecole Polytechnique, 91128 Palaiseau cedex, France

GHERGHETTA Tony, School of Physics and Astronomy, University of Minnesota, Minneapolis, MN 55455, USA

GROJEAN Christophe, SPhT, CEA/Saclay - Orme des Merisiers, 91191 Gif-sur-Yvette cedex, France

HEWETT JoAnne, Stanford Linear Accelerator Center, 2575 Sand Hill Road, Menlo Park, CA 94025, USA

MASIERO Antonio, Dipartimento di Fisica G. Galilei, Universita di Padova and INFN, Sezione di Padova, Via Marzolo 8, 35131 Padua, Italy

NIR Yosef, Department of Particle Physics, Weizmann Institute of Science, Rehovot 76100, Israel

ROMANINO Andrea, SISSA/ISAS, via Beirut 4, 34014 Trieste, Italy

RUBAKOV Valery, Institute for Nuclear Research of the Russian Academy of Sciences, 60th October Anniversary Prospect 7a, Moscow, 117312, Russia

SHADMI Yael, Physics Department, Technion - Israel Institute of Technology, Haifa 32000, Israel

SMIRNOV Alexei, The Abdus Salam International Centre for Theoretical Physics, Strada Costiera 11, 34100 Trieste, Italy

STRUMIA Alessandro, Dipartimento di Fisica, Universita di Pisa, Via Buonarroti 2, 56127 Pisa, Italy

TINYAKOV Peter, Université Libre de Bruxelles, CP 225, Bvd. du Triomphe, 1050 Bruxelles, Belgium

WAGNER Carlos, 9700 South Cass Ave., HEP 362, Argonne National Laboratory, Argonne IL 60439-4815, USA

Participants

ARVANITAKI Asimina, Stanford University, Department of Physics, 382 Via Pueblo Mall, room 365, Stanford, CA 94305, USA

BALLESTEROS Guillermo, Instituto de Fisica Teorica, modulo C-XVI 3a planta, Facultad de Ciencias, UAM, Cantoblanco, 28049 Madrid, Spain

BEDNYAKOV Alexander, Joint Institute for Nuclear research, Joliot Curie, 6, 141980, Dubna, Moscow region, Russia

BOGACHEV Dmitry, Institute of Theoretical and Experimental Physics, Bolshaya Cheremushkinskaya, 25, 117218 Moscow, Russia

BOGDANOS Charalampos, University of Ioannina, Physics Department, Theoretical Physics division, 45110 Ioannina, Greece

BOURJAILY Jacob, University of Michigan, Michigan Center for Theoretical Physics, 3484 Randall Laboratory, 500E University, Ann Arbor, MI 48109-1040, USA

BRÜMMER Felix, Institut für Theoretische Physik, Universität Heidelberg, Philosophenweg 19, 69120 Heidelberg, Germany

DASGUPTA Basudeb, Tata Institute of Fundamental Research, Dept. of Theoretical Physics, Homi Bhabha Road, Colaba, Mumbai 400005, India

DEMARIA Alison, School of Physics, The University of Melbourne, Parkville, 3010, Australia

ENGELHARD Guy, Faculty of Physics, Weizmann Institute of Sciences, PO Box 26, Rehovot, 76100, Israel

EYTON-WILLIAMS Oliver, School of Physics and Astronomy, University of Southampton, Highfield, Southampton SO17 1BJ, UK

FLACKE Thomas, Rudolf Peierls Centre for Theoretical Physics, 1 Keble Road, Oxford, OX 3NP, UK

GHERSON David, Institut de Physique Nucléaire de Lyon, Groupe de Physique Théorique, Bâtiment P. Dirac, Domaine Scientifique de la Doua, 4 rue Enrico Fermi, 69622 Villeurbanne cedex, France

GOUZEVITCH Maxime, Ecole Polytechnique, 91128 Palaiseau cedex, France

GRIPAIOS Ben, Rudolf Peierls Centre for Theoretical Physics, 1 Keble Road, Oxford, OX1 3NP, UK

GRISA Luca, New York University, Physics Department, 4 Washington Place, Mailbox 93, New York, NY 10003, USA

HIRN Johannes, IFIC, Universitat de Valencia, Edificio Institutos de Investigacion, Apartado de Correos 22085, 46071 Valencia, Spain

HOSTEINS Pierre, SPhT, CEA/Saclay - Orme des Merisiers, 91191 Gif-sur-Yvette cedex, France

ITOU Etsuko, Institute of Theoretical Physics, Kanazawa University, Kakumamachi, Kanazawa, 920-1192, Japan

JENKINS Alejandro, CALTECH, Mail code 452-48, Pasadena, CA 91 125, USA

JEONG Kwang Sik, Department of Physics, KAIST, 373-1, Kuseong-dong, Yuseong-gu, Daejeon 305-701, Republic of Korea

KAMENIK Jernej, Jozef Stefan Institute, Jamova 39, 1000 Ljubljana, Slovenia

KASHTI Tamar, Faculty of Physics, Weizmann Institute of Science, PO Box 26, Rehovot, 76100, Israel

KATZ Andrey, Technion, Israel Institute of Technology, Department of Physics, Haifa 32000, Israel

LABONNE Benjamin, Laboratoire de Physique Corpusculaire de Clermont-Ferrand, 24 avenue des Landais, 63177 Aubière, France

LEE Seung, Cornell University, Department of Physics, 109 Clark Hall, Ithaca, NY 14853, USA

LENNEK Michael, University of Arizona, Department of Physics, 1118 E 4th St., Tucson AZ 85721, USA

MALINSKY Michal, SISSA, Via Beirut 2, 34014 Trieste, Italy

MARRA Valerio, Padua University, Department of Physics "Galileo Galilei", Via Marzolo 8, 35131 Padova, Italy

MAYBURY David, University of Alberta, Department of Physics, P-412, Avadh Bhatia Physics Laboratory, University of Alberta, Edmonton, Alberta, T6G 2J1, Canada

MEDINA Anibal, University of Chicago, Department of Astronomy and Astrophysics, 5640 S. Ellis Ave., Chicago, Illinois 60637, USA

MEHTA Poonam, Dept. of Particle Physics, Weizmann Institute of Science, Rehovot 76100, Israel

NARDINI Germano, Istitut de Fisica d'Altes Energies, Edifici CN, Facultat Ciencies UAB, 08193 Bellaterra, Spain

NECHAEV Alexey, Institute of Theoretical and Experimental Physics, Bolshaya Cheremushkinskaya, 25, 117218 Moscow, Russia

NEMEVSEK Miha, Jozef Stefan Institute, Jamova 39, 1000 Ljubljana, Slovenia

NOTH David, Paul Scherrer Institut, Theory Group LTP, WHGA/120, 5232 Villigen PSI, Switzerland

PAPINEAU Chloé, LPT, Bâtiment 210, Université Orsay-Paris Sud, 91405 Orsay, France

PARK Jae-Hyeon, Department of Physics, Tohoku University, Sendaï 980-8578, Japan

PASSEMAR Emilie, I P N , 15 rue Georges Clémenceau, 91406 Orsay cedex, France

PEINADO-RODRIGUEZ Eduardo, Departamento de Fisica Teorica, Inst. de Fisica UNAM, Apartado postal 20-364, Ciudad Universitaria, 01000 Mexico DF, Mexico

POSCHENRIEDER Anton, Munich Technical University, Physics Department T31, James Franck Strasse, 85748 Garching, Germany

SCHERBAKOVA Elena, Joint Institute for Nuclear research, Joliot Curie, 6, 141980, Dubna, Moscow region, Russia

SHINDO Tetsuo, SISSA/ISAS, Via Beirut 2-4, 34014 Beirut, Italy

SOMOGYI Gabor, Institute of Nuclear research, H-Debrecen, bem ter18/c PO BOX 51, Hungary

TU Huitzu, Dept. of Physics and Astronomy, University of Aarhus, Ny Munkegade, Building 520, 8000 Aarhus C, Denmark

TURCHETTI Paolo, Physics Department, University of Roma "La Sapienza", 00185 Roma, Italy

VERNAZZA Leonardo, Dipartimento di Fisica, Universita di Genova, via Dodecaneso 33, 16146 Genova, Italy

VOLANSKY Tomer, Faculty of Physics, Weizmann Institute of Science, Rehovot, 76100 Israel

WELZEL Julien, IPN Lyon, Groupe Théorie, Bâtiment P. Dirac, Université Claude Bernard, 43 Boulevard du 11 Novembre 1918, 69622 Villeurbanne cedex, France

WEST Stephen, R. Peierls Centre for Theoretical Physics, 1 Keble road, Oxford, OX1 3NP, UK

WYSZOMIRSKI Elmar, Munich Technical University, Physics Department T31, James Franck Strasse, 85748 Garching, Germany

YANG Bai, Yale University, Physics Department, Sloane Physics Lab., 217 prospect St., New Haven, CT, 06511-8499, USA

ZANZI Andrea, Dipartimento di Fisica, Universita di Padova, Via Marzolo 8, 35131 Padova, Italy

Preface

The Standard Model of elementary particles and interactions is one of the best tested theories in physics. It has been found to be in remarkable agreement with experiment, and its validity at the quantum level has been successfully probed in the electroweak sector. In spite of its experimental successes, though, the Standard Model suffers from a number of limitations, and is likely to be an incomplete theory. It contains many arbitrary parameters; it does not include gravity, the fourth elementary interaction; it does not provide an explanation for the hierarchy between the scale of electroweak interactions and the Planck scale, characteristic of gravitational interactions; and finally, it fails to account for the dark matter and the baryon asymmetry of the universe. This led particle theorists to develop and study various extensions of the Standard Model, such as supersymmetric theories, Grand Unified Theories or theories with extra space-time dimensions – most of which have been proposed well before the experimental verification of the Standard Model. The coming generation of experimental facilities (high-energy colliders, B-physics experiments, neutrino superbeams, as well as astrophysical and cosmological observational facilities) will allow us to test the predictions of these theories and to deepen our understanding of the fundamental laws of nature.

A previous Les Houches Summer School (August 1997 session) has been devoted to the Standard Model. Two years before the start of the CERN Large Hadron Collider, a key instrument to search for new physics around the electroweak scale, it was appropriate to organize a new session on particle physics.

Particle Physics beyond the Standard Model is a wide subject, not only due to the existence of several candidate extensions of the Standard Model, but also because of the large variety of domains in which new physics may manifest itself, from collider physics to cosmology via flavour and neutrino physics. The August 2005 session was therefore divided into three parts and 14 lectures covering the main aspects of particle physics beyond the Standard Model.

The first part of the school was devoted to what is considered by many particle theorists as the most plausible extension of the Standard Model, namely supersymmetry. The phenomenology of supersymmetry is very rich and covers many topics: flavour physics and CP violation, Higgs and collider physics, astroparticle physics. Flavour physics and CP violation were the main subjects of Yossi

Nir and Antonio Masiero's lectures. Yossi Nir gave a comprehensive course on CP violation in meson decays and argued that, while the Kobayashi-Maskawa mechanism has been successfully tested by experiment, CP violation remains a powerful probe of new physics. Antonio Masiero discussed the supersymmetric contributions to flavour and CP violating processes, both in the hadron and in the lepton sectors, and showed that experimental data provide strong hints about the flavour structure of supersymmetry breaking. Antonio Masiero's lectures also reviewed the main predictions and problems of Grand Unification, from the Georgi-Glashow minimal $SU(5)$ model to recent attempts to build realistic supersymmetric Grand Unified Theories. Higgs physics and supersymmetry phenomenology at colliders were covered by Marcela Carena and Carlos Wagner (unfortunately, their written lectures notes were not available at the time of printing this book). Finally, the issue of supersymmetry breaking was discussed by Yael Shadmi, who reviewed dynamical supersymmetry breaking and the main mechanisms for mediating supersymmetry breaking to the observable sector.

The second part of the school dealt with a completely different class of extensions of the Standard Model, which rely on the existence of extra space-time dimensions, small (Kaluza-Klein picture) or large (brane-world picture), "factorizable" or "warped" (in which case the extra dimensions may even be infinite). Valery Rubakov gave an introduction to theories with large or infinite extra dimensions, focusing on theoretical ideas and on the gravity aspects, and stressed the relevance of brane worlds as theoretical laboratories. JoAnne Hewett reviewed the gravitational tests and collider phenomenology of different classes of extra-dimensional models, as well as the astrophysical and cosmological considerations that can constrain or confirm these scenarios. Tony Gherghetta presented the theoretical tools needed to build particle physics models with warped extra dimensions, and showed that this framework provides new solutions to long-standing problems such as the hierarchy of Yukawa couplings or supersymmetry breaking. He also introduced the holographic interpretation based on the AdS/CFT correspondence, which can be used to describe strongly coupled four-dimensional gauge theories in terms of five-dimensional warped models. Christophe Grojean surveyed recent approaches to the issue of electroweak symmetry breaking: little Higgs models, in which the Higgs field is a pseudo-Goldstone boson; gauge-Higgs unification models, which identify the Higgs boson with the component of a gauge field along an extra dimension; and Higgsless models, in which electroweak symmetry is broken by boundary conditions in extra dimensions. Finally, Emilian Dudas gave an introduction to string theory with emphasis on recent developments that are of interest for phenomenological applications: large extra dimensions and brane worlds from string constructions, string-theoretical mechanisms for supersymmetry breaking, moduli stabilization and de Sitter vacua.

The third and last part of the school was devoted to neutrino physics and to particle astrophysics and cosmology, a rapidly developing field where extensions of the Standard Model can lead to observable consequences. Pierre Binétruy gave a broad overview of particle astrophysics and cosmology. He first reviewed the basic equations of cosmology, black holes, gravitational waves and stellar evolution; he further described violent phenomena in the universe, such as supernovae explosions and gamma ray bursts; then he discussed the fluctuations of the Cosmic Microwave Background and the determination of the cosmological parameters; finally, he moved to the early universe and discussed particle physics models for inflation and dark energy. Peter Tinyakov introduced the subject of ultra-high energy cosmic rays, in which many questions are still open, most notably the apparent absence of the theoretically predicted (GZK) cutoff in the energy spectrum. Alexei Smirnov's lectures on neutrino physics raised the question of what kind of new physics could be at the origin of the observed neutrino masses and mixing. After a description of neutrino oscillations in vacuum and of MSW conversion in matter, he addressed the issue of reconstructing the neutrino mass and mixing pattern from present and future experimental data. Then he discussed various attempts to explain the peculiarities of lepton mixing, and reviewed a large variety of mechanisms and new physics scenarios for generating neutrino masses. Finally, Alessandro Strumia's lectures presented the basic ingredients for a precise computation of thermal leptogenesis, and discussed the conditions for a successful generation of the observed baryon asymmetry of the universe through the decays of heavy right-handed neutrinos.

The summer school and the present volume have been made possible by the financial support of the following institutions, whose contribution is gratefully acknowledged: the "Lifelong learning" programme of the Centre National de la Recherche Scientifique (France); the Université Joseph Fourier, the French Ministry of Research and the Commissariat à l'Energie Atomique, through their constant support to the Physics School. We also acknowledge support from the ECONET programme of the French Ministry of Foreign Affairs.

The photographs reproduced in this volume have been taken by Alexey Nechaev.

The staff of the School, especially Brigitte Rousset and Isabelle Lelièvre, have been of great help for the preparation and development of the school, and we would like to thank them warmly on behalf of all students and lecturers.

D. Kazakov, S. Lavignac and J. Dalibard

CONTENTS

Course 5. Phenomenology of extra dimensions, by JoAnne L. Hewett — *229*

Course 6. Warped models and holography, by Tony Gherghetta — *263*

Contents

Contents

Course 1

FLAVOUR PHYSICS AND GRAND UNIFICATION

A. Masiero[1], S.K. Vempati[2] and O. Vives[3]

[1] *Dip. di Fisica 'G. Galilei', Univ. di Padova and INFN, Sezione di Padova, Via Marzolo 8, I-35131, Padua, Italy*
[2] *Centre de Physique Théorique, Ecole Polytechnique-CPHT, 91128 Palaiseau Cedex, France*
[3] *Departament de Física Teòrica and IFIC, Univ. de València-CSIC, E-46100 Burjassot, Spain*

D. Kazakov, S. Lavignac and J. Dalibard, eds.
Les Houches, Session LXXXIV, 2005
Particle Physics Beyond the Standard Model

1

Contents

3

1. Introduction

The success of the standard model predictions is remarkably high and, indeed, to some extent, even beyond what one would have expected. As a matter of fact, a common view before LEP started operating was that some new physics related to the electroweak symmetry breaking should be present at the TeV scale. In that case, one could reasonably expect such new physics to show up when precisions at the percent level on some electroweak observable could be reached. As we know, on the contrary, even reaching sensitivities better than the percent has not given rise to any firm indication of departure from the SM predictions. To be fair, one has to recognise that in the almost four decades of existence of the SM we have witnessed a long series of "temporary diseases" of it, with effects exhibiting discrepancies from the SM reaching even more than four standard deviations. However, such diseases represented only "colds" of the SM, all following the same destiny: disappearance after some time (few months, a year) leaving the SM absolutely unscathed and, if possible, even stronger than before. Also presently we do not lack such possible "diseases" of the SM. The electroweak fit is not equally good for all observables: for instance the forward-backward asymmetry in the decay of $Z \to b\bar{b}$; some of the penguin $b \to s$ decays, the anomalous magnetic moment of the muon, etc. exhibit discrepancies from the SM expectations. As important as all these hints may be, undoubtedly we are far from any *firm* signal of insufficiency of the SM.

All what we said above can be summarised in a powerful statement about the "low-energy" limit of any kind of new physics beyond the SM: no matter which new physics may lie beyond the SM, it has to reproduce the SM with great accuracy when we consider its limit at energy scales of the order of the electroweak scale.

The fact that with the SM we have a knowledge of fundamental interactions up to energies of $O(100)$ GeV should not be underestimated: it represents a tremendous and astonishing success of our gauge theory approach in particle physics and it is clear that it represents one of the greatest achievements in a century of major conquests in physics. Having said that, we are now confronting ourselves with an embarrassing question: if the SM is so extraordinarily good, does it make sense do go beyond it? The answer, in our view, is certainly positive. This "yes" is not only motivated by what we could define "philosophical" reasons (for instance, the fact that we should not have a "big desert" with many orders of magnitude in energy scale without any new physics, etc.), but there are specific motivations pushing us beyond the SM. We will group them into two broad categories: theoretical and "observational" reasons.

1.1. Theoretical reasons for new physics

There are three questions which "we" consider fundamental and yet do not find any satisfactory answer within the SM: the flavor problem, the unification of the fundamental interactions and the gauge hierarchy problem. The reason why "we" is put in quotes is because it is debatable whether the three above issues (or at least some of them) are really to be taken as questions that the SM should address, but fails to do. Let us first briefly go over them and then we'll comment about alternative views.

Flavor problem. All the masses and mixings of fermions are just free (un-predicted) parameters in the SM. To be sure, there is not even any hint in the SM about the number and rationale of fermion families. Leaving aside predictions for individual masses, not even any even rough relation among fermion masses within the same generation or among different generations is present. Moreover, what really constitutes a *problem*, is the huge variety of fermion masses which is present. From the MeV region, where the electron mass sits, we move to the almost two hundred GeV of the top quark mass, i.e. fermion masses span at least five orders of magnitude, even letting aside the extreme smallness of the neutrino masses. If one has in mind the usual Higgs mechanism to give rise to fermion masses, it is puzzling to insert Yukawa couplings (which are free parameters of the theory) ranging from $O(1)$ to $O(10^{-6})$ or so without any justification what-soever. Saying it concisely, we can state that a "Flavor Theory" is completely missing in the SM. To be fair, we'll see that even when we proceed to BSM new physics, the situation does not improve much in this respect. This important issue is thoroughly addressed at this school: in Yossi Nir's lectures [1] you find an ample discussion of the flavor and CP aspects mainly within the SM, but with some insights on some of its extensions. In these lectures we'll deal with the flavor issue in the context of supersymmetric and grand unified extensions of the SM.

Unification of forces. At the time of the Fermi theory we had two couplings to describe the electromagnetic and the weak interactions (the electric constant and the Fermi constant, respectively). In the SM we are trading off those two couplings with two new couplings, the gauge couplings of $SU(2)$ and $U(1)$. Moreover, the gauge coupling of the strong interactions is very different from the other two. We cannot say that the SM represents a true unification of funda-mental interactions, even leaving aside the problem that gravity is not considered at all by the model. Together with the flavor issue, the unification of fundamental interactions constitutes the main focus of the present lectures. First, also respecting the chronological evolution, we'll consider grand unified theories without an underlying supersymmetry, while then we'll move to spontaneously broken supergravity theories with a unifying gauge symmetry encompassing electroweak and strong interactions.

Gauge hierarchy. Fermion and vector boson masses are "protected" by symmetries in the SM (i.e., their mass can arise only when we break certain symmetries). On the contrary the Higgs scalar mass does not enjoy such a symmetry protection. We would expect such mass to naturally jump to some higher scale where new physics sets in (this new energy scale could be some grand unification scale or the Planck mass, for instance). The only way to keep the Higgs mass at the electroweak scale is to perform incredibly accurate fine tunings of the parameters of the scalar sector. Moreover such fine tunings are unstable under radiative corrections, i.e. they should be repeated at any subsequent order in perturbation theory (this is the so-called "technical" aspect of the gauge hierarchy problem).

We close this Section coming back to the question about how fundamental the above problems actually are, a *caveat* that we mentioned at the beginning of the Section. Do we really need a flavor theory, or can we simply consider that fermion masses as fundamental parameters which just take the values that we observe in our Universe? Analogously, for the gauge hierarchy, is it really something that we have to *explain*, or could we take the view that just the way our Universe is requires that the W mass is 17 orders of magnitude smaller than the Planck mass, i.e. taking M_W as a fundamental input much in the same way we "accept" a cosmological constant as incredibly small as it is? And, finally, why should all fundamental interactions unify, is it just an aesthetical criterion that "we" try to impose after the success of the electro-magnetic unification? The majority of particle physicists (including the authors of the present contribution) consider the above three issues as genuine problems that a *fundamental* theory should address. In this view, the SM could be considered only as a low-energy limit of such deeper theory. Obviously, the relevant question becomes: at which energy scale should such alleged new physics set in? Out of the above three issues, only that referring to the gauge hierarchy problem requires a modification of the SM physics at scales close to the electroweak scale, i.e. at the TeV scale. On the other hand, the absence of clear signals of new physics at LEP, in FCNC and CP violating processes, etc. has certainly contributed to cast doubts in some researchers about the actual existence of a gauge hierarchy *problem*. Here we'll take the point of view that the electroweak symmetry breaking calls for new physics close to the electroweak scale itself and we'll explore its implications for FCNC and CP violation in particular.

1.2. "Observational" reasons for new physics

We have already said that all the experimental particle physics results of these last years have marked one success after the other of the SM. What do we mean then by "observational" difficulties for the SM? It is curious that such difficulties do not arise from observations within the strict high energy particle physics domain, but rather they originate from astroparticle physics, in particular from possible "clashes" of the particle physics SM with the standard model of cosmology (i.e., the Hot Big Bang) or the standard model of the Sun.

Neutrino masses and mixings. The statement that non-vanishing neutrino masses imply new physics beyond the SM is almost tautological. We built the SM in such a way that neutrinos *had* to be massless (linking such property to the V-A character of weak interactions), namely we avoided Dirac neutrino masses by banning the presence of the right-handed neutrino from the fermionic spectrum, while Majorana masses for the left-handed neutrinos were avoided by limiting the Higgs spectrum to isospin doublets. Then we can say that a massive neutrino is a signal of new physics *"by construction"*. However, there is something deeper in the link massive neutrino – new physics than just the obvious correlation we mentioned. Indeed, the easiest way to make neutrinos massive is the introduction of a right-handed neutrino which can combine with the left-handed one to give rise to a (Dirac) mass term through the VEV of the usual Higgs doublet. However, once such right-handed neutrino appears, one faces the question of its possible Majorana mass. Indeed, while a Majorana mass for the left-handed neutrino is forbidden by the electroweak gauge symmetry, no gauge symmetry is able to ban a Majorana mass for the right-handed neutrino given that such particle is sterile with respect to the whole gauge group of the strong and electroweak symmetries. If we write a Majorana mass of the same order as an ordinary Dirac fermion mass we end up with unbearably heavy neutrinos. To keep neutrinos light we need to invoke a large Majorana mass for the right-handed neutrinos, i.e. we have to introduce a scale larger than the electroweak scale. At this scale the right-handed neutrinos should be no longer (gauge) sterile particles and, hence, we expect new physics to set in at such new scale. Alternatively, we could avoid the introduction of right-handed neutrinos providing (left-handed) neutrino masses via the VEV of a new Higgs scalar transforming as the highest component of an $SU(2)_L$ triplet. Once again the extreme smallness of neutrino masses would force us to introduce a new scale; this time it would be a scale much lower than the electroweak scale (i.e., the VEV of the Higgs triplet has to be much smaller than that of the usual Higgs doublet) and, consequently, new physics at a new physical mass scale would emerge.

Although, needless to say, neutrino masses and mixings play a role, and, indeed, a major one, in the vast realm of flavor physics, given the specificity of the subject, there is an entire set of independent lectures devoted to neutrino physics at this school [2]. In our lectures we'll have a chance to touch now and then aspects of neutrino physics related to grand unification, although we recommend the readers more specifically interested in the neutrino aspects to refer to the thorough discussion in Alexei Smirnov's lectures at this school. But at least a point should be emphasised here: together with the issue of dark matter that we are going to present next, massive neutrinos witness that new physics beyond the SM *is* present together with a new physical scale different from that is linked to the SM electroweak physics. Obviously, new physics can be (and probably is) associated to different scales; as we said above, we think that the gauge hierarchy problem is strongly suggesting that (some) new physics should be present close to the electroweak scale. It could be that such new physics related to the electroweak scale is *not* that which causes neutrino masses (just to provide an example, consider supersymmetric versions of the seesaw mechanism: in such schemes, low-energy SUSY would be related to the gauge hierarchy problem with a typical scale of SUSY masses close to the electroweak scale, while the lightness of the neutrino masses would result from a large Majorana mass of the right-handed neutrinos).

Clashes of the SM of particle physics and cosmology: dark matter, baryogenesis and inflation.

Astroparticle physics represents a major road to access new physics BSM. This important issue is amply covered by Pierre Binetruy's lectures at this school [3]. Here we simply point out the three main "clashes" between the SM of particle physics and cosmology.

Dark Matter. There exists an impressive evidence that not only most of the matter in the Universe is *dark*, i.e. it doesn't emit radiation, but what is really crucial for a particle physicist is that (almost all) such dark matter (DM) has to be provided by particles other than the usual baryons. Combining the WMAP data on the cosmic microwave background radiation (CMB) together with all the other evidences for DM on one side, and the relevant bounds on the amount of baryons present in the Universe from Big Bang nucleosynthesis and the CMB information on the other side, we obtain the astonishing result that at something like 10 standard deviations DM has to be of non-baryonic nature. Since the SM does not provide any viable non-baryonic DM candidate, we conclude that together with the evidence for neutrino masses and oscillations, DM represents the most impressive observational evidence we have so far for new physics beyond the standard model. Notice also that it has been repeatedly shown that massive neutrinos cannot account for such non-baryonic DM, hence implying that we need wilder new physics beyond the SM rather than the obvious possibility of providing neutrinos a mass to have a weakly interactive massive particle (WIMP) for DM candidate. Thus, the existence of a (large) amount of non-baryonic DM push us to introduce new particles in addition to those of the SM.

Baryogenesis. Given that we have strong evidence that the Universe is vastly matter-antimatter asymmetric (i.e. no sizable amount of primordial antimatter has survived), it is appealing to have a dynamical mechanism to give rise to such large baryon-antibaryon asymmetry starting from a symmetric situation. In the SM it is not possible to have such an efficient mechanism for baryogenesis. In spite of the fact that at the quantum level sphaleronic interactions violate baryon number in the SM, such violation cannot lead to the observed large matter-antimatter asymmetry (both CP violation is too tiny in the SM and also the present experimental lower bounds on the Higgs mass do not allow for a conveniently strong electroweak phase transition). Hence a dynamical baryogenesis calls for the presence of new particles and interactions beyond the SM (successful mechanisms for baryogenesis in the context of new physics beyond the SM are well known).

Inflation. Several serious cosmological problems (flatness, causality, age of the Universe, ...) are beautifully solved if the early Universe underwent some period of exponential expansion (inflation). The SM with its Higgs doublet does not succeed to originate such an inflationary stage. Again some extensions of the SM, where in particular new scalar fields are introduced, are able to produce a temporary inflation of the early Universe.

As we discussed for the case of theoretical reasons to go beyond the SM, also for the above mentioned observational reasons one has to wonder which scales might be preferred by the corresponding new physics which is called for. Obviously, neutrino masses, dark matter, baryogenesis and inflation are likely to refer to *different* kinds of new physics with some possible interesting correlations. Just to provide an explicit example of what we mean, baryogenesis could occur through leptogenesis linked to the decay of heavy right-handed neutrinos in a see-saw context. At the same time, neutrino masses could arise through the same see-saw mechanism, hence establishing a potentially tantalising and fascinating link between neutrino masses and the cosmic matter-antimatter asymmetry. The scale of such new physics could be much higher than the electroweak scale.

On the other hand, the dark matter issue could be linked to a much lower scale, maybe close enough to the electroweak scale. This is what occurs in one of the most appealing proposals for cold dark matter, namely the case of a WIMP in the mass range between tens to hundreds of GeV. What really makes such a WIMP a "lucky" CDM candidate is that there is an impressive quantitative "coincidence" between Big Bang cosmological SM parameters (Hubble parameter, Planck mass, Universe expansion rate, etc.) and particle physics parameters (weak interactions, annihilation cross section, etc.) leading to a surviving relic abundance of WIMPs just appropriate to provide an energy density contribution in the right ball-park to reproduce the dark matter energy density. A particularly interesting example of WIMP is represented by the lightest SUSY particle (LSP) in SUSY extensions of the SM with a discrete symmetry called R parity (see below more about it). Once again, and in a completely independent way, we are led to consider low-energy SUSY as a viable candidate for new physics providing some answer to open questions in the SM.

As exciting as the above considerations on dark matter and unification are in suggesting us the presence of new physics at the weak scale, we should not forget that they are just *strong suggestions*, but alternative solutions to both the unification and dark matter puzzles could come from (two kinds of) new physics at scales much larger than M_W.

1.3. The SM as an effective low-energy theory

The above theoretical and "observational" arguments strongly motivate us to go beyond the SM. On the other hand, the clear success of the SM in reproducing all the known phenomenology up to energies of the order of the electroweak scale is telling us that the SM has to be recovered as the low-energy limit of such new physics. Indeed, it may even well be the case that we have a "tower" of underlying theories which show up at different energy scales.

If we accept the above point of view we may try to find signals of new physics considering the SM as a truncation to renormalisable operators of an effective low-energy theory which respects the $SU(3) \otimes SU(2) \otimes U(1)$ symmetry and whose fields are just those of the SM. The renormalisable (i.e. of canonical dimension less or equal to four) operators giving rise to the SM enjoy three crucial properties which have no reason to be shared by generic operators of dimension larger than four. They are the conservation (at any order in perturbation theory) of Baryon (B) and Lepton (L) numbers and an adequate suppression of Flavour Changing Neutral Current (FCNC) processes through the GIM mechanism.

Now consider the new physics (directly above the SM in the "tower" of new physics theories) to have a typical energy scale Λ. In the low-energy effective Lagrangian such scale appears with a positive power only in the quadratic scalar term (scalar mass) and in the dimension zero operator which can be considered a cosmological constant. Notice that Λ cannot appear in dimension three operators related to fermion masses because chirality forbids direct fermion mass terms in the Lagrangian. Then in all operators of dimension larger than four Λ will show up in the denominator with powers increasing with the dimension of the corresponding operator.

The crucial question that all of us, theorists and experimentalists, ask ourselves is: where is Λ? Namely is it close to the electroweak scale (i.e. not much above 100 GeV) or is Λ of the order of the grand unification scale or the Planck scale? B- and L-violating processes and FCNC phenomena represent a potentially interesting clue to answer this fundamental question.

Take Λ to be close to the electroweak scale. Then we may expect non-renormalisable operators with B, L and flavour violations not to be largely suppressed by the presence of powers of Λ in the denominator. Actually this constitutes in general a formidable challenge for any model builder who wants to envisage new physics close to M_W. Theories with dynamical breaking of the electroweak symmetry (technicolour) and low-energy supersymmetry constitute examples of new physics with a "small" Λ. In these lectures we will only focus on a particularly interesting "ultra-violet completion" of the Standard Model, namely low energy supersymmetry (SUSY). Other possibilities are considered in other lectures.

Alternatively, given the above-mentioned potential danger of having a small Λ, one may feel it safer to send Λ to super-large values. Apart from kind of "philosophical" objections related to the unprecedented gap of many orders of magnitude without any new physics, the above discussion points out a typical problem of this approach. Since the quadratic scalar terms have a coefficient in front scaling with Λ^2 we expect all scalar masses to be of the order of the super-large scale Λ. This is the gauge hierarchy problem and it constitutes the main (if not only) reason to believe that SUSY should be a low-energy symmetry.

Notice that the fact that SUSY should be a fundamental symmetry of Nature (something of which we have little doubt given the "beauty" of this symmetry) does not imply by any means that SUSY should be a low-energy symmetry, namely that it should hold unbroken down to the electroweak scale. SUSY may well be present in Nature but be broken at some very large scale (Planck scale or string compactification scale). In that case SUSY would be of no use in tackling the gauge hierarchy problem and its phenomenological relevance would be practically zero. On the other hand if we invoke SUSY to tame the growth of the scalar mass terms with the scale Λ, then we are forced to take the view that SUSY should hold as a good symmetry down to a scale Λ close to the electroweak scale. Then B, L and FCNC may be useful for us to shed some light on the properties of the underlying theory from which the low-energy SUSY Lagrangian resulted. Let us add that there is an independent argument in favour of this view that SUSY should be a low-energy symmetry. The presence of SUSY partners at low energy creates the conditions to have a correct unification of the strong and electroweak interactions. If they were at M_{Planck} and the SM were all the physics up to super-large scales, the program of achieving such a unification would largely fail, unless one complicates the non-SUSY GUT scheme with a large number of Higgs representations and/or a breaking chain with intermediate mass scales is invoked.

In the above discussion we stressed that we are not only insisting on the fact that SUSY should be present at some stage in Nature, but we are asking for something much more ambitious: we are asking for SUSY to be a low-energy symmetry, namely it should be broken at an energy scale as low as the electroweak symmetry breaking scale. This fact can never be overestimated. There are indeed several reasons pushing us to introduce SUSY: it is the most general symmetry compatible with a local, relativistic quantum field theory, it softens the degree of divergence of the theory, it looks promising for a consistent quantum description of gravity together with the other fundamental interactions. However, all these reasons are not telling us where we should expect SUSY to be broken. for that matter we could even envisage the maybe "natural" possibility that SUSY is broken at the Planck scale. What is relevant for phenomenology is that the gauge hierarchy problem and, to some extent, the unification of the gauge couplings are actually forcing us to ask for SUSY to be unbroken down to the electroweak scale, hence implying that the SUSY copy of all the known particles, the so-called s–particles should have a mass in the 100–1000 GeV mass range. If Tevatron is not going to see any SUSY particle, at least the advent of LHC will be decisive in establishing whether low-energy SUSY actually exists or it is just a fruit of our (ingenious) speculations. Although even after LHC, in case of a negative result for the search of SUSY particles, we will not be able to "mathematically" exclude all the points of the SUSY parameter space, we will certainly be able to very reasonably assess whether the low-energy SUSY proposal makes sense or not.

Before the LHC (and maybe Tevatron) direct searches for SUSY signals we should ask ourselves whether we can hope to have some indirect manifestation of SUSY through virtual effects of the SUSY particles.

We know that in the past virtual effects (i.e. effects due to the exchange of yet unseen particles in the loops) were precious in leading us to major discoveries, like the prediction of the existence of the charm quark or the heaviness of the top quark long before its direct experimental observation. Here we focus on the potentialities of SUSY virtual effects in processes which are particularly suppressed (or sometime even forbidden) in the SM; the flavour changing neutral current phenomena and the processes where CP violation is violated.

However, the above role of the studies of FCNC and CP violation in relation to the *discovery* of new physics should not make us forget they are equally important for another crucial task: this is the step going from *discovery* of new physics to its *understanding*. Much in the same way that discovering quarks, leptons or electroweak gauge bosons (but without any information about quark mixings and CP violation) would not allow us to *reconstruct* the theory that we call the GWS

Standard Model, in case LHC finds, say, a squark or a gluino we would not be able to *reconstruct* the correct SUSY theory. Flavour and CP physics would play a fundamental role in helping us in such effort. In this sense, we can firmly state that the study of FCNC and CP violating processes is *complementary* to the direct searches of new physics at LHC.

1.4. Flavor, CP and new physics

The generation of fermion masses and mixings ("flavour problem") gives rise to a first and important distinction among theories of new physics beyond the electroweak standard model.

One may conceive a kind of new physics which is completely "flavour blind", i.e. new interactions which have nothing to do with the flavour structure. To provide an example of such a situation, consider a scheme where flavour arises at a very large scale (for instance the Planck mass) while new physics is represented by a supersymmetric extension of the SM with supersymmetry broken at a much lower scale and with the SUSY breaking transmitted to the observable sector by flavour-blind gauge interactions. In this case one may think that the new physics does not cause any major change to the original flavour structure of the SM, namely that the pattern of fermion masses and mixings is compatible with the numerous and demanding tests of flavour changing neutral currents.

Alternatively, one can conceive a new physics which is entangled with the flavour problem. As an example consider a technicolour scheme where fermion masses and mixings arise through the exchange of new gauge bosons which mix together ordinary and technifermions. Here we expect (correctly enough) new physics to have potential problems in accommodating the usual fermion spectrum with the adequate suppression of FCNC. As another example of new physics which is not flavour blind, take a more conventional SUSY model which is derived from a spontaneously broken $N = 1$ supergravity and where the SUSY breaking information is conveyed to the ordinary sector of the theory through gravitational interactions. In this case we may expect that the scale at which flavour arises and the scale of SUSY breaking are not so different and possibly the mechanism itself of SUSY breaking and transmission is flavour-dependent. Under these circumstances we may expect a potential flavour problem to arise, namely that SUSY contributions to FCNC processes are too large.

1.4.1. The flavor problem in SUSY

The potentiality of probing SUSY in FCNC phenomena was readily realised when the era of SUSY phenomenology started in the early 80's [4–10]. In particular, the major implication that the scalar partners of quarks of the same electric charge but belonging to different generations had to share a remarkably high mass degeneracy was emphasised.

Throughout the large amount of work in this last decade it became clearer and clearer that generically talking of the implications of low-energy SUSY on FCNC may be rather misleading. We have a minimal SUSY extension of the SM, the so-called Minimal Supersymmetric Standard Model (MSSM) [11–17] where the FCNC contributions can be computed in terms of a very limited set of unknown new SUSY parameters. Remarkably enough, this minimal model succeeds to pass all the set of FCNC tests unscathed. To be sure, it is possible to severely constrain the SUSY parameter space, for instance using $b \rightarrow s\gamma$, in a way which is complementary to what is achieved by direct SUSY searches at colliders.

However, the MSSM is by no means equivalent to low-energy SUSY. A first sharp distinction concerns the mechanism of SUSY breaking and transmission to the observable sector which is chosen. As we mentioned above, in models with gauge-mediated SUSY breaking (GMSB models) [18] it may be possible to avoid the FCNC threat "ab initio" (notice that this is not an automatic feature of this class of models, but it depends on the specific choice of the sector which transmits the SUSY breaking information, the so-called messenger sector). The other more "canonical" class of SUSY theories that was mentioned above has gravitational messengers and a very large scale at which SUSY breaking occurs. In this talk we will focus only on this class of gravity-mediated SUSY breaking models. Even sticking to this more limited choice we have a variety of options with very different implications for the flavour problem.

First, there exists an interesting large class of SUSY realisations where the customary R-parity (which is invoked to suppress proton decay) is replaced by other discrete symmetries which allow either baryon or lepton violating terms in the superpotential. But, even sticking to the more orthodox view of imposing R-parity, we are still left with a large variety of extensions of the MSSM at low energy. The point is that low-energy SUSY "feels" the new physics at the super-large scale at which supergravity (i.e., local supersymmetry) broke down. In this last couple of years we have witnessed an increasing interest in supergravity realisations without the so-called flavour universality of the terms which break SUSY explicitly. Another class of low-energy SUSY realisations which differ from the MSSM in the FCNC sector is obtained from SUSY-GUT's. The interactions involving super-heavy particles in the energy range between the GUT and the Planck scale bear important implications for the amount and kind of FCNC that we expect at low energy.

Even when R parity is imposed the FCNC challenge is not over. It is true that in this case, analogously to what happens in the SM, no tree level FCNC contributions arise. However, it is well-known that this is a necessary but not sufficient condition to consider the FCNC problem overcome. The loop contributions to FCNC in the SM exhibit the presence of the GIM mechanism and we have to make sure that in the SUSY case with R parity some analog of the GIM mechanism is active.

To give a qualitative idea of what we mean by an effective super-GIM mechanism, let us consider the following simplified situation where the main features emerge clearly. Consider the SM box diagram responsible for the $K^0 - \bar{K}^0$ mixing and take only two generations, i.e. only the up and charm quarks run in the loop. In this case the GIM mechanism yields a suppression factor of $O((m_c^2 - m_u^2)/M_W^2)$. If we replace the W boson and the up quarks in the loop with their SUSY partners and we take, for simplicity, all SUSY masses of the same order, we obtain a super-GIM factor which looks like the GIM one with the masses of the superparticles instead of those of the corresponding particles. The problem is that the up and charm squarks have masses which are much larger than those of the corresponding quarks. Hence the super-GIM factor tends to be of $O(1)$ instead of being $O(10^{-3})$ as it is in the SM case. To obtain this small number we would need a high degeneracy between the mass of the charm and up squarks. It is difficult to think that such a degeneracy may be accidental. After all, since we invoked SUSY for a naturalness problem (the gauge hierarchy issue), we should avoid invoking a fine-tuning to solve its problems! Then one can turn to some symmetry reason. For instance, just sticking to this simple example that we are considering, one may think that the main bulk of the charm and up squark masses is the same, i.e. the mechanism of SUSY breaking should have some universality in providing the mass to these two squarks with the same electric charge. Flavour universality is by no means a prediction of low-energy SUSY. The absence of flavour universality of soft-breaking terms may result from radiative effects at the GUT scale or from effective supergravities derived from string theory. Indeed, from the point of view of effective supergravity theories derived from superstrings it may appear more natural not to have such flavor universality. To obtain it one has to invoke particular circumstances, like, for instance, strong dilaton over moduli dominance in the breaking of supersymmetry, something which is certainly not expected on general ground.

Another possibility one may envisage is that the masses of the squarks are quite high, say above few TeV's. Then even if they are not so degenerate in mass, the overall factor in front of the four-fermion operator responsible for the kaon mixing becomes smaller and smaller (it decreases quadratically with the mass of the squarks) and, consequently, one can respect the observational result. We see from this simple example that the issue of FCNC may be closely linked to the crucial problem of the way we break SUSY.

We now turn to some general remarks about the worries and hopes that CP violation arises in the SUSY context.

1.4.2. CP violation in SUSY

CP violation has major potentialities to exhibit manifestations of new physics beyond the standard model. Indeed, the reason behind this statement is at least twofold: CP violation is a "rare" phenomenon and hence it constitutes an ideal ground for NP to fight on equal footing with the (small) SM contributions; generically any NP present in the neighbourhood of the electroweak scale is characterised by the presence of new "visible" sources of CP violation in addition to the usual CKM phase of the SM. A nice introduction to this subject by R.N. Mohapatra can be found in the book "CP violation", Jarlskog, C. (Ed.), Singapore: World Scientific (1989) [19].

Our choice of low energy SUSY for NP is due on one side to the usual reasons related to the gauge hierarchy problem, gauge coupling unification and the possibility of having an interesting cold dark matter candidate and on the other hand to the fact that it provides the only example of a completely defined extension of the SM where the phenomenological implications can be fully detailed [13–17]. SUSY fully respects the above statement about NP and new sources of CP violation: indeed a generic SUSY extension of the SM provides numerous new CP violating phases and in any case even going to the most restricted SUSY model at least two new flavour conserving CP violating phases are present. Moreover the relation of SUSY with the solution of the gauge hierarchy problem entails that at least some SUSY particles should have a mass close to the electroweak scale and hence the new SUSY CP phases have a good chance to produce visible effects in the coming experiments [20–23]. This sensitivity of CP violating phenomena to SUSY contributions can be seen i) in a "negative" way: the "SUSY CP problem" i.e. the fact that we have to constrain general SUSY schemes to pass the demanding experimental CP tests and ii) in a "positive" way: indirect SUSY searches in CP violating processes provide valuable information on the structure of SUSY viable realisations. Concerning this latter aspect, we emphasise that not only the study of CP violation could give a first hint for the presence of low energy SUSY before LHC, but, even after the possible discovery of SUSY at LHC, the study of indirect SUSY signals in CP violation will represent a complementary and very important source of information for many SUSY spectrum features which LHC will never be able to detail [20–23].

Given the mentioned potentiality of the relation between SUSY and CP violation and obvious first question concerns the selection of the most promising phenomena to provide such indirect SUSY hints. It is interesting to notice that SUSY CP violation can manifest itself both in flavour conserving and flavour violating processes. As for the former class we think that the electric dipole moments (EDMs) of the neutron, electron and atoms are the best place where SUSY phases, even in the most restricted scenarios, can yield large departures from the SM expectations. In the flavour changing class we think that the study of CP violation in several B decay channels can constitute an important test of the uniqueness of the SM CP violating source and of the presence of the new SUSY phases. CP violation in kaon physics remains of great interest and it will be important to explore rare decay channels ($K_L \rightarrow \pi^0 \nu \bar{\nu}$ and $K_L \rightarrow \pi^0 e^+ e^-$ for instance) which can provide complementary information on the presence of different NP SUSY phases in other flavour sectors. Finally let us remark that SUSY CP violation can play an important role in baryo- and/or lepto-genesis. In particular in the leptogenesis scenario the SUSY CP violation phases can be related to new CP phases in the neutrino sector with possible links between hadronic and leptonic CP violations.

2. Grand unification and SUSY GUTS

Unification of all the known forces in nature into a universal interaction describing all the processes on equal footing has been for a long time and keeps being nowadays a major goal for particle physics. In a sense, we witness a first, extraordinary example of a "unified explanation" of apparently different phenomena under a common fundamental interaction in Newton's "Principia", where the universality of gravitational law succeeds to link together the fall of a stone with the rotation of the Moon. But it is with Maxwell's "Treatise of Electromagnetism" at the end of the 19th century that two seemingly unlinked interactions, electricity and magnetism, merge into the common description of electromagnetism. Another amazing step along this path was completed in the second half of the last century when electromagnetic and weak interactions were unified in the electroweak interactions giving rise to the Standard Model. However, the Standard Model is by no means satisfactory because it still involves three different gauge groups with independent gauge couplings $SU(3) \times SU(2) \times U(1)$. Strictly speaking, if we intend "unification" of fundamental interactions as a reduction of the fundamental coupling constants, no much gain was achieved in the SM with respect to the time when weak and electromagnetic interactions were associated to the Fermi and electric couplings, respectively. Nevertheless, one should recognise that, even though, e and G_F are traded with g_2 and g_1 of $SU(2) \times U(1)$, in the SM electromagnetic and weak forces are no longer two separate interactions, but they are closely entangled.

Another distressing feature of the Standard Model is its strange matter content. There is no apparent reason why a family contains a doublet of quarks, a doublet of leptons, two singlets of quarks and a charged lepton singlet with quantum numbers,

$$Q\left(3, 2, \frac{1}{3}\right), \quad u_R\left(\bar{3}, 1, \frac{4}{3}\right), \quad d_R\left(\bar{3}, 1, -\frac{2}{3}\right), \quad L\,(1, 2, -1), \quad e_R\,(1, 1, -2). \quad (2.1)$$

The $U(1)$ quantum numbers are specially disturbing. In principle any charge is allowed for a $U(1)$ symmetry, but, in the SM, charges are quantised in units of $1/3$.

These three problems find an answer in Grand Unified Theories (GUTs). The first theoretical attempt to solve these questions was the Pati-Salam model, $SU(4)_C \times SU(2)_L \times SU(2)_R$ [24]. The original idea of this model was to consider quarks and leptons as different components of the same representation, extending $SU(3)$ to include *leptons as the fourth colour*. In this way the matter multiplet would be

$$F^e_{L,R} = \begin{bmatrix} u_r & u_y & u_b & v_e \\ d_r & d_y & d_b & e^- \end{bmatrix}_{L,R}, \tag{2.2}$$

with F^e_L and F^e_R transforming as $(4, 2, 1)$ and $(4, 1, 2)$, respectively under the gauge group. Thus, this theory simplifies the matter content of the SM to only two representations containing 16 states, with the sixteenth component which is missing in the SM fermion spectrum, carrying the quantum numbers of a right-handed neutrino. More importantly, it provides a very elegant answer to the problem of charge quantisation in the SM. Notice that, while the eigenvalues of Abelian groups are continuous, those corresponding to non-Abelian group are discrete. Therefore, if we embed the hypercharge interaction of the SM in a non-Abelian group, the charge will necessarily be quantised. In this case the electric charge is given by $Q_{em} = T_{3L} + T_{3R} + 1/2(B - L)$, where $SU(3)_C \times U(1)_{B-L}$ is the subgroup contained in $SU(4)_C$. Still, this group contains three independent gauge couplings and it does not really unify all the known interactions (even imposing a discrete symmetry interchanging the two $SU(2)$ subgroups, we are left with two independent gauge couplings).

The Standard Model has four diagonal generators corresponding to T_3 and T_8 of $SU(3)$, T_3 of $SU(2)$ and the hypercharge generator Y, i.e. it has rank four. If we want to unify all these interactions into a simple group it must have rank four at least. Indeed, to achieve a unification of the gauge couplings, we have to require the gauge group of such unified theory to be simple or the product of identical simple factors whose coupling constants can be set equal by a discrete symmetry.

There exist 9 simple or semi-simple groups of rank four. Imposing that the viable candidate contains an SU(3) factor and that it possesses some *complex* representations (in order to accommodate the chiral fermions), one is left with $SU(3) \times SU(3)$ and $SU(5)$. Since in the former case the quarks u, d and s should be put in the same triplet representation, one would run into evident problems with exceeding FCNC contributions in d-s transitions. Hence, we are left with $SU(5)$ as the only viable candidate of rank four for grand unification. The minimal $SU(5)$ model was originally proposed by Georgi and Glashow [25]. In this theory there is a single gauge coupling α_{GUT} defined at the grand unification scale M_{GUT}. The whole SM particle content is contained in two $SU(5)$ representations $\bar{\mathbf{5}} = (\bar{3}, 1, -\frac{2}{3}) + (1, 2, -1)$ and $\mathbf{10} = (3, 2, \frac{1}{3}) + (\bar{3}, 1, \frac{4}{3}) + (1, 1, -2)$ under $SU(3) \times SU(2) \times U(1)$. Once more the $U(1)_Y$ generator is a combination of the diagonal generators of the $SU(5)$ and electric charge is also quantised in this model. The minimal $SU(5)$ will be described below.

A perhaps more complete unification is provided by the $SO(10)$ model [26, 27]. We have also a single gauge coupling and charge quantisation but in $SO(10)$ a single representation, the $\mathbf{16}$ includes both the $\bar{\mathbf{5}}$ and $\mathbf{10}$ plus a singlet corresponding to a right handed neutrino.

2.1. SU(5) the prototype of GUT theory

A grand unified theory would require the equality of the three SM gauge couplings to a single unified coupling $g_1 = g_2 = g_3 = g_{GUT}$. However this requirement seems to be phenomenologically unacceptable: the strong coupling g_3 is much bigger than the electroweak couplings g_2 and g_1 that are also different between themselves. The key point in attempting a unification of the coupling "constants" is the observation that they are, in fact, not constant. The couplings evolve with energy, they "run". The values of the renormalised couplings depend on the energy scale at which they are measured through the renormalisation group equations (RGEs). Georgi, Quinn and Weinberg [28] realised that the equality of the gauge couplings applies only at a high scale M_{GUT} where, possibly, but not necessarily, a new "grand unified" symmetry (like $SU(5)$, for instance) sets in. The evolution of the couplings with energy is regulated by the equations of the renormalisation group (RGE):

$$\frac{d\alpha_i}{d \log \mu^2} = \beta_i \alpha_i^2 + O(\alpha_i^3), \tag{2.3}$$

where $\alpha_i = g_i^2/(4\pi)$ and $i = 1, 2, 3$ refers to the $U(1)$, $SU(2)$ and $SU(3)$ gauge couplings. The coefficients β_i receive contributions from vector-boson, fermion and scalar loops shown in figure 1. These coefficients are obtained from the 1

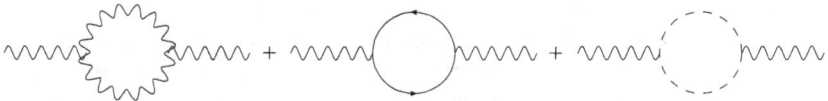

Fig. 1. One loop corrections to the gluon propagator.

loop renormalised gauge couplings,

$$\beta_i = -\frac{1}{4\pi}\left[\frac{11}{3}C_2(G_i) - \frac{2}{3}\sum_f T(R_f) - \frac{1}{3}\sum_s T(R_s)\right], \qquad (2.4)$$

with $C_2(G_i) = N$ the eigenvalue of Casimir operator of the group $SU(N)$, and $T(R) = T(S) = 1/2$ for fermions and scalars in the fundamental representation. The sums is extended over all fermions and scalars in the representations R_f and R_s.

For the particle content of the SM, the β coefficients read:

$$\beta_3 = -\frac{1}{4\pi}\left[\frac{11}{3}\cdot 3 - \frac{2}{3}\cdot 4\,n_g\cdot\frac{1}{2}\right] = -\frac{1}{4\pi}\left[11 - 4\right],$$

$$\beta_2 = -\frac{1}{4\pi}\left[\frac{11}{3}\cdot 2 - \frac{2}{3}\cdot 4\,n_g\cdot\frac{1}{2} - \frac{1}{3}\cdot n_H\cdot\frac{1}{2}\right] = -\frac{1}{4\pi}\left[\frac{22}{3} - 4 - \frac{1}{6}\right],$$

$$\beta_1 = -\frac{1}{4\pi}\frac{3}{5}\left[-\frac{2}{3}\cdot\frac{10}{3}\cdot n_g - \frac{1}{3}\cdot n_H\cdot\frac{1}{2}\right] = -\frac{1}{4\pi}\left[-4 - \frac{1}{10}\right], \qquad (2.5)$$

with $n_g = 3$ the number of generations and $n_H = 1$ the number of Higgs doublets. Care must be taken in evaluating the β_1 coefficient of $U(1)$ hypercharge. Obviously its value depends on the normalisation one chooses for the hypercharge generator (indeed, hypercharge is related to the electric charge Q and the isospin T_3 by the relation $Q = T_3 + aY$, with a being the normalisation factor of the hypercharge generator Y). Asking for a unifying gauge symmetry group G embedding the SM to set in at the scale M_{GUT} at which the SM couplings obtain a common value implies that *all* the SM generators are normalised in the same way. At this point the hypercharge normalisation is no longer arbitrary and the coefficient $\frac{3}{5}$ appearing in β_1 is readily explained (the interested reader is invited to explicitly derive this result, for instance considering one fermion family of the SM, computing $\text{Tr}(T)^2$ over such fermions and then imposing that $\text{Tr}(T)^2=\text{Tr}(Y)^2$).

Eq. (2.3) can be easily integrated and we obtain

$$\frac{1}{\alpha_i(Q^2)} = \frac{1}{\alpha_i(M^2)} + \beta_i\,\log\frac{M^2}{Q^2},$$

$$\alpha_i(Q^2) = \frac{\alpha_i(M^2)}{\left(1 + \beta_i\alpha_i(M^2)\log\frac{M^2}{Q^2}\right)}. \qquad (2.6)$$

This result is very encouraging because we can see that both α_3 and α_2 decrease with increasing energies because β_3 and β_2 are negative. Moreover α_3 decreases more rapidly because $|\beta_3| > |\beta_2|$. Finally β_1 is positive and hence α_1 increases. Now we can ask whether starting from the measured values of the gauge couplings at low energies and using the β_i parameters of the SM in Eq. (2.5) there is a scale, M_{GUT}, where the three couplings meet. To do this we have to solve the equations:

$$\frac{1}{\alpha_3(\mu^2)} = \frac{1}{\alpha_{GUT}} + \beta_3 \, \log \frac{M_{GUT}^2}{\mu^2},$$

$$\frac{1}{\alpha_2(\mu^2)} = \frac{\sin^2\theta_W(\mu^2)}{\alpha_{em}(\mu^2)} = \frac{1}{\alpha_{GUT}} + \beta_2 \, \log \frac{M_{GUT}^2}{\mu^2},$$

$$\frac{1}{\alpha_1(\mu^2)} = \frac{3}{5}\frac{\cos^2\theta_W(\mu^2)}{\alpha_{em}(\mu^2)} = \frac{1}{\alpha_{GUT}} + \beta_1 \, \log \frac{M_{GUT}^2}{\mu^2}. \tag{2.7}$$

Now, we can use α_3 and α_{em} to determine M_{GUT} and then use the remaining equation to "predict" $\sin^2\theta_W$:

$$\frac{3}{5\,\alpha_{em}(\mu^2)} - \frac{8}{5\,\alpha_3(\mu^2)} = \frac{67}{20\pi} \, \log \frac{M_{GUT}^2}{\mu^2},$$

$$\sin^2\theta_W(\mu^2) = \frac{3}{8}\left[1 - \frac{109}{36\pi}\alpha_{em} \log \frac{M_{GUT}^2}{\mu^2}\right]. \tag{2.8}$$

The result is astonishing (we say this without any exaggeration!): starting from the measured values of α_3 and α_{em} we find that all three gauge couplings would unify at a scale $M_{GUT} \simeq 2 \times 10^{15}$ GeV if $\sin^2\theta_W \simeq 0.21$. This is remarkably close to the experimental value of $\sin^2\theta_W$ and this constitutes a major triumph of the grand unification idea and the strategy we adopted to implement it. Let us finally comment that in deriving these results we have used the so-called step approximation and one loop RGE equations. The step approximation consists in using the beta parameters of the SM (the three of them different) all the way from M_W to M_{GUT} where the beta parameter would change with a step function to a common beta parameter corresponding to $SU(5)$. In reality the β-function transitions from $\mu \ll M_{GUT}$ to $\mu \gg M_{GUT}$ are smooth ones that take into account the different threshold when new particles enter the RGE evolution. This effects can be included using mass dependent beta functions [29]. Similarly, we have only used one loop RGE equations although two loop RGE equations are also available. The inclusion of these additional refinements in our RGEs would not improve substantially the agreement with the experimental results.

2.1.1. The Georgi-Glashow minimal SU(5) model

As has been discussed above, the fermions of the Standard Model can be arranged in terms of the fundamental $\bar{\mathbf{5}}$ and the anti-symmetric $\mathbf{10}$ representation of the SU(5) [30]. The appropriate particle assignments in these two representations are:

$$
\bar{\mathbf{5}} = \begin{pmatrix} d^c \\ d^c \\ d^c \\ v_e \\ e^- \end{pmatrix}_L \quad \mathbf{10} = \begin{pmatrix} 0 & u^c & u^c & u & d \\ & 0 & u^c & u & d \\ & & 0 & u & d \\ & & & 0 & e^c \\ & & & & 0 \end{pmatrix}_L , \tag{2.9}
$$

where $\bar{\mathbf{5}} = (\bar{3}, 1, -\frac{2}{3}) + (1, 2, -1)$ and $\mathbf{10} = (3, 2, \frac{1}{3}) + (\bar{3}, 1, \frac{4}{3}) + (1, 1, -2)$ under $SU(3) \times SU(2) \times U(1)$ (here, we consider Y normalised as $Q = T_3 + Y$, for simplicity). It is easy to check that this combination of the representations is anomaly free. The gauge theory of SU(5) contains 24 gauge bosons. They are decomposed in terms of the standard model gauge group SU(3) × SU(2) × U(1) as:

$$
\mathbf{24} = (\mathbf{8}, \mathbf{1}) + (\mathbf{1}, \mathbf{3}) + (\mathbf{1}, \mathbf{1}) + (\mathbf{3}, \mathbf{2}) + (\bar{\mathbf{3}}, \mathbf{2}). \tag{2.10}
$$

The first component represents the gluon fields (G) mediating the colour, the second one corresponds to the Standard Model $SU(2)$ mediators (W) and the third component corresponds to the $U(1)$ mediator (B). The fourth and fifth components carry both colour as well as the $SU(2)$ indices and are called the X and Y gauge bosons. Schematically, they can be represented in terms of the 5×5 matrix as:

$$
V = \begin{pmatrix} \left[\dfrac{(G - 2B)}{\sqrt{30}}\right]^{\alpha}_{\beta} & X_1 & Y_1 \\ & X_2 & Y_2 \\ & X_3 & Y_3 \\ X_1 \quad X_2 \quad X_3 & \dfrac{W^3}{\sqrt{2}} + \dfrac{3B}{\sqrt{30}} & W^+ \\ Y_1 \quad Y_2 \quad Y_3 & W^- & -\dfrac{W^3}{\sqrt{2}} + \dfrac{3B}{\sqrt{30}} \end{pmatrix} . \tag{2.11}
$$

As we will see later, these gauge bosons play an important role in various observational aspects of Grand Unification like proton decay, etc. Before entering such discussion, let us study the fermion masses in the prototype $SU(5)$. Given that fermions are in $\bar{\mathbf{5}}$ and $\mathbf{10}$ representations, after some simple algebra we conclude that the scalars that can form Yukawa couplings are

$$
\mathbf{10} \times \mathbf{10} = \bar{\mathbf{5}} + \overline{\mathbf{45}} + \mathbf{50}, \tag{2.12}
$$

$$
\mathbf{10} \times \bar{\mathbf{5}} = \mathbf{5} + \mathbf{45}. \tag{2.13}
$$

From the above, we see that we need at least two Higgs representations transforming as the fundamental (5_H) and the anti-fundamental ($\bar{5}_H$) to reproduce the fermion Yukawa couplings. The corresponding Yukawa terms read:

$$\mathcal{L}^{\text{yuk}}_{SU(5)} = h^u_{ij} 10_i 10_j \bar{5}_H + h^d_{ij} 10_i \bar{5}_j 5_H. \tag{2.14}$$

This simple form which we have written has some problems which we will discuss later. A Higgs in the adjoint representation can be used to break $SU(5)$ to the diagonal subgroup of the Standard Model. Denoting the adjoint as $\Phi = \sum_{i=1}^{24} \lambda_i/\sqrt{2}\phi_i$, where λ_i are generators of the SU(5), the most general renormalisable scalar potential is

$$V(\Phi) = -\frac{1}{2}m_1^2 \text{Tr}(\Phi^2) + \frac{1}{4}a(\text{Tr}(\Phi^2))^2 + \frac{1}{2}b\text{Tr}(\Phi^4) + \frac{1}{3}c\text{Tr}(\Phi^3). \tag{2.15}$$

However, to simplify the problem, we impose a discrete symmetry ($\Phi \leftrightarrow -\Phi$) which sets c to zero. The remaining potential has the following minimum when $b > 0$ and $a > -7/15\,b$:

$$< 0|\Phi|0 > = \begin{pmatrix} v & 0 & 0 & 0 & 0 \\ 0 & v & 0 & 0 & 0 \\ 0 & 0 & v & 0 & 0 \\ 0 & 0 & 0 & -3/2v & 0 \\ 0 & 0 & 0 & 0 & -3/2v \end{pmatrix}, \tag{2.16}$$

with v determined by

$$m_1^2 = \frac{15}{2}av^2 + \frac{7}{2}bv^2. \tag{2.17}$$

2.1.2. Distinctive features of GUTs and problems in building a realistic model

(i) Fermion Masses
In the previous section, we have seen that in the typical prototype SU(5) model, the fermions attain their masses through a 5_H and $\bar{5}_H$ of Higgses. A simple consequence of this approach is that there is an equality of $Y_d^T = Y_e$ at the GUT scale; which would mean equal charged lepton and down quark masses at the M_{GUT} scale. Schematically, these are given as:

$$m_e(M_{\text{GUT}}) = m_d(M_{\text{GUT}}), \tag{2.18}$$

$$m_\mu(M_{\text{GUT}}) = m_s(M_{\text{GUT}}), \tag{2.19}$$

$$m_\tau(M_{\text{GUT}}) = m_b(M_{\text{GUT}}). \tag{2.20}$$

We would have to verify these prediction by running the Yukawa couplings from the SM to the GUT scale. Let us have a more closer look at these RGEs. For the bottom mass and the τ Yukawa these are given by [31, 32]:

$$\frac{d \log Y_b(\mu)}{d \log \mu} = -3 \, C_3^b \frac{\alpha_3(\mu)}{4\pi} - 3 \, C_2^b \frac{\alpha_2(\mu)}{4\pi} - 3 \, C_1^b \frac{\alpha_1(\mu)}{4\pi}, \qquad (2.21)$$

$$\frac{d \log Y_\tau(\mu)}{d \log \mu} = -3 \, C_3^\tau \frac{\alpha_3(\mu)}{4\pi} - 3 \, C_2^\tau \frac{\alpha_2(\mu)}{4\pi} - 3 \, C_1^\tau \frac{\alpha_1(\mu)}{4\pi}, \qquad (2.22)$$

with $C_3^b = \frac{4}{3}$, $C_2^b = \frac{3}{4}$, $C_1^b = -\frac{1}{30}$, $C_3^\tau = 0$, $C_2^\tau = \frac{3}{4}$, $C_1^\tau = -\frac{3}{10}$. Knowing the scale dependence of the gauge couplings, Eq. (2.6), we can integrate this equation, neglecting the effects of other Yukawa except the top Yukawa in the RHS of the above equations. Taking the masses to be equal at M_{GUT}, we obtain

$$\frac{m_b(M_Z)}{m_\tau(M_Z)} \approx E_t^{-1/2} \left[\frac{\alpha_3(M_Z)}{\alpha_3(M_{\text{GUT}})} \right]^{\frac{-3C_3}{4\pi\beta_3}} \approx E_t^{-1/2} \left[\frac{\alpha_3(M_Z)}{\alpha_3(M_{\text{GUT}})} \right]^{\frac{4}{7}}, \qquad (2.23)$$

where $E_t = \text{Exp}[\frac{1}{2\pi} \int_{M_Z}^{M_{\text{GUT}}} Y_t(t) dt]$. Taking these masses at the weak scale, we obtain a rough relation

$$\frac{m_b(M_W)}{m_\tau(M_W)} \approx 3, \qquad (2.24)$$

which is quite in agreement with the experimental values. This can be considered as one of the major predictions of the $SU(5)$ grand unification. However, there is a caveat. If we extend similar analysis to the first two generations we end-up with relations:

$$\frac{m_\mu}{m_e} = \frac{m_s}{m_d}, \qquad (2.25)$$

which don't hold water at weak scale. The question remains how can one modify the *bad* relations of the first two generations while keeping the *good* relation of the third generation intact. Georgi and Jarlskog solved this puzzle [33] with a simple trick using an additional Higgs representation. As we have seen in Eq. (2.13), the **10** and **5̄** can couple to a **45** in addition to the **5** representation. The **45** is a completely anti-symmetric representation and a textures can be chosen such that the bad relations can be modified keeping the good relation intact.

(ii) Doublet-Triplet Splitting

We have seen that in minimal $SU(5)$ we need at least two Higgs representations transforming as **5**$_\text{H}$ and **5̄**$_\text{H}$ to accommodate the fermion Yukawa couplings. The **5** representation of $SU(5)$ contains a $(\mathbf{3}, \mathbf{1})$ and $(\mathbf{1}, \mathbf{2})$ under $(SU(3)_C, SU(2)_L)$. So, the **5**$_\text{H}$ and **5̄**$_\text{H}$ representations contain the required Higgs doublets that breaks the electroweak symmetry at low energies, but they contain also colour triplets, extremely dangerous, as we will see later, because they mediate a fast proton decay if their mass is much lower than the GUT scale. The doublet-triplet splitting problem is then the question of how one can enforce the mass of the Higgs doublet to remain at the electroweak scale, while the Higgs triplet mass should jump to M_{GUT} [34].

The $SU(5)$ symmetry is broken by the VEV of the Higgs Φ sitting in the adjoint representation, as we saw in Eq. (2.15). At the electroweak scale we need a second breaking step, $SU(3)_C \times SU(2)_L \times U(1)_Y \to SU(3)_C \times U(1)_{\text{em}}$, which is obtained by the potential

$$
V(H) = -\frac{\mu^2}{2} \mathbf{5_H}^\dagger \mathbf{5_H} + \frac{\lambda}{4} \left(\mathbf{5_H}^\dagger \mathbf{5_H} \right)^2 ,
\tag{2.26}
$$

with a VEV

$$
\langle \mathbf{5_H} \rangle = \begin{pmatrix} 0 \\ 0 \\ 0 \\ 0 \\ \frac{v_0}{\sqrt{2}} \end{pmatrix} , \qquad v_0^2 = \frac{2\mu^2}{\lambda} .
\tag{2.27}
$$

However, the potential $V = V(\Phi) + V(H)$ does not give rise to a viable model. Clearly both the Higgs doublet and triplet fields remain with masses at the M_W scale which is catastrophic for proton decay.

This problem can find a solution if we consider also the following Φ–$\mathbf{5_H}$ cross terms which are allowed by $SU(5)$

$$
V(\Phi, H) = \alpha \mathbf{5_H}^\dagger \mathbf{5_H} \text{Tr}(\Phi^2) + \beta \mathbf{5_H}^\dagger \Phi^2 \mathbf{5_H} .
\tag{2.28}
$$

Notice that even if one does not introduce the above mixed term at the tree level, one expects it to arise at higher order given that the underlying $SU(5)$ symmetry does not prevent its appearance.

Let's turn to the minimisation of the full potential $V = V(\Phi) + V(H) + V(\Phi, H)$. Now that Φ and $\mathbf{5_H}$ are coupled, $\langle \Phi \rangle$ may also break $SU(2)_L$ whilst $SU(3)_C$ must be rigorously unbroken. Therefore, we look for solutions with $\langle \Phi \rangle = \text{Diag.}\left(v, v, v, \left(-\frac{3}{2} - \frac{\varepsilon}{2} \right) v, \left(-\frac{3}{2} - \frac{\varepsilon}{2} \right) v \right)$. In the absence of Φ–$\mathbf{5_H}$ mixing, i.e. $\alpha = \beta = 0$, ε must vanish. The solution with this properties has

$$
\varepsilon = \frac{3}{20} \frac{\beta v_0^2}{b v^2} + O\left(\frac{v_0^4}{v^4} \right) .
\tag{2.29}
$$

As $v \sim O(M_{GUT})$ and $v_0 \sim O(M_W)$, we have that the breaking of $SU(2)$ due to $\langle \Phi \rangle$ is much smaller than that due to $\langle H \rangle$. Now, the expressions for m_1^2 (corresponding to Eq. (2.17)) and μ_5^2 (corresponding to Eq. (2.27)) are more complicated

$$m_1^2 = \frac{15}{2}av^2 + \frac{7}{2}15bv^2 + +\alpha v_0^2 + \frac{9}{30}\beta v_0^2, \tag{2.30}$$

and

$$\mu^2 = \frac{1}{2}\lambda v_0^2 + 15\alpha v^2 + \frac{9}{2}\beta v^2 - 3\epsilon\beta v^2. \tag{2.31}$$

We can see that Eq. (2.30) shows only a very small modification from Eq. (2.17) being $v_0 \ll v$. What is very worrying is the result of Eq. (2.31). Since the parameter in the Lagrangian $\mu \sim O(M_W)$, i.e. $\mu \ll v$, the natural thing to happen would be that v_0 takes a value order v to reduce the right-hand side of this equation (remember that in this equation v and v_0 are our unknowns). In other words, without putting any particular constraint on α and β, we would expect $v_0 \sim O(v)$. However, this would completely spoil the hierarchy between M_W and M_{GUT}. If we want to avoid such a disaster, we have to fine-tune α and β to one part in $\left(\frac{v^2}{v_0^2}\right) \sim 10^{24}$!!! Even more, such an adjustment must be repeated at any order in perturbation theory, since radiative correction will displace α and β for more than one part in 10^{24}. This is our first glimpse in the so-called hierarchy problem.

(iii) Nucleon Decay

As we saw in the previous section, perhaps the most prominent feature of GUT theories is the non-conservation of baryon (and lepton) number [35]. In the minimal $SU(5)$ model this is due to the tree-level exchange of X and Y gauge bosons in the adjoint of $SU(5)$ with $(\mathbf{3}, \mathbf{2})$ quantum numbers under $SU(3) \times SU(2)_L$. The couplings of these gauge bosons to fermions are

$$\mathcal{L}_X = \sqrt{\frac{1}{2}}g X_{\mu\alpha}^a \left[\epsilon^{\alpha\beta\gamma} \, \bar{u}_\gamma^c \gamma^\mu q_{\beta a} + \epsilon^{ab} \left(\bar{q}_{\alpha b}\gamma^\mu e^+ - \bar{l}_b\gamma^\mu d_\alpha^c \right) \right], \tag{2.32}$$

where $\epsilon_{\alpha\beta\gamma}$ and ϵ_{ab} are the totally antisymmetric tensors, (α, β, γ) are $SU(3)$ and a, b are $SU(2)_L$ indices. Thus the $SU(2)_L$ doublets are

$$X_{\alpha a} = (X_\alpha, Y_\alpha), \quad q_{\alpha a} = (u_\alpha, u_\alpha), \quad l_a = (\nu_e, e). \tag{2.33}$$

We can see in Eq. (2.32) that the (X, Y) bosons have two couplings to fermions with different baryon numbers. They have a leptoquark coupling with $B_1 = -1/3$ and a diquark couplings with $B_2 = 2/3$. Therefore, through the coupling

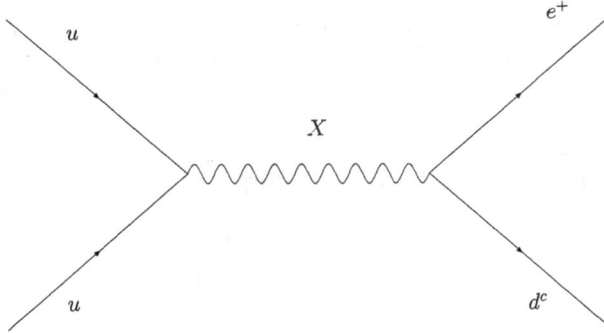

Fig. 2. Baryon number violating couplings of the X boson.

of an X boson we can change a $B = -1/3$ channel into a $B = 2/3$ channel and a $\Delta B = 1$ process occurs at tree level as shown in Figure 2. If the mass of the X boson, M_X is large compared to the other masses, we can obtain the effective four-fermion interactions [36–39]

$$\mathcal{L}^{\text{eff}}_{\Delta B=1} = \frac{g^2}{2M_X^2} \, \epsilon^{\alpha\beta\gamma} \epsilon^{ab} \left(\overline{u}^c_\gamma \gamma^\mu q_{\beta a} \right) \left(\overline{d}^c_\alpha \gamma_\mu l_b + \overline{e}^+ \gamma_\mu q_{\alpha b} \right). \tag{2.34}$$

From this effective Lagrangian we can see that although baryon number is violated, $(B - L)$ is still conserved, thus the decay $p \to e^+ \pi^0$ is allowed but a decay $n \to e^- \pi^+$ is forbidden. From this effective Lagrangian we can obtain the proton decay rate and we have $\Gamma_p \sim 10^{-3} m_p^5 / M_X^4$ and therefore from the present bound on the proton lifetime $\tau_p \geq 10^{33}$ yrs, we have that $M_X \geq 4 \times 10^{15}$ GeV. From this simple dimensional estimate of the proton decay lifetime, we can already see that the minimal non-supersymmetric $SU(5)$ can easily get into trouble because of matter stability. Indeed, performing an accurate analysis of proton decay, even taking into account the relevant theoretical uncertainty factors, like the evaluation of the hadronic matrix element, one can safely conclude that the minimal grand unified extension of the SM is ruled out because of the exceedingly high matter instability. Analogously, the high precision achieved on electroweak observables (in particular thanks to LEP physics) allows us to further exclude the minimal SU(5) model: indeed, the low-energy quantity one can *predict* solving the RGE's for the gauge coupling evolution (be it the electroweak angle θ_W, or the strong coupling α_s) exhibits a large discrepancy with respect to its measured value. The precise $SU(5)$ prediction for $\sin^2 \theta_W$ is [40]:

$$\sin^2 \theta_W(M_W) = 0.214^{+0.004}_{-0.003}, \tag{2.35}$$

while the experimental value obtained from LEP data is:

$$\sin^2 \theta_W(M_W) = 0.23108 \pm 0.00005, \tag{2.36}$$

and both values only agree at 5 standard deviations.

The fate of the minimal $SU(5)$ should not induce the reader to conclude "tout-court" that non-supersymmetric grand unification is killed by proton decay and $\sin^2 \theta_W$. Once one abandons the minimality criterion, for instance enlarging the Higgs spectrum or changing the grand unified gauge group, it is possible to rescue some GUT models. The price to pay for it is that we lose the simplicity and predictivity of minimal SU(5) ending up into more and more complicate grand unified realisations.

2.2. Supersymmetric grand unification

2.2.1. The hierarchy problem and supersymmetry

The Standard Model as a $SU(3) \otimes SU(2) \otimes U(1)$ gauge theory with three generations of quarks and leptons and a Higgs doublet provides an accurate description of all known experimental results. However, as we have discussed, the SM cannot be the final theory, and instead we consider the SM as a low energy effective theory of some more fundamental theory at higher energies. Typically we have a Grand Unification (GUT) Scale around 10^{16} GeV where the strong and electroweak interactions unify in a simple group like $SU(5)$ or $SO(10)$ [24, 25] and the Plank scale of 10^{19} GeV where these gauge interactions unify with gravity. The presence of such different scales in our theory gives rise to the so–called *hierarchy problem* (see a nice discussion in [41]). This problem refers to the difficulty to stabilise the large gap between the electroweak scale and the GUT or Plank scales under radiative corrections. Such difficulty arises from a general property of the scalar fields in a gauge theory, namely their tendency of scalar to get their masses in the neighbourhood of the largest available energy scale in the theory. In the previous section, when dealing with the scalar potential of the minimal $SU(5)$ model, we have directly witnessed the existence of such problem. From such a particular example, let us move to more general considerations about what distinguishes the behaviour of scalar fields from that of fermion and vector fields in gauge theories.

To understand this problem let us compare the one loop corrections to the electron mass and the Higgs mass. These one loop corrections are given by the diagrams in Fig. 3. The self-energy contribution to the electron mass can be calculated from this diagram to be,

$$\delta m_e = 2 \frac{\alpha_{em}}{\pi} m_e \log \frac{\Lambda}{m_e}, \tag{2.37}$$

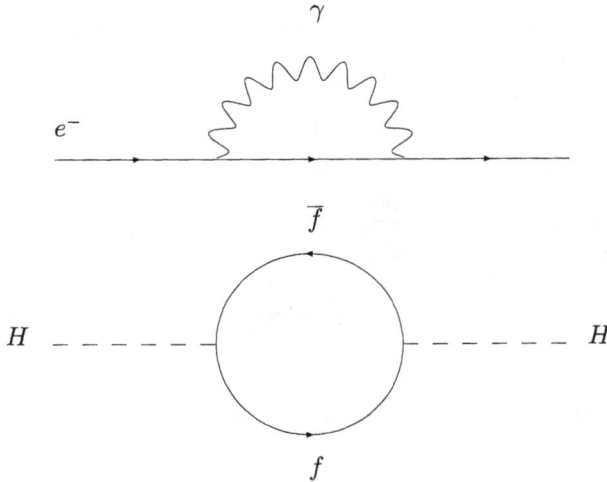

Fig. 3. One loop correction to fermion and scalar masses.

and it is logarithmically divergent. Here we have regulated the integral with an ultraviolet cutoff Λ. However, it is important to notice that this correction is proportional to the electron mass itself. This can be understood in terms of symmetry. In the limit where $m_e \to 0$, our theory acquires a new chiral symmetry where right-handed and left-handed electrons are decoupled. Were such a symmetry exact, the one loop corrections to the mass would have to vanish. This chiral symmetry is only broken by the electron mass itself and therefore any loop correction breaking this symmetry must be proportional to m_e, the only source of chiral symmetry breaking in the theory. This has important implications. If we replace the cutoff Λ by the largest possible scale, the Planck mass we get,

$$\delta m_e = 2\frac{\alpha_{em}}{\pi} m_e \log \frac{M_{Plank}}{m_e} \simeq 0.24\, m_e, \tag{2.38}$$

which is only a small correction to the electron mass.

Analogously, for the gauge vector bosons there is the gauge symmetry itself which constitutes the "natural barrier" preventing their masses to become arbitrarily large. Indeed, if a vector boson V is associated to the generator of a certain symmetry G, as long as G is unbroken the vector V has to remain massless. Its mass will be of the order of the scale at which the symmetry G is (spontaneously) broken. Hence, once again, we have a symmetry protecting the mass of vector bosons.

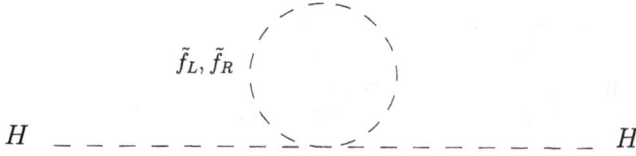

Fig. 4. Additional Supersymmetric contribution to the scalar mass.

On the other hand, the situation is very different in the case of the Higgs boson,

$$\delta m_H^2(f) = -2N_f \frac{|\lambda_f|^2}{16\pi^2}\left[\Lambda^2 - 2m_f^2 \ln \frac{\Lambda}{m_f} + \dots\right]. \tag{2.39}$$

But, in this case, the one loop contribution is quadratically divergent !!. This is due to the fact that no symmetry protects the scalar mass and in the limit $m_H^2 \to 0$ the symmetry of our model is not increased. The combination $HH^\dagger$ is always neutral under any symmetry independently of the charges of the field H. So, the scalar mass should naturally be of the order of the largest scale of the theory, as either at tree level or at loop level this scale feeds into the scalar mass.

So, if now we repeat the exercise we made with the electron mass and replace the cutoff by the Plank mass, we obtain $\delta m_H^2 \simeq 10^{30}$ GeV^2. In fact we could cancel these large correction with a bare mass of the same order and opposite sign. However, these two contributions should cancel with a precision of one part in 10^{26} and even then we should worry about the two loop contribution and so on. This is the so-called hierarchy problem and Supersymmetry constitutes so far the most interesting answer to it (later on, we'll briefly comment on the existence of other approaches tackling the hierarchy problem, although, in our view, not as effectively as low-energy supersymmetry does).

As we have seen in the previous section, Supersymmetry associates a fermion with every scalar in the theory with, in principle, identical masses and gauge quantum numbers. Therefore, in a Supersymmetric theory we would have a new contribution to the Higgs mass at one loop. Now this graph gives a contribution to the Higgs mass as,

$$\delta m_H^2(\tilde{f}) = -2N_{\tilde{f}} \frac{\lambda_{\tilde{f}}}{16\pi^2}\left[\Lambda^2 - 2m_{\tilde{f}}^2 \ln \frac{\Lambda}{m_{\tilde{f}}} + \dots\right]. \tag{2.40}$$

If we compare Eqs. (2.39) and (2.40) we see that with $N_f = N_{\tilde{f}}$, $|\lambda_f|^2 = -\lambda_{\tilde{f}}$ and $m_f = m_{\tilde{f}}$ we obtain a total correction $\delta m_H^2(f) + \delta m_H^2(\tilde{f}) = 0$!!. This means we need a symmetry that associates a bosonic partner to every fermion with equal mass and related couplings and this symmetry is **Supersymmetry**.

Still, we have not found scalars exactly degenerate with the SM fermions in our experiments. In fact, it would have been very easy to find a scalar partner of the electron if it existed. Thus, Supersymmetry can not be an exact symmetry of nature, it must be broken. Fortunately, we can break Supersymmetry while at the same time preserving to an acceptable extent the Supersymmetric solution of the hierarchy problem. To do that, we want to ensure the cancellation of quadratic divergences and comparing Eq. (2.39) and Eq. (2.40) we can see that we must still require equal number of scalar and fermionic degrees of freedom, $N_f = N_{\tilde{f}}$, and supersymmetric dimensionless couplings $|\lambda_f|^2 = -\lambda_{\tilde{f}}$. Supersymmetry can be broken only in couplings with positive mass dimension, as for instance the masses. This is called soft breaking [42]. Now if we take $m_{\tilde{f}}^2 = m_f^2 + \delta^2$ we obtain a correction to the Higgs mass,

$$\delta m_H^2(f) + \delta m_H^2(\tilde{f}) \simeq 2N_f \frac{|\lambda_f|^2}{16\pi^2} \delta^2 \ln \frac{\Lambda}{m_{\tilde{f}}} + \dots, \qquad (2.41)$$

and this is only logarithmically divergent and proportional to the mass difference between the fermion and its scalar partner. Still we must require this correction to be smaller than the Higgs mass itself (around the electroweak scale) implies that this mass difference, δ, can not be too large, in fact $\delta \lesssim 1$ TeV. If Supersymmetry is the solution to the hierarchy problem it must be softly broken and the SUSY partners must be roughly below 1 TeV. The rich SUSY phenomenology is thoroughly discussed in Marcela Carena and Carlos Wagner's lectures at this School [43].

What is relevant for our discussion on grand unification is the effect of the presence of new SUSY particles at 1 TeV in the evolution of the gauge couplings. We saw in the previous section that the RGE equations in the SM predict that the gauge couplings get very close at a large scale $\simeq 2 \times 10^{15}$ GeV. Nevertheless this unification was not perfect and, using the precise determination of the gauge couplings at LEP we see that the SM couplings do not unify at seven standard deviations. If we have new SUSY particles around 1 TeV, these RGE equations are modified. Using Eq. (2.4), it is straightforward to obtain the new β_i parameters in the MSSM. We have to take into account that for every gauge boson we have to add a fermion, called gaugino, both in the adjoint representation. Therefore from gauge bosons and gauginos we have

$$\beta_i(V) = -\frac{1}{4\pi}\left[\frac{11}{3}C_2(G_i) - \frac{2}{3}C_2(G_i)\right] = -\frac{1}{4\pi}3\,C_2(G_i). \qquad (2.42)$$

While for every fermion we have a corresponding scalar partner in the same representation. Thus we have

$$\beta_i(F) = -\frac{1}{4\pi} \sum_F \left[-\frac{2}{3} T(R_F) - \frac{1}{3} T(R_F) \right] = \frac{1}{4\pi} \sum_F T(R_F), \qquad (2.43)$$

summed over all the chiral supermultiplet (fermion plus scalar) representations. Therefore the total β_i coefficient in a supersymmetric model is

$$\beta_i = -\frac{1}{4\pi} \left[3\, C_2(G_i) - \sum_F T(R_F) \right]. \qquad (2.44)$$

And for the MSSM

$$\begin{aligned}
\beta_3 &= -\frac{1}{4\pi} \left[9 - 2\, n_g \right] = -\frac{3}{4\pi}, \\
\beta_2 &= -\frac{1}{4\pi} \left[6 - 2\, n_g - \frac{1}{2}\, n_H \right] = +\frac{1}{4\pi}, \\
\beta_1 &= -\frac{1}{4\pi} \left[-\frac{10}{3}\, n_g - \frac{1}{2}\, n_H \right] = +\frac{11}{4\pi}.
\end{aligned} \qquad (2.45)$$

From the comparison of Eq. (2.5) and Eq. (2.45) we see that the evolution of the gauge couplings is significantly modified. We can easily calculate the grand unification scale and the "predicted" value of $\sin^2 \theta_W$ as done in Eqs. (2.7) and (2.8) and we obtain

$$M_{\text{GUT}}^{\text{MSSM}} = 1.5 \times 10^{16} \text{GeV}, \qquad \sin^2 \theta_W(M_Z) = 0.234, \qquad (2.46)$$

which is remarkably close to the experimental vale $\sin^2 \theta_W^{\text{exp}}(M_Z) = 0.23149 \pm 0.00017$. And we obtain easily the grand unified coupling constant

$$\frac{5}{3}\alpha_1(M_{\text{GUT}}) = \alpha_2(M_{\text{GUT}}) = \alpha_3(M_{\text{GUT}}) \approx \frac{1}{24}. \qquad (2.47)$$

In fact, the actual analysis, including two loop RGEs and threshold effects predicts $\alpha_3(M_Z) = 0.129$ which is slightly higher than the observed value (such discrepancy could be justified by the presence of threshold effects when approaching the GUT scale in the running). The couplings meet at the value $M_X = 2 \times 10^{16}$ GeV [44–46]. The "exact" unification of the gauge couplings within the MSSM may or may not be an accident. But it provides enough reasons to consider supersymmetric standard models seriously as it links supersymmetry and grand unification in an inseparable manner [47].

2.2.2. SUSY GUT predictions and problems

(i) Doublet-Triplet Splitting

As we saw in the non-supersymmetric case, a very accurate fine-tuning in the parameters of the scalar potential was required to reproduce the hierarchy between the electroweak and the GUT scale. In a supersymmetric grand unified theory the problem is very similar. The relevant terms in the superpotential are,

$$W = \alpha \bar{\mathbf{5}}_\mathbf{H} \Phi^2 \mathbf{5}_\mathbf{H} + \mu \bar{\mathbf{5}}_\mathbf{H} \mathbf{5}_\mathbf{H}. \tag{2.48}$$

The breaking of $SU(5)$ in the $SU(3) \times SU(2) \times U(1)$ direction via

$$\langle \Phi \rangle = \frac{2\,m'}{3\,\alpha} \text{ Diag. } \left(1, 1, 1, -\frac{3}{2}, -\frac{3}{2} \right), \tag{2.49}$$

leads to

$$W = \bar{\mathbf{3}}_\mathbf{H} \mathbf{3}_\mathbf{H} \left(\mu + \frac{2}{3}m' \right) + \bar{\mathbf{2}}_\mathbf{H} \mathbf{2}_\mathbf{H} \left(\mu - m' \right). \tag{2.50}$$

Choosing $\mu = m'$ (both them $\sim O(M_{\text{GUT}})$) renders the Higgs doublets massless. However, although due to supersymmetry this equality is stable under radiative corrections, this extremely accurate adjustment is extremely unnatural.

There are several mechanisms in supersymmetric theories to render doublet-triplet splitting natural. Here we will briefly discus the "missing partner mechanism" [48]. From Eq. (2.50) we see that if the direct mass term for the Higgses, μ, was absent the doublets would obtain super-heavy masses from the vacuum expectation value of the adjoint Higgs Φ. The strategy we will use to solve the doublet-triplet splitting problem is to introduce representations that contain Higgs triplets but no doublets. We can choose the **50** that is decomposed under $SU(3) \times SU(2)$ as:

$$\mathbf{50} = (\mathbf{8}, \mathbf{2}) + (\mathbf{6}, \mathbf{3}) + (\bar{\mathbf{6}}, \mathbf{1}) + (\mathbf{3}, \mathbf{2}) + (\bar{\mathbf{3}}, \mathbf{1}) + (\mathbf{1}, \mathbf{1}). \tag{2.51}$$

We need both the **50** and $\bar{\mathbf{50}}$ to get an anomaly-free model. In order to write mixing terms between **5**, $\bar{\mathbf{5}}$ and **50**, $\bar{\mathbf{50}}$ we need a field Σ in the **75** instead of the **24** to break $SU(5)$. The relevant part of the superpotential is then

$$W = \frac{M}{2} \text{Tr}(\Sigma^2) + \frac{a}{3} \text{ Tr}(\Sigma^3) + b \, \mathbf{50} \, \Sigma \, \mathbf{5}_\mathbf{H} + c \, \bar{\mathbf{50}} \, \Sigma \, \bar{\mathbf{5}}_\mathbf{H} + \tilde{M} \, \bar{\mathbf{50}} \, \mathbf{50}, \tag{2.52}$$

where no mass term $\bar{\mathbf{5}}_\mathbf{H} \, \mathbf{5}_\mathbf{H}$ is present. Σ gets a VEV, $\langle \Sigma \rangle \sim \frac{M}{a}$, breaking $SU(5)$ to $SU(3) \times SU(2) \times U(1)$. The resulting $SU(3) \times SU(2) \times U(1)$ superpotential is,

$$W = \mathbf{50}_3 \, \frac{M\,b}{a} \, H_3 + \bar{\mathbf{50}}_3 \, \frac{M\,c}{a} \, \bar{H}_3 + \tilde{M} \, \bar{\mathbf{50}}_3 \, \mathbf{50}_3, \tag{2.53}$$

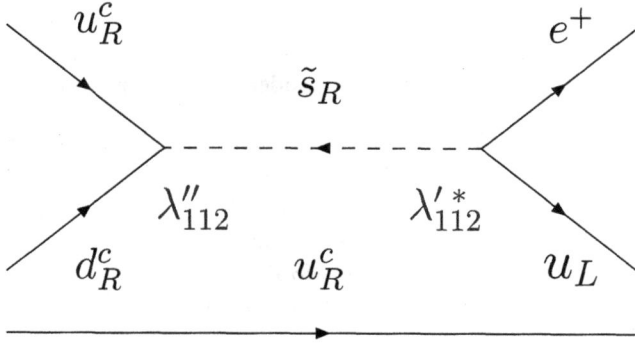

Fig. 5. Proton decay through R-parity violating couplings.

with H_3 and $\mathbf{50_3}$ the Higgs triplets in the $\mathbf{5_H}$ and $\mathbf{50}$ representations respectively. Therefore the Higgs triplets get a mass of the order of $M \sim \tilde{M} \sim M_{\text{GUT}}$ and the Higgs doublets remain massless because there is no mass term for the doublets. In this way we solve the doublet-triplet splitting problem without unnatural fine-tuning of the parameters.

(ii) Proton Decay
In the non-supersymmetric $SU(5)$ proton decay arises four fermion operators, hence from operators of canonical dimension 6. In addition to such dim=6 operators, in the supersymmetric case we encounter also dim=5 and even dim=4 operators leading to proton decay.

Dimension 4 operators are not suppressed by any power of the GUT scale. In fact, these terms are gauge invariant and in principle are allowed to appear in the superpotential,

$$W_{\Delta L=1} = \lambda^{ijk} L_i L_j e^c_{Rk} + \lambda'^{ijk} L_i Q_j d^c_{Rk} + \epsilon^i L_i H_2,$$
$$W_{\Delta B=1} = \lambda''^{ijk} u^c_{Ri} d^c_{Rj} d^c_{Rk}. \tag{2.54}$$

However, these terms violate baryon or lepton number by 1 unit. So, these terms are very dangerous. Indeed, if λ' and λ'' are simultaneously present, a very fast proton decay arises through the diagram in Figure 5. Clearly, the major difference is that in the non-SUSY case the mediation of proton decay occurs through the exchange of super-heavy (vector or scalar) bosons whose masses are at the GU scale. On the contrary, in Figure 5 the mediator is a SUSY particle and, hence, at least if we insist in invoking low-energy SUSY to tackle the hierarchy problem, its mass is at the electroweak scale instead of being at M_{GUT}! From the bounds to the decay $p^+ \to e^+ \pi^0$ we obtain $\lambda'^*_{112} \cdot \lambda''_{112} \leq 2 \times 10^{-27}$. Clearly this product is too small and it is more natural to consider it as exactly zero. Other couplings from Eq. (2.54) are not so stringently bounded but in general all of them must be very small from phenomenological considerations (in particular, from FCNC constraints).

One possibility is to introduce a new discrete symmetry, called R-parity to forbid these terms. R-parity is defined as $R_P = (-1)^{3B+L+2S}$ such that the SM particles and Higgs bosons have $R_P = +1$ and all superpartners have $R_P = -1$. In the MSSM R_P is conserved and this has some interesting consequences.

• $W_{\Delta L=1}$ and $W_{\Delta B=1}$ are absent in the MSSM.

• The Lightest Supersymmetric Particle (LSP) is completely stable and it provides a (cold) dark matter candidate.

• Any sparticle produced in laboratory experiments decays into a final state with an odd number of LSP.

• In colliders, Supersymmetric particles can only be produced (or destroyed) in pairs.

A second contribution to proton decay, already present in non-SUSY GUTs, comes from dimension 6 operators. The discussion is analogous to the analysis in non-SUSY GUTS. Here we will only recall that a generic four-fermion operator of the form $1/\Lambda^2\, q\, q\, q\, l$ results in a proton decay rate of the order $\Gamma_p \sim 10^{-3} m_p^5/\Lambda^4$. Given the bound on the proton lifetime $\tau_p > 5 \times 10^{33}$ yrs, this constrains the scale Λ to be $\Lambda > 4 \times 10^{15}$ GeV. Therefore we can see that with $\Lambda \simeq M_{\text{GUT}} \simeq 2 \times 10^{16}$ GeV, dimension 6 operators are still in agreement with the experimental bound.

Dimension 5 operators are new in supersymmetric grand unified theories. They are generated by the exchange of the coloured Higgs multiplet and are of the form

$$W_5 = \frac{c_L^{ijkl}}{M_T}(Q_k Q_l Q_i L_j) + \frac{c_R^{ijkl}}{M_T}(u_i^c u_k^c d_j^c e_l^c), \qquad (2.55)$$

commonly called LLLL and RRRR operators respectively, with M_T the mass of the coloured Higgs triplet. The coefficients c_L^2 and c_R^2 are model dependent factors depending on the Yukawa couplings. For instance in Reference [49, 50] they are

$$\begin{aligned} c_L^{ijkl} &= (Y_D)_{ij} \left(V^T P Y_U V \right)_{kl}, \\ c_R^{ijkl} &= \left(P^* V^* Y_D \right)_{ij} (Y_U V)_{kl}, \end{aligned} \qquad (2.56)$$

where Y_D and Y_U are diagonal Yukawa matrices, V is the CKM mixing matrix and P is a diagonal phase matrix. The RRRR dimension 5 operator contributes

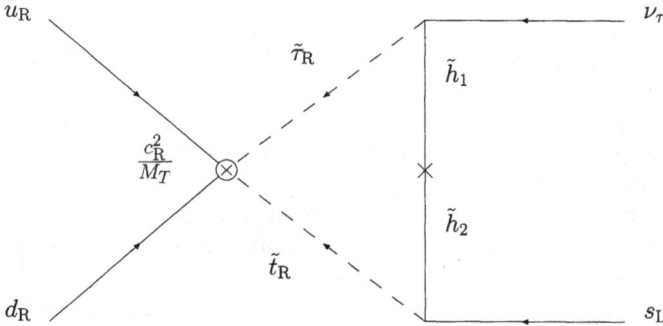

Fig. 6. Proton decay, $p \to K^+ \bar{\nu}_\tau$, through dimension 5 RRRR operator.

to the decay $p \to K^+ \bar{\nu}_\tau$ through the diagram of Figure 6. The corresponding amplitude is roughly given by

$$A_\tau(t_R) \propto g^2 Y_d Y_t^2 Y_\tau V_{tb}^* V_{ud} V_{ts} \frac{\mu}{M_T m_{\tilde{f}}^2},\tag{2.57}$$

with μ the Higgs mass parameter in the superpotential and $m_{\tilde{f}}^2$ a typical squark or slepton mass. Notice that this amplitude is proportional to $\tan^2 \beta$.

In fact these contributions from dimension 5 operators are extremely dangerous. From the bound on the proton lifetime we have that, for $\tan \beta = 2.5$ and $m_{\tilde{f}} \lesssim 1$ TeV

$$M_T \geq 6.5 \times 10^{16} \text{GeV},\tag{2.58}$$

and this bound becomes more severe for larger values of $\tan \beta$ given that the RRRR amplitude scales as $\tan^2 \beta / M_T$. On the other hand, in minimal $SU(5)$, there is an upper bound on the Higgs triplet mass if we require correct gauge coupling unification, $M_T \leq 2.5 \times 10^{16}$ GeV at 90% C.L.. This implies that the minimal SUSY SU(5) model would be excluded by proton decay if the sfermion masses are smaller than 1 TeV. Obviously, much in the same way that non-SUSY GUTs can be complicated enough to avoid the too fast proton decay present in minimal $SU(5)$, also in the SUSY case it is possible to avoid the mentioned problem in the minimal $SU(5)$ realisation by going to non-minimal SU(5) realisations or changing the gauge group altogether. How "realistic" such non-minimal SUSY-GUTs are is what we shortly discuss in the next subsection.

2.2.3. "Realistic" supersymmetric $SU(5)$ models

Gauge coupling unification in supersymmetric grand unified theories is a big quantitative success. However, minimal $SU(5)$ models, face a series of other problems like proton decay or doublet-triplet splitting. A sufficiently "realistic" model should be able to address and solve all these problems simultaneously [51]. The problems we would like this model to solve are: i) gauge coupling unification with an acceptable value of $\alpha_s(M_Z)$ given α and $\sin^2\theta_W$ at M_Z, ii) compatibility with the very stringent bounds on proton decay and iii) natural doublet-triplet splitting.

To solve the doublet-triplet problem we use the missing partner mechanism presented above. The superpotential of this model will be that of Eq. (2.52) with the addition of the Yukawa couplings. Now the $SU(5)$ symmetry is broken to the SM by a VEV of the representation **75**. This provides a mass for the Higgs triplets while the doublets remain massless. Later a μ-term for the Higgs doublets of the order of the electroweak scale is generated through the Giudice-Masiero mechanism [52].

Regarding gauge coupling unification, it is well known that in minimal supersymmetric $SU(5)$ the central value of $\alpha_3(M_Z)$ required by gauge coupling unification is too large: $\alpha_3(M_Z) \simeq 0.13$ to be compared with the experimental value $\alpha_3^{\exp}(M_Z) \simeq 0.1187 \pm 0.002$. Using two loop RGE equations and taking into account the threshold effects we can write the corrected value of $\alpha_3(M_Z)$ as

$$
\alpha_3(M_Z) = \frac{\alpha_3^{(0)}(M_Z)}{1 + \alpha_3^{(0)}(M_Z)\,\delta},
$$

$$
\delta = k + \frac{1}{2\pi}\log\frac{M_{\text{SUSY}}}{M_Z} - \frac{3}{5\pi}\log\frac{M_T}{M_{\text{GUT}}}, \tag{2.59}
$$

with $\alpha_3^{(0)}$ the leading log value of this coupling equal to the minimal $SU(5)$ value and k contains the contribution from two loop running, SUSY and GUT thresholds. M_T is an effective mass defined as

$$
m_T = \frac{M_{T_1} M_{T_2}}{\tilde{M}}, \tag{2.60}
$$

with M_{T_1} and M_{T_1} the two eigenvalues of the Higgs triplet mass matrix and $\tilde{M}$ the mass of the **50** in the superpotential, Eq. (2.53). The value of the parameter k is different in the minimal $SU(5)$ model and in the realistic model with a **75** breaking the $SU(5)$ symmetry:

$$
k^{\text{minimal}} = -1.243, \quad k^{\text{realistic}} = 0.614. \tag{2.61}
$$

This difference is very important and improves substantially the comparison of the prediction with the experimental value of $\alpha_3(M_Z)$. In fact, for k large and negative we need to take M_{SUSY} as large as possible and M_T as small as possible, but this runs into problems with proton decay. On the other hand if k is positive and large, we can take $M_T > M_{GUT}$. For instance, with $M_T = 6 \times 10^{17}$GeV $\simeq 30 M_{GUT}$ and $M_{SUSY} = 0.25$ TeV we obtain $\alpha_3(M_Z) \simeq 0.116$ which is acceptable.

Regarding proton decay the main contribution comes again from dimension five operators when the Higgs triplets are integrated out. Clearly these operators depend on M_T, but we have seen above that a large M_T is preferred in this model. Typical values would be

$$M_{GUT} = 2.9 \times 10^{16}\text{GeV}, \quad \tilde{M} = 2.0 \times 10^{16}\text{GeV},$$
$$M_{T_1} = 1.2 \times 10^{17}\text{GeV}, \quad M_{T_2} = 1.0 \times 10^{17}\text{GeV},$$
$$M_T = 6 \times 10^{17}\text{GeV}. \tag{2.62}$$

Notice that in this case the couplings of the triplets to the fermions is **not** related to the fermion masses as the Higgs triplets are now a mixing between the triplets in the $\mathbf{5_H}$ and the triplets in the $\mathbf{50}$. Therefore we have some unknown Yukawa coupling $Y_{\bar{50}}$. Assuming a hierarchical structure in these couplings somewhat analogous to the doublet Yukawa couplings [51] we would obtain a proton decay rate in the range 8×10^{31}–3×10^{34} yrs for the channel $p \to K^+\bar{\nu}$ and 2×10^{32}–8×10^{34} yrs for the channel $p \to \pi^+\bar{\nu}$. The present bound at 90% C.L. on $\tau/\text{BR}(p \to K^+\bar{\nu})$ is 1.9×10^{33} yrs. Thus we see that agreement with the stringent proton decay bounds is possible in this model.

2.3. Other GUT Models

So, far we have discussed $SU(5)$, the prototype Grand Unified theory in both supersymmetric and non-supersymmetric versions. Given that supersymmetric Grand Unification ensures gauge coupling unification, most models of Grand Unification which have been studied in recent years have been supersymmetric. Other than the SUSY $SU(5)$, historically, one of the first unified models constructed was the Pati-Salam model [24]. The gauge group was given by, $SU(4)_c \times SU(2)_L \times SU(2)_R$. The fermion representations, as explained above, require the presence of a right-handed neutrino. Realistic models can be built incorporating bi-doublets of Higgs giving rise to fermion masses and suitable representations for the breaking of the gauge group. However the Pati-Salam Model is not truly a unified model in a strict sense. For this reason, one needs to go for a larger group of which the Pati-Salam gauge group would be a subgroup. The simplest gauge group in this category is an orthogonal group $SO(10)$ of rank 5.

2.3.1. The seesaw mechanism

There are several reasons to consider models beyond the simple $SU(5)$ gauge group we have considered here. One of the major reasons is the question of neutrino masses. This can be elegantly be solved through a mechanism which goes by the name *seesaw* mechanism [53–57]. The seesaw mechanism requires an additional standard model singlet fermion, which could be the right handed neutrino. Given that it is electrically neutral, this particle can have a Majorana mass (violating lepton number by two units) in addition to the standard *Dirac* mass that couples it to the SM left handed neutrino. Representing the three left-handed fields by a column vector ν_L and the three right handed fields by ν_R, the Dirac mass terms are given by

$$-\mathcal{L}^D = \bar{\nu}_L \mathcal{M}_D \nu_R + \text{h.c.}, \tag{2.63}$$

where $\mathcal{M}^D$ represents the Dirac mass matrix. The Majorana masses for the right handed neutrinos are given by

$$-\mathcal{L}^R = \frac{1}{2} \bar{\nu}_R^c \mathcal{M}_R \nu_R + \text{h.c.}. \tag{2.64}$$

The total mass matrix is given as

$$-\mathcal{L}^{total} = \frac{1}{2} \bar{\nu}_p \mathcal{M} \nu_p, \tag{2.65}$$

where the column vector ν_p is

$$\nu_p = \begin{pmatrix} \nu_L \\ \nu_R^c \end{pmatrix}. \tag{2.66}$$

And the matrix $\mathcal{M}$ is

$$\mathcal{M} = \begin{pmatrix} 0 & \mathcal{M}_D^T \\ \mathcal{M}_D & \mathcal{M}_R \end{pmatrix}. \tag{2.67}$$

Diagonalising the above matrix, one sees that the left handed neutrinos attain Majorana masses of order,

$$\mathcal{M}^\nu = -\mathcal{M}_D^T \mathcal{M}_R^{-1} \mathcal{M}_D. \tag{2.68}$$

This is called the seesaw mechanism. Choosing for example the Dirac mass of the neutrinos to be typically of the order of charged lepton masses or down quark masses, we see that for a heavy right handed neutrino mass scale, (left-handed) neutrinos masses are suppressed. In this way, the smallness of neutrino masses can be explained naturally by the seesaw mechanism. While the seesaw mechanism is elegant, as mentioned in the Introduction, by construction we do not have right handed neutrinos in the SM particle spectrum. The $SU(5)$ representations do not contain a right handed singlet particle either, as we have seen above. However, in larger GUT groups like $SO(10)$ these additional particles are naturally present.

2.3.2. SO(10)

The group theory of $SO(10)$ and its spinorial representations can be simplified by using the $SU(N)$ basis for the $SO(2N)$ generators or the tensorial approach. The spinorial representation of the $SO(10)$ is given by a 16-dimensional spinor, which could accommodate all the SM model particles as well as the right handed neutrino. Let's now see how fermions attain their masses in this model. The product of two **16** matter representations can only couple to **10**, **120** or **126** representations, which can be formed by either a single Higgs field or a non-renormalisable product of representations of several Higgs fields. In either case, the Yukawa matrices resulting from the couplings to **10** and **126** are complex-symmetric, whereas they are antisymmetric when the couplings are to the **120**. Thus, the most general $SO(10)$ superpotential relevant to fermion masses can be written as

$$W_{SO(10)} = h_{ij}^{10} \mathbf{16_i} \ \mathbf{16_j} \ \mathbf{10} + h_{ij}^{126} \mathbf{16_i} \ \mathbf{16_j} \ \mathbf{126} + h_{ij}^{120} \mathbf{16_i} \ \mathbf{16_j} \ \mathbf{120}, \tag{2.69}$$

where i, j refer to the generation indices. In terms of the SM fields, the Yukawa couplings relevant for fermion masses are given by [30, 58][1]:

$$\mathbf{16\ 16\ 10} \supset \mathbf{5}\ (uu^c + vv^c) + \bar{\mathbf{5}}\ (dd^c + ee^c), \tag{2.70}$$

$$\mathbf{16\ 16\ 126} \supset \mathbf{1}\ v^c v^c + \mathbf{15}\ vv + \mathbf{5}\ (uu^c - 3\ vv^c) + \overline{\mathbf{45}}\ (dd^c - 3\ ee^c),$$

$$\mathbf{16\ 16\ 120} \supset \mathbf{5}\ vv^c + \mathbf{45}\ uu^c + \bar{\mathbf{5}}\ (dd^c + ee^c) + \overline{\mathbf{45}}\ (dd^c - 3\ ee^c),$$

where we have specified the corresponding $SU(5)$ Higgs representations for each of the couplings and all the fermions are left handed fields. The resulting mass matrices can be written as

$$M^u = M_{10}^5 + M_{126}^5 + M_{120}^{45}, \tag{2.71}$$

$$M_{LR}^v = M_{10}^5 - 3\ M_{126}^5 + M_{120}^5, \tag{2.72}$$

$$M^d = M_{10}^{\bar{5}} + M_{126}^{\overline{45}} + M_{120}^{\bar{5}} + M_{120}^{\overline{45}}, \tag{2.73}$$

$$M^e = M_{10}^{\bar{5}} - 3M_{126}^{\overline{45}} + M_{120}^{\bar{5}} - 3M_{120}^{\overline{45}}, \tag{2.74}$$

$$M_{LL}^v = M_{126}^{15}, \tag{2.75}$$

$$M_R^v = M_{126}^1. \tag{2.76}$$

---

[1]Recently, SO(10) couplings have also been evaluated for various renormalisable and non-renormalisable couplings in [59].

We can see here the relations between the different fermionic species. In particular, notice the relation between up-quarks and neutrino (Dirac) mass matrices, Eqs. (2.71) and (2.72).

The breaking of $SO(10)$ to the Standard Model group on the other hand can be quite complex compared to that of the $SU(5)$ model we have studied so far. In particular, the gauge group offers the possibility of the existence of an intermediate scale where another "gauge symmetry", a subgroup of $SO(10)$, can exist. Some of the popular ones are summarised in the figure below: Each of these

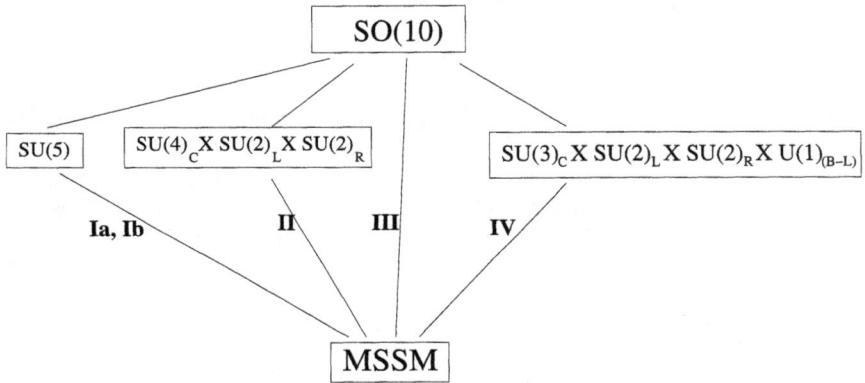

Fig. 7. The various breaking chains of $SO(10)$ are summarised in this figure.

breaking chains would have its own RG scaling which can in principle lead to different results at the weak scale even though the initial conditions at the $SO(10)$ scale are the same. Different Higgs representations are used for the breaking in each of these cases. In the recent years, attempts have been made to construct complete (renormalisable) models of $SO(10)$ where it could be possible to have precision studies of $SO(10)$ models. This studies do give a good handle on the predictions on proton life time, gauge coupling unification, and, to some extent, fermion masses. However, for processes involving the supersymmetric spectra like flavour changing neutral currents, etc, the situation is more model dependent.

Apart from $SU(5)$ and $SO(10)$, there are other GUT models in the literature based on gauge groups E_6, $SU(6)$ etc, which we have not touched in this set of lectures.

3. Flavour and CP violation in SUSY

The simplest Supersymmetric version of the Standard Model that we can build is the so-called Minimal Supersymmetric Standard Model (MSSM). Clearly, this model must include all the SM interactions and particle spectrum together with their Supersymmetric partners. This means that to every quark and lepton in the SM we add a scalar Supersymmetric partner, called "squark" or "slepton" respectively, with identical gauge quantum numbers and, in principle, identical mass, forming a "chiral supermultiplet". In the same way, to the SM Higgs or more exactly to the Higgses in a 2 Higgs doublet version of the SM[2] we add fermionic partners called "higgsinos" with the same quantum numbers and masses in another "chiral Supermultiplet". Then every gauge boson is also joined by a gaugino ("gluino", "wino", "bino"...) with spin 1/2 in the adjoint representation in a "vector supermultiplet" (for a complete formulation of Supersymmetric theories in superfield notation see Ref. [60]).

The gauge interactions in our MSSM are completely fixed by the gauge quantum numbers of the different particles in the usual way. However, we still need the Yukawa interactions of the Standard Model that give masses to the fermions once we break the electroweak symmetry. These interactions are included in the MSSM Superpotential, which is a gauge invariant analytic function of the MSSM superfields (i.e. a function of fields ϕ_i but not of complex conjugate fields ϕ_i^*) with dimensions of mass cube. If we include all possible terms invariant under the gauge symmetry then it turns out that some of these terms violate either baryon or lepton number. As we have seen in the previous section, this endangers proton stability; hence one usually imposes a discrete symmetry called R-parity under which the ordinary particles are even while their SUSY partners are odd [61][3]. The MSSM Superpotential (using standard notation) is then,

$$W = Y_d^{ij} Q_i H_1 d_{Rj}^c + Y_e^{ij} L_i H_1 e_{Rj}^c + Y_u^{ij} Q_i H_2 u_{Rj}^c + \mu H_1 H_2, \tag{3.1}$$

and this gives rise to the interactions,

$$\mathcal{L}_W = \left| \frac{\partial W}{\partial \phi_i} \right|^2 + \psi_i \psi_j \frac{\partial^2 W}{\partial \phi_i \partial \phi_j}, \tag{3.2}$$

with ϕ_i any scalar in the MSSM and ψ_i its corresponding fermionic partner.

Still, we know that Supersymmetry is not an exact symmetry in nature and it must be broken. If Supersymmetry is the solution to the hierarchy problem, the breaking of Supersymmetry must be soft, i.e. should not reintroduce the quadratic divergences which are forbidden in the SUSY invariant case, and the scale of SUSY breaking must be close to the electroweak scale. The most general set of possible Soft SUSY breaking terms (SBT) [42] under these conditions are,

---

[2] As it is well-known, Supersymmetry requires two different Higgs doublets to give mass to fermions of weak isospin $+1/2$ and $-1/2$ [13–17].

[3] Since these terms violate either lepton or baryon number, it is also possible to forbid only lepton number or baryon number violation to ensure proton stability [62].

1. Gaugino masses

$$\mathcal{L}_{\text{soft}}^{(1)} = \tfrac{1}{2}\left(M_1\,\tilde{B}\tilde{B} + M_2\,\tilde{W}\tilde{W} + M_3\,\tilde{g}\tilde{g}\right) + \text{h.c.},$$

2. Scalar masses

$$\mathcal{L}_{\text{soft}}^{(2)} = (M_{\tilde{Q}}^2)_{ij}\,\tilde{Q}_i\,\tilde{Q}_j^* + (M_{\tilde{u}}^2)_{ij}\,\tilde{u}_{Ri}^c\,\tilde{u}_{Rj}^{c*} + (M_{\tilde{d}}^2)_{ij}\,\tilde{d}_{Ri}^c\,\tilde{d}_{Rj}^{c*} + (M_{\tilde{L}}^2)_{ij}\,\tilde{L}_i\,\tilde{L}_j^* +$$
$$(M_{\tilde{e}}^2)_{ij}\,\tilde{e}_{Ri}^c\,\tilde{e}_{Rj}^{c*} + (m_{H_1}^2)H_1 H_1^* + (m_{H_2}^2)H_2 H_2^*,$$

3. Trilinear couplings and B–term

$$\mathcal{L}_{\text{soft}}^{(3)} = (Y_d^A)^{ij}\,\tilde{Q}_i H_1 \tilde{d}_{Rj} + (Y_e^A)^{ij}\,\tilde{L}_i H_1 \tilde{e}_{Rj}^c + (Y_u^A)^{ij}\,\tilde{Q}_i H_2 \tilde{u}_{Rj}^c + B\mu H_1 H_2,$$

where, $M_{\tilde{Q}}^2$, $M_{\tilde{u}}^2$, $M_{\tilde{d}}^2$, $M_{\tilde{L}}^2$ and $M_{\tilde{e}}^2$ are hermitian 3×3 matrices in flavour space, while (Y_d^A), (Y_u^A) and (Y_u^A) are complex 3×3 matrices and M_1, M_2, M_3 denote the Majorana gaugino masses for the $U(1)$, $SU(2)$, $SU(3)$ gauge symmetries respectively.

This completes the definition of the MSSM. However, these conditions include a huge variety of models with very different phenomenology specially in the flavour and CP violation sectors.

It is instructive to identify all the observable parameters in a general MSSM [63]. Here we distinguish the flavour independent sector which includes the gauge and Higgs sectors and the flavour sector involving the three generations of chiral multiplets containing the SM fermions and their Supersymmetric partners.

In the flavour independent sector, we have three real gauge couplings, g_i, and three complex gaugino masses, M_i. In the Higgs sector, also flavour independent, we have a complex μ parameter in the superpotential, a complex $B\mu$ soft term and two real squared soft masses $m_{H_1}^2$ and $m_{H_2}^2$. However, not all the phases in these parameters are physical [64]. In the limit of $\mu = B\mu = 0$, vanishing gaugino masses and zero trilinear couplings, Y^A, (we will discuss trilinear terms in the flavour dependent sector), our theory has two global $U(1)$ symmetries: $U(1)_R$ and $U(1)_{PQ}$. This implies that we can use these two global symmetries to remove two of the phases of these parameters. For instance, we can choose a real $B\mu$ and a real gluino mass M_3. Then, in the flavour independent sector, we have 10 real parameters (g_i, $|M_i|$, $|\mu|$, $B\mu$, $m_{H_1}^2$ and $m_{H_2}^2$) and 3 phases (arg(μ), arg(M_1) and arg(M_2)).

Next, we have to analyse the flavour dependent sector. Here we do not take into account neutrino mass matrices. Then, in the superpotential we have the up quark, down quark and charged lepton Yukawa couplings, Y_u, Y_d and Y_e, that are complex 3×3 matrices. In the soft breaking sector we have 5 hermitian mass squared matrices, $M_{\tilde{Q}}^2$, $M_{\tilde{U}}^2$, $M_{\tilde{D}}^2$, $M_{\tilde{L}}^2$ and $M_{\tilde{E}}^2$ and three complex trilinear matrices, Y_u^A, Y_d^A and Y_e^A. This implies we have 6×9 moduli and 6×9 phases from the 6 complex matrices (Y_u, Y_d, Y_e, Y_u^A, Y_d^A and Y_e^A) and 5×6 moduli and 5×3 phases from the 5 hermitian matrices. Therefore, in the flavour sector we have 84 moduli and 69 phases. However, it is well-known that not all these parameters are observable. In the absence of these flavour matrices the theory has a global $U(3)_{Q_L} \otimes U(3)_{u_R} \otimes U(3)_{d_R} \otimes U(3)_{L_L} \otimes U(3)_{e_R}$ flavour symmetry under exchange of the different particles of the three generations. The number of observable parameters is easily determined using the method in Ref. [65] as,

$$N = N_{fl} - N_G - N_{G'}, \tag{3.3}$$

where N_{fl} is the number of parameters in the flavour matrices. N_G is the number of parameters of the group of invariance of the theory in the absence of the flavour matrices $G = U(3)_{Q_L} \otimes U(3)_{u_R} \otimes U(3)_{d_R} \otimes U(3)_{L_L} \otimes U(3)_{e_R}$. Finally $N_{G'}$ is the number of parameters of the group G', the subgroup of G still unbroken by the flavour matrices. In this case, G' corresponds to two $U(1)$ symmetries, baryon number conservation and lepton number conservation and therefore $N_{G'} = 2$. Furthermore Eq. (3.3) can be applied separately to phases and moduli. In this way, and taking into account that a $U(N)$ matrix contains $n(n-1)/2$ moduli and $n(n+1)/2$ phases, it is straightforward to obtain that we have, $N_{ph} = 69 - 5 \times 6 + 2 = 41$ phases and $N_{mod} = 84 - 5 \times 3 = 69$ moduli in the flavour sector. This amounts to a total of 123 parameters in the model[4], out of which 44 are CP violating phases!! As we know, in the SM, there is only one observable CP violating phase, the CKM phase, and therefore we have here 43 new phases, 40 in the flavour sector and three in the flavour independent sector.

Clearly, to explore completely the flavour and CP violating phenomena in a generic MSSM is a formidable task as we have to determine a huge number of unknown parameters [66]. However, this parameter counting corresponds to a completely general MSSM at the electroweak scale but the number of parameters is largely reduced in most of the theory motivated models defined at high energies. In these models most of the parameters at M_W are fixed as a function of a handful of parameters at the scale of the transmission of SUSY breaking, for instance M_{Pl} in the case of supergravity mediation, and therefore there are relations among the parameters at M_W. So, our task will be to determine as many as possible of the CP violating and flavour parameters at M_W to look for possible relations among them that will allow us to explore the physics of SUSY and CP breaking at high energies.

The so-called Constrained MSSM (CMSSM), or Sugra-MSSM, (for an early version of these models see, [11, 12]) is the simplest version we can build of

---

[4]Notice that we did not include the parameter θ_{QCD} which was also present in the 124 parameters MSSM of H. Haber [63].

the MSSM. For instance a realisation of this model is obtained in string models
with dilaton dominated SUSY breaking [67, 68]. Here all the SBT are universal.
The soft masses are all proportional to the identity matrix and the trilinear couplings are directly proportional to the corresponding Yukawa matrix. Moreover
the gaugino masses are all unified at the high scale. So, we have at M_{GUT},

$$M_{\tilde{Q}}^2 = M_{\tilde{U}}^2 = M_{\tilde{D}}^2 = M_{\tilde{L}}^2 = M_{\tilde{E}}^2 = m_0^2 \, \mathbb{1},$$

$$Y_u^A = A_0 \, Y_u, \qquad Y_d^A = A_0 \, Y_d, \qquad Y_e^A = A_0 \, Y_e,$$

$$m_{H_1}^2 = M_{H_2}^2 = m_0^2, \qquad M_3 = M_2 = M_1 = M_{1/2}. \tag{3.4}$$

In this way the number of parameters is strongly reduced. If we repeat the counting of parameters in this case we have only 27 complex parameters in the Yukawa
matrices, out of which only 12 moduli and 1 phase are observable. In the soft
breaking sector we have only a real mass square, m_0^2, and a complex trilinear
term, A_0. We have a single unified gauge coupling, g_U, and a complex universal
gaugino mass $M_{1/2}$ in the gauge sector. Finally in the Higgs sector there are two
complex parameters μ and $B\mu$. Again two of these phases can be reabsorbed
through the $U(1)_R$ and $U(1)_{PQ}$ symmetries. Therefore, we have only 21 parameters, 18 moduli and 3 phases. In fact, 14 of these parameters are already
known in the Standard Model and we are left with only 7 unknown parameters
from SUSY: $(m_0^2, |M_{1/2}|, |\mu|, \arg(\mu), |A_0|, \arg(A_0)$ and $|B|)$. If we require radia-
tive electroweak symmetry breaking [69] we get an additional constraint which
is used to relate $|B|$ to M_W. In the literature it is also customary to exchange
$|\mu|$ by $\tan\beta = v_2/v_1$, the ratio of the two Higgs vacuum expectation values, so
that the set of parameters usually considered in the MSSM with radiative sym-
metry breaking is $(m_0^2, |M_{1/2}|, \tan\beta, |A_0|, \arg(A_0)$ and $\arg(\mu))$. Regarding CP
violation, we see that even in the simplest MSSM version we have two new CP
violating phases, which we have chosen to be $\varphi_\mu \equiv \arg(\mu)$ and $\varphi_A \equiv \arg(A_0)$.
These phases will have a very strong effect on CP violating observables, mainly
the Electric Dipole Moments (EDMs) of the electron and the neutron as we will
show in the next section. We must remember that a generic MSSM will always
include at least these two phases and therefore the constraints from EDMs are
always applicable in any MSSM.

   All these flavour parameters and phases are encoded at the electroweak scale
in the different mass matrices of sfermions and gauginos/higgsinos. For instance,
after breaking the $SU(2)_L$ symmetry, the superpartners of $W^\pm$ and $H^\pm$ have the
same unbroken quantum number and thus can mix through a matrix,

$$-\frac{1}{2} \begin{pmatrix} \tilde{W}^- & \tilde{H}_1^- \end{pmatrix} \begin{pmatrix} M_2 & \sqrt{2}M_W \sin\beta \\ \sqrt{2}M_W \cos\beta & \mu \end{pmatrix} \begin{pmatrix} \tilde{W}^+ \\ \tilde{H}_2^+ \end{pmatrix}, \tag{3.5}$$

This non-symmetric (non-hermitian) matrix is diagonalised with two unitary matrices, $U^* \cdot M_{\chi^+} \cdot V^\dagger = \text{Diag.}(m_{\chi_1^+}, m_{\chi_2^+})$.

In the same way, once we break the electroweak symmetry, neutral higgsinos and neutral gauginos mix. In the basis $(\tilde{B}\ \tilde{W}^0 \tilde{H}_1^0 \tilde{H}_2^0)$, the mass matrix is,

$$
\begin{pmatrix}
M_1 & 0 & -M_Z c\beta s\theta_W & M_Z s\beta s\theta_W \\
0 & M_2 & M_Z c\beta c\theta_W & M_Z s\beta c\theta_W \\
-M_Z c\beta s\theta_W & M_Z c\beta c\theta_W & 0 & -\mu \\
M_Z s\beta s\theta_W & -M_Z s\beta c\theta_W & -\mu & 0
\end{pmatrix},
\tag{3.6}
$$

with $c\beta(s\beta)$ and $c\theta_W(s\theta_W)$, $\cos(\sin)\beta$ and $\cos(\sin)\theta_W$ respectively. This is diagonalised by a unitary matrix N,

$$
N^* \cdot M_{\tilde{N}} \cdot N^\dagger = \text{Diag.}(m_{\chi_1^0}, m_{\chi_2^0}, m_{\chi_3^0}, m_{\chi_4^0}),
\tag{3.7}
$$

Finally, the different sfermions, as $\tilde{f}_L$ and $\tilde{f}_R$, mix after EW breaking. In fact they can also mix with fermions of different generations and in general we have a 6×6 mixing matrix.

$$
M_{\tilde{f}}^2 = \begin{pmatrix} m_{\tilde{f}_{LL}}^2 & m_{\tilde{f}_{LR}}^2 \\ m_{\tilde{f}_{LR}}^{2\dagger} & m_{\tilde{f}_{RR}}^2 \end{pmatrix}, \quad m_{\tilde{f}_{LR}}^2 = \left(Y_f^A \cdot \frac{v_2}{v_1} - m_f \mu \frac{\tan\beta}{\cot\beta}\right) \text{ for } f = \frac{e,d}{u},
$$

$$
m_{\tilde{f}_{LL}}^2 = M_{\tilde{f}_L}^2 + m_{\tilde{f}}^2 + M_Z^2 \cos 2\beta (I_3 + \sin^2\theta_W Q_{em}),
$$

$$
m_{\tilde{f}_{RR}}^2 = M_{\tilde{f}_R}^2 + m_{\tilde{f}}^2 + M_Z^2 \cos 2\beta \sin^2\theta_W Q_{em}.
\tag{3.8}
$$

These hermitian sfermion mass matrices are diagonalised by a unitary rotation, $R_{\tilde{f}} \cdot M_{\tilde{f}} \cdot R_{\tilde{f}}^\dagger = \text{Diag.}(m_{\tilde{f}_1}, m_{\tilde{f}_2}, \ldots, m_{\tilde{f}_6})$.

Therefore all the new SUSY phases are kept in these gaugino and sfermion mixing matrices. However, it is not necessary to know the full mass matrices to estimate the CP violation effects. We have some powerful tools as the Mass Insertion (MI) approximation [70–73] to analyse FCNCs and CP violation. In this approximation, we use flavour diagonal gaugino vertices and the flavour changing is encoded in non-diagonal sfermion propagators. These propagators are then expanded assuming that the flavour changing parts are much smaller than the flavour diagonal ones. In this way we can isolate the relevant elements of the sfermion mass matrix for a given flavour changing process and it is not necessary to analyse the full 6×6 sfermion mass matrix. Using this method, the experimental limits lead to upper bounds on the parameters (or combinations of) $\delta_{ij}^f \equiv \Delta_{ij}^f/m_{\tilde{f}}^2$, known as mass insertions; where Δ_{ij}^f is the flavour-violating off-

A. *Masiero et al.*

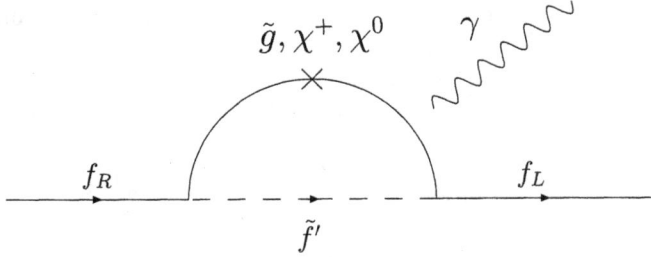

Fig. 8. 1 loop contributing to a fermion EDM.

diagonal entry appearing in the $f = (u, d, l)$ sfermion mass matrices and $m_{\tilde{f}}^2$ is the average sfermion mass. In addition, the mass-insertions are further sub-divided into LL/LR/RL/RR types, labelled by the chirality of the corresponding SM fermions. In the following sections we will use both the full mass matrix di-agonalisation and this MI formalism to analyse flavour changing and CP violation processes. Now, we will start by studying the EDM calculations and constraints which are common to all Supersymmetric models.

3.1. Electric Dipole Moments in the MSSM

The large SUSY contributions to the electric dipole moments of the electron and the neutron are the main source of the so-called "Supersymmetric CP problem". This "problem" is present in any MSSM due to the presence of the flavour inde-pendent phases φ_μ and φ_A. Basically Supersymmetry gives rise to contributions to the EDMs at 1 loop order with no suppression associated to flavour as these phases are flavour diagonal [74–79]. Taking into account these facts, this contri-bution can be expected to be much larger than the SM contribution which appears only at three loops and is further suppressed by CKM angles and fermion masses. In fact the SM contribution to the neutron EDM is expected to be of the order of 10^{-32} e cm, while the present experimental bounds are $d_n \leq 6.3 \times 10^{-26}$ e cm (90% C.L.) [80] and $d_e \leq 1.6 \times 10^{-27}$ e cm (90% C.L.) [81]. As we will show here the Supersymmetric 1 loop contributions to the EDM for SUSY masses be-low several TeV can easily exceed the present experimental bounds. Therefore, these experiments impose very stringent bounds on φ_μ and φ_A.

The typical diagram giving rise to a fermion EDM is shown in Figure 8. In the case of a quark EDM, the dominant contribution typically corresponds to the diagram with internal gluino and squark states. Here all the phases appear only in the squark mass matrix. If we neglect intergenerational mixing (that can be expected to be small), we have a 2×2 squark mass matrix. For instance the down squark mass matrix $M_{\tilde{d}}$ is, in the basis $(\tilde{d}_L, \tilde{d}_R)$,

$$
\begin{pmatrix}
m_{\tilde{d}_L}^2 + m_d^2 - (\frac{1}{2} - \frac{1}{3} s^2\theta_W) \, c2\beta M_Z^2 & Y_d^{A*} \, v \, c\beta - m_d \, \mu \, tg\beta \\
Y_d^A \, v \, c\beta - m_d \, \mu^* \, tg\beta & m_{\tilde{d}_R}^2 + m_d^2 - \frac{1}{3} s^2\theta_W \, c2\beta M_Z^2
\end{pmatrix}
\tag{3.9}
$$

with $Y_d^A \simeq A_0 Y_d$ except 1 loop correction in the RGE evolution from M_{GUT} to M_W. Therefore we have both φ_μ and φ_A in the left-right squark mixing and these phases appear then in the down squark mixing matrix, R^d, $R^{\tilde{d}} M_{\tilde{d}} R^{\tilde{d}\dagger} = \text{Diag.}(m_{\tilde{d}_1}, m_{\tilde{d}_2})$. In terms of this mixing matrix the 1 loop gluino contribution to the EDM of the down quark is (in a similar way we would obtain the gluino contribution to the up quark EDM),

$$
d_{\tilde{g}}^d = \frac{2\alpha_s e}{9\pi} \sum_{k=1}^{2} \text{Im}\big[R_{k2}^{\tilde{d}} R_{k1}^{\tilde{d}*}\big] \frac{1}{m_{\tilde{g}}} B\left(\frac{m_{\tilde{g}}^2}{m_{\tilde{d}_k}^2}\right),
\tag{3.10}
$$

with,

$$
B(r) = \frac{r}{2(1-r)^2}\left(1 + r + \frac{2r \log r}{1-r}\right).
\tag{3.11}
$$

It is interesting to obtain the corresponding formula in terms of the $(\delta_{11}^f)_{\text{LR}}$ mass insertion. To do this we observe that given a $n \times n$ hermitian matrix $A = A^0 + A^1$ diagonalised by $U \cdot A \cdot U^\dagger = \text{Diag}(a_1, \ldots, a_n)$, with $A^0 = \text{Diag}(a_1^0, \ldots, a_n^0)$ and A^1 completely off-diagonal, we have at first order in A^1 [82,83],

$$
U_{ki}^* f(a_k) U_{kj} \simeq \delta_{ij} f(a_i^0) + A_{ij}^1 \frac{f(a_i^0) - f(a_j^0)}{a_i^0 - a_j^0}.
\tag{3.12}
$$

Therefore, for small off-diagonal entries A^1 and taking into account that for approximately degenerate squarks we can replace the finite differences by the derivative of the function, $B'(x)$, Eq. (3.10) is converted into,

$$
\begin{aligned}
d_{\tilde{g}}^d &\simeq \frac{2\alpha_s e}{9\pi} \frac{m_{\tilde{g}}}{m_{\tilde{d}}^2} B'\left(\frac{m_{\tilde{g}}^2}{m_{\tilde{d}}^2}\right) \text{Im}\left[\frac{Y_d^{A*} \, v \cos\beta - m_d \, \mu \tan\beta}{m_{\tilde{d}}^2}\right] \\
&\equiv \frac{2\alpha_s e}{9\pi} \frac{m_{\tilde{g}}}{m_{\tilde{d}}^2} B'\left(\frac{m_{\tilde{g}}^2}{m_{\tilde{d}}^2}\right) \text{Im}\left[(\delta_{11}^d)_{\text{LR}}\right].
\end{aligned}
\tag{3.13}
$$

with $m_{\tilde{d}}^2$ the average down squark mass. From this equation it is straightforward to obtain a simple numerical estimate. Taking $m_{\tilde{g}} = m_{\tilde{d}} = 500$ GeV, $Y_d^A = A_0 Y_d$

and $\mu \simeq A_0 \simeq 500$ GeV, we have,

$$d_{\tilde{g}}^d \simeq 2.8 \times 10^{-20} \; \mathrm{Im}\left[(\delta_{11}^d)_{\mathrm{LR}}\right] \; \mathrm{e \; cm}$$
$$= 2.8 \times 10^{-25} \; (\sin \varphi_A - \tan \beta \sin \varphi_\mu) \; \mathrm{e \; cm}, \qquad (3.14)$$

where we used $\alpha_s = 0.12$ and $m_d = 5$ MeV. Comparing with the experimental bound on the neutron EDM and using, for simplicity the quark model relation $d_n = \frac{1}{3}(4d_d - d_u)$, we see immediately that φ_A and $(\tan \beta \; \varphi_\mu) \leq 0.16$. This is a simple aspect of the "supersymmetric CP problem". As we will see the constraints from the electron EDM give rise to even stronger bounds on these phases.

In addition, we have also contributions from chargino and neutralino loops which are usually subdominant in the quark EDMs but are the leading contribution in the electron EDM. A simple example is the chargino contribution to the electron EDM. The corresponding diagram is shown in Figure 8 with the chargino and sneutrino in the internal lines,

$$d_{\chi^+}^e = -\frac{\alpha e}{4\pi \sin^2 \theta_W} \frac{m_e}{\sqrt{2} M_W \cos \beta} \sum_{j=1}^{2} \mathrm{Im}[U_{j2} V_{j1}] \frac{m_{\chi_j^+}}{m_{\tilde{\nu}_e}^2} A\left(\frac{m_{\chi_j^+}^2}{m_{\tilde{\nu}_e}^2}\right), \qquad (3.15)$$

with,

$$A(r) = \frac{1}{2(1-r)^2} \left(3 - r + \frac{2 \log r}{1-r}\right). \qquad (3.16)$$

It is also useful to use a technique similar to Eq. (3.12) to expand the chargino mass matrix. In this case we have to be careful because the chargino mass matrix is not hermitian. However due to the necessary chirality flip in the chargino line we know that the EDM is a function of odd powers of M_{χ^+} [84],

$$\sum_{j=1}^{2} U_{j2} V_{j1} m_{\chi_j^+} A(m_{\chi_j^+}^2) = \sum_{j,k,l=1}^{2} U_{lk} m_{\chi_l^+} V_{l1} \; U_{j2} A(m_{\chi_j^+}^2) U_{jk}^*. \qquad (3.17)$$

where we have simply introduced an identity $\delta_{lj} = \sum_k U_{lk} U_{jk}^*$. Now, assuming $M_W \ll M_2, \mu$, we can use Eq. (3.12) to develop the loop function A(x) as a function of the hermitian matrix $M_{\chi^+} M_{\chi^+}^\dagger$ and we get,

$$d_{\chi^+}^e \simeq \frac{-\alpha \, e \, m_e}{4\pi \sin^2 \theta_W} \frac{\mathrm{Im}\left[\sum_k \left(M_{\chi^+} M_{\chi^+}^\dagger\right)_{2k} (M_{\chi^+})_{k1}\right]}{\sqrt{2} M_W \cos \beta \, m_{\tilde{\nu}_e}^2} \frac{A(r_1) - A(r_2)}{m_{\chi_1^+}^2 - m_{\chi_2^+}^2}$$
$$= \frac{-\alpha \, e \, m_e \tan \beta}{4\pi \sin^2 \theta_W} \frac{\mathrm{Im}[M_2 \, \mu]}{m_{\tilde{\nu}_e}^2} \frac{A(r_1) - A(r_2)}{m_{\chi_1^+}^2 - m_{\chi_2^+}^2}. \qquad (3.18)$$

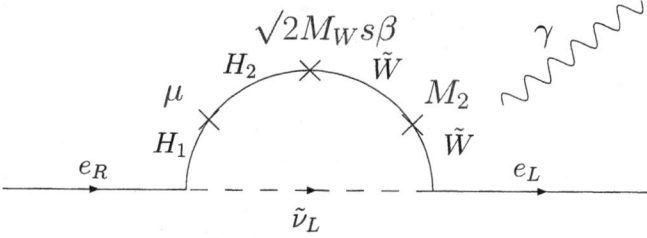

Fig. 9. 1 loop chargino contribution to the electron EDM at leading order in chargino mass insertions.

with $r_i = m^2_{\chi^+_i}/m^2_{\tilde{\nu}_e}$. This structure with three chargino MIs is shown in figure 9. Here we can see that only φ_μ enters in the chargino contribution. In fact $\arg(M_2\,\mu)$ is the rephasing invariant expression of the observable phase that we usually call φ_μ. Again we can make a rough estimate with $\mu \simeq M_2 \simeq m_{\tilde{\nu}} \simeq 200$ GeV (taking the derivative of $A(r)$),

$$d^e_{\chi^+} \simeq 1.5 \times 10^{-25} \ \tan\beta \sin\varphi_\mu \text{ e cm.} \tag{3.19}$$

Now, comparing with the experimental bound on the electron EDM, we obtain a much stronger bound, $(\tan\beta\ \varphi_\mu) \leq 0.01$. These two examples give a clear idea of the strength of the "SUSY CP problem".

As we have seen in these examples typically the bound on φ_μ is stronger than the bound on φ_A. There are several reason for this, as we can see φ_μ enters the down-type sfermion mass matrix together with $\tan\beta$ while φ_A is not enhanced by this factor. Furthermore, φ_μ appears also in the chargino and neutralino mass matrices. This difference is increased if we consider the bounds on the original parameters at M_{GUT}. The μ phase is unchanged in the RGE evolution, but $\varphi_A = \arg(M_{1/2}A_0)$ (where $M_{1/2}$ is the gaugino mass) is reduced due to large gaugino contributions to the trilinear couplings in the running from M_{GUT} to M_W. The bounds we typically find in the literature [85, 86] are,

$$\varphi_\mu \leq 10^{-2} - 10^{-3}, \qquad \varphi_A \leq 10^{-1} - 10^{-2}. \tag{3.20}$$

Nevertheless, a full computation should take into account all the different contributions to the electron and neutron EDM. In the case of the electron, we have both chargino and neutralino contributions at 1 loop. For the neutron EDM, we have to include also the gluino contribution, the quark chromoelectric dipole moments and the dimension six gluonic operator [85, 87–91]. When all these contributions are taken into account our estimates above may not be accurate enough and the bound can be loosened.

In fact, there can be regions on the parameter space where different contributions to the neutron or electron EDM have opposite signs and similar size. Thus the complete result for these EDM can be smaller than the individual contributions. In this way, it is possible to reduce the stringent constraints on these phases and $\varphi_A = \mathcal{O}(1)$ and $\varphi_\mu = \mathcal{O}(0.1)$ can be still allowed [92–98]. However, when all the EDM constraints, namely electron, neutron and also mercury atom EDM, are considered simultaneously the cancellation regions practically disappear and the bounds in Eq. (3.20) remain basically valid [99, 100].

3.2. *Flavour changing neutral currents in the MSSM*

In the previous section we have analysed the effects of the "flavour independent" SUSY phases, φ_μ and φ_A, on the EDMs of the electron and the neutron. However, we have seen that a generic MSSM contains many other observable phases and flavour changing parameters. This huge number of new parameters in the SUSY soft breaking sector can easily generate dangerous contributions in FCNC and flavour changing CP violation processes.

Given the large number of unknown parameters involved in FC processes, it is particularly helpful to make use of the Mass Insertion formalism. The mass insertions are defined in the so-called Super CKM (SCKM) basis. This is the basis where the Yukawa couplings for the down or up quarks are diagonal and we keep the neutral gaugino couplings flavour diagonal. In this basis squark mass matrices are not diagonal and therefore the flavour changing is exhibited by the non-diagonality of the sfermion propagators. Denoting by Δ_{ij}^f the flavour-violating off-diagonal entry appearing in the $f = (u_L, d_L, u_R, d_R, u_{LR}, d_{LR})$ sfermion mass matrices, the sfermion propagators are expanded as a series in terms of $(\delta^f)_{ij} = \Delta_{ij}^f / m_{\tilde{f}}$, which are known as mass insertions (MI). Clearly the goodness of this approximation depends on the smallness of the expansion parameter δ_{ij}^f. As we will see, indeed the phenomenological constraints require these parameters to be small and it is usually enough to keep the first terms in this expansion. The use of the MI approximation presents the major advantage that it is not necessary to know and diagonalise the full squark mass matrix to perform an analysis of FCNC in a given MSSM. It is enough to know the single entry contributing to a given process and in this way it is easy to isolate the relevant phases.

In terms of the MI, and taking all diagonal elements approximately equal to $m_{\tilde{d}}^2$, the down squark mass matrix is,

$$
\frac{M_{\tilde{d}}^2}{m_{\tilde{d}}^2} \simeq
\begin{pmatrix}
1 & (\delta_{12}^d)_{LL} & (\delta_{13}^d)_{LL} & (\delta_{11}^d)_{LR} & (\delta_{12}^d)_{LR} & (\delta_{13}^d)_{LR} \\
(\delta_{12}^d)_{LL}^* & 1 & (\delta_{23}^d)_{LL} & (\delta_{21}^d)_{LR} & (\delta_{22}^d)_{LR} & (\delta_{23}^d)_{LR} \\
(\delta_{13}^d)_{LL}^* & (\delta_{23}^d)_{LL}^* & 1 & (\delta_{31}^d)_{LR} & (\delta_{32}^d)_{LR} & (\delta_{33}^d)_{LR} \\
(\delta_{11}^d)_{LR}^* & (\delta_{21}^d)_{LR}^* & (\delta_{31}^d)_{LR}^* & 1 & (\delta_{12}^d)_{RR} & (\delta_{13}^d)_{RR} \\
(\delta_{12}^d)_{LR}^* & (\delta_{22}^d)_{LR}^* & (\delta_{32}^d)_{LR}^* & (\delta_{12}^d)_{RR}^* & 1 & (\delta_{23}^d)_{RR} \\
(\delta_{13}^d)_{LR}^* & (\delta_{23}^d)_{LR}^* & (\delta_{33}^d)_{LR}^* & (\delta_{13}^d)_{RR}^* & (\delta_{23}^d)_{RR}^* & 1
\end{pmatrix}
\tag{3.21}
$$

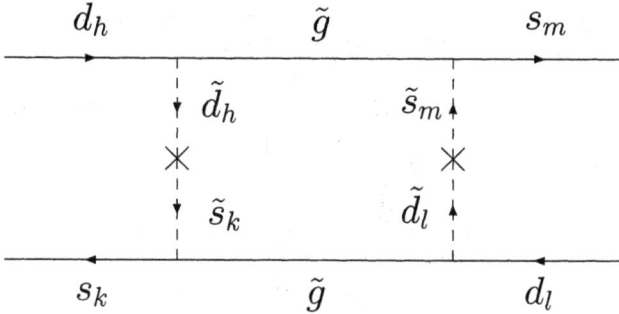

Fig. 10. 1 loop contribution to K–$\bar{K}$ mixing.

with all the off-diagonal elements complex which means we have 15 new moduli and 15 phases. The same would be true for the up squark mass matrix, although the $\left(\delta_{ij}^u\right)_{LL}$ would be related to $\left(\delta_{ij}^d\right)_{LL}$ by a CKM rotation. Therefore there would be a total of 27 moduli and 27 phases in the squark sector.

An illustrative example of the usage of the MI formalism is provided by the SUSY contribution to K–$\bar{K}$ [70–73] mixing. The relevant diagram at leading order in the MI approximation is shown in Fig. 10. Here the MI are treated as new vertices in our theory. We have to compute the contribution to the Wilson coefficients of the different four–fermion operators in the $\Delta S = 2$ effective Hamiltonian [73, 101]. For example the Wilson coefficient associated with the operator $Q_1 = \bar{d}_L^\alpha \gamma^\mu s_L^\alpha\, \bar{d}_L^\beta \gamma_\mu s_L^\beta$ would be,

$$C_1 = -\frac{\alpha_s^2}{216 m_{\tilde{q}}^2}\left(24 x f_6(x) + 66 \tilde{f}_6(x)\right)\left(\delta_{12}^d\right)_{LL}^2 , \tag{3.22}$$

with $x = m_{\tilde{g}}^2 / m_{\tilde{q}}^2$ and the functions $f_6(x)$ and $\tilde{f}_6(x)$ given by,

$$f_6(x) = \frac{6(1 + 3x)\log x + x^3 - 9x^2 - 9x + 17}{6(x - 1)^5},$$

$$\tilde{f}_6(x) = \frac{6x(1 + x)\log x - x^3 - 9x^2 + 9x + 1}{3(x - 1)^5}. \tag{3.23}$$

It is straightforward to understand the different factors in this formula: we have four flavour diagonal gluino vertices providing a factor g_s^4 and the two MI which supply the necessary flavour transition. The remainder corresponds only to the loop functions. A full computation of the whole set of Wilson coefficients can be found in Refs. [73, 101].

The complete leading order expression for K^0–$\bar{K}^0$ mixing, using the Vacuum Insertion Approximation (VIA) for the matrix elements of the different operators, is [73],

$$
\langle K^0 | \mathcal{H}_{\text{eff}}^{\Delta S=2} | \bar{K}^0 \rangle = -\frac{\alpha_s^2}{216 m_{\tilde{q}}^2} \frac{1}{3} m_K f_K^2 \Bigg\{ \tag{3.24}
$$

$$
\left(\left(\delta_{12}^d \right)_{\text{LL}}^2 + \left(\delta_{12}^d \right)_{\text{RR}}^2 \right) \left(24 x f_6(x) + 66 \tilde{f}_6(x) \right)
$$

$$
+ \left(\delta_{12}^d \right)_{\text{LL}} \left(\delta_{12}^d \right)_{\text{RR}} \left[\left(84 \left(\frac{m_K}{m_s + m_d} \right)^2 + 72 \right) x f_6(x) \right.
$$

$$
\left. + \left(-24 \left(\frac{m_K}{m_s + m_d} \right)^2 + 36 \right) \tilde{f}_6(x) \right]
$$

$$
+ \left(\left(\delta_{12}^d \right)_{\text{LR}}^2 + \left(\delta_{12}^d \right)_{\text{RL}}^2 \right) \left(-132 \left(\frac{m_K}{m_s + m_d} \right)^2 \right) x f_6(x)
$$

$$
+ \left(\delta_{12}^d \right)_{\text{LR}} \left(\delta_{12}^d \right)_{\text{RL}} \left[-144 \left(\frac{m_K}{m_s + m_d} \right)^2 - 84 \right] \tilde{f}_6(x) \Bigg\}.
$$

The neutral kaon mass difference and the mixing CP violating parameter, ε_K, are given by,

$$
\Delta M_K = 2 \Re \langle K^0 | \mathcal{H}_{\text{eff}}^{\Delta S=2} | \bar{K}^0 \rangle,
$$

$$
\varepsilon_K = \frac{1}{\sqrt{2} \Delta M_K} \Im \langle K^0 | \mathcal{H}_{\text{eff}}^{\Delta S=2} | \bar{K}^0 \rangle. \tag{3.25}
$$

To obtain a model independent bound on the different MI, we assume that each time only one of these MI is different from zero neglecting accidental cancellations between different MIs. Moreover, it is customary to consider only the gluino contributions leaving aside other SUSY contributions as chargino, charged Higgs or neutralino. In fact, in the presence of sizable MI, the gluino contribution provides typically a large part of the full SUSY contribution. Barring sizable accidental cancellations between the SM and SUSY contributions a conservative limit on the δs is obtained by requiring the SUSY contribution by itself not to exceed the experimental value of the observable under consideration.

The different MI bounds for the $\left(\delta_{12}^d \right)_a$ (a = LL,RR,LR) are presented in Table 1. As can be seen explicitly in Eq. (3.24) gluino contributions are completely symmetrical under the interchange $L \leftrightarrow R$ and therefore the limits on $\left(\delta_{12}^d \right)_{\text{RR}}$ are equal to those on $\left(\delta_{12}^d \right)_{\text{LL}}$ and the limits on $\left(\delta_{12}^d \right)_{\text{RL}}$ to those on $\left(\delta_{12}^d \right)_{\text{LR}}$. In this table we present the bounds at tree level in the four fermion effective Hamiltonian

Table 1

Maximum allowed values for $|\Re(\delta^d_{12})_{AB}|$ and $|\Im(\delta^d_{12})_{AB}|$, with $A, B = (L, R)$ for an average squark mass $m_{\tilde{q}} = 500$ GeV and for different values of $x = m^2_{\tilde{g}}/m^2_{\tilde{q}}$. The bounds are given at tree level in the effective Hamiltonian and at NLO in QCD corrections as explained in the text. For different values of $m_{\tilde{q}}$ the bounds scale roughly as $m_{\tilde{q}}/500$ GeV.

	$\sqrt{|\Re(\delta^d_{12})^2_{LL}|}$		$\sqrt{|\Im(\delta^d_{12})^2_{LL}|}$	
x	TREE	NLO	TREE	NLO
0.3	1.4×10^{-2}	2.2×10^{-2}	1.8×10^{-3}	2.9×10^{-3}
1.0	3.0×10^{-2}	4.6×10^{-2}	3.9×10^{-3}	6.1×10^{-3}
4.0	7.0×10^{-2}	1.1×10^{-1}	9.2×10^{-3}	1.4×10^{-2}

	$\sqrt{|\Re(\delta^d_{12})_{LL}(\delta^d_{12})_{RR}|}$		$\sqrt{|\Im(\delta^d_{12})_{LL}(\delta^d_{12})_{RR}|}$	
x	TREE	NLO	TREE	NLO
0.3	1.8×10^{-3}	8.6×10^{-4}	2.3×10^{-4}	1.1×10^{-4}
1.0	2.0×10^{-3}	9.6×10^{-4}	2.6×10^{-4}	1.3×10^{-4}
4.0	2.8×10^{-3}	1.3×10^{-3}	3.7×10^{-4}	1.8×10^{-4}

	$\sqrt{|\Re(\delta^d_{12})^2_{LR}|}$		$\sqrt{|\Im(\delta^d_{12})^2_{LR}|}$	
x	TREE	NLO	TREE	NLO
0.3	3.1×10^{-3}	2.6×10^{-3}	4.1×10^{-4}	3.4×10^{-4}
1.0	3.4×10^{-3}	2.8×10^{-3}	4.6×10^{-4}	3.7×10^{-4}
4.0	4.9×10^{-3}	3.9×10^{-3}	6.5×10^{-4}	5.2×10^{-4}

(TREE), i.e. using directly Eq. (3.24) without any further QCD corrections and we compare them with bounds obtained using the NLO QCD evolution with lattice B parameters in the matrix elements [101]. As we can see, although QCD corrections may change the bounds even a factor 2, the tree level estimates remain valid as order of magnitude bounds. The main conclusion we can draw from this table is that MI bounds in $s \to d$ transitions are very tight and this is specially true on the imaginary parts. This poses a very stringent constraint in most attempts to build a viable MSSM or any realistic supersymmetric flavour model [102–104]. Conversely we can say that $s \to d$ transitions are very sensitive to the presence of relatively small SUSY contributions and a deviation from SM predictions here could provide the first indirect sign of SUSY [105, 106].

CP violating supersymmetric contributions can also be very interesting in the B system [107, 108]. Similarly to the previous case, we can build the $\Delta B = 2$ effective Hamiltonian to obtain the bounds from B_d–$\bar{B}_d$ mixing. A full calculation is presented in Ref. [109], in Table 2 we present the results. As we can see here, the constraints in the B_d system are less stringent than in the K sector specially in the imaginary parts of the MI which come from ε_K and $\sin 2\beta$ [110–113]. At first sight this may be surprising as it is well-known that CP violation is more promi-

Table 2

Maximum allowed values for $|\Re(\delta^d_{13})_{AB}|$ and $|\Im(\delta^d_{13})_{AB}|$, with $A, B = (L, R)$ for an average squark mass $m_{\tilde{q}} = 500$ GeV and different values of $x = m^2_{\tilde{g}}/m^2_{\tilde{q}}$. with NLO evolution and lattice B parameters, denoted by NLO. The missing entries correspond to cases in which no constraint was found for $|(\delta^d_{ij})_{AB}| < 0.9$.

x	$\|\Re(\delta^d_{13})_{LL}\|$		$\|\Re(\delta^d_{13})_{LL=RR}\|$	
	TREE	NLO	TREE	NLO
0.25	4.9×10^{-2}	6.2×10^{-2}	3.1×10^{-2}	1.9×10^{-2}
1.0	1.1×10^{-1}	1.4×10^{-1}	3.4×10^{-2}	2.1×10^{-2}
4.0	6.0×10^{-1}	7.0×10^{-1}	4.7×10^{-2}	2.8×10^{-2}

x	$\|\Im(\delta^d_{13})_{LL}\|$		$\|\Im(\delta^d_{13})_{LL=RR}\|$	
	TREE	NLO	TREE	NLO
0.25	1.1×10^{-1}	1.3×10^{-1}	1.3×10^{-2}	8.0×10^{-3}
1.0	2.6×10^{-1}	3.0×10^{-1}	1.5×10^{-2}	9.0×10^{-3}
4.0	2.6×10^{-1}	3.4×10^{-1}	2.0×10^{-2}	1.2×10^{-2}

x	$\|\Re(\delta^d_{13})_{LR}\|$		$\|\Re(\delta^d_{13})_{LR=RL}\|$	
	TREE	NLO	TREE	NLO
0.25	3.4×10^{-2}	3.0×10^{-2}	3.8×10^{-2}	2.6×10^{-2}
1.0	3.9×10^{-2}	3.3×10^{-2}	8.3×10^{-2}	5.2×10^{-2}
4.0	5.3×10^{-2}	4.5×10^{-2}	1.2×10^{-1}	–

x	$\|\Im(\delta^d_{13})_{LR}\|$		$\|\Im(\delta^d_{13})_{LR=RL}\|$	
	TREE	NLO	TREE	NLO
0.25	7.6×10^{-2}	6.6×10^{-2}	1.5×10^{-2}	9.0×10^{-3}
1.0	8.7×10^{-2}	7.4×10^{-2}	3.6×10^{-2}	2.3×10^{-2}
4.0	1.2×10^{-1}	1.0×10^{-1}	2.7×10^{-1}	–

nent in the B system. To understand this difference we analyse more closely these two observables.

Let us assume that the imaginary part of K^0–$\bar{K}^0$ and B^0_d–$\bar{B}^0_d$ is entirely provided by SUSY from a single $(\delta^d_{ij})_{LL}$ MI, while the real part is mostly given by SM loops. The Standard Model contribution to K^0–$\bar{K}^0$ mixing is given by,

$$\langle K^0|\mathcal{H}^{\Delta S=2}_{\text{eff}}|\bar{K}^0\rangle = -\frac{\alpha^2_{\text{em}}}{8M^2_W \sin^4\theta_W} \frac{m^2_c}{M^2_W} \frac{f^2_K m_K}{3} (V_{cs}V^*_{cd})^2. \qquad (3.26)$$

Replacing this expression and Eq. (3.24) in Eq. (3.25) we have,

$$\varepsilon^{\text{SUSY}}_K = \frac{\text{Im } M_{12}|_{\text{SUSY}}}{\sqrt{2}\,\Delta M_K}\bigg|_{\text{SM}} \simeq \frac{\alpha^2_s \sin^2\theta_W}{\alpha^2_{\text{em}}} \frac{M^4_W}{M^2_{\text{SUSY}}\, m^2_c} \frac{\text{Im}\left\{(\delta^d_{12})^2_{\text{LL}}\right\}}{\left(V_{cd}V^*_{cs}\right)^2},$$

$$\frac{8(24xf_6(x) + 66\tilde{f}_6(x))}{216\sqrt{2}} \simeq 12.5 \times 84 \times \frac{\text{Im}\left\{(\delta_{12}^d)_{\text{LL}}^2\right\}}{0.05} \times 0.026,$$

$$\varepsilon_K^{\text{SUSY}} \leq 2.3 \times 10^{-3} \Rightarrow \sqrt{\text{Im}\left\{(\delta_{12}^d)_{\text{LL}}^2\right\}} \leq 2.0 \times 10^{-3}, \qquad (3.27)$$

where we used $x = 1$ and $M_{\text{SUSY}} = 500$ GeV. In the same way, we can obtain an estimate of the MI bound from the B^0 CP asymmetries. The gluino and SM contributions to B^0–$\bar{B}^0$ mixing are analogous to Eq. (3.24) and Eq. (3.26) respectively changing $f_K^2 m_K \rightarrow f_B^2 m_B$, $m_s \rightarrow m_b$, $m_c \rightarrow m_t$ and $(V_{cs} V_{cd}^*)$ by $(V_{tb} V_{td}^*)$. Then we have,

$$a_{J/\psi}\big|_{\text{SUSY}} = \frac{\text{Im}\, M_{12}|_{\text{SUSY}}}{|M_{12}|_{\text{SM}}} \simeq \frac{\alpha_s^2 \sin^2 \theta_W}{\alpha_{\text{em}}^2} \frac{M_W^4}{M_{\text{SUSY}}^2 m_t^2} \frac{\text{Im}\left\{(\delta_{13}^d)_{\text{LL}}^2\right\}}{(V_{tb} V_{td}^*)^2},$$

$$\frac{8(24xf_6(x) + 66\tilde{f}_6(x))}{216} \simeq 12.5 \times 0.005 \times \frac{\text{Im}\left\{(\delta_{13}^d)_{\text{LL}}^2\right\}}{(0.008)^2} \times 0.037,$$

$$a_{J/\psi}\big|_{\text{SUSY}} \leq 0.74 \Rightarrow \sqrt{\text{Im}\left\{(\delta_{13}^d)_{\text{LL}}^2\right\}} \leq 0.14. \qquad (3.28)$$

From here we see that, although there is a difference due to masses and mixings, $m_c^2 (V_{cs} V_{cd}^*)^2$ versus $m_t^2 (V_{tb} V_{td}^*)^2$, the main reason for the difference in the MI bounds is the experimental sensitivity to CP violation observables. In the kaon system we can measure imaginary contributions to K–$\bar{K}$ mixings three orders of magnitude smaller than the real part while in the B system we can only distinguish imaginary contributions if they are of the same order as the mass difference. It is clear that we need much larger MI in the B system that in the K system to have observable effects [105]. On the other hand, as we will show in the next section, in realistic flavour models we expect larger MI in b transitions that in s transitions. Whether the B–system or K–system is more sensitive to SUSY will finally depend on the particular model considered.

Similarly, $b \rightarrow s$ transitions can be very interesting in SUSY models [114–125]. In fact, the only phenomenological constraints in this sector come from the $b \rightarrow s\gamma$ process. As we can see in Table 3, the bounds are stringent only for the $(\delta_{23}^d)_{\text{LR}}$ while they are very weak for $(\delta_{23}^d)_{\text{LL,RR}}$. A large $(\delta_{23}^d)_{\text{LL,RR,LR}}$ could have observable effects in several decays like $B \rightarrow \Phi K_S$ that can still differ from the SM predictions [126, 127].

Another interesting CP violating process in SUSY is ε'/ε [128–134]. We present the corresponding MI bounds from $\varepsilon'/\varepsilon < 2.7 \times 10^{-3}$ [135, 136] in Table 4. This observable is more sensitive to chirality changing MI due to the dominance of the gluonic and electroweak penguin operators. The bounds on $\Im (\delta_{12}^d)_{\text{LR}}$ look really tight and in fact these are the strongest bounds attainable on this MI. However, it is important to remember that these off-diagonal LR mass

Table 3

Limits on $|(\delta_{13}^d)|$, from the $b \to s\gamma$ decay, for an average squark mass $m_{\tilde{q}} = 500\text{GeV}$ and for different values of $x = m_{\tilde{g}}^2/m_{\tilde{q}}^2$. For different values of $m_{\tilde{q}}$, the limits can be obtained multiplying the ones in the table by $(m_{\tilde{q}}(\text{GeV})/500)^2$.

x	$|(\delta_{23}^d)_{\text{LL}}|$	$|(\delta_{23}^d)_{\text{LR}}|$
0.3	4.4	1.3×10^{-2}
1.0	8.2	1.6×10^{-2}
4.0	26	3.2×10^{-2}

Table 4

Limits from $\varepsilon'/\varepsilon < 2.7 \times 10^{-3}$ on $\Im(\delta_{12}^d)$, for an average squark mass $m_{\tilde{q}} = 500\text{GeV}$ and for different values of $x = m_{\tilde{g}}^2/m_{\tilde{q}}^2$. For different values of $m_{\tilde{q}}$, the limits can be obtained multiplying the ones in the table by $(m_{\tilde{q}}(\text{GeV})/500)^2$.

x	$|\Im(\delta_{12}^d)_{\text{LL}}|$	$|\Im(\delta_{12}^d)_{\text{LR}}|$
0.3	1.0×10^{-1}	1.1×10^{-5}
1.0	4.8×10^{-1}	2.0×10^{-5}
4.0	2.6×10^{-1}	6.3×10^{-5}

insertions come from the trilinear soft breaking terms which in realistic models are always proportional to fermion masses. Thus this MI typically contains a suppression $m_s/M_{\text{SUSY}} \simeq 2 \times 10^{-4}$ for $M_{\text{SUSY}} = 500$ GeV. So, if we consider this "intrinsic" suppression the bounds are less impressive.

In summary, these MI bounds show the present sensitivity of CP violation experiments to the presence of new phases and flavour structures in the SUSY soft breaking terms. An important lesson we can draw from the stringent bounds in the tables is that, in fact we already posses a crucial information on the enormous (123-dimensional) parameter space of a generic MSSM: most of this parameter space is already now excluded by flavour physics, and indeed the "realistic" MSSM realisation should not depart too strongly from the CMSSM, at least barring significant accidental cancellations.

3.3. *Grand unification of FCNCs*

As we have been discussing, in a SUSY-GUT, quarks and leptons sit in same multiplets and are transformed ones into the others through GU symmetry transformations. If the supergravity Lagrangian, and, in particular, its Kähler function are present at a scale larger than the GUT breaking scale, they have to fully respect the underlying gauge symmetry which is the GU symmetry itself. The subsequent SUSY breaking will give rise to the usual soft breaking terms in the

Lagrangian. In particular, the sfermion mass matrices, originating from the Kähler potential, will have to respect the underlying GU symmetry. Hence we expect hadron-lepton correlations among entries of the sfermion mass matrices. In other words, the quark-lepton unification seeps also into the SUSY breaking soft sector [137].

Imposition of a GU symmetry on the $\mathcal{L}_{soft}$ entails relevant implications at the weak scale. This is because the flavour violating (FV) mass-insertions do not get strongly renormalised through RG scaling from the GUT scale to the weak scale in the absence of new sources of flavor violation. On the other hand, if such new sources are present, for instance due to the presence of new neutrino Yukawa couplings in SUSY GUTs with a seesaw mechanism for neutrino masses, then one can compute the RG-induced effects in terms of these new parameters. Hence, the correlations between hadronic and leptonic flavor violating MIs survive at the weak scale to a good approximation. As for the flavor conserving (FC) mass insertions (i.e., the diagonal entries of the sfermion mass matrices), they get strongly renormalised, but in a way which is RG computable.

To summarise, in SUSY GUTs where the soft SUSY breaking terms respect boundary conditions which are subject to the GU symmetry to start with, we generally expect the presence of relations among the (bilinear and trilinear) scalar terms in the hadronic and leptonic sectors. Such relations hold true at the (superlarge) energy scale where the correct symmetry of the theory is the GU symmetry. After its breaking, the mentioned relations will undergo corrections which are computable through the appropriate RGE's which are related to the specific structure of the theory between the GU and the electroweak scale (for instance, new Yukawa couplings due to the presence of right-handed (RH) neutrinos acting down to the RH neutrino mass scale, presence of a symmetry breaking chain with the appearance of new symmetries at intermediate scales, etc.). As a result of such a computable running, we can infer the correlations between the softly SUSY breaking hadronic and leptonic δ terms at the low scale where we perform our FCNC tests.

Given that a common SUSY soft-breaking scalar term of $\mathcal{L}_{soft}$ at scales close to M_{Planck} can give rise to RG-induced δ^q's and δ^l's at the weak scale, one may envisage the possibility to make use of the FCNC constraints on such low-energy δ's to infer bounds on the soft breaking parameters of the original supergravity Lagrangian ($\mathcal{L}_{sugra}$). Indeed, for each scalar soft parameter of $\mathcal{L}_{sugra}$ one can ascertain whether the hadronic or the leptonic corresponding bound at the weak scale yields the stronger constraint at the large scale. One can then go through an exhaustive list of the low-energy constraints on the various δ^q's and δ^l's and, then, after RG evolving such δ's up to M_{Planck}, we will establish for each δ of $\mathcal{L}_{sugra}$ which one between the hadronic and leptonic constraints is going to win, namely which provides the strongest constraint on the corresponding δ_{sugra} [138].

Consider for example the scalar soft breaking sector of the MSSM:

$$-\mathcal{L}_{soft} = m^2_{\tilde{Q}_{ii}} \tilde{Q}^\dagger_i \tilde{Q}_i + m^2_{u^c_{ii}} \tilde{u}^{c\star}_i \tilde{u}^c{}_i + m^2_{e^c_{ii}} \tilde{e}^{c\star}_i \tilde{e}^c{}_i + m^2_{d^c_{ii}} \tilde{d}^{c\star}_i \tilde{d}^c{}_i$$

$$+ m^2_{\tilde{L}_{ii}} \tilde{L}^\dagger_i \tilde{L}_i + m^2_{H_1} H^\dagger_1 H_1 + m^2_{H_2} H^\dagger_2 H_2 + A^u_{ij} \tilde{Q}_i \tilde{u}^c{}_j H_2$$

$$+ A^d_{ij} \tilde{Q}_i \tilde{d}^c{}_j H_1 + A^e_{ij} \tilde{L}_i \tilde{e}^c{}_j H_1 + (\Delta^l_{ij})_{LL} \tilde{L}^\dagger_i \tilde{L}_j + (\Delta^e_{ij})_{RR} \tilde{e}^{c\star}_i \tilde{e}^c{}_j$$

$$+ (\Delta^q_{ij})_{LL} \tilde{Q}^\dagger_i \tilde{Q}_j + (\Delta^u_{ij})_{RR} \tilde{u}^{c\star}_i \tilde{u}^c{}_j + (\Delta^d_{ij})_{RR} \tilde{d}^{c\star}_i \tilde{d}^c{}_j$$

$$+ (\Delta^e_{ij})_{LR} \tilde{e}^{\star}_{L_i} \tilde{e}^c{}_j + (\Delta^u_{ij})_{LR} \tilde{u}^{\star}_{L_i} \tilde{u}^c{}_j + (\Delta^d_{ij})_{LR} \tilde{d}^{\star}_{L_i} \tilde{d}^c{}_j, \qquad (3.29)$$

where we have explicitly written down the various Δ parameters.

Consider now that $SU(5)$ is the relevant symmetry at the scale where the above soft terms firstly show up. Then, taking into account that matter is organised into the SU(5) representations $\mathbf{10} = (q, u^c, e^c)$ and $\bar{\mathbf{5}} = (l, d^c)$, one obtains the following relations

$$m^2_{\tilde{Q}} = m^2_{\tilde{e}^c} = m^2_{\tilde{u}^c} = m^2_{\mathbf{10}}, \qquad (3.30)$$

$$m^2_{\tilde{d}^c} = m^2_{\tilde{L}} = m^2_{\bar{\mathbf{5}}}, \qquad (3.31)$$

$$A^e_{ij} = A^d_{ji}. \qquad (3.32)$$

Eqs. (3.30, 3.31, 3.32) are matrices in flavor space. These equations lead to relations between the slepton and squark flavor violating off-diagonal entries Δ_{ij}. These are:

$$(\Delta^u_{ij})_{LL} = (\Delta^d_{ij})_{RR} = (\Delta^d_{ij})_{LL} = (\Delta^l_{ij})_{RR}, \qquad (3.33)$$

$$(\Delta^d_{ij})_{RR} = (\Delta^l_{ij})_{LL}, \qquad (3.34)$$

$$(\Delta^d_{ij})_{LR} = (\Delta^l_{ji})_{LR} = (\Delta^l_{ij})^{\star}_{RL}. \qquad (3.35)$$

These GUT correlations among hadronic and leptonic scalar soft terms that are summarised in the second column of Table 5. Assuming that no new sources of flavor structure are present from the $SU(5)$ scale down to the electroweak scale, apart from the usual SM CKM one, one infers the relations in the first column of Table 5 at low scale. Here we have taken into account that due to their different gauge couplings "average" (diagonal) squark and slepton masses acquire different values at the electroweak scale.

Two comments are in order when looking at Table 5. First, the boundary conditions on the sfermion masses at the GUT scale (last column in Table 5) imply that the squark masses are *always* going to be larger at the weak scale compared to the slepton masses due to the participation of the QCD coupling in

Table 5

Links between various transitions between up-type, down-type quarks and charged leptons for SU(5). $m_{\tilde{f}}^2$ refers to the average mass for the sfermion f, $m_{\tilde{Q}_{\text{avg}}}^2 = \sqrt{m_{\tilde{Q}}^2 m_{\tilde{d}^c}^2}$ and $m_{\tilde{L}_{\text{avg}}}^2 = \sqrt{m_{\tilde{L}}^2 m_{\tilde{e}^c}^2}$.

	Relations at weak-scale	Boundary conditions at M_{GUT}
(1)	$(\delta_{ij}^u)_{RR} \approx (m_{\tilde{e}^c}^2/m_{\tilde{u}^c}^2)\,(\delta_{ij}^l)_{RR}$	$m_{\tilde{u}^c}^2(0) = m_{\tilde{e}^c}^2(0)$
(2)	$(\delta_{ij}^q)_{LL} \approx (m_{\tilde{e}^c}^2/m_{\tilde{Q}}^2)\,(\delta_{ij}^l)_{RR}$	$m_{\tilde{Q}}^2(0) = m_{\tilde{e}^c}^2(0)$
(3)	$(\delta_{ij}^d)_{RR} \approx (m_{\tilde{L}}^2/m_{\tilde{d}^c}^2)\,(\delta_{ij}^l)_{LL}$	$m_{\tilde{d}^c}^2(0) = m_{\tilde{L}}^2(0)$
(4)	$(\delta_{ij}^d)_{LR} \approx (m_{\tilde{L}_{\text{avg}}}^2/m_{\tilde{Q}_{\text{avg}}}^2)\,(m_b/m_\tau)\,(\delta_{ij}^l)_{LR}^\star$	$A_{ij}^e = A_{ji}^d$

Table 6

Links between various transitions between up-type, down-type quarks and charged leptons for PS/SO(10) type models.

	Relations at weak-scale	Boundary conditions at M_{GUT}
(1)	$(\delta_{ij}^u)_{RR} \approx (m_{\tilde{e}^c}^2/m_{\tilde{u}^c}^2)\,(\delta_{ij}^l)_{RR}$	$m_{\tilde{u}^c}^2(0) = m_{\tilde{e}^c}^2(0)$
(2)	$(\delta_{ij}^q)_{LL} \approx (m_{\tilde{L}}^2/m_{\tilde{Q}}^2)\,(\delta_{ij}^l)_{LL}$	$m_{\tilde{Q}}^2(0) = m_{\tilde{L}}^2(0)$

the RGEs. As a second remark, notice that the relations between hadronic and leptonic δ MI in Table 5 always exhibit opposite "chiralities", i.e. LL insertions are related to RR ones and vice-versa. This stems from the arrangement of the different fermion chiralities in $SU(5)$ five- and ten-plets (as it clearly appears from the final column in Table 5). This restriction can easily be overcome if we move from $SU(5)$ to left-right symmetric unified models like SO(10) or the Pati-Salam (PS) case (we exhibit the corresponding GUT boundary conditions and δ MI at the electroweak scale in Table 6).

So far we have confined our discussion within the simple $SU(5)$ model, without the presence of any extra particles like right handed (RH) neutrinos. In the presence of RH neutrinos, one can envisage of two scenarios [139]: (a) with either very small neutrino Dirac Yukawa couplings and/or very small mixing present in the neutrino Dirac Yukawa matrix, (b) Large Yukawa and large mixing in the neutrino sector. In the latter case, Eqs. (3.33–3.35) are not valid at all scales in general, as large RGE effects can significantly modify the sleptonic flavour structure while keeping the squark sector essentially unmodified; thus essentially breaking the GUT symmetric relations. In the former case where the neutrino Dirac Yukawa couplings are tiny and do not significantly modify the sleptonic flavour structure, the GUT symmetric relations are expected to be valid at the weak scale. However, in both cases it is possible to say that there exists a upper bound on the hadronic δ parameters of the form [137]:

$$|(\delta_{ij}^d)_{\text{RR}}| \geq \frac{m_{\tilde{L}}^2}{m_{\tilde{d}^c}^2}|(\delta_{ij}^l)_{\text{LL}}|. \tag{3.36}$$

As an example of these GUT relations, let us compute the bounds on $\left(\delta_{ij}^d\right)_{AB}$ parameters, with $A, B = L, R$, from Lepton Flavour Violation (LFV) rare decays $l_j \to l_i, \gamma$, using the relations described above. First, we will analyse the 23 sector, that has been recently of much interest due to the discrepancy with SM expectations in the measurements of the CP asymmetry $A_{CP}(B \to \phi K_s)$, which can be attributed to the presence of large neutrino mixing within $SO(10)$ models [83, 116, 140, 141]. Subsequently, a detailed analysis has been presented [118, 119] within the context of MSSM. It has been shown that [119] the presence of a large $\sim \mathcal{O}(1)$ δ_{23}^d of LL or RR type could lead to significant discrepancies from the SM expectations and in particular one could reach the present central value for the measurement of $A_{CP}(B \to \phi K_s)$. Similar statements hold for a relatively small $\sim \mathcal{O}(10^{-2})$ LR and RL type MI.

Now, we would like to analyse the impact of LFV bounds on these hadronic δ parameters and its effect on B-physics observables. In table 7, we present up-

Table 7

Bounds on (δ_{23}^d) from $\tau \to \mu, \gamma$ for three different values of the branching ratios for $\tan \beta = 10$.

Type	$< 1.1\ 10^{-6}$	$< 6\ 10^{-7}$	$< 1.\ 10^{-7}$
LL	–	–	–
RR	0.105	0.075	0.03
RL	0.108	0.08	0.035
LR	0.108	0.08	0.035

per bounds on $\left(\delta_{23}^d\right)_{\text{RR}}$ with squark masses in the range 350–500 GeVs and for three different upper bounds on $\text{Br}(\tau \to \mu, \gamma)$. There are no bounds on $\left(\delta_{23}^d\right)_{\text{LL}}$ because large values of $\left(\delta_{23}^l\right)_{\text{RR}}$ are still allowed due to possible cancellations of bino and higgsino contributions for the decay amplitudes [83, 140, 142]. In Fig. 11 we present the allowed ranges of $\left(\delta_{23}^d\right)_{\text{RR}}$ and its effects on the CP asymmetry, $A_{CP}(B \to \phi K_s)$, taking into account only hadronic constraints (left) or hadronic and leptonic constraints simultaneously (right). Thus, we can see that in a $SU(5)$ GUT model where SUSY-breaking terms have a supergravity origin, LFV constraints are indeed very relevant for $(\delta_{23}^d)_{RR}$ and it is not possible to generate large effects on $A_{CP}(B \to \phi K_s)$. Naturally we have to take into account that the leptonic bounds and their effects on hadronic MIs scale as $10/\tan \beta$ for

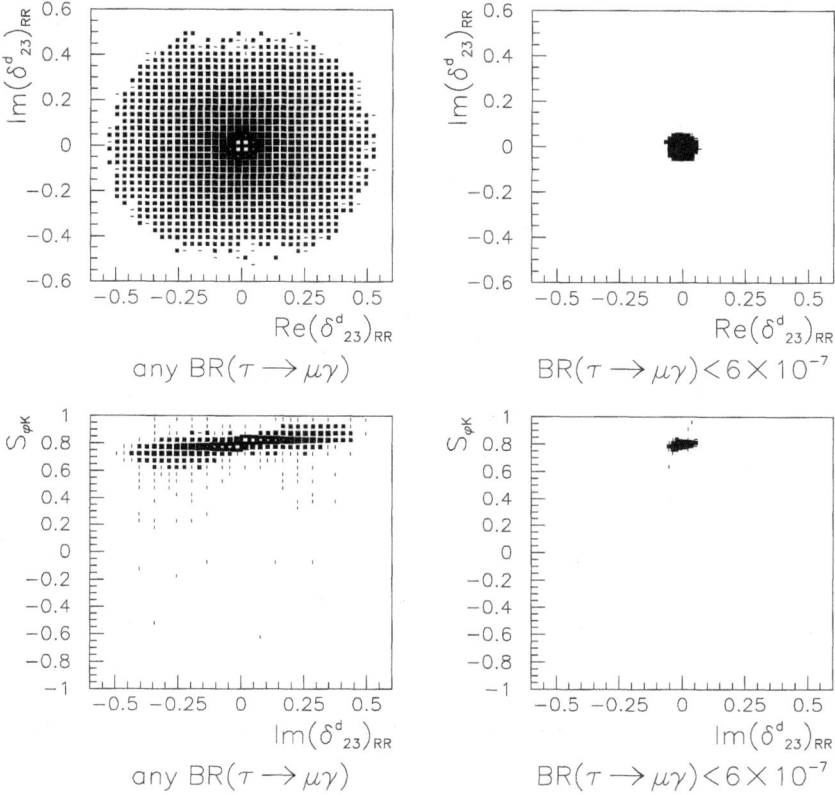

Fig. 11. Allowed regions in the $\text{Re}(\delta^d_{23})_{RR}$–$\text{Im}(\delta^d_{23})_{RR}$ plane (top) and in the $S_{K\phi}$–$\text{Im}(\delta^d_{23})_{RR}$ plane (bottom). Constraints from $B \to X_s\gamma$, $BR(B \to X_s\ell^+\ell^-)$, and the lower bound on ΔM_s have been used.

different values of $\tan\beta$. However, even for $\tan\beta \leq 5$ the leptonic bounds would be very relevant on this MI.

Finally, we will also analyse the effects of leptonic constraints in the 12 sector. In Fig. 12 we present the allowed values of $\text{Re}\left(\delta^d_{12}\right)_{RR}$ and $\text{Im}\left(\delta^d_{12}\right)_{RR}$. The upper left plot corresponds to the values that satisfy the hadronic bounds, coming mainly from $\varepsilon_K = (2.284 \pm 0.014) \times 10^{-3}$. The upper right plot takes also into account the present $\mu \to e\gamma$ bound, $BR(\mu \to e, \gamma) < 1.1 \times 10^{-11}$, and the plots in the second row correspond to projected bounds from the proposed experiments, $BR(\mu \to e, \gamma) < 10^{-13}$ and $BR(\mu \to e, \gamma) < 10^{-14}$ respectively. Now the GUT symmetry relates $\left(\delta^d_{12}\right)_{RR}$ to $\left(\delta^l_{12}\right)_{LL}$ and in this case leptonic bounds

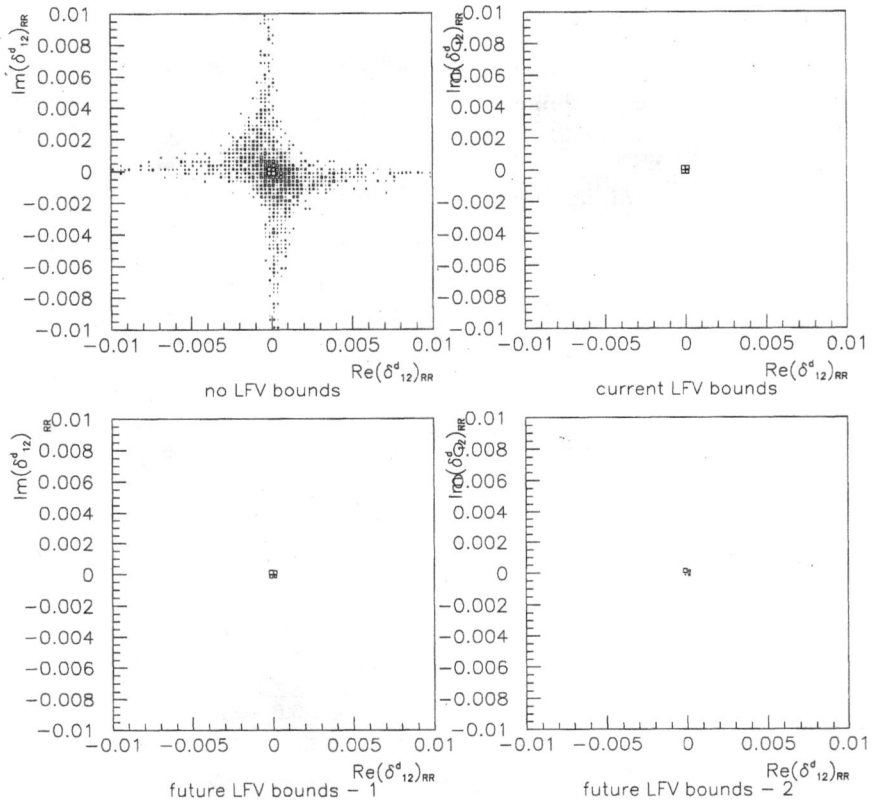

Fig. 12. Allowed regions in the $\text{Re}(\delta^d_{12})_{\text{RR}}$–$\text{Im}(\delta^d_{12})_{\text{RR}}$ plane from hadronic and leptonic constraints. Upper left plot takes into account only hadronic bounds, upper right plot includes the present bound on the $\mu \to e\gamma$ decay, $\text{BR}(\mu \to e, \gamma) < 1.1 \times 10^{-11}$. The second row correspond to the projected bounds from the proposed LFV experiments, $\text{BR}(\mu \to e, \gamma) < 10^{-13}$ and $\text{BR}(\mu \to e, \gamma) < 10^{-14}$ respectively. We have to take into account that we use $\tan\beta = 10$ and leptonic bounds scale as $10/\tan\beta$.

(already the present bounds) are very stringent and reduce the allowed values of $(\delta^d_{12})_{\text{RR}}$ by more than one order of magnitude to a value $(\delta^d_{12})_{\text{RR}} \leq 4 \times 10^{-4}$ for $\tan\beta = 10$.

In the case of $(\delta^d_{12})_{\text{LL}}$ the $\mu \to e\gamma$ decay does not provide a bound to this MI due to the presence of cancellations between different contributions. We can only obtain a relatively mild bound, $(\delta^l_{12})_{\text{RR}} \leq 0.09$ for $\tan\beta = 10$, if we take into account $\mu \to eee$ and μ–e conversion in nuclei. After rescaling this bound by the factor $\frac{\tilde{m}^2_{e^c}}{\tilde{m}^2_{d_L}}$ the leptonic bound is still able to reduce the maximum values of Re $\left(\delta^d_{12}\right)_{\text{LL}}$ and Im $\left(\delta^d_{12}\right)_{\text{LL}}$ by a factor of 2, although the hadronic bound is still more constraining in a big part of the parameter space.

In summary, Supersymmetric Grand Unification predicts links between various leptonic and hadronic FCNC Observables. Though such relations can be constructed for any GUT group, we have concentrated on SU(5) and quantitatively studied the implications for the 23 and 12 sectors. In particular we have shown that the present limit on $BR(\tau \to \mu, \gamma)$ is sufficient to significantly constrain the observability of supersymmetry in CP violating B-decays.

3.4. Supersymmetric seesaw and lepton flavour violation

As discussed in the above, flavour violation can also be generated through renormalisation group running even if one starts with flavour-blind soft masses at the scale where supersymmetry is mediated to the visible sector. A classic example of this is the supersymmetric seesaw mechanism and the generation of lepton flavour violation at the weak scale.

The seesaw mechanism can be incorporated in the Minimal Supersymmetric Standard Model in a manner similar to what is done in the Standard Model by adding right-handed neutrino superfields to the MSSM superpotential:

$$
\begin{aligned}
W = {}& h^u_{ij} Q_i u^c_j H_2 + h^d_{ii} Q_i d^c_i H_1 + h^e_{ii} L_i e^c_i H_1 + h^\nu_{ij} L_i \nu^c_j H_2 \\
& + M_{R_{ii}} \nu^c_i \nu^c_i + \mu H_1 H_2,
\end{aligned} \tag{3.37}
$$

where we are in the basis of diagonal charged lepton, down quark and right-handed Majorana mass matrices. M_R represents the (heavy) Majorana mass matrix for the right-handed neutrinos. Eq. (3.37) leads to the standard seesaw formula for the (light) neutrino mass matrix

$$
\mathcal{M}_\nu = -h^\nu M_R^{-1} h^{\nu\,T} v_2^2, \tag{3.38}
$$

where v_2 is the vacuum expectation value (VEV) of the up-type Higgs field, H_2. Under suitable conditions on h^ν and M_R, the correct mass splittings and mixing angles in $\mathcal{M}_\nu$ can be obtained. Detailed analyses deriving these conditions are already present in the literature [143–150].

Following the discussion in the previous section, we will assume that the mechanism that breaks supersymmetry and conveys it to the observable sector at the high scale $\sim M_P$ is flavour-blind, as in the CMSSM (also called mSUGRA). However, this flavour blindness is not protected down to the weak scale [151][5].

---

[5]This is always true in a gravity mediated supersymmetry breaking model, but it also applies to other mechanisms under some specific conditions [152, 153].

The slepton mass matrices are no longer invariant under RG evolution from the super-large scale where supersymmetry is mediated to the visible sector down to the seesaw scale. The flavour violation present in the neutrino Dirac Yukawa couplings h^ν is now "felt" by the slepton mass matrices in the presence of heavy right-handed neutrinos [154, 155].

The weak-scale flavour violation so generated can be obtained by solving the RGEs for the slepton mass matrices from the high scale to the scale of the right-handed neutrinos. Below this scale, the running of the FV slepton mass terms is RG-invariant as the right-handed neutrinos decouple from the theory. For the purpose of illustration, a leading-log estimate can easily be obtained for these equations[6]. Assuming the flavour blind mSUGRA specified by the high-scale parameters, m_0, the common scalar mass, A_0, the common trilinear coupling, and $M_{1/2}$, the universal gaugino mass, the flavour violating entries in these mass matrices at the weak scale are given as:

$$(\Delta_{ij}^l)_{\mathrm{LL}} \approx -\frac{3m_0^2 + A_0^2}{8\pi^2} \sum_k (h_{ik}^\nu h_{jk}^{\nu*}) \ln \frac{M_X}{M_{R_k}}, \tag{3.39}$$

where h^ν are given in the basis of diagonal charged lepton masses and diagonal Majorana right-handed neutrino mass matrix M_R, and M_X is the scale at which soft terms appear in the Lagrangian. Given this, the branching ratios for LFV rare decays $l_j \rightarrow l_i, \gamma$ can be roughly estimated using

$$\mathrm{BR}(l_j \rightarrow l_i \gamma) \approx \frac{\alpha^3 \ |\delta_{ij}^l|^2}{G_F^2 \ m_{\mathrm{SUSY}}^4} \tan^2 \beta. \tag{3.40}$$

From above it is obvious that the amount of lepton flavour violation generated by the SUSY seesaw at the weak scale crucially depends on the flavour structure of h^ν and M_R, the "new" sources of flavour violation not present in the MSSM, Eq. (3.37). If either the neutrino Yukawa couplings or the flavour mixings present in h^ν are very tiny, the strength of LFV will be significantly reduced. Further, if the right-handed neutrino masses were heavier than the supersymmetry breaking scale (as in GMSB models) they would decouple from the theory before the SUSY soft breaking matrices enter into play and hence these effects would vanish.

---

[6]Within mSUGRA, the leading-log approximation works very well for most of the parameter space, except for regions of large $M_{1/2}$ and low m_0. The discrepancy with the exact result increases with low $\tan \beta$ [156].

3.5. Seesaw in GUTs: SO(10) and LFV

A simple analysis of the fermion mass matrices in the $SO(10)$ model, as detailed in the Eq. (2.72) leads us to the following result: *At least one of the Yukawa couplings in $h^\nu = v_u^{-1} M_{LR}^\nu$ has to be as large as the top Yukawa coupling* [139]. This result holds true in general, independently of the choice of the Higgses responsible for the masses in Eqs. (2.71), (2.72), provided that no accidental fine-tuned cancellations of the different contributions in Eq. (2.72) are present. If contributions from the **10**'s solely dominate, h^ν and h^u would be equal. If this occurs for the **126**'s, then $h^\nu = -3\, h^u$ [157]. In case both of them have dominant entries, barring a rather precisely fine-tuned cancellation between M_{10}^5 and M_{126}^5 in Eq. (2.72), we expect at least one large entry to be present in h^ν. A dominant antisymmetric contribution to top quark mass due to the **120** Higgs is phenomenologically excluded, since it would lead to at least a pair of heavy degenerate up quarks.

Apart from sharing the property that at least one eigenvalue of both M^u and M_{LR}^ν has to be large, for the rest it is clear from Eqs. (2.71) and (2.72) that these two matrices are not aligned in general, and hence we may expect different mixing angles appearing from their diagonalisation. This freedom is removed if one sticks to particularly simple choices of the Higgses responsible for up quark and neutrino masses. A couple of remarks are in order here. Firstly, note that in general there can be an additional contribution, Eq. (2.75), to the light neutrino mass matrix, independent of the canonical seesaw mechanism. Taking into consideration also this contribution leads to the so-called Type-II seesaw formula [158, 159]. Secondly, the correlation between neutrino Dirac Yukawa coupling and the top Yukawa is in general independent of the type of seesaw mechanism, and thus holds true irrespective of the light-neutrino mass structure.

Therefore, we see that the $SO(10)$ model with only two ten-plets would inevitably lead to small mixing in h^ν. In fact, with two Higgs fields in symmetric representations, giving masses to the up-sector and the down-sector separately, it would be difficult to avoid the small CKM-like mixing in h^ν. We will call this case the CKM case. From here, the following mass relations hold between the quark and leptonic mass matrices at the GUT scale[7]:

$$h^u = h^\nu, \quad h^d = h^e. \tag{3.41}$$

In the basis where charged lepton masses are diagonal, we have

$$h^\nu = V_{\text{CKM}}^T\, h_{Diag}^u\, V_{\text{CKM}}. \tag{3.42}$$

---

[7] Clearly this relation cannot hold for the first two generations of down quarks and charged leptons. One expects, small corrections due to non-renormalisable operators or suppressed renormalisable operators [33] to be invoked.

The large couplings in $h^\nu \sim \mathcal{O}(h_t)$ induce significant off-diagonal entries in $m_{\tilde{L}}^2$ through the RG evolution between M_{GUT} and the scale of the right-handed Majorana neutrinos[8], M_{R_i}. The induced off-diagonal entries relevant to $l_j \to l_i, \gamma$ are of the order of:

$$(m_{\tilde{L}}^2)_{21} \approx -\frac{3m_0^2 + A_0^2}{8\pi^2} h_t^2 V_{td} V_{ts} \ln \frac{M_{\text{GUT}}}{M_{R_3}} + \mathcal{O}(h_c^2), \qquad (3.43)$$

$$(m_{\tilde{L}}^2)_{32} \approx -\frac{3m_0^2 + A_0^2}{8\pi^2} h_t^2 V_{tb} V_{ts} \ln \frac{M_{\text{GUT}}}{M_{R_3}} + \mathcal{O}(h_c^2), \qquad (3.44)$$

$$(m_{\tilde{L}}^2)_{31} \approx -\frac{3m_0^2 + A_0^2}{8\pi^2} h_t^2 V_{tb} V_{td} \ln \frac{M_{\text{GUT}}}{M_{R_3}} + \mathcal{O}(h_c^2). \qquad (3.45)$$

In these expressions, the CKM angles are small but one would expect the presence of the large top Yukawa coupling to compensate such a suppression. The required right-handed neutrino Majorana mass matrix, consistent with both the observed low energy neutrino masses and mixings as well as with CKM-like mixings in h^ν is easily determined from the seesaw formula defined at the scale of right-handed neutrinos[9].

The Br($l_i \to l_j\gamma$) are now predictable in this case. Considering mSUGRA boundary conditions and taking $\tan\beta = 40$, we obtain that reaching a sensitivity of 10^{-14} for BR($\mu \to e\gamma$) would allow us to probe the SUSY spectrum completely up to $M_{1/2} = 300$ GeV (notice that this corresponds to gluino and squark masses of order 750 GeV) and would still probe large regions of the parameter space up to $M_{1/2} = 700$ GeV. Thus, in summary, though the present limits on BR($\mu \to e, \gamma$) would not induce any significant constraints on the supersymmetry-breaking parameter space, an improvement in the limit to $\sim \mathcal{O}(10^{-14})$, as foreseen, would start imposing non-trivial constraints especially for the large $\tan\beta$ region.

To obtain mixing angles larger than CKM angles, asymmetric mass matrices have to be considered. In general, it is sufficient to introduce asymmetric textures either in the up-sector or in the down-sector. In the present case, we assume that the down-sector couples to a combination of Higgs representations (symmetric

---

[8]Typically one has different mass scales associated with different right-handed neutrino masses.

[9]The neutrino masses and mixings here are defined at M_R. Radiative corrections can significantly modify the neutrino spectrum from that of the weak scale [160]. This is more true for the degenerate spectrum of neutrino masses [161–163] and for some specific forms of h^ν [164]. For our present discussion, with hierarchical neutrino masses and up-quark like neutrino Yukawa matrices, we expect these effects not to play a very significant role.

and antisymmetric)[10] Φ, leading to an asymmetric mass matrix in the basis where the up-sector is diagonal. As we will see below, this would also require that the right-handed Majorana mass matrix be diagonal in this basis. We have:

$$W_{SO(10)} = \frac{1}{2} h_{ii}^{u,v} \, \mathbf{16_i 16_i 10^u} + \frac{1}{2} h_{ij}^{d,e} \, \mathbf{16_i 16_j} \Phi + \frac{1}{2} h_{ii}^R \, \mathbf{16_i 16_i 126}, \qquad (3.46)$$

where the **126**, as before, generates only the right-handed neutrino mass matrix. To study the consequences of these assumptions, we see that at the level of $SU(5)$, we have

$$W_{SU(5)} = \frac{1}{2} h_{ii}^u \, \mathbf{10_i 10_i 5_u} + h_{ii}^v \, \mathbf{\bar{5}_i 1_i 5_u} + h_{ij}^d \, \mathbf{10_i \bar{5}_j \bar{5}_d} + \frac{1}{2} M_{ii}^R \, \mathbf{1_i 1_i}, \qquad (3.47)$$

where we have decomposed the **16** into $\mathbf{10 + \bar{5} + 1}$ and $\mathbf{5_u}$ and $\mathbf{\bar{5}_d}$ are components of $\mathbf{10_u}$ and Φ respectively. To have large mixing $\sim U_{\text{PMNS}}$ in h^v we see that the asymmetric matrix h^d should now give rise to both the CKM mixing as well as PMNS mixing. This is possible if

$$V_{\text{CKM}}^T \, h^d \, U_{\text{PMNS}}^T = h_{Diag}^d. \qquad (3.48)$$

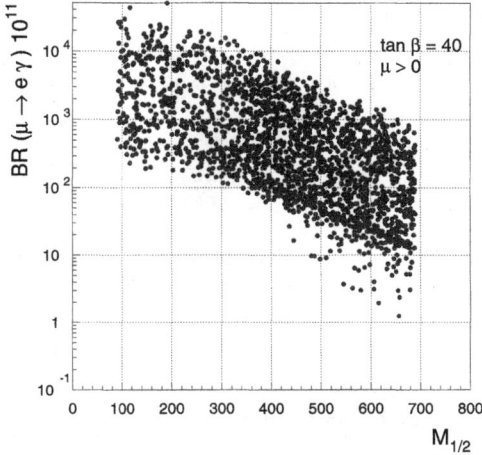

Fig. 13. The scatter plots of branching ratios of $\mu \rightarrow e, \gamma$ decays as a function of $M_{1/2}$ are shown for the (maximal) PMNS case for $\tan \beta = 40$. The results do not alter significantly with the change of sign(μ).

---

[10]The couplings of the Higgs fields in the superpotential can be either renormalisable or non-renormalisable. See [116] for a non-renormalisable example.

Therefore the **10** that contains the left-handed down-quarks would be rotated by the CKM matrix whereas the **5̄** that contains the left-handed charged leptons would be rotated by the U_{PMNS} matrix to go into their respective mass bases [116, 165–167]. Thus we have, in analogy with the previous subsection, the following relations in the basis where charged leptons and down quarks are diagonal:

$$h^u = V_{\text{CKM}} \, h^u_{Diag} \, V^T_{\text{CKM}}, \tag{3.49}$$

$$h^\nu = U_{\text{PMNS}} \, h^u_{Diag}. \tag{3.50}$$

Using the seesaw formula of Eqs. (3.38) and (3.50), we have

$$M_R = Diag\left\{ \frac{m_u^2}{m_{\nu_1}}, \ \frac{m_c^2}{m_{\nu_2}}, \ \frac{m_t^2}{m_{\nu_3}} \right\}. \tag{3.51}$$

We now turn our attention to lepton flavour violation in this case. The branching ratio, $BR(\mu \to e, \gamma)$ would now depend on

$$[h^\nu h^{\nu \, T}]_{21} = h_t^2 \, U_{\mu 3} \, U_{e3} + h_c^2 \, U_{\mu 2} \, U_{e2} + \mathcal{O}(h_u^2). \tag{3.52}$$

It is clear from the above that in contrast to the CKM case, the dominant contribution to the off-diagonal entries depends on the unknown magnitude of the element U_{e3} [168]. If U_{e3} is very close to its present limit ~ 0.2 [169], the first term on the RHS of the Eq. (3.52) would dominate. Moreover, this would lead to large contributions to the off-diagonal entries in the slepton masses with $U_{\mu 3}$ of $\mathcal{O}(1)$. From Eq. (3.39) we have

$$(m_{\tilde{L}}^2)_{21} \approx -\frac{3m_0^2 + A_0^2}{8\pi^2} \, h_t^2 U_{e3} U_{\mu 3} \ln \frac{M_{\text{GUT}}}{M_{R_3}} + \mathcal{O}(h_c^2). \tag{3.53}$$

This contribution is larger than the CKM case by a factor of $(U_{\mu 3}U_{e3})/(V_{td}V_{ts}) \sim 140$. From Eq. (3.40) we see that it would mean about a factor 10^4 times larger than the CKM case in $BR(\mu \to e, \gamma)$. In case U_{e3} is very small, *i.e* either zero or $\lesssim (h_c^2/h_t^2) \, U_{e2} \sim 4 \times 10^{-5}$, the second term $\propto h_c^2$ in Eq. (3.52) would dominate. However the off-diagonal contribution in slepton masses, now being proportional to charm Yukawa could be much smaller, even smaller than the CKM contribution by a factor

$$\frac{h_c^2 \, U_{\mu 2} \, U_{e2}}{h_t^2 \, V_{td} \, V_{ts}} \sim 7 \times 10^{-2}. \tag{3.54}$$

If U_{e3} is close to its present limit, the current bound on $R(\mu \to e, \gamma)$ would already be sufficient to produce stringent limits on the SUSY mass spectrum.

Similar U_{e3} dependence can be expected in the $\tau \to e$ transitions where the off-diagonal entries are given by:

$$(m_{\tilde{L}}^2)_{31} \approx -\frac{3m_0^2 + A_0^2}{8\pi^2} h_t^2 U_{e3} U_{\tau 3} \ln \frac{M_{\text{GUT}}}{M_{R_3}} + \mathcal{O}(h_c^2). \tag{3.55}$$

The $\tau \to \mu$ transitions are instead U_{e3}-independent probes of SUSY, whose importance was first pointed out in Ref. [170]. The off-diagonal entry in this case is given by:

$$(m_{\tilde{L}}^2)_{32} \approx -\frac{3m_0^2 + A_0^2}{8\pi^2} h_t^2 U_{\mu 3} U_{\tau 3} \ln \frac{M_{\text{GUT}}}{M_{R_3}} \mathcal{O}(h_c^2). \tag{3.56}$$

In the PMNS scenario, Fig. 3 shows the plot for BR$(\mu \to e, \gamma)$ for $\tan \beta = 40$. In this plot, the value of U_{e3} chosen is very close to the present experimental upper limit [169]. As long as $U_{e3} \gtrsim 4 \times 10^{-5}$, the plots scale as U_{e3}^2, while for $U_{e3} \lesssim 4 \times 10^{-5}$ the term proportional to m_c^2 in Eq. (3.53) starts dominating; the result is then insensitive to the choice of U_{e3}. For instance, a value of $U_{e3} = 0.01$ would reduce the BR by a factor of 225 and still a significant amount of the parameter space for $\tan \beta = 40$ would be excluded. We further find that with the present limit on BR$(\mu \to e, \gamma)$, all the parameter space would be completely excluded up to $M_{1/2} = 300$ GeV for $U_{e3} = 0.15$, for any vale of $\tan \beta$ (not shown in the figure).

In the $\tau \to \mu\gamma$ decay the situation is similarly constrained. For $\tan \beta = 2$, the present bound of 3×10^{-7} starts probing the parameter space up to $M_{1/2} \lesssim 150$ GeV. The main difference is that this does not depend on the value of U_{e3}, and therefore it is already a very important constraint on the parameter space of the model. In fact, for large $\tan \beta = 40$, as shown in Fig. 4, reaching the expected limit of 1×10^{-8} would be able to rule out completely this scenario up to gaugino masses of 400 GeV, and only a small portion of the parameter space with heavier gauginos would survive. In the limit $U_{e3} = 0$, this decay mode would provide a constraint on the model stronger than $\mu \to e, \gamma$, which would now be suppressed as it would contain only contributions proportional to h_c^2, as shown in Eq. (3.53).

In summary, in the PMNS/maximal mixing case, even the present limits from BR$(\mu \to e, \gamma)$ can rule out large portions of the supersymmetry-breaking parameter space if U_{e3} is either close to its present limit or within an order of magnitude of it (as the planned experiments might find out soon [171]). These limits are more severe for large $\tan \beta$. In the extreme situation of U_{e3} being zero or very small $\sim \mathcal{O}(10^{-4} - 10^{-5})$, BR$(\tau \to \mu\gamma)$ will start playing an important role with its present constraints already disallowing large regions of the parameter space at large $\tan \beta$. While the above example concentrated on the hierarchical light

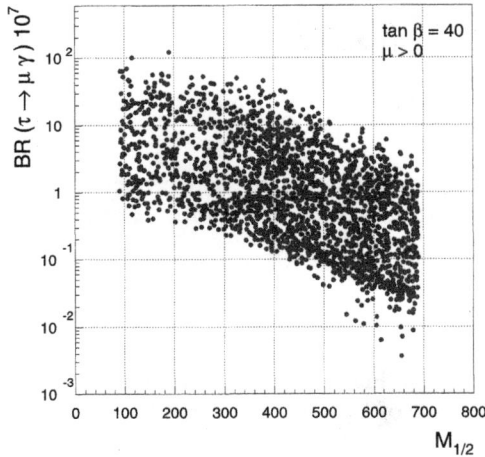

Fig. 14. The scatter plots of branching ratios of $\tau \to \mu, \gamma$ decays as a function of $M_{1/2}$ are shown for the (maximal) PMNS case for n the PMNS scenario, Fig. 3 shows the plot for BR($\mu \tan \beta = 40$. The results do not alter significantly with the change of sign(μ).

neutrinos, similar 'benchmark' mixing scenarios have been explored in great detail, for degenerate spectra of light neutrinos, by Ref. [172], taking also in to consideration running between the Planck scale and the GUT scale.

Conclusions

The ideas of the Grand Unification and Supersymmetry are closely connected and represent the main avenue to explore in the search of physics beyond the Standard Model. In these lectures we have presented the reasons that make us believe in the existence of new physics beyond the SM. We have presented the (non-supersymmetric) Grand Unification idea and analysed its achievements and failures. Supersymmetric grand unification was shown to cure some of these problems and make the construction of "realistic" models possible. The phenomenology of low-energy supersymmetry has been discussed in the second part of these lectures with special emphasis on the SUSY flavour and CP problems. We have seen that, quite generally, SUSY extensions of the SM lead to the presence of a host of new flavour and CP violation parameters. The solution of the "SUSY flavour problem" and the "SUSY CP problem" are intimately linked. However, there is an "intrinsic" CP problem in SUSY which goes beyond the flavour issue and requires a deeper comprehension of the link between CP violation and breaking of SUSY. We tried to emphasise that the these two problems have not only a dark and worrying side, but also they provide promising tools to obtain indirect SUSY hints. We have also seen that the presence of a grand unified symmetry and/or new particles, like right-handed neutrinos, at super-large scales has observable consequences in the structure of soft masses at the electroweak scale. Thus the discovery of low energy SUSY at the LHC or low energy FCNC experiments and the measurement of the SUSY spectrum may provide a fundamental clue for the assessment of SUSY GUTs and SUSY seesaw in nature.

Acknowledgements

Although some part of the material of these lectures reflects the personal views of the authors, much of it results from works done in collaboration with several friends, in particular M. Ciuchini, P. Paradisi and L. Silvestrini. We thank them very much. A.M. is very grateful to the organisers and participants of this school for the pleasant and stimulating environment they succeeded to create all along the school itself.

A.M. acknowledges partial support from the MIUR PRIN "Fisica Astroparticellare" 2004-2006. O.V. acknowledges partial support from the Spanish MCYT FPA2005-01678. S.K.V. acknowledges support from Indo-French Centre for Promotion of Advanced Research (CEFIPRA) project No:2904-2 "Brane world phenomenology". This work was partaly supported by RTN contract MRTN-CT-2004-503369. All the figures have been plotted with Jaxodraw [173].

References

[1] Y. Nir, "CP violation in meson decays", lectures at the Les Houches summer school: Particle Physics beyond the Standard Model, Aug. 2005 (this volume).

[2] A. Y. Smirnov, "Neutrino mass and mixing: toward the underlying physics", lectures at the Les Houches summer school: Particle Physics beyond the Standard Model, Aug. 2005 (this volume).

[3] P. Binétruy, "Particle astrophysics and cosmology", lectures at the Les Houches summer school: Particle Physics beyond the Standard Model, Aug. 2005 (this volume).

[4] J. R. Ellis and D. V. Nanopoulos, Phys. Lett. B **110** (1982) 44.

[5] R. Barbieri and R. Gatto, Phys. Lett. B **110** (1982) 211.

[6] M. J. Duncan and J. Trampetic, Phys. Lett. B **134** (1984) 439.

[7] J. M. Gerard, W. Grimus, A. Raychaudhuri, and G. Zoupanos, Phys. Lett. B **140** (1984) 349.

[8] J. M. Gerard, W. Grimus, A. Masiero, D. V. Nanopoulos, and A. Raychaudhuri, Phys. Lett. B **141** (1984) 79.

[9] P. Langacker and B. Sathiapalan, Phys. Lett. B **144** (1984) 401.

[10] J. M. Gerard, W. Grimus, A. Masiero, D. V. Nanopoulos, and A. Raychaudhuri, Nucl. Phys. B **253** (1985) 93.

[11] R. Barbieri, S. Ferrara, and C. A. Savoy, Phys. Lett. B **119** (1982) 343.

[12] A. H. Chamseddine, R. Arnowitt, and P. Nath, Phys. Rev. Lett. **49** (1982) 970.

[13] H. P. Nilles, Phys. Rept. **110** (1984) 1.

[14] H. E. Haber and G. L. Kane, Phys. Rept. **117** (1985) 75.

[15] G. G. Ross, *Grand Unified Theories*. Benjamin/cummings, Reading, Usa, 1984 (Frontiers in Physics, 60).

[16] H. E. Haber, hep-ph/9306207.

[17] S. P. Martin, hep-ph/9709356.

[18] G. F. Giudice and R. Rattazzi, Phys. Rept. **322** (1999) 419 [hep-ph/9801271].

[19] R. N. Mohapatra, Adv. Ser. Direct. High Energy Phys. **3** (1989) 436.

[20] Y. Grossman, Y. Nir, and R. Rattazzi, Adv. Ser. Direct. High Energy Phys. **15** (1998) 755 [hep-ph/9701231].

[21] M. Misiak, S. Pokorski, and J. Rosiek, Adv. Ser. Direct. High Energy Phys. **15** (1998) 795 [hep-ph/9703442].

[22] A. Masiero and O. Vives, Ann. Rev. Nucl. Part. Sci. **51** (2001) 161 [hep-ph/0104027].

[23] A. Masiero and O. Vives, New J. Phys. **4** (2002) 4.

[24] J. C. Pati and A. Salam, Phys. Rev. D **10** (1974) 275.

[25] H. Georgi and S. L. Glashow, Phys. Rev. Lett. **32** (1974) 438.

[26] H. Fritzsch and P. Minkowski, Ann. Phys. **93** (1975) 193.

[27] H. Georgi and D. V. Nanopoulos, Nucl. Phys. B **155** (1979) 52.

[28] H. Georgi, H. R. Quinn, and S. Weinberg, Phys. Rev. Lett. **33** (1974) 451.

[29] I. Antoniadis, C. Kounnas, and C. Roiesnel, Nucl. Phys. B **198** (1982) 317.

[30] P. Langacker, Phys. Rept. **72** (1981) 185.

[31] D. V. Nanopoulos and D. A. Ross, Nucl. Phys. B **157** (1979) 273.

[32] D. V. Nanopoulos and D. A. Ross, Phys. Lett. B **108** (1982) 351.

[33] H. Georgi and C. Jarlskog, Phys. Lett. B **86** (1979) 297.

[34] A. J. Buras, J. R. Ellis, M. K. Gaillard, and D. V. Nanopoulos, Nucl. Phys. B **135** (1978) 66.

[35] P. Nath and P. F. Perez, hep-ph/0601023.

[36] S. Weinberg, Phys. Rev. Lett. **43** (1979) 1566.

[37] S. Weinberg, Phys. Rev. D **22** (1980) 1694.

[38] F. Wilczek and A. Zee, Phys. Rev. Lett. **43** (1979) 1571.

[39] H. A. Weldon and A. Zee, Nucl. Phys. B **173** (1980) 269.

[40] W. J. Marciano, talk at the 4th Workshop on Grand Unification, Philadelphia, Pa., Apr 21-23, 1983.

[41] M. Drees, hep-ph/9611409.

[42] L. Girardello and M. T. Grisaru, Nucl. Phys. B **194** (1982) 65.

[43] M. Carena and C. Wagner, "Higgs physics and supersymmetry phenomenology", lectures at the Les Houches summer school: Particle Physics beyond the Standard Model, Aug. 2005.

[44] S. Dimopoulos, S. Raby, and F. Wilczek, Phys. Rev. D **24** (1981) 1681.

[45] W. J. Marciano and G. Senjanovic, Phys. Rev. D **25** (1982) 3092.

[46] U. Amaldi, W. de Boer, P. H. Frampton, H. Furstenau, and J. T. Liu, Phys. Lett. B **281** (1992) 374.

[47] R. N. Mohapatra, hep-ph/9911272.

[48] A. Masiero, D. V. Nanopoulos, K. Tamvakis, and T. Yanagida, Phys. Lett. B **115** (1982) 380.

[49] R. Arnowitt, A. H. Chamseddine, and P. Nath, Phys. Lett. B **156** (1985) 215.

[50] P. Nath, A. H. Chamseddine, and R. Arnowitt, Phys. Rev. D **32** (1985) 2348.

[51] G. Altarelli, F. Feruglio, and I. Masina, JHEP **11** (2000) 040 [hep-ph/0007254].

[52] G. F. Giudice and A. Masiero, Phys. Lett. B **206** (1988) 480.

[53] P. Minkowski, Phys. Lett. B **67** (1977) 421.

[54] T. Yanagida, in: Proceedings of the Workshop on the Baryon Number of the Universe and Unified Theories, Tsukuba, Japan, 13-14 Feb 1979.

[55] M. Gell-Mann, P. Ramond, and R. Slansky, Print-80-0576 (CERN).

[56] R. N. Mohapatra and G. Senjanovic, Phys. Rev. Lett. **44** (1980) 912.

[57] J. Schechter and J. W. F. Valle, Phys. Rev. D **22** (1980) 2227.

[58] R. Barbieri, D. V. Nanopoulos, G. Morchio, and F. Strocchi, Phys. Lett. B **90** (1980) 91.

[59] P. Nath and R. M. Syed, Phys. Lett. B **506** (2001) 68 [hep-ph/0103165].

[60] J. Wess and J. Bagger, *Supersymmetry and supergravity.* Princeton University Press, 1992.

[61] G. R. Farrar and P. Fayet, Phys. Lett. B **76** (1978) 575.

[62] H. K. Dreiner, hep-ph/9707435.

[63] H. E. Haber, Nucl. Phys. Proc. Suppl. **62** (1998) 469 [hep-ph/9709450].

[64] S. Dimopoulos and S. Thomas, Nucl. Phys. B **465** (1996) 23 [hep-ph/9510220].

[65] A. Santamaria, Phys. Lett. B **305** (1993) 90 [hep-ph/9302301].

[66] F. J. Botella, M. Nebot, and O. Vives, JHEP **01** (2006) 106 [hep-ph/0407349].

[67] V. S. Kaplunovsky and J. Louis, Phys. Lett. B **306** (1993) 269 [hep-th/9303040].

[68] A. Brignole, L. E. Ibanez, and C. Munoz, Nucl. Phys. B **422** (1994) 125 [hep-ph/9308271].

[69] L. E. Ibanez and G. G. Ross, hep-ph/9204201.

[70] L. J. Hall, V. A. Kostelecky, and S. Raby, Nucl. Phys. B **267** (1986) 415.

[71] F. Gabbiani and A. Masiero, Nucl. Phys. B **322** (1989) 235.

[72] J. S. Hagelin, S. Kelley, and T. Tanaka, Nucl. Phys. B **415** (1994) 293.

[73] F. Gabbiani, E. Gabrielli, A. Masiero, and L. Silvestrini, Nucl. Phys. B **477** (1996) 321 [hep-ph/9604387].

[74] J. R. Ellis, S. Ferrara, and D. V. Nanopoulos, Phys. Lett. B **114** (1982) 231.

[75] W. Buchmuller and D. Wyler, Phys. Lett. B **121** (1983) 321.

[76] J. Polchinski and M. B. Wise, Phys. Lett. B **125** (1983) 393.

[77] E. Franco and M. L. Mangano, Phys. Lett. B **135** (1984) 445.

[78] M. Dugan, B. Grinstein, and L. J. Hall, Nucl. Phys. B **255** (1985) 413.

[79] W. Fischler, S. Paban, and S. Thomas, Phys. Lett. B **289** (1992) 373 [hep-ph/9205233].

[80] P. G. Harris et al., Phys. Rev. Lett. **82** (1999) 904.

[81] B. C. Regan, E. D. Commins, C. J. Schmidt, and D. DeMille, Phys. Rev. Lett. **88** (2002) 071805.

[82] A. J. Buras, A. Romanino, and L. Silvestrini, Nucl. Phys. B **520** (1998) 3 [hep-ph/9712398].

[83] J. Hisano and D. Nomura, Phys. Rev. D **59** (1999) 116005 [hep-ph/9810479].

[84] L. Clavelli, T. Gajdosik, and W. Majerotto, Phys. Lett. B **494** (2000) 287 [hep-ph/0007342].

[85] S. Abel, S. Khalil, and O. Lebedev, Nucl. Phys. B **606** (2001) 151 [hep-ph/0103320].

[86] V. D. Barger et al., Phys. Rev. D **64** (2001) 056007 [hep-ph/0101106].

[87] S. Weinberg, Phys. Rev. Lett. **63** (1989) 2333.

[88] T. Ibrahim and P. Nath, Phys. Lett. B **418** (1998) 98 [hep-ph/9707409].

[89] T. Ibrahim and P. Nath, Phys. Rev. D **57** (1998) 478 [hep-ph/9708456].

[90] D. Chang, W.-Y. Keung, and A. Pilaftsis, Phys. Rev. Lett. **82** (1999) 900 [hep-ph/9811202].

[91] M. Pospelov and A. Ritz, Phys. Rev. D **63** (2001) 073015 [hep-ph/0010037].

[92] T. Ibrahim and P. Nath, Phys. Rev. D **58** (1998) 111301 [hep-ph/9807501].

[93] M. Brhlik, G. J. Good, and G. L. Kane, Phys. Rev. D **59** (1999) 115004 [hep-ph/9810457].

[94] A. Bartl, T. Gajdosik, W. Porod, P. Stockinger, and H. Stremnitzer, Phys. Rev. D **60** (1999) 073003 [hep-ph/9903402].

[95] M. Brhlik, L. L. Everett, G. L. Kane, and J. Lykken, Phys. Rev. Lett. **83** (1999) 2124 [hep-ph/9905215].

[96] M. Brhlik, L. L. Everett, G. L. Kane, and J. Lykken, Phys. Rev. D **62** (2000) 035005 [hep-ph/9908326].

[97] T. Ibrahim and P. Nath, Phys. Rev. D **61** (2000) 093004 [hep-ph/9910553].

[98] A. Bartl et al., Phys. Rev. D **64** (2001) 076009 [hep-ph/0103324].

[99] S. Abel, S. Khalil, and O. Lebedev, Phys. Rev. Lett. **86** (2001) 5850 [hep-ph/0103031].

[100] O. Lebedev, K. A. Olive, M. Pospelov, and A. Ritz, Phys. Rev. D **70** (2004) 016003 [hep-ph/0402023].

[101] M. Ciuchini et al., JHEP **10** (1998) 008 [hep-ph/9808328].

[102] G. G. Ross, L. Velasco-Sevilla, and O. Vives, Nucl. Phys. B **692** (2004) 50 [hep-ph/0401064].

[103] K. S. Babu, J. C. Pati, and P. Rastogi, hep-ph/0410200.

[104] K. S. Babu, J. C. Pati, and P. Rastogi, Phys. Lett. B **621** (2005) 160 [hep-ph/0502152].

[105] A. Masiero and O. Vives, Phys. Rev. Lett. **86** (2001) 26 [hep-ph/0007320].

[106] A. Masiero, M. Piai, and O. Vives, Phys. Rev. D **64** (2001) 055008 [hep-ph/0012096].

[107] S. Bertolini, F. Borzumati, A. Masiero, and G. Ridolfi, Nucl. Phys. B **353** (1991) 591.

[108] M. Ciuchini, E. Franco, G. Martinelli, A. Masiero, and L. Silvestrini, Phys. Rev. Lett. **79** (1997) 978 [hep-ph/9704274].

[109] D. Becirevic et al., Nucl. Phys. B **634** (2002) 105 [hep-ph/0112303].

[110] B. Aubert et al., [BABAR Collaboration], Phys. Rev. Lett. **89** (2002) 201802 [hep-ex/0207042].

[111] K. Abe et al., [Belle Collaboration], Phys. Rev. Lett. **87** (2001) 091802 [hep-ex/0107061].

[112] K. Abe et al., [Belle Collaboration], hep-ex/0308036.

[113] T. Affolder et al., [CDF Collaboration], Phys. Rev. D **61** (2000) 072005 [hep-ex/9909003].

[114] E. Lunghi and D. Wyler, Phys. Lett. B **521** (2001) 320 [hep-ph/0109149].

[115] T. Goto, Y. Okada, Y. Shimizu, T. Shindou, and M. Tanaka, Phys. Rev. D **66** (2002) 035009 [hep-ph/0204081].

[116] D. Chang, A. Masiero, and H. Murayama, Phys. Rev. D **67** (2003) 075013 [hep-ph/0205111].

[117] S. Khalil and E. Kou, Phys. Rev. D **67** (2003) 055009 [hep-ph/0212023].

[118] R. Harnik, D. T. Larson, H. Murayama, and A. Pierce, Phys. Rev. D **69** (2004) 094024 [hep-ph/0212180].

[119] M. Ciuchini, E. Franco, A. Masiero, and L. Silvestrini, Phys. Rev. D **67** (2003) 075016 [hep-ph/0212397].

[120] S. Baek, Phys. Rev. D **67** (2003) 096004 [hep-ph/0301269].

[121] K. Agashe and C. D. Carone, Phys. Rev. D **68** (2003) 035017 [hep-ph/0304229].

[122] G. L. Kane et al., Phys. Rev. Lett. **90** (2003) 141803 [hep-ph/0304239].

[123] S. Mishima and A. I. Sanda, Phys. Rev. D **69** (2004) 054005 [hep-ph/0311068].

[124] M. Endo, M. Kakizaki, and M. Yamaguchi, Phys. Lett. B **594** (2004) 205 [hep-ph/0403260].

[125] M. Endo, S. Mishima, and M. Yamaguchi, Phys. Lett. B **609** (2005) 95 [hep-ph/0409245].

[126] B. Aubert et al., [BABAR Collaboration], hep-ex/0207070.

[127] K. Abe et al., [Belle Collaboration], Phys. Rev. Lett. **91** (2003) 261602 [hep-ex/0308035].

[128] E. Gabrielli and G. F. Giudice, Nucl. Phys. B **433** (1995) 3 [hep-lat/9407029].

[129] A. Masiero and H. Murayama, Phys. Rev. Lett. **83** (1999) 907 [hep-ph/9903363].

[130] G. Eyal, A. Masiero, Y. Nir, and L. Silvestrini, JHEP **11** (1999) 032 [hep-ph/9908382].

[131] R. Barbieri, R. Contino, and A. Strumia, Nucl. Phys. B **578** (2000) 153 [hep-ph/9908255].

[132] K. S. Babu, B. Dutta, and R. N. Mohapatra, Phys. Rev. D **61** (2000) 091701 [hep-ph/9905464].

[133] S. Khalil, T. Kobayashi, and O. Vives, Nucl. Phys. B **580** (2000) 275 [hep-ph/0003086].

[134] A. J. Buras, P. Gambino, M. Gorbahn, S. Jager, and L. Silvestrini, Nucl. Phys. B **592** (2001) 55 [hep-ph/0007313].

[135] G. D. Barr et al., [NA31 Collaboration], Phys. Lett. B **317** (1993) 233.

[136] A. Alavi-Harati et al., [KTeV Collaboration], Phys. Rev. Lett. **83** (1999) 22 [hep-ex/9905060].

[137] M. Ciuchini, A. Masiero, L. Silvestrini, S. K. Vempati, and O. Vives, Phys. Rev. Lett. **92** (2004) 071801 [hep-ph/0307191].

[138] M. Ciuchini, A. Masiero, P. Paradisi, L. Silvestrini, S. K. Vempati, and O. Vives, "Flavour violating constraints at the weak and GUT scales" (Work in progress, 2006).

[139] A. Masiero, S. K. Vempati, and O. Vives, Nucl. Phys. B **649** (2003) 189 [hep-ph/0209303].

[140] J. Hisano, T. Moroi, K. Tobe, and M. Yamaguchi, Phys. Rev. D **53** (1996) 2442 [hep-ph/9510309].

[141] J. Hisano and Y. Shimizu, Phys. Lett. B **565** (2003) 183 [hep-ph/0303071].

[142] I. Masina and C. A. Savoy, Nucl. Phys. B **661** (2003) 365 [hep-ph/0211283].

[143] G. Altarelli and F. Feruglio, hep-ph/0206077.

[144] G. Altarelli and F. Feruglio, hep-ph/0306265.

[145] G. Altarelli and F. Feruglio, New J. Phys. **6** (2004) 106 [hep-ph/0405048].

[146] I. Masina, Int. J. Mod. Phys. A **16** (2001) 5101 [hep-ph/0107220].

[147] R. N. Mohapatra, hep-ph/0211252.

[148] R. N. Mohapatra, hep-ph/0306016.

[149] S. F. King, Rept. Prog. Phys. **67** (2004) 107 [hep-ph/0310204].

[150] A. Y. Smirnov, Int. J. Mod. Phys. A **19** (2004) 1180 [hep-ph/0311259].

[151] F. Borzumati and A. Masiero, Phys. Rev. Lett. **57** (1986) 961.

[152] K. Tobe, J. D. Wells, and T. Yanagida, Phys. Rev. D **69** (2004) 035010 [hep-ph/0310148].

[153] M. Ibe, R. Kitano, H. Murayama, and T. Yanagida, Phys. Rev. D **70** (2004) 075012 [hep-ph/0403198].

[154] J. A. Casas and A. Ibarra, Nucl. Phys. B **618** (2001) 171 [hep-ph/0103065].

[155] A. Masiero, S. K. Vempati, and O. Vives, New J. Phys. **6** (2004) 202 [hep-ph/0407325].

[156] S. T. Petcov, S. Profumo, Y. Takanishi, and C. E. Yaguna, Nucl. Phys. B **676** (2004) 453 [hep-ph/0306195].

[157] R. N. Mohapatra and B. Sakita, Phys. Rev. D **21** (1980) 1062.

[158] G. Lazarides, Q. Shafi, and C. Wetterich, Nucl. Phys. B **181** (1981) 287.

[159] R. N. Mohapatra and G. Senjanovic, Phys. Rev. D **23** (1981) 165.

[160] P. H. Chankowski and S. Pokorski, Int. J. Mod. Phys. A **17** (2002) 575 [hep-ph/0110249].

[161] J. R. Ellis and S. Lola, Phys. Lett. B **458** (1999) 310 [hep-ph/9904279].

[162] J. A. Casas, J. R. Espinosa, A. Ibarra, and I. Navarro, Nucl. Phys. B **556** (1999) 3 [hep-ph/9904395].

[163] N. Haba and N. Okamura, Eur. Phys. J. C **14** (2000) 347 [hep-ph/9906481].

[164] S. Antusch, J. Kersten, M. Lindner, and M. Ratz, Phys. Lett. B **544** (2002) 1 [hep-ph/0206078].

[165] T. Moroi, JHEP **03** (2000) 019 [hep-ph/0002208].

[166] T. Moroi, Phys. Lett. B **493** (2000) 366 [hep-ph/0007328].

[167] N. Akama, Y. Kiyo, S. Komine, and T. Moroi, Phys. Rev. D **64** (2001) 095012 [hep-ph/0104263].

[168] J. Sato, K. Tobe, and T. Yanagida, Phys. Lett. B **498** (2001) 189 [hep-ph/0010348].

[169] M. Apollonio et al., [CHOOZ Collaboration], Phys. Lett. B **466** (1999) 415 [hep-ex/9907037].

[170] T. Blazek and S. F. King, Phys. Lett. B **518** (2001) 109 [hep-ph/0105005].

[171] M. Goodman, hep-ex/0404031.

[172] J. I. Illana and M. Masip, Eur. Phys. J. C **35** (2004) 365 [hep-ph/0307393].

[173] D. Binosi and L. Theussl, Comput. Phys. Commun. **161** (2004) 76 [hep-ph/0309015].

Course 2

CP VIOLATION IN MESON DECAYS

Yosef Nir

Department of Particle Physics, Weizmann Institute of Science, Rehovot 76100, Israel

D. Kazakov, S. Lavignac and J. Dalibard, eds.
Les Houches, Session LXXXIV, 2005
Particle Physics Beyond the Standard Model
© *2006 Elsevier B.V. All rights reserved*

Contents

1. Introduction

The Standard Model predicts that the only way that CP is violated is through the Kobayashi-Maskawa mechanism [1]. Specifically, the source of CP violation is a *single* phase in the mixing matrix that describes the charged current weak interactions of quarks. In the introductory chapter, we briefly review the present evidence that supports the Kobayashi-Maskawa picture of CP violation, as well as the various arguments against this picture.

1.1. Why believe the Kobayashi-Maskawa mechanism?

Experiments have measured to date nine independent CP violating observables:[1]

1. Indirect CP violation in $K \rightarrow \pi\pi$ decays [2] and in $K \rightarrow \pi\ell\nu$ decays is given by

$$\varepsilon_K = (2.28 \pm 0.02) \times 10^{-3} \, e^{i\pi/4}. \tag{1.1}$$

2. Direct CP violation in $K \rightarrow \pi\pi$ decays [3–5] is given by

$$\varepsilon'/\varepsilon = (1.72 \pm 0.18) \times 10^{-3}. \tag{1.2}$$

3. CP violation in the interference of mixing and decay in the $B \rightarrow \psi K_S$ and other, related modes is given by [6, 7]:

$$S_{\psi K_S} = +0.69 \pm 0.03. \tag{1.3}$$

4. CP violation in the interference of mixing and decay in the $B \rightarrow K^+ K^- K_S$ mode is given by [8, 9]

$$S_{K^+ K^- K_S} = -0.45 \pm 0.13. \tag{1.4}$$

5. CP violation in the interference of mixing and decay in the $B \rightarrow D^{*+} D^{*-}$ mode is given by [10, 11]

$$S_{D^{*+} D^{*-}} = -0.75 \pm 0.23. \tag{1.5}$$

---

[1]The list of measured observables in B decays is somewhat conservative. I include only observables where the combined significance of Babar and Belle measurements (taking an inflated error in case of inconsistencies) is above 3σ.

6. CP violation in the interference of mixing and decay in the $B \to \eta' K^0$ modes is given by [12–14]

$$S_{\eta' K_S} = +0.50 \pm 0.09(0.13).$$ (1.6)

7. CP violation in the interference of mixing and decay in the $B \to f_0 K_S$ mode is given by [13, 15]

$$S_{f_0 K_S} = -0.75 \pm 0.24.$$ (1.7)

8. Direct CP violation in the $\overline{B}^0 \to K^- \pi^+$ mode is given by [16, 17]

$$\mathcal{A}_{K^\mp \pi^\pm} = -0.115 \pm 0.018.$$ (1.8)

9. Direct CP violation in the $B \to \rho\pi$ mode is given by [18, 19]

$$\mathcal{A}_{\rho\pi}^{-+} = -0.48 \pm 0.14.$$ (1.9)

All nine measurements – as well as many other, where CP violation is not (yet) observed at a level higher than 3σ – are consistent with the Kobayashi-Maskawa picture of CP violation. In particular, the measurement of the phase β from the CP asymmetry $B \to \psi K$ and the measurement of the phase α from CP asymmetries and decay rates in the $B \to \pi\pi$, $\rho\pi$ and $\rho\rho$ modes have provided the first two precision tests of CP violation in the Standard Model. Since the model has passed these tests successfully, we are able, for the first time, to make the following statement: *The Kobayashi-Maskawa phase is, very likely, the dominant source of CP violation in low-energy flavor-changing processes.*

In contrast, various alternative scenarios of CP violation that have been phenomenologically viable for many years are now unambiguously excluded. Two important examples are the following:

• The superweak framework [20], that is, the idea that CP violation is purely indirect, is excluded by the evidence that $\varepsilon'/\varepsilon \neq 0$.

• Approximate CP, that is, the idea that all CP violating phases are small (see, for example, [21]), is excluded by the evidence that $S_{\psi K_S} = \mathcal{O}(1)$.

Indeed, I am not aware of any viable, reasonably motivated, scenario which provides a complete alternative to the KM mechanism, that is of a framework where the KM phase plays no significant role in the observed CP violation.

The experimental results from the B-factories, such as those in Eqs. (1.3–1.9), and their implications for theory signify a new era in the study of CP violation. In this series of lectures we explain these recent developments and their significance.

1.2. Why doubt the Kobayashi-Maskawa mechanism?

1.2.1. The baryon asymmetry of the universe

Baryogenesis is a consequence of CP violating processes [22]. Therefore the present baryon number, which is accurately deduced from nucleosynthesis and CMBR constraints,

$$Y_B \equiv \frac{n_B - n_{\overline{B}}}{s} \simeq 9 \times 10^{-11}, \tag{1.10}$$

is essentially a CP violating observable! It can be added to the list of known CP violating observables, Eqs. (1.1–1.9). Within a given model of CP violation, one can check for consistency between the data from cosmology, Eq. (1.10), and those from laboratory experiments.

The surprising point is that the Kobayashi-Maskawa mechanism for CP violation fails to account for (1.10). It predicts present baryon number density that is many orders of magnitude below the observed value [23–25]. This failure is independent of other aspects of the Standard Model: the suppression of Y_B from CP violation is much too strong, even if the departure from thermal equilibrium is induced by mechanisms beyond the Standard Model. This situation allows us to make the following statement: *There must exist sources of CP violation beyond the Kobayashi-Maskawa phase.*

Two important examples of viable models of baryogenesis are the following:

1. Leptogenesis [26]: a lepton asymmetry is induced by CP violating decays of heavy fermions that are singlets of the Standard Model gauge group (sterile neutrinos). Departure from thermal equilibrium is provided if the lifetime of the heavy neutrino is long enough that it decays when the temperature is below its mass. Processes that violate $B + L$ are fast before the electroweak phase transition and partially convert the lepton asymmetry into a baryon asymmetry. The CP violating parameters may be related to CP violation in the mixing matrix for the light neutrinos (but this is a model dependent issue [27]).

2. Electroweak baryogenesis (for a review see [28]): the source of the baryon asymmetry is the interactions of top (anti)quarks with the Higgs field during the electroweak phase transition. CP violation is induced, for example, by supersymmetric interactions. Sphaleron configurations provide baryon number violating interactions. Departure from thermal equilibrium is provided by the wall between the false vacuum ($\langle \phi \rangle = 0$) and the expanding bubble with the true vacuum, where electroweak symmetry is broken.

1.2.2. The strong CP problem

Nonperturbative QCD effects induce an additional term in the SM Lagrangian,

$$\mathcal{L}_\theta = \frac{\theta_{\mathrm{QCD}}}{32\pi^2} \epsilon_{\mu\nu\rho\sigma} F^{\mu\nu a} F^{\rho\sigma a}. \tag{1.11}$$

This term violates CP. In particular, it induces an electric dipole moment (EDM) to the neutron. The leading contribution in the chiral limit is given by [29]

$$d_N = \frac{g_{\pi NN}\bar{g}_{\pi NN}}{4\pi^2 M_N} \ln \frac{M_N}{m_\pi} \approx 5 \times 10^{-16} \, \theta_{QCD} \, e \, \text{cm}, \qquad (1.12)$$

where M_N is the nucleon mass, and $g_{\pi NN}$ ($\bar{g}_{\pi NN}$) is the pseudoscalar coupling (CP-violating scalar coupling) of the pion to the nucleon. (The leading contribution in the large N_c limit was calculated in the Skyrme model [30] and leads to a similar estimate.) The experimental bound on d_N is given by [31]

$$d_N \leq 6.3 \times 10^{-26} \, e \, \text{cm}. \qquad (1.13)$$

It leads to the following bound on θ_{QCD}:

$$\theta_{QCD} \lesssim 10^{-10}. \qquad (1.14)$$

Since θ_{QCD} arises from nonperturbative QCD effects, it is impossible to calculate it. Yet, there are good reasons to expect that these effects should yield $\theta_{QCD} = \mathcal{O}(1)$ (for a review, see [32]). Within the SM, a value as small as in Eq. (1.14) is unnatural, since setting θ_{QCD} to zero does not add symmetry to the model. [In particular, as we will see below, CP is violated by $\delta_{KM} = \mathcal{O}(1)$.] Understanding why CP is so small in the strong interactions is the strong CP problem.

It seems then that the strong CP problem is a clue to new physics. Among the solutions that have been proposed are a massless u-quark (for a review, see [33]), the Peccei-Quinn mechanism [34, 35] and spontaneous CP violation.

1.2.3. New physics

Almost any extension of the Standard Model provides new sources of CP violation. For example, in the supersymmetric extension of the Standard Model (with R-parity), there are 44 independent phases, most of them in flavor changing couplings. If there is new physics at or below the TeV scale, it is quite likely that the KM phase is not the only source of CP violation that is playing a role in meson decays.

1.3. Will new CP violation be observed in experiments?

The SM picture of CP violation is testable because the Kobayashi-Maskawa mechanism is unique and predictive. These features are mainly related to the fact that there is a single phase that is responsible to all CP violation. As a consequence of this situation, one finds two classes of tests:

(i) Correlations: many independent CP violating observables are correlated within the SM. For example, the SM predicts that the CP asymmetries in $B \rightarrow$

ψK_S and in $B \rightarrow \phi K_S$, which proceed through different quark transitions, are equal to each other (to a few percent accuracy) [36, 37]. Another important example is the strong SM correlation between CP violation in $B \rightarrow \psi K_S$ and in $K \rightarrow \pi \nu \bar{\nu}$ [38–40]. It is a significant fact, in this context, that several CP violating observables can be calculated with very small hadronic uncertainties. To search for violations of the correlations, precise measurements are important.

(ii) Zeros: since the KM phase appears in flavor-changing, weak-interaction couplings of quarks, and only if all three generations are involved, many CP violating observables are predicted to be negligibly small. For example, the transverse lepton polarization in semileptonic meson decays, CP violation in $t\bar{t}$ production, tree level D decays, and (assuming $\theta_{QCD} = 0$) the electric dipole moment of the neutron are all predicted to be orders of magnitude below the (present and near future) experimental sensitivity. To search for lifted zeros, measurements of CP violation in many different systems should be performed.

The strongest argument that new sources of CP violation must exist in Nature comes from baryogenesis. Whether the CP violation that is responsible for baryogenesis would be manifest in measurements of CP asymmetries in B decays depends on two issues:

(i) The scale of the new CP violation: if the relevant scale is very high, such as in leptogenesis, the effects cannot be signalled in these measurements. To estimate the limit on the scale, the following three facts are relevant: First, the Standard Model contributions to CP asymmetries in B decays are $\mathcal{O}(1)$. Second, the expected experimental accuracy would reach in some cases the few percent level. Third, the contributions from new physics are expected to be suppressed by $(\Lambda_{EW}/\Lambda_{NP})^2$. The conclusion is that, if the new source of CP violation is related to physics at $\Lambda_{NP} \gg 1\ TeV$, it cannot be signalled in B decays. Only if the true mechanism is electroweak baryogenesis, it can potentially affect B decays.

(ii) The flavor dependence of the new CP violation: if it is flavor diagonal, its effects on B decays would be highly suppressed. It can still manifest itself in other, flavor diagonal CP violating observables, such as electric dipole moments.

We conclude that new measurements of CP asymmetries in meson decays are particularly sensitive to new sources of CP violation that come from physics at (or below) the few TeV scale and that are related to flavor changing couplings. This is, for example, the case, in certain supersymmetric models of baryogenesis [41, 42]. The search for electric dipole moments can reveal the existence of new flavor diagonal CP violation.

Of course, there could be new flavor physics at the TeV scale that is not related to the baryon asymmetry and may give signals in B decays. The best motivated extension of the SM where this situation is likely is that of supersymmetry.

Finally, we would like to mention that, in the past, flavor physics and the physics of CP violation led indeed to the discovery of new physics or to probing

it before it was directly observed in experiments:
- The smallness of $\frac{\Gamma(K_L \to \mu^+\mu^-)}{\Gamma(K^+ \to \mu^+\nu)}$ led to predicting a fourth (the charm) quark;
- The size of Δm_K led to a successful prediction of the charm mass;
- The size of Δm_B led to a successful prediction of the top mass;
- The measurement of ε_K led to predicting the third generation.

2. The Kobayashi-Maskawa mechanism

2.1. Yukawa interactions are the source of CP violation

A model of elementary particles and their interactions is defined by three ingredients:

1. The symmetries of the Lagrangian;

2. The representations of fermions and scalars;

3. The pattern of spontaneous symmetry breaking.

The Standard Model (SM) is defined as follows:
   1. The gauge symmetry is

$$G_{SM} = SU(3)_C \times SU(2)_L \times U(1)_Y. \tag{2.1}$$

   2. There are three fermion generations, each consisting of five representations of G_{SM}:

$$Q_{Li}^I(3,2)_{+1/6}, \quad U_{Ri}^I(3,1)_{+2/3}, \quad D_{Ri}^I(3,1)_{-1/3},$$
$$L_{Li}^I(1,2)_{-1/2}, \quad E_{Ri}^I(1,1)_{-1}. \tag{2.2}$$

Our notations mean that, for example, left-handed quarks, Q_L^I, are triplets of $SU(3)_C$, doublets of $SU(2)_L$ and carry hypercharge $Y = +1/6$. The super-index I denotes interaction eigenstates. The sub-index $i = 1,2,3$ is the flavor (or generation) index. There is a single scalar representation,

$$\phi(1,2)_{+1/2}. \tag{2.3}$$

   3. The scalar ϕ assumes a VEV,

$$\langle\phi\rangle = \begin{pmatrix} 0 \\ \frac{v}{\sqrt{2}} \end{pmatrix}, \tag{2.4}$$

so that the gauge group is spontaneously broken,

$$G_{SM} \to SU(3)_C \times U(1)_{EM}. \tag{2.5}$$

The Standard Model Lagrangian, $\mathcal{L}_{SM}$, is the most general renormalizable Lagrangian that is consistent with the gauge symmetry (2.1), the particle content (2.2, 2.3) and the pattern of spontaneous symmetry breaking (2.4). It can be divided to three parts:

$$\mathcal{L}_{SM} = \mathcal{L}_{kinetic} + \mathcal{L}_{Higgs} + \mathcal{L}_{Yukawa}. \qquad (2.6)$$

As concerns the kinetic terms, to maintain gauge invariance, one has to replace the derivative with a covariant derivative:

$$D^{\mu} = \partial^{\mu} + ig_s G_a^{\mu} L_a + ig W_b^{\mu} T_b + ig' B^{\mu} Y. \qquad (2.7)$$

Here G_a^{μ} are the eight gluon fields, W_b^{μ} the three weak interaction bosons and B^{μ} the single hypercharge boson. The L_a's are $SU(3)_C$ generators (the 3×3 Gell-Mann matrices $\frac{1}{2}\lambda_a$ for triplets, 0 for singlets), the T_b's are $SU(2)_L$ generators (the 2×2 Pauli matrices $\frac{1}{2}\tau_b$ for doublets, 0 for singlets), and the Y's are the $U(1)_Y$ charges. For example, for the left-handed quarks Q_L^I, we have

$$\mathcal{L}_{kinetic}(Q_L) = i\overline{Q_{Li}^I}\gamma_{\mu}\left(\partial^{\mu} + \frac{i}{2}g_s G_a^{\mu}\lambda_a + \frac{i}{2}g W_b^{\mu}\tau_b + \frac{i}{6}g' B^{\mu}\right)Q_{Li}^I, \qquad (2.8)$$

while for the left-handed leptons L_L^I, we have

$$\mathcal{L}_{kinetic}(L_L) = i\overline{L_{Li}^I}\gamma_{\mu}\left(\partial^{\mu} + \frac{i}{2}g W_b^{\mu}\tau_b - ig' B^{\mu}\right)L_{Li}^I. \qquad (2.9)$$

These parts of the interaction Lagrangian are always CP conserving.

The Higgs potential, which describes the scalar self interactions, is given by:

$$\mathcal{L}_{Higgs} = \mu^2 \phi^{\dagger}\phi - \lambda(\phi^{\dagger}\phi)^2. \qquad (2.10)$$

For the Standard Model scalar sector, where there is a single doublet, this part of the Lagrangian is also CP conserving. For extended scalar sectors, such as that of a two Higgs doublet model, $\mathcal{L}_{Higgs}$ can be CP violating. Even in case that it is CP symmetric, it may lead to spontaneous CP violation.

The quark Yukawa interactions are given by

$$-\mathcal{L}_{Yukawa}^{quarks} = Y_{ij}^d \overline{Q_{Li}^I}\phi D_{Rj}^I + Y_{ij}^u \overline{Q_{Li}^I}\tilde{\phi}U_{Rj}^I + h.c.. \qquad (2.11)$$

This part of the Lagrangian is, in general, CP violating. More precisely, CP is violated if and only if [43]

$$\mathcal{I}m(\det[Y^d Y^{d\dagger}, Y^u Y^{u\dagger}]) \neq 0. \qquad (2.12)$$

An intuitive explanation of why CP violation is related to *complex* Yukawa couplings goes as follows. The hermiticity of the Lagrangian implies that $\mathcal{L}_{\text{Yukawa}}$ has its terms in pairs of the form

$$Y_{ij}\overline{\psi_{Li}}\phi\psi_{Rj} + Y_{ij}^*\overline{\psi_{Rj}}\phi^\dagger\psi_{Li}. \tag{2.13}$$

A CP transformation exchanges the operators

$$\overline{\psi_{Li}}\phi\psi_{Rj} \leftrightarrow \overline{\psi_{Rj}}\phi^\dagger\psi_{Li}, \tag{2.14}$$

but leaves their coefficients, Y_{ij} and Y_{ij}^*, unchanged. This means that CP is a symmetry of $\mathcal{L}_{\text{Yukawa}}$ if $Y_{ij} = Y_{ij}^*$.

The lepton Yukawa interactions are given by

$$-\mathcal{L}_{\text{Yukawa}}^{\text{leptons}} = Y_{ij}^e \overline{L_{Li}^I}\phi E_{Rj}^I + \text{h.c.}. \tag{2.15}$$

It leads, as we will see in the next section, to charged lepton masses but predicts massless neutrinos. Recent measurements of the fluxes of atmospheric and solar neutrinos provide evidence for neutrino masses (for a review, see [44]). That means that $\mathcal{L}_{\text{SM}}$ cannot be a complete description of Nature. The simplest way to allow for neutrino masses is to add dimension-five (and, therefore, non-renormalizable) terms, consistent with the SM symmetry and particle content:

$$-\mathcal{L}_{\text{Yukawa}}^{\text{dim}-5} = \frac{Y_{ij}^\nu}{M} L_i L_j \phi\phi + \text{h.c.}. \tag{2.16}$$

The parameter M has dimension of mass. The dimensionless couplings Y_{ij}^ν are symmetric ($Y_{ij}^\nu = Y_{ji}^\nu$). We refer to the SM extended to include the terms $\mathcal{L}_{\text{Yukawa}}^{\text{dim}-5}$ of Eq. (2.16) as the "extended SM" (ESM):

$$\mathcal{L}_{\text{ESM}} = \mathcal{L}_{\text{kinetic}} + \mathcal{L}_{\text{Higgs}} + \mathcal{L}_{\text{Yukawa}} + \mathcal{L}_{\text{Yukawa}}^{\text{dim}-5}. \tag{2.17}$$

The inclusion of non-renormalizable terms is equivalent to postulating that the SM is only a low energy effective theory, and that new physics appears at the scale M.

How many independent CP violating parameters are there in $\mathcal{L}_{\text{Yukawa}}^{\text{quarks}}$? Each of the two Yukawa matrices Y^q ($q = u, d$) is 3×3 and complex. Consequently, there are 18 real and 18 imaginary parameters in these matrices. Not all of them are, however, physical. One can think of the quark Yukawa couplings as spurions that break a global symmetry,

$$U(3)_Q \times U(3)_D \times U(3)_U \rightarrow U(1)_B. \tag{2.18}$$

This means that there is freedom to remove 9 real and 17 imaginary parameters [the number of parameters in three 3×3 unitary matrices minus the phase related to $U(1)_B$]. We conclude that there are 10 quark flavor parameters: 9 real ones and a single phase. This single phase is the source of CP violation in the quark sector.

How many independent CP violating parameters are there in the lepton Yukawa interactions? The matrix Y^e is a general complex 3×3 matrix and depends, therefore, on 9 real and 9 imaginary parameters. The matrix Y^ν is symmetric and depends on 6 real and 6 imaginary parameters. Not all of these 15 real and 15 imaginary parameters are physical. One can think of the lepton Yukawa couplings as spurions that break (completely) a global symmetry,

$$U(3)_L \times U(3)_E. \tag{2.19}$$

This means that 6 real and 12 imaginary parameters are not physical. We conclude that there are 12 lepton flavor parameters: 9 real ones and three phases. These three phases induce CP violation in the lepton sector.

2.2. CKM mixing is the (only!) source of CP violation in the quark sector

Upon the replacement $\mathcal{R}e(\phi^0) \rightarrow \frac{v+H^0}{\sqrt{2}}$ [see Eq. (2.4)], the Yukawa interactions (2.11) give rise to mass terms:

$$-\mathcal{L}_M^q = (M_d)_{ij}\, \overline{D_{Li}^I} D_{Rj}^I + (M_u)_{ij}\, \overline{U_{Li}^I} U_{Rj}^I + \text{h.c.}, \tag{2.20}$$

where

$$M_q = \frac{v}{\sqrt{2}} Y^q, \tag{2.21}$$

and we decomposed the $SU(2)_L$ quark doublets into their components:

$$Q_{Li}^I = \begin{pmatrix} U_{Li}^I \\ D_{Li}^I \end{pmatrix}. \tag{2.22}$$

The mass basis corresponds, by definition, to diagonal mass matrices. We can always find unitary matrices V_{qL} and V_{qR} such that

$$V_{qL} M_q V_{qR}^\dagger = M_q^{\text{diag}} \quad (q = u, d), \tag{2.23}$$

with M_q^{diag} diagonal and real. The quark mass eigenstates are then identified as

$$q_{Li} = (V_{qL})_{ij} q_{Lj}^I, \quad q_{Ri} = (V_{qR})_{ij} q_{Rj}^I \quad (q = u, d). \tag{2.24}$$

The charged current interactions for quarks [that is the interactions of the charged $SU(2)_L$ gauge bosons $W_\mu^\pm = \frac{1}{\sqrt{2}}(W_\mu^1 \mp i W_\mu^2)$], which in the interaction basis are described by (2.8), have a complicated form in the mass basis:

$$-\mathcal{L}_{W^\pm}^q = \frac{g}{\sqrt{2}} \overline{u}_{Li} \gamma^\mu (V_{uL} V_{dL}^\dagger)_{ij} d_{Lj} W_\mu^+ + \text{h.c.}. \tag{2.25}$$

The unitary 3×3 matrix,

$$V = V_{uL} V_{dL}^\dagger, \quad (VV^\dagger = 1), \tag{2.26}$$

is the Cabibbo-Kobayashi-Maskawa (CKM) *mixing matrix* for quarks [1, 45]. A unitary 3×3 matrix depends on nine parameters: three real angles and six phases.

The form of the matrix is not unique:

(i) There is freedom in defining V in that we can permute between the various generations. This freedom is fixed by ordering the up quarks and the down quarks by their masses, *i.e.* $(u_1, u_2, u_3) \to (u, c, t)$ and $(d_1, d_2, d_3) \to (d, s, b)$. The elements of V are written as follows:

$$V = \begin{pmatrix} V_{ud} & V_{us} & V_{ub} \\ V_{cd} & V_{cs} & V_{cb} \\ V_{td} & V_{ts} & V_{tb} \end{pmatrix}. \tag{2.27}$$

(ii) There is further freedom in the phase structure of V. Let us define P_q ($q = u, d$) to be diagonal unitary (phase) matrices. Then, if instead of using V_{qL} and V_{qR} for the rotation (2.24) to the mass basis we use $\tilde{V}_{qL}$ and $\tilde{V}_{qR}$, defined by $\tilde{V}_{qL} = P_q V_{qL}$ and $\tilde{V}_{qR} = P_q V_{qR}$, we still maintain a legitimate mass basis since M_q^{diag} remains unchanged by such transformations. However, V does change:

$$V \to P_u V P_d^*. \tag{2.28}$$

This freedom is fixed by demanding that V has the minimal number of phases. In the three generation case V has a single phase. (There are five phase differences between the elements of P_u and P_d and, therefore, five of the six phases in the CKM matrix can be removed.) This is the Kobayashi-Maskawa phase δ_{KM} which is the single source of CP violation in the quark sector of the Standard Model [1].

As a result of the fact that V is not diagonal, the $W^\pm$ gauge bosons couple to quark (mass eigenstates) of different generations. Within the Standard Model, this is the only source of *flavor changing* quark interactions.

2.3. *The three phases in the lepton mixing matrix*

The leptonic Yukawa interactions (2.15) and (2.16) give rise to mass terms:

$$-\mathcal{L}_M^\ell = (M_e)_{ij}\overline{e_{Li}^I}e_{Rj}^I + (M_\nu)_{ij}\nu_{Li}^I\nu_{Lj}^I + \text{h.c.}, \tag{2.29}$$

where

$$M_e = \frac{v}{\sqrt{2}}Y^e, \quad M_\nu = \frac{v^2}{2M}Y^\nu, \tag{2.30}$$

and we decomposed the $SU(2)_L$ lepton doublets into their components:

$$L_{Li}^I = \begin{pmatrix} \nu_{Li}^I \\ e_{Li}^I \end{pmatrix}. \tag{2.31}$$

We can always find unitary matrices V_{eL} and V_ν such that

$$\begin{aligned} V_{eL}M_e M_e^\dagger V_{eL}^\dagger &= \text{diag}(m_e^2, m_\mu^2, m_\tau^2), \\ V_\nu M_\nu^\dagger M_\nu V_\nu^\dagger &= \text{diag}(m_1^2, m_1^2, m_3^2). \end{aligned} \tag{2.32}$$

The charged current interactions for leptons, which in the interaction basis are described by (2.9), have the following form in the mass basis:

$$-\mathcal{L}_{W^\pm}^\ell = \frac{g}{\sqrt{2}}\overline{e_{Li}}\gamma^\mu(V_{eL}V_\nu^\dagger)_{ij}\nu_{Lj}W_\mu^- + \text{h.c.}. \tag{2.33}$$

The unitary 3×3 matrix,

$$U = V_{eL}V_\nu^\dagger, \tag{2.34}$$

is the *lepton mixing matrix* [46]. Similarly to the CKM matrix, the form of the lepton mixing matrix is not unique. But there are differences in choosing conventions:

(*i*) We can permute between the various generations. This freedom is usually fixed in the following way. We order the charged leptons by their masses, *i.e.* $(e_1, e_2, e_3) \rightarrow (e, \mu, \tau)$. As concerns the neutrinos, one takes into account that the atmospheric and solar neutrino data imply that $\Delta m_{\text{atm}}^2 \gg \Delta m_{\text{sol}}^2$. It follows that one of the neutrino mass eigenstates is separated in its mass from the other two, which have a smaller mass difference. The convention is to denote this separated state by ν_3. For the remaining two neutrinos, ν_1 and ν_2, the convention is to call the heavier state ν_2. In other words, the three mass eigenstates are defined by the following conventions:

$$|\Delta m_{3i}^2| \gg |\Delta m_{21}^2|, \quad \Delta m_{21}^2 > 0. \tag{2.35}$$

Note in particular that ν_3 can be either heavier ('normal hierarchy') or lighter ('inverted hierarchy') than $\nu_{1,2}$. The elements of U are written as follows:

$$U = \begin{pmatrix} U_{e1} & U_{e2} & U_{e3} \\ U_{\mu 1} & U_{\mu 2} & U_{\mu 3} \\ U_{\tau 1} & U_{\tau 2} & U_{\tau 3} \end{pmatrix}. \tag{2.36}$$

(*ii*) There is further freedom in the phase structure of U. One can change the charged lepton mass basis by the transformation $e_{(L,R)i} \rightarrow e'_{(L,R)i} = (P_e)_{ii} \times e_{(L,R)i}$, where P_e is a phase matrix. There is, however, no similar freedom to redefine the neutrino mass eigenstates: From Eq. (2.29) one learns that a transformation $\nu_L \rightarrow P_\nu \nu_L$ will introduce phases into the diagonal mass matrix. This is related to the Majorana nature of neutrino masses, assumed in Eq. (2.16). The allowed transformation modifies U:

$$U \rightarrow P_e U. \tag{2.37}$$

This freedom is fixed by demanding that U will have the minimal number of phases. Out of six phases of a generic unitary 3×3 matrix, the multiplication by P_e can be used to remove three. We conclude that the three generation U matrix has three phases. One of these is the analog of the Kobayashi-Maskawa phase. It is the only source of CP violation in processes that conserve lepton number, such as neutrino flavor oscillations. The other two phases can affect lepton number changing processes.

With $U \neq \mathbf{1}$, the $W^\pm$ gauge bosons couple to lepton (mass eigenstates) of different generations. Within the ESM, this is the only source of *flavor changing* lepton interactions.

2.4. The flavor parameters

Examining the quark mass basis, one can easily identify the flavor parameters. In the quark sector, we have six quark masses and four mixing parameters: three mixing angles and a single phase.

The fact that there are only three real and one imaginary physical parameters in V can be made manifest by choosing an explicit parameterization. For example, the standard parameterization [47], used by the particle data group, is given by

$$V = \begin{pmatrix} c_{12}c_{13} & s_{12}c_{13} & s_{13}e^{-i\delta} \\ -s_{12}c_{23} - c_{12}s_{23}s_{13}e^{i\delta} & c_{12}c_{23} - s_{12}s_{23}s_{13}e^{i\delta} & s_{23}c_{13} \\ s_{12}s_{23} - c_{12}c_{23}s_{13}e^{i\delta} & -c_{12}s_{23} - s_{12}c_{23}s_{13}e^{i\delta} & c_{23}c_{13} \end{pmatrix} \tag{2.38}$$

where $c_{ij} \equiv \cos\theta_{ij}$ and $s_{ij} \equiv \sin\theta_{ij}$. The three $\sin\theta_{ij}$ are the three real mixing parameters while δ is the Kobayashi-Maskawa phase. Another, very useful, example is the Wolfenstein parametrization, where the four mixing parameters are

(λ, A, ρ, η) with $\lambda = |V_{us}| = 0.22$ playing the role of an expansion parameter and η representing the CP violating phase [48, 49]:

$$V = \begin{pmatrix} 1 - \frac{1}{2}\lambda^2 - \frac{1}{8}\lambda^4 & \lambda & A\lambda^3(\rho - i\eta) \\ -\lambda + \frac{1}{2}A^2\lambda^5[1 - 2(\rho + i\eta)] & 1 - \frac{1}{2}\lambda^2 - \frac{1}{8}\lambda^4(1 + 4A^2) & A\lambda^2 \\ A\lambda^3[1 - (1 - \frac{1}{2}\lambda^2)(\rho + i\eta)] & -A\lambda^2 + \frac{1}{2}A\lambda^4[1 - 2(\rho + i\eta)] & 1 - \frac{1}{2}A^2\lambda^4 \end{pmatrix}. \quad (2.39)$$

Various parametrizations differ in the way that the freedom of phase rotation, Eq. (2.28), is used to leave a single phase in V. One can define, however, a CP violating quantity in V_{CKM} that is independent of the parametrization [43]. This quantity, J_{CKM}, is defined through

$$\mathcal{I}m(V_{ij}V_{kl}V_{il}^*V_{kj}^*) = J_{\text{CKM}} \sum_{m,n=1}^{3} \epsilon_{ikm}\epsilon_{jln}, \quad (i, j, k, l = 1, 2, 3). \quad (2.40)$$

In terms of the explicit parametrizations given above, we have

$$J_{\text{CKM}} = c_{12}c_{23}c_{13}^2 s_{12}s_{23}s_{13} \sin\delta \simeq \lambda^6 A^2 \eta. \quad (2.41)$$

It is interesting to translate the condition (2.12) to the language of the flavor parameters in the mass basis. One finds that the following is a necessary and sufficient condition for CP violation in the quark sector of the SM (we define $\Delta m_{ij}^2 \equiv m_i^2 - m_j^2$):

$$\Delta m_{tc}^2 \Delta m_{tu}^2 \Delta m_{cu}^2 \Delta m_{bs}^2 \Delta m_{bd}^2 \Delta m_{sd}^2 J_{\text{CKM}} \neq 0. \quad (2.42)$$

Equation (2.42) puts the following requirements on the SM in order that it violates CP:

(i) Within each quark sector, there should be no mass degeneracy;
(ii) None of the three mixing angles should be zero or $\pi/2$;
(iii) The phase should be neither 0 nor π.

As concerns the lepton sector of the ESM, the flavor parameters are the six lepton masses, and six mixing parameters: three mixing angles and three phases. One can parameterize U in a convenient way by factorizing it into $U = \hat{U}P$. Here P is a diagonal unitary matrix that depends on two phases, *e.g.* $P = \text{diag}(e^{i\phi_1}, e^{i\phi_2}, 1)$, while $\hat{U}$ can be parametrized in the same way as (2.38). The advantage of this parametrization is that for the purpose of analyzing lepton number conserving processes and, in particular, neutrino flavor oscillations, the parameters of P are usually irrelevant and one can use the same Chau-Keung parametrization as is being used for V. (An alternative way to understand these statements is to use a single-phase mixing matrix and put the extra two phases in the

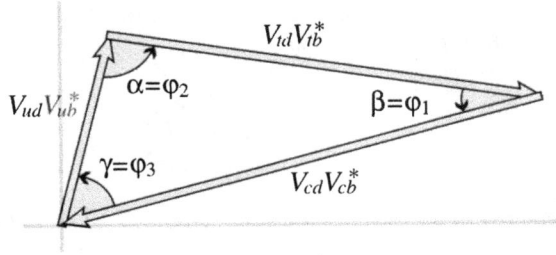

Fig. 1. Graphical representation of the unitarity constraint $V_{ud}V_{ub}^* + V_{cd}V_{cb}^* + V_{td}V_{tb}^* = 0$ as a triangle in the complex plane.

neutrino mass matrix. Then it is obvious that the effects of these 'Majorana-phases' always appear in conjunction with a factor of the Majorana mass that is lepton number violating parameter.) On the other hand, the Wolfenstein parametrization [Eq. (2.39)] is inappropriate for the lepton sector: it assumes $|V_{23}| \ll |V_{12}| \ll 1$, which does not hold here.

In order that the CP violating phase δ in $\hat{U}$ would be physically meaningful, *i.e.* there would be CP violation that is not related to lepton number violation, a condition similar to Eq. (2.42) should hold:

$$\Delta m_{\tau\mu}^2 \Delta m_{\tau e}^2 \Delta m_{\mu e}^2 \Delta m_{32}^2 \Delta m_{31}^2 \Delta m_{21}^2 J_\ell \neq 0. \tag{2.43}$$

2.5. The unitarity triangles

A very useful concept is that of the *unitarity triangles*. We focus on the quark sector, but analogous triangles can be defined in the lepton sector. The unitarity of the CKM matrix leads to various relations among the matrix elements, *e.g.*

$$V_{ud}V_{us}^* + V_{cd}V_{cs}^* + V_{td}V_{ts}^* = 0, \tag{2.44}$$

$$V_{us}V_{ub}^* + V_{cs}V_{cb}^* + V_{ts}V_{tb}^* = 0, \tag{2.45}$$

$$V_{ud}V_{ub}^* + V_{cd}V_{cb}^* + V_{td}V_{tb}^* = 0. \tag{2.46}$$

Each of these three relations requires the sum of three complex quantities to vanish and so can be geometrically represented in the complex plane as a triangle. These are "the unitarity triangles", though the term "unitarity triangle" is usually reserved for the relation (2.46) only. The unitarity triangle related to Eq. (2.46) is depicted in Fig. 1.

It is a surprising feature of the CKM matrix that all unitarity triangles are equal in area: the area of each unitarity triangle equals $|J_{CKM}|/2$ while the sign of J_{CKM} gives the direction of the complex vectors around the triangles.

The rescaled unitarity triangle is derived from (2.46) by (a) choosing a phase convention such that $(V_{cd}V_{cb}^*)$ is real, and (b) dividing the lengths of all sides by $|V_{cd}V_{cb}^*|$. Step (a) aligns one side of the triangle with the real axis, and step (b) makes the length of this side 1. The form of the triangle is unchanged. Two vertices of the rescaled unitarity triangle are thus fixed at $(0,0)$ and $(1,0)$. The coordinates of the remaining vertex correspond to the Wolfenstein parameters (ρ, η). The area of the rescaled unitarity triangle is $|\eta|/2$.

Depicting the rescaled unitarity triangle in the (ρ, η) plane, the lengths of the two complex sides are

$$R_u \equiv \left| \frac{V_{ud}V_{ub}}{V_{cd}V_{cb}} \right| = \sqrt{\rho^2 + \eta^2}, \quad R_t \equiv \left| \frac{V_{td}V_{tb}}{V_{cd}V_{cb}} \right| = \sqrt{(1-\rho)^2 + \eta^2}. \quad (2.47)$$

The three angles of the unitarity triangle are defined as follows [50, 51]:

$$\alpha \equiv \arg\left[-\frac{V_{td}V_{tb}^*}{V_{ud}V_{ub}^*} \right], \quad \beta \equiv \arg\left[-\frac{V_{cd}V_{cb}^*}{V_{td}V_{tb}^*} \right], \quad \gamma \equiv \arg\left[-\frac{V_{ud}V_{ub}^*}{V_{cd}V_{cb}^*} \right]. \quad (2.48)$$

They are physical quantities and can be independently measured by CP asymmetries in B decays. It is also useful to define the two small angles of the unitarity triangles (2.45) and (2.44):

$$\beta_s \equiv \arg\left[-\frac{V_{ts}V_{tb}^*}{V_{cs}V_{cb}^*} \right], \quad \beta_K \equiv \arg\left[-\frac{V_{cs}V_{cd}^*}{V_{us}V_{ud}^*} \right]. \quad (2.49)$$

To make predictions for CP violating observables, we need to find the allowed ranges for the CKM phases. There are three ways to determine the CKM parameters (see *e.g.* [52]):

(i) **Direct measurements** are related to SM tree level processes. At present, we have direct measurements of $|V_{ud}|$, $|V_{us}|$, $|V_{ub}|$, $|V_{cd}|$, $|V_{cs}|$, $|V_{cb}|$ and $|V_{tb}|$.

(ii) **CKM Unitarity** ($V^\dagger V = 1$) relates the various matrix elements. At present, these relations are useful to constrain $|V_{td}|$, $|V_{ts}|$, $|V_{tb}|$ and $|V_{cs}|$.

(iii) **Indirect measurements** are related to SM loop processes. At present, we constrain in this way $|V_{tb}V_{td}|$ (from Δm_B and Δm_{B_s}) and the phase structure of the matrix (for example, from ε_K and $S_{\psi K_S}$).

Direct measurements are expected to hold almost model independently. Most extensions of the SM have a special flavor structure that suppresses flavor changing couplings and, in addition, have a mass scale Λ_{NP}, that is higher than the electroweak breaking scale. Consequently, new physics contributions to tree level processes are suppressed, compared to the SM ones, by at least $\mathcal{O}(m_Z^2/\Lambda_{NP}^2) \ll 1$.

Unitarity holds if the only quarks (that is fermions in color triplets with electric charge $+2/3$ or $-1/3$) are those of the three generations of the SM. This is the situation in many extensions of the SM, including the supersymmetric SM (SSM).

Using tree level constraints and unitarity, the 90% confidence limits on the magnitude of the elements are [53]

$$
\begin{pmatrix}
0.9739 - 0.9751 & 0.221 - 0.227 & 0.0029 - 0.0045 \\
0.221 - 0.227 & 0.9730 - 0.9744 & 0.039 - 0.044 \\
0.0048 - 0.014 & 0.037 - 0.043 & 0.9990 - 0.9992
\end{pmatrix}. \tag{2.50}
$$

Note that $|V_{ub}|$ and $|V_{td}|$ are the only elements with uncertainties of order one.

Indirect measurements are sensitive to new physics. Take, for example, the $B^0 - \overline{B}^0$ mixing amplitude. Within the SM, the leading contribution comes from an electroweak box diagram and is therefore $\mathcal{O}(g^4)$ and depends on small mixing angles, $(V_{td}^* V_{tb})^2$. (It is this dependence on the CKM elements that makes the relevant indirect measurements, particularly Δm_B and $S_{\psi K_S}$, very significant in improving our knowledge of the CKM matrix.) These suppression factors do not necessarily persist in extensions of the SM. For example, in the SSM there are (gluino-mediated) contributions of $\mathcal{O}(g_s^4)$ and the mixing angles could be comparable to, or even larger than the SM ones. The validity of indirect measurements is then model dependent. Conversely, inconsistencies among indirect measurements (or between indirect and direct measurements) can give evidence for new physics.

When all available data are taken into account, one finds [54]:

$$
\lambda = 0.226 \pm 0.001, \quad A = 0.83 \pm 0.03, \tag{2.51}
$$
$$
\bar{\rho} = 0.21 \pm 0.04, \quad \bar{\eta} = 0.33 \pm 0.02, \tag{2.52}
$$
$$
\sin 2\beta = 0.720 \pm 0.025, \quad \alpha = (99 \pm 7)^o, \tag{2.53}
$$
$$
\gamma = (58 \pm 7)^o, \quad \beta_s = (1.03 \pm 0.08)^o, \tag{2.54}
$$
$$
R_u = 0.40 \pm 0.02, \quad R_t = 0.86 \pm 0.04. \tag{2.55}
$$

Of course, there are correlations between the various parameters. The present constraints on the shape of the unitarity triangle or, equivalently, the allowed region in the $\rho - \eta$ plane, are presented in Fig. 2.

2.6. The uniqueness of the Standard Model picture of CP violation

In the previous subsections, we have learnt several features of CP violation as explained by the Standard Model. It is important to understand that various reasonable (and often well-motivated) extensions of the SM provide examples where some or all of these features do not hold. Furthermore, until a few years ago, none of the special features of the Kobayashi-Maskawa mechanism of CP violation has been experimentally tested. This situation has dramatically changed recently. Let

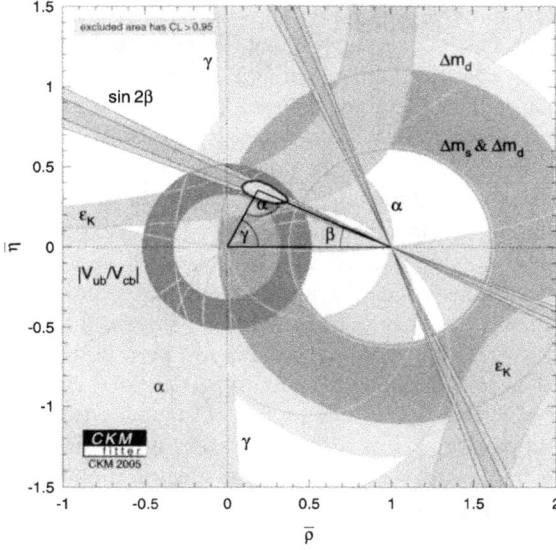

Fig. 2. Allowed region in the ρ, η plane. Superimposed are the individual constraints from charmless semileptonic B decays ($|V_{ub}/V_{cb}|$), mass differences in the B^0 (Δm_d) and B_s (Δm_s) neutral meson systems, and CP violation in $K \to \pi\pi$ (ε_K), $B \to \psi K$ ($\sin 2\beta$), $B \to \pi\pi$, $\rho\pi$, $\rho\rho$ (α), and $B \to DK$ (γ). Taken from [54].

us survey some of the SM features, how they can be modified with new physics, and whether experiment has shed light on these questions.

(i) δ_{KM} *is the only source of CP violation in meson decays.* This is arguably the most unique feature of the SM and gives the model a strong predictive power. It is violated in almost any low-energy extension. For example, in the supersymmetric extension of the SM there are 44 physical CP violating phases, many of which affect meson decays. The measured value of $S_{\psi K_S}$ is consistent with the correlation between K and B decays that is predicted by the SM. The value of $S_{\phi K_S}$ is equal (within the present experimental accuracy) with $S_{\psi K_S}$, consistent with the SM correlation between the asymmetries in $b \to s\bar{s}s$ and $b \to c\bar{c}s$ transitions. It is therefore very likely that δ_{KM} is indeed the dominant source of CP violation in meson decays.

(ii) *CP violation is small in $K \to \pi\pi$ decays because of flavor suppression and not because CP is an approximate symmetry.* In many (though certainly not all) supersymmetric models, the flavor suppression is too mild, or entirely ineffective, requiring approximate CP to hold. The measurement of $S_{\psi K_S} = \mathcal{O}(1)$ confirms that not all CP violating phases are small.

(iii) *CP violation appears in both $\Delta F = 1$ (decay) and $\Delta F = 2$ (mixing) amplitudes.* Superweak models suggest that CP is violated only in mixing amplitudes. The measurements of non-vanishing ε'/ε, $\mathcal{A}_{K^{\mp}\pi^{\pm}}$ and $\mathcal{A}_{\rho\pi}^{-+}$ confirm that there is CP violation in $\Delta S = 1$ and $\Delta B = 1$ processes.

(iv) *CP is not violated in the lepton sector.* Models that allow for neutrino masses, such as the ESM framework presented above, predict CP violation in leptonic charged current interactions. Thus, while there is no measurement of leptonic CP violation, the data from neutrino oscillation experiments, which give evidence that neutrinos are massive and mix, make it very likely that charged current weak interactions violate CP also in the lepton sector.

(v) *CP violation appears only in the charged current weak interactions and in conjunction with flavor changing processes.* Here both various extensions of the SM (such as supersymmetry) and non-perturbative effects within the SM (θ_{QCD}) allow for CP violation in other types of interactions and in flavor diagonal processes. In particular, it is difficult to avoid flavor-diagonal phases in the supersymmetric framework. The fact that no electric dipole moment has been measured yet poses difficulties to many models with diagonal CP violation (and, of course, is responsible to the strong CP problem within the SM).

(vi) *CP is explicitly broken.* In various extensions of the scalar sector, it is possible to achieve spontaneous CP violation. It is very difficult to test this question experimentally.

This situation, where the Standard Model has a very unique and predictive description of CP violation, is the basis for the strong interest, experimental and theoretical, in CP violation.

3. Meson decays

The phenomenology of CP violation is superficially different in K, D, B, and B_s decays. This is primarily because each of these systems is governed by a different balance between decay rates, oscillations, and lifetime splitting. However, the underlying mechanisms of CP violation are identical for all pseudoscalar mesons.

In this section we present a general formalism for, and classification of, CP violation in the decay of a pseudoscalar meson P that might be a charged or neutral K, D, B, or B_s meson. Subsequent sections describe the CP-violating phenomenology, approximations, and alternate formalisms that are specific to each system. We follow here closely the discussion in [55].

3.1. Charged and neutral meson decays

We define decay amplitudes of a pseudoscalar meson P (which could be charged or neutral) and its CP conjugate $\overline{P}$ to a multi-particle final state f and its CP conjugate $\overline{f}$ as

$$
\begin{aligned}
A_f &= \langle f|\mathcal{H}|P\rangle, & \overline{A}_f &= \langle f|\mathcal{H}|\overline{P}\rangle, \\
A_{\overline{f}} &= \langle \overline{f}|\mathcal{H}|P\rangle, & \overline{A}_{\overline{f}} &= \langle \overline{f}|\mathcal{H}|\overline{P}\rangle,
\end{aligned}
\tag{3.1}
$$

where $\mathcal{H}$ is the Hamiltonian governing weak interactions. The action of CP on these states introduces phases ξ_P and ξ_f that depend on their flavor content, according to

$$
\begin{aligned}
CP|P\rangle &= e^{+i\xi_P}\,|\overline{P}\rangle, & CP|f\rangle &= e^{+i\xi_f}\,|\overline{f}\rangle, \\
CP|\overline{P}\rangle &= e^{-i\xi_P}\,|P\rangle, & CP|\overline{f}\rangle &= e^{-i\xi_f}\,|f\rangle,
\end{aligned}
\tag{3.2}
$$

so that $(CP)^2 = 1$. The phases ξ_P and ξ_f are arbitrary and unphysical because of the flavor symmetry of the strong interaction. If CP is conserved by the dynamics, $[CP, \mathcal{H}] = 0$, then A_f and $\overline{A}_{\overline{f}}$ have the same magnitude and an arbitrary unphysical relative phase

$$
\overline{A}_{\overline{f}} = e^{i(\xi_f - \xi_P)}\, A_f.
\tag{3.3}
$$

3.2. Neutral meson mixing

A state that is initially a superposition of P^0 and $\overline{P}^0$, say

$$
|\psi(0)\rangle = a(0)|P^0\rangle + b(0)|\overline{P}^0\rangle,
\tag{3.4}
$$

will evolve in time acquiring components that describe all possible decay final states $\{f_1, f_2, \ldots\}$, that is,

$$
|\psi(t)\rangle = a(t)|P^0\rangle + b(t)|\overline{P}^0\rangle + c_1(t)|f_1\rangle + c_2(t)|f_2\rangle + \cdots.
\tag{3.5}
$$

If we are interested in computing only the values of $a(t)$ and $b(t)$ (and not the values of all $c_i(t)$), and if the times t in which we are interested are much larger than the typical strong interaction scale, then we can use a much simplified formalism [56]. The simplified time evolution is determined by a 2×2 effective Hamiltonian $\mathcal{H}$ that is not Hermitian, since otherwise the mesons would only oscillate and not decay. Any complex matrix, such as $\mathcal{H}$, can be written in terms of Hermitian matrices M and Γ as

$$
\mathcal{H} = M - \frac{i}{2}\Gamma.
\tag{3.6}
$$

M and Γ are associated with $(P^0, \overline{P}^0) \leftrightarrow (P^0, \overline{P}^0)$ transitions via off-shell (dispersive) and on-shell (absorptive) intermediate states, respectively. Diagonal elements of M and Γ are associated with the flavor-conserving transitions $P^0 \to P^0$

and $\overline{P}^0 \to \overline{P}^0$ while off-diagonal elements are associated with flavor-changing transitions $P^0 \leftrightarrow \overline{P}^0$.

The eigenvectors of $\mathcal{H}$ have well defined masses and decay widths. We introduce complex parameters $p_{L,H}$ and $q_{L,H}$ to specify the components of the strong interaction eigenstates, P^0 and $\overline{P}^0$, in the light (P_L) and heavy (P_H) mass eigenstates:

$$|P_{L,H}\rangle = p_{L,H}|P^0\rangle \pm q_{L,H}|\overline{P}^0\rangle, \tag{3.7}$$

with the normalization $|p_{L,H}|^2 + |q_{L,H}|^2 = 1$. (Another possible choice, which is in standard usage for K mesons, defines the mass eigenstates according to their lifetimes: K_S for the short-lived and K_L for the long-lived state. The K_L is the heavier state.) If either CP or CPT is a symmetry of $\mathcal{H}$ (independently of whether T is conserved or violated) then $M_{11} = M_{22}$ and $\Gamma_{11} = \Gamma_{22}$, and solving the eigenvalue problem for $\mathcal{H}$ yields $p_L = p_H \equiv p$ and $q_L = q_H \equiv q$ with

$$\left(\frac{q}{p}\right)^2 = \frac{M_{12}^* - (i/2)\Gamma_{12}^*}{M_{12} - (i/2)\Gamma_{12}}. \tag{3.8}$$

If either CP or T is a symmetry of $\mathcal{H}$ (independently of whether CPT is conserved or violated), then M_{12} and Γ_{12} are relatively real, leading to

$$\left(\frac{q}{p}\right)^2 = e^{2i\xi_P} \quad \Rightarrow \quad \left|\frac{q}{p}\right| = 1, \tag{3.9}$$

where ξ_P is the arbitrary unphysical phase introduced in Eq. (3.2). If, and only if, CP is a symmetry of $\mathcal{H}$ (independently of CPT and T) then both of the above conditions hold, with the result that the mass eigenstates are orthogonal

$$\langle P_H | P_L \rangle = |p|^2 - |q|^2 = 0. \tag{3.10}$$

From now on we assume that CPT is conserved.

The real and imaginary parts of the eigenvalues of $\mathcal{H}$ corresponding to $|P_{L,H}\rangle$ represent their masses and decay-widths, respectively. The mass difference Δm and the width difference $\Delta\Gamma$ are defined as follows:

$$\Delta m \equiv M_H - M_L, \quad \Delta\Gamma \equiv \Gamma_H - \Gamma_L. \tag{3.11}$$

Note that here Δm is positive by definition, while the sign of $\Delta\Gamma$ is to be experimentally determined. (Alternatively, one can use the states defined by their lifetimes to have $\Delta\Gamma \equiv \Gamma_S - \Gamma_L$ positive by definition.) The average mass and width are given by

$$m \equiv \frac{M_H + M_L}{2}, \quad \Gamma \equiv \frac{\Gamma_H + \Gamma_L}{2}. \tag{3.12}$$

It is useful to define dimensionless ratios x and y:

$$x \equiv \frac{\Delta m}{\Gamma}, \quad y \equiv \frac{\Delta \Gamma}{2\Gamma}. \tag{3.13}$$

Solving the eigenvalue equation gives

$$(\Delta m)^2 - \frac{1}{4}(\Delta \Gamma)^2 = (4|M_{12}|^2 - |\Gamma_{12}|^2), \quad \Delta m \Delta \Gamma = 4\mathcal{R}e(M_{12}\Gamma_{12}^*). \tag{3.14}$$

3.3. CP-violating observables

All CP-violating observables in P and $\overline{P}$ decays to final states f and $\overline{f}$ can be expressed in terms of phase-convention-independent combinations of A_f, $\overline{A}_f$, $A_{\overline{f}}$ and $\overline{A}_{\overline{f}}$, together with, for neutral-meson decays only, q/p. CP violation in charged-meson decays depends only on the combination $|\overline{A}_{\overline{f}}/A_f|$, while CP violation in neutral-meson decays is complicated by $P^0 \leftrightarrow \overline{P}^0$ oscillations and depends, additionally, on $|q/p|$ and on $\lambda_f \equiv (q/p)(\overline{A}_f/A_f)$.

The decay-rates of the two neutral K mass eigenstates, K_S and K_L, are different enough ($\Gamma_S/\Gamma_L \sim 500$) that one can, in most cases, actually study their decays independently. For neutral D, B, and B_s mesons, however, values of $\Delta \Gamma / \Gamma$ are relatively small and so both mass eigenstates must be considered in their evolution. We denote the state of an initially pure $|P^0\rangle$ or $|\overline{P}^0\rangle$ after an elapsed proper time t as $|P^0_{\text{phys}}(t)\rangle$ or $|\overline{P}^0_{\text{phys}}(t)\rangle$, respectively. Using the effective Hamiltonian approximation, we obtain

$$\begin{aligned}
|P^0_{\text{phys}}(t)\rangle &= g_+(t)\,|P^0\rangle - (q/p)\,g_-(t)|\overline{P}^0\rangle, \\
|\overline{P}^0_{\text{phys}}(t)\rangle &= g_+(t)\,|\overline{P}^0\rangle - (p/q)\,g_-(t)|P^0\rangle,
\end{aligned} \tag{3.15}$$

where

$$g_\pm(t) \equiv \frac{1}{2}\left(e^{-im_H t - \frac{1}{2}\Gamma_H t} \pm e^{-im_L t - \frac{1}{2}\Gamma_L t}\right). \tag{3.16}$$

One obtains the following time-dependent decay rates:

$$\begin{aligned}
\frac{d\Gamma[P^0_{\text{phys}}(t) \to f]/dt}{e^{-\Gamma t}\mathcal{N}_f} &= \left(|A_f|^2 + |(q/p)\overline{A}_f|^2\right)\cosh(y\Gamma t) \\
&+ \left(|A_f|^2 - |(q/p)\overline{A}_f|^2\right)\cos(x\Gamma t) \\
&+ 2\,\mathcal{R}e((q/p)A_f^*\overline{A}_f)\sinh(y\Gamma t) \\
&- 2\,\mathcal{I}m((q/p)A_f^*\overline{A}_f)\sin(x\Gamma t),
\end{aligned} \tag{3.17}$$

$$\frac{d\Gamma[\overline{P}^0_{\text{phys}}(t) \to f]/dt}{e^{-\Gamma t}\mathcal{N}_f} = \left(|(p/q)A_f|^2 + |\overline{A}_f|^2\right)\cosh(y\Gamma t)$$

$$- \left(|(p/q)A_f|^2 - |\overline{A}_f|^2\right)\cos(x\Gamma t)$$

$$+ 2\mathcal{R}e((p/q)A_f\overline{A}_f^*)\sinh(y\Gamma t)$$

$$- 2\mathcal{I}m((p/q)A_f\overline{A}_f^*)\sin(x\Gamma t), \tag{3.18}$$

where $\mathcal{N}_f$ is a common normalization factor. Decay rates to the CP-conjugate final state $\overline{f}$ are obtained analogously, with $\mathcal{N}_f = \mathcal{N}_{\overline{f}}$ and the substitutions $A_f \to A_{\overline{f}}$ and $\overline{A}_f \to \overline{A}_{\overline{f}}$ in Eqs. (3.17, 3.18). Terms proportional to $|A_f|^2$ or $|\overline{A}_f|^2$ are associated with decays that occur without any net $P \leftrightarrow \overline{P}$ oscillation, while terms proportional to $|(q/p)\overline{A}_f|^2$ or $|(p/q)A_f|^2$ are associated with decays following a net oscillation. The $\sinh(y\Gamma t)$ and $\sin(x\Gamma t)$ terms of Eqs. (3.17, 3.18) are associated with the interference between these two cases. Note that, in multi-body decays, amplitudes are functions of phase-space variables. Interference may be present in some regions but not others, and is strongly influenced by resonant substructure.

3.4. Classification of CP-violating effects

We distinguish three types of CP-violating effects in meson decays [57]:

[I] CP violation in decay is defined by

$$|\overline{A}_{\overline{f}}/A_f| \neq 1. \tag{3.19}$$

In charged meson decays, where mixing effects are absent, this is the only possible source of CP asymmetries:

$$\mathcal{A}_{f^\pm} \equiv \frac{\Gamma(P^- \to f^-) - \Gamma(P^+ \to f^+)}{\Gamma(P^- \to f^-) + \Gamma(P^+ \to f^+)}$$

$$= \frac{|\overline{A}_{f^-}/A_{f^+}|^2 - 1}{|\overline{A}_{f^-}/A_{f^+}|^2 + 1}. \tag{3.20}$$

[II] CP violation in mixing is defined by

$$|q/p| \neq 1. \tag{3.21}$$

In charged-current semileptonic neutral meson decays $P, \overline{P} \to \ell^\pm X$ (taking $|A_{\ell^+ X}| = |\overline{A}_{\ell^- X}|$ and $A_{\ell^- X} = \overline{A}_{\ell^+ X} = 0$, as is the case in the Standard Model, to lowest order, and in most of its reasonable extensions), this is the only source of CP violation, and can be measured via the asymmetry of "wrong-sign" decays induced by oscillations:

$$\mathcal{A}_{SL}(t) \equiv \frac{d\Gamma/dt[\overline{P}^0_{\text{phys}}(t) \to \ell^+ X] - d\Gamma/dt[P^0_{\text{phys}}(t) \to \ell^- X]}{d\Gamma/dt[\overline{P}^0_{\text{phys}}(t) \to \ell^+ X] + d\Gamma/dt[P^0_{\text{phys}}(t) \to \ell^- X]}$$
$$= \frac{1 - |q/p|^4}{1 + |q/p|^4}. \tag{3.22}$$

Note that this asymmetry of time-dependent decay rates is actually time independent.

[III] CP violation in interference between a decay without mixing, $P^0 \to f$, and a decay with mixing, $P^0 \to \overline{P}^0 \to f$ (such an effect occurs only in decays to final states that are common to P^0 and $\overline{P}^0$, including all CP eigenstates), is defined by

$$\mathcal{I}m(\lambda_f) \neq 0, \tag{3.23}$$

with

$$\lambda_f \equiv \frac{q}{p}\frac{\overline{A}_f}{A_f}. \tag{3.24}$$

This form of CP violation can be observed, for example, using the asymmetry of neutral meson decays into final CP eigenstates f_{CP}

$$\mathcal{A}_{f_{CP}}(t) \equiv \frac{d\Gamma/dt[\overline{P}^0_{\text{phys}}(t) \to f_{CP}] - d\Gamma/dt[P^0_{\text{phys}}(t) \to f_{CP}]}{d\Gamma/dt[\overline{P}^0_{\text{phys}}(t) \to f_{CP}] + d\Gamma/dt[P^0_{\text{phys}}(t) \to f_{CP}]}. \tag{3.25}$$

If $\Delta\Gamma = 0$ and $|q/p| = 1$, as expected to a good approximation for B mesons but not for K mesons, then $\mathcal{A}_{f_{CP}}$ has a particularly simple form [58–60]:

$$\mathcal{A}_f(t) = S_f \sin(\Delta mt) - C_f \cos(\Delta mt),$$
$$S_f \equiv \frac{2\mathcal{I}m(\lambda_f)}{1 + |\lambda_f|^2}, \quad C_f \equiv \frac{1 - |\lambda_f|^2}{1 + |\lambda_f|^2}, \tag{3.26}$$

If, in addition, the decay amplitudes fulfill $|\overline{A}_{f_{CP}}| = |A_{f_{CP}}|$, the interference between decays with and without mixing is the only source of the asymmetry and

$$\mathcal{A}_{f_{CP}}(t) = \mathcal{I}m(\lambda_{f_{CP}}) \sin(x\Gamma t). \tag{3.27}$$

4. Theoretical interpretation: general considerations

Consider the $P \to f$ decay amplitude A_f, and the CP conjugate process, $\overline{P} \to \overline{f}$, with decay amplitude $\overline{A}_{\overline{f}}$. There are two types of phases that may appear in these decay amplitudes. Complex parameters in any Lagrangian term that contributes to the amplitude will appear in complex conjugate form in the CP-conjugate amplitude. Thus their phases appear in A_f and $\overline{A}_{\overline{f}}$ with opposite signs. In the Standard Model, these phases occur only in the couplings of the $W^\pm$ bosons and hence are often called "weak phases". The weak phase of any single term is convention dependent. However, the difference between the weak phases in two different terms in A_f is convention independent. A second type of phase can appear in scattering or decay amplitudes even when the Lagrangian is real. Their origin is the possible contribution from intermediate on-shell states in the decay process. Since these phases are generated by CP-invariant interactions, they are the same in A_f and $\overline{A}_{\overline{f}}$. Usually the dominant rescattering is due to strong interactions and hence the designation "strong phases" for the phase shifts so induced. Again, only the relative strong phases between different terms in the amplitude are physically meaningful.

The 'weak' and 'strong' phases discussed here appear in addition to the 'spurious' CP-transformation phases of Eq. (3.3). Those spurious phases are due to an arbitrary choice of phase convention, and do not originate from any dynamics or induce any CP violation. For simplicity, we set them to zero from here on.

It is useful to write each contribution a_i to A_f in three parts: its magnitude $|a_i|$, its weak phase ϕ_i, and its strong phase δ_i. If, for example, there are two such contributions, $A_f = a_1 + a_2$, we have

$$A_f = |a_1|e^{i(\delta_1 + \phi_1)} + |a_2|e^{i(\delta_2 + \phi_2)},$$
$$\overline{A}_{\overline{f}} = |a_1|e^{i(\delta_1 - \phi_1)} + |a_2|e^{i(\delta_2 - \phi_2)}. \tag{4.1}$$

Similarly, for neutral meson decays, it is useful to write

$$M_{12} = |M_{12}|e^{i\phi_M}, \quad \Gamma_{12} = |\Gamma_{12}|e^{i\phi_\Gamma}. \tag{4.2}$$

Each of the phases appearing in Eqs. (4.1, 4.2) is convention dependent, but combinations such as $\delta_1 - \delta_2$, $\phi_1 - \phi_2$, $\phi_M - \phi_\Gamma$ and $\phi_M + \phi_1 - \overline{\phi}_1$ (where $\overline{\phi}_1$ is a weak phase contributing to $\overline{A}_f$) are physical.

It is now straightforward to evaluate the various asymmetries in terms of the theoretical parameters introduced here. We will do so with approximations that are often relevant to the most interesting measured asymmetries.

1. The CP asymmetry in charged meson decays [Eq. (3.20)] is given by

$$\mathcal{A}_{f^\pm} = -\frac{2|a_1 a_2| \sin(\delta_2 - \delta_1) \sin(\phi_2 - \phi_1)}{|a_1|^2 + |a_2|^2 + 2|a_1 a_2| \cos(\delta_2 - \delta_1) \cos(\phi_2 - \phi_1)}. \tag{4.3}$$

The quantity of most interest to theory is the weak phase difference $\phi_2 - \phi_1$. Its extraction from the asymmetry requires, however, that the amplitude ratio and the strong phase are known. Both quantities depend on non-perturbative hadronic parameters that are difficult to calculate.

2. In the approximation that $|\Gamma_{12}/M_{12}| \ll 1$ (valid for B and B_s mesons), the CP asymmetry in semileptonic neutral-meson decays [Eq. (3.22)] is given by

$$
\mathcal{A}_{\mathrm{SL}} = - \left| \frac{\Gamma_{12}}{M_{12}} \right| \sin(\phi_M - \phi_\Gamma). \tag{4.4}
$$

The quantity of most interest to theory is the weak phase $\phi_M - \phi_\Gamma$. Its extraction from the asymmetry requires, however, that $|\Gamma_{12}/M_{12}|$ is known. This quantity depends on long distance physics that is difficult to calculate.

3. In the approximations that only a single weak phase contributes to decay, $A_f = |a_f| e^{i(\delta_f + \phi_f)}$, and that $|\Gamma_{12}/M_{12}| = 0$, we obtain $|\lambda_f| = 1$ and the CP asymmetries in decays to a final CP eigenstate f [Eq. (3.25)] with eigenvalue $\eta_f = \pm 1$ are given by

$$
\mathcal{A}_{fCP}(t) = \mathcal{I}m(\lambda_f) \, \sin(\Delta m t) \ \text{ with } \ \mathcal{I}m(\lambda_f) = \eta_f \sin(\phi_M + 2\phi_f). \tag{4.5}
$$

Note that the phase so measured is purely a weak phase, and no hadronic parameters are involved in the extraction of its value from $\mathcal{I}m(\lambda_f)$.

The discussion above allows us to introduce another classification:

1. **Direct CP violation** is one that cannot be accounted for by just $\phi_M \neq 0$. CP violation in decay (type I) belongs to this class.

2. **Indirect CP violation** is consistent with taking $\phi_M \neq 0$ and setting all other CP violating phases to zero. CP violation in mixing (type II) belongs to this class.

As concerns type III CP violation, observing $\eta_{f_1} \mathcal{I}m(\lambda_{f_1}) \neq \eta_{f_2} \mathcal{I}m(\lambda_{f_2})$ (for the same decaying meson and two different final CP eigenstates f_1 and f_2) would establish direct CP violation. The significance of this classification is related to theory. In superweak models [20], CP violation appears only in diagrams that contribute to M_{12}, hence they predict that there is no direct CP violation. In most models and, in particular, in the Standard Model, CP violation is both direct and indirect. The experimental observation of $\epsilon' \neq 0$ (see Section 5) excluded the superweak scenario.

5. K decays

CP violation was discovered in $K \to \pi\pi$ decays in 1964 [2]. The same mode provided the first evidence for direct CP violation [3–5].

The decay amplitudes actually measured in neutral K decays refer to the mass eigenstates K_L and K_S rather than to the K and $\overline{K}$ states referred to in Eq. (3.1). We define CP-violating amplitude ratios for two-pion final states,

$$\eta_{00} \equiv \frac{\langle \pi^0 \pi^0 | \mathcal{H} | K_L \rangle}{\langle \pi^0 \pi^0 | \mathcal{H} | K_S \rangle}, \quad \eta_{+-} \equiv \frac{\langle \pi^+ \pi^- | \mathcal{H} | K_L \rangle}{\langle \pi^+ \pi^- | \mathcal{H} | K_S \rangle}. \tag{5.1}$$

Another important observable is the asymmetry of time-integrated semileptonic decay rates:

$$\delta_L \equiv \frac{\Gamma(K_L \to \ell^+ \nu_\ell \pi^-) - \Gamma(K_L \to \ell^- \bar{\nu}_\ell \pi^+)}{\Gamma(K_L \to \ell^+ \nu_\ell \pi^-) + \Gamma(K_L \to \ell^- \bar{\nu}_\ell \pi^+)}. \tag{5.2}$$

CP violation has been observed as an appearance of K_L decays to two-pion final states [53],

$$
\begin{aligned}
|\eta_{00}| &= (2.275 \pm 0.017) \times 10^{-3} & \phi_{00} &= 43.6° \pm 0.8° \\
|\eta_{+-}| &= (2.286 \pm 0.017) \times 10^{-3} & \phi_{+-} &= 43.2° \pm 0.6° \\
|\eta_{00}/\eta_{+-}| &= 0.9950 \pm 0.0008 & \phi_{00} - \phi_{+-} &= 0.4° \pm 0.5°,
\end{aligned}
\tag{5.3}
$$

where ϕ_{ij} is the phase of the amplitude ratio η_{ij} determined without assuming CPT invariance. (A fit that assumes CPT gives [53] $\phi_{+-} = \phi_{00} = 43.49° \pm 0.07°$.) CP violation has also been observed in semileptonic K_L decays [53]

$$\delta_L = (3.27 \pm 0.12) \times 10^{-3}, \tag{5.4}$$

where δ_L is a weighted average of muon and electron measurements, as well as in K_L decays to $\pi^+ \pi^- \gamma$ and $\pi^+ \pi^- e^+ e^-$ [53].

Historically, CP violation in neutral K decays has been described in terms of parameters ϵ and ϵ'. The observables η_{00}, η_{+-}, and δ_L are related to these parameters, and to those of Section 3, by

$$
\begin{aligned}
\eta_{00} &= \frac{1 - \lambda_{\pi^0 \pi^0}}{1 + \lambda_{\pi^0 \pi^0}} &= \epsilon - 2\epsilon', \\
\eta_{+-} &= \frac{1 - \lambda_{\pi^+ \pi^-}}{1 + \lambda_{\pi^+ \pi^-}} &= \epsilon + \epsilon', \\
\delta_L &= \frac{1 - |q/p|^2}{1 + |q/p|^2} &= \frac{2\mathcal{R}e(\epsilon)}{1 + |\epsilon|^2},
\end{aligned}
\tag{5.5}
$$

where, in the last line, we have assumed that $|A_{\ell^+ \nu_\ell \pi^-}| = |\overline{A}_{\ell^- \bar{\nu}_\ell \pi^+}|$ and $|A_{\ell^- \bar{\nu}_\ell \pi^+}|$ $= |\overline{A}_{\ell^+ \nu_\ell \pi^-}| = 0$. A fit to the $K \to \pi\pi$ data yields [53]

$$
\begin{aligned}
|\epsilon| &= (2.283 \pm 0.017) \times 10^{-3}, \\
\mathcal{R}e(\epsilon'/\epsilon) &= (1.67 \pm 0.26) \times 10^{-3}.
\end{aligned}
\tag{5.6}
$$

In discussing two-pion final states, it is useful to express the amplitudes $A_{\pi^0 \pi^0}$ and $A_{\pi^+ \pi^-}$ in terms of their isospin components via

$$
\begin{aligned}
A_{\pi^0 \pi^0} &= \sqrt{\frac{1}{3}} |A_0| e^{i(\delta_0 + \phi_0)} - \sqrt{\frac{2}{3}} |A_2| e^{i(\delta_2 + \phi_2)}, \\
A_{\pi^+ \pi^-} &= \sqrt{\frac{2}{3}} |A_0| e^{i(\delta_0 + \phi_0)} + \sqrt{\frac{1}{3}} |A_2| e^{i(\delta_2 + \phi_2)},
\end{aligned}
\tag{5.7}
$$

where we parameterize the amplitude $A_I (\overline{A}_I)$ for $K^0 (\overline{K}^0)$ decay into two pions with total isospin $I = 0$ or 2 as

$$
\begin{aligned}
A_I &\equiv \langle (\pi\pi)_I | \mathcal{H} | K^0 \rangle = |A_I| e^{i(\delta_I + \phi_I)}, \\
\overline{A}_I &\equiv \langle (\pi\pi)_I | \mathcal{H} | \overline{K}^0 \rangle = |A_I| e^{i(\delta_I - \phi_I)}.
\end{aligned}
\tag{5.8}
$$

The smallness of $|\eta_{00}|$ and $|\eta_{+-}|$ allows us to approximate

$$
\epsilon \simeq \frac{1}{2} (1 - \lambda_{(\pi\pi)_{I=0}}), \qquad \epsilon' \simeq \frac{1}{6} \left(\lambda_{\pi^0 \pi^0} - \lambda_{\pi^+ \pi^-} \right).
\tag{5.9}
$$

The parameter ϵ represents indirect CP violation, while ϵ' parameterizes direct CP violation: $\mathcal{R}e(\epsilon')$ measures CP violation in decay (type I), $\mathcal{R}e(\epsilon)$ measures CP violation in mixing (type II), and $\mathcal{I}m(\epsilon)$ and $\mathcal{I}m(\epsilon')$ measure the interference between decays with and without mixing (type III).

The following expressions for ϵ and ϵ' are useful for theoretical evaluations:

$$
\epsilon \simeq \frac{e^{i\pi/4}}{\sqrt{2}} \frac{\mathcal{I}m(M_{12})}{\Delta m}, \qquad \epsilon' = \frac{i}{\sqrt{2}} \left| \frac{A_2}{A_0} \right| e^{i(\delta_2 - \delta_0)} \sin(\phi_2 - \phi_0).
\tag{5.10}
$$

The expression for ϵ is only valid in a phase convention where $\phi_2 = 0$, corresponding to a real $V_{ud} V_{us}^*$, and in the approximation that also $\phi_0 = 0$. The phase of $\pi/4$ is approximate, and determined by hadronic parameters, $\arg \epsilon \approx \arctan(-2\Delta m/\Delta\Gamma)$, independently of the electroweak model. The calculation of ϵ benefits from the fact that $\mathcal{I}m(M_{12})$ is dominated by short distance physics. Consequently, the main source of uncertainty in theoretical interpretations of ϵ are the values of matrix elements such as $\langle K^0 | (\bar{s}d)_{V-A} (\bar{s}d)_{V-A} | \overline{K}^0 \rangle$. The expression for ϵ' is valid to first order in $|A_2/A_0| \sim 1/20$. The phase of ϵ' is experimentally determined, $\pi/2 + \delta_2 - \delta_0 \approx \pi/4$ and is independent of the electroweak model. Note that, accidentally, ϵ'/ϵ is real to a good approximation.

A future measurement of much interest is that of CP violation in the rare $K \to \pi \nu \bar{\nu}$ decays. The signal for CP violation is simply observing the $K_L \to \pi^0 \nu \bar{\nu}$ decay. The effect here is that of interference between decays with and without mixing (type III) [61]:

$$\frac{\Gamma(K_L \to \pi^0 \nu \bar{\nu})}{\Gamma(K^+ \to \pi^+ \nu \bar{\nu})} = \frac{1}{2} \left[1 + |\lambda_{\pi \nu \bar{\nu}}|^2 - 2 \mathcal{R}e(\lambda_{\pi \nu \bar{\nu}}) \right] \simeq 1 - \mathcal{R}e(\lambda_{\pi \nu \bar{\nu}}), \quad (5.11)$$

where in the last equation we neglect CP violation in decay and in mixing (expected, model independently, to be of order 10^{-5} and 10^{-3}, respectively). Such a measurement would be experimentally very challenging and theoretically very rewarding [62]. Similar to the CP asymmetry in $B \to J/\psi K_S$, the CP violation in $K \to \pi \nu \bar{\nu}$ decay is predicted to be large and can be very cleanly interpreted.

Within the Standard Model, the $K_L \to \pi^0 \nu \bar{\nu}$ decay is dominated by an intermediate top quark contribution and, consequently, can be cleanly interpreted in terms of CKM parameters [63]. (For the charged mode, $K^+ \to \pi^+ \nu \bar{\nu}$, the contribution from an intermediate charm quark is not negligible and constitutes a source of hadronic uncertainty.) In particular, $\mathcal{B}(K_L \to \pi^0 \nu \bar{\nu})$ provides a theoretically clean way to determine the Wolfenstein parameter η [64]:

$$\mathcal{B}(K_L \to \pi^0 \nu \bar{\nu}) = \kappa_L X^2(m_t^2/m_W^2) A^4 \eta^2, \tag{5.12}$$

where $\kappa_L = 1.80 \times 10^{-10}$ incorporates the value of the four-fermion matrix element which is deduced, using isospin relations, from $\mathcal{B}(K^+ \to \pi^0 e^+ \nu)$, and $X(m_t^2/m_W^2)$ is a known function of the top mass.

5.1. Implications of ε_K

The measurement of ε_K has had (and still has) important implications. Two implications of historical importance are the following:

(i) CP violation was discovered through the measurement of ε_K. Hence this measurement played a significant role in the history of particle physics.

(ii) The observation of $\varepsilon_K \neq 0$ led to the prediction that a third generation must exist, so that CP is violated in the Standard Model. This provides an excellent example of how precision measurements at low energy can lead to the discovery of new physics (even if, at present, this new physics is old...)

Within the Standard Model, $\mathcal{I}m(M_{12})$ is accounted for by box diagrams:

$$\begin{aligned} \varepsilon_K = \ & e^{i\pi/4} C_\varepsilon B_K \mathcal{I}m(V_{ts}^* V_{td}) \left\{ \mathcal{R}e(V_{cs}^* V_{cd})[\eta_1 S_0(x_c) \right. \\ & \left. - \eta_3 S_0(x_c, x_t)] - \mathcal{R}e(V_{ts}^* V_{td}) \eta_2 S_0(x_t) \right\}, \end{aligned} \tag{5.13}$$

where $C_\varepsilon \equiv \frac{G_F^2 f_K^2 m_K m_W^2}{6\sqrt{2}\pi^2 \Delta m_K}$ is a well known parameter, the η_i are QCD correction factors, S_0 is a kinematic factor, and B_K is the ratio between the matrix element of the four quark operator and its value in the vacuum insertion approximation. The measurement of ε_K has the following implications within the SM:

• This measurement allowed one to set the value of δ_{KM}. Furthermore, by implying that $\delta_{\text{KM}} = \mathcal{O}(1)$, it made the KM mechanism plausible. Having been the single measured CP violating parameter it could not, however, serve as a test of the KM mechanism. More precisely, a value of $|\varepsilon_K| \gg 10^{-3}$ would have invalidated the KM mechanism, but any value $|\varepsilon_K| \lesssim 10^{-3}$ was acceptable. It is only the combination of the new measurements in B decays (particularly $S_{\psi K_S}$) with ε_K that provides the first precision test of the KM mechanism.

• Within the SM, the smallness of ε_K is not related to suppression of CP violation but rather to suppression of flavor violation. Specifically, it is the smallness of the ratio $|(V_{td}V_{ts})/(V_{ud}V_{us})| \sim \lambda^4$ that explains $|\varepsilon_K| \sim 10^{-3}$.

• Until recently, the measured value of ε_K provided a unique type of information on the KM phase. For example, the measurement of $\mathcal{R}e(\varepsilon_K) > 0$ tells us that $\eta > 0$ and excludes the lower half of the $\rho - \eta$ plane. Such information cannot be obtained from any CP conserving observable.

• The ε_K constraint in Eq. (5.13) gives hyperbole in the $\rho - \eta$ plane. It is shown in Fig. 2. The measured value is consistent with all other CKM-related measurements and further narrows the allowed region.

Beyond the SM, ε_K is an extremely powerful probe of new physics. This aspect will be discussed later.

6. *D* decays

Unlike the case of neutral K, B, and B_s mixing, $D^0 - \overline{D}^0$ mixing has not yet been observed. Long-distance contributions make it difficult to calculate the Standard Model prediction for the $D^0 - \overline{D}^0$ mixing parameters. Therefore, the goal of the search for $D^0 - \overline{D}^0$ mixing is not to constrain the CKM parameters but rather to probe new physics. Here CP violation plays an important role [65]. Within the Standard Model, the CP-violating effects are predicted to be negligibly small since the mixing and the relevant decays are described, to an excellent approximation, by physics of the first two generations. Observation of CP violation in $D^0 - \overline{D}^0$ mixing (at a level much higher than $\mathcal{O}(10^{-3})$) will constitute an unambiguous signal of new physics.[2] At present, the most sensitive searches involve the $D \to K^+ K^-$ and $D \to K^\pm \pi^\mp$ modes.

---

[2] In contrast, neither $y_D \sim 10^{-2}$ [66], nor $x_D \sim 10^{-2}$ [67] can be considered as evidence for new physics.

The neutral D mesons decay via a singly-Cabibbo-suppressed transition to the CP eigenstate K^+K^-. Since the decay proceeds via a Standard-Model tree diagram, it is very likely unaffected by new physics and, furthermore, dominated by a single weak phase. It is safe then to assume that direct CP violation plays no role here [68, 69]. In addition, given the experimental bounds [53], $x \equiv \Delta m/\Gamma \lesssim 0.03$ and $y \equiv \Delta\Gamma/(2\Gamma) = 0.008 \pm 0.005$, we can expand the decay rates to first order in these parameters. Using Eq. (3.17) with these assumptions and approximations yields, for $xt, yt \lesssim \Gamma^{-1}$,

$$
\begin{aligned}
\Gamma[D^0_{\text{phys}}(t) \to K^+K^-] &= e^{-\Gamma t}|A_{KK}|^2 \\
&\quad \times [1 - |q/p|(y\cos\phi_D - x\sin\phi_D)\Gamma t], \\
\Gamma[\overline{D}^0_{\text{phys}}(t) \to K^+K^-] &= e^{-\Gamma t}|A_{KK}|^2 \\
&\quad \times [1 - |p/q|(y\cos\phi_D + x\sin\phi_D)\Gamma t],
\end{aligned} \tag{6.1}
$$

where ϕ_D is defined via $\lambda_{K^+K^-} = -|q/p|e^{i\phi_D}$. (In the limit of CP conservation, choosing $\phi_D = 0$ is equivalent to defining the mass eigenstates by their CP eigenvalue: $|D_\mp\rangle = p|D^0\rangle \pm q|\overline{D}^0\rangle$, with $D_-(D_+)$ being the CP-odd (CP-even) state; that is, the state that does not (does) decay into K^+K^-.) Given the small values of x and y, the time dependences of the rates in Eq. (6.1) can be recast into purely exponential forms, but with modified decay-rate parameters [70]:

$$
\begin{aligned}
\Gamma_{D^0 \to K^+K^-} &= \Gamma \times [1 + |q/p|(y\cos\phi_D - x\sin\phi_D)], \\
\Gamma_{\overline{D}^0 \to K^+K^-} &= \Gamma \times [1 + |p/q|(y\cos\phi_D + x\sin\phi_D)].
\end{aligned} \tag{6.2}
$$

One can define CP-conserving and CP-violating combinations of these two observables (normalized to the true width Γ):

$$
\begin{aligned}
Y &\equiv \frac{\Gamma_{\overline{D}^0 \to K^+K^-} + \Gamma_{D^0 \to K^+K^-}}{2\Gamma} - 1 \\
&= \frac{|q/p| + |p/q|}{2} y\cos\phi_D - \frac{|q/p| - |p/q|}{2} x\sin\phi_D, \\
\Delta Y &\equiv \frac{\Gamma_{\overline{D}^0 \to K^+K^-} - \Gamma_{D^0 \to K^+K^-}}{2\Gamma} \\
&= \frac{|q/p| + |p/q|}{2} x\sin\phi_D - \frac{|q/p| - |p/q|}{2} y\cos\phi_D.
\end{aligned} \tag{6.3}
$$

In the limit of CP conservation (and, in particular, within the Standard Model), $Y = y$ and $\Delta Y = 0$.

The $K^\pm\pi^\mp$ states are not CP eigenstates but they are still common final states for D^0 and $\overline{D}^0$ decays. Since $D^0(\overline{D}^0) \to K^-\pi^+$ is a Cabibbo-favored (doubly-Cabibbo-suppressed) process, these processes are particularly sensitive to x and/or $y = \mathcal{O}(\lambda^2)$. Taking into account that $|\lambda_{K^-\pi^+}|$, $|\lambda^{-1}_{K^+\pi^-}| \ll 1$ and $x, y \ll 1$, assuming that there is no direct CP violation (again, these are Standard Model tree level decays dominated by a single weak phase) and expanding the time dependent rates for $xt, yt \lesssim \Gamma^{-1}$, one obtains

$$\frac{\Gamma[D^0_{\text{phys}}(t) \to K^+\pi^-]}{\Gamma[\overline{D}^0_{\text{phys}}(t) \to K^+\pi^-]} = r_d^2 + r_d \left|\frac{q}{p}\right| (y' \cos\phi_D - x' \sin\phi_D)\Gamma t$$

$$+ \left|\frac{q}{p}\right|^2 \frac{y^2 + x^2}{4}(\Gamma t)^2,$$

$$\frac{\Gamma[\overline{D}^0_{\text{phys}}(t) \to K^-\pi^+]}{\Gamma[D^0_{\text{phys}}(t) \to K^-\pi^+]} = r_d^2 + r_d \left|\frac{p}{q}\right| (y' \cos\phi_D + x' \sin\phi_D)\Gamma t$$

$$+ \left|\frac{p}{q}\right|^2 \frac{y^2 + x^2}{4}(\Gamma t)^2, \tag{6.4}$$

where

$$y' \equiv y \cos\delta - x \sin\delta,$$
$$x' \equiv x \cos\delta + y \sin\delta. \tag{6.5}$$

The weak phase ϕ_D is the same as that of Eq. (6.1) (a consequence of the absence of direct CP violation), δ is a strong phase difference for these processes, and $r_d = \mathcal{O}(\tan^2\theta_c)$ is the amplitude ratio, $r_d = |\overline{A}_{K^-\pi^+}/A_{K^-\pi^+}| = |A_{K^+\pi^-}/\overline{A}_{K^+\pi^-}|$, that is, $\lambda_{K^-\pi^+} = r_d(q/p)e^{-i(\delta-\phi_D)}$ and $\lambda^{-1}_{K^+\pi^-} = r_d(p/q)e^{-i(\delta+\phi_D)}$. By fitting to the six coefficients of the various times dependences, one can extract r_d, $|q/p|$, $(x^2 + y^2)$, $y'\cos\phi_D$, and $x'\sin\phi_D$. In particular, finding CP violation, that is, $|q/p| \neq 1$ and/or $\sin\phi_D \neq 0$, would constitute evidence for new physics.

7. B decays

The upper bound on the CP asymmetry in semileptonic B decays [53] implies that CP violation in $B^0 - \overline{B}^0$ mixing is a small effect [we use $\mathcal{A}_{\text{SL}}/2 \approx 1 - |q/p|$, see Eq. (3.22)]:

$$\mathcal{A}_{\text{SL}} = (0.3 \pm 1.3) \times 10^{-2} \implies |q/p| = 0.998 \pm 0.007. \tag{7.1}$$

The Standard Model prediction is

$$\mathcal{A}_{\text{SL}} = \mathcal{O}\left(\frac{m_c^2}{m_t^2} \sin\beta\right) \lesssim 0.001. \tag{7.2}$$

In models where Γ_{12}/M_{12} is approximately real, such as the Standard Model, an upper bound on $\Delta\Gamma/\Delta m \approx \mathcal{R}e(\Gamma_{12}/M_{12})$ provides yet another upper bound on the deviation of $|q/p|$ from one. This constraint does not hold if Γ_{12}/M_{12} is approximately imaginary.

The small deviation (less than one percent) of $|q/p|$ from 1 implies that, at the present level of experimental precision, CP violation in B mixing is a negligible effect. Thus, for the purpose of analyzing CP asymmetries in hadronic B decays, we can use

$$\lambda_f = e^{-i\phi_B}(\overline{A}_f/A_f), \tag{7.3}$$

where ϕ_B refers to the phase of M_{12} [see Eq. (4.2)]. Within the Standard Model, the corresponding phase factor is given by

$$e^{-i\phi_B} = (V_{tb}^* V_{td})/(V_{tb} V_{td}^*). \tag{7.4}$$

Some of the most interesting decays involve final states that are common to B^0 and $\overline{B}^0$ [71–73]. Here Eq. (3.26) applies [58–60]. The processes of interest proceed via quark transitions of the form $\bar{b} \rightarrow \bar{q}q\bar{q}'$ with $q' = s$ or d. For $q = c$ or u, there are contributions from both tree (t) and penguin (p^{q_u}, where $q_u = u, c, t$ is the quark in the loop) diagrams (see Fig. 3) which carry different weak phases:

$$A_f = \left(V_{qb}^* V_{qq'}\right)t_f + \sum_{q_u=u,c,t} \left(V_{q_ub}^* V_{q_uq'}\right)p_f^{q_u}. \tag{7.5}$$

(The distinction between tree and penguin contributions is a heuristic one, the separation by the operator that enters is more precise. For a detailed discussion of the more complete operator product approach, which also includes higher order QCD corrections, see, for example, ref. [74].) Using CKM unitarity, these decay amplitudes can always be written in terms of just two CKM combinations. For example, for $f = \pi\pi$, which proceeds via $\bar{b} \rightarrow \bar{u}u\bar{d}$ transition, we can write

$$A_{\pi\pi} = \left(V_{ub}^* V_{ud}\right) T_{\pi\pi} + \left(V_{tb}^* V_{td}\right) P_{\pi\pi}^t, \tag{7.6}$$

where $T_{\pi\pi} = t_{\pi\pi} + p_{\pi\pi}^u - p_{\pi\pi}^c$ and $P_{\pi\pi}^t = p_{\pi\pi}^t - p_{\pi\pi}^c$. CP violating phases in Eq. (7.6) appear only in the CKM elements, so that

$$\frac{\overline{A}_{\pi\pi}}{A_{\pi\pi}} = \frac{\left(V_{ub} V_{ud}^*\right) T_{\pi\pi} + \left(V_{tb} V_{td}^*\right) P_{\pi\pi}^t}{\left(V_{ub}^* V_{ud}\right) T_{\pi\pi} + \left(V_{tb}^* V_{td}\right) P_{\pi\pi}^t}. \tag{7.7}$$

For $f = J/\psi K$, which proceeds via $\bar{b} \rightarrow \bar{c}c\bar{s}$ transition, we can write

$$A_{\psi K} = \left(V_{cb}^* V_{cs}\right) T_{\psi K} + \left(V_{ub}^* V_{us}\right) P_{\psi K}^u, \tag{7.8}$$

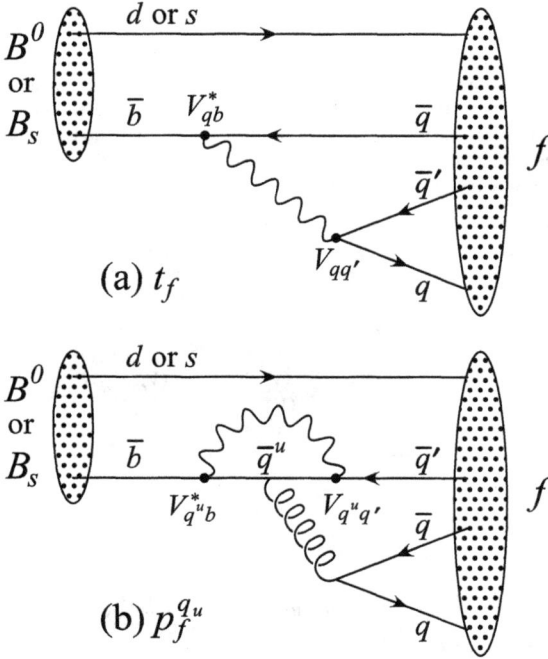

Fig. 3. Feynman diagrams for (a) tree and (b) penguin amplitudes contributing to $B^0 \to f$ or $B_s \to f$ via a $\bar{b} \to \bar{q}q\bar{q}'$ quark-level process.

where $T_{\psi K} = t_{\psi K} + p^c_{\psi K} - p^t_{\psi K}$ and $P^u_{\psi K} = p^u_{\psi K} - p^t_{\psi K}$. A subtlety arises in this decay that is related to the fact that $B^0 \to J/\psi K^0$ and $\overline{B}^0 \to J/\psi \overline{K}^0$. A common final state, e.g. $J/\psi K_S$, is reached only via $K^0 - \overline{K}^0$ mixing. Consequently, the phase factor corresponding to neutral K mixing, $e^{-i\phi_K} = (V^*_{cd} V_{cs})/(V_{cd} V^*_{cs})$, plays a role:

$$\frac{\overline{A}_{\psi K_S}}{A_{\psi K_S}} = -\frac{\left(V_{cb}V^*_{cs}\right) T_{\psi K} + \left(V_{ub}V^*_{us}\right) P^u_{\psi K}}{\left(V^*_{cb}V_{cs}\right) T_{\psi K} + \left(V^*_{ub}V_{us}\right) P^u_{\psi K}} \times \frac{V^*_{cd}V_{cs}}{V_{cd}V^*_{cs}}. \tag{7.9}$$

For $q = s$ or d, there are only penguin contributions to A_f, that is, $t_f = 0$ in Eq. (7.5). (The tree $\bar{b} \to \bar{u}u\bar{q}'$ transition followed by $\bar{u}u \to \bar{q}q$ rescattering is included below in the P^u terms.) Again, CKM unitarity allows us to write A_f in terms of two CKM combinations. For example, for $f = \phi K_S$, which proceeds

Table 1

Summary of $\bar{b} \to \bar{q}q\bar{q}'$ modes with $q' = s$ or d. The second and third columns give examples of final hadronic states. The fourth column gives the CKM dependence of the amplitude A_f, using the notation of Eqs. (7.6, 7.8, 7.10), with the dominant term first and the sub-dominant second. The suppression factor of the second term compared to the first is given in the last column. "Loop" refers to a penguin versus tree suppression factor (it is mode-dependent and roughly $\mathcal{O}(0.2 - 0.3)$) and $\lambda = 0.22$ is the expansion parameter of Eq. (2.39).

$\bar{b} \to q\bar{q}\bar{q}'$	$B^0 \to f$	$B_s \to f$	CKM dependence of A_f	Suppression
$\bar{b} \to \bar{c}c\bar{s}$	ψK_S	$\psi\phi$	$(V_{cb}^* V_{cs})T + (V_{ub}^* V_{us})P^u$	loop $\times \lambda^2$
$\bar{b} \to \bar{s}s\bar{s}$	ϕK_S	$\phi\phi$	$(V_{cb}^* V_{cs})P^c + (V_{ub}^* V_{us})P^u$	λ^2
$\bar{b} \to \bar{u}u\bar{s}$	$\pi^0 K_S$	$K^+ K^-$	$(V_{cb}^* V_{cs})P^c + (V_{ub}^* V_{us})T$	λ^2/loop
$\bar{b} \to \bar{c}c\bar{d}$	$D^+ D^-$	ψK_S	$(V_{cb}^* V_{cd})T + (V_{tb}^* V_{td})P^t$	loop
$\bar{b} \to \bar{s}s\bar{d}$	$\phi\pi$	ϕK_S	$(V_{tb}^* V_{td})P^t + (V_{cb}^* V_{cd})P^c$	$\lesssim 1$
$\bar{b} \to \bar{u}u\bar{d}$	$\pi^+\pi^-$	$\pi^0 K_S$	$(V_{ub}^* V_{ud})T + (V_{tb}^* V_{td})P^t$	loop

via $\bar{b} \to \bar{s}s\bar{s}$ transition, we can write

$$\frac{\overline{A}_{\phi K_S}}{A_{\phi K_S}} = -\frac{\left(V_{cb}V_{cs}^*\right) P_{\phi K}^c + \left(V_{ub}V_{us}^*\right) P_{\phi K}^u}{\left(V_{cb}^* V_{cs}\right) P_{\phi K}^c + \left(V_{ub}^* V_{us}\right) P_{\phi K}^u} \times \frac{V_{cd}^* V_{cs}}{V_{cd} V_{cs}^*}, \qquad (7.10)$$

where $P_{\phi K}^c = p_{\phi K}^c - p_{\phi K}^t$ and $P_{\phi K}^u = p_{\phi K}^u - p_{\phi K}^t$.

Since the amplitude A_f involves two different weak phases, the corresponding decays can exhibit both CP violation in the interference of decays with and without mixing, $S_f \neq 0$, and CP violation in decays, $C_f \neq 0$. (At the present level of experimental precision, the contribution to C_f from CP violation in mixing is negligible, see Eq. (7.1).) If the contribution from a second weak phase is suppressed, then the interpretation of S_f in terms of Lagrangian CP-violating parameters is clean, while C_f is small. If such a second contribution is not suppressed, S_f depends on hadronic parameters and, if the relevant strong phase is large, C_f is large.

A summary of $\bar{b} \to \bar{q}q\bar{q}'$ modes with $q' = s$ or d is given in Table 1. The $\bar{b} \to \bar{d}d\bar{q}$ transitions lead to final states that are similar to the $\bar{b} \to \bar{u}u\bar{q}$ transitions and have similar phase dependence. Final states that consist of two vector-mesons ($\psi\phi$ and $\phi\phi$) are not CP eigenstates, and angular analysis is needed to separate the CP-even from the CP-odd contributions.

The cleanliness of the theoretical interpretation of S_f can be assessed from the information in the last column of Table 1. In case of small uncertainties, the expression for S_f in terms of CKM phases can be deduced from the fourth column of Table 1 in combination with Eq. (7.4) (and, for $b \to q\bar{q}s$ decays, the example in Eq. (7.9)). In the next three sections, we consider three interesting classes.

For B_s decays, one has to replace Eq. (7.4) with

$$e^{-i\phi_{B_s}} = (V_{tb}^* V_{ts})/(V_{tb} V_{ts}^*). \tag{7.11}$$

Note that one expects $\Delta\Gamma_s/\Gamma_s = \mathcal{O}(0.1)$, and therefore y_{B_s} should not be put to zero in the expressions for the time dependent decay rates, but $|q/p| = 1$ is expected to hold to an even better approximation than for B mesons. The CP asymmetry in $B_s \to D_s^+ D_s^-$ (or in $B_s \to \psi\phi$ with angular analysis to disentangle the CP-even and CP-odd components of the final state) will determine $\sin 2\beta_s$, where β_s is defined in Eq. (2.49). Since the SM prediction is that this asymmetry is small [see Eq. (2.53)], $\sin 2\beta_s \sim 0.036$, an observation of a $S_{B_s \to D_s^+ D_s^-} \gg 0.04$ will provide evidence for new physics.

8. $b \to c\bar{c}s$ transitions

For $B \to J/\psi K_S$ and other $\bar{b} \to \bar{c}c\bar{s}$ processes, we can neglect the P^u contribution to $A_{\psi K}$, in the SM, to an approximation that is better than one percent:

$$\lambda_{\psi K_S} = -e^{-2i\beta} \implies S_{\psi K_S} = \sin 2\beta, \quad C_{\psi K_S} = 0. \tag{8.1}$$

(Below the percent level, several effects modify this equation [75, 76].) The experimental measurements give the following ranges [77]:

$$S_{\psi K_S} = 0.69 \pm 0.03, \quad C_{\psi K_S} = 0.02 \pm 0.05. \tag{8.2}$$

The consistency of the experimental results (8.2) with the SM predictions means that the KM mechanism of CP violation has successfully passed its first precision test. For the first time, we can make the following statement based on experimental evidence:

Very likely, the Kobayashi-Maskawa mechanism is the dominant source of CP violation in flavor changing processes.

There are three qualifications implicit in this statement, and we now explain them in little more detail [78].

• *'Very likely'*: It could be that the success is accidental. It could happen, for example, that $\sin 2\beta$ is significantly different from the SM value and that, at the same time, there is a significant CP violating contribution to the $B^0 - \overline{B}^0$ mixing amplitude, and the sum of $M_{12}^{\text{SM}} + M_{12}^{\text{NP}}$ accidentally carries the same phase as the one predicted by the SM alone. It could also happen that the size of NP contributions to $b \to d$ transitions is small, or that its phase is similar to the SM one, but that in $b \to s$ transitions the deviation is significant.

• *'Dominant'*: While $S_{\psi K}$ is measured with an accuracy of order 0.04, the accuracy of the SM prediction for $\sin 2\beta$ is only at the level of 0.2. Thus, it is quite possible that there is a new physics contribution at the level of $|M_{12}^{NP}/M_{12}^{SM}| \lesssim \mathcal{O}(0.2)$.

• *'Flavor changing'*: It may well happen that the KM phase, which is closely related to flavor violation through the CKM matrix, dominates meson decays while new, flavor diagonal phases (such as the two unavoidable phases in the universal version of the MSSM) dominate observables such as electric dipole moments by many orders of magnitude.

The measurement of $S_{\psi K}$ provides a significant constraint on the unitarity triangle. In the $\rho - \eta$ plane, it reads:

$$\sin 2\beta = \frac{2\eta(1 - \rho)}{\eta^2 + (1 - \rho)^2} = 0.69 \pm 0.03. \tag{8.3}$$

One can get an impression of the impact of this constraint by looking at Fig. 2, where the blue region represents $\sin 2\beta = 0.69 \pm 0.03$. An impression of the KM test can be achieved by observing that the blue region has an excellent overlap with the region allowed by all other measurements. A comparison between the constraints in the $\rho - \eta$ plane from CP conserving and CP violating processes is provided in Fig. 4. The impressive consistency between the two allowed regions is the basis for our statement that the KM mechanism has passed its first precision tests. The fact that the allowed region from the CP violating processes is more strongly constrained is related to the fact that CP is a good symmetry of the strong interactions and that, therefore, various CP violating observables – in particular $S_{\psi K}$ – can be cleanly interpreted.

9. Penguin dominated $b \to s$ transitions

9.1. General considerations

The present experimental situation concerning CP asymmetries in decays to final CP eigenstates dominated by $b \to s$ penguins is summarized in Table 2.

For $B \to \phi K_S$ and other $\bar{b} \to \bar{s}s\bar{s}$ processes, we can neglect the P^u contribution to A_f, in the Standard Model, to an approximation that is good to order of a few percent:

$$\lambda_{\phi K_S} \approx -e^{-2i\beta} \implies S_{\phi K_S} \approx \sin 2\beta, \quad C_{\phi K_S} \approx 0. \tag{9.1}$$

In the presence of new physics, both A_f and M_{12} can get contributions that are comparable in size to those of the Standard Model and carry new weak phases [36]. Such a situation gives several interesting consequences for $\bar{b} \to \bar{s}s\bar{s}$ decays:

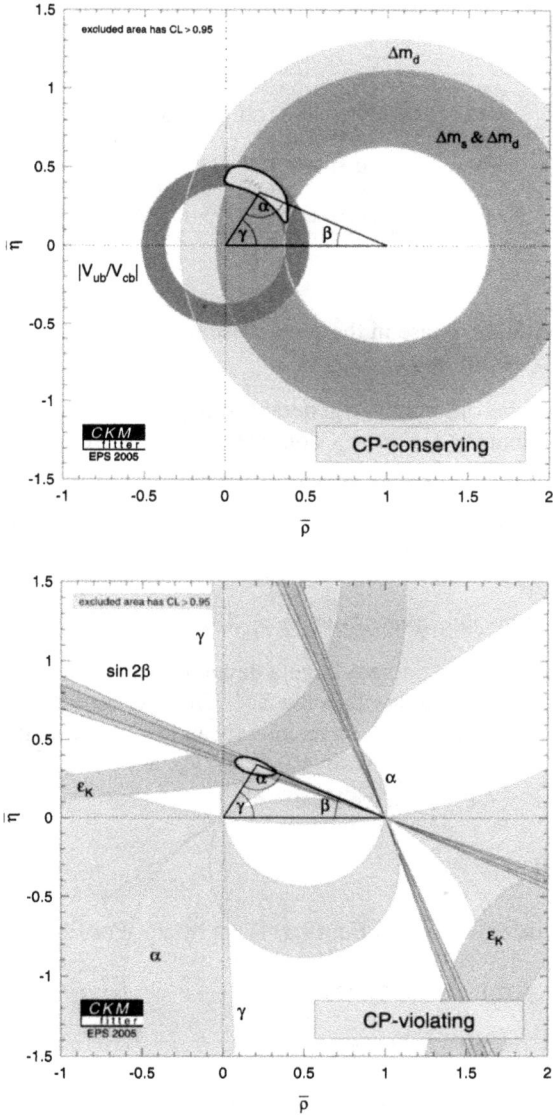

Fig. 4. Constraints in the $\rho - \eta$ plane from (a) CP conserving or (b) CP violating loop processes.

Table 2

CP asymmetries in $b \to s$ penguin dominated modes.

f_{CP}	$-\eta_{f_{CP}} S_{f_{CP}}$	$C_{f_{CP}}$
ϕK_S	$+0.47 \pm 0.19$	-0.09 ± 0.15
$\eta' K_S$	$+0.50 \pm 0.09(0.13)$	$-0.07 \pm 0.07(0.10)$
$f_0 K_S$	$+0.75 \pm 0.24$	$+0.06 \pm 0.21(0.23)$
$\pi^0 K_S$	$+0.31 \pm 0.26$	-0.02 ± 0.13
ωK_S	$+0.63 \pm 0.30$	-0.44 ± 0.24
$K_S K_S K_S$	$+0.61 \pm 0.23$	$-0.31 \pm 0.17(0.20)$

1. A new CP violating phase in the $b \to s$ decay amplitude will lead to a deviation of $-\eta_f S_f$ from $S_{\psi K}$.

2. The S_f's will be different, in general, among the various f's. Only if the new physics contribution to A_f dominates over the SM we should expect a universal S_f.

3. A new CP violating phase in the $b \to s$ decay amplitude in combination with a strong phase will lead to $C_f \neq 0$.

9.2. Calculating the deviations from $S_f = S_{\psi K}$

It is important to understand how large a deviation from the approximate equalities in Eq. (9.1) is expected within the SM. The SM contribution to the decay amplitudes, related to $\bar{b} \to \bar{q} q \bar{s}$ transitions, can always be written as a sum of two terms, $A_f^{SM} = A_f^c + A_f^u$, with $A_f^c \propto V_{cb}^* V_{cs}$ and $A_f^u \propto V_{ub}^* V_{us}$. Defining the ratio $a_f^u \equiv e^{-i\gamma}(A_f^u / A_f^c)$, we have

$$A_f^{SM} = A_f^c (1 + a_f^u e^{i\gamma}). \tag{9.2}$$

The size of the deviations from Eq. (9.1) is set by a_f^u. For $|a_f^u| \ll 1$, we obtain

$$-\eta_f S_f \simeq \sin 2\beta + 2 \cos 2\beta \, \mathcal{R}e(a_f^u) \sin \gamma,$$
$$C_f \simeq -2\mathcal{I}m(a_f^u) \sin \gamma. \tag{9.3}$$

For charmless modes, the effects of the a_f^u terms (often called 'the SM pollution') are at least of order $|(V_{ub}^* V_{us})/(V_{cb}^* V_{cs})| \sim$ a few percent.

To calculate them explicitly, we use the operator product expansion (OPE). We follow the notations of ref. [79]. We consider the following effective Hamiltonian for $\Delta B = \pm 1$ decays:

$$\mathcal{H}_{\text{eff}} = \frac{G_F}{\sqrt{2}} \sum_{p=u,c} V_{ps}^* V_{pb} \tag{9.4}$$

$$\times \left(C_1 O_1^p + C_2 O_2^p + \sum_{i=3}^{10} C_i O_i + C_{7\gamma} O_{7\gamma} + C_{8g} O_{8g} \right) + \text{h.c.},$$

with

$$O_1^p = (\bar{p}b)_{V-A}(\bar{s}p)_{V-A},$$

$$O_2^p = (\bar{p}_\beta b_\alpha)_{V-A}(\bar{s}_\alpha p_\beta)_{V-A},$$

$$O_3 = (\bar{s}b)_{V-A} \sum_q (\bar{q}q)_{V-A},$$

$$O_4 = (\bar{s}_\alpha b_\beta)_{V-A} \sum_q (\bar{q}_\beta q_\alpha)_{V-A},$$

$$O_5 = (\bar{s}b)_{V-A} \sum_q (\bar{q}q)_{V+A},$$

$$O_6 = (\bar{s}_\alpha b_\beta)_{V-A} \sum_q (\bar{q}_\beta q_\alpha)_{V+A},$$

$$O_7 = \frac{3}{2}(\bar{s}b)_{V-A} \sum_q e_q (\bar{q}q)_{V+A},$$

$$O_8 = \frac{3}{2}(\bar{s}_\alpha b_\beta)_{V-A} \sum_q e_q (\bar{q}_\beta q_\alpha)_{V+A},$$

$$O_9 = \frac{3}{2}(\bar{s}b)_{V-A} \sum_q e_q (\bar{q}q)_{V-A},$$

$$O_{10} = \frac{3}{2}(\bar{s}_\alpha b_\beta)_{V-A} \sum_q e_q (\bar{q}_\beta q_\alpha)_{V-A},$$

$$O_{7\gamma} = -\frac{em_b}{8\pi^2}\bar{s}\sigma^{\mu\nu}(1+\gamma_5)F_{\mu\nu}b,$$

$$O_{8g} = -\frac{g_s m_b}{8\pi^2}\bar{s}\sigma^{\mu\nu}(1+\gamma_5)G_{\mu\nu}b, \tag{9.5}$$

where $(\bar{q}_1 q_2)_{V\pm A} = \bar{q}_1 \gamma_\mu (1 \pm \gamma_5) q_2$, the sum is over active quarks, with e_q denoting their electric charge in fractions of $|e|$ and α, β are color indices. The decay amplitudes can be calculated from this effective Hamiltonian:

$$A_f = \langle f | \mathcal{H}_{\text{eff}} | B^0 \rangle, \quad \overline{A}_f = \langle f | \mathcal{H}_{\text{eff}} | \overline{B}^0 \rangle. \tag{9.6}$$

The electroweak model determines the Wilson coefficients while QCD (or, more practically, a calculational method such as QCD factorization) determines the matrix elements $\langle f|O_i|B^0(\overline{B}^0)\rangle$.

Take, for example, the $B^0 \to K^0\pi^0$ decay amplitude. It can be written as follows (for simplicity, we omit the contributions from O_{7-10}):

$$A^c_{K^0\pi^0} \approx i V^*_{cb}V_{cs}\frac{G_F}{2} f_K F^{B\to\pi}(m^2_K)(m^2_B - m^2_\pi)\left(a_4 + r_\chi a_6\right), \tag{9.7}$$

$$A^u_{K^0\pi^0} \approx i V^*_{ub}V_{us}\frac{G_F}{2}\left[f_K F^{B\to\pi}(m^2_K)(m^2_B - m^2_\pi)\left(a_4 + r_\chi a_6\right)\right.$$
$$\left. - f_\pi F^{B\to K}(m^2_\pi)(m^2_B - m^2_K)a_2\right], \tag{}$$

where $r_\chi = 2m^2_K/[m_b(m_s + m_d)]$. The a_i parameters are related to the Wilson coefficients as follows:

$$a_i \equiv C_i + \frac{1}{N_c}C_{i\pm1} \text{ for } i = \text{even, odd.} \tag{9.8}$$

Within the SM, at leading order,

$$C_1(m_W) = 1, \quad C_{i\neq1}(m_W) = 0. \tag{9.9}$$

(Strictly speaking, $C_{7\gamma}(m_W)$ and $C_{8g}(m_W)$ are also different from zero. Their contributions to the decay processes of interest occur, however, at next-to-leading order which we neglect here for simplicity.) To run the Wilson coefficients from the weak scale m_W to the low scale of order m_b, we use

$$\vec{C}(\mu) = [\alpha_s(m_W)/\alpha_s(\mu)]^{\gamma/2\beta_0}, \tag{9.10}$$

where $\beta_0 = (33 - 2f)/3$, with $f = 5$ for $m_b \leq \mu \leq m_W$, and γ is the 12-dimensional leading-log anomalous dimension matrix which can be found, for example, in ref. [80]. The bottom line is the following set of values for the relevant a_i parameters at the scale $\mu = m_b$:

$$a_1 = 1.028, \quad a_2 = 0.105, \quad a_4 = -0.0233, \quad a_6 = -0.0314. \tag{9.11}$$

We use the following values for the relevant hadronic parameters:

$$f_\pi = 131 \text{ MeV}, \quad f_K = 160 \text{ MeV},$$
$$F^{B\to\pi}(0) = 0.28, \quad F^{B\to K}(0) = 0.34, \quad r_\chi(m_b) = 1.170. \tag{9.12}$$

Thus we can estimate $a^u_{\pi K}$:

$$a^u_{\pi K} \approx \lambda^2 R_u \left(1 - \frac{f_\pi}{f_K}\frac{F^{B\to K}}{F^{B\to\pi}}\frac{a_2}{a_4 + r_\chi a_6}\right) \approx 2.75\lambda^2 R_u \approx 0.052. \tag{9.13}$$

We learn that the SM and factorization predict that $-S_{\pi^0 K_S} - S_{\psi K_S} \approx +0.05$.

Table 3

The a^u_f parameters, calculated in QCD factorization at leading log and to zeroth order in Λ/m_b (except for chirally enhanced corrections), and the SM values of S_f for $\mu = m_b$ and in parentheses the respective values for $\mu = 2m_b$ (first) and $\mu = m_b/2$ (second) if different from the central one. In the last column, the results of ref. [83], using QCD factorization at NLO, are given. Taken from [80].

f	a^u_f [80]	$-\eta_{CP}S_f$ [80]	$-\eta_{CP}S_f$ [83]
ψK_S	0	0.69	0.69
ϕK_S	0.019	0.71	0.71 ± 0.01
$\pi^0 K_S$	0.052 [0.094, 0.021]	0.75 [0.79, 0.72]	$0.76^{+0.05}_{-0.04}$
ηK_S	0.08 [0.16, 0.02]	0.78 [0.84, 0.72]	$0.79^{+0.11}_{-0.07}$
$\eta' K_S$	0.007 [−0.006, 0.019]	0.70 [0.68, 0.71]	0.70 ± 0.01
ωK_S	0.22 [0.37, 0.04]	0.88 [0.94, 0.74]	0.82 ± 0.08
$\rho^0 K_S$	−0.16 [−0.32, 0.005]	0.45 [0.15, 0.70]	$0.61^{+0.08}_{-0.12}$

In Table 3 we give the values of the a^u_f parameter (obtained in ref. [80] by using factorization [79,81,82]) for all relevant modes.

An examination of Table 3 shows that the SM pollution is small (that is, at the naively expected level of $|(V_{ub}V^*_{us})/(V_{cb}V^*_{cs})| \sim$ a few percent) for $f = \phi K_S$, $\eta' K_S$ and $\pi^0 K_S$. It is larger for $f = \eta K_S$, ωK_S and $\rho^0 K_S$. In these modes, a^u_f is enhanced because, within the QCD factorization approach, there is an accidental cancellation between the leading contributions to A^c_f. The reason for the suppression of the leading A^c_f piece in $f = \rho K$, ωK versus $f = \pi^0 K$ is that the dominant QCD-penguin coefficients a_4 and a_6 appear in $A^c_{(\rho,\omega)K}$ as $(a_4 - r_\chi a_6)$ and in $A^c_{\pi^0 K}$ as $(a_4 + r_\chi a_6)$. Since $r_\chi \simeq 1$ and, within the Standard Model, $a_4 \sim a_6$, there is a cancellation in $A^c_{(\rho,\omega)K}$ while there isn't one in $A^c_{\pi^0 K}$. The suppression for $A^c_{\eta K}$ with respect to $A^c_{\eta' K}$ has a different reason: it is due to the octet-singlet mixing, which causes destructive (constructive) interference in the $\eta(\eta')K$ penguin amplitude [84].

10. $b \to u\bar{u}d$ transitions

The present experimental situation concerning CP asymmetries in decays to final CP eigenstates via $b \to d$ transitions is summarized in Table 4.

For $B \to \pi\pi$ and other $\bar{b} \to \bar{u}ud$ processes, the penguin-to-tree ratio can be estimated using SU(3) relations and experimental data on related $B \to K\pi$ decays. The result is that the suppression is of order $0.2 - 0.3$ and so cannot

Table 4

CP asymmetries in $b \to c\bar{c}d$ (above line) or $b \to u\bar{u}d$ (below line) modes.

f_{CP}	$-\eta_{f_{CP}} S_{f_{CP}}$	$C_{f_{CP}}$
$\psi\pi^0$	$+0.69 \pm 0.25$	-0.11 ± 0.20
$D^+ D^-$	$+0.29 \pm 0.63$	$+0.11 \pm 0.35$
$D^{*+} D^{*-}$	$+0.75 \pm 0.23$	-0.04 ± 0.14
$\pi^+\pi^-$	$+0.50 \pm 0.12(0.18)$	$-0.37 \pm 0.10(0.23)$
$\pi^0\pi^0$		-0.28 ± 0.39
$\rho^+\rho^-$	$+0.22 \pm 0.22$	-0.02 ± 0.17

be neglected. The expressions for $S_{\pi\pi}$ and $C_{\pi\pi}$ to leading order in $R_{PT} \equiv (|V_{tb}V_{td}|P_{\pi\pi}^t)/(|V_{ub}V_{ud}|T_{\pi\pi})$ are:

$$\lambda_{\pi\pi} = e^{2i\alpha} \left[(1 - R_{PT}e^{-i\alpha})/(1 - R_{PT}e^{+i\alpha}) \right] \Rightarrow \tag{10.1}$$
$$S_{\pi\pi} \approx \sin 2\alpha + 2\mathcal{R}e(R_{PT}) \cos 2\alpha \sin\alpha, \quad C_{\pi\pi} \approx 2\mathcal{I}m(R_{PT}) \sin\alpha.$$

Note that R_{PT} is mode-dependent and, in particular, could be different for $\pi^+\pi^-$ and $\pi^0\pi^0$. If strong phases can be neglected then R_{PT} is real, resulting in $C_{\pi\pi} = 0$. The size of $C_{\pi\pi}$ is an indicator of how large the strong phase is. As concerns $S_{\pi\pi}$, it is clear from (10.1) that the relative size and strong phase of the penguin contribution must be known to extract α. This is the problem of penguin pollution.

The cleanest solution involves isospin relations among the $B \to \pi\pi$ amplitudes. Let us derive this relation step by step. The $SU(2)$-isospin representations of the $\pi\pi$ states are as follows:

$$\langle \pi^+\pi^-| = \sqrt{\frac{1}{2}} \langle (1,+1)(1,-1) + (1,-1)(1,+1)|$$

$$= \sqrt{\frac{1}{3}} \langle 2,0| + \sqrt{\frac{2}{3}} \langle 0,0|,$$

$$\langle \pi^0\pi^0| = \langle (1,0)(1,0)| = \sqrt{\frac{2}{3}} \langle 2,0| - \sqrt{\frac{1}{3}} \langle 0,0|,$$

$$\langle \pi^+\pi^0| = \sqrt{\frac{1}{2}} \langle (1,+1)(1,0) + (1,0)(1,+1)| = \langle 2,+1|. \tag{10.2}$$

The Hamiltonian, with its four quark operators, has two features that are important for our purposes:

1. There are $\Delta I = 1/2$ and $\Delta I = 3/2$ pieces, but no $\Delta I = 5/2$ one. The absence of the latter gives isospin relations among the $B \to \pi\pi$ amplitdues.

2. The penguin operatores are purely $\Delta I = 1/2$. Thus we will find that they do not contribute to the $\pi^{\pm}\pi^0$ modes.

We contract the Hamiltonian with with the $(B^+, B^0) = (1/2, \pm 1/2)$ states:

$$H_{3/2,+1/2}|1/2, -1/2\rangle \propto \sqrt{\frac{1}{2}}\,|2, 0\rangle + \sqrt{\frac{1}{2}}\,|1, 0\rangle,$$

$$H_{3/2,+1/2}|1/2, +1/2\rangle \propto \sqrt{\frac{3}{4}}\,|2, 1\rangle - \sqrt{\frac{1}{4}}\,|1, 1\rangle,$$

$$H_{1/2,+1/2}|1/2, -1/2\rangle \propto \sqrt{\frac{1}{2}}\,|1, 0\rangle - \sqrt{\frac{1}{2}}\,|0, 0\rangle,$$

$$H_{1/2,+1/2}|1/2, +1/2\rangle \propto |1, 0\rangle. \tag{10.3}$$

Combining (10.2) and (10.3), we obtain:

$$A_{\pi^+\pi^-} = \sqrt{1/6}\,A_{3/2} - \sqrt{1/3}\,A_{1/2},$$
$$A_{\pi^0\pi^0} = \sqrt{1/3}\,A_{3/2} + \sqrt{1/6}\,A_{1/2},$$
$$A_{\pi^+\pi^0} = \sqrt{3/4}\,A_{3/2}. \tag{10.4}$$

Analogous relation hold for the CP-conjugate amplitudes, $\overline{A}_{\pi^i\pi^j}$. These isospin decompositions lead to the Gronau-London triangle relations [85]:

$$\frac{1}{\sqrt{2}}A_{\pi^+\pi^-} + A_{\pi^0\pi^0} = A_{\pi^+\pi^0},$$

$$\frac{1}{\sqrt{2}}\overline{A}_{\pi^+\pi^-} + \overline{A}_{\pi^0\pi^0} = \overline{A}_{\pi^-\pi^0}. \tag{10.5}$$

The method further exploits the fact that the penguin contribution to $P_{\pi\pi}$ is pure $\Delta I = \frac{1}{2}$ (this is not true for the electroweak penguins which, however, are expected to be small), while the tree contribution to $T_{\pi\pi}$ contains pieces which are both $\Delta I = \frac{1}{2}$ and $\Delta I = \frac{3}{2}$. A simple geometric construction then allows one to find R_{PT} and extract α cleanly from $S_{\pi^+\pi^-}$. Explicitly, one notes that, since $A_{3/2}$ comes purely from tree contributions, we have

$$\frac{q}{p}\frac{\overline{A}_{3/2}}{A_{3/2}} = -e^{2i\alpha}. \tag{10.6}$$

The branching ratios of the various modes determine $|A_{\pi^i \pi^j}|$ and $|\overline{A}_{\pi^i \pi^j}|$ (with $|A_{\pi^+ \pi^0}| = |\overline{A}_{\pi^- \pi^0}|$). This would determine the shape of each of the triangles (10.5). Defining

$$A_0 \equiv (1/\sqrt{6}) A_{1/2}, \quad A_2 \equiv (1/\sqrt{12}) A_{3/2}, \tag{10.7}$$

we can obtain $A_2 = (1/3) A_{\pi^+ \pi^0}$ and $A_0 = (1/\sqrt{2}) A_{\pi^+ \pi^-} - A_2$. Similarly, we can obtain $\overline{A}_2$ and $\overline{A}_0$. Next, we define (and obtain)

$$\theta \equiv \arg(A_0 A_2^*), \quad \overline{\theta} \equiv \arg(\overline{A}_0 \overline{A}_2^*). \tag{10.8}$$

Then we have

$$\mathcal{I}m \lambda_{\pi^+ \pi^-} = \mathcal{I}m \left(-e^{-2i\alpha} \frac{|\overline{A}_2| - |\overline{A}_0| e^{i\overline{\theta}}}{|A_2| - |A_0| e^{i\theta}} \right). \tag{10.9}$$

On the other hand, we can use the experimentally measured quantities to extract $\mathcal{I}m \lambda_{\pi^+ \pi^-}$:

$$\mathcal{I}m \lambda_{\pi^+ \pi^-} = \frac{S_{\pi^+ \pi^-}}{1 + C_{\pi^+ \pi^-}}. \tag{10.10}$$

The key experimental difficulty is that one must measure accurately the separate rates for $B^0, \overline{B}^0 \to \pi^0 \pi^0$. It has been noted that an upper bound on the average rate allows one to put a useful upper bound on the deviation of $S_{\pi^+ \pi^-}$ from $\sin 2\alpha$ [86–88]. Parametrizing the asymmetry by $S_{\pi^+ \pi^-}/\sqrt{1 - (C_{\pi^+ \pi^-})^2} = \sin(2\alpha_{\text{eff}})$, the bound reads

$$\cos(2\alpha_{\text{eff}} - 2\alpha) \geq \frac{1}{\sqrt{1 - (C_{\pi^+ \pi^-})^2}}$$
$$\times \left[1 - \frac{2\mathcal{B}_{00}}{\mathcal{B}_{+0}} + \frac{(\mathcal{B}_{+-} - 2\mathcal{B}_{+0} + 2\mathcal{B}_{00})^2}{4\mathcal{B}_{+-}\mathcal{B}_{+0}} \right],$$

where $\mathcal{B}_{ij}$ are the averages over CP-conjugate branching ratios; *e.g.*, $\mathcal{B}_{00} = \frac{1}{2}[\mathcal{B}(B^0 \to \pi^0 \pi^0) + \mathcal{B}(\overline{B}^0 \to \pi^0 \pi^0)]$. CP asymmetries in $B \to \rho\pi$ and, in particular, in $B \to \rho\rho$ can also be used to determine α [89–93]. At present, the constraints read [54]

$$|\alpha_{\text{eff}}^{\pi^+ \pi^-} - \alpha| < 38^o, \quad R_{PT}^{\pi^+ \pi^-} = 0.37 \pm 0.17,$$
$$|\alpha_{\text{eff}}^{\rho^+ \rho^-} - \alpha| < 14^o, \quad R_{PT}^{\rho^+ \rho^-} = 0.07^{+0.14}_{-0.07}. \tag{10.11}$$

Using isospin analyses for all three systems ($\pi\pi$, $\rho\pi$ and $\rho\rho$), one obtains [54]

$$\alpha(\pi\pi, \pi\rho, \rho\rho) = \left[101^{+16}_{-9} \right]^o, \tag{10.12}$$

to be compared with the result of the CKM fit,

$$\alpha(\text{CKM fit}) = 96 \pm 16^o. \tag{10.13}$$

We would like to emphasize the following points:
• The consistency of (10.12) with (10.13) means that **the KM mechanism of CP violation has successfully passed a second precision test.**
• The α measurement via the $b \to u\bar{u}d$ transitions provides a significant constraint on the unitarity triangle.
• The isospin analysis determines the relative phase between the $B^0 - \overline{B}^0$ mixing amplitude and the tree decay amplitude $A_{3/2}$, independent of the electroweak model. The tree decay amplitude is not affected by new physics. Any new physics modification of the mixing amplitude is measured by $S_{\psi k}$. Thus, the combination of $S_{\psi K}$ and the isospin analysis of $S_{\pi\pi, \rho\pi, \rho\rho}$ constrains α even in the presence of new physics in $B^0 - \overline{B}^0$ mixing.

11. $b \to c\bar{u}s, u\bar{c}s$ transitions

An interesting set of measurements is that of $B \to DK$ which proceed via the quark transitions $\bar{b} \to \bar{c}u\bar{s}$ or $\bar{b} \to \bar{u}c\bar{s}$ (and their CP conjugates). Given the quark processes, it is clear that there is no penguin contribution here. Thus, the quark transitions are purely tree processes. The interference between the two quark transitions (if they lead to the same final states – see below) is sensitive to $\arg[(V_{ub}^* V_{us})/(V_{cb}^* V_{cs})] \approx \gamma$.

There are three variants on this method: GLW [94, 95], ADS [96] and GGSZ [97]. The simplest one to explain involves branching ratios of charged B decays, and thus $B^0 - \overline{B}^0$ mixing plays no role. Consider the decay $B^\pm \to D_1^0 K^\pm$, where $D_{1,2}^0 = \frac{1}{\sqrt{2}}(D^0 \pm \overline{D}^0)$ are the CP eigenstates. Taking into account that

$$A(B^+ \to D^0 K^+) \times A(D^0 \to D_1^0) \propto (V_{ub}^* V_{cs}) \times (V_{cs}^* V_{us}),$$
$$A(B^+ \to \overline{D}^0 K^+) \times A(\overline{D}^0 \to D_1^0) \propto (V_{cb}^* V_{us}) \times (V_{us}^* V_{cs}), \tag{11.1}$$

we can write the relevant decay amplitudes as follows:

$$\sqrt{2}A_{D_1^0 K^+} = |A_{D^0 K^+}|e^{i(\delta + \gamma)} + |A_{\overline{D}^0 K^+}| = A_{D^0 K^+} + A_{\overline{D}^0 K^+},$$
$$\sqrt{2}A_{D_1^0 K^-} = |A_{D^0 K^-}|e^{i(\delta - \gamma)} + |A_{\overline{D}^0 K^-}| = A_{\overline{D}^0 K^-} + A_{D^0 K^-}. \tag{11.2}$$

Measuring the rates for the six relevant decay modes ($D_1^0 K^+$, $D^0 K^+$, $\overline{D}^0 K^+$ and the CP conjugate modes), one can construct an amplitude triangle for each of the two relations in Eq. (11.2). We can choose a phase convention where $A_{\overline{D}^0 K^+} = A_{D^0 K^-}$. Then, the relative angle between $A_{D^0 K^+}$ and $A_{\overline{D}^0 K^-}$ is 2γ.

The method of [97] gives, at present, the most significant constraints. It allows one to determine the amplitude ratios, $r_B = 0.12^{+0.03}_{-0.04}$ and $r_B^* = 0.09^{+0.03}_{-0.04}$, and the weak phase γ [54]:

$$\gamma(DK) = (63^{+15}_{-13})^o. \tag{11.3}$$

This range is to be compared with the range of γ derived from the CKM fit (not including the direct γ measurements):

$$\gamma(\text{CKM fit}) = (57^{+7}_{-14})^o. \tag{11.4}$$

We would like to emphasize the following points:

• The consistency of (11.3) with (11.4) means that **the KM mechanism of CP violation has successfully passed a third precision test.**

• The γ measurement via the $b \rightarrow c\bar{u}s, u\bar{c}s$ transitions provides yet another constraint on the unitarity triangle. The constraint will become more significant when the experimental precision improves.

• The determination of γ here relies on tree decay amplitudes. Thus, the analysis of $B \rightarrow DK$ decays constrains γ even in the presence of new physics in loop processes.

12. CP violation as a probe of new physics

We have argued that the Standard Model picture of CP violation is unique and highly predictive. We have also stated that reasonable extensions of the Standard Model have a very different picture of CP violation. Experimental results are now starting to decide between the various possibilities. Our discussion of CP violation in the presence of new physics is aimed to demonstrate that, indeed, models of new physics can significantly modify the Standard Model predictions and that present and near future measurements have therefore a strong impact on the theoretical understanding of CP violation.

To understand how the Standard Model predictions could be modified by New Physics, we focus on CP violation in the interference between decays with and without mixing. As explained above, this type of CP violation may give, due to its theoretical cleanliness, unambiguous evidence for New Physics most easily. We now demonstrate what type of questions can be (or have already been) answered when many such observables are measured.

I. Consider $S_{\psi K_S}$, the CP asymmetry in $B \rightarrow \psi K_S$. This measurement cleanly determines the relative phase between the $B^0 - \overline{B}^0$ mixing amplitude and the $b \to c\bar{c}s$ decay amplitude ($\sin 2\beta$ in the SM). The $b \to c\bar{c}s$ decay has Standard Model tree contributions and therefore is very unlikely to be significantly affected by new physics. On the other hand, the mixing amplitude can be easily modified by new physics. We parametrize such a modification as follows:

$$r_d^2 \, e^{2i\theta_d} = \frac{M_{12}}{M_{12}^{\text{SM}}}. \tag{12.1}$$

Then the following observables provide constraints on r_d^2 and $2\theta_d$:

$$\begin{aligned} S_{\psi K_S} &= \sin(2\beta + 2\theta_d), \\ \Delta m_B &= r_d^2 (\Delta m_B)^{\text{SM}}, \\ \mathcal{A}_{\text{SL}} &= -\mathcal{R}e \left(\frac{\Gamma_{12}}{M_{12}} \right)^{\text{SM}} \frac{\sin 2\theta_d}{r_d^2} + \mathcal{I}m \left(\frac{\Gamma_{12}}{M_{12}} \right)^{\text{SM}} \frac{\cos 2\theta_d}{r_d^2}. \end{aligned} \tag{12.2}$$

Examining whether $S_{\psi K_S}$, Δm_B and $\mathcal{A}_{\text{SL}}$ fit the SM prediction, that is, whether $\theta_d \neq 0$ and/or $r_d^2 \neq 1$, we can answer the following question (see *e.g.* [98]):

(i) *Is there new physics in $B^0 - \overline{B}^0$ mixing?*

Thanks to the fact that quite a few observables that are related to SM tree level processes have already been measured, we are able to refer to this question in a quantitative way. The tree level processes are insensitive to new physics and can be used to constrain ρ and η even in the presence of new physics contributions to loop processes, such as Δm_B. Among these observables we have $|V_{cb}|$ and $|V_{ub}|$ from semileptonic B decays, the phase γ from $B \to DK$ decays, and the phase α from $B \to \rho\rho$ decays (in combination with $S_{\psi K}$). One can fit these observables, and the ones in Eq. (12.2) to the four parameters ρ, η, r_d^2 and $2\theta_d$. The resulting constraints are shown in Fig. 5.

A long list of models that require a significant modification of the $B^0 - \overline{B}^0$ mixing amplitude are excluded. We can further conclude from Fig. 5 that a new physics contribution to the $B^0 - \overline{B}^0$ mixing amplitude at a level higher than 20% is now disfavored. Yet, it is still possible that ρ and η are well outside their SM range and that NP gives $2\theta_d$ very different from zero and/or r_d^2 very different from one. In this case, the SM and the NP 'conspire' to mimic the SM values of the observables (12.2). This is what we meant concretely in our statement that the KM dominance of the observed CP violation is now very likely but not guaranteed.

II. Consider $S_{\phi K_S}$, the CP asymmetry in $B \to \phi K_S$. This measurement is sensitive to the relative phase between the $B - \bar{B}$ mixing amplitude and the $b \to s\bar{s}s$ decay amplitude ($\sin 2\beta$ in the SM). The $b \to s\bar{s}s$ decay has only Standard Model penguin contributions and therefore is sensitive to new physics. For the

Y. Nir

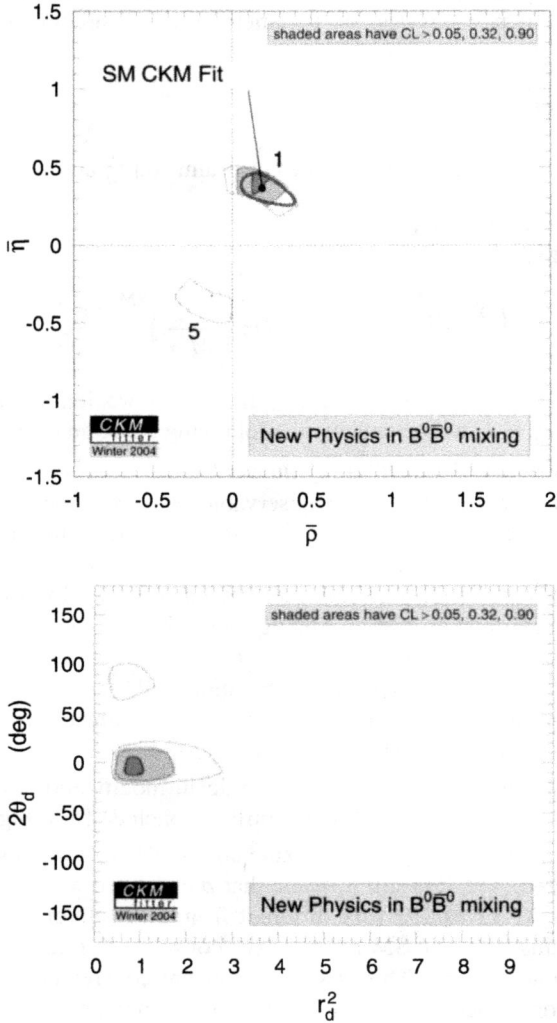

Fig. 5. Constraints in the (a) $\rho - \eta$ plane (b) $r_d^2 - 2\theta_d$ plane, assuming that NP contributions to tree level processes are negligible [54].

simple case that the NP contribution depends on a single CP violating phase ϕ_{bs}, we parametrize the modification of the decay amplitude as follows (for simplicity, we neglect here the a_f^u terms of Eq. (9.2)):

$$A_f = A_f^c \left(1 + b_f \, e^{i\phi_{bs}}\right). \tag{12.3}$$

Here b_f is complex only if it carries a strong phase. The effects of this new physics contribution are simple to understand in two limits:

1. The new physics contribution is dominant, $|b_f| \gg 1$. The shift in all modes where this condition is valid is universal and depends only on ϕ_{bs}:

$$-\eta_f S_f \simeq \sin(2\beta + 2\theta_d) \cos 2\phi_{bs} + \cos(2\beta + 2\theta_d) \sin 2\phi_{bs},$$
$$C_f \simeq 0. \tag{12.4}$$

2. The new physics contribution is small. Explicitly, $|b_f| \ll 1$. The shift is mode dependent and depends on both b_f and $\sin \phi_{bs}$:

$$-\eta_f S_f \simeq \sin(2\beta + 2\theta_d) + 2\cos(2\beta + 2\theta_d)\mathcal{R}e(b_f)\sin\phi_{bs},$$
$$C_f \simeq -2\mathcal{I}m(b_f^c)\sin\phi_{bs}. \tag{12.5}$$

Note that the effect of the NP is similar to that of the SM a_f^u terms (with $b_f \leftrightarrow a_f^u$ and $\phi_{bs} \leftrightarrow \gamma$), so that the latter have to be known in order to probe the b_f terms. Once that is done, the value of $S_{\psi K}$ determines $2\beta + 2\theta_d$ and one can examine whether $\phi_{bs} \neq 0$ and answer the following questions:

(ii) *Is there new physics in $b \to s$ transitions?*

So far, the experimental data – see Table 2 – do not provide any evidence for $\phi_{bs} \neq 0$. Yet, the experimental accuracy is still not sufficient to make qualitative statements such as we made for $b \to d$ transitions ($B^0 - \overline{B}^0$ mixing). To see this, we compare the constraints in the $\rho - \eta$ plane that arise from tree plus $b \to d$ loops (Δm_B, $S_{\psi K_S}$, $S_{\rho\rho}$, etc.) to those from tree plus $b \to s$ loops ($S_{\phi K_S}$, $S_{\eta'K_S}$, Δm_s). This is done in Fig. 6.

III. Together with a future measurement of $B_s - \overline{B}_s$ mixing, we may also try to answer the following question:

(iii) *Is there new physics in $\Delta B = 1$ processes? in $\Delta B = 2$? in both?*

IV. Consider $a_{\pi\nu\bar{\nu}} \equiv \Gamma_{K_L \to \pi^0 \nu\bar{\nu}}/\Gamma_{K^+ \to \pi^+\nu\bar{\nu}}$, see Eq. (5.11). This measurement will cleanly determine the relative phase between the $K^0 - \overline{K}^0$ mixing amplitude and the $s \to d\nu\bar{\nu}$ decay amplitude (of order $\sin^2\beta$ in the SM). The experimentally measured small value of ε_K requires that the phase of the $K^0 - \overline{K}^0$ mixing amplitude is not modified from the SM prediction. (More precisely, it requires that the phase of the mixing amplitude is very close to twice the phase of the $s \to d\bar{u}u$ decay amplitude [99].) On the other hand, the decay, which in

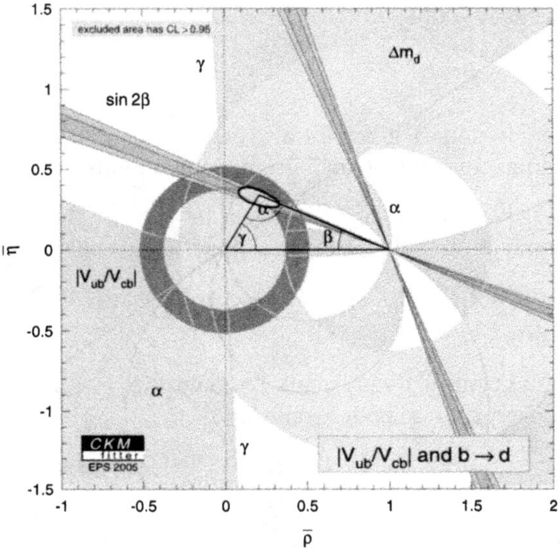

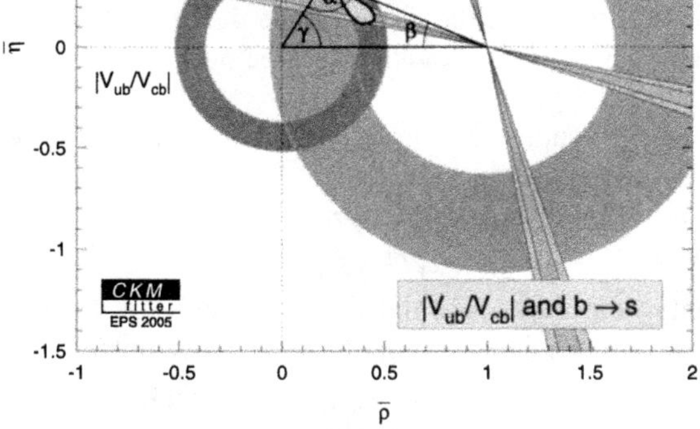

Fig. 6. Constraints in the $\rho - \eta$ plane from tree processes and (a) $b \to d$ or (b) $b \to s$ loop processes.

the SM is a loop process with small mixing angles, can be easily modified by new physics. Examining whether the SM correlation between $a_{\pi\nu\bar{\nu}}$ and $S_{\psi K_S}$ is fulfilled, we can answer the following question:

(iv) *Is there new physics related solely to the third generation? to all generations?*

To understand the present situation, we present in Fig. 7 the constraints in the $\rho - \eta$ plane from tree plus loop processes that do not involve external third generation quarks, namely $s \rightarrow d$ transitions only (ϵ and $\mathcal{B}(K^+ \rightarrow \pi^+\nu\bar{\nu})$). This can be compared with the constraints from tree plus loop processes that do involve the third generation, namely $b \rightarrow d$ and $b \rightarrow s$ transitions. Again, one can see that there is a lot to be learnt from future measurements. (For a recent, comprehensive analysis of this question, see ref. [100].)

V. Consider ϕ_D, defined in Eq. (6.4), which is the relative phase between the $D^0 - \overline{D}^0$ mixing amplitude and the $c \rightarrow d\bar{s}u$ and $c \rightarrow s\bar{d}u$ decay amplitudes. Within the Standard Model, the two decay channels are tree level. It is unlikely that they are affected by new physics. On the other hand, the mixing amplitude can be easily modified by new physics. Examining whether $\phi_D \neq 0$, we can answer the following question:

(v) *Is there new physics in the down sector? in the up sector? in both?*

VI. Consider d_N, the electric dipole moment of the neutron. We did not discuss this quantity so far because, unlike CP violation in meson decays, flavor changing couplings are not necessary for d_N. In other words, the CP violation that induces d_N is *flavor diagonal*. It does in general get contributions from flavor changing physics, but it could be induced by sectors that are flavor blind. Within the SM (and ignoring θ_{QCD}), the contribution from δ_{KM} arises at the three loop level and is at least six orders of magnitude below the experimental bound (1.13). If the bound is further improved (or a signal observed), we can answer the following question:

(vi) *Are there new sources of CP violation that are flavor changing? flavor diagonal? both?*

It is no wonder then that with such rich information, flavor and CP violation provide an excellent probe of new physics. We next demonstrate this situation more concretely by discussing CP violation in supersymmetry.

13. Supersymmetric CP violation

Supersymmetry solves the fine-tuning problem of the Standard Model and has many other virtues. But at the same time, it leads to new problems: baryon number violation, lepton number violation, large flavor changing neutral current processes and large CP violation. The first two problems can be solved by imposing R-parity on supersymmetric models. There is no such simple, symmetry-

Y. Nir

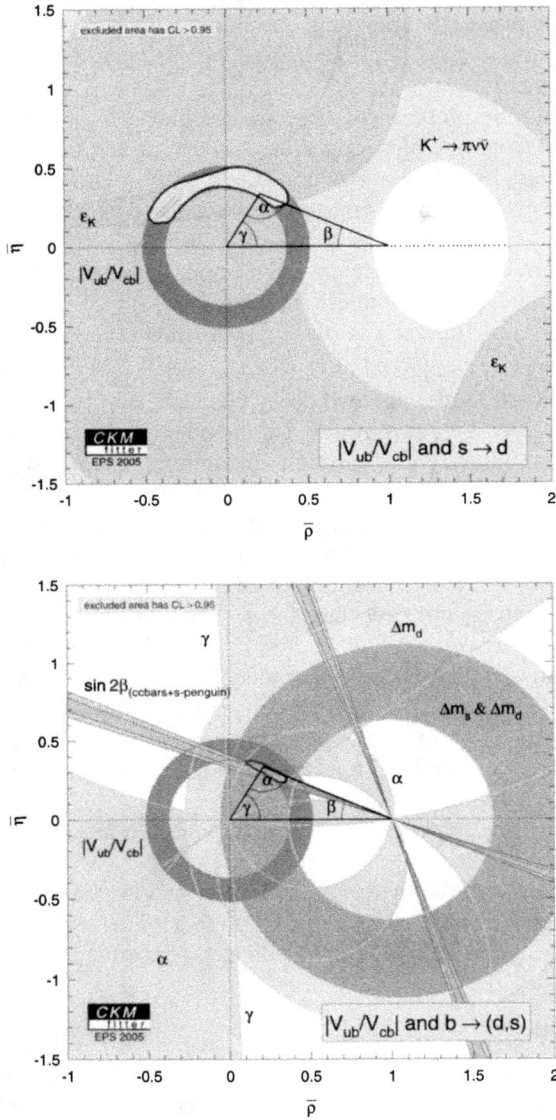

Fig. 7. Constraints in the $\rho - \eta$ plane from tree processes and (a) $s \to d$ or (b) $b \to d$ and $b \to s$ loop processes.

related solution to the problems of flavor and CP violation. Instead, suppression of the relevant couplings can be achieved by demanding very constrained structures of the soft supersymmetry breaking terms. There are two important questions here: First, can theories of dynamical supersymmetry breaking naturally induce such structures? Second, can measurements of flavor changing and/or CP violating processes shed light on the structure of the soft supersymmetry breaking terms? Since the answer to both questions is in the affirmative, we conclude that flavor changing neutral current processes and, in particular, CP violating observables will provide clues to the crucial question of how supersymmetry breaks.

13.1. CP-violating parameters

A generic supersymmetric extension of the Standard Model contains a host of new flavor and CP violating parameters. (For a review of CP violation in supersymmetry see [101, 102].) It is an amusing exercise to count the number of parameters [103]. The supersymmetric part of the Lagrangian depends, in addition to the three gauge couplings of G_{SM}, on the parameters of the superpotential W:

$$
\begin{aligned}
W = {} & \sum_{i,j} \left(Y_{ij}^u H_u Q_{Li} \overline{U}_{Lj} + Y_{ij}^d H_d Q_{Li} \overline{D}_{Lj} + Y_{ij}^\ell H_d L_{Li} \overline{E}_{Lj} \right) \\
& + \mu H_u H_d.
\end{aligned}
\tag{13.1}
$$

In addition, we have to add soft supersymmetry breaking terms:

$$
\begin{aligned}
\mathcal{L}_{\mathrm{soft}} = {} & -\sum_S (m_S^2)_{ij} A_i \bar{A}_j - \frac{1}{2} \sum_{(a)=1}^3 \left(\tilde{m}_{(a)} (\lambda\lambda)_{(a)} + \text{h.c.} \right) + (B H_u H_d \\
& - (A_{ij}^u H_u \tilde{Q}_{Li} \overline{\tilde{U}}_{Lj} + A_{ij}^d H_d \tilde{Q}_{Li} \overline{\tilde{D}}_{Lj} + A_{ij}^\ell H_d \tilde{L}_{Li} \overline{\tilde{E}}_{Lj} + \text{h.c.}),
\end{aligned}
\tag{13.2}
$$

where S stands for the scalar components of $Q_L, \overline{D}_L, \overline{U}_L, L_L, \overline{E}_L$. The three Yukawa matrices Y^f depend on 27 real and 27 imaginary parameters. Similarly, the three A^f-matrices depend on 27 real and 27 imaginary parameters. The five m_S^2 hermitian 3×3 mass-squared matrices for sfermions have 30 real parameters and 15 phases. The gauge and Higgs sectors depend on

$$
\theta_{\mathrm{QCD}}, \tilde{m}_{(1)}, \tilde{m}_{(2)}, \tilde{m}_{(3)}, g_1, g_2, g_3, \mu, B, m_{h_u}^2, m_{h_d}^2,
\tag{13.3}
$$

that is 11 real and 5 imaginary parameters. Summing over all sectors, we get 95 real and 74 imaginary parameters. The various couplings (other than the gauge couplings) can be thought of as spurions that break a global symmetry,

$$
U(3)^5 \times U(1)_{\mathrm{PQ}} \times U(1)_R \;\rightarrow\; U(1)_B \times U(1)_L.
\tag{13.4}
$$

The $U(1)_{PQ} \times U(1)_R$ charge assignments are:

$$
\begin{array}{cccccc}
 & H_u & H_d & Q\overline{U} & Q\overline{D} & L\overline{E} \\
U(1)_{PQ} & 1 & 1 & -1 & -1 & -1 \\
U(1)_R & 1 & 1 & 1 & 1 & 1
\end{array}
\tag{13.5}
$$

Consequently, we can remove 15 real and 30 imaginary parameters, which leaves

$$
124 = \begin{cases} 80 & \text{real} \\ 44 & \text{imaginary} \end{cases} \text{physical parameters.}
\tag{13.6}
$$

In particular, there are 43 new CP violating phases! In addition to the single Kobayashi-Maskawa of the SM, we can put 3 phases in M_1, M_2, μ (we used the $U(1)_{PQ}$ and $U(1)_R$ to remove the phases from μB^* and M_3, respectively) and the other 40 phases appear in the mixing matrices of the fermion-sfermion-gaugino couplings. (Of the 80 real parameters, there are 11 absolute values of the parameters in (13.3), 9 fermion masses, 21 sfermion masses, 3 CKM angles and 36 SCKM angles.) Supersymmetry provides a nice example to our statement that reasonable extensions of the Standard Model may have more than one source of CP violation.

The requirement of consistency with experimental data provides strong constraints on many of these parameters. For this reason, the physics of flavor and CP violation has had a profound impact on supersymmetric model building. A discussion of CP violation in this context can hardly avoid addressing the flavor problem itself. Indeed, many of the supersymmetric models that we analyze below were originally aimed at solving flavor problems.

As concerns CP violation, one can distinguish two classes of experimental constraints. First, bounds on nuclear and atomic electric dipole moments determine what is usually called the *supersymmetric CP problem*. Second, the physics of neutral mesons and, most importantly, the small experimental value of ε_K pose the *supersymmetric ε_K problem*. In the next two subsections we describe the two problems.

13.2. The supersymmetric CP problem

One aspect of supersymmetric CP violation involves effects that are flavor preserving. Then, for simplicity, we describe this aspect in a supersymmetric model without additional flavor mixings, *i.e.* the minimal supersymmetric standard model (MSSM) with universal sfermion masses and with the trilinear SUSY-breaking scalar couplings proportional to the corresponding Yukawa couplings. (The generalization to the case of non-universal soft terms is straightforward.) In such a constrained framework, there are four new phases beyond the two phases of the SM (δ_{KM} and θ_{QCD}). One arises in the bilinear μ-term of the superpotential (13.1), while the other three arise in the soft supersymmetry breaking parameters of (13.2): $\tilde{m}$ (the gaugino mass), A (the trilinear scalar coupling) and B (the bilinear scalar coupling). Only two combinations of the four phases are physical [104, 105]:

$$\phi_A = \arg(A^*\tilde{m}), \quad \phi_B = \arg(\tilde{m}\mu B^*). \qquad (13.7)$$

In the more general case of non-universal soft terms there is one independent phase ϕ_{A_i} for each quark and lepton flavor. Moreover, complex off-diagonal entries in the sfermion mass-squared matrices represent additional sources of CP violation.

The most significant effect of ϕ_A and ϕ_B is their contribution to electric dipole moments (EDMs). For example, the contribution from one-loop gluino diagrams to the down quark EDM is given by [106, 107]:

$$d_d = m_d \frac{e\alpha_3}{18\pi\tilde{m}^3} \left(|A| \sin \phi_A + \tan \beta |\mu| \sin \phi_B\right), \qquad (13.8)$$

where we have taken $m_Q^2 \sim m_D^2 \sim m_{\tilde{g}}^2 \sim \tilde{m}^2$, for left- and right-handed squark and gluino masses. We define, as usual, $\tan \beta = \langle H_u \rangle / \langle H_d \rangle$. Similar one-loop diagrams give rise to chromoelectric dipole moments. The electric and chromoelectric dipole moments of the light quarks (u, d, s) are the main source of d_N (the EDM of the neutron), giving [108]

$$d_N \sim 2 \left(\frac{100\, GeV}{\tilde{m}}\right)^2 \sin \phi_{A,B} \times 10^{-23}\, e\, \text{cm}, \qquad (13.9)$$

where, as above, $\tilde{m}$ represents the overall SUSY scale. In a generic supersymmetric framework, we expect $\tilde{m} = \mathcal{O}(m_Z)$ and $\sin \phi_{A,B} = \mathcal{O}(1)$. Then the constraint (1.13) is generically violated by about two orders of magnitude. This is *the Supersymmetric CP Problem*.

Eq. (13.9) shows two possible ways to solve the supersymmetric CP problem:
(i) Heavy squarks: $\tilde{m} \gtrsim 1\, TeV$;
(ii) Approximate CP: $\sin \phi_{A,B} \ll 1$.

13.3. The supersymmetric ε_K problem

The supersymmetric contribution to the ε_K parameter is dominated by diagrams involving Q and $\bar{d}$ squarks in the same loop. For $\tilde{m} = m_{\tilde{g}} \simeq m_Q \simeq m_D$ (our results depend only weakly on this assumption) and focusing on the contribution from the first two squark families, one gets (see, for example, [109]):

$$\varepsilon_K = \frac{5\,\alpha_3^2}{162\sqrt{2}} \frac{f_K^2 m_K}{\tilde{m}^2 \Delta m_K} \left[\left(\frac{m_K}{m_s + m_d} \right)^2 + \frac{3}{25} \right] \mathcal{I}m((\delta_{12}^d)_{LL}(\delta_{12}^d)_{RR}). \quad (13.10)$$

Here

$$(\delta_{12}^d)_{LL} = \left(\frac{m_{\tilde{Q}_2}^2 - m_{\tilde{Q}_1}^2}{m_{\tilde{Q}}^2} \right) K_{12}^{dL},$$

$$(\delta_{12}^d)_{RR} = \left(\frac{m_{\tilde{D}_2}^2 - m_{\tilde{D}_1}^2}{m_{\tilde{D}}^2} \right) K_{12}^{dR}, \quad (13.11)$$

where K_{12}^{dL} (K_{12}^{dR}) are the mixing angles in the gluino couplings to left-handed (right-handed) down quarks and their scalar partners. Note that CP would be violated even if there were two families only [110]. Using the experimental value of ε_K, we get

$$\frac{(\Delta m_K \varepsilon_K)^{\text{SUSY}}}{(\Delta m_K \varepsilon_K)^{\text{EXP}}} \sim 10^7 \left(\frac{300\,GeV}{\tilde{m}} \right)^2 \quad (13.12)$$

$$\times \left(\frac{m_{\tilde{Q}_2}^2 - m_{\tilde{Q}_1}^2}{m_{\tilde{Q}}^2} \right) \left(\frac{m_{\tilde{D}_2}^2 - m_{\tilde{D}_1}^2}{m_{\tilde{D}}^2} \right) |K_{12}^{dL} K_{12}^{dR}| \sin \phi,$$

where ϕ is the CP violating phase. In a generic supersymmetric framework, we expect $\tilde{m} = \mathcal{O}(m_Z)$, $\delta m_{Q,D}^2 / m_{Q,D}^2 = \mathcal{O}(1)$, $K_{ij}^{Q,D} = \mathcal{O}(1)$ and $\sin \phi = \mathcal{O}(1)$. Then the constraint (13.12) is generically violated by about seven orders of magnitude.

The Δm_K constraint on $\mathcal{R}e((\delta_{12}^d)_{LL}(\delta_{12}^d)_{RR})$ is about two orders of magnitude weaker. One can distinguish then three interesting regions for $\langle \delta_{12}^d \rangle = \sqrt{(\delta_{12}^d)_{LL}(\delta_{12}^d)_{RR}}$:

$$\langle \delta_{12}^d \rangle \begin{cases} \gg 0.003 & \text{excluded;} \\ \in [0.0002, 0.003] & \text{viable with small phases;} \\ \ll 0.0002 & \text{viable with } \mathcal{O}(1) \text{ phases.} \end{cases} \quad (13.13)$$

The first bound comes from the Δm_K constraint (assuming that the relevant phase is not particularly close to $\pi/2$). The bounds here apply to squark masses of order 500 GeV and scale like $\tilde{m}$. There is also dependence on $m_{\tilde{g}}/\tilde{m}$, which is here taken to be one.

Table 5

Theoretical predictions for supersymmetric flavor changing couplings in viable models of alignment, and the experimental constraints.

$(\delta^q_{MN})_{ij}$	Prediction	Upper bound	$(\delta^d_{MN})_{ij}$	Prediction	Upper bound
$(\delta^d_{LL})_{12}$	$\lambda^5 - \lambda^3$	λ^3	$(\delta^d_{LR})_{12}$	$\lambda^7(m_b/\tilde{m})$	$\lambda^7(\mathcal{I}m)$
$(\delta^d_{RR})_{12}$	$\lambda^7 - \lambda^3$	$\lambda^{10}/(\delta^d_{LL})_{12}$	$(\delta^d_{RL})_{12}$	$\lambda^9(m_b/\tilde{m})$	$\lambda^7(\mathcal{I}m)$
$(\delta^d_{LL})_{13}$	λ^3	λ	$(\delta^d_{LR})_{13}$	$\lambda^3(m_b/\tilde{m})$	λ^2
$(\delta^d_{RR})_{13}$	$\lambda^7 - \lambda^3$	$\lambda^4/(\delta^d_{LL})_{13}$	$(\delta^d_{RL})_{13}$	$\lambda^7(m_b/\tilde{m})$	λ^2
$(\delta^d_{LL})_{23}$	λ^2	$\lambda^2(\mathcal{R}e) - \lambda(\mathcal{I}m)$	$(\delta^d_{LR})_{23}$	$\lambda^2(m_b/\tilde{m})$	$\lambda^4(\mathcal{R}e) - \lambda^3(\mathcal{I}m)$
$(\delta^d_{RR})_{23}$	$\lambda^4 - \lambda^2$	1	$(\delta^d_{RL})_{23}$	$\lambda^4(m_b/\tilde{m})$	λ^3
$(\delta^u_{LL})_{12}$	λ	λ			
$(\delta^u_{RR})_{12}$	$\lambda^4 - \lambda^2$	$\lambda^4/(\delta^u_{LL})_{12}$			

Eq. (13.12) also shows what are the possible ways to solve the supersymmetric ε_K problem:

(i) Heavy squarks: $\tilde{m} \gg 300\ GeV$;

(ii) Universality: $\delta m^2_{Q,D} \ll m^2_{Q,D}$;

(iii) Alignment: $|K^d_{12}| \ll 1$;

(iv) Approximate CP: $\sin\phi \ll 1$.

13.4. More on supersymmetric flavor and CP violation

The flavor and CP constraints on supersymmetric models apply to almost all flavor changing couplings. The size of supersymmetric flavor violation depends on the overall scale of the soft supersymmetry breaking terms, on mass degeneracies between sfermion generations, and on the mixing angles in gaugino couplings. One can choose a representative scale (say, $\tilde{m} \sim 300$ GeV) and then conveniently present the constraints in terns of the $(\delta^q_{ij})_{MN}$ parameters [see Eq. (13.11)]. In a given supersymmetric flavor model, one can find predictions for the $(\delta^q_{ij})_{MN}$ and test the model.

A summary of upper bounds on the supersymmetric flavor changing couplings is given in Table 5. The bounds on the $\mathcal{I}m(\delta^d_{12})_{LR,RL}$ parameters are taken from [111], on δ^d_{13} from [112] and on δ^d_{23} from [113, 114]. The bounds are expressed in powers of the Wolfenstein parameter λ, which makes it easy to compare with model predictions. As an example, we give the range of these parameters that is expected in a large class of viable models of alignment [115–117].

Until some time ago, the δ^d_{23} parameters have been only weakly constrained (the improving accuracy of the measurements of $\mathcal{B}(B \to X\ell^+\ell^-)$ have strengthened the constraints considerably). Furthermore, measurements of various CP asymmetries in penguin dominated modes (particularly $S_{\phi K}$ and $S_{\eta' K}$) gave central values that were far off the expected value $\sim S_{\psi K}$ (at present the strongest discrepancy is down to the 2σ level). One may still ask whether effects of order 0.1, which is the order of the expected experimental accuracy and probably above the theoretical error on $S_{\phi K}$ and $S_{\eta' K}$, are still possible within supersymmetric flavor models and, in particular, alignment models.

To answer this question, we use the results of ref. [113]. From their Fig. 3, we make the following estimates:

$$\frac{\Delta S_{\phi K}}{\Delta \mathcal{I}m(\delta^d_{LL})_{23}} \sim \frac{\Delta S_{\phi K}}{\Delta \mathcal{I}m(\delta^d_{RR})_{23}} \sim 0.3,$$

$$\frac{\Delta S_{\phi K}}{\Delta \mathcal{I}m(\delta^d_{LR})_{23}} \sim \frac{\Delta S_{\phi K}}{\Delta \mathcal{I}m(\delta^d_{RL})_{23}} \sim 100. \tag{13.14}$$

Thus, for $S_{\phi K}$ to be shifted by $\mathcal{O}(0.1)$, we need at least one of the following four options:

$$\mathcal{I}m(\delta^d_{LL})_{23} \sim \lambda, \quad \mathcal{I}m(\delta^d_{RR})_{23} \sim \lambda,$$

$$\mathcal{I}m(\delta^d_{LR})_{23} \sim \lambda^4, \quad \mathcal{I}m(\delta^d_{RL})_{23} \sim \lambda^4. \tag{13.15}$$

Examining Table 5, we learn that in alignment models $\mathcal{I}m(\delta^d_{LR})_{23} \sim 7 \times 10^{-4}$ $(350 \text{ GeV}/\tilde{m})$ is the closest to satisfying the condition in Eq. (13.15), though the unknown numbers of order one should be on the large side to give an observable effect.

13.5. Discussion

We define two scales that play an important role in supersymmetry: Λ_S, where the soft supersymmetry breaking terms are generated, and Λ_F, where flavor dynamics takes place. When $\Lambda_F \gg \Lambda_S$, it is possible that there are no genuinely new sources of flavor and CP violation. This class of models, where the Yukawa couplings (or, in the mass basis, the CKM matrix) are the only source of flavor and CP breaking, are often called 'minimal flavor violation.' The most important features of the supersymmetry breaking terms are universality of the scalar masses-squared and proportionality of the A-terms. When $\Lambda_F \lesssim \Lambda_S$, we do not expect, in general, that flavor and CP violation are limited to the Yukawa matrices. One way to suppress CP violation would be to assume that, similarly to the Standard Model, CP violating phases are large, but their effects are screened, possibly by the same physics that explains the various flavor puzzles, such as models with Abelian or non-Abelian horizontal symmetries. It is also possible that CP violating effects are suppressed because squarks are heavy. Another option, which is now excluded, was to assume that CP is an approximate symmetry of the full theory (namely, CP violating phases are all small).

We would like to emphasize the following points:

(i) For supersymmetry to be established, a direct observation of supersymmetric particles is necessary. Once it is discovered, then measurements of CP violating observables will be a very sensitive probe of its flavor structure and, consequently, of the mechanism of dynamical supersymmetry breaking.

(ii) It seems possible to distinguish between models of exact universality and models with genuine supersymmetric flavor and CP violation. The former tend to give $d_N \lesssim 10^{-31}$ e cm while the latter usually predict $d_N \gtrsim 10^{-28}$ e cm.

(iii) The proximity of $S_{\psi K_S}$ to the SM predictions is obviously consistent with models of exact universality. It disfavors models of heavy squarks such as that of ref. [118]. Models of flavor symmetries allow deviations of order 20% (or smaller) from the SM predictions. To be convincingly signalled, an improvement in the theoretical calculations that lead to the SM predictions for $S_{\psi K_S}$ will be required [119].

(iv) The fact that $K \to \pi \nu \bar{\nu}$ decays are not affected by most supersymmetric flavor models [120–122] is an advantage here. The SM correlation between $a_{\pi \nu \bar{\nu}}$ and $S_{\psi K_S}$ is a much cleaner test than a comparison of $S_{\psi K_S}$ to the CKM constraints.

(v) $D^0 - \overline{D}^0$ mixing provides a stringent test of alignment. Observation of CP violation in the $D \to K \pi$ decays will make a convincing case for new physics.

(vi) CP violation in $b \to s$ transition remains an interesting probe of supersymmetry. Deviations of order 0.1 from the SM predictions are possible if one of the conditions in Eq. (13.15) is satisfied.

14. Lessons from the B factories

Let us summarize the main lessons that have been learned from the measurements of CP violation in B decays:
- The KM phase is different from zero, that is, the SM violates CP.
- The KM mechanism is the dominant source of CP violation in meson decays.
- The size and the phase of new physics contributions to $b \to d$ transitions ($B^0 - \overline{B}^0$ mixing) is severely constrained ($\leq \mathcal{O}(0.2)$).
- Complete alternatives to the KM mechanism (the superweak mechanism and approximate CP) are excluded.
- Corrections to the KM mechanism are possible, particularly for $b \to s$ transitions, but there is no evidence at present for such corrections.
- There is still a lot to be learned from future measurements.

Acknowledgments

I am grateful to Andreas Höcker, Sandrine Laplace and, in particular, Stephane T'Jampens for providing me with beautiful plots of CKM constraints. Their work has helped me to understand and, hopefully, to explain the significance of the B-factory measurements of CP violating asymmetries to our understanding of flavor and CP violation. I am grateful to Guy Raz and to Zoltan Ligeti for their contributions to the basic ideas and to the details of this review. I thank David Kirkby for collaboration on the PDG review on CP violation in meson decays [55] which is the basis of some sections in these lecture notes. This work was supported by a grant from the G.I.F., the German–Israeli Foundation for Scientific Research and Development, by the Israel Science Foundation founded by the Israel Academy of Sciences and Humanities, by EEC RTN contract HPRN-CT-00292-2002, by the Minerva Foundation (München), and by the United States-Israel Binational Science Foundation (BSF), Jerusalem, Israel.

References

[1] M. Kobayashi and T. Maskawa, Prog. Theor. Phys. **49**, 652 (1973).
[2] J. H. Christenson, J. W. Cronin, V. L. Fitch and R. Turlay, Phys. Rev. Lett. **13**, 138 (1964).
[3] H. Burkhardt et al. [NA31 Collaboration], Phys. Lett. B **206**, 169 (1988).
[4] V. Fanti et al. [NA48 Collaboration], Phys. Lett. B **465**, 335 (1999) [hep-ex/9909022].
[5] A. Alavi-Harati et al. [KTeV Collaboration], Phys. Rev. Lett. **83**, 22 (1999) [hep-ex/9905060].
[6] B. Aubert et al. [BaBar Collaboration], Phys. Rev. Lett. **87**, 091801 (2001) [hep-ex/0107013].
[7] K. Abe et al. [Belle Collaboration], Phys. Rev. Lett. **87**, 091802 (2001) [hep-ex/0107061].
[8] K. Abe et al. [Belle Collaboration], hep-ex/0409049.
[9] B. Aubert et al. [BaBar Collaboration], Phys. Rev. D **71**, 091102 (2005) [hep-ex/0502019].
[10] T. Aushev et al. [Belle Collaboration], hep-ex/0411021.
[11] B. Aubert et al. [BaBar Collaboration], Phys. Rev. Lett. **95**, 151804 (2005) [hep-ex/0506082].
[12] B. Aubert et al. [BaBar Collaboration], Phys. Rev. Lett. **94**, 191802 (2005) [hep-ex/0502017].
[13] K. Abe et al. [Belle Collaboration], hep-ex/0507037.
[14] B. Aubert et al. [BaBar Collaboration], hep-ex/0507087.
[15] B. Aubert et al. [BaBar Collaboration], hep-ex/0408095.
[16] B. Aubert et al. [BaBar Collaboration], Phys. Rev. Lett. **93**, 131801 (2004) [hep-ex/0407057].
[17] K. Abe et al. [Belle Collaboration], hep-ex/0507045.
[18] C. C. Wang et al. [Belle Collaboration], Phys. Rev. Lett. **94**, 121801 (2005) [hep-ex/0408003].
[19] B. Aubert et al. [BaBar Collaboration], hep-ex/0408099.
[20] L. Wolfenstein, Phys. Rev. Lett. **13**, 562 (1964).
[21] G. Eyal and Y. Nir, Nucl. Phys. B **528**, 21 (1998) [hep-ph/9801411].
[22] A. D. Sakharov, Pisma Zh. Eksp. Teor. Fiz. **5**, 32 (1967) [JETP Lett. **5**, 24 (1967)].
[23] G. R. Farrar and M. E. Shaposhnikov, Phys. Rev. D **50**, 774 (1994) [hep-ph/9305275].

[24] P. Huet and E. Sather, Phys. Rev. D **51**, 379 (1995) [hep-ph/9404302].

[25] M. B. Gavela, M. Lozano, J. Orloff and O. Pene, Nucl. Phys. B **430**, 345 (1994) [hep-ph/9406288].

[26] M. Fukugita and T. Yanagida, Phys. Lett. B **174**, 45 (1986).

[27] G. C. Branco, T. Morozumi, B. M. Nobre and M. N. Rebelo, hep-ph/0107164.

[28] A. G. Cohen, D. B. Kaplan and A. E. Nelson, Ann. Rev. Nucl. Part. Sci. **43**, 27 (1993) [hep-ph/9302210].

[29] R. J. Crewther, P. Di Vecchia, G. Veneziano and E. Witten, Phys. Lett. B **88**, 123 (1979) [Erratum-ibid. B **91**, 487 (1979)].

[30] L. J. Dixon, A. Langnau, Y. Nir and B. Warr, Phys. Lett. B **253**, 459 (1991).

[31] P. G. Harris et al., Phys. Rev. Lett. **82**, 904 (1999).

[32] M. Dine, hep-ph/0011376.

[33] T. Banks, Y. Nir and N. Seiberg, hep-ph/9403203.

[34] R. D. Peccei and H. R. Quinn, Phys. Rev. Lett. **38**, 1440 (1977).

[35] R. D. Peccei and H. R. Quinn, Phys. Rev. D **16**, 1791 (1977).

[36] Y. Grossman and M. P. Worah, Phys. Lett. B **395**, 241 (1997) [hep-ph/9612269].

[37] Y. Grossman, G. Isidori and M. P. Worah, Phys. Rev. D **58**, 057504 (1998) [hep-ph/9708305].

[38] G. Buchalla and A. J. Buras, Phys. Lett. B **333**, 221 (1994) [hep-ph/9405259].

[39] G. Buchalla and A. J. Buras, Phys. Rev. D **54**, 6782 (1996) [hep-ph/9607447].

[40] S. Bergmann and G. Perez, JHEP **0008**, 034 (2000) [hep-ph/0007170].

[41] M. P. Worah, Phys. Rev. Lett. **79**, 3810 (1997) [hep-ph/9704389].

[42] M. P. Worah, Phys. Rev. D **56**, 2010 (1997) [hep-ph/9702423].

[43] C. Jarlskog, Phys. Rev. Lett. **55**, 1039 (1985).

[44] M. C. Gonzalez-Garcia and Y. Nir, Rev. Mod. Phys. **75**, 345 (2003) [hep-ph/0202058].

[45] N. Cabibbo, Phys. Rev. Lett. **10**, 531 (1963).

[46] Z. Maki, M. Nakagawa and S. Sakata, Prog. Theor. Phys. **28**, 870 (1962).

[47] L. Chau and W. Keung, Phys. Rev. Lett. **53**, 1802 (1984).

[48] L. Wolfenstein, Phys. Rev. Lett. **51**, 1945 (1983).

[49] A. J. Buras, M. E. Lautenbacher, and G. Ostermaier, Phys. Rev. D **50**, 3433 (1994) [hep-ph/9403384].

[50] C. Dib, I. Dunietz, F. J. Gilman and Y. Nir, Phys. Rev. D **41**, 1522 (1990).

[51] J. L. Rosner, A. I. Sanda and M. P. Schmidt, EFI-88-12-CHICAGO [Presented at Workshop on High Sensitivity Beauty Physics, Batavia, IL, Nov 11-14, 1987].

[52] H. Harari and Y. Nir, Phys. Lett. B **195**, 586 (1987).

[53] S. Eidelman et al. [Particle Data Group], Phys. Lett. B **592**, 1 (2004).

[54] CKMfitter Group (J. Charles et al.), Eur. Phys. J. C41, 1-131 (2005), [hep-ph/0406184], updated results and plots available at: http://ckmfitter.in2p3.fr

[55] D. Kirkby and Y. Nir, review of 'CP violation in meson decays' in ref. [53].

[56] V. Weisskopf and E. P. Wigner, Z. Phys. **63**, 54 (1930); Z. Phys. **65**, 18 (1930). [See Appendix A of P. K. Kabir, "The CP Puzzle: Strange Decays of the Neutral Kaon", Academic Press (1968).]

[57] Y. Nir, SLAC-PUB-5874 [Lectures given at 20th Annual SLAC Summer Institute on Particle Physics (Stanford, CA, 1992)].

[58] I. Dunietz and J. L. Rosner, Phys. Rev. D **34**, 1404 (1986).

144

Y. Nir

[59] Ya. I. Azimov, N. G. Uraltsev, and V. A. Khoze, Sov. J. Nucl. Phys. **45**, 878 (1987) [Yad. Fiz. **45**, 1412 (1987)].

[60] I. I. Bigi and A. I. Sanda, Nucl. Phys. B **281**, 41 (1987).

[61] Y. Grossman and Y. Nir, Phys. Lett. B **398**, 163 (1997) [hep-ph/9701313].

[62] L. S. Littenberg, Phys. Rev. D **39**, 3322 (1989).

[63] A. J. Buras, Phys. Lett. B **333**, 476 (1994) [hep-ph/9405368].

[64] G. Buchalla and A. J. Buras, Nucl. Phys. B **400**, 225 (1993).

[65] G. Blaylock, A. Seiden and Y. Nir, Phys. Lett. B **355**, 555 (1995) [hep-ph/9504306].

[66] A. F. Falk, Y. Grossman, Z. Ligeti and A. A. Petrov, Phys. Rev. D **65**, 054034 (2002) [hep-ph/0110317].

[67] A. F. Falk, Y. Grossman, Z. Ligeti, Y. Nir and A. A. Petrov, Phys. Rev. D **69**, 114021 (2004) [hep-ph/0402204].

[68] S. Bergmann and Y. Nir, JHEP **9909**, 031 (1999) [hep-ph/9909391].

[69] G. D'Ambrosio and D. Gao, Phys. Lett. B **513**, 123 (2001) [hep-ph/0105078].

[70] S. Bergmann, Y. Grossman, Z. Ligeti, Y. Nir and A. A. Petrov, Phys. Lett. B **486**, 418 (2000) [hep-ph/0005181].

[71] A. B. Carter and A. I. Sanda, Phys. Rev. Lett. **45**, 952 (1980).

[72] A. B. Carter and A. I. Sanda, Phys. Rev. D **23**, 1567 (1981).

[73] I. I. Bigi and A. I. Sanda, Nucl. Phys. B **193**, 85 (1981).

[74] G. Buchalla, A. J. Buras, and M. E. Lautenbacher, Rev. Mod. Phys. **68**, 1125 (1996) [hep-ph/9512380].

[75] Y. Grossman, A. L. Kagan and Z. Ligeti, Phys. Lett. B **538**, 327 (2002) [hep-ph/0204212].

[76] H. Boos, T. Mannel and J. Reuter, Phys. Rev. D **70**, 036006 (2004) [hep-ph/0403085].

[77] K. Abe, talk given at Lepton-Photon 2005.

[78] Y. Nir, Nucl. Phys. Proc. Suppl. **117**, 111 (2003) [hep-ph/0208080].

[79] M. Beneke, G. Buchalla, M. Neubert and C. T. Sachrajda, Nucl. Phys. B **591**, 313 (2000) [hep-ph/0006124].

[80] G. Buchalla, G. Hiller, Y. Nir and G. Raz, JHEP **0509**, 074 (2005) [hep-ph/0503151].

[81] A. Ali, G. Kramer and C. D. Lü, Phys. Rev. D **58**, 094009 (1998) [hep-ph/9804363].

[82] M. Beneke and M. Neubert, Nucl. Phys. B **651**, 225 (2003) [hep-ph/0210085].

[83] M. Beneke, Phys. Lett. B **620**, 143 (2005) [hep-ph/0505075].

[84] H. J. Lipkin, Phys. Rev. Lett. **46**, 1307 (1981).

[85] M. Gronau and D. London, Phys. Rev. Lett. **65**, 3381 (1990).

[86] Y. Grossman and H. R. Quinn, Phys. Rev. D **58**, 017504 (1998) [hep-ph/9712306].

[87] J. Charles, Phys. Rev. D **59**, 054007 (1999) [hep-ph/9806468].

[88] M. Gronau, D. London, N. Sinha, and R. Sinha, Phys. Lett. B **514**, 315 (2001) [hep-ph/0105308].

[89] H. J. Lipkin, Y. Nir, H. R. Quinn and A. Snyder, Phys. Rev. D **44**, 1454 (1991).

[90] M. Gronau, Phys. Lett. B **265**, 389 (1991).

[91] A. E. Snyder and H. R. Quinn, Phys. Rev. D **48**, 2139 (1993).

[92] H. R. Quinn and J. P. Silva, Phys. Rev. D **62**, 054002 (2000) [hep-ph/0001290].

[93] A. F. Falk, Z. Ligeti, Y. Nir and H. Quinn, Phys. Rev. D **69**, 011502 (2004) [hep-ph/0310242].

[94] M. Gronau and D. London., Phys. Lett. B **253**, 483 (1991).

[95] M. Gronau and D. Wyler, Phys. Lett. B **265**, 172 (1991).

[96] D. Atwood, I. Dunietz and A. Soni, Phys. Rev. Lett. **78**, 3257 (1997) [hep-ph/9612433].

[97] A. Giri, Y. Grossman, A. Soffer and J. Zupan, Phys. Rev. D **68**, 054018 (2003) [hep-ph/0303187].

[98] Y. Grossman, Y. Nir and M. P. Worah, Phys. Lett. B **407**, 307 (1997) [hep-ph/9704287].

[99] Y. Nir and D. J. Silverman, Nucl. Phys. B **345**, 301 (1990).

[100] K. Agashe, M. Papucci, G. Perez and D. Pirjol, hep-ph/0509117.

[101] Y. Grossman, Y. Nir and R. Rattazzi, hep-ph/9701231.

[102] M. Dine, E. Kramer, Y. Nir and Y. Shadmi, Phys. Rev. D **63**, 116005 (2001) [hep-ph/0101092].

[103] H. E. Haber, Nucl. Phys. Proc. Suppl. **62**, 469 (1998) [hep-ph/9709450].

[104] M. Dugan, B. Grinstein and L. Hall, Nucl. Phys. B **255**, 413 (1985).

[105] S. Dimopoulos and S. Thomas, Nucl. Phys. B **465**, 23 (1996) [hep-ph/9510220].

[106] W. Buchmuller and D. Wyler, Phys. Lett. B **121**, 321 (1983).

[107] J. Polchinski and M. B. Wise, Phys. Lett. B **125**, 393 (1983).

[108] W. Fischler, S. Paban and S. Thomas, Phys. Lett. B **289**, 373 (1992) [hep-ph/9205233].

[109] F. Gabbiani, E. Gabrielli, A. Masiero and L. Silvestrini, Nucl. Phys. B **477**, 321 (1996) [hep-ph/9604387].

[110] Y. Nir, Nucl. Phys. B **273**, 567 (1986).

[111] G. Eyal, A. Masiero, Y. Nir and L. Silvestrini, JHEP **9911**, 032 (1999) [hep-ph/9908382].

[112] D. Becirevic et al., Nucl. Phys. B **634**, 105 (2002) [hep-ph/0112303].

[113] M. Ciuchini, E. Franco, G. Martinelli, A. Masiero, M. Pierini and L. Silvestrini, hep-ph/0407073.

[114] L. Silvestrini, hep-ph/0510077.

[115] Y. Nir and N. Seiberg, Phys. Lett. B **309**, 337 (1993) [hep-ph/9304307].

[116] M. Leurer, Y. Nir and N. Seiberg, Nucl. Phys. B **420**, 468 (1994) [hep-ph/9310320].

[117] Y. Nir and G. Raz, Phys. Rev. D **66**, 035007 (2002) [hep-ph/0206064].

[118] A. G. Cohen, D. B. Kaplan, F. Lepeintre and A. E. Nelson, Phys. Rev. Lett. **78**, 2300 (1997) [hep-ph/9610252].

[119] G. Eyal, Y. Nir and G. Perez, JHEP **0008**, 028 (2000) [hep-ph/0008009].

[120] Y. Nir and M. P. Worah, Phys. Lett. B **423**, 319 (1998) [hep-ph/9711215].

[121] A. J. Buras, A. Romanino and L. Silvestrini, Nucl. Phys. B **520**, 3 (1998) [hep-ph/9712398].

[122] G. Colangelo and G. Isidori, JHEP **9809**, 009 (1998) [hep-ph/9808487].

Course 3

SUPERSYMMETRY BREAKING

Yael Shadmi

Physics Department, Technion,
Haifa, Israel

D. Kazakov, S. Lavignac and J. Dalibard, eds.
Les Houches, Session LXXXIV, 2005
Particle Physics Beyond the Standard Model
© *2006 Elsevier B.V. All rights reserved*

Contents

1. Preamble

Need we motivate lectures on supersymmetry breaking? Not really. If there is supersymmetry in Nature, it must be broken. But it's worth emphasizing that the breaking of supersymmetry, namely, the masses of superpartners, determines the way supersymmetry would manifest itself in experiment.

From a purely theoretical point of view, supersymmetry breaking is a very beautiful subject, and I hope these lectures will convey some of this beauty.

1.1. Structure and further reading

It is very hard to cover supersymmetry-breaking in three lectures. In the first lecture, section 2, we will describe the essentials of supersymmetry breaking. In the second lecture, section 3, we will study dynamical supersymmetry breaking. In the last lecture, section 4, we will describe several mechanisms for generating supersymmetry-breaking terms for the standard-model superpartners. This section can be read independently of section 3.

For lack of time, we will not cover supersymmetry-breaking mechanisms, or mechanisms for mediating the breaking, that rely on extra dimensions (we will discuss anomaly-mediation, because it is always present in four dimensions). Supersymmetry-breaking in extra dimensions will be mentioned in the lectures by Gherghetta.

These lectures assume basic knowledge of supersymmetry (essentially the first seven chapters of Wess and Bagger [1], whose notations we will use). Supersymmetry basics were covered in Wagner's lectures, and we will review some definitions in the appendix to establish common notations.

I tried to make section 3 self-contained, but a serious treatment of non-perturbative effects in supersymmetric gauge theories is beyond the scope of these lectures. For excellent reviews of the subject see, e.g., [2–4]. For more details and examples of dynamical supersymmetry breaking, see [5,6].

Finally, ref. [7] is a comprehensive review of gauge-mediation models.

2. Basic features of supersymmetry breaking

In this section, we will discuss the fundamentals of supersymmetry breaking: the order parameters for the breaking, the Goldstone fermion, F-type and D-type tree-level breaking, and some general criteria for determining when supersymmetry is broken. The discussion will mostly be in the framework of $\mathcal{N} = 1$ global supersymmetry, but we will end this section by commenting on how things are modified for local supersymmetry.

2.1. Order parameters for supersymmetry breaking

When looking for spontaneous supersymmetry breaking, we are asking whether the variation of some field under the supersymmetry transformation is non-zero in the ground state,

$$\langle 0|\delta(\text{field})|0\rangle \neq 0. \tag{2.1}$$

For a chiral superfield ϕ, with scalar component $\tilde{\phi}$, fermion component ψ, and auxiliary component F, the supersymmetry variation are roughly (omitting numerical coefficients),

$$
\begin{aligned}
\delta_\xi \tilde{\phi}(x) &\sim \xi \psi(x), \\
\delta_\xi \psi(x) &\sim i\sigma^\mu \bar{\xi}\, \partial_\mu \tilde{\phi}(x) + \xi F(x), \\
\delta_\xi F(x) &\sim i\bar{\xi}\bar{\sigma}^\mu \partial_\mu \psi(x),
\end{aligned}
\tag{2.2}
$$

where ξ parameterizes the supersymmetry variation. Clearly, the only Lorentz invariant on the RHS of eqn. (2.2) is F, so supersymmetry is broken if

$$< F > \neq 0, \tag{2.3}$$

and the field whose variation is non-zero in this case is the fermion, $\langle 0|\delta_\xi \psi(x)|0\rangle \neq 0$.

Similarly, for the vector superfield, only the gaugino variation can be non-zero

$$\langle 0|\delta_\xi \lambda(x)|0\rangle \propto \langle 0|D|0\rangle \neq 0, \tag{2.4}$$

so a non-zero $\langle D \rangle$ signals supersymmetry breaking.

A much more physical order parameter for global supersymmetry breaking is the vacuum energy. The supersymmetry algebra contains the translation operator P_μ

$$\{Q_\alpha, \bar{Q}_{\dot{\alpha}}\} = 2\sigma^\mu_{\alpha\dot{\alpha}} P_\mu, \tag{2.5}$$

where Q is the supersymmetry generator. Therefore the Hamiltonian H can be written as

$$H = \frac{1}{4} (\bar{Q}_1 Q_1 + \bar{Q}_2 Q_2 + \text{h.c.}). \tag{2.6}$$

Since this is a positive operator, the energy of a supersymmetric system is either positive or zero. Furthermore, if supersymmetry is unbroken, the vacuum is annihilated by the supersymmetry generators, and

$$E_{\text{vacuum}} = \langle 0|H|0 \rangle = 0. \tag{2.7}$$

Thus, a non-zero vacuum energy signals spontaneous supersymmetry breaking.

In order to know whether global supersymmetry is spontaneously broken, we therefore need to study the minima of the scalar potential, and to see whether there is a minimum with zero energy.

2.2. The scalar potential and flat directions

In a theory with chiral superfields ϕ_i, superpotential $W(\phi_i)$ and Kähler potential $K(\phi_i, \phi_i^\dagger)$, the scalar potential is given by

$$V_F = K_{i*j}^{-1} \frac{\partial W^*}{\partial \phi_i^*} \frac{\partial W}{\partial \phi_j} = K_{ij}^{-1} F_i^* F_j, \tag{2.8}$$

where

$$K_{ij*} = \frac{\partial^2 K}{\partial \phi_i \partial \phi_{j*}}. \tag{2.9}$$

In eqn. (2.8) we used the fact that, on-shell, the auxiliary fields are given by

$$F_i = \frac{\partial W}{\partial \phi_i}. \tag{2.10}$$

If there are gauge interactions in the theory the scalar potential has additional contributions and is given by

$$V = V_F + V_D = V_F + \frac{1}{2} g^2 \sum_a (D^a)^2, \tag{2.11}$$

where $D^a = \sum_i \phi_i^\dagger T^a \phi_i$. As expected, the scalar potential is non-negative, and again we see that supersymmetry is broken by a non-zero F and /or D vacuum expectation value (VEV). Only then is the ground state energy non-zero.

To look for the zeros of the scalar potential (in field space) in a theory with gauge interactions, we need to do the following:

1. Find the sub-(field)space for which $D^a = 0$. This is often called the space of "D-flat directions". Note that along these directions, the potential is not merely flat, but rather zero[1]. The space of D-flat directions can be parametrized by the VEVs of the chiral gauge invariants that one can construct from the fundamental chiral fields of the theory. This is an extremely useful result and we will often use it in the following.

2. If for a subspace of the D-flat directions we also have $F_i = 0$ (for all F_i's) then the potential is zero. The sub-(field) space for which this happens is often called the "moduli space".

To look for supersymmetry breaking, we will be interested then in the moduli-space of the theory. If there is no moduli space, supersymmetry is broken.

Exercise: D-flat directions: Consider an $SU(N)$ gauge theory with chiral fields $Q_i \sim N$, $\bar{Q}^A \sim \bar{N}$, with $i, A = 1, \ldots, F$. (This theory is usually called $SU(N)$ with F flavors.) Assume $F < N$. Denote the $SU(N)$ gauge index by α. Show that

$$Q_{i\alpha} = \bar{Q}_{i\alpha} = v_i \delta_{i\alpha}, \tag{2.12}$$

are D-flat. The D-flat directions of the theory are then given by (2.12) up to global $SU(F)_L \times SU(F)_R$ and gauge rotations.

As mentioned above, an alternative parameterization of the D-flat directions is in terms of the VEVs of the gauge invariants of the theory. In this case, the only chiral gauge invariants are the "mesons" $M_i^A = Q_i \cdot \bar{Q}^A$. Indeed, using the global symmetry we can always write the meson VEVs as

$$M_i^A = \text{diag}(V_1, V_2, \ldots, V_N). \tag{2.13}$$

and the two parameterizations are clearly equivalent $V_i \leftrightarrow v_i^2$.

2.3. The Goldstino

With broken supersymmetry $Q_\alpha |0\rangle$ is non-zero. What is it then? The generator of a broken bosonic global symmetry gives the Goldstone boson. Likewise, $Q_\alpha |0\rangle$ gives the Goldstone fermion of supersymmetry breaking, or "Goldstino", which we denote by $\psi_\alpha^G(x)$.

---

[1]The reason why these directions are called "flat" will become clear once we discuss radiative corrections. Typically, in non-supersymmetric theories, if we have a flat potential at tree level, the degeneracy is lifted by radiative corrections. As we will see, in supersymmetric theories, if the ground state energy is zero at tree-level, it remains zero to all orders in perturbation theory. Therefore, the directions in field space for which $V = 0$ are the only ones that are truly flat—they remain zero to all orders in perturbation theory.

To see the Goldstino concretely, we should examine the supersymmetry current, and look for a piece that is linear in the fields. The supersymmetry current is of the form

$$J_\alpha^\mu \sim \sum_\phi \frac{\delta \mathcal{L}}{\delta(\partial_\mu \phi)} (\delta \phi)_\alpha, \tag{2.14}$$

where $\delta\phi$ is the supersymmetry variation of the the field ϕ. Since $\frac{\delta\mathcal{L}}{\delta(\partial_\mu\phi)}$ cannot get a VEV, a term that is linear in the fields can only occur when $\delta\phi$ gets a non-zero VEV. As we saw before, the only fields whose supersymmetry variations can have non-zero VEVS are the fermion of the chiral superfield, ψ (the VEV of whose variation is F), and the fermion of the vector superfield, λ (the VEV of whose variation is D). Thus,

$$J_\mu^\alpha \sim \sum_i \frac{\delta \mathcal{L}}{\delta(\partial_\mu \psi_{i\alpha})} \langle F_i \rangle + \frac{1}{\sqrt{2}} \sum_a \frac{\delta \mathcal{L}}{\delta(\partial_\mu \lambda_\alpha^a)} \langle D^a \rangle, \tag{2.15}$$

so that

$$\psi_\mu^G \sim \sum_i \langle F_i \rangle \psi_i + \sum_a \langle D^a \rangle \lambda^a. \tag{2.16}$$

We see that the Goldstino is a combination of the fermions that correspond to non-zero auxiliary field VEVs.

To demonstrate the basics we have seen so far, let us now turn to two examples of supersymmetry breaking. These examples will also illustrate some other general features of supersymmetry breaking.

2.4. Tree-level breaking: F-type

In this section we will study a variation of the O'Raifeartaigh model [8], with chiral fields Y_i, Z_i, and X with $i = 1, 2$, with the superpotential

$$W = X(Y_1 Y_2 - M^2) + m_1 Z_1 Y_1 + m_2 Z_2 Y_2, \tag{2.17}$$

where M and m_i are parameters with the dimension of mass. Note that the superpotential has a term that is linear in one of the fields (X). This is crucial for breaking supersymmetry at tree-level.

The original O'Raifeartaigh model is obtained by identifying $Y_1 = Y_2 = Y$, and $X_1 = X_2 = X$. We are complicating the model in order to illustrate the interplay between broken global symmetries and supersymmetry breaking, which we will get to later. But let's postpone that, and see whether the model breaks supersymmetry.

Since there are no gauge interactions in the model, we don't have to worry about D-terms, and we can turn directly to finding whether there are F-flat directions for which the potential vanishes. Equating all the F-terms to zero we have the following equations:

1 $\quad Y_1 Y_2 = M^2 \quad (F_X)$
2 $\quad XY_2 + m_1 Z_1 = 0 \quad (F_{Y_1})$
3 $\quad XY_1 + m_2 Z_2 = 0 \quad (F_{Y_2})$
4 $\quad m_1 Y_1 = 0 \quad (F_{Z_1})$
5 $\quad m_2 Y_2 = 0 \quad (F_{Z_2})$

Clearly, equations 4 and 5 clash with equation 1. There is no point for which the potential vanishes, and supersymmetry is broken. Note that it is crucial that M, m_1 and m_2 are all non-zero. If $M = 0$, there is no linear term in the superpotential, and the origin of field space is always a supersymmetric point[2]. If for example, $m_2 = 0$, we can have a solution with $Y_1 \to 0$ and $Y_2 \to \infty$, such that their product is M^2.

You may be gasping with disbelief at how simple supersymmetry breaking is. And it's true: given a superpotential, finding out whether supersymmetry is broken simply amounts to solving a system of equations. The tricky part, as we will see, is to derive the superpotential, which usually involves understanding the dynamics of the theory.

As we saw above, supersymmetry is broken in this model. Very often, this is all one can say about a model. There are many other questions one can ask, such as: Where is the minimum of the potential? Which global symmetries are preserved in this minimum? What is the ground state energy? What is the light spectrum? To answer these questions, we need to know the Kähler potential of the theory.

In fact, we have already made an implicit assumption about the Kähler potential when we determined that supersymmetry is broken. We found that some F terms are non-zero in the model, but inspecting (2.8), we see that the potential can still vanish if K_{ij} blows up. So we are assuming that the Kähler potential is well behaved. For the simple chiral model we wrote above, this is a completely innocent assumption. But in general, when we study gauge theories with complicated dynamics, this is an important caveat to keep in mind.

But let's take the tree-level Kähler potential of our toy model to be canonical. The potential is then

$$V = \left| Y_1 Y_2 - M^2 \right|^2 + \left[\left| XY_2 + m_1 Z_1 \right|^2 + m_1^2 \left| Y_1^2 \right|^2 + 1 \leftrightarrow 2 \right]. \tag{2.18}$$

---

[2]This will no longer hold when we discuss non-perturbative effects, which can give superpotential terms with negative powers of the fields.

Exercise: Show that for $m, m_2 \ll M$, the potential is minimized along

$$\langle Y_1 \rangle = v_1 \equiv \sqrt{\frac{m_2}{m_1}} \sqrt{M^2 - m_1 m_2},$$

$$Z_1 = -\frac{1}{m_1} X Y_2, \tag{2.19}$$

and similarly for $1 \leftrightarrow 2$.

Instead of an isolated minimum, there is a direction in field space for which V is constant and non-zero. This is typical of O'Raifeartaigh like models. The degeneracy is removed at the loop level. For example, in our toy model the true minimum will occur at $X = Z_i = 0$.

We now turn to a useful criterion for supersymmetry breaking [10, 11]. Suppose a theory has

1. A spontaneously broken global symmetry

2. No classical flat directions

then supersymmetry is broken.

Let us illustrate this in our toy model. As we saw above, the model has no flat directions.

The model has a $U(1)$ global symmetry, under which we can choose the charges to be

$$X(0) \;\; Y_1(1) \;\; Y_2(-1) \;\; Z_1(-1) \;\; Z_2(1). \tag{2.20}$$

Take for simplicity $m_1 = m_2 = m << M$. The ground state is at

$$\langle Y_1 \rangle = \langle Y_2 \rangle = v = \sqrt{M^2 - m^2}. \tag{2.21}$$

So the $U(1)$ is broken and there is a massless Goldstone boson, which we can parameterize as ϕ_R with

$$Y_1 = v e^{i(\phi_R + i\phi_I)},$$

$$Y_2 = v e^{-i(\phi_R + i\phi_I)}. \tag{2.22}$$

Consider the potential at $X = Z_i = 0$,

$$V = \left| Y_1 Y_2 - M^2 \right|^2 + m^2 \left(|Y_1|^2 + |Y_2|^2 \right). \tag{2.23}$$

As expected, ϕ_R drops out, but ϕ_I doesn't. However, for the supersymmetric theory with $m = 0$, ϕ_I drops out too. What we are seeing of course is that the supersymmetric theory is invariant under the "complexified" $U(1)$. With unbroken supersymmetry, the massless Goldstone boson is part of a massless chiral superfield, so there must be an *additional* massless real scalar, and together they form a complex scalar. In our example, the Goldstone is ϕ_R, and it corresponds to a compact flat direction. In the supersymmetric theory ($m = 0$), the Goldstone is accompanied by another massless scalar, ϕ_I, which corresponds to a non-compact flat direction. When $m \neq 0$, there is no non-compact flat direction and therefore no other massless scalar. Thus, the Goldstone cannot be part of a supersymmetric multiplet and supersymmetry must be broken.

In our toy example, it was easy to verify directly that supersymmetry is broken. But in some examples, where a direct analysis is impossible, it is still possible to show that there are no classical flat directions, and that the global symmetry of the model is broken, and thus to conclude that supersymmetry is broken. We will see such an example in the next lecture.

2.5. Tree-level breaking: D-type

In this section we will study the Fayet-Iliopulos model [9], in which supersymmetry is broken by a non-zero D term (and/or F term). The model has a $U(1)$ gauge symmetry. The important observation is that the auxiliary field of the $U(1)$ vector field is gauge invariant, and therefore can appear in the Lagrangian. (From the point of view of supersymmetry, we can always add an auxiliary field to the Lagrangian because its supersymmetry variation is a total derivative.) Consider then a model with chiral fields Q and $\bar{Q}$, whose $U(1)$ charges are 1 and -1 respectively, and with the Kähler potential

$$K = Q^\dagger e^V Q + \bar{Q}^\dagger e^{-V} \bar{Q} + \xi_{\text{FI}} V, \tag{2.24}$$

and superpotential

$$W = m Q \bar{Q}, \tag{2.25}$$

where V is the vector superfield.

The potential is

$$V = \frac{1}{2} g^2 \left[|Q|^2 - |\bar{Q}|^2 + \xi_{\text{FI}} \right]^2 + m^2 \left[|Q|^2 + |\bar{Q}|^2 \right]. \tag{2.26}$$

We see that the potential is never zero. For the D-part to vanish we need $\langle \bar{Q} \rangle \neq 0$, but then the F_Q term

$$\frac{\partial W}{\partial Q} = m \bar{Q} \neq 0. \tag{2.27}$$

Exercise: Show that the minimum is at

1. $\langle Q \rangle = \langle \bar{Q} \rangle = 0$ for $g^2 \xi_{FI} < m^2$. In this case the $U(1)$ is unbroken, the D-term is non-zero, but all F-terms vanish.

2. $\langle Q \rangle = 0$, $\langle \bar{Q} \rangle = v = \sqrt{2}\sqrt{\xi_{FI} - m^2/g^2}$ for $g^2 \xi_{FI}^2 > m^2$. In this case the $U(1)$ is broken, the D-term is non-zero, and one F-term is non-zero.

Exercise: Show that the Goldstino is

$$\psi_G \sim m\lambda + \frac{i}{2} g v \psi_Q, \tag{2.28}$$

where λ is the gaugino and ψ_Q is the Q-fermion. We explicitly see that the Goldstino is a combination of fermion fields whose F- or D-terms are non-zero.

2.6. Going local

So far we only discussed global supersymmetry, so let us briefly mention which parts of our discussion above are modified when we promote supersymmetry to a local symmetry. For lack of time and space, we just present here the results. Although we can't see the origin of these results, they are still useful in order to understand, at least qualitatively, what Nature looks like if it has spontaneously broken supersymmetry.

• The order parameter for F-type breaking now becomes

$$D_\phi W = \frac{\partial W}{\partial \phi} + \frac{1}{M_{Pl}^2} \frac{\partial K}{\partial \phi} W. \tag{2.29}$$

If we decouple gravity by taking the Planck scale M_{Pl} to infinity, this reduces to (2.3).

• The vacuum energy is no-longer an order parameter for supersymmetry breaking. This is very fortunate, because we certainly don't want the cosmological constant to be of the order of the supersymmetry breaking scale. The scalar potential is now (omitting D-terms, which are not modified)

$$V = e^{K/M_{Pl}^2} \left[(D_i W)^* K_{ij}^{-1} (D_j W) - \frac{3}{M_{Pl}^2} |W|^2 \right]. \tag{2.30}$$

We can always shift the superpotential by a constant, $W(\phi) \to W(\phi) + W_0$ so that $V = 0$ even when $D_i W = 0$.

• When supersymmetry is broken, the gravitino gets a mass. The Goldstino is eaten by the gravitino, and supplies the extra two degrees of freedom required for a massive gravitino.

• The supergravity multiplet contains the graviton, gravitino, and auxiliary fields. When supersymmetry is broken, the scalar auxiliary field of the supergravity multiplet acquires a VEV.

In the last lecture, when we discuss how supersymmetry breaking terms are generated for the minimal supersymmetric standard model (MSSM), we will need to know how a non-zero VEV of the supergravity scalar auxiliary field affects the MSSM fields. So we need to know how this auxiliary field couples to chiral and vector fields. It is convenient to parameterize this auxiliary field as the F-component of a *non-dynamical* chiral superfield[3]

$$\Phi = 1 + F_\Phi \theta^2. \qquad (2.31)$$

The supergravity auxiliary field then couples to chiral and vector superfields through the following rescaling of the usual Lagrangian.

$$\mathcal{L} = \int d^2\theta\, \Phi^3 W(Q) + \int d^4\theta\, \Phi^\dagger \Phi K(Q^\dagger, e^V Q) + \int d^2\theta\, \tau W^\alpha W_\alpha. \qquad (2.32)$$

It is easy to see from this that Φ is related to scale transformations. We can also see from the Lagrangian (2.32) that when supersymmetry is broken, F_Φ becomes non-zero. Equation (2.32) will be our starting point when we discuss anomaly mediated supersymmetry breaking in section 4.2.

3. Beyond tree level: dynamical supersymmetry breaking

Consider a supersymmetric gauge theory with some tree-level superpotential W_{tree}, and with a minimum at zero energy, $V_{\text{tree}} = 0$. Then the ground state energy remains zero to all orders in perturbation theory [14–16]. This follows from the "non-renormalization" of the superpotential— the tree-level superpotential is not corrected in perturbation theory, which in turn, follows from the fact that the superpotential is a holomorphic function of the fields [17]. We will not prove here this non-renormalization theorem, but we will see in detail two examples of how holomorphy and global symmetries dictate the form of the superpotential in section 3.2. It will be clear in these examples that the tree-level superpotential is not corrected radiatively.

---

[3] This field is called the chiral compensator, because it is often introduced in order to write down a superspace Lagrangian for supergravity that is manifestly invariant under Weyl-rescaling. Note that the lowest component of Φ breaks the "fake" Weyl invariance. Non-dynamical fields of this type, which are introduced in order to make the Lagrangian look invariant under some fake symmetry, are called spurions.

This leads to one of the most important results about supersymmetry breaking: If supersymmetry is unbroken at tree-level, it can only be broken by non-perturbative effects. Only the dynamics of the theory can generate a non-zero potential. This makes the study of supersymmetry breaking hard (and interesting!). However, holomorphy, which forces us to consider non-perturbative phenomena when studying supersymmetry breaking, also comes to our aid. As we will see, we can say a lot about the dynamics of supersymmetric theories based on holomorphy.

Before going on, let us pause to say a word about one kind of non-perturbative phenomenon—instantons[4], which we will encounter in the following. Instantons are classical solutions of the Euclidean Yang-Mills action that approach pure gauge for $|x| \to \infty$. Therefore, the field strength for these solutions goes to zero at infinity, and the instanton action is finite. The one-instanton action is

$$S_{\text{inst}} = \frac{1}{2g^2} \int d^4x \, F_{\mu\nu}^2 \sim \frac{8\pi^2}{g^2}, \tag{3.1}$$

where g is the gauge coupling. If there are fermions charged under the gauge group, instantons can generate a fermion interaction with strength proportional to the instanton action, $\exp(-8\pi^2/g^2)$. The gauge coupling is of course scale-dependent, and obeys at one-loop

$$\mu \frac{dg}{d\mu} = -\frac{b}{16\pi^2} g^3 \tag{3.2}$$

(in our conventions, $\mathcal{N} = 1$ $SU(N)$ with F flavors has $b = 3N - F$). So the instanton-generated interactions involve

$$\exp\left(-\frac{8\pi^2}{g^2(\mu)}\right) = \frac{\Lambda^b}{\mu^b}, \tag{3.3}$$

where Λ is the strong coupling scale of the theory. In an $SU(N)$ theory with $F = N - 1$ flavors, instantons generate fermion-scalar interactions that can be encoded by the superpotential [12]

$$W_{np} = \left(\frac{\Lambda^{3N-F}}{\det(Q \cdot \bar{Q})}\right)^{\frac{1}{N-F}} \tag{3.4}$$

(we will study this example in detail below). So instanton effects, and more generally, dynamical effects, involve the strong coupling scale of the theory.

---

[4]This is intended for students who have never heard about instantons, and would still like to follow these lectures. It is by no means a serious introduction to instantons, and I refer you to [4] for an introduction to instantons in supersymmetric gauge theories.

Going back to supersymmetry breaking, we see that if the ground state energy is zero at tree-level (unbroken supersymmetry), only dynamical effects can alter that, and therefore the full ground state energy, or supersymmetry-breaking scale, is proportional to some strong coupling scale Λ. This has a profound implication: If a theory breaks supersymmetry spontaneously, with supersymmetry unbroken at tree-level, then the supersymmetry breaking scale, or the ground state energy, is proportional to some strong coupling scale Λ,

$$E_{\text{vac}} \sim \Lambda \sim M_{UV}\, e^{-\frac{8\pi^2}{g^2(M_{UV})}}, \tag{3.5}$$

where M_{UV} is the cutoff scale of the theory, say, M_{Pl}. Thus, supersymmetry can do much more than *stabilize* the Planck-electroweak scale hierarchy. It can actually *generate* this hierarchy if it's broken dynamically [13], because the factor $e^{-\frac{8\pi^2}{g^2(M_{UV})}}$ can easily be 10^{-17}.

In general, there are three types of (dynamical) supersymmetry-breaking models.

1. In some models we can only tell that supersymmetry is broken based on indirect arguments. In particular, we have no information about the potential of the theory, and all we know is that the supersymmetry-breaking scale is of the order of the relevant strong-coupling scale.

2. In some models, we can derive the superpotential at low-energies (in variables such that the Kähler potential is non-singular), and conclude that some F-terms are non-zero. Such models are often called "non-calculable", because apart from determining that supersymmetry is broken, we cannot calculate any of the properties of the ground state (including the supersymmetry breaking scale).

   How do we determine the superpotential in these models? There are many methods, some of which we will see today. These typically involve holomorphy, global symmetries, known exact results and even Seiberg duality.

3. In some models, we can calculate the superpotential as above, but for certain ranges of parameters, the theory is weakly coupled and we can also calculate the Kähler potential. Then we can compute the supersymmetry breaking scale, the light spectrum, and other properties of the ground state.

   Roughly, these models have the following behavior. There is a tree-level superpotential W_{tree}, with some couplings λ, that lifts all flat directions (classically). Because W_{tree} is a polynomial in the fields, it vanishes at the origin, and grows for large field VEVs. On the other hand, non-perturbative effects generate a potential that is strong in the origin of field space, but decreases for large field VEVs (because the gauge symmetry is Higgsed with large scalar VEVs, so the low energy is weakly coupled). The interplay between the tree-level potential and the non-perturbatively generated potential may give a supersymmetry breaking ground state. Clearly, if we decrease the tree-level coupling λ, V_{tree} becomes smaller, so that the ground state is obtained at larger values of the field VEVs, where the theory is weakly coupled.

In the remainder of this section, we will demonstrate this through two examples out of the many known supersymmetry breaking models. We will spend most of our time studying the $3 - 2$ model. This example will illustrate how holomorphy, symmetries and known results about the superpotentials of various theories, completely determine the superpotential of the model.

3.1. Indirect analysis—SU(5) with single antisymmetric

We will now see an example of the first type of models discussed above, where there is only indirect evidence for supersymmetry breaking. We will apply here the criterion explained in section 2.4: If a theory has broken global symmetries and no flat directions, supersymmetry is broken. Our example is an $SU(5)$ gauge theory with fields $T \sim 10$, $\bar{F} \sim \bar{5}$ [10, 18]. As explained above, the D-flat directions of a gauge theory can be parametrized by the chiral gauge invariants. Since we cannot form any gauge invariants out of T and $\bar{F}$, there are no flat directions.

The global anomaly-free symmetry of the model is $G = U(1) \times U(1)_R$, with charges $T(1, 1)$ and $\bar{F}(-3, -9)$. We can now argue, based on 't Hooft anomaly matching, that G is spontaneously broken.

So let's show that G is (most likely) broken. First, the $SU(5)$ theory probably confines. (We stress that we cannot prove this, but since this $SU(5)$ is asymptotically free, with few matter fields, this is a very likely possibility.) Suppose then that the global symmetry is unbroken. Then the $SU(5)$-invariant composite fields of the confined theory should reproduce the global anomalies, $U(1)^3$, $U(1)^2U(1)_R$, etc of the original theory. Denoting the fields of the confined theory by X_i, and their charges under G by (q_i, r_i), we obtain four equations for the q_i's and r_i's. There is no simple solution to these equations. Allowing only charges below 50, we need at least 5 fields to obtain a solution. We conclude then that the global symmetry is (probably) broken. Since there are no classical flat directions, supersymmetry is (probably) broken.

The supersymmetry breaking scale is proportional to the only scale in the problem, which is the strong coupling scale of $SU(5)$.

3.2. Direct analysis: the 3 − 2 model

The $3 - 2$ model is probably the canonical example of supersymmetry breaking [11]. It is certainly one of the simplest models in the sense that it has a small gauge group $SU(3) \times SU(2)$, and relatively small field content. But it is actually not the simplest model to analyze. Still, this makes it an interesting example, and we will use it to demonstrate several important points. We will see how the superpotential is determined by holomorphy and symmetries. The basic observation we will use is that the parameters of the theory can be thought of as the VEVs of background fields. The notion of holomorphy can then be extended to these parameters.

Furthermore, this model will also demonstrate the three types of analysis detailed in the beginning of this section. We will first establish supersymmetry breaking by the indirect argument we saw in section 2.4: we will show that the model has no flat directions and a broken global symmetry. We will then derive the exact superpotential of the theory and show that it gives at least one non-zero F-term. Finally, we will choose parameters such that the minimum is calculable.

3.2.1. Classical theory

The field content of the model is $Q \sim (3, 2)$, $\bar{Q}_A \sim (\bar{3}, 1)$, $L \sim (1, 2)$ with $A = 1, 2$. We add the superpotential

$$W_{\text{tree}} = \lambda Q \cdot \bar{Q}_2 \cdot L. \tag{3.6}$$

As explained in section 2.2, we should first find the D-flat directions, and these can be parametrized by the classical gauge-invariants that we can make out of the chiral fields

$$X_A = Q \cdot \bar{Q}_A \cdot L = Q_{i\alpha} \bar{Q}_A^i L_\beta \epsilon^{\alpha\beta}, \tag{3.7}$$

$$Y = \det(Q \cdot \bar{Q}) = \epsilon^{\alpha\beta} \epsilon^{AB} (Q_{i\alpha} \bar{Q}_A^i) Q_{j\beta} \bar{Q}_B^j, \tag{3.8}$$

where i (α) is the $SU(3)$ ($SU(2)$) gauge index. To see this, it is easy to start by making $SU(3)$ invariants: $Q_\alpha \cdot \bar{Q}_A$. These are $SU(2)$ doublets, and together with the remaining doublet L_α, they can be combined into the $SU(2)$ invariants X_A and Y.

Next, we should find the subspace of the D-flat directions for which all F-terms vanish. Consider for example the requirement that the L F-term vanishes,

$$\frac{\partial W}{\partial L_\alpha} = \lambda Q_\alpha \cdot \bar{Q}_2 = 0. \tag{3.9}$$

Contracting this equation with L_α we see that $X_2 = 0$. Similarly, you can show that $X_1 = Y = 0$. Thus, there are no flat directions classically–only the origin is a supersymmetric point.

Remembering our indirect criterion of section 2.4, let's consider the global symmetry of the model. The only anomaly-free symmetry that's preserved by the superpotential (3.6), is $U(1) \times U(1)_R$, with charges $Q(1/3, 1)$, $\bar{Q}_1(-4/3, -8)$,

$\bar{Q}_2(2/3, 4)$, and $L(-1, -3)$. If we can show that this global symmetry is broken, we'll know that supersymmetry is broken.

3.2.2. Exact superpotential

So let's turn to the quantum theory. We already know that only non-perturbative effects can change the potential (and in particular "lift" the classical zero potential at the origin). We also mentioned that the tricky part is to find the proper variables, for which the Kähler potential is well behaved. Our first task is then to find such variables and derive the superpotential [19]. Let's first see if we missed any gauge invariants. The way we constructed the gauge invariants above was to contract $SU(3)$ indices first. What happens if we do it the other way around? We find one new gauge invariant

$$Z = (Q^2) \cdot (Q \cdot L) = \epsilon^{ijk} Q_{i\alpha} Q_{j\beta} \epsilon^{\alpha\beta} Q_{k\gamma} L_\delta \epsilon^{\gamma\delta}. \tag{3.10}$$

Note that $Z = 0$ classically.

We turn now to deriving the superpotential. Beyond tree level, there can be contributions to the superpotential generated by the $SU(3)$ and $SU(2)$ dynamics. To analyze these, it is useful to consider various limits.

Take first $\Lambda_3 >> \Lambda_2$, and λ much smaller than the gauge couplings. Then we have an $SU(3)$ theory with two flavors. An $SU(3)$ instanton then gives rise to the superpotential

$$W_3 = \frac{\Lambda_3^7}{Y}. \tag{3.11}$$

Below we will see that (3.11) is the most general superpotential allowed by the symmetries of the theory.

But before doing that, let's note that we can already conclude that supersymmetry is broken! As a result of the the superpotential (3.11), the ground state is at non-zero Y. But Y appears in the superpotential, so its R-charge must be non-zero (you can check that it is indeed -2). Therefore the global R-symmetry is broken, and since there are no flat directions, supersymmetry must be broken too.

Note the difference between an R- and non-R symmetry in this respect. We were able to conclude that the R symmetry is broken because a certain superpotential term is non-zero at the ground state, and any superpotential term is charged under the R-symmetry (assuming of course that there is an R symmetry that the superpotential preserves). Since the superpotential is neutral under non-R symmetries, we cannot conclude analogously that a non-R symmetry is broken.

Let us now show that the $SU(3)$ superpotential must be of the form (3.11). In fact, we will show this more generally for an $SU(N)$ gauge theory with $F < N$

flavors Q and $\bar{Q}$. The global symmetry of this theory is $SU(F)_L \times SU(F)_R \times U(1)_B \times U(1)_R$, with $Q \sim (F, 1, 1, (F - N)/F)$, and $\bar{Q} \sim (1, \bar{F}, -1, (F - N)/F)$. The superpotential must be gauge invariant, so it can only depend on the "mesons", $M_{ij} = Q_i \cdot \bar{Q}_j$ (with a slight abuse of notation, we are using now Latin indices to denote both $SU(F)_L$ and $SU(F)_R$ indices, with $i, j = 1, \ldots, F$). So $W = W(M_{ij})$.

Furthermore, the superpotential better be invariant under $SU(F)_L \times SU(F)_R$, so $W = W(\det M)$, where M stands for the meson matrix. Now $\det M$ is neutral under $U(1)_B$, but has $U(1)_R$ charge $2(F - N)$. Therefore

$$W \propto \left(\frac{1}{\det \bar{Q} \cdot Q} \right)^{\frac{1}{N-F}}. \tag{3.12}$$

The only other thing W can depend on is the $SU(N)$ scale Λ^{3N-f}, so on dimensional grounds it is of the form

$$W = \text{const} \left(\frac{\Lambda^{3N-F}}{\det \bar{Q} \cdot Q} \right)^{\frac{1}{N-F}}. \tag{3.13}$$

Note that holomorphy was crucial in this argument—without it we could make invariants such as $Q^\dagger Q$. Also note that we have just proven the non-renormalization theorem for this theory. We did not put in any tree-level superpotential, so $W_{tree} = 0$. We argued that (3.13) is the most general form of the superpotential in the quantum theory. But radiative corrections can only produce positive powers of the fields. So indeed the tree-level superpotential is not corrected radiatively.

Of course, we have only shown that the superpotential (3.13) is allowed. We haven't shown that it is actually generated, because that's much harder [12, 20]. But it is generated, by an instanton for $F = N - 1$, and by gaugino condensation for other $F < N$. Going back to the $3 - 2$ model, an $SU(3)$ instanton generates the superpotential (3.11).

Finally, we get to the $SU(2)$ dynamics. In the limit $\Lambda_2 \gg \Lambda_3$, we have $SU(2)$ with two flavors. The classical moduli space of this theory is parametrized by the "mesons" $V_{ij} = Q_i \cdot Q_j$, $V_{i4} = Q_i \cdot L$. An $SU(2)$ instanton modifies this moduli space, so that, at the quantum level, the moduli space is given by the V's subject to the constraint

$$W = A \, (\epsilon^{i_1 i_2 i_3 i_4} V_{i_1 i_2} V_{i_3 i_4} - \Lambda_2^4) = A(Z - \Lambda_2^4). \tag{3.14}$$

where A is a Lagrange multiplier.

We can now use these different limits to obtain the full superpotential of the model, which is a function

$$W = W(X_A, Y, Z, \lambda, \Lambda_3^7, \Lambda_2^4). \tag{3.15}$$

As in the $SU(N)$ example above, we want to use the global symmetry, which in this case is $U(1) \times U(1)_R$ to constrain this function. However λ, Λ_i are of course neutral under this symmetry, so that wouldn't work. Note that in our $SU(N)$ example this was not a problem, because there was only one parameter in the theory, Λ, and at the last step we could constrain the way Λ enters on dimensional grounds. So we need symmetries under which λ, Λ_i are charged, i.e., global symmetries that are broken by the tree-level superpotential, and/or have global anomalies. In particular, we want to treat λ as a background field, or spurion, and use the fact that the superpotential cannot depend on $\lambda^\dagger$.

The simplest symmetries to consider are the following: Introduce $U(1)_Q$ under which Q has charge 1, with all other fields neutral. Under this symmetry, λ has charge -1, Λ_3^7 has charge 2, and Λ_2^4 has charge 3. It is probably clear why λ has charge -1. We are introducing a "fake" symmetry and treating λ as a background field charged under this symmetry. For the superpotential to be invariant under $U(1)_Q$, λ must have charge -1. Let's now see why we can think of Λ_3^7 as having charge 2. The $U(1)_Q$ symmetry is anomalous. Therefore, if we rotate Q by this symmetry, we will shift the $SU(3)$ θ-angle. The shift is proportional to the number of $SU(3)$ fermion zero modes charged under the global symmetry. This number is 2, because Q also has an $SU(2)$ index. Finally, recall that

$$\Lambda^b = \mu^b e^{-\frac{8\pi^2}{g^2(\mu)} + i\theta},$$

(3.16)

so under the anomalous rotation, Λ_3^7 has charge 2.

Exercise: Introduce similarly $U(1)_{\bar{Q}_1}$, $U(1)_{\bar{Q}_2}$, and $U(1)_L$, and compute the charges of λ, Λ_3^7, Λ_2^4 under these symmetries. Then use these symmetries, together with $U(1)_R$, to show that the superpotential is of the form

$$W = \frac{\Lambda_3^7}{Y} f(t_1, t_2) + A(Z - \Lambda_2^4) g(t_1, t_2),$$

(3.17)

where f and g are general functions of

$$t_1 = \frac{\lambda X_2 Y}{\Lambda_3^7}, \quad t_2 = \frac{Z}{\Lambda_2^4}.$$

(3.18)

Now consider the limit

$$X_2, \Lambda_3, \lambda \to 0.$$

(3.19)

In this limit, t_1 and t_2 can take any value, and we know

$$W \to A(Z - \Lambda_2^4).$$

(3.20)

Therefore $g(t_1, t_2) \equiv 1$. Now take

$$Y \to \infty, \quad \lambda \to 0. \tag{3.21}$$

Again t_1 and t_2 can take any value. But for large Y VEVs, the gauge symmetry is completely Higgsed with the gauge bosons very heavy. The low-energy theory is therefore weakly coupled, and the superpotential is given by

$$W = \frac{\Lambda_3^7}{Y} + \lambda X_2, \tag{3.22}$$

so that $f(t_1, t_2) = 1 + t_1$. We then have the full superpotential

$$W = \frac{\Lambda_3^7}{Y} + A(Z - \Lambda_2^4) + \lambda X_2. \tag{3.23}$$

Since

$$\frac{\partial W}{\partial X_2} \neq 0, \tag{3.24}$$

supersymmetry is broken.

We assumed here that the Kähler potential is non-singular in X_2. This is justified because the theory is driven away from the origin by the first term of (3.22), so that the gauge symmetry is completely broken. We can then integrate out the heavy gauge bosons, and the low energy theory can be described in terms of the gauge invariants X_A, Y and Z. Note that, as a result, the tree-level superpotential becomes linear in the fields, just as in the O'Raifeartaigh model.

Finally, we note that we derived the non-renormalization theorem once again. The tree-level superpotential is not modified by perturbative corrections.

3.2.3. Calculable minimum

We established supersymmetry breaking by deriving the full superpotential of the theory. We can now choose parameters for which the minimum is calculable. For $\Lambda_3 \gg \Lambda_2$, $\lambda \ll 1$, Y gets a large VEV, and the gauge symmetry is completely broken. Because of the superpotential (3.22), Z gets mass and we can integrate it out[5], to get

$$W = \frac{\Lambda_3^7}{Y} + \lambda X_2. \tag{3.25}$$

---

[5]Because Z vanishes classically, the term $Z^\dagger Z$ in the Kähler potential is suppressed by some power of Λ_2/v, where v is the typical VEV. Therefore the Z mass is enhanced by v/Λ_2, and we can indeed integrate it out.

Since the theory is weakly coupled in this limit, the Kähler potential is just the canonical Kähler potential

$$Q^\dagger Q + \bar{Q}_A^\dagger \bar{Q}_A + L^\dagger L, \qquad (3.26)$$

and we can calculate the potential, either in terms of the elementary fields or in terms of the classical gauge invariants X_A and Y (to use the latter, one needs to project (3.26) on the classical moduli space). In particular, it is easy to show that in terms of elementary fields, the typical VEV is $v \sim \lambda^{-1/7}\Lambda_3$ and $E_{vac} \sim \lambda^{5/14}\Lambda_3$. This demonstrates the general features of calculable minima mentioned at the beginning of this section. As we lower the superpotential coupling λ, the ground state is driven to large VEVs, for which the theory is weakly coupled. Note also that, as expected, the supersymmetry breaking scale is proportional to to the relevant strong coupling scale, (Λ_3 in this limit) and to some positive power of the Yukawa coupling λ.

We end this section with a few comments.

First, in this example, we were able to derive the exact superpotential of the theory and conclude from it that supersymmetry is broken. It would have been much easier to just consider the limit $\Lambda_3 \gg \Lambda_2$, $\lambda \ll 1$, and show that supersymmetry is broken as we did in section 3.2.3. In general, even if we can only establish supersymmetry breaking for some range of parameters, (say $\Lambda_3 \gg \Lambda_2$, $\lambda \ll 1$), we expect this to hold generally, because there should not be any phase transition as we vary the parameters of the theory. However, the details of the breaking, such as the supersymmetry-breaking scale, can be different.

Second, we used two examples to demonstrate the analysis of supersymmetry breaking. There is a long list of models that we know break supersymmetry [5]. The analysis of some of these involves many interesting ingredients and phenomena: quantum removal of flat directions, supersymmetry breaking without R symmetry, and the use of a Seiberg-dual theory to establish supersymmetry breaking, to name but a few. Unfortunately, there is no fundamental organizing principle that would allow us to systematically classify known models, or to guide us in the quest for new ones.

4. Mediating the breaking

We now know that supersymmetry can be broken, and that if broken dynamically, its scale is proportional to some strong coupling scale, Λ, which can be much lower than the Planck scale. In fact, this is all we need from the previous sections in order to discuss the mediation of supersymmetry breaking to the MSSM.

The MSSM contains many soft supersymmetry-breaking terms: scalar masses, gaugino masses, A-terms etc. This is often cited as a drawback of supersymmetry. But in any sensible theory, the soft terms must be generated by some underlying theory, and this underlying theory may have very few parameters. In fact, as we will see, if the soft terms are generated by anomaly-mediation, they are controlled by a *single* new parameter—the overall supersymmetry breaking scale.

The MSSM soft terms were discussed in detail in the lectures of Wagner, Masiero and Nir. As we saw in these lectures, the soft terms determine the way we would observe supersymmetry in collider experiments, and are severely constrained by flavor changing processes. Here we will discuss several mechanisms for generating the soft terms

• Mediation by Planck-suppressed higher-dimension operators (a.k.a. "gravity mediation")

• Anomaly mediated supersymmetry breaking (AMSB)

• Gauge mediated supersymmetry breaking (GMSB)

We will focus on AMSB, because it is always present, and because it is probably the most tricky. In addition, there are fewer reviews of AMSB than of GMSB.

Suppose then that the fundamental theory contains, in addition to the MSSM, some fields and interactions that break supersymmetry (these are usually referred to as a supersymmetry breaking "sector", and the MSSM is sometimes referred to as the "visible sector"). We can think of the supersymmetry breaking sector as the $3-2$ model, or the $SU(5)$ model we saw above, or even as a model with tree-level breaking, if we don't mind having very small parameters in the Lagrangian. The question is then: What do we need to do in order to communicate supersymmetry breaking to the MSSM, namely, generate the MSSM soft terms?

4.1. Mediating supersymmetry-breaking by Planck-suppressed operators

The short answer to this question is—nothing. The effective field theory below the Planck scale generically contains higher dimension operators that are generated when heavy states with masses of order the Planck scale are integrated out. These higher dimension terms couple the MSSM fields to the fields of the supersymmetry breaking sector. Denoting the MSSM matter superfields by Q_i, where i is a generation index, and a field of the supersymmetry breaking sector by X, the Kähler potential is then of the form

$$Q_i^\dagger Q_i + X^\dagger X + c_{ij} \frac{1}{M_{Pl}^2} X^\dagger X Q_i^\dagger Q_j + \cdots, \tag{4.1}$$

where c_{ij} are order-one coefficients. If X has a non-zero F-term, the last term of (4.1) gives rise to scalar masses for the Q's:

$$\left(m_{\tilde{Q}}^2 \right)_{ij} = c_{ij} \left(\frac{F_x}{M_{Pl}} \right)^2. \tag{4.2}$$

For the scalar masses to be around the electroweak scale we need

$$\frac{F_x}{M_{Pl}} \sim 100 \text{GeV}, \tag{4.3}$$

or $\sqrt{F_x} \sim 10^{11}\text{GeV}$. So it is very easy to generate the required scalar masses. However, there is no reason for the coefficients c_{ij} to be flavor blind. The fundamental theory above the Planck scale is certainly not flavor blind, because it must generate the fermion masses we observe. Generically then, this mechanism, which is usually referred to as "gravity mediation", leads to large flavor changing neutral currents. There are some solutions to this problem. One solution, which we heard about in Nir's lecture, uses flavor symmetries, with different generation fields transforming differently under the symmetry, leading to "alignment" of the fermion and sfermion mass matrices [21].

In fact, the name "gravity-mediation" is misleading, because the mass terms are not generated by purely gravitational interactions. As we saw above, they are mediated by heavy string states which couple to the MSSM and to supersymmetry-breaking fields with unknown couplings.

Can we suppress these dangerous contributions to the masses? One way to do this, is to suppress the coefficients c_{ij}. It is easy to do this if there are extra-dimensions [22]. For example, if the MSSM is confined to a 3-brane, and the supersymmetry breaking sector lives on a different 3-brane, separated by an extra dimension, then tree-level couplings of the the two sectors are exponentially suppressed, $c_{ij} \sim \exp(-MR)$, where R is the distance between the branes, and M is the mass of the heavy state that mediates the coupling. Such models are called sequestered models[6].

Assume then that tree-level couplings of the MSSM and supersymmetry breaking sector are negligible. As it turns out, gravity automatically generates soft masses for the MSSM fields through the scale anomaly of the standard model. This time, the mediation of supersymmetry breaking is purely gravitational.

4.2. Anomaly mediated supersymmetry breaking

As we said above, we are assuming that apart from the MSSM, the theory contains a supersymmetry-breaking sector. Therefore, as mentioned in section 2.6, the scalar auxiliary field of the supergravity multiplet develops a non-zero VEV F_ϕ. The couplings of this auxiliary field to the MSSM are contained in eqn (2.32) which we repeat here for convenience

$$\mathcal{L} = \int d^2\theta \, \Phi^3 W(Q) + \int d^4\theta \, \Phi^\dagger \Phi K(Q^\dagger, e^V Q) + \int d^2\theta \, \tau \, W^\alpha W_\alpha. \tag{4.4}$$

---

[6]The c_{ij}'s can be suppressed in 4d theories too, using "conformal sequestering" [23].

Here Q denotes collectively the MSSM matter fields, and V the gauge fields. Note that because

$$\Phi = 1 + F_\Phi \theta^2, \tag{4.5}$$

F_ϕ has dimension one. We could instead write it as $F_\Phi = F/M_{Pl}$, where F has the usual dimension-2.

At first sight, it seems that the non-zero F_Φ has no effect on the MSSM fields, because we can rotate it away by rescaling

$$Q \to \Phi^{-1} Q. \tag{4.6}$$

Note however that this assumes that the superpotential is trilinear in the fields, as is true for the MSSM apart from the μ term. If the superpotential contains a quadratic term then the rescaling gives, schematically,

$$\int d^2\theta \, \Phi^3 [Q^3 + M Q^2] \to \int d^2\theta [Q^3 + M\Phi Q^2]. \tag{4.7}$$

Thus, an explicit mass parameter would pick up one power of Φ

$$M \to M\Phi = M(1 + F_\Phi \theta^2). \tag{4.8}$$

We will come back to this point often in the following. But as we said above, the MSSM classical Lagrangian is scale invariant—no mass parameter appears, and therefore the non-zero F_Φ has no effect.

This scale invariance is lost of course when we include quantum effects. The gauge and Yukawa couplings become scale dependent, and the dependence is controlled by the relevant β functions. We now have an explicit mass scale—the cut-off scale Λ_{UV}. As we saw above, this mass scale will pick up powers of Φ. Since the latter has a non-zero θ^2 component, we will obtain supersymmetry breaking masses for the MSSM fields [22, 24].

Consider first gaugino masses. These will come from

$$\int d^2\theta \, \frac{1}{4g^2(\frac{\mu}{\Lambda_{UV}})} W^\alpha W_\alpha. \tag{4.9}$$

Since Λ_{UV} is rescaled by Φ^7, (4.9) becomes

$$\int d^2\theta \, \frac{1}{4g^2(\frac{\mu}{\Lambda_{UV}\Phi})} W^\alpha W_\alpha = \int d^2\theta \left[\frac{1}{4g_{UV}^2} + \frac{b}{32\pi^2} \ln \frac{\mu}{\Lambda_{UV}\Phi} \right] W^\alpha W_\alpha, \tag{4.10}$$

---

[7] The simplest way to see this is to think of Λ_{UV} as the mass of regulator fields.

where b is the one-loop β function coefficient for the gauge coupling. Substituting (4.4) and expanding in θ, we get

$$\frac{1}{4g^2(\mu)} W^\alpha W_\alpha \Big|_{\theta^2} - \frac{b}{32\pi^2} F_\Phi \lambda^\alpha \lambda_\alpha. \tag{4.11}$$

The last term is a mass term for the gaugino. Going to canonical normalization for the gaugino,

$$m_\lambda(\mu) = \frac{b}{2\pi} \alpha(\mu) F_\Phi. \tag{4.12}$$

Exercise: scalar masses. Repeat this analysis for the scalars. Start from the Kähler potential

$$\int d^4\theta Z \left(\frac{\mu}{\Lambda_{UV}} \right) Q^\dagger Q, \tag{4.13}$$

where Z is the wave-function renormalization. After the rescaling this becomes

$$\int d^4\theta Z \left(\frac{\mu}{\Lambda_{UV}(\Phi^\dagger \Phi)^{1/2}} \right) Q^\dagger Q, \tag{4.14}$$

(The combination $(\Phi^\dagger \Phi)^{1/2}$ appears because Z is real). Expand this to obtain

$$\begin{aligned} m_0^2(\mu) &= -\frac{1}{4} \frac{\partial \gamma(\mu)}{\partial \ln \mu} |F_\Phi|^2 \\ &= \frac{1}{4} \left[\frac{b_g}{2\pi} \alpha_g^2 \frac{\partial \gamma}{\partial \alpha_g} + \frac{b_\lambda}{2\pi} \alpha_\lambda^2 \frac{\partial \gamma}{\partial \alpha_\lambda} \right] |F_\Phi|^2, \end{aligned} \tag{4.15}$$

where

$$\gamma(\mu) = \frac{\partial \ln Z(\mu)}{\partial \ln(\mu)} \tag{4.16}$$

is the anomalous dimension, $\alpha_g = g^2/(4\pi)$, $\alpha_\lambda = \lambda^2/(4\pi)$, λ is a Yukawa coupling, and β_λ is its one-loop β function coefficient.

The AMSB masses (and A-terms, which can be derived similarly) are determined by the MSSM couplings, beta functions, and anomalous dimensions. The only new parameter that appears is F_Φ, which sets the overall scale. Since gaugino masses are generated at one-loop, and scalar masses squared are generated at two loops, the masses are comparable, of order a loop factor times F_Φ. Furthermore, the scalar masses are largely generation blind. Apart from third generation fields, for which flavor-changing constraints are rather weak, the masses are dominated by gauge contributions. Thus, FCNC's are not a problem. Finally, the expressions (4.12) and (4.15) are valid at any scale, and in particular, at low energies. Thus, AMSB is extremely elegant. Unfortunately, it predicts negative masses-squared for the sleptons, because $\beta_g < 0$ for $SU(2)$ and $SU(3)$.

So minimal AMSB does not work. Furthermore, we can already guess, from the fact that the soft terms can be calculated directly at low energies, that it will not be easy to modify them by introducing new physics at some high scale. We will explain this in detail in section 4.4. But before doing that, let's pause to consider gauge mediation. We will then use the results of this section together with the results of the next section to tackle the question of "fixing" anomaly mediation in section 4.4.

4.3. Gauge mediated supersymmetry breaking

In the last two sections, we assumed that the supersymmetry breaking sector and the MSSM only couple indirectly, either through higher-dimension operators, or through the supersymmetry breaking VEV of the supergravity multiplet. In this section, we will instead extend the MSSM, and couple it, mainly through gauge (but typically also through Yukawa) interactions, to the supersymmetry breaking sector. The main ingredient of gauge mediation are new fields, that are charged under the standard-model gauge group, and couple directly to the supersymmetry breaking sector, so that they get supersymmetry-breaking mass splittings at tree level. These fields are usually called the "messengers" of supersymmetry breaking. The MSSM scalars and gauginos then obtain supersymmetry-breaking mass splittings at the loop level, from diagrams with messengers running in the loop.

We cannot go into detailed model building here. Instead, we will concentrate on the simplest set of messenger fields. Furthermore, to simplify the discussion, we will focus on the $SU(3)$ gauge interactions, and ignore $SU(2) \times U(1)$. Our discussion can be trivially extended to include these.

We then consider a "vector-like" pair of messengers, chiral superfields Q_3 and $\bar{Q}_3$, transforming as a 3 and $\bar{3}$ of $SU(3)$ respectively [25, 26]. The messengers couple to the supersymmetry-breaking sector through the superpotential

$$W_{mess} = X Q_3 \bar{Q}_3, \tag{4.17}$$

where X is a standard-model singlet, with a non-zero VEV, $\langle X \rangle = M$ and F-term VEV, $\langle F_X \rangle = F \neq 0$. The Q_3, $\bar{Q}_3$ fermions then get mass M. The scalar mass terms are of the form

$$M^2|\tilde{Q}_3|^2 + M^2|\tilde{\bar{Q}}_3|^2 + \left(F \tilde{Q}_3 \tilde{\bar{Q}}_3 + h.c. \right), \tag{4.18}$$

so that

$$m_{Q_3}^2 = M^2 \pm F. \tag{4.19}$$

The gluinos then get mass at one loop (with the Q_3 scalar and fermion running in the loop)

$$m_\lambda = \frac{\alpha_3}{4\pi}\frac{F}{M} + \mathcal{O}\left(\frac{F}{M^2}\right)^2,$$ (4.20)

and the squarks get masses at two loops,

$$m_0^2 \sim \left(\frac{\alpha_3}{4\pi}\right)^2 \left(\frac{F}{M}\right)^2 + \mathcal{O}\left(\frac{F}{M^2}\right)^4,$$ (4.21)

We will see how to calculate these masses in the following.

The masses only depend on the SM gauge couplings, and are therefore flavor blind, so that there are no FCNC's. The gaugino and scalar masses are again comparable, and given by a loop factor times F/M. We therefore want F/M to be around $10^4 - 10^5$GeV. For M lower than, roughly, 10^{16}GeV, the M_{Pl} suppressed contributions we saw in section 4.1 are negligible. They would contribute soft masses of the order of F/M_{Pl}, at least two orders of magnitude below the gauge-mediated masses. (The AMBS masses are smaller by a loop factor.)

We will now see a nice trick [27] for calculating the GMSB soft masses, to leading order in the supersymmetry breaking, F/M^2. In the model we considered above, the masses are generated when the messengers are integrated out at $\langle X \rangle = M$. The effective theory for the gluinos below the messenger scale depends on M through the gauge coupling,

$$\mathcal{L} = \int d^2\theta \, \frac{1}{4g^2(\mu)} W^\alpha W_\alpha,$$ (4.22)

with

$$\frac{1}{g^2(\mu)} = \frac{b_H}{8\pi^2}\ln\frac{X}{\Lambda_{\rm UV}} + \frac{b_L}{8\pi^2}\ln\frac{\mu}{X},$$ (4.23)

where b_H is the one-loop beta-function coefficient above M (MSSM $+Q_3 + \bar{Q}_3$) and b_L is the one-loop beta function coefficient below M (MSSM). The key point is that we promoted the VEV of X to the field X. Since $X = M + \theta^2 F$ the situation is completely analogous to what we had in the previous section. We can Taylor expand in θ to get the gaugino mass

$$m_\lambda(\mu) = \frac{\alpha(\mu)}{4\pi}(b_L - b_H)\frac{F}{M}.$$ (4.24)

In our case, $b_L - b_H = 1$, so we recover (4.20). Similarly, starting with the quark kinetic term we can essentially repeat the derivation of scalar masses in AMSB, to get (4.21).

This concludes our short review of gauge mediation.

4.4. How NOT to fix AMSB

As we saw above, minimal AMSB gives rise to tachyonic sleptons. One might try to modify the slepton masses by adding some new physics at a high scale. We will now show that this has no effect on the masses at low scales. We assume here that the only source of supersymmetry breaking in the visible sector is anomaly mediation. For simplicity let us take the new fields to be the vector-like pair Q_3, $\bar{Q}_3$ of the previous section. We also add the superpotential

$$W = M Q_3 \bar{Q}_3. \tag{4.25}$$

Now let us calculate the AMSB masses at low energies below M. For simplicity, we will consider gaugino masses only, but a similar discussion applies for scalar masses and A terms. Just above the scale M, the gaugino masses are given by the usual AMSB prediction (4.12)

$$m_\lambda(\mu) = \frac{b_H}{2\pi} \alpha(\mu) F_\Phi \qquad \text{for} \quad \mu > M, \tag{4.26}$$

where b_H is the beta function coefficient for the MSSM $+ Q_3, \bar{Q}_3$, $b_H = b_{\text{MSSM}} + 1$.

At the scale M, we need to integrate out the heavy fields. But because the superpotential (4.25) contains an explicit mass parameter, these fields get super-symmetry-breaking mass splittings at tree-level

$$W = M Q_3 \bar{Q}_3 \rightarrow \Phi M Q_3 \bar{Q}_3 = (1 + F_\Phi \theta^2) M Q_3 \bar{Q}_3, \tag{4.27}$$

with the fermions at M, and the scalars at $m^2 = M^2 \pm M F_\Phi$. So Q_3 and $\bar{Q}_3$ behave just like the messengers of gauge mediation! We can calculate their contribution to the gaugino masses just as we did in the previous section. Clearly, the effect of this contribution is to precisely cancel the $Q_3 \bar{Q}_3$ part in b_H, so that below M, the gaugino mass is

$$m_\lambda(\mu) = \frac{b_{\text{MSSM}}}{2\pi} \alpha(\mu) F_\Phi \quad \text{for} \quad \mu < M, \tag{4.28}$$

as in minimal AMSB. The heavy fields decouple completely and have no effect on the soft masses [22, 28, 29].

Note that it was crucial here that the new fields get mass in a supersymmetric manner. To emphasize this, let's give an even simpler argument for the decoupling. Consider the low-energy theory below M,

$$\int d^2\theta \, \tau(\mu, M, \Lambda_{\text{UV}}) \, W^\alpha W_\alpha. \tag{4.29}$$

On dimensional grounds

$$\tau = \tau \left(\frac{\mu}{\Lambda_{UV}}, \frac{M}{\Lambda_{UV}} \right). \tag{4.30}$$

Rescaling explicit mass scales by Φ

$$\tau \left(\frac{\mu}{\Lambda_{UV}}, \frac{M}{\Lambda_{UV}} \right) \rightarrow \tau \left(\frac{\mu}{\Lambda_{UV}\Phi}, \frac{M\Phi}{\Lambda_{UV}\Phi} \right) = \tau \left(\frac{\mu}{\Lambda_{UV}\Phi}, \frac{M}{\Lambda_{UV}} \right). \tag{4.31}$$

The Φ dependence cancels out completely in M so we recover the minimal AMSB prediction. Note that the cancellation only holds to leading order in F_Φ. The reason is that the AMSB masses are given fully by (4.12) and (4.15), with no corrections at higher order in F_Φ. In contrast, the "GMSB" contributions from integrating out Q_3 and $\bar{Q}_3$, do contain higher order corrections, that are not captured by the trick we saw in the previous section.

The same discussion applies to different heavy thresholds, and in particular those associated with D terms, which have attracted some attention lately [30]. The basic idea [29] is to get slepton masses by adding a new $U(1)$ symmetry, under which the MSSM matter fields are charged. Probably the simplest model [29] involves new fields $h_\pm$, $\xi_\pm$, with charges ± 1 under the $U(1)$, as well as gauge singlets n_i, $i = 1, 2$, and S, with the superpotential

$$W = S(\lambda h_+ h_- - M^2) + y_1 n_1 h_+ \xi_- + y_2 n_2 h_- \xi_+. \tag{4.32}$$

Because of the first term, h_+ and h_- obtain VEVs and break the $U(1)$. All new fields get mass either by the Higgs mechanism or through the superpotential. With no supersymmetry breaking, h_+ and h_- get equal VEVs. However, assuming that there is some supersymmetry breaking sector, all fields get supersymmetry breaking masses through AMSB. In particular, for $y_1 \neq y_2$, h_1 and h_2 have different soft masses and therefore different VEVs, so that the $U(1)$ D-term is non-zero. If the sleptons are charged under the $U(1)$, one might naively think that the D term affects the slepton masses. But as explained in [29], this is not the case. The model described above has no effect on the soft masses at low energy, to leading order in the supersymmetry breaking, F/M^2. In [29], the surviving F^4/M^2 contributions were used in order to generate acceptable slepton masses, using the fact that these enter scalar masses-squared at *one*-loop. The scale M was generated dynamically from F, so that it was roughly two orders of magnitude (an inverse loop-factor) above F.

To conclude, we see that we cannot modify the AMSB predictions at leading order in F_Φ using new heavy thresholds that get mass in the limit of unbroken supersymmetry. Clearly then, there are two possible approaches to fixing AMSB models: One is to use higher order terms in the supersymmetry breaking F_Φ. The second is to introduce thresholds for which some fields remain light in the limit of unbroken supersymmetry.

Acknowledgements

I thank the organizers, Stephane Lavignac and Dmitri Kazakov, for running such a smooth and enjoyable school. And I thank the students, who asked many good questions, and made giving these lectures fun.

Appendix A. Supersymmetry basics

It is convenient to write the Lagrangian of a 4d, $\mathcal{N} = 1$ supersymmetric gauge theory using superspace, with anti-commuting coordinates θ_α, $\bar{\theta}^{\dot\alpha}$, $\alpha = 1, 2$. Matter fermions and scalars combine into chiral superfields,

$$\phi = \tilde{\phi} + \sqrt{2}\theta\psi + \theta^2 F, \tag{A.1}$$

and gauge bosons and gauginos combine into vector superfields

$$V = -\theta\sigma^\mu\bar{\theta}A_\mu + i\theta^2\bar{\theta}\bar{\lambda} - i\bar{\theta}^2\theta\lambda + \frac{1}{2}\theta^2\bar{\theta}^2 D, \tag{A.2}$$

in Wess-Zumino gauge.

The Lagrangian is given by

$$\mathcal{L} = \int d^4\theta \, K(\phi^\dagger, e^V\phi) + \int d^2\theta \, W(\phi) + \text{h.c.} + \frac{1}{4g^2}\int d^2\theta \, W^\alpha W_\alpha + \text{h.c.} \tag{A.3}$$

Here $W_\alpha = -\frac{1}{4}\bar{D}\bar{D}D_\alpha V$, where D_α is the superspace derivative, contains the gauge field strength $F_{\mu\nu}$. The first term in (A.3) contains the Kähler potential K, and gives the matter kinetic terms, and the second term contains the superpotential W, which is holomorphic in the fields: it is a function of ϕ but not $\phi^\dagger$. The auxiliary fields are given by

$$F_i = \frac{\partial W}{\partial \phi_i}, \tag{A.4}$$

$$D^a = \sum_i \tilde{\phi}_i^\dagger T^a \tilde{\phi}_i. \tag{A.5}$$

We will often use R symmetries. Under a $U(1)_R$ transformation with parameter α, a chiral superfield of charge q rotates as

$$\phi(x, \theta, \bar{\theta}) \rightarrow e^{-iq\alpha}\phi(x, e^{i\alpha}\theta, e^{-i\alpha}\bar{\theta}), \tag{A.6}$$

so that different components transform differently. It is easy to see from (A.3) that the superpotential W has R-charge 2 in these conventions.

When discussing non-perturbative effects, we will also be interested in the $F^{\mu\nu}\tilde{F}_{\mu\nu}$ term, so it's useful to replace the $W^\alpha W_\alpha$ term above by

$$\int d^2\theta\, \tau\, W^\alpha W_\alpha + \text{h.c.}, \tag{A.7}$$

where τ is the "holomorphic coupling",

$$\tau = \frac{1}{4g^2} - \frac{i\,\theta_{YM}}{\pi}. \tag{A.8}$$

References

[1] J. Wess and J. Bagger, *Supersymmetry and supergravity* (Princeton University Press, 1992).
[2] K. A. Intriligator and N. Seiberg, Nucl. Phys. Proc. Suppl. **45BC** (1996) 1 [hep-th/9509066].
[3] M. A. Shifman and A. I. Vainshtein, hep-th/9902018.
[4] J. Terning, hep-th/0306119.
[5] Y. Shadmi and Y. Shirman, Rev. Mod. Phys. **72** (2000) 25 [hep-th/9907225].
[6] E. Poppitz and S. P. Trivedi, Ann. Rev. Nucl. Part. Sci. **48** (1998) 307 [hep-th/9803107].
[7] G. F. Giudice and R. Rattazzi, Phys. Rept. **322** (1999) 419 [hep-ph/9801271].
[8] L. O'Raifeartaigh, Nucl. Phys. B **96** (1975) 331.
[9] P. Fayet and J. Iliopoulos, Phys. Lett. B **51** (1974) 461.
[10] I. Affleck, M. Dine and N. Seiberg, Phys. Lett. B **137** (1984) 187.
[11] I. Affleck, M. Dine and N. Seiberg, Nucl. Phys. B **256** (1985) 557.
[12] I. Affleck, M. Dine and N. Seiberg, Nucl. Phys. B **241** (1984) 493.
[13] E. Witten, Nucl. Phys. B **188** (1981) 513.
[14] S. Ferrara, J. Iliopoulos and B. Zumino, Nucl. Phys. B **77** (1974) 413.
[15] J. Wess and B. Zumino, Phys. Lett. B **49** (1974) 52.
[16] M. T. Grisaru, W. Siegel and M. Rocek, Nucl. Phys. B **159** (1979) 429.
[17] N. Seiberg, Phys. Lett. B **318** (1993) 469 [hep-ph/9309335].
[18] Y. Meurice and G. Veneziano, Phys. Lett. B **141** (1984) 69.
[19] K. A. Intriligator and S. D. Thomas, Nucl. Phys. B **473** (1996) 121 [hep-th/9603158].
[20] S. F. Cordes, Nucl. Phys. B **273** (1986) 629.
[21] Y. Nir and N. Seiberg, Phys. Lett. B **309** (1993) 337 [hep-ph/9304307].
[22] L. Randall and R. Sundrum, Nucl. Phys. B **557** (1999) 79 [hep-th/9810155].
[23] M. Luty and R. Sundrum, Phys. Rev. D **67** (2003) 045007 [hep-th/0111231].
[24] G. F. Giudice, M. A. Luty, H. Murayama and R. Rattazzi, JHEP **9812** (1998) 027 [hep-ph/9810442].
[25] M. Dine, A. E. Nelson and Y. Shirman, Phys. Rev. D **51** (1995) 1362 [hep-ph/9408384].
[26] M. Dine, A. E. Nelson, Y. Nir and Y. Shirman, Phys. Rev. D **53** (1996) 2658 [hep-ph/9507378].
[27] G. F. Giudice and R. Rattazzi, Nucl. Phys. B **511** (1998) 25 [hep-ph/9706540].

[28] A. Pomarol and R. Rattazzi, JHEP **9905** (1999) 013 [hep-ph/9903448].

[29] E. Katz, Y. Shadmi and Y. Shirman, JHEP **9908** (1999) 015 [hep-ph/9906296].

[30] R. Harnik, H. Murayama and A. Pierce, JHEP **0208** (2002) 034 [hep-ph/0204122].

Course 4

EXTRA DIMENSIONS: A PRIMER

Valery A. Rubakov

Institute for Nuclear Research of the Russian Academy of Sciences, 60th October Anniversary Prospect, 7a, Moscow, 117312, Russia

D. Kazakov, S. Lavignac and J. Dalibard, eds.
Les Houches, Session LXXXIV, 2005
Particle Physics Beyond the Standard Model

Contents

183

1. Introduction

The possibility that our space has more than three spatial dimensions has been attracting continuing interest for many years. Strong motivation for considering space as multi-dimensional comes from theories which incorporate gravity in a reliable manner — string theory and M-theory: almost all their versions are naturally and/or consistently formulated in space-time of more than four dimensions. In parallel to developments in the fundamental theory, studies along more phenomenological lines have lead to new insights on whether and how extra dimensions may manifest themselves, and whether and how they may help to solve long-standing problems of particle theory (notably, the hierarchy problem).

An important issue in multi-dimensional theories is the mechnaism by which extra dimensions are hidden, so that the space-time is effectively four-dimensional for us. Until fairly recently, the main emphasis was put on theories of Kaluza–Klein type, where extra dimensions are compact and essentially homogeneous. In this picture, it is the compactness of extra dimensions that ensures that space-time is effectively four-dimensional at distances exceeding the compactification scale (size of extra dimensions). Hence, the size of extra dimensions must be microscopic; a "common wisdom" was that this size was roughly of the order of the Planck scale (although compactifications at the electroweak scale were also considered since quite some time ago, see, e.g., [1–3]). With the Planck length $l_{Pl} \sim 10^{-33}$ cm and the corresponding energy scale $M_{Pl} \sim 10^{19}$ GeV, probing extra dimensions directly appeared hopeless.

More recently, however, the emphasis has shifted towards "brane world" picture which assumes that ordinary matter is trapped in one way or another to a three-dimensional submanifold — brane — embedded in fundamental multi-dimensional space. In the brane world scenario, extra dimensions may be large, and even infinite; we shall see that they may then have experimentally observable effects.

Yet another possibility is a combination of the Kaluza–Klein and brane-world scenarios. This is particularly interesting for warped extra dimensions [4, 5].

Certainly, the potential detectability of large and infinite extra dimensions is one of the reasons of why they are interesting. Another reason is that lower-dimensional manifolds, p-branes, are inherent in string/M-theory. Some kinds of p-branes are capable of carrying matter fields; for example, D-branes have gauge

185

fields residing on them (for a review, see Ref. [6]). Hence, the general idea of brane world appears naturally in M-theory context, and, indeed, numerous realistic brane-world models based on M-theory have been proposed since [7,8]. Even though the phenomenological models to be discussed in these lectures may have nothing to do with M-theory p-branes, one hopes that some of their properties will have counterparts in the fundamental theory. We note in this regard that the term "brane" has quite different meaning in different contexts; we shall use this term for any three-dimensional submanifold to which ordinary matter is trapped, irrespectively of the trapping mechanism.

The purpose of this lectures is to expose, on the basis of simple models, some ideas and results related to large and infinite extra dimensions. We do not attempt at a comprehensive discussion of such constructions; our choice of topics will thus be very personal, and the list of references very incomplete. Neither are we going to present historical overview; a view on history of the brane world scenario is presented, e.g., in Ref. [9].

2. Homogeneous and small extra dimensions: Kaluza–Klein

To begin with, let us outline the basic idea of the Kaluza–Klein scenario; this will serve as a point of reference for further discussions. The simplest case is one extra spatial dimension z, so that the complete set of coordinates in (4+1)-dimensional space-time is $X^A = (x^\mu, z)$, $\mu = 0, 1, 2, 3$. The low energy physics will be effectively four-dimensional if the coordinate z is compact with a certain compactification radius R. This means that z runs from 0 to $2\pi R$, and points $z = 0$ and $z = 2\pi R$ are identified. In other words, the four-dimensional space is a cylider whose three dimensions x^1, x^2, x^3 are infinite, and the fourth dimension z is a circle of radius R. Assuming that this cylinder is homogeneous, and that the metric is flat, one writes a complete set of wave functions of a free massless particle on this cylinder (e.g., solutions to five-dimensional Klein–Gordon equation),

$$\phi_{\mathbf{p},n} = e^{ip_\mu x^\mu} e^{inz/R}, \quad n = 0, \pm 1, \pm 2, \dots.$$

Here p_μ is the (3+1)-dimensional momentum and n is the eigenvalue of (one-dimensional) angular momentum. Since $\phi(x, z)$ obeys $\partial_A \partial^A \phi = 0$, these quantities are related,

$$p_\mu p^\mu - \frac{n^2}{R^2} = 0. \tag{2.1}$$

Hence, inhomogeneous modes with $n \neq 0$ carry energy of order $1/R$, and they cannot be excited in low energy processes. Below the energy scale $1/R$, only homogeneous modes with $n = 0$ are relevant, and low energy physics is effectively four-dimensional.

From (3+1)-dimensional point of view, each Kaluza–Klein (KK) mode can be interpreted as a separate type of particle with mass $m_n = |n|/R$, according to eq. (2.1). Every multi-dimensional field corresponds to a Kaluza–Klein tower of four-dimensional particles with increasing masses. At low energies, only massless (on the scale $1/R$) particles can be produced, whereas at $E \sim 1/R$ extra dimensions will show up. Since the KK partners of ordinary particles (electrons, photons, etc.) have not been observed, the energy scale $1/R$ must be at least in a few hundred GeV range, so in the Kaluza–Klein scenario, the size of extra dimensions must be microscopic ($R \lesssim 10^{-17}$ cm). These properties are inherent in all models of Kaluza–Klein type (with larger number of extra dimensions, compactifications on non-trivial manifolds or orbifolds instead of a circle, etc.).

Let us make use of the Kaluza–Klein picture to illustrate one property which is generic to models with extra dimensions, namely, the fact that (some of the) effective coupling constants that one measures at low energies in our four-dimensional world are related to the "fundamental" couplings of a multi-dimensional theory through volume factors characterizing the geometry of extra dimensions. Consider, e.g., a scalar theory in the above simple setting. One wishes to derive the effective four-dimensional theory from the "fundamental" multi-dimensional action (we still consider massless scalar field in five dimensions)

$$S^{(5)} = \int d^4 dz \left(\frac{1}{2} (\partial_A \varphi)^2 - \lambda_{(5)} \varphi^4 \right).$$

One performs the KK decomposition

$$\varphi(x, z) = \sum_{n=0,\pm 1,\ldots} \phi_n(x) e^{inz/R},$$

and plugs it back into $S^{(5)}$. One obtains the action in four-dimensional terms,

$$S^{(4)} = V \int d^4 x \left(\sum \frac{1}{2} (\partial_\mu \phi_n)^2 + \lambda_{(5)} \phi_0^4 + \lambda_{(5)} \phi_0^2 \sum_{n \neq 0} \phi_n \phi_{-n} + \ldots \right), \quad (2.2)$$

where $V = 2\pi R$ is the volume of extra dimension. One point to note is that the fields ϕ_n are not canonically normalized; the canonically normalized ones are

$$\phi_n^{can} = \sqrt{V} \phi_n.$$

Their couplings are governed by the action

$$S_{int}^{(4)} = \int d^4x \left[\frac{1}{V}\lambda_{(5)}(\phi_0^{can})^4 + \frac{1}{V}\lambda_5(\phi_0^{can})^2 \sum_{n\neq 0} \phi_n^{can}\phi_{-n}^{can} + \cdots \right]. \quad (2.3)$$

Thus, the effective four-dimensional coupling constant is related to the "fundamental" coupling $\lambda_{(5)}$ as follows,

$$\lambda_{(4)} = \frac{1}{V}\lambda_{(5)}. \quad (2.4)$$

Another point is that the coupling constant $\lambda_{(5)}$ has dimension of $(\text{mass})^{-1}$, so the five-dimensional theory (and its four-dimensional counterpart too) makes sense up to energies $M \sim \lambda_{(5)}^{-1}$, i.e., M may be considered as the UV cutoff scale. The whole classical picture is valid provided the size of extra dimension(s) is large,

$$R \gg M^{-1},$$

otherwise KK modes would have masses above the UV cutoff scale M. This ensures that the four-dimensional theory is weakly coupled at low energies,

$$\lambda_{(4)} \ll 1,$$

and that the properties of sufficiently low-lying KK states may be reliably calculated.

Yet another point is that the angular momentum along the extra dimension is conserved, which is reflected in the last term in (2.3). Because of that, KK modes can be created only in pairs. This property does *not* hold in brane-world models where translational symmetry along extra dimension(s) is broken by the brane(s).

Finally, it is worth noting that the self-coupling of the zero mode and its coupling to KK modes, as well as couplings between the KK modes, are all of the same strength: they are all equal to $\lambda_{(4)}$. These couplings appear weak, so one may wonder how the UV cutoff M shows up in the four-dimensional theory. The reason is that above M, the KK modes become numerous, so, say, scattering amplitudes mediated by these modes, become large. We will see this explicitly in various contexts in these lectures.

3. Localized matter

To see that it is indeed conceivable that ordinary matter may be trapped to a brane, we present in this section simple field-theoretic models exhibiting this property. Throughout this section we neglect gravity; new possibilities emerging when gravitational interactions are included will be considered later on.

3.1. Simplest brane: domain wall

It is fairly straightforward to construct field-theoretic models with localized scalars and fermions. The simplest model of this sort has one extra dimension z, with the brane being a domain wall [10] (see also Ref. [11]). Namely, let us consider a theory of one real scalar field φ whose action is

$$S_\varphi = \int d^4x \, dz \left[\frac{1}{2} (\partial_A \varphi)^2 - V(\varphi) \right]. \tag{3.1}$$

Let the scalar potential $V(\varphi)$ have a double-well shape with two degenerate minima at $\varphi = \pm v$, as shown in Fig. 1.

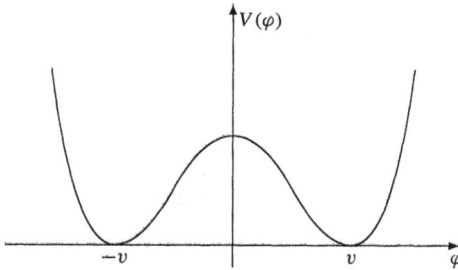

Fig. 1. Scalar potential in the model (3.1).

There exists a classical solution $\varphi_c(z)$, kink, depending on one coordinate only. This solution is sketched in Fig. 2. It has asymptotics

$$\varphi_c(z \to +\infty) = +v,$$

$$\varphi_c(z \to -\infty) = -v,$$

and describes a domain wall separating two classical vacua of the model. Obviously, the field $\varphi_c(z)$ breaks translational invariance along the extra dimension, but leaves the four-dimensional Poincaré invariance intact.

3.2. Scalars

Due to breaking of translational invariance along extra dimension, there must be a gapless Goldstone mode among perturbations of the field φ about the classical solution φ_c. Since this breaking occurs only near the domain wall (brane), this mode must have a wave function concentrating near the brane. Furthermore, since four-dimensional Lorentz-invariance remains intact, the fact that the mode

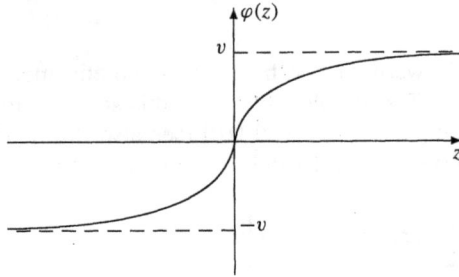

Fig. 2. Domain wall solution.

is gapless (i.e., its energy can be arbitrarily close to zero), implies that the dispersion relation for this mode is

$$p^2 \equiv p_\mu p^\mu = 0.$$

This general argument shows that there must be a four-dimensionally massless mode, localized on the brane (and traveling along the brane with speed of light).

To see this explicitly, one writes the field equation linearized about the kink solution, $\varphi(x, z) = \varphi_c(z) + \phi(x, z)$. In four-dimensional Fourier representation one has

$$p^2\phi + \partial_z^2\phi - \left(\frac{\partial^2 V}{\partial\varphi^2}\right)_{\varphi=\varphi_c} \phi = 0. \tag{3.2}$$

The last term here acts as a potential binding the scalar particle to the brane; its effect is preclslely to ensure the existence of the zero mode. The shape of this mode is

$$\phi_0(z) = \partial_z\varphi_c.$$

Indeed, this function obeys eq. (3.2) with $p^2 = 0$. Since the kink configuration tends to a constant at infinity, this mode is normalizable.

This is the simplest example of a field localized on a brane. Less trivial example emerges if one introduces another complex scalar field $\chi(x, z)$ whose scalar potential is

$$V_\chi = -M^2|\chi|^2 + \lambda_1\varphi^2|\chi|^2 + \lambda_2|\chi|^4.$$

Neglecting for the sake of argument the back reaction of this scalar field on the kink itself (this can be achieved by an appropriate choice of parameters), and choosing $\lambda_1 v^2 > M^2$, one finds that away from the kink, the expectation value of χ vanishes, while near the kink it does not (again with an appropriate choice of parameters): in the kink background the field χ develops real expectation value $\chi_c(z)$ with $\chi_c \to 0$ as $y \to \pm\infty$. Since $U(1)$ global symmetry of the phase rotations of χ is broken *near the brane*, there is four-dimesnionally massless Goldsone mode among perturbations of χ about χ_c,

$$\delta\chi_0 = i\alpha(x)\chi_c(z),$$

where α is a massless four-dimensional field, and $\chi_c(z)$ serves as the wave function of this mode in the transverse direction.

3.3. Fermions

Let us now introduce fermions into this model. We recall that fermions in five-dimensional space-time are four-component columns, and that the five-dimensional gamma-matrices can be chosen as follows,

$$\Gamma^\mu = \gamma^\mu, \quad \mu = 0, 1, 2, 3; \qquad \Gamma^z = -i\gamma^5,$$

where γ^μ and γ^5 are the standard Dirac matrices of four-dimensional theory. Introducing the Yukawa interaction of fermions with the scalar field φ, we write the five-dimensional action for fermions,

$$S_\Psi = \int d^4x \, dz \left(i\bar{\Psi}\Gamma^A \partial_A \Psi - h\varphi\bar{\Psi}\Psi \right). \tag{3.3}$$

Note that in each of the scalar field vacua, $\varphi = \pm v$, five-dimensional fermions acquire a mass

$$m_5 = hv.$$

Let us consider fermions in the domain wall background. The corresponding Dirac equation is

$$i\Gamma^A \partial_A \Psi - h\varphi_c(z)\Psi = 0. \tag{3.4}$$

Due to unbroken four-dimensional Poincaré invariance, the fermion wave functions may be characterized by four-momentum p_μ, and we are interested in the spectrum of four-dimensional masses $m^2 = p_\mu p^\mu$. A key point is that there exists a zero mode [12], a solution to eq. (3.4) with $m = 0$. For this mode one has $\gamma^\mu p_\mu \Psi_0 = 0$, and the Dirac equation (3.4) becomes

$$\gamma^5 \partial_z \Psi_0 = h\varphi_c(z)\Psi_0.$$

The zero mode is left-handed from the four-dimensional point of view,

$$\gamma_5 \Psi_0 = -\Psi_0,$$

and has the form

$$\Psi_0 = e^{-\int_0^z dz' \, h\varphi_c(z')} \psi_L(p), \tag{3.5}$$

where $\psi_L(p)$ is the usual solution of the four-dimensional Weyl equation. The zero mode (3.5) is localized near $z = 0$, i.e., at the domain wall, and at large $|z|$ it decays exponentially, $\Psi_0 \propto \exp(-m_5|z|)$.

The spectrum of four-dimensional masses is shown in Fig. 3. Besides the chiral zero mode, there may or may not exist bound states, but in any case the masses of the latter are proportional to v and are large for large v. There is also a continuum part of the spectrum starting at $m = m_5$; the continuum states correspond to five-dimensional fermions which are not bound to the domain wall and escape to $|z| = \infty$.

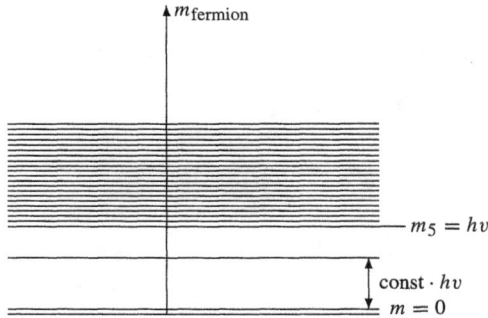

Fig. 3. Spectrum of four-dimensional masses of fermions in domain wall background. The gap between zero mode (with $m = 0$) and non-zero modes is of order hv. Continuum starts at $m = m_5 \equiv hv$.

Massless four-dimensional fermions and scalars localized on the domain wall, zero modes, are meant to mimic our matter. They propagate with the speed of light along the domain wall, but do not move along z. Of course, in realistic theories they should acquire small masses by one or another mechanism. At low energies, their interactions can produce only zero modes again, so physics is effectively four-dimensional. Zero modes interacting at high energies, however, will produce continuum modes, the extra dimension will open up, and particles will be able to leave the brane, escape to $|z| = \infty$ (if the size of extra dimension is infinite) and literally disappear from our world. For a four-dimensional observer (composed of particles trapped to the brane), these high energy processes will look like $e^+e^- \rightarrow$ nothing or $e^+e^- \rightarrow \gamma +$ nothing. We shall discuss later on whether these and similar processes are indeed possible when gravitational and gauge interactions are taken into account, whether they may lead to apparent non-conservation of energy, electric charge, etc.

The above constructions are straightforwardly generalized to more than one extra dimensions. This is done by considering, instead of the domain wall, topological defects of higher codimensions: the Abrikosov–Nielsen–Olesen vortex in six-dimensional space-time ($D = 6$, number of extra dimensions $d = 2$), 't Hooft–Polyakov monopole in $D = 7$ ($d = 3$), etc. Scalars may then be localized in analogy with above examples. In many cases, the existence of fermion zero modes in the background of the topological defect is guaranteed by the corresponding index theorem. Explicit expressions for fermion zero modes in various backgrounds are given in Refs. [12–14]. As a bonus, the four-dimensional massless fermions localized on topological defects are usually chiral. Furthermore, the number of fermion zero modes may be greater than one, so from one family of multi-dimensional fermions one can obtain several four-dimensional families. This possibility of explaining the origin of three Standard Model generations has been explored in Refs. [15–18] where it has been found that reasonable pattern of masses and mixings can be obtained in a fairly natural way. Noteworthy, an interesting pattern of flavor violation is inherent in models of this sort [19].

3.4. Gauge fields

Localizing gauge fields on a brane is more difficult. The mechanism just described does not have chance to work, at least for massless non-Abelian fields. The reason is as follows. If the gauge field had a zero mode whose wave function $A(z)$ is localized near the brane, the four-dimensional effective interaction between this localized field and other localized fields (say, fermions) would involve overlap integrals of the form

$$\int dz \, \Psi_0^\dagger(z) A(z) \Psi_0(z), \tag{3.6}$$

where Ψ_0 is the fermion zero mode. We have seen that fermion zero modes may depend on various parameters (e.g., the coupling constant h in the example of the previous subsection: the width of the zero mode explicitly depends on h, see eq. (3.5)). Therefore, the gauge charges in effective four-dimensional theory would be different, at least in principle, for different types of particles, and they would take arbitrary values depending on the overlap integrals like (3.6). This is impossible in non-Abelian gauge theories where the gauge charge is necessarily quantized, i.e., it depends only on representation to which a matter field belongs, up to a factor common to all fields.

Any mechanism of the localization of (non-Abelian) massless gauge fields must automatically preserve charge universality, i.e., ensure that gauge charges of all four-dimensional particles are the same (up to group representation factors) irrespectively of the structure of their wave functions in transverse directions or other details of a mechanism that binds these particles to our brane. A field-theoretic mechanism of gauge field localization in the absence of gravity, which ensures charge universality, has been proposed in Ref. [20]. It invokes a gauge theory which is in confinement phase in the bulk (outside the brane), whereas there is no confinement on the brane. Then the electric field of a charge residing on the brane will not penetrate into the bulk, the multi-dimensional Gauss' law will reduce to the four-dimensional Gauss' law, and the electric field on the brane will fall off according to the four-dimensional Coulomb law, $E \propto 1/r^2$.

It is worth noting that confinement in the bulk may imply that all states propagating in the bulk are heavy. If the corresponding mass gap is large enough, light particles carrying gauge charges will be bound to the brane, and bulk modes will not be excited at low energies. Hence, the mechanism of Ref. [20] is simultaneously a mechanism of trapping matter (fermions, Higgs bosons, etc.) to the brane, alternative to that discussed in Section 3.1.

We would like to warn the reader, however, that it is not known whether non-Abelian field theories, exhibiting confinement, exist at all in more than four space-time dimensions (see Ref. [21] for discussion). So, in field theory context, the mechanism of Ref. [20] in somewhat up in the air. On the other hand, the picture with confinement in the bulk and no confinement on the brane has a lot of similarities with the localization of gauge fields on D-branes of string/M-theory.

4. Large extra dimensions: ADD

4.1. KK picture for gravitons

Localization of matter on a brane explains why low energy physics is effectively four-dimensional insofar as all interactions except gravity are concerned. To include gravity, one may proceed in different ways. One approach [22, 23], called ADD, is to neglect the brane tension (energy density per unit three-volume of the brane) and consider compact extra dimensions. In this way the Kaluza–Klein picture is reintroduced. The size of extra dimensions R need not, however, be microscopic (we assume for simplicity that the sizes of all extra dimensions are of the same order). Indeed, the distances at which non-gravitational interactions cease to be four-dimensional are determined by the dynamics on the brane, and may be much smaller than R. Only gravity becomes multi-dimensional at scales just below R. The four-dimensional law of gravitational attraction has been established experimentally down to distances[1] of about 0.2 mm [24], so the size of extra dimensions is allowed to be as large as 0.1 mm.

This possibility opens up a new way to address the hierarchy problem [22, 23], the problem of why the elctroweak scale (of order $M_{EW} \sim 1$ TeV) is so different from the Planck scale ($M_{Pl} \sim 10^{16}$ TeV). According to the discussion in Section 2, in multi-dimensional theories, the four-dimensional Planck scale is not a fundamental parameter. Rather, the mass scale of multi-dimensional gravity, which we denote simly by M, is fundamental, as it is this latter scale that enters the full multi-dimensional gravitational action,

$$S = -\frac{1}{16\pi G_{(D)}} \int d^D X \sqrt{g^{(D)}} R^{(D)}, \tag{4.1}$$

where

$$G_{(D)} = \frac{1}{M^{D-2}} \equiv \frac{1}{M^{d+2}}$$

is the fundamental D-dimensional Newton's constant, $d = D - 4$ is the number of extra dimensions, and $d^D X = d^4 x \, d^d z$.

In ADD picture, the long-distance four-dimensional gravity is mediated by the graviton zero mode (cf. Section 2) whose wave function is homogeneous over extra dimensions. Hence, the four-dimensional effective action describing long-distance gravity is obtained from eq. (4.1) by taking the metric to be independent of extra coorinates z. The integration over z is then trivial, and the effective four-dimensional gravitational action is (cf. (2.2))

$$S_{eff} = -\frac{V_d}{16\pi G_{(D)}} \int d^4 x \sqrt{g^{(4)}} R^{(4)},$$

where $V_d \sim R^d$ is the volume of extra dimensions. We see that the four-dimensional Planck mass is, up to a numerical factor of order one, equal to (cf. (2.4))

$$M_{Pl} = M(MR)^{\frac{d}{2}}. \tag{4.2}$$

If the size of extra dimensions is large compared to the fundamental length M^{-1}, the Planck mass is much larger than the fundamental gravity scale M.

---

[1]Until recently, the distance down to which the Newton law was established experimentally was in the several millimeter range [25, 26], for a review Ref. [27]. The new round of experiments was stimulated precicely by the idea that extra dimensions may be large. For a review of recent progress see Ref. [28].

One may push this line of reasoning to extreme and suppose that the fundamental gravity scale is of the same order as the electroweak scale, $M \sim 1$ TeV. Then the hierarchy between M_{Pl} and M_{EW} is entirely due to the large size of extra dimensions. The hierarchy problem becomes now the problem of explaining why R is large. This is certainly an interesting reformulation.

Assuming that $M \sim 1$ TeV, one calculates from eq. (4.2) the value of R,

$$R \sim M^{-1} \left(\frac{M_{Pl}}{M} \right)^{\frac{2}{d}} \sim 10^{\frac{32}{d}} \cdot 10^{-17} \text{ cm.} \qquad (4.3)$$

For one extra dimension one obtains unacceptably large value of R. An interesting case is $d = 2$ when roughly $R \sim 1$ mm. This observation [22] stimulated recent activity in experimental search for deviations from Newton's gravity law at sub-millimeter distances. As we shall discuss later, the mass scale as low as $M \sim 1$ TeV is in fact excluded, for $d = 2$, by astrophysics and cosmology; a more realistic value $M \sim 30$ TeV implies $R \sim 1 - 10$ μm. This motivates search for deviations from Newton's law in a micro-meter range, which is difficult but not impossible [28–30].

For $d > 2$, eq. (4.3) results in smaller values of R. For example, for $d = 3$ and $M \sim 1$ TeV one obtains $R \sim 10^{-6}$ cm. Search for violation of Newton's law at these scales appears hopeless. For $d = 6$ (full dimensionality of space time $D = 10$, as suggested by superstring theory), one has $R \sim 10^{-12}$ cm, which is still much larger than the elctroweak scale, (1 TeV)$^{-1} \sim 10^{-17}$ cm. We note, however, that the compactification scales of different extra dimensions are not guaranteed to be of the same order; if some of these are much smaller than the others, the situation with deviations from Newton's gravity in spaces with $d > 2$ may be similar to that of $d = 2$. In other words, deviations from Newton's gravity law may occur in micro-meter range even for $d > 2$.

4.2. Potential conflict with cosmology and astrophysics

If the fundamental gravity scale is indeed in the TeV range, one expects that extra dimensions should start to show up in collider experiments at energies approaching this scale [31]. One also notes that they may have important effects in cosmology and astrophysics. In the picture described in this section, extra dimensions are felt exclusively by gravitons capable of propagating in the bulk. Hence, the most distinctive feature of this scenario is the possibility to emit gravitons into the bulk; this process has strong dependence on the center-of-mass energy of particles colliding on the brane and has large probablity at energies comparable to the fundamental gravity scale.

From the four-dimensional viewpoint, emission of gravitons into extra dimensions corresponds to the production of Kaluza–Klein gravitons. One process of

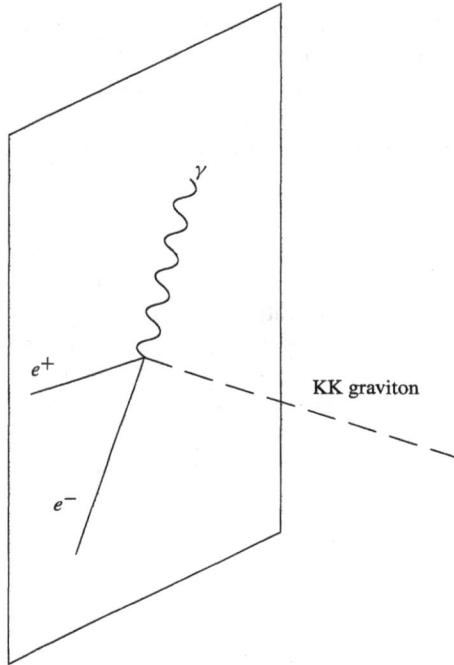

Fig. 4. Emission of a graviton into extra dimensions (or, equivalently, creation of a KK graviton) in the process $e^+e^- \rightarrow \gamma +$ graviton. Electron, positron and photon propagate along the brane, the graviton escapes into the bulk.

this type is shown in Fig. 4. Each of the KK graviton states interacts with matter on the brane with four-dimensional gravitational strength. Indeed, the quadratic action for each type of KK graviton, and its interaction with matter on the brane are schematically (omitting all indices, tensor structure, etc.) written as

$$S_{\mathbf{n}} = \frac{1}{16\pi G_{(D)}} \int d^D X [\partial h(x) e^{i \mathbf{q_n z}}]^* [\partial h(x) e^{i \mathbf{q_n z}}] + \int d^4 x \, h(x) T(x),$$

where $\mathbf{q_n}$ are the discrete momenta along extra dimensions (in the case of toroidal compactification with equal sizes of extra dimensions one has $\mathbf{q_n} = \mathbf{n}/R$, $\mathbf{n} = (n_1, \ldots n_d)$) and $T_{\mu\nu}$ is the energy-momentum tensor of matter on the brane. The integration over z again gives the volume factor V_d in front of the first term, so the coupling of each type of KK graviton is determined by the four-dimensional Planck mass.

Even though the coupling of every KK graviton is weak, the total emission rate of KK gravitons may be large due to large number of KK graviton states. It is straightforward to find the number $N(E)$ of KK graviton spieces with masses below E. Since the momenta along extra dimensions, $\mathbf{q_n}$, are quantized in units of $1/R$, and the KK graviton masses are $m_n = |\mathbf{q_n}|$, one has

$$N(E) \sim (ER)^d. \tag{4.4}$$

Obiously, this number gets large at large E, so the emission rate of KK gravitons rapidly increases with energy. In the early Universe, KK gravitons can be produced at high enough temperatures, and therefore may destroy the standard Big Bang picture [32]. Consistency with the Big Bang nucleosynthesis, as well as the present composition of the Universe impose strong bounds on the maximum temperature of the Universe [32], as we shall now see.

At high enough temperature, $T \gg 1/R$, the creation rate (per unit time per unit volume) of one KK graviton spieces of mass $m_{\mathbf{n}} \lesssim T$ is estimated as

$$\Gamma \sim \frac{T^6}{M_{Pl}^2},$$

where the factor M_{Pl}^{-2} comes from the strength of the graviton–matter interaction, and the dependence on temperature is restored on dimensional grounds. Taking into account the number of KK states, cf. eq. (4.4), one obtains the estimate of the total rate of creation of KK gravitons,

$$\frac{dn}{dt} \sim \frac{T^6}{M_{Pl}^2}(TR)^d \sim T^4 \left(\frac{T}{M}\right)^{2+d}, \tag{4.5}$$

where the latter relation comes from eq. (4.2).

Let us pause here to notice that the latter formula has a natural multi-dimensional interpretation. At $T \gg R^{-1}$, the wavelengths $\lambda_z \sim T^{-1}$ of gravitons along extra dimensions are typically much smaller than the size of these dimensions. Thus, their emission proceeds as if extra dimensions were infinite. Then the emission rate is determined by the "fundamental" gravitational constant, and is proportional to $G_{(D)} = M^{-(d+2)}$. With temperature dependence restored on dimensional grounds, this gives precisely the result (4.5).

Assuming that the Universe expands in the standard way[2],

$$H = \frac{T^2}{M_{Pl}^*}, \tag{4.6}$$

---

[2]If KK gravitons dominated the expansion of the Universe, eq. (4.6) would not hold. It is straightforward to see, however, that such a scenario is not viable anyway.

with $M_{Pl}^* = M_{Pl}/(1.66g_*^{1/2}) \sim$ (a few) $\cdot 10^{18}$ GeV, where g_* is the effective number of degrees of freedom, one finds the total number density of KK gravitons created in the Hubble time H^{-1},

$$n(T) \sim T^2 M_{Pl}^* \left(\frac{T}{M}\right)^{2+d}.$$

Even though the creation rate (4.5) is fairly small at $T \ll M$, the total number of gravitons may be large because of slow expansion rate (4.6).

A stringent bound on the maximum temperature T_*, that ever occurred in the Universe after inflation, emerges if one takes the ADD picture literally, i.e., assumes that KK gravitons survive in the bulk. Most of the gravitons created at temperature T_* have masses of order T_*. Below this temperature they are non-relativistic, and their number density scales as T^3. Hence, at the nucleosynthesis epoch ($T_{NS} \sim 1$ MeV) the mass density of KK gravitons is of order

$$\rho_{grav}(T_{NS}) \sim \left(\frac{T_{NS}}{T_*}\right)^3 \cdot T_* n(T_*) \sim T_{NS}^3 M_{Pl}^* \left(\frac{T_*}{M}\right)^{2+d}. \tag{4.7}$$

Requiring that this energy density is lower than the energy density of one mass-less spieces (otherwise the standard Big Bang nucleosynthesis would fail), i.e., that $\rho_{grav}(T_{NS}) \lesssim T_{NS}^4$, one obtains

$$T_* \lesssim M \left(\frac{T_{NS}}{M_{Pl}^*}\right)^{\frac{1}{2+d}} \sim M \cdot 10^{-\frac{21}{2+d}}.$$

For $d = 2$ and $M = 1$ TeV one finds that the maximum temperature should not exceed 10 MeV, and even for $d = 6$ one obtains fairly low maximum temperature, $T_* \lesssim 1$ GeV.

Even stronger bounds on T_* are obtained by requiring that the present mass density of KK gravitons does not exceed the actual present energy density, which is equal to the critical density, and that decaying KK gravitons do not produce too much diffuse photon background [32]. For $d = 2$, the very fact that the Universe underwent the nucleosynthesis epoch (i.e., that $T_* \gtrsim$ (a few) MeV) pushes the fundamental gravity scale up to $M \gtrsim$ (a few) $\cdot 10$ TeV.

Low maximum temperature of the Universe (say, in the range 10 MeV – 1 GeV) does not directly contradict cosmological data: we know for sure that the Universe underwent the standard hot Big Bang evolution at the nucleosynthesis epoch, but have no observational handle on higher temperature epochs[3].

---

[3]This has been explored in detail in Ref. [33].

With low T_*, however, one has to invoke fairly exotic mechanisms of baryogenesis and inflation, which are not impossible but not very appealing either, to say the least.

We note in passing that we have assumed in the above discussion that KK gravitons do not decay before the nucleosynthesis epoch. This is correct if nothing happens to gravitons emitted into the bulk: the width of a graviton with mass of order T_* with respect to decay into ordinary particles (photons, e^+e^--pairs, etc.) is of order T_*^3/M_{Pl}^2, which, for $T_* \lesssim 1$ GeV (and even for substantially larger T_*), is much smaller than the expansion rate before and at nucleosynthesis, $H \sim T_{NS}^2/M_{Pl}^*$. One might invent mechanisms of faster decay of KK gravitons into something massless in the bulk (or on a different brane). Then the energy density of the latter would scale as T^4, and the cosmological constraints on T_* would be weaker. Still, T_* is required to be rather low. Indeed, instead of eq. (4.7) one would have

$$\rho_{extra}(T_{NS}) \sim \left(\frac{T_{NS}}{T_*}\right)^4 \cdot T_* n(T_*) \sim \frac{T_{NS}^4}{T_*} M_{Pl}^* \left(\frac{T_*}{M}\right)^{2+d}.$$

Requiring again that $\rho_{extra}(T_{NS}) \lesssim T_{NS}^4$, one obtains

$$T_* \lesssim M \left(\frac{M}{M_{Pl}^*}\right)^{\frac{1}{2+d}}.$$

For $M \sim 1$ TeV and $d = 2$ one again obtains $T_* \lesssim 10$ MeV, whereas for $d = 6$ one has $T_* \lesssim 10$ GeV. This model-independent estimate shows that the maximum temperature must be quite low irrespectively of the fate of emitted KK gravitons.

Light KK gravitons are potentially dangerous for astrophysics as well, as they may be produced by stars or supernovae, take away energy, and hence contradict observational data [32]. Strong bounds on the fundamental scale M are obtained in this way for $d = 2$ only, as for larger number of extra dimensions, the number of KK graviton states with small masses is suppressed, cf. (4.4). As an example, by requiring that the emission of gravitons during the collapse of SN1987a is not the dominant cooling process (otherwise no neutrinos would be produced, in contradiction to observations), one obtains [32, 34–36]

$$M > 30 \text{ TeV},$$

which is comparable to the cosmological bounds. We note, however, that this bound is obtained without any assumptions concerning the lifetime of KK gravitons. As we alredy mentioned, with so high scale M, the deviations from Newton's law are only allowed well below 1 mm — more realistically, at distances at most in 1 to 10 μm range. Even stronger bound is obtained in Ref. [37] under the assumption that KK gravitons produced during suprenovae collapses decay into usual photons (and not, say, into particles residing on other branes).

To conclude this section, cosmology in the ADD scenario is not very appealing: the maximum temperature of the Universe must be below 10 GeV, so one has to rely upon fairly exotic mechanisms of baryogenesis and inflation.

5. Warped extra dimension

5.1. Non-factorizable geometry

Until now we have ignored the energy density of the brane itself, i.e., the gravitational field that the brane produces. Here we shall see that a gravitating brane induces an interesting geometry in multi-dimensional space, and that a number of novel properties emerge.

When considering distance scales much larger than the brane thickness, one may view the brane as a delta-function source of the gravitational field. In the simplest case, the gravitating brane is characterized by just one parameter, the energy density per unit three-volume σ. This quantity is also called brane tension. We shall mostly discuss the case of one extra dimension, so the five-dimensional gravitational action in the presence of the brane is

$$
\begin{aligned}
S_g = & -\frac{1}{16\pi G_{(5)}} \int d^4x \, dz \, \sqrt{g^{(5)}} R^{(5)} \\
& - \Lambda \int d^4x \, dz \, \sqrt{g^{(5)}} - \sigma \int d^4x \, \sqrt{g^{(4)}},
\end{aligned} \tag{5.1}
$$

where Λ is the five-dimensional cosmological constant, and the integral in the last term is evaluated along the world surface of the 3-brane with $g_{\mu\nu}^{(4)}$ being the induced metric.

The resulting field equations are straightforward to obtain. In the bulk, these are the standard five-dimensional Einstein equations with the cosmological constant Λ, while the last term in eq. (5.1) gives rise to the Israel junction conditions [38] on the brane surface (for pedagogical presentation of the Israel conditions see, e.g., Ref. [39]). Notably, this set of equations allows for a solution preserving four-dimensional Poincaré invariance. This fact was extensively discussed in the D-brane context (see, e.g, Ref. [40] and references therein) and its relevance for phenomenological models has been stressed in Refs. [41–44]. The existence of four-dimensionally flat solution requires fine-tuning between Λ and σ: the five-dimensional cosmological constant must be negative and equal to [43]

$$
\Lambda = -\frac{4\pi}{3} G_{(5)} \sigma^2 \tag{5.2}
$$

(note that the parameters here have dimensions $[\Lambda] = M^5$, $[\sigma] = M^4$, $[G_{(5)}] = M^{-3}$). This fine-tuning is very similar to fine-tuning of the cosmological constant to zero in conventional four-dimensional gravity; indeed, if eq. (5.2) does not hold, the intrinsic geometry on the brane is (anti-)de Sitter rather than flat.

With the relation (5.2) satisfied, the four-dimensionally flat solution has the form known[4] as Randall–Sundrum geometry [43]

$$ds^2 = a^2(z)\eta_{\mu\nu}dx^\mu dx^\nu - dz^2, \tag{5.3}$$

where $\eta_{\mu\nu}$ is the four-dimensional Minkowski metric and the "warp factor" has the form

$$a(z) = e^{-k|z|}, \tag{5.4}$$

where

$$k = \frac{4\pi}{3}G_{(5)}\sigma. \tag{5.5}$$

The brane is located at $z = 0$.

To see that this is indeed a solution to the complete system of the Einstein equations, we write for the metric (5.3)

$$G_{\mu\nu} \equiv R_{\mu\nu}^{(5)} - \frac{1}{2}g_{\mu\nu}^{(5)}R^{(5)} = g_{\mu\nu}^{(5)}\left[-3\frac{a''}{a} - 3\left(\frac{a'}{a}\right)^2\right],$$

$$G_{z\mu} = 0,$$

$$G_{zz} = g_{zz}^{(5)}\left[-6\left(\frac{a'}{a}\right)^2\right], \tag{5.6}$$

where prime denotes the derivative with respect to z. In is then straightforward to see that metric (5.3) is a solution to the Einstein equations

$$G_{\mu\nu} = 8\pi G_{(5)}\Lambda g_{\mu\nu}^{(5)} + 8\pi G_{(5)}\sigma g_{\mu\nu}^{(5)}\delta(z),$$

$$G_{z\mu} = 0,$$

$$G_{zz} = 8\pi G_{(5)}\Lambda g_{zz}^{(5)}, \tag{5.7}$$

provided eq. (5.5) is satisfied, and

$$k^2 = -\frac{4\pi}{3}G_{(5)}\Lambda,$$

---

[4]In fact, non-factorizable geometries of the type (5.3) (but not with (5.4)) have been discussed long ago, see, e.g., Refs. [9,45–48].

which is equivalent to eq. (5.2). Note that eq. (5.5) comes from the $\mu\nu$ component of the Einstein equations across the brane,

$$-3\frac{[a']}{a}g^{(5)}_{\mu\nu} = 8\pi G_{(5)}\sigma g^{(5)}_{\mu\nu}, \quad z = 0,$$

where $[a']$ denotes the jump of a' at $z = 0$. The latter equation is essentially the Israel condition in the case considered.

The metric (5.3) is non-factorizable: unlike the metrics appearing in the usual Kaluza–Klein scenarios, it does not correspond to a product of the four-dimensional Minkowski space and a (compact) manifold of extra dimensions. This metric rather corresponds to two patches of anti-de Sitter space of radius $1/k$ glued together along $z = 0$, i.e., along the brane. The four-dimensional hypersurfaces $z = \text{const}$ are flat; in particular, the metric induced on the brane is the Minkowski metric $\eta_{\mu\nu}$.

At this point it is worth mentioning one property of the metric (5.3). Due to four-dimensional Poincaré invariance, every field in this background can be decomposed into four-dimensional plane waves,

$$\phi \propto e^{ip_\mu x^\mu}\phi_p(z).$$

The coordinate four-momentum p_μ coincides with the physical momentum *on the brane*, but from the point of view of an observer residing at $z \neq 0$, the physical four-momentum is larger,

$$p_\mu^{phys}(z) = \frac{1}{a(z)}p_\mu = e^{k|z|}p_\mu. \tag{5.8}$$

The modes which are soft on the brane become harder away from the brane. This scaling property is behind many peculiarities of physics in the background (5.3).

5.2. Slice of adS_5

There are several approaches which make use of the solution (5.3). One of them [43] is to make extra dimension compact by introducing two branes: one with positive tension σ at $z = 0$, and the other with negative tension $(-\sigma)$ located at distance z_c, see Fig. 5. Allowing the negative tension brane to vibrate freely is dangerous, as this would give rise to physical excitations of arbitrarily large negative energy (see Ref. [49] for detailed discussion). To circumvent this problem, the branes are placed at fixed points of an orbifold; in our case, this means that all bosonic fields, including gravity, are required to be symmetric under reflections with respect to both z_c, the position of the negative tension brane, and $z = 0$, the position of positive tension one (fermion fields may have more complicated symmetry properties). The metric (5.3) is still a solution of the complete set of

Einstein equations in the presence of the two branes; extra dimension is compact, as the coordinate z runs now from $z = 0$ to $z = z_c$,

$$z \in [0, z_c].$$

The orbifold boundary conditions (reflection symmetry) project the undesirable negative energy modes out, and there remain positive energy excitations only.

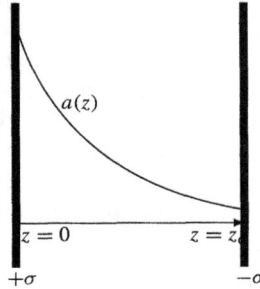

Fig. 5. Two-brane set up: branes of positive and negative tensions and warp factor between them.

Let us consider small perturbations about the metric (5.3). To make the long story short, one can use the gauge

$$g_{55} = -1, \quad g_{5\mu} = 0,$$

i.e., consider perturbed metric of the form

$$ds^2 = [a^2(z)\eta_{\mu\nu} + h_{\mu\nu}(x, z)]dx^\mu dx^\nu - dz^2.$$

If there are no sources of the gravitational field except for the bulk cosmological constant and the two branes, one can further specify the coordinate frame (i.e., fix the gauge), so that $h_{\mu\nu}$ are transverse and trace-free in the bulk,

$$\partial_\mu h^\mu_\nu = 0, \quad h^\mu_\mu = 0.$$

For all types of perturbations but one, this frame is at the same time Gaussian normal with respect to both branes. This means that the positions of the branes are still $z = 0$ and $z = z_c$. Then all components of $h_{\mu\nu}$ obey the same equation (we omit the subscripts)

$$h'' - 4k^2 h - \frac{m^2}{a^2(z)}h = 0, \tag{5.9}$$

where

$$m^2 = \eta^{\mu\nu} p_\mu p_\nu,$$

is the four-dimensional mass of the perturbation. The junction conditions on the branes are (assuming the orbifold symmetry)

$$h' + 2kh = 0 \quad \text{at } z = 0 \text{ and } z = z_c. \tag{5.10}$$

Equations (5.9) and (5.10) determine the mass spectrum of the KK gravitons, where the mass is defined with respect to the positive tension brane, cf. eq. (5.8).

Before considering this spectrum, let us point out that there exists one scalar mode which cannot be treated in the above way. This mode — the radion — is massless and corresponds to oscillations of the relative distance between the two branes. Its properties are discussed, e.g., in Refs. [49–51]. In many phenomenological models based on this set up, massless radion is unacceptable (we shall briefly discuss this point below). Giving the radion a mass corresponds to stabilizing the distance between the branes; field theory mechanisms for this stabilization are suggested, e.g., in Ref. [52, 53]. We shall not consider the radion mode in what follows, assuming that the distance between the branes is stabilized in one or another way.

Let us now turn to the graviton spectrum, i.e., solutions to eqs (5.9) and (5.10). There exists a zero mode, $m^2 = 0$, whose wave function, up to normalization, is

$$h_0(z) = e^{-kz}. \tag{5.11}$$

This mode describes the usual four-dimensional gravity. Unlike in the Kaluza–Klein theories with factorizable geometry, the zero-mode wave function depends on z non-trivially, and decreases towards $z = z_c$. This suggests that the gravitational coupling between particles residing on the negative tension brane is weak as compared to the positive tension brane. We shall discuss this feature in more detail later on.

Solutions to eq. (5.9) obeying the boundary condition (5.10) at $z = 0$ (not yet at $z = z_c$) are, again up to normalization,

$$h_m(z) = N_1\left(\frac{m}{k}\right) J_2\left(\frac{m}{k}e^{kz}\right) - J_1\left(\frac{m}{k}\right) N_2\left(\frac{m}{k}e^{kz}\right), \tag{5.12}$$

where N and J are the Bessel functions. The mass spectrum is determined by the boundary condition (5.10) at $z = z_c$. One obtains that the mass splitting between the KK modes is of order

$$\Delta m \sim k e^{-kz_c}. \tag{5.13}$$

The phenomenological interpretation of these results depends on whether the Standard Model particles are bound to the brane of positive or negative tension.

5.3. Matter on negative tension brane: RS1

Let us first consider the possibility that the ordinary matter resides on the negative tension brane (RS1 scenario [43]). We are interested in gravitational interactions of this matter at large distances, then the dominant contribution to the gravitational attraction is due to the zero graviton mode. It is convenient to rescale the four-dimensional coordinates in such a way that the warp factor is equal to 1 at the negative tension brane (i.e., at $z = z_c$),

$$a(z) = e^{k(z_c - z)}.$$

Similarly, it is convenient to normalize the zero mode so that it is equal to 1 at the negative tension brane. Then the massless gravitational perturbations are described by the field

$$h_{\mu\nu}(x, z) = e^{2k(z_c - z)} h_{\mu\nu}^{(4)}(x). \tag{5.14}$$

The coordinates x^μ are now the physical coordinates on the negative tension brane, and the four-dimensional graviton field $h_{\mu\nu}^{(4)}(x)$ couples to energy-momentum of the ordinary matter in the usual way,

$$S_{int} = \int d^4x \, h_{\mu\nu}^{(4)} T^{\mu\nu}. \tag{5.15}$$

The strength of gravitational interactions is read off from the quadratic part of the action for $h_{\mu\nu}^{(4)}$. This is obtained by plugging the expression (5.14) in the five-dimensional gravitational action (5.1). Schematically, one has

$$
\begin{aligned}
S_g &= \frac{1}{16\pi G_{(5)}} \int_{-z_c}^{z_c} \frac{dz}{a^2(z)} \, d^4x (\partial_\mu h)^2 \\
&= \frac{1}{8\pi G_{(5)}} \int_0^{z_c} dz \, e^{2k(z_c - z)} \int d^4x (\partial_\mu h^{(4)})^2 \\
&= \frac{(e^{2kz_c} - 1)}{16\pi G_{(5)}k} \int d^4x (\partial_\mu h^{(4)})^2. \tag{5.16}
\end{aligned}
$$

The integral in the last expression is the quadratic action for four-dimensional gravitons. Hence, the four-dimensional Newton's constant is

$$G_{(4)} = G_{(5)}k \frac{1}{e^{2kz_c} - 1}, \tag{5.17}$$

which means that at relatively large z_c, the gravitational interactions of matter residing on the negative tension brane are weak.

This observation opens up a novel possibility to address the hierarchy problem. Indeed, one can take the fundamental five-dimensional gravity scale, as well as the inverse anti-de Sitter radius k to be of the order of the weak scale, $M_{EW} \sim 1$ TeV. As is clear from eq. (5.17), the effective four-dimensional Planck mass is then of order

$$M_{Pl} \sim e^{k z_c} M_{EW}, \qquad (5.18)$$

which means the exponential hierarchy between the Planck mass and the weak scale: for z_c only about 37 times larger than the anti-de Sitter radius k^{-1}, the value of M_{Pl}/M_{EW} is of the right order of magnitude.

One may wonder whether the low energy theory of gravity obtained in this set up is indeed the conventional four-dimensional General Relativity. It has been shown explicitly in Refs. [54, 55] that this is indeed the case, *provided* the radion is given a mass (see also [50]). Otherwise the radion would act as a Brans–Dicke field with unacceptably strong (TeV scale) coupling to matter.

Let us now turn to KK gravitons. The mass splitting (5.13) refers to the masses measured by an observer on the positive tension brane. According to eq. (5.8), the physical masses measured by an observer on the negative tension brane are of order

$$m_{grav} \sim k.$$

Hence, KK gravitons have masses in TeV range, in clear distinction to ADD scenario.

Unlike the zero mode, the coupling of KK gravitons to matter residing on the negative tension brane is characterized by the fundamental mass scale (of order M_{EW}). To see this, we write the massive modes as follows (for m somewhat larger than $k e^{-k z_c}$)

$$h_m(x, z) = e^{k(z_c - z)/2} \sin\left(\frac{m}{k} e^{kz} - \varphi_m\right) h_m^{(4)}(x). \qquad (5.19)$$

This expression is valid at $(m/k)e^{kz} \gg 1$, while the KK wave functions decrease towards $z = 0$. The pre-factor in eq. (5.19) has been chosen in such a way that the four-dimensional fields $h_m^{(4)}(x)$ couple to matter at $z = z_c$ with unit strength, cf. eq. (5.15). In the same way as eq. (5.16) one obtains the quadratic action

$$
\begin{aligned}
S_{g,m} &= \frac{1}{16\pi G_{(5)}} \int \frac{dz}{a^2(z)} \, d^4x (\partial_\mu h_m)^2 \\
&= \frac{1}{8\pi G_{(5)}} \int_0^{z_c} dz \, e^{-k(z_c - z)} \int d^4x (\partial_\mu h_m^{(4)})^2 \\
&= \frac{(1 - e^{-k z_c})}{8\pi G_{(5)} k} \int d^4x (\partial_\mu h_m^{(4)})^2.
\end{aligned}
\qquad (5.20)
$$

Hence, the mass scale determining the interactions of KK gravitons with matter is of order

$$M_m \sim \frac{1}{\sqrt{G_{(5)}k}},$$

which is of order M_{EW}. The interaction of matter and KK gravitons becomes strong in the TeV energy range.

Thus, RS1 scenario leads to exponential hierarchy between the weak and Planck scales. Similarly to ADD, gravity becomes strong at TeV energies; manifestations of this phenomenon in collider experiments are quite different in RS1 model as compared to ADD [31]. The reason is of course that the graviton spectra are entirely different; a distinctive feature of RS1 collider phenomenology is TeV scale graviton resonances quite strongly coupled to ordinary particles. The cosmology in RS1 is much more plausible than in ADD: the maximum temperature in the Universe can be in 100 GeV – TeV range without contradicting cosmological data, since KK gravitons have masses in the TeV range.

5.4. Matter on positive tension brane

Another option is that the conventional matter resides on the positive tension brane. The analysis similar to that leading to eq. (5.17) shows that in this case the effective four-dimensional Newton's constant is

$$G_{(4)} = G_{(5)}k\frac{1}{1 - e^{-2kz_c}}. \tag{5.21}$$

If one does not introduce huge hierarchy between the fundamental five-dimensional gravity scale and the inverse anti-de Sitter radius k, the fundamental scale must be of order of M_{Pl}. This does not mean, however, that the exponential hierarchy between the scales cannot be generated. As an example, there is a possibility that the electroweak symmetry breaking and/or supersymmetry breaking occur due to physics on the negative tension brane, and are transfered to "our" brane by one or another mechanism [56]. In that case the electroweak scale and/or supersymmetry breaking scale in our world are naturally exponentially smaller than the Planck scale, essentially because of the scaling relation (5.8), and the exponential hierarchy (5.18) is again generated. The development of this set of ideas is discussed by T. Gherghetta in these Proceedings.

6. Infinite extra dimension: RS2

6.1. Localized graviton

The graviton zero mode (5.11) that appeared in the set up of Section 5, is normalizable for $z_c \to \infty$, i.e., for negative tension brane moved away. This means that gravity is still localized if there exists a single positive tension brane only, and extra dimension is infinite [44]. Hence, one is lead to consider a set up, called RS2 [44], with matter residing on the positive tension brane and experiencing four-dimensional gravity law at large distances due to the exchange of the graviton zero mode. The fact that gravity is four-dimensional at large distances is clear also from eq. (5.21): in the limit $z_c \to \infty$, the four-dimensional Newton's constant tends to a finite value

$$G_{(4)} = G_{(5)}k.$$

Obviously, in this simplest set up, unlike in constructions of Sections 4 and 5, the hierarchy between the Planck and weak scales is not explained by physics of extra dimension, and one has to rely upon other, more conventional mechanisms (we shall mention another option later on).

One property of the anti-de Sitter geometry in the bulk is worth mentioning. Although the distance from the brane to $z = \infty$, measured along the z-axis is infinite, it is straightforward to see that $z = \infty$ is in fact a particle horizon. Indeed, let us consider, as an example, a massive particle that starts from the brane at $t = 0$ and $\mathbf{x} = 0$ with zero velocity, and then freely moves along the z-axis. The corresponding solution to the geodesic equation is [57,58]

$$z_c(t) = \frac{1}{2k} \ln(1 + k^2 t^2).$$

The particle accelerates towards $z \to \infty$, its velocity tends to the speed of light. According to eq. (5.3), the proper-time interval is determined by

$$d\tau^2 = a^2(z_c(t))dt^2 - \left(\frac{dz_c}{dt}\right)^2 dt^2.$$

The particle reaches $z = \infty$ at infinite time t, but finite proper time

$$\tau = \int_0^\infty \frac{dt}{1 + k^2 t^2} = \frac{\pi}{2k}.$$

Hence, $z = \infty$ is indeed the particle horizon. When considering physics in the background (5.3), one has to impose certain boundary conditions at the horizon $z = \infty$ (which in principle may affect physics on the brane); it is usually assumed that nothing comes in from "behind the horizon".

To substantiate the claim that gravity experienced by matter residing on the (positive tension) brane is effectively four-dimensional at large distances, let us consider the Kaluza–Klein gravitons. According to eq. (5.12), the spectrum of KK gravitons *is continuous* and *starts from zero m^2*. In this situation, the wave functions of KK gravitons are to be normalized to delta-function (again with the measure $a^{-2} dz$, see eq. (5.16)),

$$\int \frac{dz}{a^2(z)} h_m(z) h_{m'}(z) = \delta(m - m').$$

Making use of the asymptotics of the Bessel functions, one obtains that the properly normalized KK graviton wave functions are

$$h_m(z) = \sqrt{\frac{m}{k}} \frac{J_1\left(\frac{m}{k}\right) N_2\left(\frac{m}{k}e^{kz}\right) - N_1\left(\frac{m}{k}\right) J_2\left(\frac{m}{k}e^{kz}\right)}{\sqrt{\left[J_1\left(\frac{m}{k}\right)\right]^2 + \left[N_1\left(\frac{m}{k}\right)\right]^2}}. \qquad (6.1)$$

At large z, these wave functions oscillate,

$$h_m(z) = \text{const} \cdot \sqrt{a(z)} \sin\left(\frac{m}{k}e^{kz} + \varphi_m\right),$$

whereas they decrease towards small z and are suppressed at $z = 0$,

$$h_m(0) = \text{const} \cdot \sqrt{\frac{m}{k}}.$$

The wave functions (6.1) correspond to gravitons escaping into extra dimension, i.e., towards $z \to \infty$ (or coming towards the brane from $z = \infty$). The coupling of these KK gravitons to matter, residing on the brane, is fairly weak at small m, so their production at relatively low energies (and/or temperatures) is unimportant (for details, see Ref. [60] and references therein). Likewise, the contribution of virtual KK gravitons into low energy processes is small.

As an example, let us consider the contribution of KK graviton exchange into gravitational potential between two unit point masses placed on the brane. Each KK graviton produces the potential of Yukawa type, so the total contribution is

$$\begin{aligned}
\Delta V_{KK}(r) &= -G_{(5)} \int_0^\infty dm |h_m(0)|^2 \frac{e^{-mr}}{r} \\
&= -\frac{G_{(5)}k}{r} \cdot \text{const} \cdot \int_0^\infty \frac{m\,dm}{k^2} e^{-mr} \\
&= -\frac{G_{(4)}}{r} \cdot \frac{\text{const}}{k^2 r^2}. \qquad (6.2)
\end{aligned}$$

Hence, the gravitational potential, including the contribution of the graviton zero mode, is [44]

$$V(r) = -\frac{G_{(4)}}{r}\left(1 + \frac{\text{const}}{k^2 r^2}\right).$$

The correction to Newton's law has power law behaviour at large r, in contrast to theories with compact extra dimensions where the corrections are suppressed exponentially at large distances. However, this correction is negligible at distances exceeding the anti-de Sitter radius k^{-1}. It has been explicitly shown in Refs. [54, 55] that the tensor structure of the gravitational interactions at large distances indeed corresponds to (the weak field limit of) the four-dimensional General Relativity.

Note that the radion excitation is absent in RS2 set up. This property may appear counter-intuitive, as the massless radion should be restored in, e.g., domain wall model of the brane in the limit $G_{(5)} \to 0$, in which gravity is switched off. The way this occurs is discussed in Ref. [61].

We have already mentioned that in RS2 set up with one brane, extra dimension does not help to solve the hierarchy problem. It was pointed out, however, that modest extension of this set up leads to exponential hierarchy even if extra dimension is infinite [59]. Instead of assuming that our matter is bound to the "central" brane, one may introduce one more, "probe" brane which is placed at a position z_c in extra dimension and, for simplicity, has zero tension. Metric (5.3) on this probe brane still has four-dimensional Poincaré invariance. If our matter is put on the probe brane, the exponential hierarchy (5.18) is generated in much the same way as in RS1 set up discussed in Section 5.

6.2. Escape into extra dimension

If one or more extra dimensions are infinite, one naturally expects that particles may eventually leave our brane and escape into extra dimension. In RS2 set up, this process is certainly possible for gravitons, as the excitation of a KK mode is interpreted precisely as escape of a graviton towards $z \to \infty$. If other fields have bulk modes, the corresponding particles may also leave our brane. As an example, even in the absence of gravity, fermions bound to the brane by the mechanism presented in subsection 3.1, are capable of leaving the brane provided they are given enough energy. As we discussed in subsection 3.1, this would show up as a process like $e^+ e^- \to$ nothing which would be possible at *high* energies.

A novelty of the bulk with anti-de Sitter metric is that such processes become possible also at *low* energies [62]. The ultimate reason is again the scaling property (5.8): energies which are low if measured at the brane position, become high if measured at large z. Low energy physics on the brane is high energy physics away from the brane.

Quantitatively, this feature is manifest in a peculiar property of KK continuum for fields having bulk modes: the continuum starts from *zero* m^2 irrespectively of the dynamics near the brane. Suppose now, that in the absence of gravity, a field has a bound state of a non-zero mass, whose wave function concentrates near the brane and hence corresponds to a four-dimensional particle. When gravity is turned on, this would-be bound state becomes embedded in a continuum of KK modes, which describe particles capable of escaping to $z \to \infty$. Hence, this would-be bound state becomes quasi-localized (there are no true bound states embedded in continuum, unless the potential is very contrived): its energy obtains an imaginary part which determines (finite) probability of tunneling to large z. Particle on the brane becomes metastable against escape into extra dimension.

To illustrate this fairly general phenomenon, let us consider real scalar field in the presence of the brane, with the action

$$S_\phi = \int d^4x \, dz \, \sqrt{g} \left[\frac{1}{2} g^{AB} \partial_A \phi \partial_B \phi - \frac{1}{2} V(z) \phi^2 \right],$$ (6.3)

where $x^A = (x^\mu, z)$ are coordinates in five-dimensional space-time. Effects of the brane are encoded in the potential $V(z)$, which is assumed to tend to a (possibly, non-zero) non-negative constant as $z \to \infty$ (we asuume the orbifold symmetry $z \to -z$ for simplicity). If gravity is switched off, the field ϕ obeys the Klein–Gordon equation

$$-\partial_\mu^2 \phi + \partial_z^2 \phi - V(z) \phi = 0.$$

The spectrum of four-dimensional masses is determined by the potential $V(z)$,

$$p_\mu p^\mu \phi \equiv m^2 \phi = [-\partial_z^2 + V(z)] \phi.$$ (6.4)

An interesting case is when the operator on the right hand side of this equation has discrete eigenmodes which correspond to particles trapped to the brane. This situation is shown in Fig. 6. The continuum starts at $m^2 = V(\infty)$.

When gravity of the brane is turned on, the situation changes. The action in the background metric (5.3) is

$$S_\phi = \int d^4x \, dz \, a^4 \left[\frac{1}{a^2} \eta^{\mu\nu} \partial_\mu \phi \partial_\nu \phi - \frac{1}{2} (\partial_z \phi)^2 - \frac{1}{2} V(z) \phi^2 \right],$$ (6.5)

where, as before, $a(z) = e^{-k|z|}$. Since a^{-2} grows at large z, the first term in the integrand of eq. (6.5) dominates away from the brane over the potential term, and continuum of KK modes starts from zero. The eigenvalue equation for four-dimensional masses reads now

$$\frac{1}{a^4} \partial_z (a^4 \partial_z \phi) - V(z) \phi + \frac{m^2}{a^2} \phi = 0.$$ (6.6)

It is useful to note that the normalization condition for the eigenfunctions $\phi_m(z)$ is

$$\int dz\, a^2(z)\phi_m(z)\phi_{m'}(z) = \delta_{mm'}. \tag{6.7}$$

Indeed, the eigenvalue equation (6.6) can be written in the form

$$-\partial_z(a^4\partial_z\phi_m) + a^4 V(z)\phi_m = m^2 a^2\phi_m.$$

The operator entering the left hand side is Hermitean, so the eigenfunctions are orthogonal with the measure $a^2 dz$. This is precisely the same measure as multiplies the kinetic term in the action, i.e., the first term in the integrand of eq. (6.5).

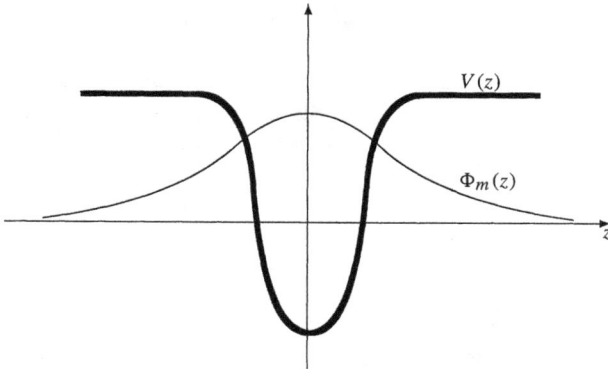

Fig. 6. Binding potential $V(z)$ and bound state with $m^2 \neq 0$ in the absence of the warp factor.

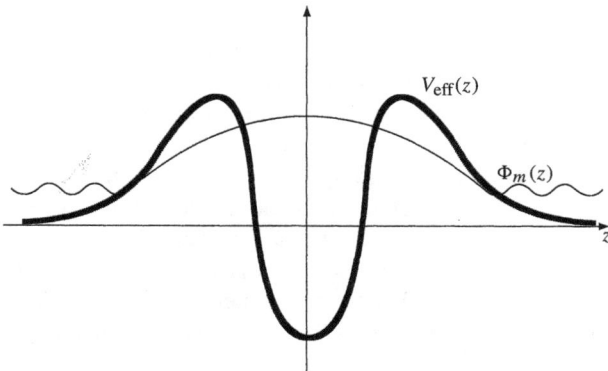

Fig. 7. In anti-de Sitter space, the effective binding potential gets modified, and would-be bound state with $m^2 \neq 0$ becomes quasi-localized. Particle has finite probability to escape the brane via tunneling.

At large z, the second term in eq. (6.6) is negligible as compared to the third one. Effectively, this means that the binding potential gets modified and tends to zero as $|z| \to \infty$, see Fig. 7. The wave functions at $z \to \infty$ are

$$\phi(z) = \text{const} \cdot e^{3kz/2} \sin\left(\frac{m}{k}e^{kz} + \varphi_m\right)$$

(these are normalized to delta-function with weight $a^2 dz$). The point is that the continuum spectrum is determined by the large-z asymptotics of eq. (6.6), and it starts from $m^2 = 0$ irrespectively of the form of $V(z)$ (provided that $V(z)$ does not rapidly increase as $z \to \infty$). As there are no bound states embedded in the continuum, the massive bound states of eq. (6.4) become now resonances, i.e., quasi-localized states having finite widths of decay (finite probability of escape from the brane to $z = \infty$). These widths depend on the potential $V(z)$ binding the particles to the brane.

It is worth noting that in RS2 set up, massless scalar field has a zero mode even without the potential $V(z)$ [63]. Its wave function is $\phi = \text{const}$, and it is normalizable with the appropriate weight $a^2 dz$. Once this field is given a mass, the would-be bound state becomes metastable against escape into extra dimension. For example, for constant $V(z) \equiv \mu^2$, the mass of the four-dimensional particle is equal to [62]

$$m = \frac{\mu}{\sqrt{2}}, \tag{6.8}$$

and its width of escape to $z = \infty$ is

$$\Gamma = m\frac{\pi}{16}\left(\frac{m}{k}\right)^2.$$

The latter formula illustrates a general feature of these decays: for small mass of the would-be bound state, the probability of its decay into extra dimension is small, the reason being that the decay occurs through tunneling, and hence it is naturally suppressed.

The above general arguments imply that the metastability of massive particles against escape into extra dimension should be characteristic to all kinds of matter, including fermions, provided these have bulk modes. The calculation of the life-time of a fermion bound to the brane by the mechanism of subsection 3.1, has been performed in Ref. [62]. This life-time depends not only on the fermion mass and anti-de Sitter radius, but also on other parameters, so quantitative estimates are premature at this stage.

Yet another interesting property of anti-de Sitter bulk has to do with virtual KK states. With massive four-dimensional bosons only, the potential between sources is of the Yukawa type. Once there exist arbitrarily light KK states, one expects the potential to have long ranged tail. For example, in the scalar field model (6.5) with constant $V(z) \equiv \mu^2$, one finds the potential between two distant sources of the scalar field, q_1 and q_2, located on the brane [62]

$$V(r) = -\pi q_1 q_2 k \frac{e^{-mr}}{r} - 60\pi q_1 q_2 \cdot \frac{1}{km^4} \cdot \frac{1}{r^7}, \qquad (6.9)$$

where m is given by eq. (6.8). The first (Yukawa) term here is due to the exchange by massive quasi-localized mode, whereas the second one is due to the exchange by the KK continuum states. In a model meant to describe a massive four-dimensional particle, the potential has power-law behaviour at large distances! The assumption that the field has bulk modes, which has been crucial for the above discussion, is not, in fact, an innocent one, especially if particles are charged. Indeed, if gauge fields are localized by the mechanism of subsection 3.2, charged particles are confined in the bulk, so they *do not* have bulk modes. In this scenario, escape of a charged particle from the brane into extra dimension is impossible.

6.3. Holographic interpretation

Is it possible to describe the low energy physics on the brane entirely in four-dimensional language? Certainly not, if one thinks in terms of usual weakly coupled theories allowing for particle interpretation: continuum KK modes do not correspond to particles travelling in four-dimensional space-time along the brane. On the basis of adS/CFT correspondence [64–66] it has been argued [67, 68], however, that RS2 scenario may be described by a strongly coupled four-dimensional conformal field theory (CFT) with an ultraviolet cut-off, interacting with conventional gravitational field. The correction (6.2) to Newton's gravity law is then interpreted as coming from "one loop" contribution of conformal matter to the graviton propagator [67–69]. Indeed, in conformal field theory language, this correction to the graviton propagator has the following form (indices are omitted),

$$\delta G(x - y) = \text{const} \int d^4u\, d^4v\, D(x - u)\langle T(u)T(v)\rangle D(v - y),$$

where $D(x - y)$ is a free four-dimensional graviton propagator, and $T(u)$ is the energy-momentum tensor of conformal fields. Hence,

$$\partial_x^2 \partial_y^2 (\delta G(x - y)) = \text{const} \cdot \langle T(x)T(y)\rangle.$$

Now, the correlator of the energy-momentum tensor of a conformal field theory is

$$\langle T(x)T(y)\rangle = \frac{\text{const}}{(x - y)^8},$$

so that

$$\delta G(x - y) = \frac{\text{const}}{(x - y)^4}.$$

The contribution to the gravitational potential is the time integral of this expression, which immediately gives

$$\Delta V(r) = \frac{\text{const}}{r^3},$$

in agreement with eq. (6.2). This is the most straightforward check of the interpretation of KK gravitons in terms of four-dimensional conformal field theory; there are a number of other checks.

Likewise, escape of a particle into extra dimension has a CFT interpretation as a decay into conformal modes, now interacting with matter fields, whereas the power-law correction to the Yukawa potential, eq. (6.9), is again due to the exchange by these modes.

Another way to see how effective conformal matter shows up in four-dimensional theory is to consider gravitational field of a massive point-like particle which sits on the brane until some moment of time (say, $t = 0$) and then leaves the brane and escapes into extra dimension along the geodesic normal to the brane [58,70]. The four-dimensional gravitational field induced by this particle is straightforward to calculate in linearized five-dimensional theory. One finds that outside the light cone, i.e., at $(\mathbf{x}^2 - t^2) > 0$, the four-dimensional gravitational field induced on the brane is still described by the linearized Schwarzschild metric (in other words, four-dimensional gravitational field does not change outside the light cone, in accord with causality). Inside the light cone, the induced four-dimensional metric is flat. If one *defines* the effective four-dimensional energy-momentum tensor by

$$8\pi G_{(4)} T_{\mu\nu}^{(\textit{eff})} \equiv R_{\mu\nu}^{(4)} - \frac{1}{2} g_{\mu\nu}^{(4)} R^{(4)},$$

where $g_{\mu\nu}^{(4)}$ is the four-dimensional metric induced on the brane, then $T_{\mu\nu}^{(\textit{eff})}$ corresponds to a thin shell of matter expanding along the four-dimensional light cone and dissipating as $1/r^2$. This is precisely the behaviour of energy-momentum expected in any conformal field theory [71].

The holographic approach is useful for analysing aspects of phenomenology of RS2 model [55, 69, 70] and sheds new light on cosmology with infinite extra dimesnions (see, e.g., Refs. [72–74] and references therein). The holographic approach to RS1 model and, in general, to models in a slice of adS$_5$ is also very interesting; it is discussed by T. Gherghetta in these Proceedings.

7. Brane worlds as a theoretical laboratory

Among other things, the brane world scenario provides a framework for addressing a variety of issues in field theory, particle phenomenology and cosmology which are hard to even address in conventional four-dimensional theories. Some examples may be found in these Proceedings [4, 5]. Other examples include:

– Apparent electric charge non-conservation for brane-based observer [75,76]. It may occur if electric charge may escape from the brane to bulk. A potential problem is that electrodynamics must be four-dimensional at least for charges on the brane, while electric charge non-conservation in four dimensions contradicts Maxwell's equations. Nevertheless, there exists at least one way to localize electromagnetic fields on the brane [75, 77], such that electric charges can indeed move into bulk. This mechanism is very much the same as the Randall–Sundrum mechanism of the gravity localization, and charges can escape into bulk in much the same way as masses in RS2 setup do. Unlike in the gravitational case, however, a holographic interpretation of the charge escape remains obscure.

– Infrared modification of gravity. There is still a hope that modification of gravity at cosmological distances may provide an explanation of the accelerated expansion of the Universe, which would be an alternative to the dark energy explanation. The idea behind is that the graviton itself may be *quasi*-localized, rather than perfectly localized on the brane. However, models of this sort which preserve 4d Lorentz-invariance, except possibly [78] for the DGP model [79], have either ghosts (fields with negative kinetic terms) or unacceptably low UV strong coupling scale. Models violating Lorentz-invariance are more promising, but none of the existing ones leads to accelerating Universe.

Here we discuss in some detail yet another example [80, 81], which has to do with "trans-Planckian problem" in inflationary cosmology [82–84]. Namely, the standard inflationary mechanism of the generation of cosmological perturbations assumes Lorentz-invariance down to very short distances. This assumption is worth questioning, especially in view of the fact that relevant perturbations initially have wavelengths much shorter than the Planck length. In Lorentz-invariant theory these perturbations are always very close to mass shell, so it does not matter for their description what their spatial wavelength is; one never deals with sub-Planckian physics (in Lorentz-invariant theory it is not three-momentum, but rather momentum transfer — virtuality — that actually probes "short distance physics"). If Lorentz-invariance does not hold at very short spatial distances, the latter argument does not work, so one might expect observable physical effects originating from sub-Planckian physics. Let us note further, that there is no guarantee that Lorentz-invariance is valid even at the length scale characteristic of inflation itself, the horizon size towards the end of inflation.

In this context, there are two possible viwepoints. The more conservative one is that the three-momentum scale of Lorentz-violation, P_{LV}, is much larger than the inflationary Hubble parameter H. It has been argued [85, 86] that for $P_{LV} \gg H$, the standard inflationary predictions should remain almost intact, although there are fairly exotic four-dimensional counterexamples [87]. Barring exotica, there is still some debate [85, 88, 89] on whether the effects of Lorentz-violation, and "heavy physics" in general, are at best of order $(H/P_{LV})^2$, or weaker suppression is possible. It is this conservative case $P_{LV} \gg H$ that we discuss below.

Another option is that $P_{LV} \lesssim H$, that is Lorentz-violation at inflation is strong. In that case the outcome for the spectrum of cosmological perturbations generated at inflation may be even more interesting [81].

A particularly fascinating possibility is "mode generation": one may try to invent a model where the number of modes of a field increases in time, as the Universe expands [90–94]; this happens, e.g., if the physical momentum is effectively bounded from above. One may have in mind a possibility that there is a physical lattice in space with fixed lattice spacing l_{lat} (then the maximum spatial momentum is of order l_{lat}^{-1}, i.e., it is indeed finite). As the Universe expands, more and more lattice sites must somehow appear in comoving volume, thus more and more modes (degrees of freedom) of any field must somehow be created. The problem is that in four-dimensional theories, the result for the spectrum of perturbations strongly depends on the choice of a state in which a newborn mode emerges. If this is the adiabatic vacuum, the outcome is similar to the standard scenario, at least for pure de Sitter inflation [91, 92, 94], but there does not seem to be a particular reason to choose the adiabatic vacuum for the newborn mode. We will see shortly that brane worlds offer an unambiguous and calculable framework for mode generation.

In brane-world scenarios, it is not inconceivable that four-dimensional Lorentz-invariance is violated in the bulk [9, 95–100]. An example is given by $(4 + 1)$-dimensional setup with the background five-dimensional metric

$$ds^2 = [\alpha^2(z)dt^2 - \beta^2(z)a^2(t)d\mathbf{x}^2] - \alpha^2(z)dz^2, \tag{7.1}$$

where the coordinate choice for z is made for convenience. There is a single brane at $z = z_B$. The warp factors $\alpha(z)$ and $\beta(z)$ are continuous across the brane, while their derivatives $\partial_z\alpha \equiv \alpha'$ and $\partial_z\beta \equiv \beta'$ are not. The static background, $a(t) = \text{const}$, is not four-dimensinally Lorentz-invariant; it is this case that has been considered in Refs. [98, 99], where it has been shown that, with appropriate choice of the warp factors, four-dimensional Lorenz-invariance still holds for brane modes. An equivalent form of the metric is

$$ds^2 = a^2(\eta)[\alpha^2(z)d\eta^2 - \beta^2(z)d\mathbf{x}^2] - \alpha^2(z)dz^2, \tag{7.2}$$

where the conformal time η is related to time t in the usual way, $dt = a(\eta)d\eta$.

Let us choose the warp factors in such a way that both $\alpha(z)$ and $\beta(z)$ are Z_2-symmetric across the brane, monotonically decrease towards large z with $\beta'' > 0$ and decay away from the brane,

$$\alpha(z), \beta(z) \to 0, \quad \text{as } z \to \infty.$$

Furthermore, we assume that the ratio of warp factors tends to a small constant,

$$\frac{\alpha(z)}{\beta(z)} \to \epsilon, \quad \text{as } z \to \infty, \tag{7.3}$$

where $\epsilon \ll 1$. The parameter ϵ is a free small parameter of our model. We are interested in inflating background,

$$a(t) = \exp\left(\int H(t)dt\right), \tag{7.4}$$

where $H(t)$ is a slowly varying function of time.

Finally, we assume that $\beta(z)$ decays sufficiently slowly as $z \to \infty$, so that

$$\frac{\beta'(z)}{\beta(z)}, \frac{\beta''(z)}{\beta(z)} \to 0, \quad \text{as } z \to \infty. \tag{7.5}$$

We should stress that this bulk geometry is completely *ad hoc*; neither the source of Lorentz-violation nor mechanism of inflation are specified. An advantage, however, is that the behavior of quantum fields, in particular, their initial state, is well under control, so the calculation of the spectrum of perturbations generated at inflation is unambiguous.

Let us consider a real massless scalar field Φ in the background (7.1), meant to model perturbations of gravitational and/or inflaton field. It is consistent with Z_2 symmetry across the brane to impose the Neumann boundary condition,

$$\Phi'(z = z_B) = 0.$$

Indeed, this boundary condition applies to appropriately defined gravitational perturbations, and also to free scalar field in the RS2 background. It is convenient to introduce the field

$$\chi(\eta, x^i, z) = a(\eta)\beta^{3/2}(z) \cdot \Phi(X^A),$$

which obeys the field equation, in terms of three-dimensional Fourier harmonics,

$$\ddot{\chi} - \frac{\ddot{a}}{a}\chi + U(z)k^2\chi + a^2(\eta)L_z\chi = 0. \tag{7.6}$$

Hereafter dot denotes $d/d\eta$, k is time-independent conformal 3-momentum, while $P(t) = \frac{k}{a(t)}$ is physical 3-momentum. The quantities entering (7.6) are

$$U(z) = \frac{\alpha^2(z)}{\beta^2(z)}, \tag{7.7}$$

and $L_z = -\partial_z^2 + V(z)$ with

$$V(z) = \frac{3}{2}\frac{\beta''}{\beta} + \frac{3}{4}\frac{\beta'^2}{\beta^2}.$$

The boundary condition for χ is

$$\left(\chi' - \frac{3}{2}\frac{\beta'}{\beta}\chi\right)_{z=z_B} = 0. \tag{7.8}$$

The field χ is canonically normalized: its action is

$$S_\chi = \int d\eta\, d^3x\, dz \left(\frac{1}{2}\dot{\chi}^2 + \dots\right),$$

where dots denote terms independent of $\dot{\chi}$.

Let us first discuss bulk modes, assuming for the time being that the background is static, $H = 0$. According to our assumptions, the potential $V(z)$ is positive and vanishes as $z \to \infty$, so the spectrum of the operator L_z is continuous and starts from zero. Recalling eq. (7.3), we find that the dispersion relation for these modes is

$$\omega^2 = \epsilon^2 P^2 + \lambda^2, \tag{7.9}$$

where λ^2 are eigenvalues of L_z, which can be arbitrarily small.

Let us now consider brane modes, still in static background. At $P^2 = 0$ there is a normalizable zero mode, $\omega = 0$, whose shape is

$$\chi_0(z) = \beta^{3/2}(z).$$

It is worth noting that in terms of the original filed Φ this mode is constant along extra dimension. Now, at finite but small P, this mode gets lifted; the third term in eq. (7.6) can be treated as perturbation, and for the (real part of) energy one finds

$$\omega^2 = P^2 \cdot \frac{\int_{z_B}^{\infty} dz\, U(z)|\chi_0|^2}{\int_{z_B}^{\infty} dz|\chi_0|^2}.$$

By rescaling of t and x^i one obtains the Lorentz-invariant dispersion relation, $\omega^2 = P^2$. Thus, the theory on the brane is (almost) Lorentz-invariant at low 3-momenta.

This is not the whole story, however. At small but finite P, the would be zero mode is embedded in the continuum of bulk modes: its energy is larger than the lowest energy $\omega = \epsilon P$ of continuum modes, see eq. (7.9). Therefore, the brane mode is *quasi*-localized even at low 3-momenta, i.e., it has finite width against escape into extra dimension. This effect was found in Ref. [99]. The quasi-localization is due to mixing with bulk modes, which comes from the third term in eq. (7.6). Introducing the overlap between the (normalized) zero mode χ_0 and continuum modes χ_λ,

$$I_\lambda = \int dz U(z) \chi_0^*(z) \chi_\lambda(z), \qquad (7.10)$$

we estimate the width of the quasilocalized state as

$$\Gamma(P) \sim P^2 \cdot |I_{\lambda=P}|^2. \qquad (7.11)$$

The whole picture of weak Lorentz-violation on the brane at low 3-momenta and strong Lorentz-violation at high 3-momenta works provided that the warp factors are chosen in such a way that

$$\lambda |I_\lambda|^2 \to 0, \quad \text{as} \quad \frac{\lambda}{P_{LV}} \to 0, \qquad (7.12)$$

and

$$\lambda |I_\lambda|^2 \sim 1, \quad \text{at} \quad \lambda \sim P_{LV}.$$

It is this case that we consider in what follows. Then, the width of the brane mode is small compared to its energy at low P, but it becomes comparable to energy $\omega = P$ at $P \gtrsim P_{LV}$. At $P > P_{LV}$ even quasi-localized brane mode ceases to exist, so four-dimensional Lorentz-invariance is indeed completely destroyed.

Thus, the model indeed has a pre-requisite property for mode generation, understood here as generation of a brane mode from bulk modes: maximum three-momentum for which the brane mode exists is finite and equal to P_{LV}. In the cosmological context, this means for a given conformal momentum k that at early times, when physical momentum k/a exceeds P_{LV}, there are bulk modes only, while at the time when the physical momentum gets redshifted to $k/a \sim P_{LV}$ the brane mode gets formed.

Let us now consider the scalar field in inflating setup, eq. (7.4). It is convenient to work with the field $\chi(\eta, z)$ and decompose it in eigenfunctions of the operator L_z,

$$\chi(\eta, z) = \psi_0(\eta)\chi_0(z) + \int_0^\infty d\lambda \psi_\lambda(\eta)\chi_\lambda(z).$$

From eq. (7.6) one obtains the system of equations

$$\ddot{\psi}_0 - \frac{\ddot{a}}{a}\psi_0 + k^2\psi_0 = -k^2 \int d\lambda \, \psi_\lambda I_\lambda, \tag{7.13}$$

$$\ddot{\psi}_\lambda - \frac{\ddot{a}}{a}\psi_\lambda + \epsilon^2 k^2 \psi_\lambda + a^2\lambda^2\psi_\lambda = -k^2 I_\lambda^* \psi_0 - k^2 \int d\lambda' \, \psi_{\lambda'} I_{\lambda,\lambda'}, \tag{7.14}$$

where

$$I_{\lambda\lambda'} = \int dz \, \chi_\lambda^*[U(z) - \epsilon^2]\chi_{\lambda'}. \tag{7.15}$$

We are interested in the brane mode, ψ_0, towards the end of inflation. To this end, we will treat the right hand sides of Eqs. (7.13) and (7.14) as perturbations. This approximation is certainly *not* valid at very early times, when $P(\eta) \equiv k/a(\eta) \gtrsim P_{LV}$: at those times the very notion of the brane mode does not make sense. As we discussed above, this is precisely the brane-world version of mode generation. However, for $P_{LV} \gg H$ (the case we consider here) the relevant modes are in the adiabatic regime at those times, mixing between positive- and negative-frequency components of the field is negligible, and the field remains in its adiabatic vacuum state. Thus, mode generation *per se* does not introduce any interesting features into the spectrum of cosmological perturbations. At later times, perturbation theory *is* justified by the fact that $U(z)$ peaks near the brane, where $\chi_\lambda(z)$ are suppressed; hence the overlap integrals (7.10) and (7.15) are small. In the zeroth approximation, with overlaps neglected, equations for the brane mode and bulk modes decouple and can be straighforwardly solved. At this level, equation for the brane mode is

$$\ddot{\psi}_0 - \frac{\ddot{a}}{a}\psi_0 + k^2\psi_0 = 0.$$

It exactly coincides with the corresponding equation in four-dimensional theory, so the zeroth order result for the spectrum is exactly the same as in four dimensions. Namely, towards the end of inflation, $\psi_0^{(0)}(\mathbf{k}, \eta)$ (the superscript here refers to the zeroth order approximation) is a Gaussian field with the correlation function

$$\langle \psi_0^{(0)}(\mathbf{k}), \psi_0^{(0)}(\mathbf{k}')\rangle = a^2(\eta)\frac{2\pi^2}{k^3}\mathcal{P}^{(0)}(k)\delta^3(\mathbf{k} - \mathbf{k}'),$$

with

$$\mathcal{P}^{(0)} = \frac{H_k^2}{4\pi^2},$$

and $H_k = H(\eta_k)$, where η_k is the time at which the mode of momentum k crosses out the horizon, $H(\eta_k) = \frac{k}{a(\eta_k)}$. Thus, four-dimensional behavior of the field in the brane mode is trivially obtained in the zeroth approximation.

The first non-trivial contribution to the field in the brane mode occurs at the first order of perturbation theory, and is due to the overlap with bulk modes in the right hand side of eq. (7.13). To evaluate this contribution, let us first solve eq. (7.14) at the zeroth order.

At the zeroth order, eq. (7.14) becomes

$$\ddot{\psi}_\lambda - \frac{\ddot{a}}{a}\psi_\lambda + \epsilon^2 k^2 \psi_\lambda + a^2 \lambda^2 \psi_\lambda = 0.$$

In the asymptotic past, the second and fourth terms in this equation are negligible, the field is in the adiabatic regime, and we immediately write its decomposition in creation and annihilation operators,

$$\psi_\lambda^{(0)} = \frac{1}{\sqrt{2\epsilon k}}(\psi_\lambda^+(\eta) A_{\lambda,k}^+ + \text{h.c.}), \tag{7.16}$$

where in the asymptotic past

$$\psi_\lambda^+ = e^{i\epsilon k\eta}, \quad \eta \to -\infty.$$

The main effect on the brane mode comes from the bulk modes with $\lambda \ll H$, where H is, roughly speaking, the inflationary Hubble parameter. These modes get out from the adiabatic regime at the time $\eta_{\epsilon k}$ such that

$$H(\eta_{\epsilon k}) \equiv H_{\epsilon k} \sim \frac{\epsilon k}{a(\eta_{\epsilon k})}.$$

Let us call this moment of time "ϵ-horizon crossing". One of the key observations is that for small ϵ, this moment occurs much earlier than the horizon crossing by the brane mode. In terms of the original field Φ, the bulk modes thus freeze out at much larger amplitudes than the brane mode, and their effect on brane mode is enhanced.

Immediately after ϵ-horizon crossing, the bulk mode behaves as

$$\psi_\lambda^+(\eta) = a(\eta)\frac{H_{\epsilon k}}{\epsilon k}, \quad \eta \gtrsim \eta_{\epsilon k}. \tag{7.17}$$

We need this mode at the moment η_k at which the brane mode crosses out the horizon. For small ϵ this moment occurs much later than ϵ-horizon crossing, so some care must be taken at this point. One can see that at $\eta \sim \eta_k$ the amplitude is

$$\psi_\lambda^+(\eta) = a(\eta) \frac{H_{\epsilon k}}{\epsilon k} \exp\left(-\frac{\lambda^2}{3\hat{H}^2}|\log \epsilon|\right), \quad \eta \sim \eta_k, \tag{7.18}$$

where $\hat{H}$ is a sort of average value of the Hubble parameter in the time interval $(\eta_{\epsilon k}, \eta_k)$.

We now wish to calculate the effect of the bulk modes on the brane mode, in the lowest non-trivial order of perturbation theory. To this end, we insert the zeroth order expression for the bulk modes, eq. (7.16) into eq. (7.13) and obtain the equation for the first correction to ψ_0,

$$\ddot{\psi}_0^{(1)} - \frac{\ddot{a}}{a}\psi_0^{(1)} + k^2 \psi_0^{(1)} = -k^2 \int d\lambda\, \psi_\lambda^{(0)} I_\lambda. \tag{7.19}$$

We are interested in the solution to this equation with zero initial condition at infinite past: all modes oscillate as $\eta \to -\infty$, and mixing between the modes is negligible at that time. This solution can be obtained by standard Green's function technique. One obtains for late times, $\eta \gg \eta_k$,

$$\psi_0^{(1)}(\eta) = -a(\eta)\frac{1}{\sqrt{2}}\frac{H_{\epsilon k}}{\epsilon^{\frac{3}{2}}k^{\frac{3}{2}}}\int d\lambda\, I_\lambda e^{-\frac{\lambda^2}{3\hat{H}^2}|\log \epsilon|} A_\lambda^+ + \text{h.c.}. \tag{7.20}$$

This is again a Gaussian field whose spectrum is

$$\mathcal{P}^{(1)} = \frac{H_{\epsilon k}^2}{4\pi^2\epsilon^3}\int d\lambda\, |I_\lambda|^2 e^{-\frac{2\lambda^2}{3\hat{H}^2}|\log \epsilon|}.$$

Because of the relation (7.12), the integral here is convergent at small λ, and therefore it is saturated at

$$\lambda \sim \lambda_c = \frac{\hat{H}}{|\log \epsilon|^{1/2}} \ll H.$$

One obtains finally

$$\mathcal{P}^{(1)} = \text{const} \cdot \frac{H_{\epsilon k}^2}{\epsilon^3}\lambda_c|I_{\lambda_c}|^2, \tag{7.21}$$

where the constant is of order 1.

Several remarks are in order. First, the Gaussian field (7.20) is independent of the field $\psi_0^{(0)}$ considered in the previous subsection, as the latter contains creation and annihilation operators in the incoming brane modes. Thus, towards the end of inflation, perturbations on the brane are the sum of two independent Gaussian fields with spectra $\mathcal{P}^{(0)}$ and $\mathcal{P}^{(1)}$. Second, because of the relation (7.12), the contribution (7.21) indeed tends to zero as $P_{LV} \to \infty$. However, the suppression at small H/P_{LV} may be quite weak, depending on the form of the warp factors. In general

$$\lambda_c I_{\lambda_c}^2 \sim \left(\frac{\hat{H}}{P_{LV}} \right)^{2\nu - 2}$$

(up to $\log \epsilon$), while the only restriction on the parameter ν is $\nu > 1$. Third, the bulk-induced contribution to the spectrum is enhanced by ϵ^{-3}; we have already discussed the origin of this enhancement. Fourth, the overall magnitude of $\mathcal{P}^{(1)}$ is determined by $H_{\epsilon k}$, the value of the Hubble parameter at ϵ-horizon crossing, which occurs earlier than the usual horizon crossing by the brane mode, and also by $\hat{H}$, which is a certain average of the Hubble parameter over rather long time. Thus, the two contributions to the spectrum, $\mathcal{P}^{(0)}$ and $\mathcal{P}^{(1)}$ generically have different tilts. Finally, for exactly de Sitter inflation with $H(t) = const$, both spectra $\mathcal{P}^{(0)}$ and $\mathcal{P}^{(1)}$ are exactly flat. This appears to be a very general property of the de Sitter background even in Lorentz-violating theories. This property originates from the symmetries of de Sitter space-time.

To summarize, in Lorentz-violating brane-world models, the properties of cosmological perturbations generated at inflation may strongly depend on dynamics in the bulk, even if 3-momentum scale of Lorentz-violation on the brane largely exceeds the inflationary Hubble parameter. Needless to say, this dynamics may naturally be entirely different for, e.g., inflaton and graviton modes, so that the standard relations between the scalar and tensor perturbations may be completely (or partially) destroyed.

References

[1] I. P. Volobuev and Y. A. Kubyshin, Theor. Math. Phys. **68** (1986) 788; Theor. Math. Phys. **68** (1986) 885; JETP Lett. **45** (1987) 581.

[2] I. Antoniadis, Phys. Lett. B **246** (1990) 377.

[3] J. D. Lykken, Phys. Rev. D **54** (1996) 3693 [hep-th/9603133].

[4] T. Gherghetta, these proceedings.

[5] C. Grojean, these proceedings.

[6] J. Polchinski, hep-th/9611050.

[7] P. Horava and E. Witten, Nucl. Phys. B **460** (1996) 506 [hep-th/9510209].

[8] A. Lukas, B. A. Ovrut, K. S. Stelle and D. Waldram, Phys. Rev. D **59** (1999) 086001 [hep-th/9803235].

[9] M. Visser, Phys. Lett. B **159** (1985) 22 [hep-th/9910093].

[10] V. A. Rubakov and M. E. Shaposhnikov, Phys. Lett. B **125** (1983) 136.

[11] K. Akama, Lect. Notes Phys. **176** (1982) 267 [hep-th/0001113].

[12] R. Jackiw and C. Rebbi, Phys. Rev. D **13** (1976) 3398.

[13] R. Jackiw and P. Rossi, Nucl. Phys. B **190** (1981) 681.

[14] G. 't Hooft, Phys. Rev. D **14** (1976) 3432.

[15] M. V. Libanov and S. V. Troitsky, Nucl. Phys. B **599** (2001) 319 [hep-ph/0011095].

[16] J. M. Frere, M. V. Libanov and S. V. Troitsky, Phys. Lett. B **512** (2001) 169 [hep-ph/0012306].

[17] M. V. Libanov and E. Y. Nougaev, JHEP **0204** (2002) 055 [hep-ph/0201162].

[18] J. M. Frere, M. V. Libanov, E. Y. Nugaev and S. V. Troitsky, JHEP **0306** (2003) 009 [hep-ph/0304117].

[19] J. M. Frere, M. V. Libanov, E. Y. Nugaev and S. V. Troitsky, JHEP **0403** (2004) 001 [hep-ph/0309014]; JETP Lett. **79** (2004) 598 [Pisma Zh. Eksp. Teor. Fiz. **79** (2004) 734] [hep-ph/0404139].

[20] G. Dvali and M. Shifman, Phys. Lett. B **396** (1997) 64 [hep-th/9612128].

[21] M. Laine, H. B. Meyer, K. Rummukainen and M. Shaposhnikov, JHEP **0404** (2004) 027 [hep-ph/0404058].

[22] N. Arkani-Hamed, S. Dimopoulos and G. Dvali, Phys. Lett. B **429** (1998) 263 [hep-ph/9803315].

[23] I. Antoniadis, N. Arkani-Hamed, S. Dimopoulos and G. Dvali, Phys. Lett. B **436** (1998) 257 [hep-ph/9804398].

[24] C. D. Hoyle, U. Schmidt, B. R. Heckel, E. G. Adelberger, J. H. Gundlach, D. J. Kapner and H. E. Swanson, Phys. Rev. Lett. **86** (2001) 1418 [hep-ph/0011014].

[25] V. P. Mitrofanov and O. I. Ponomareva, JETP **67** (1988) 1963.

[26] Y. Su, B. R. Heckel, E. G. Adelberger, J. H. Gundlach, M. Harris, G. L. Smith and H. E. Swanson, Phys. Rev. D **50** (1994) 3614.

[27] J. C. Long, H. W. Chan and J. C. Price, Nucl. Phys. B **539** (1999) 23 [hep-ph/9805217].

[28] E. G. Adelberger, B. R. Heckel and A. E. Nelson, Ann. Rev. Nucl. Part. Sci. **53** (2003) 77 [hep-ph/0307284].

[29] J. C. Long, A. B. Churnside and J. C. Price, hep-ph/0009062.

[30] V. B. Braginsky, private communication.

[31] J. Hewett, these proceedings.

[32] N. Arkani-Hamed, S. Dimopoulos and G. Dvali, Phys. Rev. D **59** (1999) 086004 [hep-ph/9807344].

[33] G. F. Giudice, E. W. Kolb and A. Riotto, Phys. Rev. D **64** (2001) 023508 [hep-ph/0005123].

[34] S. Cullen and M. Perelstein, Phys. Rev. Lett. **83** (1999) 268 [hep-ph/9903422].

[35] V. Barger, T. Han, C. Kao and R. J. Zhang, Phys. Lett. B **461** (1999) 34 [hep-ph/9905474]

[36] C. Hanhart, D. R. Phillips, S. Reddy and M. J. Savage, Nucl. Phys. B **595** (2001) 335 [nucl-th/0007016].

[37] S. Hannestad and G. Raffelt, Phys. Rev. Lett. **87** (2001) 051301 [hep-ph/0103201].

[38] W. Israel, Nuovo Cim. B **44S10** (1966) 1.

[39] V. A. Berezin, V. A. Kuzmin and I. I. Tkachev, Phys. Rev. D **36** (1987) 2919.

[40] D. Z. Freedman, S. S. Gubser, K. Pilch and N. P. Warner, JHEP **0007** (2000) 038 [hep-th/9906194].

[41] M. Gogberashvili, Europhys. Lett. **49** (2000) 396 [hep-ph/9812365].

[42] M. Gogberashvili, Mod. Phys. Lett. A **14** (1999) 2025 [hep-ph/9904383].

[43] L. Randall and R. Sundrum, Phys. Rev. Lett. **83** (1999) 3370 [hep-ph/9905221].

[44] L. Randall and R. Sundrum, Phys. Rev. Lett. **83** (1999) 4690 [hep-th/9906064].

[45] V. A. Rubakov and M. E. Shaposhnikov, Phys. Lett. B **125** (1983) 139.

[46] T. Maehara, T. Muta, J. Saito and K. Shimizu, Phys. Rev. D **30** (1984) 1397.

[47] C. Wetterich, Nucl. Phys. B **253** (1985) 366.

[48] H. Nicolai and C. Wetterich, Phys. Lett. B **150** (1985) 347.

[49] L. Pilo, R. Rattazzi and A. Zaffaroni, JHEP **0007** (2000) 056 [hep-th/0004028].

[50] C. Csaki, M. Graesser, L. Randall and J. Terning, Phys. Rev. D **62** (2000) 045015 [hep-ph/9911406].

[51] C. Charmousis, R. Gregory and V. A. Rubakov, Phys. Rev. D **62** (2000) 067505 [hep-th/9912160].

[52] W. D. Goldberger and M. B. Wise, Phys. Rev. Lett. **83** (1999) 4922 [hep-ph/9907447].

[53] M. A. Luty and R. Sundrum, Phys. Rev. D **62** (2000) 035008 [hep-th/9910202].

[54] J. Garriga and T. Tanaka, Phys. Rev. Lett. **84** (2000) 2778 [hep-th/9911055].

[55] S. B. Giddings, E. Katz and L. Randall, JHEP **0003** (2000) 023 [hep-th/0002091].

[56] T. Gherghetta and A. Pomarol, Nucl. Phys. B **602** (2001) 3 [hep-ph/0012378].

[57] W. Muck, K. S. Viswanathan and I. V. Volovich, Phys. Rev. D **62** (2000) 105019 [hep-th/0002132].

[58] R. Gregory, V. A. Rubakov and S. M. Sibiryakov, Class. Quant. Grav. **17** (2000) 4437 [hep-th/0003109].

[59] J. Lykken and L. Randall, JHEP **0006** (2000) 014 [hep-th/9908076].

[60] A. Hebecker and J. March-Russell, Nucl. Phys. B **608** (2001) 375 [hep-ph/0103214].

[61] M. Shaposhnikov, P. Tinyakov and K. Zuleta, JHEP **0509** (2005) 062 [hep-th/0508102].

[62] S. L. Dubovsky, V. A. Rubakov and P. G. Tinyakov, Phys. Rev. D **62** (2000) 105011 [hep-th/0006046].

[63] B. Bajc and G. Gabadadze, Phys. Lett. B **474** (2000) 282 [hep-th/9912232].

[64] J. Maldacena, Adv. Theor. Math. Phys. **2** (1998) 231 [hep-th/9711200].

[65] S. S. Gubser, I. R. Klebanov and A. M. Polyakov, Phys. Lett. B **428** (1998) 105 [hep-th/9802109].

[66] E. Witten, Adv. Theor. Math. Phys. **2** (1998) 505 [hep-th/9803131].

[67] H. Verlinde, talk at the ITP Santa Barbara Conference on "New Dimensions in Field Theory and String Theory", http://www.itp.ucsb.edu/online/susy_c99/verlinde;
E. Witten, ibid., http://www.itp.ucsb.edu/online/susy_c99/discussion.

[68] S. S. Gubser, Phys. Rev. D **63** (2001) 084017 [hep-th/9912001].

[69] N. Arkani-Hamed, M. Porrati and L. Randall, JHEP **0108** (2001) 017 [hep-th/0012148].

[70] S. B. Giddings and E. Katz, J. Math. Phys. **42** (2001) 3082 [hep-th/0009176].

[71] S. Coleman and L. Smarr, Commun. Math. Phys. **56** (1977) 1.

[72] S. W. Hawking, T. Hertog and H. S. Reall, Phys. Rev. D **62** (2000) 043501 [hep-th/0003052].

[73] S. Nojiri and S. D. Odintsov, Phys. Lett. B **484** (2000) 119 [hep-th/0004097].

[74] L. Anchordoqui, C. Nunez and K. Olsen, JHEP **0010** (2000) 050 [hep-th/0007064].

[75] S. L. Dubovsky, V. A. Rubakov and P. G. Tinyakov, JHEP **0008** (2000) 041 [hep-ph/0007179].

[76] S. L. Dubovsky and V. A. Rubakov, hep-th/0204205.

[77] I. Oda, Phys. Lett. B **496** (2000) 113 [hep-th/0006203].

[78] A. Nicolis and R. Rattazzi, JHEP **0406** (2004) 059 [hep-th/0404159].

[79] G. R. Dvali, G. Gabadadze and M. Porrati, Phys. Lett. B **485** (2000) 208 [hep-th/0005016].

[80] M. V. Libanov and V. A. Rubakov, JCAP **0509** (2005) 005 [astro-ph/0504249].

[81] M. V. Libanov and V. A. Rubakov, hep-ph/0509148.

[82] J. Martin and R. H. Brandenberger Phys. Rev. D **63** (2001) 123501 [hep-th/0005209].

[83] J. Martin and R. H. Brandenberger Mod. Phys. Lett. A **16** (2001) 999 [astro-ph/0005432].

[84] J. C. Niemeyer Phys. Rev. D **63** (2001) 123502 [astro-ph/0005533].

[85] N. Kaloper, M. Kleban, A. E. Lawrence and S. Shenker, Phys. Rev. D **66** (2002) 123510 [hep-th/0201158].

[86] K. Schalm, G. Shiu and J. P. van der Schaar, JHEP **0404** (2004) 076 [hep-th/0401164].

[87] J. Martin and R. H. Brandenberger, Phys. Rev. D **65** (2002) 103514 [hep-th/0201189].

[88] C. P. Burgess, J. M. Cline, F. Lemieux and R. Holman, JHEP **0302** (2003) 048 [hep-th/0210233].

[89] C. P. Burgess, J. M. Cline and R. Holman, JCAP **0310** (2003) 004 [hep-th/0306079].

[90] A. Kempf Phys. Rev. D **63** (2001) 083514 [astro-ph/0009209].

[91] A. Kempf and J. C. Niemeyer Phys. Rev. D **64** (2001) 103501 [astro-ph/0103225].

[92] R. Easther, B. R. Greene, W. H. Kinney and G. Shiu Phys. Rev. D **64** (2001) 103502 [hep-th/0104102].

[93] R. Easther, B. R. Greene, W. H. Kinney and G. Shiu Phys. Rev. D **67** (2003) 063508 [hep-th/0110226].

[94] R. Brandenberger R and P. M. Ho Phys. Rev. D **66** (2002) 023517 [hep-th/0203119].

[95] D. J. H. Chung and K. Freese, Phys. Rev. D **61** (2000) 023511 [hep-ph/9906542].

[96] D. J. H. Chung and K. Freese, Phys. Rev. D **62** (2000) 063513 [hep-ph/9910235].

[97] D. J. H. Chung, E. W. Kolb and A. Riotto, Phys. Rev. D **65** (2002) 083516 [hep-ph/0008126].

[98] C. Csaki, J. Erlich and C. Grojean, Nucl. Phys. B **604** (2001) 312 [hep-th/0012143].

[99] S. L. Dubovsky, JHEP **0201** (2002) 012 [hep-th/0103205].

[100] D. S. Gorbunov and S. M. Sibiryakov, hep-th/0506067.

Course 5

PHENOMENOLOGY OF EXTRA DIMENSIONS

JoAnne L. Hewett

Stanford Linear Accelerator Center, Stanford University, Stanford, CA 94309 USA

D. Kazakov, S. Lavignac and J. Dalibard, eds.
Les Houches, Session LXXXIV, 2005
Particle Physics Beyond the Standard Model
© *2006 Elsevier B.V. All rights reserved*

Contents

1. Introduction

In particle physics, 4-dimensional Minkowski spacetime is the underlying fundamental framework under which the laws of nature are formulated and interpreted. Relativistic quantum fields exist in spacetime, interactions occur at spacetime points and the laws governing these fields and their interactions are constructed using weighted averages over their spacetime histories. According to the general theory of relativity, fluctuations of the spacetime curvature provide gravitational dynamics. Indeed, experiments show evidence for the predictions of general relativity and hence that spacetime is dynamical at very long length scales. However, gravitational dynamics have yet to be probed at short distances, and it is possible that they are quite different from that implied by a simple extrapolation of the long range theory.

Early attempts to extend general relativity in order to unify gravity and electromagnetism within a common geometrical framework trace back to Gunnar Nordström (1914) [1], Theodor Kaluza (1921) and Oscar Klein (1926) [2]. They proposed that unification of the two forces occurred when spacetime was extended to a five dimensional manifold and imposed the condition that the fields should not depend on the extra dimension. A difficulty with the acceptance of these ideas at the time was a lack of both experimental implications and a quantum description of gravitational dynamics.

Today, one of the most striking requirements of modern string theory, which incorporates both gauge theories and gravitation, is that there must be six or seven extra spatial dimensions. Otherwise the theory is anomalous. Recently, concepts developed within string theory have led to new phenomenological ideas which relate the physics of extra dimensions to observables in a variety of physics experiments.

These new theories have been developed to address the hierarchy problem, *i.e.*, the large disparity between the electroweak scale ($\sim 10^3$ GeV) where electroweak symmetry breaking occurs and the traditional scale of gravity defined by the Planck scale (10^{19} GeV). The source of physics which generates and stabilizes this sixteen order of magnitude difference between the two scales is unknown and represents one of the most puzzling aspects of nature. The novel approach to this long-standing problem proposed in these recent theories is that the geometry of extra spatial dimensions may be responsible for the hierarchy:

233

the gravitational field lines spread throughout the full higher dimensional space and modify the behavior of gravity. Indeed, the fact that gravity has yet to be measured at energy scales much above 10^{-3} eV in laboratory experiments admits for the possibility that at higher energies gravity behaves quite differently than expected. The first scenario of this type to be proposed [3] suggested that the apparent hierarchy between these two important scales of nature is generated by a large volume of the extra dimensions, while in a later theoretical framework [4, 5] the observed hierarchy results from a strong curvature of the extra dimensional space. If new dimensions are indeed relevant to the source of the hierarchy, then they should provide detectable signatures at the electroweak scale. These physics scenarios with additional dimensions hence afford concrete and distinctive phenomenological predictions for high energy colliders, as well as producing observable consequences for astrophysics and short-range gravity experiments.

Theoretical frameworks with extra dimensions have some general features. In most scenarios, our observed 3-dimensional space is a 3-brane (sometimes called a wall), where the terminology is derived from a generalization of a 2-dimensional membrane. This 3-brane is embedded in a higher D-dimensional spacetime, $D = 3 + \delta + 1$, with δ extra spatial dimensions which are orthogonal to our 3-brane. The higher D-dimensional space is known as the "bulk". The branes provide a mechanism to hide the existence of extra dimensions in that an observer trapped on a brane can not directly probe the dimensions transverse to the brane without overcoming the brane tension. String theory contains branes upon which particles can be naturally confined or localized [6]. In a general picture, branes carry the Standard Model gauge charges and the ends of open strings are stuck to the branes and represent the Standard Model fields. Fields, such as gravitons, which do not carry Standard Model gauge charges correspond to closed strings and may pop off the brane and propagate throughout the bulk.

The picture is thus one where matter and gauge forces are confined to our 3-dimensional subspace, while gravity propagates in a higher dimensional volume. In this case, the Standard Model fields maintain their usual behavior, however, the gravitational field spreads throughout the full $3 + \delta$ spatial volume. Conventional wisdom dictates that if the additional dimensions are too large, this would result in observable deviations from Newtonian gravity. The extra dimensional space must then be *compactified*, *i.e*, made finite. However, in some alternative theories [5, 7], the extra dimensions are infinite and the gravitational deviations are suppressed by other means.

If the additional dimensions are small enough, the Standard Model fields are phenomenologically allowed to propagate in the bulk. This possibility allows for new model-building techniques to address gauge coupling unification [8], supersymmetry breaking [9, 10], the neutrino mass spectrum [11], and the fermion mass hierarchy [12]. Indeed, the field content which is allowed to propagate in the bulk, as well as the size and geometry of the bulk itself, varies between different models.

As a result of compactification, fields propagating in the bulk expand into a series of states known as a Kaluza-Klein (KK) tower, with the individual KK excitations being labeled by mode numbers. Similar to a particle in a box, the momentum of the bulk field is then quantized in the compactified dimensions. For an observer trapped on the brane, each quanta of momentum in the compactified volume appears as a KK excited state with mass $m^2 = \vec{p}_\delta^{\,2}$. This builds a KK tower of states, where each state carries identical spin and gauge quantum numbers. If the additional dimensions are infinite instead of being compactified, the δ-dimensional momentum and resulting KK spectrum is continuous.

More technically, in the case where gravity propagates in a compactified bulk, one starts from a D-dimensional Einstein-Hilbert action and performs a KK expansion about the metric field of the higher dimensional spacetime. The graviton KK towers arise as a solution to the linearized equation of motion of the metric field in this background [13]. The resulting 4-dimensional fields are the Kaluza-Klein modes. Counting the degrees of freedom within the original higher dimensional metric, the reduction of a spin-2 bulk field results in three distinct classes of towers of KK modes: symmetric tensor, vector fields and scalar fields. The KK zero-mode fields are massless, while the excitation states acquire mass by 'eating' lower spin degrees of freedom. This results in a single 5-component tensor KK tower of massive graviton states, $\delta - 1$ gauge KK towers of massive vector states, and $\delta(\delta - 1)/2$ scalar towers. The zero-mode scalar states are radius moduli fields associated with the size of the additional dimensions.

A generalized calculation of the action for linearized gravity in D dimensions can be used to compute the effective 4-dimensional theory. The spin-2 tower of KK states couples to Standard Model fields on the brane via the conserved symmetric stress-energy tensor. The spin-1 KK tower does not induce interactions on the 3-brane. The scalar KK states couple to the Standard Model fields on the brane via the trace of the stress-energy tensor.

The possible experimental signals for the existence of extra dimensions are: (i) the direct or indirect observation of a KK tower of states, or (ii) the observation of deviations in the inverse-square law of gravity in short-range experiments. The detailed properties of the KK states are determined by the geometry of the compactified space and their measurement would reveal the underlying geometry of the bulk.

We now discuss each of the principal scenarios and how they may be probed in experiment.

2. Large extra dimensions

The large extra dimensions scenario postulated by Arkani-Hamed, Dimopoulos and Dvali (ADD) [3] makes use of the string inspired braneworld hypothesis. In this model, the Standard Model gauge and matter fields are confined to a 3-dimensional brane that exists within a higher dimensional bulk. Gravity alone propagates in the δ extra spatial dimensions which are compactified. Gauss' Law relates the Planck scale of the effective 4-d low-energy theory, M_{Pl}, to the scale where gravity becomes strong in the $4 + \delta$-dimensional spacetime, M_D, through the volume of the compactified dimensions V_δ via

$$M_{\text{Pl}}^2 = V_\delta M_D^{2+\delta}. \tag{2.1}$$

Taking $M_D \sim$ TeV, as assumed by ADD, eliminates the hierarchy between M_{Pl} and the electroweak scale. M_{Pl} is generated by the large volume of the higher dimensional space and is thus no longer a fundamental scale. The hierarchy problem is now translated to the possibly more tractable question of why the compactification scale of the extra dimensions is large.

If the compactified dimensions are flat, of equal size, and of toroidal form, then $V_\delta = (2\pi R_c)^\delta$. For $M_D \sim$ TeV, the radius R_c of the extra dimensions ranges from a fraction of a millimeter to ~ 10 fermi for δ varying between 2 and 6. The compactification scale $(1/R_c)$ associated with these parameters then ranges from $\sim 10^{-4}$ eV to tens of MeV. The case of one extra dimension is excluded as the corresponding dimension (of size $R_c \approx 10^{11}$ m) would directly alter Newton's law at solar-system distances. Our knowledge of the electroweak and strong forces extends with great precision down to distances of order 10^{-15} mm, which corresponds to $\sim (100 \text{ GeV})^{-1}$. Thus the Standard Model fields do not feel the effects of the large extra dimensions present in this scenario and must be confined to the 3-brane. Therefore in this model only gravity probes the existence of the extra dimensions[1].

We now discuss the derivation of the 4-dimensional effective theory, which is computed within linearized quantum gravity. The flat metric is expanded via

$$G_{AB} = \eta_{AB} + \frac{h_{AB}}{M_D^{\delta/2+1}}, \tag{2.2}$$

where the upper case indices extend over the full D-dimensional spacetime and h_{AB} represents the bulk graviton fluctuation. The interactions of the graviton are

---

[1] Any Standard Model singlet field, *e.g.*, right handed neutrinos, could also be in the bulk in this scenario.

then described by the action

$$S_{int} = -\frac{1}{M_D^{\delta/2+1}} \int d^4x \, d^\delta y_i \, h_{AB}(x_\mu, y_i) T_{AB}(x_\mu, y_i), \qquad (2.3)$$

with T_{AB} being the symmetric conserved stress-energy tensor. Upon compactification, the bulk graviton decomposes into the various spin states as described in the introduction and Fourier expands into Kaluza-Klein towers of spin-0, 1, and 2 states which have equally spaced masses of $m_{\vec{n}} = \sqrt{\vec{n}^2/R_c^2}$, where $\vec{n} = (n_1, n_2, \ldots, n_\delta)$ labels the KK excitation level. The spin-1 states do not interact with fields on the 3-brane, the spin-0 states couple to the trace of the stress-energy tensor and will not be considered here. Their phenomenology is described in [14]. Performing the KK expansion for the spin-2 tower, setting $T_{AB} = \eta_A^\mu \eta_B^\nu T_{\mu\nu} \delta(y_i)$ for the Standard Model fields confined to the brane, and integrating the action over the extra dimensional coordinates y_i gives the interactions of the graviton KK states with the Standard Model fields. All the states in the KK tower, including the $\vec{n} = 0$ massless state, couple in an identical manner with universal strength of M_{Pl}^{-1}. The corresponding Feynman rules are catalogued in [15, 16].

The existence of large extra dimenions would affect a broad range of physical processes. Their presence may be detected in tests of short range gravity, astrophysical considerations, and collider experiments. We now review each of these in turn.

2.1. Short range tests of gravity

Until very recently, the inverse square force law of Newtonian gravity had been precisely tested only down to distances of order a centimeter [17, 18]. Such tests are performed by short range gravity experiments that probe new interactions by searching for deviations from Newtonian gravity at small distances. There are several parameterizations which describe these potential deviations [19]; the one most widely used by experiments is that where the classical gravitational potential is expanded to include a Yukawa interaction:

$$V(r) = -\frac{1}{M_{\text{Pl}}^2} \frac{m_1 m_2}{r} (1 + \alpha e^{-r/\lambda}). \qquad (2.4)$$

Here, r is the distance between two masses m_1 and m_2 and is fixed by the experimental apparatus, α is a dimensionless parameter relating the strength of the additional Yukawa interaction to that of gravity, and λ is the range of the new interaction. The best experimental sensitivity is achieved for the case $\lambda \approx r$, with the sensitivity decreasing rapidly for smaller distances. The experimental results are presented in the $\alpha - \lambda$ plane in the form of convex curves that are centered around the distance at which a particular experiment operates. These short range tests are performed by Van-der-Waals (probing $1/r^3$ deviations) and Casimir (probing $1/r^4$ terms) force experiments, as well as Cavendish-type detectors which directly measure the gravitational force.

The two-body potential given by Gauss' Law in the presence of additional dimensions (for distances $r < R_c$) is expressed as [3]:

$$V(r) = -\frac{1}{8\pi M_D^{2+\delta}} \frac{m_1 m_2}{r^{\delta+1}} \tag{2.5}$$

in the conventions of [15], which we employ throughout these lectures. When the two masses are separated by a distance $r > R_c$ and the dimensions are assumed to be compactified on a torus of radius R_c the potential becomes:

$$V(r) = -\frac{1}{8\pi M_D^{2+\delta}} \frac{m_1 m_2}{R_c^\delta} \frac{1}{r}, \tag{2.6}$$

i.e., the usual $1/r$ Newtonian potential is recovered using Gauss' Law. The parameters in the general form of the two-body potential in Eq. (2.4), *i.e.*, α and λ, depend on the number of extra dimensions and the type of compactification [20]; for the simple case of compactifying on a torus, the range λ of the new interaction is the compactification radius R_c, and $\alpha = 2\delta$. It should be noted that the dependence of Eq. (2.6) on M_D is related to the compactification scheme through the precise form of the volume factor.

The most recent Cavendish-type experiment [21] used a torsion pendulum and a rotating attractor. These results are displayed as the curve labeled Eöt-Wash in Figure 1. Interpreted within the framework of two large additional spatial dimensions, the results imply that $R_c < 130\ \mu$m. The relation of this bound to the fundamental scale M_D depends on the compactification scheme. For $\delta > 2$, R_c is too small for the effects of extra dimensions to be probed in mechanical experiments. Results from other searches are also shown in the $\alpha - \lambda$ plane in this figure. The predictions and allowed regions in the $\alpha - \lambda$ parameter space from other theoretical considerations are also presented in the figure; they include scenarios with axions, dilatons and scalar moduli fields from string theory, and attempted solutions to the cosmological constant problem.

2.2. Astrophysical and cosmological constraints

Astrophysical and cosmological considerations impose strict constraints on some theories of extra dimensions; in particular, early universe cosmology can be drastically altered from the standard picture. The typical energy scale associated with

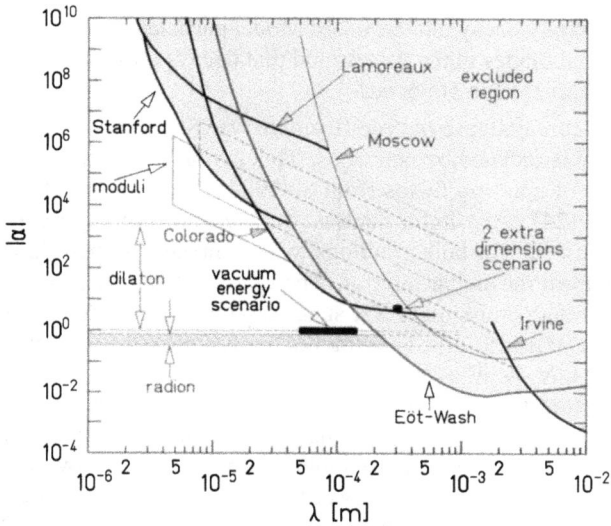

Fig. 1. 95% confidence level upper limits on the strength α (relative to Newtonian gravity) as a function of the range λ of additional Yukawa interactions. The region excluded by previous experiments [18] lies above the curves labeled Irvine, Moscow and Lamoreaux. The most recent results [21] correspond to the curves labeled Eöt-Wash.

such considerations is of order 100 MeV, and models with KK states that can be produced in this energy regime are highly restricted.

For the case of large extra dimensions of flat and toroidal form, the astrophysical bounds far surpass those from collider or short range gravity experiments for $\delta = 2$. If these large additional dimensions are compactified on a hyperbolic manifold instead, then the astrophysical constraints are avoided [22] as the modified spectrum of KK graviton states admits for a first excitation mass of order several GeV. Alternatively, these bounds are also weakened if a Ricci term is present on the brane since that serves to suppress graviton emission rates [23].

We now describe the various astrophysical and cosmological considerations that restrict the scenario with large flat extra dimensions. These processes include graviton emission during the core collapse of supernovae, the heating of neutron stars from graviton decays, considerations of the cosmic diffuse γ-ray background, overclosure of the universe, matter dominated cooling of the universe, and reheating of the universe. The restrictions obtained from processes that include effects from the decays of KK states rely on the assumption that the

KK modes can only decay into Standard Model particles on one brane, *i.e.*, there are no additional branes in the theory, and that decays into other KK modes with smaller bulk momenta do not occur.

During the core collapse of type II supernova (SN), most of the gravitational binding energy is radiated by neutrinos. This hypothesis has been confirmed by measurements of neutrino fluxes from SN1987A by the Kamiokande and IMB collaborations [24]. Any light, neutral, weakly interacting particle which couples to nucleons, such as bulk gravitons, will compete with neutrinos in carrying energy away from the stellar interior. The rate at which the supernova core can lose energy through emission of KK states can then be used to constrain the fundamental scale M_D [25, 26]. The graviton emission process is nucleon 'gravisstrahlung', $N + N \rightarrow N + N + X$, where N can be a proton or neutron, and X represents the contributions from massive KK graviton states, ordinary gravitons, and the KK dilaton (scalar) modes which are a remnant of the bulk graviton decomposition. If present, this gravisstrahlung process would provide an additional heat sink and accelerate the supernova cooling in violation with the observations of SN1987A. This process is highly dependent on the temperature of the core at collapse, which is estimated to be $T \approx 30 - 70$ MeV, and on the core density, $\rho \approx (3 - 10) \times 10^{14} \text{gcm}^{-3}$. Several additional uncertainties, such as the form of the nucleon scattering matrix, the specific heat of matter at high density, and the neutrino transport mean-free path at high density, also enter the calculation. These uncertainties are computed using the well studied nucleon-nucleon axion bremsstrahlung process. The most conservative constraint on KK emission [26] yields $R_c \leq 7.1 \times 10^{-4}$ mm for $\delta = 2$ and $R_c \leq 8.5 \times 10^{-7}$ mm for $\delta = 3$, taking $T_{SN1987A} = 30$ MeV.

A complementary bound arises from the radiative decay of the Kaluza Klein gravitons produced by the core collapse of all supernovae that have exploded during the history of the universe (SNe). The two photon decay mode is kinematically favored for the lower mass KK modes [16] with this lifetime being $\tau_{\gamma\gamma} \approx 3 \times 10^9 \text{yr} \left(\frac{100 \text{ MeV}}{m_{KK}} \right)^3$. Over the age of the universe, a significant fraction of the KK states emitted from supernovae cores will have decayed into photons, contributing to the cosmic diffuse γ-ray background. This is estimated using the present day supernova rate and the gravisstrahlung rate discussed above. A bound on the size of the additional dimensions is then imposed from the measured cosmic γ-ray background. For a choice of cosmological parameters the predicted γ-flux exceeds the observations by EGRET or COMPTEL [28] unless the fraction of the SN energy released via gravisstrahlung is less than about 0.5-1% of the total. For two extra dimensions, the limit on the compactification radius is $R_c \leq 0.9 \ 10^{-4}$ mm and for three extra dimensions the bound is $R_c \leq 0.19 \ 10^{-6}$ mm [27]. Additional contributions to the cosmic diffuse γ-ray background arise when the KK gravitons are produced from other sources such as neutrino annihilation, $\nu\bar{\nu} \rightarrow G_n \rightarrow \gamma\gamma$. These were considered in [29], and by placing a bound on the normalcy temperature required by Big Bang Nucleosynthesis the limit $R_c \leq 5.1 \times 10^{-5}$ mm for $\delta = 2$ is obtained.

In [29], it is assumed that the universe enters the radiation dominated epoch instantaneously at the reheating temperature. However, it is plausible that the universe enters the radiation epoch after being reheated by the decay of a massive scalar field or by some other means of entropy production. If a large number of KK states are produced during reheating, they are non-relativistic and hence are not diluted by entropy production. Their subsequent decays contribute to the diffuse γ-ray background. Using data from COMPTEL and EGRET [28], the constraints on M_D are tightened and are 167, 21.7, 4.75, and 1.55 TeV for $\delta = 2, 3, 4$, and 5, respectively, assuming that a 1 GeV maximum temperature is reached during reheating [30].

The escape velocity of a neutron star is similar to the average speed of thermally produced KK states in a SN core collapse, and hence a large fraction of the KK states can become trapped within the core halo. The decays of these states will continue to be a source of γ-rays long after the SN explosion. Comparisons of the expected contributions to the γ-ray flux rate from this source with EGRET data [28] from nearby neutron stars and pulsars constrains [31] the fraction of the SN energy released via gravisstrahlung to be less than about 10^{-5} of the total. For two extra dimensions this yields the bound $M_D \gtrsim 450$ TeV and for $\delta = 3$ the constraint is $M_D \gtrsim 30$ TeV. The expected sensitivity from GLAST [32] will increase these limits by a factor of 2 to 3.

The Hubble space telescope has observed that the surface temperature of several older neutron stars is higher than that expected in standard cooling models. A possible source for this excess heat is the decays of the KK graviton states trapped in the halo surrounding the star. The γ's, electrons, and neutrinos from the KK decays then hit the star and heat it. For the estimated heating rate from this mechanism not to exceed the observed luminosity, the fraction of the SN energy released via gravisstrahlung must be $\lesssim 5 \times 10^{-8}$ of the total [31], with the exact number being uncertain by a factor of a few due to theoretical and experimental uncertainties. This is by far the most stringent constraint yielding $M_D \gtrsim 1700, 60$ TeV for $\delta = 2, 3$, respectively. Although the calculations for SN emissions have not been performed for $\delta > 4$, simple scaling suggests that this mechanism results in $M_D \gtrsim 4, 0.8$ TeV for $\delta = 4, 5$, respectively.

Once produced, the massive KK gravitons are sufficiently long-lived as to potentially overclose the universe. Comparisons of KK graviton production rates from photon, as well as neutrino, annihilation to the critical density of the universe results [29] in $R_c < 1.5h \times 10^{-5}$ m for 2 additional dimensions, where h is the current Hubble parameter in units of 100 km/sMpc. While this constraint is

Table 1

Summary of constraints on the fundamental scale M_D in TeV from astrophysical and cosmological considerations as discussed in the text.

	δ			
	2	3	4	5
Supernova Cooling [26]	30	2.5		
Cosmic Diffuse γ-Rays:				
Cosmic SNe [27]	80	7		
$\nu\bar{\nu}$ Annihilation [29]	110	5		
Re-heating [30]	170	20	5	1.5
Neutron Star Halo [31]	450	30		
Overclosure of Universe [29]	$6.5/\sqrt{h}$			
Matter Dominated Early Universe [33]	85	7	1.5	
Neutron Star Heat Excess [31]	1700	60	4	1

milder than those obtained above, it is less dependent on assumptions regarding the existence of additional branes.

Overproduction of Kaluza Klein modes in the early universe could initiate an early epoch of matter radiation equality which would lead to a too low value for the age of the universe. For temperatures below ~ 100 MeV, the cooling of the universe can be accelerated by KK mode production and evaporation into the bulk, as opposed to the normal cosmological expansion. Using the present temperature of the cosmic microwave background of 2.73 K ($= 2.35 \times 10^{-10}$ MeV) and taking the minimum age of the universe to be 12.8 Gyrs ($= 6.2 \times 10^{39}$ MeV^{-1}), as determined by the mean observed age of globular clusters, a maximum temperature can be imposed at radiation-matter equality which cannot be exceeded by the overproduction of KK modes at early times. The resulting lower bounds are M_D are 86, 7.4, and 1.5 TeV for $\delta = 2, 3$, and 4 respectively [33]. Further considerations of the effects from overproduction of KK states on the characteristic scale of the turn-over of the matter power spectra at the epoch of matter radiation equality show that the period of inflation must be extended down to very low temperatures in order to be consistent with the latest data from galaxy surveys [34].

We collect the constraints from these considerations in Table 1, where we state the restrictions in terms of bounds on the fundamental scale M_D. We note that the relation of the above constraints to M_D is tricky as numerical conventions, as well as assumptions regarding the compactification scheme, explicitly enter some of the computations; in particular, that of gravisstrahlung production during supernova collapse. In addition, all of these bounds assume that all of the additional dimensions are of the same size. The constraints in the table are thus merely indicative and should not be taken as exact.

To conclude this section, we discuss the possible contribution of graviton KK states to the production of high-energy cosmic rays beyond the GZK cut-off of 10^{20} eV. About 20 super-GZK events have been observed and their origin is presently unknown. In the case of large extra dimensions, KK graviton exchange can contribute to high-energy ν-nucleon scattering and produce hadronic sized cross sections above the GZK cut-off for M_D in the range of 1 to 10 TeV [35].

2.3. Collider probes

If such additional dimensions are present and quantum gravity becomes strong at the TeV scale, then observable signatures at colliders operating at TeV energies must be induced.

One may wonder how interactions of this type can be observable at colliders since the coupling strength is so weak. In the ADD scenario, there are $(ER_c)^\delta$ massive Kaluza-Klein modes that are kinematically accessible in a collider process with energy E. For $\delta = 2$ and $E = 1$ TeV, that totals 10^{30} graviton KK states which may individually contribute to a process. It is the sum over the contribution from each KK state which removes the Planck scale suppression in a process and replaces it by powers of the fundamental scale $M_D \sim$ TeV. The interactions of the massive Kaluza-Klein graviton modes can then be observed in collider experiments either through missing energy signatures or through their virtual exchange in Standard Model processes. At future colliders with very high energies, it is possible that quantum gravity phenomena are accessible resulting in explicit signals for string or brane effects; these will be discussed briefly in Section 5. We now discuss in detail the two classes of collider signatures for large extra dimensions.

The first class of collider processes involves the real emission of Kaluza-Klein graviton states in the scattering processes $e^+e^- \rightarrow \gamma(Z) + G_n$, and $p\bar{p} \rightarrow g + G_n$, or in $Z \rightarrow f\bar{f} + G_n$. The produced graviton behaves as if it were a massive, non-interacting, stable particle and thus appears as missing energy in the detector. The cross section is computed for the production of a single massive KK excitation and then summed over the full tower of KK states. Since the mass splittings between the KK states is so small, the sum over the states may be replaced by an integral weighted by the density of KK states. The specific process kinematics cut off this integral, rendering a finite and model independent result. The expected suppression from the M_{Pl}^{-1} strength of the graviton KK couplings is exactly compensated by a M_{Pl}^2 enhancement in the phase space integration. The cross section for on-shell production of massive Kaluza Klein graviton modes then scales as simple powers of $\sqrt{s}/M_D$,

$$\sigma_{KK} \sim \frac{1}{M_{\mathrm{Pl}}^2}(\sqrt{s}R_c)^\delta \sim \frac{1}{M_D^2}\left(\frac{\sqrt{s}}{M_D}\right)^\delta. \qquad (2.7)$$

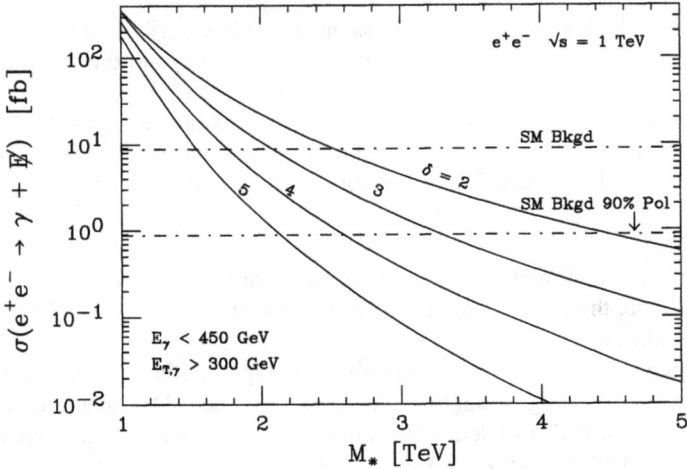

Fig. 2. The cross section for $e^+e^- \to \gamma G_n$ for $\sqrt{s} = 1$ TeV as a function of the fundamental Planck scale for various values of δ as indicated. The cross section for the SM background, with and without 90% beam polarization correspond to the horizontal lines as labeled. From [15].

The exact expression may be found in [15, 36]. It is important to note that due to integrating over the effective density of states, the radiated graviton appears to have a continuous mass distribution; this corresponds to the probability of emitting gravitons with different extra dimensional momenta. The observables for graviton production, such as the γ/Z angular and energy distributions in e^+e^- collisions, are then distinct from those of other physics processes involving fixed masses for the undetectable particles. In particular, the SM background is given by the 3-body production $e^+e^- \to \nu\bar{\nu}\gamma$.

The cross section for $e^+e^- \to \gamma G_n$ as a function of the fundamental Planck scale is presented in Fig. 2 for $\sqrt{s} = 1$ TeV. The level of SM background is also shown, with and without electron beam polarization set at 90%. We note that the signal(background) increases(decreases) with increasing $\sqrt{s}$. Details of the various distributions associated with this process can be found in Ref. [15].

Searches for direct KK graviton production in the reaction $e^+e^- \to G_n + \gamma(Z)$ at LEP II, using the characteristic final states of missing energy plus a single photon or Z boson, have excluded [37] fundamental scales up to ~ 1.45 TeV for two extra compactified dimensions and ~ 0.6 TeV for six extra dimensions. These analyses use both total cross section measurements and fits to angular distributions to set a limit on the graviton production rates as a function of the number of extra dimensions. The expected discovery reach from this process has been computed in Ref. [38] at a high energy linear e^+e^- collider with $\sqrt{s} = 800$ GeV, 1000 fb^{-1} of integrated luminosity, and various configurations for the beam polarization. These results are displayed in Table 2 and include kinematic acceptance cuts, initial state radiation, and beamsstrahlung.

The emission process at hadron colliders, for example, $q\bar{q} \rightarrow g + G_n$, results in a monojet plus missing transverse energy signature. For larger numbers of extra dimenisons the density of the KK states increases rapidly and the KK mass distribution is shifted to higher values. This is not reflected in the missing energy distribution: although the heavier KK gravitons are more likely to carry larger energy, they are also more likely to be produced at threshold due to the rapidly decreasing parton distribution functions. These two effects compensate each other, leaving nearly identical missing energy distributions. In addition, the effective low-energy theory breaks down for some regions of the parameter space as the parton-level center of mass energy can exceed the value of M_D. Experiments are then sensitive to the new physics appearing above M_D that is associated with the low-scale quantum gravity.

Searches from the Tevatron Run I yield similar results as those from LEP II, and it is anticipated that Run II at the Tevatron will have a higher sensitivity [39]. An ATLAS simulation [40] of the missing transverse energy in signal and background events at the LHC with 100 fb^{-1} is presented in Fig. 3 for various values of M_D and δ. This study results in the discovery range displayed in Table 2. The lower end of the range corresponds to where the ultraviolet physics sets in and the effective theory fails, while the upper end represents the boundary where the signal is observable above background.

If an emission signal is observed, one would like to determine the values of the fundamental parameters, M_D and δ. The measurement of the cross section at a linear collider at two different values of $\sqrt{s}$ can be used to determine these parameters [38] and test the consistency of the data with the large extra dimensions hypothesis. This is illustrated in Fig. 4.

Lastly, we note that the cross section for the emission process can be reduced somewhat if the 3-brane is flexible, or soft, instead of being rigid [41]. In this case, the brane is allowed to recoil when the KK graviton is radiated; this can be parameterized as an exponential suppression of the cross section, with the exponential being a function of the brane tension Δ_τ

$$\frac{d\sigma}{dx_\gamma d\cos\theta}(\text{soft}) = \frac{d\sigma}{dx_\gamma d\cos\theta}(\text{stiff})e^{s(1-x_\gamma)/\Delta_\tau^2}, \tag{2.8}$$

where x_γ is the scaled energy of the photon. For reasonable values of the brane tension, this suppression is not numerically large. The brane tension can also be determined by mapping out the cross section as a function of $\sqrt{s}$ as shown in Fig. 4.

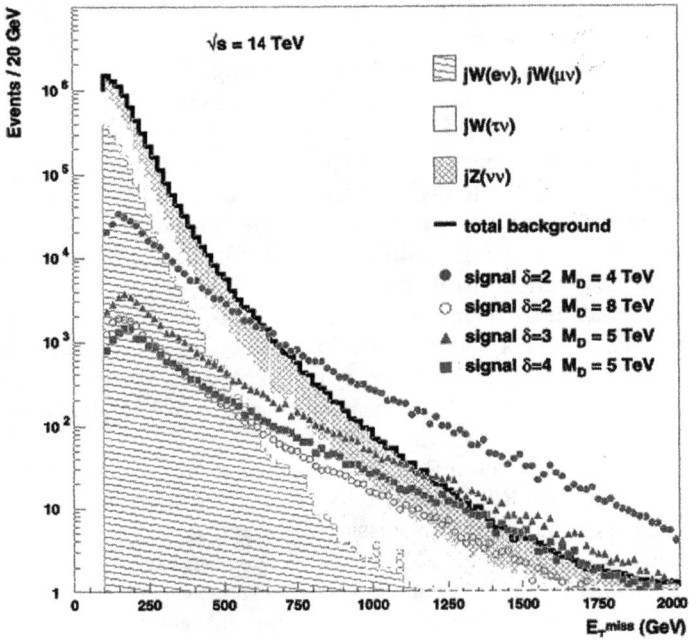

Fig. 3. Distribution of the missing transverse energy in background events and signal events for $100 \, \text{fb}^{-1}$. The contribution of the three principal Standard Model background processes is shown as well as the distribution of the signal for several values of δ and M_D. From [40].

Table 2

95% CL sensitivity to the fundamental scale M_D in TeV for different values of δ, from the emission process for various polarization configurations and different colliders as discussed in the text. $\sqrt{s} = 800 \, \text{GeV}$ and $1 \, \text{ab}^{-1}$ has been assumed for the LC and $100 \, \text{fb}^{-1}$ for the LHC. Note that the LHC only probes M_D within the stated range.

$e^+e^- \to \gamma + G_n$		2	4	6
LC	$P_{-,+} = 0$	5.9	3.5	2.5
LC	$P_- = 0.8$	8.3	4.4	2.9
LC	$P_- = 0.8, P_+ = 0.6$	10.4	5.1	3.3
$pp \to g + G_n$		2	3	4
LHC		$4 - 8.9$	$4.5 - 6.8$	$5.0 - 5.8$

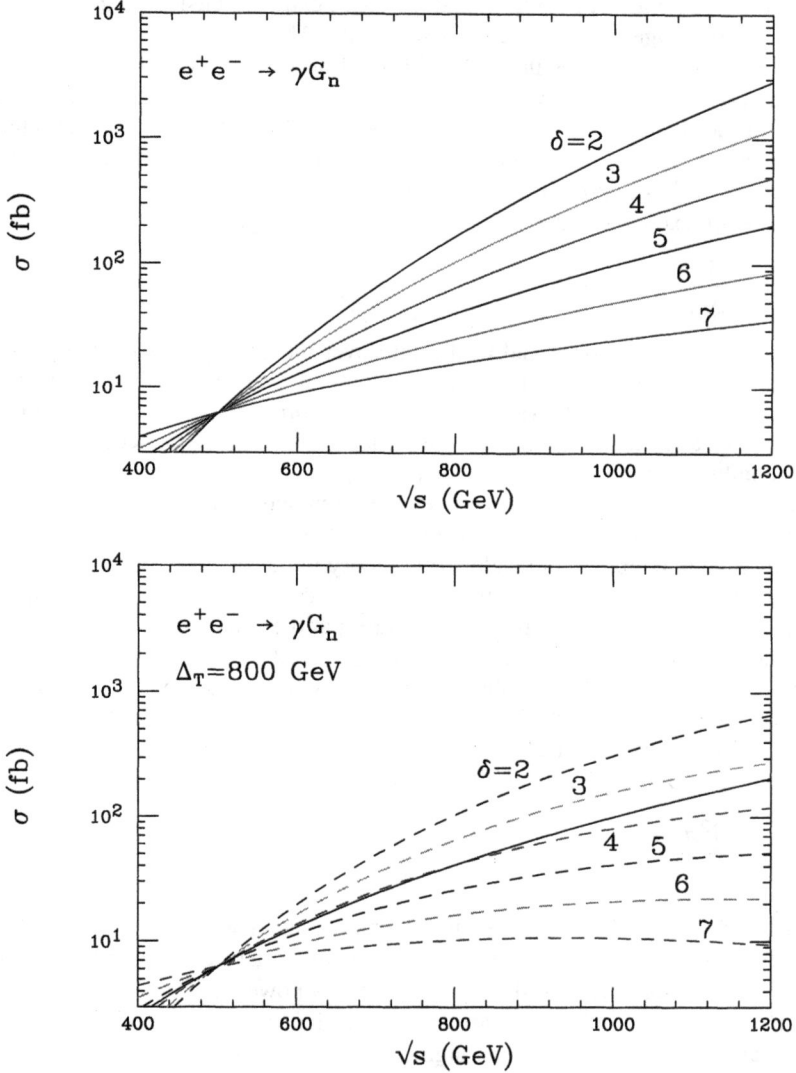

Fig. 4. Emission cross section in e^+e^- annihilation as a function of $\sqrt{s}$ for $\delta = 2 - 7$ from bottom to top on the right-hand side. The cross sections are normalized to $M_D = 5$ TeV and $\delta = 2$ at $\sqrt{s} = 500$ GeV. Top" Brane terms are not included. Bottom: The effects of finite brane tension are included, taking the relevant tension parameter to be $\Delta_\tau = 800$ GeV. Here, the lone solid curve represents the case without brane term effects for $\delta = 5$.

The second class of collider signals for large extra dimensions is that of virtual graviton exchange [15, 42] in $2 \to 2$ scattering. This leads to deviations in cross sections and asymmetries in Standard Model processes, such as $e^+e^- \to f\bar{f}$. It may also give rise to new production processes which are not present at tree-level in the Standard Model, such as $gg \to \ell^+\ell^-$. The signature is similar to that expected in composite theories and provides a good experimental tool for searching for large extra dimensions for the case $\sqrt{s} < M_D$.

Graviton exchange is governed by the effective Lagrangian

$$\mathcal{L} = i\frac{4\lambda}{\Lambda_H^4}T_{\mu\nu}T^{\mu\nu} + h.c.. \tag{2.9}$$

The amplitude is proportional to the sum over the propagators for the graviton KK tower which may be converted to an integral over the density of KK states. However, in this case, there is no specific cut-off associated with the process kinematics and the integral is divergent for $\delta > 1$. This introduces a sensitivity to the unknown ultraviolet physics which appears at the fundamental scale. This integral needs to be regulated and several approaches have been proposed: (i) a naive cut-off scheme [15, 42] (ii) brane fluctuations [43], or (iii) the inclusion of full weakly coupled TeV-scale string theory in the scattering process [39, 44]. The most model independent approach which does not make any assumptions as to the nature of the new physics appearing at the fundamental scale is that of the naive cut-off. Here, the cut-off is set to $\Lambda_H \neq M_D$; the exact relationship between Λ_H and M_D is not calculable without knowledge of the full theory. The parameter $\lambda = \pm 1$ is also usually incorporated in direct analogy with the standard parameterization for contact interactions [45] and accounts for uncertainties associated with the ultraviolet physics. The substitution

$$\mathcal{M} \sim \frac{i^2\pi}{M_{\text{Pl}}^2} \sum_{\vec{n}=1}^{\infty} \frac{1}{s - m_{\vec{n}}^2} \to \frac{\lambda}{\Lambda_H^4} \tag{2.10}$$

is then performed in the matrix element for s-channel KK graviton exchange with corresponding replacements for t- and u-channel scattering. As above, the Planck scale suppression is removed and superseded by powers of $\Lambda_H \sim$ TeV.

The resulting angular distributions for fermion pair production are quartic in $\cos\theta$ and thus provide a unique signal for spin-2 exchange. An illustration of this is given in Fig. 5 which displays the angular dependence of the polarized Left-Right asymmetry in $e^+e^- \to b\bar{b}$.

The experimental analyses also make use of the cut-off approach. Using virtual Kaluza-Klein graviton exchange in reactions with diphoton, diboson and dilepton final states, $e^+e^- \to G_n \to \gamma\gamma, VV, \ell\ell$, LEP experiments [46] exclude $\Lambda_H \lesssim 0.5 - 1.0$ TeV independent of the number of extra dimensions. At

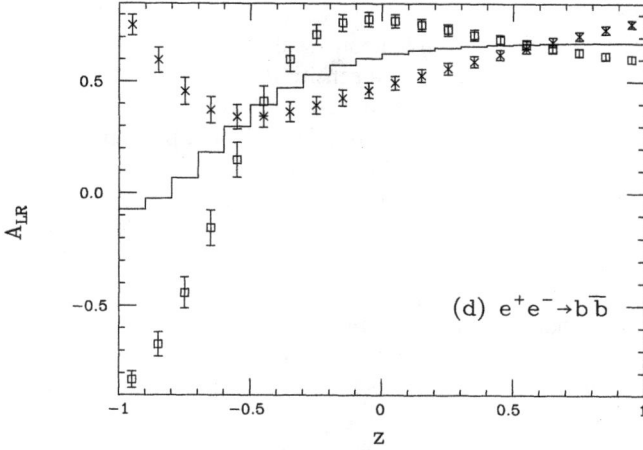

Fig. 5. Distribution of the angular dependence ($z = \cos\theta$) of the polarized Left-Right asymmetry in $e^+e^- \rightarrow b\bar{b}$ with $\sqrt{s} = 500$ GeV, taking $M_D = 1.5$ TeV and $\lambda = \pm1$. The solid histogram is the Standard Model expectation. The two sets of data points correspond to the two choices of sign for λ, and the error bars represent the statistics in each bin for an integrated luminosity of 75 fb^{-1}. From [42].

Table 3

The estimated 95% CL search reach for M_H from various processes at future accelerators.

		$\sqrt{s}$ (TeV)	M_H (TeV)
LC	$e^+e^- \rightarrow f\bar{f}$	0.5	4.1
	$e^+e^- \rightarrow f\bar{f}$	1.0	7.2
	$\gamma\gamma \rightarrow \gamma\gamma$	1.0	3.5
	$\gamma\gamma \rightarrow WW$	1.0	13.0
	$e\gamma \rightarrow e\gamma$	1.0	8.0
LHC	$pp \rightarrow \ell^+\ell^-$	14.0	7.5
	$pp \rightarrow \gamma\gamma$	14.0	7.1

the Tevatron [47], the combined Drell-Yan and diphoton channels exclude exchange scales up to ~ 1.1 TeV. In addition, H1 and ZEUS at HERA [48] have both placed the bound $\Lambda_H \gtrsim 800$ GeV.

The potential search reach for virtual KK graviton exchange in processes at future accelerators are listed in Table 3. These sensitivities are estimated for the LHC [49], a high energy e^+e^- linear collider [42], as well as for a $\gamma\gamma$ collider [50], where the initial photon beams originate from Compton laser backscattering. Note that the $\gamma\gamma \rightarrow WW$ process has the highest sensitivity to graviton exchange.

In summary, present facilities have searched for large extra dimensions and excluded their existence for fundamental scales up to $\sim$ TeV. The reach of future facilities will extend this reach to a sensitivity of $\sim$ 10 TeV. If this scenario is indeed relevant to the hierarchy, then it should be discovered in the next round of experiments. In addition, future experiments will have the capability to determine the geometry of the higher dimensional space, such as the size and number of extra dimensions, as well as the degree of the brane tension.

If the fundamental scale of gravity is at roughly a TeV, then future colliders will directly probe new exotic degrees of freedom in addition to the Kaluza Klein modes of extra dimensions, including the effects of quantum gravity itself. We do not yet have unambiguous predictions for this new and unknown physics, but it could take the form of new strongly interacting gauge sectors or string or brane excitations. For example, the exchange of string Regge excitations of Standard Model particles in $2 \rightarrow 2$ scattering [39, 44] would appear as a contact-like interaction, similar to that of graviton KK exchange, but with a large strength.

It is possible that inelastic scattering at energies $\gg$ TeV could be dominated by the production of strongly coupled objects such as microscopic black holes [51]. Assuming that these decay via Hawking radiation, they would then be observable in future very high-energy colliders [52].

3. TeV^{-1}-sized extra dimensions

The possibility of TeV^{-1}-sized extra dimensions naturally arises in braneworld theories [9, 53]. By themselves, they do not allow for a reformulation of the hierarchy problem, but they may be incorporated into a larger structure in which this problem is solved. In these scenarios, the Standard Model fields are phenomenologically allowed to propagate in the bulk. This presents a wide variety of choices for model building: (i) all, or only some, of the Standard Model gauge fields exist in the bulk; (ii) the Higgs field may lie on the brane or in the bulk; (iii) the Standard Model fermions may be confined to the brane or to specific locales in the extra dimension. The phenomenological consequences of this scenario strongly depend on the location of the fermion fields. Unless otherwise noted, our discussion assumes that all of the Standard Model gauge fields propagate in the bulk.

The masses of the excitation states in the gauge boson KK towers depend on where the Higgs boson is located. If the Higgs field propagates in the bulk, the zero-mode state of the Higgs KK tower receives a vacuum expectation value (vev) which is responsible for the spontaneous breaking of the electroweak gauge symmetry. In this case, the resulting mass matrix for the states in the gauge boson

KK towers is diagonal and the excitation masses are shifted by the mass of the gauge zero-mode, which corresponds to the Standard Model gauge field, giving

$$m_{\vec{n}} = (m_0^2 + \vec{n} \cdot \vec{n}/R_c^2)^{1/2}. \tag{3.1}$$

However, if the Higgs is confined to the brane, its vev induces mixing, amongst the gauge KK states of order $(m_0 R_c)^2$. The KK mass matrix must then be diagonalized in order to determine the excitation masses. For the case of 1 extra TeV^{-1}-sized dimension, the coupling strength of the gauge KK states to the Standard Model fermions on the brane is $\sqrt{2}g$, where g is the corresponding Standard Model gauge coupling.

We first discuss the case where the Standard Model fermions are rigidly fixed to the brane and do not feel the effects of the additional dimensions. For models in this class, precision electroweak data place strong constraints [54] on the mass of the first gauge KK excitation. Contributions to electroweak observables arise from the virtual exchange of gauge KK states and a summation over the contributions from the entire KK tower must be performed. For $D > 5$, this sum is divergent. In the full higher dimensional theory, some new, as of yet unknown, physics would regularize this sum and render it finite. An example of this is given by the possibility that the brane is flexible or non-rigid [43], which has the effect of exponentially damping the sum over KK states. Due to our present lack of knowledge of the full underlying theory, the KK sum is usually terminated by an explicit cut-off, which provides a naive estimate of the magnitude of the effects.

Since the $D = 5$ theory is finite, it is the scenario that is most often discussed and is sometimes referred to as the 5-dimensional Standard Model (5DSM). In this case, a global fit to the precision electroweak data including the contributions from KK gauge interactions yields [54] $m_1 \sim R_c^{-1} \gtrsim 4$ TeV. In addition, the KK contributions to the precision observables allow for the mass of the Higgs boson to be somewhat heavier than the value obtained in the Standard Model global fit. Given the constraint on R_c from the precision data set, the gauge KK contributions to the anomalous magnetic moment of the muon are small [55].

Such a large mass for the first gauge KK state is beyond the direct reach at present accelerators, as well as a future e^+e^- linear collider. However, they can be produced as resonances at the LHC in the Drell-Yan channel provided $m_1 \lesssim 6$ TeV. Lepton colliders can indirectly observe the existence of heavy gauge KK states in the contact interaction limit via their s-channel exchanges. In this case the contribution of the entire KK tower must be summed, and suffers the same problems with divergences discussed above. The resulting sensitivities to the gauge KK tower in the 5DSM from direct and indirect searches at various facilities is displayed in Table 4.

Table 4
95% CL search reach for the mass m_1 of the first KK gauge boson excitation [54].

	m_1 Reach (TeV)
Tevatron Run II 2 fb^{-1}	1.1
LHC 100 fb^{-1}	6.3
LEP II	3.1
LC $\sqrt{s} = 0.5$ TeV 500 fb^{-1}	13.0
LC $\sqrt{s} = 1.0$ TeV 500 fb^{-1}	23.0
LC $\sqrt{s} = 1.5$ TeV 500 fb^{-1}	31.0

We now discuss the scenario where the Standard Model fermions are localized at specific points in the extra TeV^{-1}-sized dimensions. In this case, the fermions have narrow gaussian-like wave functions in the extra dimensions with the width of their wave function being much smaller than R_c^{-1}. The placement of the different fermions at distinct locations in the additional dimensions, along with the narrowness of their wavefunctions, can then naturally suppress [12] operators mediating dangerous processes such as proton decay. The exchange of gauge KK states in $2 \to 2$ scattering processes involving initial and final state fermions is sensitive to the placement of the fermions and can be used to perform a cartography of the localized fermions [56], *i.e.*, measure the wavefunctions and locations of the fermions. At very large energies, it is possible that the cross section for such scattering will tend rapidly to zero since the fermions' wavefunctions will not overlap and hence they may completely miss each other in the extra dimensions [57].

Lastly, we discuss the case of universal extra dimensions [58], where all Standard Model fields propagate in the bulk, and branes need not be present. Translational invariance in the higher dimensional space is thus preserved. This results in the tree-level conservation of the δ-dimensional momentum of the bulk fields, which implies that KK parity, $(-1)^n$, is conserved to all orders. The phenomenology of this scenario is quite different from the cases discussed above. Since KK parity is conserved, KK excitations can no longer be produced as s-channel resonances; they can now only be produced in pairs. This results in a drastic reduction of the collider sensitivity to such states, with searches at the Tevatron yielding the bounds [58,59] $m_1 \gtrsim 400$ GeV for two universal extra dimensions. The constraints from electroweak precision data are also lowered and yield similar bounds. Since the KK states are allowed to be relatively light, they can produce observable effects [59,60] in loop-mediated processes, such as $b \to s\gamma$, $g - 2$ of the muon, and rare Higgs decays.

4. Warped extra dimensions

In this scenario, the hierarchy between the Planck and electroweak scales is generated by a large curvature of the extra dimensions [4, 5]. The simplest such framework is comprised of just one additional spatial dimension of finite size, in which gravity propagates. The geometry is that of a 5-dimensional Anti-de-Sitter space (AdS$_5$), which is a space of constant negative curvature. The extent of the 5th dimension is $y = \pi R_c$. Every slice of the 5th dimension corresponds to a 4-d Minkowski metric. Two 3-branes, with equal and opposite tension, sit at the boundaries of this slice of AdS$_5$ space. The Standard Model fields are constrained to the 3-brane located at the boundary $y = \pi R_c$, known as the TeV-brane, while gravity is localized about the opposite brane at the other boundary $y = 0$. This is referred to as the Planck brane.

The metric for this scenario preserves 4-d Poincare invariance and is

$$ds^2 = e^{-2ky}\eta_{\mu\nu}dx^\mu dx^\nu - dy^2, \tag{4.1}$$

where the exponential function of the 5th dimensional coordinate multiplying the usual 4-d Minkowski term indicates a non-factorizable geometry. This exponential is known as a warp factor. Here, the parameter k governs the degree of curvature of the AdS$_5$ space; it is assumed to be of order the Planck scale. Consistency of the low-energy theory sets $k/\overline{M}_{\rm Pl} \lesssim 0.1$, with $\overline{M}_{\rm Pl} = M_{\rm Pl}/\sqrt{8\pi} = 2.4 \times 10^{18}$ being the reduced 4-d Planck scale. The relation

$$\overline{M}_{\rm Pl}^2 = \frac{\overline{M}_5^3}{k} \tag{4.2}$$

is derived from the 5-dimensional action and indicates that the (reduced) 5-dimensional fundamental scale $\overline{M}_5$ is of order $\overline{M}_{\rm Pl}$. Since $k \sim \overline{M}_5 \sim \overline{M}_{\rm Pl}$, there are no additional hierarchies present in this model.

The scale of physical phenomena as realized by a 4-dimensional flat metric transverse to the 5th dimension is specified by the exponential warp factor. The scale $\Lambda_\pi \equiv \overline{M}_{\rm Pl}e^{-kR_c\pi}$ then describes the scale of all physical processes on the TeV-brane. With the gravitational wavefunction being localized on the Planck brane, Λ_π takes on the value ~ 1 TeV providing $kR_c \simeq 11 - 12$. It has been demonstrated [61] that this value of kR_c can be stabilized within this configuration without the fine tuning of parameters. The hierarchy is thus naturally established by the warp factor. Note that since $kR_c \simeq 10$ and it is assumed that $k \sim 10^{18}$ GeV, this is not a model with a large extra dimension.

Two parameters govern the 4-d effective theory of this scenario [62]: Λ_π and the ratio $k/\overline{M}_{Pl}$. Note that the approximate values of these parameters are known due to the relation of this model to the hierarchy problem. As in the case of large extra dimensions, the Feynman rules are obtained by a linear expansion of the flat metric,

$$G_{\alpha\beta} = e^{-2ky}(\eta_{\alpha\beta} + 2h_{\alpha\beta}/M_5^{3/2}), \tag{4.3}$$

which for this scenario includes the warp factor multiplying the linear expansion. After compactification, the resulting KK tower states are the coefficients of a Bessel expansion with the Bessel functions replacing the Fourier series of a flat geometry due to the strongly curved space and the presence of the warp factor. Here, the masses of the KK states are $m_n = x_n k e^{-kR_c\pi} = x_n \Lambda_\pi k/\overline{M}_{Pl}$ with the x_n being the roots of the first-order Bessel function, i.e., $J_1(x_n) = 0$. The first excitation is then naturally of order a TeV and the KK states are not evenly spaced. The interactions of the graviton KK tower with the Standard Model fields on the TeV-brane are given by

$$\mathcal{L} = \frac{-1}{\overline{M}_{Pl}} T^{\mu\nu}(x) h_{\mu\nu}^{(0)}(x) - \frac{1}{\Lambda_\pi} T^{\mu\nu}(x) \sum_{n=1}^{\infty} h_{\mu\nu}^{(n)}(x). \tag{4.4}$$

Note that the zero-mode decouples and that the couplings of the excitation states are inverse TeV strength. This results in a strikingly different phenomenology than in the case of large extra dimensions.

In this scenario, the principal collider signature is the direct resonant production of the spin-2 states in the graviton KK tower. To exhibit how this may appear at a collider, Figure 6 displays the cross section for $e^+e^- \to \mu^+\mu^-$ as a function of $\sqrt{s}$, assuming $m_1 = 500$ GeV and varying $k/\overline{M}_{Pl}$ in the range $0.01-0.05$. The height of the third resonance is greatly reduced as the higher KK excitations prefer to decay to the lighter graviton states, once it is kinematically allowed [63]. In this case, high energy colliders may become graviton factories! If the first graviton KK state is observed, then the parameters of this model can be uniquely determined by measurement of the location and width of the resonance. In addition, the spin-2 nature of the graviton resonance can be determined from the shape of the angular distribution of the decay products. This is demonstrated in Figure 7, which displays the angular distribution of the final state leptons in Drell-Yan production, $pp \to \ell^+\ell^-$, at the LHC [64].

Searches for the first graviton KK resonance in Drell-Yan and dijet data from Run I at the Tevatron restrict [62] the parameter space of this model, as shown in Figure 8. These data exclude larger values of $k/\overline{M}_{Pl}$ for values of m_1 which are in kinematic reach of the accelerator.

Gravitons may also contribute to precision electroweak observables. A precise description of such contributions requires a complete understanding of the full underlying theory due to the non-renormalizability of gravity. However, naive estimates of the size of such effects can be obtained in an effective field theory

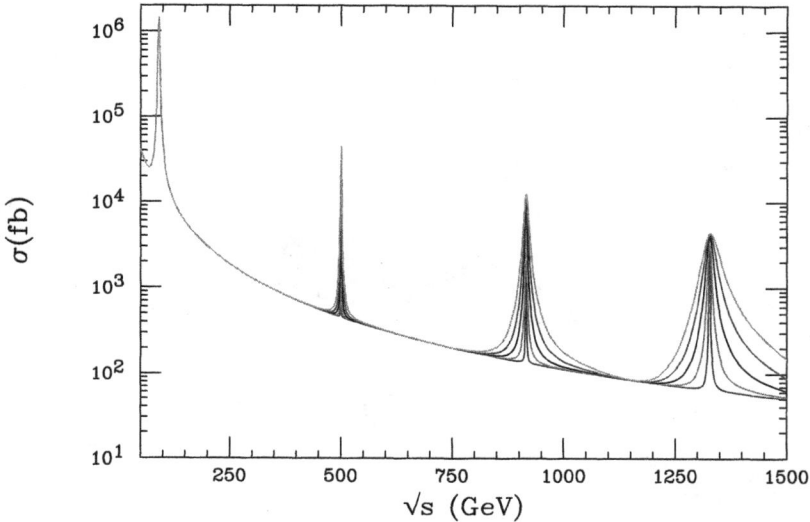

Fig. 6. The cross section for $e^+e^- \rightarrow \mu^+\mu^-$ including the exchange of a KK tower of gravitons in the Randall-Sundrum model with $m_1 = 500$ GeV. The curves correspond to $k/\overline{M}_{\rm Pl} =$ in the range $0.01 - 0.05$.

by employing a cut-off to regulate the theory [65]. The resulting cut-off dependent constraints indicate [66] that smaller values of $k/\overline{M}_{\rm Pl}$ are inconsistent with precision electroweak data, as shown in Figure 8.

These two constraints from present data, taken together with the theoretical assumptions that (i) $\Lambda_\pi \lesssim 10$ TeV, *i.e.*, the scale of physics on the TeV-brane is not far above the electroweak scale so that an additional hierarchy is not generated, and (ii) $k/\overline{M}_{\rm Pl} \lesssim 0.1$ from bounds on the curvature of the AdS$_5$, result in a closed allowed region in the two parameter space. This is displayed in Figure 8, which also shows the expected search reach for resonant graviton KK production in the Drell-Yan channel at the LHC. We see that the full allowed parameter space can be completely explored at the LHC, given our theoretical prejudices, and hence the LHC will either discover or exclude this model.

If the above theoretical assumptions are evaded, the KK gravitons may be too massive to be produced directly. However, their contributions to fermion pair production may still be felt via virtual exchange. In this case, the uncertainties associated with the introduction of a cut-off are avoided, since there is only one additional dimension and the KK states may be neatly summed. The resulting sensitivities [62] to Λ_π at current and future colliders are listed in Table 5.

The Goldberger-Wise mechanism [61] for stabilizing the separation of the two 3-branes in this configuration with $kR_c \sim 10$ leads to the existence of a

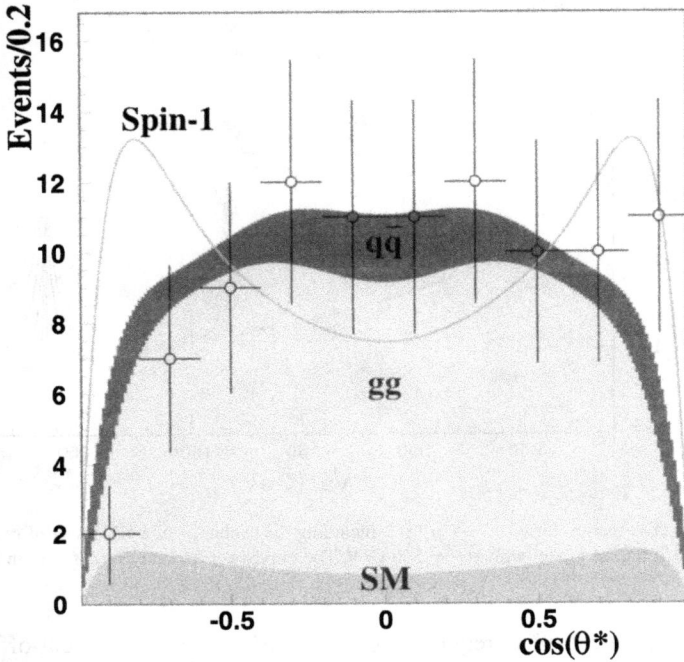

Fig. 7. The angular distribution of "data" at the LHC from Drell-Yan production of the first graviton KK excitation with $m_1 = 1.5$ TeV and 100 fb^{-1} of integrated luminosity. The stacked histograms represent the Standard Model contributions, and gg and $q\bar{q}$ initiated graviton production as labeled. The curve shows the expected distribution from a spin-1 resonance. From [64].

Table 5

95% CL search reach for Λ_π in TeV in the contact interaction regime taking 500, 2.5, 2, and 100 fb^{-1} of integrated luminosity at the LC, LEP II, Tevatron, and LHC, respectively. From [62].

	$k/\overline{M}_{Pl}$		
	0.01	0.1	1.0
LEP II	4.0	1.5	0.4
LC $\sqrt{s} = 0.5$ TeV	20.0	5.0	1.5
LC $\sqrt{s} = 1.0$ TeV	40.0	10.0	3.0
Tevatron Run II	5.0	1.5	0.5
LHC	20.0	7.0	3.0

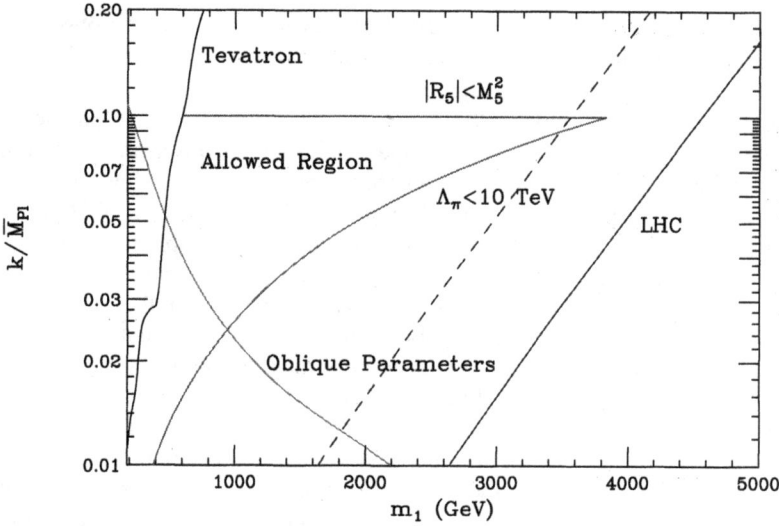

Fig. 8. Summary of experimental and theoretical constraints on the Randall-Sundrum model in the two-parameter plane $k/\overline{M}_{Pl} - m_1$, for the case where the Standard Model fields are constrained to the TeV-brane. The allowed region lies in the center as indicated. The LHC sensitivity to graviton resonances in the Drell-Yan channel is represented by the diagonal dashed and solid curves, corresponding to 10 and 100 fb^{-1} of integrated luminosity, respectively. From [66].

new, relatively light scalar field. This field is the radion and it is related to the radial fluctuations of the extra dimension, and to the scalar remnant of the bulk graviton KK decomposition. The radion couples to the Standard Model fields via the trace of the stress energy tensor with strength $\sim T_\mu^\mu/\Lambda_\pi$. These interactions are similar to those of the Higgs boson, and it is allowed to mix with the Higgs, which alters the couplings of both fields. The phenomenology of this field is detailed in [14, 67].

Astrophysical bounds are not present in this scenario since the first graviton KK state occurs at a $\sim$TeV. However, the TeV scale graviton KK states can induce high energy cosmic rays. In this case, neutrino annihilation within a GZK distance of the earth can produce a single graviton KK state on resonance which subsequently decays hadronically [68]. For neutrinos of mass $m_\nu \sim 10^{-2}$ to 10^{-1} eV, and graviton resonances of order a TeV, super-GZK events can be produced. Under the assumption that the incident neutrino spectrum extends in neutrino energy with a reasonably slow fall-off, the existence of a series of s-channel KK graviton resonances will lead to a series of ultra-GZK events. The rates for these bursts are generally at or near the present level of observability for a wide range of model parameters. The fact that such events are not as yet observed can be used to constrain the parameter space of this model once a specific form of the neutrino energy spectrum is assumed.

In a variant of this model, the Standard Model fields may propagate in the bulk. This is desirable for numerous model building reasons as mentioned in the introduction. As a first step, one can study the effect of placing the Standard Model gauge fields in the bulk and keeping the fermions on the TeV-brane. In this case, one finds [69] that the fermions on the brane couple to the KK gauge fields ~ 9 times more strongly than they couple to the Standard Model gauge fields. This results in strong bounds on gauge KK states from their contributions to electroweak precision data. A global fit to the electroweak data set yields the constraint $m_1 \gtrsim 25$ TeV on the first gauge KK mass, implying $\Lambda_\pi \gtrsim 100$ TeV.

This bound can be relaxed if the fermions also reside in the bulk [66, 70]. In this case, a third parameter is introduced, corresponding to the bulk fermion mass which is given by $m_5 = \nu k$ with ν being of order one. The parameter ν controls the shape of the fermion zero mode wavefunction. The resulting phenomenology is markedly different, and is highly dependent on the parameter ν. In particular, large mixing is induced between the zero-mode top-quark and the states in its KK tower. This results [71] in substantial shifts to the ρ-parameter and forces the third generation of fermions to be confined to the TeV-brane with only the first two generations of fermions being allowed to reside in the bulk.

An alternate scenario is possible [5] when the second brane is taken off to infinity, *i.e.*, $R_c \to \infty$, and the Standard Model fields are confined to the brane at $y = 0$ where gravity is localized. In this case, the graviton KK modes become continuous, *i.e.*, the gap between KK states disappears, and their couplings to the Standard Model fields are much weaker than M_{Pl}. This configuration no longer allows for a reformulation of the hierarchy problem, but can potentially be observable [72] in sub-mm gravitational force experiments.

Another consistent scenario of this type involves two branes, both with positive tension, separated in a five-dimensional Anti-de Sitter geometry of infinite extent [73]. The graviton is localized on one of the branes, while a gapless continuum of additional gravity modes probe the infinite fifth dimension. The phenomenological effects of this framework are similar to the process of real graviton emission in the ADD scheme with six large toroidal dimensions. The resulting cosmological constraints are also found to be very mild [74].

5. Summary

If the structure of spacetime is different than that readily observed, gravitational physics, particle physics and cosmology are all immediately affected. The physics of extra dimensions offers new insights and solutions to fundamental questions arising in these fields. Novel ideas and frameworks are continuously born and evolved. They make use of string theoretical features and tools and they may reveal if and how the 11-dimensional string theory is relevant to our four-dimensional world.

We have outlined some of the experimental observations in particle and gravitational physics as well as astrophysical and cosmological considerations thatcan constrain or confirm these scenarios. These developing ideas and the wide interdisciplinary experimental program that is charted out to investigate them mark a renewed effort to describe the dynamics behind spacetime. We look forward to the discovery of a higher dimenionsal spacetime!

References

[1] G. Nordstrom, Z. Phys. **15**, 504 (1914).

[2] T. Kaluza, Sitzungsber. Preuss. Akad. Wiss. Berlin (Math. Phys.), 966 (1921); O. Klein, Z. Phys. **37**, 895 (1926); Nature **118**, 516 (1926).

[3] N. Arkani-Hamed, S. Dimopoulos and G. R. Dvali, Phys. Lett. B **429**, 263 (1998) [hep-ph/9803315]; I. Antoniadis, N. Arkani-Hamed, S. Dimopoulos and G. R. Dvali, Phys. Lett. B **436**, 257 (1998) [hep-ph/9804398]; N. Arkani-Hamed, S. Dimopoulos and G. R. Dvali, Phys. Rev. D **59**, 086004 (1999) [hep-ph/9807344].

[4] L. Randall and R. Sundrum, Phys. Rev. Lett. **83**, 3370 (1999) [hep-ph/9905221].

[5] L. Randall and R. Sundrum, Phys. Rev. Lett. **83**, 4690 (1999) [hep-th/9906064].

[6] J. Polchinski, *String Theory* (Cambridge University Press, 1998).

[7] N. Arkani-Hamed, S. Dimopoulos, G. R. Dvali and N. Kaloper, Phys. Rev. Lett. **84**, 586 (2000) [hep-th/9907209].

[8] K. R. Dienes, E. Dudas and T. Gherghetta, Phys. Lett. B **436**, 55 (1998) [hep-ph/9803466]; Nucl. Phys. B **537**, 47 (1999) [hep-ph/9806292]; Nucl. Phys. B **567**, 111 (2000) [hep-ph/9908530].

[9] I. Antoniadis, Phys. Lett. B **246**, 377 (1990).

[10] D. E. Kaplan, G. D. Kribs and M. Schmaltz, Phys. Rev. D **62**, 035010 (2000) [hep-ph/9911293]; Z. Chacko, M. A. Luty, A. E. Nelson and E. Ponton, JHEP **0001**, 003 (2000) [hep-ph/9911323]; N. Arkani-Hamed, L. J. Hall, D. R. Smith and N. Weiner, Phys. Rev. D **63**, 056003 (2001) [hep-ph/9911421].

[11] K. R. Dienes, E. Dudas and T. Gherghetta, Nucl. Phys. B **557**, 25 (1999) [hep-ph/9811428]; N. Arkani-Hamed, S. Dimopoulos, G. R. Dvali and J. March-Russell, Phys. Rev. D **65**, 024032 (2002) [hep-ph/9811448]; Y. Grossman and M. Neubert, Phys. Lett. B **474**, 361 (2000) [hep-ph/9912408].

[12] N. Arkani-Hamed and M. Schmaltz, Phys. Rev. D **61**, 033005 (2000) [hep-ph/9903417].

[13] *Flavor Physics For The Millenium*, Proc. of the Theoretical Advanced Study Institute in Elementary Particle Physics, Boulder, Colorado, 4 - 30 June 2000, J. L. Rosner ed. (University of Chicago, 2000).

[14] G. F. Giudice, R. Rattazzi and J. D. Wells, Nucl. Phys. B **595**, 250 (2001) [hep-ph/0002178].

[15] G. F. Giudice, R. Rattazzi and J. D. Wells, Nucl. Phys. B **544**, 3 (1999) [hep-ph/9811291].

[16] T. Han, J. D. Lykken and R. J. Zhang, Phys. Rev. D **59**, 105006 (1999) [hep-ph/9811350].

[17] C. Caso et al. [Particle Data Group], Eur. Phys. J. C **3**, 781 (1998).

[18] D.E. Krause and E. Fischbach, hep-ph 9912276; G.L. Smith, C.D. Hoyle, J.H. Gundlach, E.G. Adelberger, B.R. Heckel and H.E. Swanson, Phys. Rev. D **61**, 022001 (2000); J. K. Hoskins, R. D. Newman, R. Spero, and J. Schultz, Phys. Rev D **32**, 3084 (1985); V. P. Mitrofanov and O. I. Ponomareva, Sov. Phys. JETP **67**, 1963 (1988); J.C. Long, H.W. Chan, J.C. Price, Nucl. Phys. B **539**, 23 (1999); S.K. Lamoreaux, Phys. Rev. Lett. **78**, 5 (1997); quant-ph/9907076.

[19] A. H. Cook, Contemp. Phys. **28**, 159 (1987).

[20] A. Kehagias and K. Sfetsos, Phys. Lett. B **472**, 39 (2000) [hep-ph/9905417].

[21] C. D. Hoyle, D. J. Kapner, B. R. Heckel, E. G. Adelberger, J. H. Gundlach, U. Schmidt and H. E. Swanson, Phys. Rev. D **70**, 042004 (2004) [hep-ph/0405262]; C. D. Hoyle, U. Schmidt, B. R. Heckel, E. G. Adelberger, J. H. Gundlach, D. J. Kapner and H. E. Swanson, Phys. Rev. Lett. **86**, 1418 (2001) [hep-ph/0011014]; E. G. Adelberger [EOT-WASH Group], hep-ex/0202008.

[22] N. Kaloper, J. March-Russell, G. D. Starkman and M. Trodden, Phys. Rev. Lett. **85**, 928 (2000) [hep-ph/0002001].

[23] G. R. Dvali, G. Gabadadze, M. Kolanovic and F. Nitti, Phys. Rev. D **64**, 084004 (2001) [hep-ph/0102216].

[24] K. Hirata et al., Phys. Rev. Let. **58**, 1490 (1987); R.M. Bionta et al., Phys. Rev. Let. **58**, 1494 (1987).

[25] S. Cullen and M. Perelstein, Phys. Rev. Lett. **83**, 268 (1999) [hep-ph/9903422]; V. D. Barger, T. Han, C. Kao and R. J. Zhang, Phys. Lett. B **461**, 34 (1999) [hep-ph/9905474]; C. Hanhart, J. A. Pons, D. R. Phillips and S. Reddy, Phys. Lett. B **509**, 1 (2001) [astro-ph/0102063].

[26] C. Hanhart, D. R. Phillips, S. Reddy and M. J. Savage, Nucl. Phys. B **595**, 335 (2001) [nucl-th/0007016].

[27] S. Hannestad and G. Raffelt, Phys. Rev. Lett. **87**, 051301 (2001) [hep-ph/0103201].

[28] S. C. Kappadath et al., Astron. Astrophys. Suppl. **120**, 619 (1996); D. A. Knitten et al., Astron. Astrophys. Suppl. **120**, 615 (1996); D. J. Thompson et al., Astrophys. J. Suppl. **101**, 259 (1995); *ibid.*, **107**, 227 (1996); R. C. Hartman et al., Astrophys. J. Suppl. **123**, 79 (1999).

[29] L. J. Hall and D. R. Smith, Phys. Rev. D **60**, 085008 (1999) [hep-ph/9904267].

[30] S. Hannestad, Phys. Rev. D **64**, 023515 (2001) [hep-ph/0102290].

[31] S. Hannestad and G. G. Raffelt, Phys. Rev. Lett. **88**, 071301 (2002) [hep-ph/0110067].

[32] See, for example, www-glast.stanford.edu.

[33] M. Fairbairn, Phys. Lett. B **508**, 335 (2001) [hep-ph/0101131].

[34] M. Fairbairn and L. M. Griffiths, JHEP **0202**, 024 (2002) [hep-ph/0111435].

[35] P. Jain, D. W. McKay, S. Panda and J. P. Ralston, Phys. Lett. B **484**, 267 (2000) [hep-ph/0001031]; A. Goyal, A. Gupta and N. Mahajan, Phys. Rev. D **63**, 043003 (2001) [hep-ph/0005030].

[36] E. A. Mirabelli, M. Perelstein and M. E. Peskin, Phys. Rev. Lett. **82**, 2236 (1999) [hep-ph/9811337]; K. M. Cheung and W. Y. Keung, Phys. Rev. D **60**, 112003 (1999) [hep-ph/9903294].

[37] For a summary of LEP results on graviton emission, see G. Landsberg, hep-ex/0105039. See also P. Abreu et al. [DELPHI Collaboration], Eur. Phys. J. C **17**, 53 (2000); G. Abbiendi et al. [OPAL Collaboration], Eur. Phys. J. C **18**, 253 (2000); M. Acciari et al. [L3 collaboration], Phys. Lett. B **464**, 135 (1999).

[38] J. A. Aguilar-Saavedra et al. [ECFA/DESY LC Physics Working Group Collaboration], TESLA Technical Design Report, Part III, R. D. Heuer, D. J. Miller, F. Richard and P. M. Zerwas eds. (2001), hep-ph/0106315.

[39] S. Cullen, M. Perelstein and M. E. Peskin, Phys. Rev. D **62**, 055012 (2000) [hep-ph/0001166].

[40] L. Vacavant and I. Hinchliffe, J. Phys. G **27**, 1839 (2001).

[41] H. Murayama and J. D. Wells, Phys. Rev. D **65**, 056011 (2002) [hep-ph/0109004].

[42] J. L. Hewett, Phys. Rev. Lett. **82**, 4765 (1999) [hep-ph/9811356].

[43] M. Bando, T. Kugo, T. Noguchi and K. Yoshioka, Phys. Rev. Lett. **83**, 3601 (1999) [hep-ph/9906549].

[44] E. Dudas and J. Mourad, Nucl. Phys. B **575**, 3 (2000) [hep-th/9911019]; E. Accomando, I. Antoniadis and K. Benakli, Nucl. Phys. B **579**, 3 (2000) [hep-ph/9912287].

[45] E. Eichten, K. D. Lane and M. E. Peskin, Phys. Rev. Lett. **50**, 811 (1983).

[46] ALEPH Collaboration, ALEPH Note CONF-2000-005; DELPHI Collaboration, DELPHI-CONF-464 (2001); L3 Collaboration, L3 Note 2650 (2001); OPAL Collaboration, OPAL Note PN469 (2001); D. Bourilkov, hep-ex/0103039.

[47] B. Abbott et al. [D0 Collaboration], Phys. Rev. Lett. **86**, 1156 (2001).

[48] C. Adloff et al., Phys. Lett. B **479**, 358 (2000); A. Zarnecki, talk presented at the Workshop on Higgs and Supersymmetry, Orsay, France, March 2001.

[49] F. Gianotti et al., hep-ph/0204087; A. Miagkov, talk presented at the Workshop on Higgs and Supersymmetry, Orsay, France, March 2001.

[50] T. G. Rizzo, Phys. Rev. D **60**, 115010 (1999) [hep-ph/9904380]; H. Davoudiasl, Phys. Rev. D **60**, 084022 (1999) [hep-ph/9904425]; Phys. Rev. D **61**, 044018 (2000) [hep-ph/9907347].

[51] S. B. Giddings and S. Thomas, Phys. Rev. D **65**, 056010 (2002) [hep-ph/0106219]; S. Dimopoulos and G. Landsberg, Phys. Rev. Lett. **87**, 161602 (2001) [hep-ph/0106295]; M. B. Voloshin, Phys. Lett. B **518**, 137 (2001) [hep-ph/0107119].

[52] T. G. Rizzo, JHEP **0202**, 011 (2002) [hep-ph/0201228].

[53] J. D. Lykken, Phys. Rev. D **54**, 3693 (1996) [hep-th/9603133]; I. Antoniadis and M. Quiros, Phys. Lett. B **392**, 61 (1997) [hep-th/9609209].

[54] T. G. Rizzo and J. D. Wells, Phys. Rev. D **61**, 016007 (2000) [hep-ph/9906234]; M. Masip and A. Pomarol, Phys. Rev. D **60**, 096005 (1999) [hep-ph/9902467]; W. J. Marciano, Phys. Rev. D **60**, 093006 (1999) [hep-ph/9903451].

[55] P. Nath and M. Yamaguchi, Phys. Rev. D **60**, 116006 (1999) [hep-ph/9903298].

[56] T. G. Rizzo, Phys. Rev. D **64**, 015003 (2001) [hep-ph/0101278].

[57] N. Arkani-Hamed, Y. Grossman and M. Schmaltz, Phys. Rev. D **61**, 115004 (2000) [hep-ph/9909411].

[58] T. Appelquist, H. C. Cheng and B. A. Dobrescu, Phys. Rev. D **64**, 035002 (2001) [hep-ph/0012100].

[59] T. G. Rizzo, Phys. Rev. D **64**, 095010 (2001) [hep-ph/0106336].

[60] K. Agashe, N. G. Deshpande and G. H. Wu, Phys. Lett. B **514**, 309 (2001) [hep-ph/0105084]; T. Appelquist and B. A. Dobrescu, Phys. Lett. B **516**, 85 (2001) [hep-ph/0106140]; F. J. Petriello, hep-ph/0204067.

[61] W. D. Goldberger and M. B. Wise, Phys. Rev. Lett. **83**, 4922 (1999) [hep-ph/9907447].

[62] H. Davoudiasl, J. L. Hewett and T. G. Rizzo, Phys. Rev. Lett. **84**, 2080 (2000) [hep-ph/9909255].

[63] H. Davoudiasl and T. G. Rizzo, Phys. Lett. B **512**, 100 (2001) [hep-ph/0104199].

[64] B. C. Allanach, K. Odagiri, M. A. Parker and B. R. Webber, JHEP **0009**, 019 (2000) [hep-ph/0006114].

[65] For a discussion of this point, see R. Contino, L. Pilo, R. Rattazzi and A. Strumia, JHEP **0106**, 005 (2001) [hep-ph/0103104].

[66] H. Davoudiasl, J. L. Hewett and T. G. Rizzo, Phys. Rev. D **63**, 075004 (2001) [hep-ph/0006041].

[67] W. D. Goldberger and M. B. Wise, Phys. Lett. B **475**, 275 (2000) [hep-ph/9911457]; K. M. Cheung, Phys. Rev. D **63**, 056007 (2001) [hep-ph/0009232]; J. L. Hewett and T. G. Rizzo, hep-ph/0202155.

[68] H. Davoudiasl, J. L. Hewett and T. G. Rizzo, hep-ph/0010066.

[69] H. Davoudiasl, J. L. Hewett and T. G. Rizzo, Phys. Lett. B **473**, 43 (2000) [hep-ph/9911262]; A. Pomarol, Phys. Lett. B **486**, 153 (2000) [hep-ph/9911294].

[70] S. Chang, J. Hisano, H. Nakano, N. Okada and M. Yamaguchi, Phys. Rev. D **62**, 084025 (2000) [hep-ph/9912498]; T. Gherghetta and A. Pomarol, Nucl. Phys. B **586**, 141 (2000) [hep-ph/0003129].

[71] J. L. Hewett, F. J. Petriello and T. G. Rizzo, hep-ph/0203091.

[72] D. J. Chung, L. L. Everett and H. Davoudiasl, Phys. Rev. D **64**, 065002 (2001) [hep-ph/0010103].

[73] J. Lykken and L. Randall, JHEP **0006**, 014 (2000) [hep-th/9908076].

[74] A. Hebecker and J. March-Russell, Nucl. Phys. B **608**, 375 (2001) [hep-ph/0103214].

Course 6

WARPED MODELS AND HOLOGRAPHY

Tony Gherghetta

*School of Physics and Astronomy, University of Minnesota,
Minneapolis, MN 55455, USA*

D. Kazakov, S. Lavignac and J. Dalibard, eds.
Les Houches, Session LXXXIV, 2005
Particle Physics Beyond the Standard Model
© 2006 Elsevier B.V. All rights reserved

Contents

1. Preamble

The theoretical tools required to construct models in warped extra dimensions are presented. This includes how to localise zero modes in the warped bulk and how to obtain the holographic interpretation using the AdS/CFT correspondence. Several models formulated in warped space are then discussed including non-supersymmetric and supersymmetric theories as well as their dual interpretation. Finally it is shown how grand unification occurs in warped models.

2. Introduction

Warped extra dimensions have provided a new framework for addressing the hierarchy problem in extensions of the Standard Model. The curved fifth dimension is compactified on a line segment where distance scales are measured with the nonfactorisable metric of anti-de Sitter (AdS) space. Since distance and hence energy scales are location dependent in AdS space, the hierarchy problem can be redshifted away. However this novel solution also has a more mundane explanation. By the AdS/CFT correspondence, five-dimensional (5D) AdS space has a dual four-dimensional (4D) interpretation in terms of a strongly-coupled conformal field theory (CFT). In this 4D guise the hierarchy problem is solved by having a low cutoff scale that is associated with the conformal symmetry breaking scale of the CFT. Both interpretations are equally valid and allow the infrared (IR) scale to be hierarchically smaller and stable compared to the ultraviolet (UV) scale. This is much like the photon and electron of QED compared with the baryons and mesons of QCD. The Planck scale, M_P provides the cutoff scale for QED, while in QCD the composite baryons and mesons are only valid up to the QCD scale $\Lambda_{QCD} \ll M_P$. In warped models of electroweak physics the Higgs is composite at the TeV scale, while the Standard Model fermions and gauge bosons are partly composite to varying degrees, ranging from an elementary electron to a composite top quark. The Planck scale again provides the UV cutoff for the elementary states, while the TeV scale is the compositeness or IR scale.

This hybrid framework of electroweak physics has led to a renaissance of composite Higgs models that have many desirable features. They are consistent with electroweak precision data that place strong bounds on the scales of new physics. New physics at the TeV scale generically leads to flavor problems, but these are absent in warped models, since a GIM-like mechanism operates. The warped models also explain the fermion mass hierarchy (including neutrino masses) as well as incorporate grand unification with logarithmic running and unified gauge couplings. This sophisticated level of model building is due to the complementary nature of the AdS/CFT framework. The 5D warped bulk not only provides a simple geometric picture in which to construct these models, but more importantly is also a weakly-coupled description in which calculations can be performed. In fact this theoretical tool provides a unique window into strongly-coupled 4D gauge theories and goes beyond the application to electroweak physics that will be the primary concern of these lectures. The 5D picture is complemented by the 4D description which is more intuitive, primarily from our understanding of QCD, except that the 4D gauge theory is strongly-coupled and a perturbative analysis is not possible.

Hence the plan of these lectures will be to begin in 5D warped space and describe how to localise zero modes anywhere in a slice of AdS. This will enable us to quite simply construct the Standard Model in the bulk. The more intuitive dual 4D interpretation will then be obtained by writing down the AdS/CFT dictionary. This dictionary will allow us to give either a 5D or 4D description for any type of warped model. The AdS/CFT framework need not only apply to electroweak physics. Supersymmetry will be subsequently introduced and the model-building possibilities surveyed. Lastly the novel features of grand unification in warped space will be discussed.

Since the aim of these lectures is to emphasize the dual nature of warped models a little background knowledge of warped extra dimensions will be useful. This is nicely reviewed in Refs [1, 2, 3, 4, 5]. Some theoretical aspects of warped models presented in these lectures have also been covered in Refs [3, 5, 6]. Higgless models are not covered in these lectures, but these type of warped models are discussed in Refs [6, 7]. Finally, while these lectures concentrate on the theoretical aspects of warped models, the phenomenological aspects are just as important, and these can be found in Ref [6, 8].

3. Bulk fields in a slice of AdS$_5$

3.1. A slice of AdS$_5$

Let us begin by considering a 5D spacetime with the AdS$_5$ metric

$$ds^2 = e^{-2ky}\eta_{\mu\nu}dx^{\mu}dx^{\nu} + dy^2 \equiv g_{MN}dx^M dx^N, \tag{3.1}$$

where k is the AdS curvature scale. The spacetime indices $M = (\mu, 5)$ where $\mu = 0, 1, 2, 3$ and $\eta_{\mu\nu} = \text{diag}(- + ++)$ is the Minkowski metric. The fifth dimension y is compactified on a Z_2 orbifold with a UV (IR) brane located at the orbifold fixed points $y^* = 0(\pi R)$. Between these two three-branes the metric (3.1) is a solution to Einstein's equations provided the bulk cosmological constant and the brane tensions are appropriately tuned (see, for example, the lectures by Rubakov [1]). This slice of AdS$_5$ is the Randall-Sundrum solution [9] (RS1) and is geometrically depicted in Fig. 1.

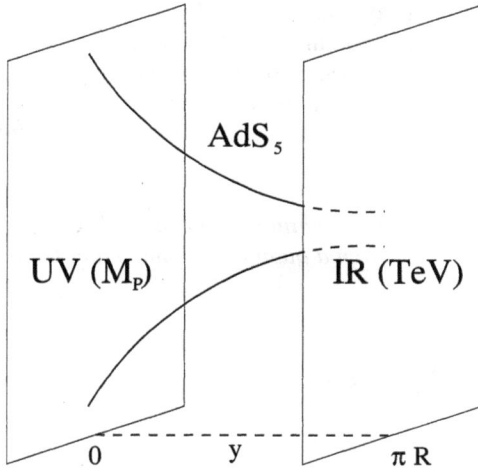

Fig. 1. A slice of AdS$_5$: The Randall-Sundrum scenario.

In RS1 the Standard Model particle states are confined to the IR brane. The hierarchy problem is then solved by noting that generic mass scales M in the 5D theory are scaled down to $Me^{-\pi kR}$ on the IR brane at $y = \pi R$. In particular since the Higgs boson H is localised on the IR brane this means that the dimension two Higgs mass term gets rescaled by an amount

$$m_H^2 |H|^2 \to (m_H e^{-\pi kR})^2 |H|^2, \tag{3.2}$$

so that a Higgs mass parameter $m_H \sim \mathcal{O}(M_5)$ in the 5D theory will naturally be associated with a hierarchically smaller scale on the IR brane (where M_5 is the 5D fundamental mass scale). However on the IR brane higher-dimension operators with dimension greater than four, such as those associated with proton decay, flavour changing neutral currents (FCNC) and neutrino masses will now be suppressed by the warped down scale

$$\frac{1}{M_5^2} \bar{\Psi}_i \Psi_j \bar{\Psi}_k \Psi_l \;\to\; \frac{1}{(M_5 e^{-\pi k R})^2} \bar{\Psi}_i \Psi_j \bar{\Psi}_k \Psi_l, \tag{3.3}$$

$$\frac{1}{M_5} \nu \nu H H \;\to\; \frac{1}{M_5 e^{-\pi k R}} \nu \nu H H, \tag{3.4}$$

where Ψ_i is a Standard Model fermion and ν is the neutrino. This leads to generic problems with proton decay and FCNC effects, and also neutrino masses are no longer consistent with experiment. Thus, while the hierarchy problem has been addressed in the Higgs sector by a classical rescaling of the Higgs field, this has come at the expense of introducing proton decay and FCNC problems from higher-dimension operators that were sufficiently suppressed in the Standard Model.

• EXERCISE: *The classical rescaling* $\Phi \to e^{d_\Phi \pi k R} \Phi$ *where* $d_\Phi = 1(\frac{3}{2})$ *for scalars (fermions), suffers from a quantum anomaly and leads to the addition of the Lagrangian term*

$$\delta \mathcal{L}_{\text{anomaly}} = \pi k R \sum_i \frac{\beta(g_i)}{4 g_i^3} \operatorname{Tr} F_{\mu\nu,i}^2, \tag{3.5}$$

where $\beta(g_i)$ *is the* β*-function for the corresponding gauge couplings* g_i. *Show that this anomaly implies that quantum mass scales, such as the gauge coupling unification scale* M_{GUT}, *are also redshifted by an amount* $M_{GUT} e^{-\pi k R}$.

Instead in the slice of AdS$_5$ with the Standard Model fields confined on the IR brane one has to resort to discrete symmetries to forbid the offending higher-dimension operators. Of course it is not adequate just to forbid the leading higher-dimension operators. Since the cutoff scale on the IR brane is low ($\mathcal{O}(\text{TeV})$), successive higher-dimension operators must also be eliminated to very high order.

This feature of RS1 stems from the fact that all Standard Model particles are localised on the IR brane. However to address the hierarchy problem, only the Higgs field needs to be localised on the IR brane. The Standard Model fermions and gauge fields do not have a hierarchy problem and therefore can be placed anywhere in the bulk [10, 11, 12]. In this way the UV brane can be used to provide a sufficiently high scale to help suppress higher-dimension operators while still solving the hierarchy problem [11].

3.2. The bulk field Lagrangian

Let us consider fermion Ψ, scalar Φ and vector A_M bulk fields. In a slice of AdS$_5$ the 5D action is given by

$$
\begin{aligned}
S_5 = -\int d^4x \, dy \, \sqrt{-g} \Bigg[& \frac{1}{2} M_5^3 R + \Lambda_5 \\
& + \frac{1}{4g_5^2} F_{MN}^2 + |D_M \Phi|^2 + i \bar{\Psi} \Gamma^M \nabla_M \Psi \\
& + m_\phi^2 |\Phi|^2 + i m_\psi \, \bar{\Psi} \Psi \Bigg],
\end{aligned}
\tag{3.6}
$$

where M_5 is the 5D fundamental scale, Λ_5 is the bulk cosmological constant and g_5 is the 5D gauge coupling. In curved space the gamma matrices are $\Gamma_M = e_M^A \gamma_A$, where e_M^A is the funfbein defined by $g_{MN} = e_M^A e_N^B \eta_{AB}$ and $\gamma_A = (\gamma_\alpha, \gamma_5)$ are the usual gamma matrices in flat space. The curved space covariant derivative $\nabla_M = D_M + \omega_M$, where ω_M is the spin connection and D_M is the gauge covariant derivative for fermion and/or scalar fields charged under some gauge symmetry. The action (3.6) includes all terms to quadratic order that are consistent with gauge symmetries and general coordinate invariance. In particular this only allows a mass term m_ϕ for the bulk scalar and a mass term m_ψ for the bulk fermion.

In general the equation of motion for the bulk fields is obtained by requiring that $\delta S_5 = 0$. This variation of the action (3.6) can be written in the form

$$
\delta S_5 = \int d^5x \, \delta\phi(\mathcal{D}\phi) + \int d^4x \, \delta\phi(\mathcal{B}\phi)\big|_{y*},
\tag{3.7}
$$

where ϕ is any bulk field. Requiring the first term in (3.7) to vanish gives the equation of motion $\mathcal{D}\phi = 0$. However the second term in (3.7) is evaluated at the boundaries y^* of the fifth dimension y. The vanishing of the second term thus leads to the boundary conditions $\delta\phi|_{y*} = 0$ or $\mathcal{B}\phi|_{y*} = 0$. Note that there are also boundary terms arising from the orthogonal directions x^μ, but these are automatically zero because ϕ is assumed to vanish at the 4D boundary $x^\mu = \pm\infty$.

3.2.1. Scalar fields

Suppose that the bulk scalar field has a mass squared $m_\phi^2 = ak^2$ where we have defined the bulk scalar mass in units of the curvature scale k with dimensionless coefficient a. The equation of motion derived from the scalar part of the variation of the action (3.6) is

$$
\partial^2 \Phi + e^{2ky} \partial_5 (e^{-4ky} \partial_5 \Phi) - m_\phi^2 e^{-2ky} \Phi = 0,
\tag{3.8}
$$

where $\partial^2 = \eta^{\mu\nu}\partial_\mu\partial_\nu$. We are interested in the zero mode solution of this equation. The solution is obtained by assuming a separation of variables

$$\Phi(x, y) = \frac{1}{\sqrt{\pi R}} \sum_n \Phi^{(n)}(x)\phi^{(n)}(y), \qquad (3.9)$$

where $\Phi^{(n)}$ are the Kaluza-Klein modes satisfying $\partial^2\Phi^{(n)}(x) = m_n^2\Phi^{(n)}(x)$ and $\phi^{(n)}(y)$ is the profile of the Kaluza-Klein mode in the bulk. The general solution for the zero mode ($m_0 = 0$) is given by

$$\phi^{(0)}(y) = c_1 e^{(2-\alpha)ky} + c_2 e^{(2+\alpha)ky}, \qquad (3.10)$$

where $\alpha \equiv \sqrt{4+a}$ and c_1, c_2 are arbitrary constants. These constants can be determined by imposing boundary conditions at the brane locations, which following from the second term of (3.7) can be either Neumann $\partial_5\phi^{(n)}|_{y^*} = 0$ or Dirichlet $\phi^{(n)}|_{y^*} = 0$. However for $a \neq 0$ imposing Neumann conditions leads to $c_1 = c_2 = 0$. Similarly Dirichlet conditions lead to $c_1 = c_2 = 0$. This implies that there is no zero mode solution with simple Neumann or Dirichlet boundary conditions.

Instead in order to obtain a zero mode we need to modify the boundary action and include boundary mass terms [11]

$$S_{bdy} = -\int d^4x\, dy\, \sqrt{-g}\, 2\, b\, k\, [\delta(y) - \delta(y - \pi R)]\, |\Phi|^2, \qquad (3.11)$$

where b is a dimensionless constant parametrising the boundary mass in units of k. The Neumann boundary conditions are now modified to

$$\left(\partial_5\phi^{(0)} - b\, k\, \phi^{(0)}\right)\Big|_{0,\pi R} = 0. \qquad (3.12)$$

• EXERCISE: *Verify that the boundary conditions (3.12) follow from (3.7) after including the boundary mass terms (3.11).*

Imposing the modified Neumann boundary conditions at $y^* = 0, \pi R$ leads to the equations

$$(2 - \alpha - b)\, c_1 + (2 + \alpha - b)\, c_2 = 0, \qquad (3.13)$$
$$(2 - \alpha - b)\, c_1\, e^{(2-\alpha)\pi kR} + (2 + \alpha - b)\, c_2\, e^{(2+\alpha)\pi kR} = 0. \qquad (3.14)$$

These equations depend on the two arbitrary mass parameters a and b. For generic values of these parameters the solution to the boundary conditions again leads to $c_1 = c_2 = 0$. However, if $b = 2 - \alpha$ then this implies that only $c_2 = 0$, while if $b = 2 + \alpha$ then we obtain $c_1 = 0$. Note that in principle there are three mass parameters if we introduce two parameters corresponding to each boundary. However in (3.11) we have chosen the mass parameters on the two boundaries to be equal and opposite. Thus a nonzero part of the general solution always survives and the zero mode solution becomes

$$\phi^{(0)}(y) \propto e^{bky}, \tag{3.15}$$

where $b = 2 \pm \alpha$. Assuming α to be real (which requires $a \geq -4$ in accord with the Breitenlohner-Freedman bound [13] for the stability of AdS space), the parameter b has a range $-\infty < b < \infty$. The localisation features of the zero mode follows from considering the kinetic term

$$
\begin{aligned}
&- \int d^5x \, \sqrt{-g} \, g^{\mu\nu} \, \partial_\mu \Phi^* \partial_\nu \Phi + \ldots \\
&= - \int d^5x \, e^{2(b-1)ky} \, \eta^{\mu\nu} \, \partial_\mu \Phi^{(0)*}(x) \partial_\nu \Phi^{(0)}(x) + \ldots .
\end{aligned} \tag{3.16}
$$

Hence, with respect to the 5D flat metric the zero mode profile is given by

$$\widetilde{\phi}^{(0)}(y) \propto e^{(b-1)ky} = e^{(1 \pm \sqrt{4+a})ky}. \tag{3.17}$$

We see that for $b < 1$ ($b > 1$) the zero mode is localised towards the UV (IR) brane and when $b = 1$ the zero mode is flat. Therefore using the one remaining free parameter b the scalar zero mode can be localised anywhere in the bulk.

The general solution of the Kaluza-Klein modes corresponding to $m_n \neq 0$ is given by

$$\phi^{(n)}(y) = e^{2ky} \left[c_1 J_\alpha \left(\frac{m_n}{ke^{-ky}} \right) + c_2 Y_\alpha \left(\frac{m_n}{ke^{-ky}} \right) \right], \tag{3.18}$$

where $c_{1,2}$ are arbitrary constants. The Kaluza-Klein masses are determined by imposing the boundary conditions and in the limit $\pi kR \gg 1$ lead to the approximate values

$$m_n \approx \left(n + \frac{1}{2}\sqrt{4+a} - \frac{3}{4} \right) \pi k \, e^{-\pi kR}. \tag{3.19}$$

The fact that the Kaluza-Klein mass scale is associated with the IR scale ($ke^{-\pi kR}$) is consistent with the fact that the Kaluza-Klein modes are localised near the IR brane, and unlike the zero mode can not be arbitrarily localised in the bulk.

3.2.2. Fermions

Let us next consider bulk fermions in a slice of AdS$_5$ [14, 11]. In five dimensions a fundmental spinor representation has four components, so fermions are described by Dirac spinors Ψ. Under the Z_2 symmetry $y \rightarrow -y$ a fermion transforms (up to a phase $\pm$) as

$$\Psi(-y) = \pm\gamma_5\Psi(y), \tag{3.20}$$

so that $\bar{\Psi}\Psi$ is odd. Since only invariant (or even) terms under the Z_2 symmetry can be added to the bulk Lagrangian the corresponding mass parameter for a fermion must necessarily be odd and given by

$$m_\psi = c\,k\,(\epsilon(y) - \epsilon(y - \pi R)), \tag{3.21}$$

where c is a dimensionless mass parameter and $\epsilon(y) = y/|y|$. We have again chosen the mass parameter c to be equal and opposite on the two boundaries. The corresponding equation of motion for fermions resulting from the action (3.6) is

$$e^{ky}\eta^{\mu\nu}\gamma_\mu\partial_\nu\widehat{\Psi}_- + \partial_5\widehat{\Psi}_+ + m_\psi\widehat{\Psi}_+ = 0,$$
$$e^{ky}\eta^{\mu\nu}\gamma_\mu\partial_\nu\widehat{\Psi}_+ - \partial_5\widehat{\Psi}_- + m_\psi\widehat{\Psi}_- = 0, \tag{3.22}$$

where $\widehat{\Psi} = e^{-2ky}\Psi$ and $\Psi_\pm$ are the components of the Dirac spinor $\Psi = \Psi_+ + \Psi_-$ with $\Psi_\pm = \pm\gamma_5\Psi_\pm$. Note that the equation of motion is now a first order coupled equation between the components of the Dirac spinor Ψ.

● EXERCISE: *Using the fact that the spin connection for the AdS$_5$ metric is given by*

$$\omega_M = \left(\frac{k}{2}\gamma_5\gamma_\mu e^{-ky},\ 0\right), \tag{3.23}$$

derive the bulk fermion equation of motion (3.22) from the action (3.6).

The solutions of the bulk fermion equation of motion (3.22) are again obtained by separating the variables

$$\Psi_\pm(x, y) = \frac{1}{\sqrt{\pi R}}\sum_n \Psi_\pm^{(n)}(x)\psi_\pm^{(n)}(y), \tag{3.24}$$

where $\Psi_\pm^{(n)}$ are the Kaluza-Klein modes satisfying $\eta^{\mu\nu}\gamma_\mu\partial_\nu\Psi_\pm^{(n)} = m_n\Psi_\pm^{(n)}$. The zero mode solutions can be obtained for $m_n = 0$ and the general solution of (3.22) is given by

$$\widehat{\psi}_\pm^{(0)}(y) = d_\pm\,e^{\mp cky}, \tag{3.25}$$

where $d_\pm$ are arbitrary constants. The Z_2 symmetry implies that one of the components $\psi_\pm$ must always be odd. If $\gamma_5 = \mathrm{diag}(1, -1)$, then (3.20) implies that $\psi_\mp$ is odd and there is no corresponding zero mode for this component of Ψ. In fact this is how 4D chirality is recovered from the vectorlike 5D bulk and is the result of compactifying on the Z_2 orbifold. For the remaining zero mode the boundary condition obtained from (3.7) with the boundary mass term (3.21) is the modified Neumann condition

$$\left(\partial_5 \widehat{\psi}^{(0)}_\pm \pm c\, k\, \widehat{\psi}^{(0)}_\pm\right)\bigg|_{0,\pi R} = 0. \tag{3.26}$$

Thus there will always be a zero mode since the boundary condition is trivially the same as the equation of motion. For concreteness let us choose ψ_- to be odd, then the only nonvanishing zero mode component of Ψ is

$$\psi^{(0)}_+(y) \propto e^{(2-c)ky}. \tag{3.27}$$

Again the localisation features of this mode are obtained by considering the kinetic term

$$-\int d^5x\, \sqrt{-g}\, g^{\mu\nu}\, i\bar{\Psi}\Gamma_\mu \partial_\nu \Psi + \dots$$
$$= -\int d^5x\, e^{2(\frac{1}{2}-c)ky}\, \eta^{\mu\nu}\, i\bar{\Psi}^{(0)}_+(x)\gamma_\mu \partial_\nu \Psi^{(0)}_+(x) + \dots. \tag{3.28}$$

Hence with respect to the 5D flat metric the fermion zero mode profile is

$$\widetilde{\psi}^{(0)}_+(y) \propto e^{(\frac{1}{2}-c)ky}. \tag{3.29}$$

When $c > 1/2$ ($c < 1/2$) the fermion zero mode is localised towards the UV (IR) brane while the zero mode fermion is flat for $c = 1/2$. So, just like the scalar field zero mode the fermion zero mode can be localised anywhere in the 5D bulk.

The nonzero Kaluza-Klein fermion modes can be obtained by solving the coupled equations of motion for the Dirac components $\psi^{(n)}_\pm$. This leads to a pair of decoupled second order equations that can be easily solved. The expressions for the corresponding wave functions and Kaluza-Klein masses are summarised in the next subsection.

3.2.3. Summary

A similar analysis can also be done for the bulk graviton [15] and bulk gauge field [16, 17]. In these cases there are no bulk or boundary masses and the corresponding graviton zero mode is localised on the UV brane while the gauge field

Table 1

The zero mode profiles of bulk fields and the corresponding CFT operator dimensions.

Field	Profile	dim $\mathcal{O}$
scalar $\phi^{(0)}(y)$	$e^{(1\pm\sqrt{4+a})ky}$	$2+\sqrt{4+a}$
fermion $\psi_+^{(0)}(y)$	$e^{\left(\frac{1}{2}-c\right)ky}$	$\frac{3}{2}+\|c+\frac{1}{2}\|$
vector $A_\mu^{(0)}(y)$	1	3
graviton $h_{\mu\nu}^{(0)}(y)$	e^{-ky}	4

zero mode is flat and not localised in the 5D bulk[1]. A summary of the bulk zero mode profiles is given in Table 1.

Similarly the Kaluza-Klein mode ($m_n \neq 0$) solutions can be obtained for all types of bulk fields and combined into one general expression [11]

$$f^{(n)}(y) = \frac{e^{\frac{s}{2}ky}}{N_n}\left[J_\alpha\left(\frac{m_n}{ke^{-ky}}\right) + b_\alpha Y_\alpha\left(\frac{m_n}{ke^{-ky}}\right)\right],\tag{3.30}$$

for $f^{(n)} = (\phi^{(n)}, \widehat{\psi}_\pm^{(n)}, A_\mu^{(n)})$ where

$$b_\alpha = -\frac{(-r+\frac{s}{2})J_\alpha(\frac{m_n}{k}) + \frac{m_n}{k}J'_\alpha(\frac{m_n}{k})}{(-r+\frac{s}{2})Y_\alpha(\frac{m_n}{k}) + \frac{m_n}{k}Y'_\alpha(\frac{m_n}{k})},\tag{3.31}$$

and

$$N_n \simeq \frac{1}{\sqrt{\pi^2 R\, m_n\, e^{-\pi kR}}},\tag{3.32}$$

with $s = (4, 1, 2)$, $r = (b, \mp c, 0)$ and $\alpha = (\sqrt{4+a}, |c \pm \frac{1}{2}|, 1)$. The graviton modes $h_{\mu\nu}^{(n)}$ are identical to the scalar modes $\phi^{(n)}$ except that $a = b = 0$. The Kaluza-Klein mass spectrum is approximately given by

$$m_n \simeq \left(n + \frac{1}{2}(\alpha-1) \mp \frac{1}{4}\right)\pi k\, e^{-\pi kR},\tag{3.33}$$

for even (odd) modes and $n = 1, 2, \ldots$. Note that the Kaluza-Klein modes for all types of bulk fields are always localised near the IR brane. Unlike the zero mode there is no freedom to delocalise the Kaluza-Klein (nonzero) modes away from the IR brane.

---

[1]Note however that bulk and boundary masses can be introduced for the graviton [18] and gauge field [19], thereby changing their localization profile and corresponding operator dimensions. But this is beyond the applications that will be considered in these lectures.

4. The Standard Model in the bulk

We can now use the freedom to localise scalar and fermion zero mode fields anywhere in the warped bulk to construct a bulk Standard Model. Recall that the hierarchy problem only affects the Higgs boson. Hence to solve the hierarchy problem the Higgs scalar field must be localised very near the TeV brane, and for simplicity we will assume that the Higgs is confined on the TeV brane (as in RS1). However we will now consider the possible effects of allowing fermions and gauge bosons to live in the warped bulk.

4.1. Yukawa couplings

One consequence of allowing fermions to be localised anywhere in the bulk, is that Yukawa coupling hierarchies are naturally generated by separating the fermions from the Higgs on the TeV brane. Each Standard Model fermion is identified with the zero mode of a corresponding 5D Dirac spinor Ψ. For example, the left-handed electron doublet e_L is identified with the zero mode of Ψ_{eL+}, which is the even component of the 5D Dirac spinor $\Psi_{eL} = \Psi_{eL+} + \Psi_{eL-}$. The odd component Ψ_{eL-} does not have a zero mode, but at the massive level it pairs up with the massive modes of Ψ_{eL+} to form a vectorlike Dirac mass. This embedding of 4D fermions into 5D fermions is repeated for each Standard Model fermion. The Standard Model Yukawa interactions, such as $\bar{\Psi}_{eL}\Psi_{eR}H$, are then promoted to 5D interactions in the warped bulk. This gives

$$\int d^4x \int dy \sqrt{-g}\,\lambda_{ij}^{(5)}\left[\bar{\Psi}_{iL}(x,y)\Psi_{jR}(x,y) + h.c.\right]H(x)\delta(y - \pi R)$$
$$\equiv \int d^4x\,\lambda_{ij}\,(\bar{\Psi}_{iL+}^{(0)}(x)\Psi_{jR+}^{(0)}(x)H(x) + h.c. + \ldots), \tag{4.1}$$

where i, j are flavour indices, $\lambda_{ij}^{(5)}$ is the (dimensionful) 5D Yukawa coupling and λ_{ij} is the (dimensionless) 4D Yukawa coupling. Given that the zero mode profile is

$$\tilde{\psi}_{iL+,R+}^{(0)}(y) \propto e^{(\frac{1}{2}-c_{iL,R})ky}, \tag{4.2}$$

this leads to an exponential hierarchy in the 4D Yukawa coupling [11]

$$\lambda_{ij} \simeq \lambda_{ij}^{(5)}k\sqrt{(c_{iL} - 1/2)(c_{iR} - 1/2)}\,e^{(1-c_{iL}-c_{jR})\pi kR}, \tag{4.3}$$

for $c_{iL,R} > 1/2$. Assuming $c_{iL} = c_{jR}$ for simplicity then the electron Yukawa coupling $\lambda_e \sim 10^{-6}$ is obtained for $c_e \simeq 0.64$. Instead when $c_{iL,R} < 1/2$, both fermions are localised near the IR brane giving

$$\lambda_{ij} \simeq \lambda_{ij}^{(5)}k\sqrt{(1/2 - c_{iL})(1/2 - c_{iR})}, \tag{4.4}$$

with no exponential suppression. Hence the top Yukawa coupling $\lambda_t \sim 1$ is obtained for $c_t \simeq -0.5$. The remaining fermion Yukawa couplings, c_f then range from $c_t < c_f < c_e$ [11, 20]. Thus, we see that for bulk mass parameters c of $\mathcal{O}(1)$ the fermion mass hierarchy is explained. The fermion mass problem is now reduced to determining the c parameters in the 5D theory. This requires a UV completion of the 5D warped bulk model with fermions, such as string theory.

Since bulk fermions naturally lead to Yukawa coupling hierarchies, the gauge bosons must also necessarily be in the bulk. From Table 1 the gauge field zero mode is flat and therefore couples with equal strength to both the UV and IR brane. Only the Higgs field is confined to the IR brane (or TeV brane). Thus the picture that emerges is a Standard Model in the warped bulk as depicted in Figure 2. The fermions are localised to varying degrees in the bulk with the electron zero mode, being the lightest fermion, furthest away from the Higgs on the TeV brane while the top, being the heaviest, is closest to the Higgs. This enables one to not only solve the hierarchy problem but also address the Yukawa coupling hierarchies.

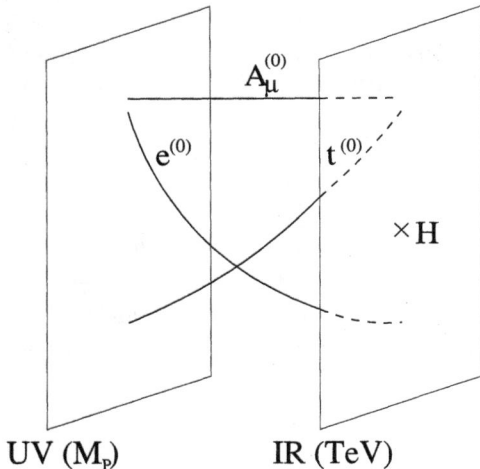

Fig. 2. The Standard Model in the warped five-dimensional bulk.

The warped bulk can also be used to obtain naturally small neutrino masses. Various scenarios are possible. If the right (left) handed neutrino is localised near the UV (IR) brane then a tiny Dirac neutrino mass is obtained [14]. However this requires that lepton number is conserved on the UV brane. Instead in the "reversed" scenario one can place the right (left) handed neutrino near the IR

(UV) brane. In this case even though lepton number is violated on the UV brane, the neutrinos will still obtain naturally tiny Dirac masses [21].

4.2. Higher-dimension operators

Let us consider the following generic four-fermion operators which are relevant for proton decay and $K - \bar{K}$ mixing

$$\int d^4x \int dy \sqrt{-g} \frac{1}{M_5^3} \bar{\Psi}_i \Psi_j \bar{\Psi}_k \Psi_l \equiv \int d^4x \frac{1}{M_4^2} \bar{\Psi}_{i+}^{(0)} \Psi_{j+}^{(0)} \bar{\Psi}_{k+}^{(0)} \Psi_{l+}^{(0)}, \quad (4.5)$$

where the effective 4D mass scale M_4 for $1/2 \lesssim c_i \lesssim 1$ is approximately given by[11]

$$\frac{1}{M_4^2} \simeq \frac{k}{M_5^3} e^{(4-c_i-c_j-c_k-c_l)\pi k R}. \quad (4.6)$$

If we want the suppression scale for higher-dimension proton decay operators to be $M_4 \sim M_P$ then (4.6) requires $c_i \simeq 1$ assuming $k \sim M_5 \sim M_P$. Unfortunately for these values of c_i the corresponding Yukawa couplings would be too small. Nevertheless, the values of c needed to explain the Yukawa coupling hierarchies still suppresses proton decay by a mass scale larger than the TeV scale [11, 22]. Thus there is no need to impose a discrete symmetry which forbids very large higher-dimension operators.

On the other hand the suppression scale for FCNC processes only needs to be $M_4 \gtrsim 1000$ TeV. This can easily be achieved for the values of c that are needed to explain the Yukawa coupling hierarchies. In fact the FCNC constraints can be used to obtain a lower bound on the Kaluza-Klein mass scale m_{KK}. For example Kaluza-Klein gluons can mediate $\Delta S = 2$ FCNC processes at tree level because the fermions are located at different points. In flat space with split fermions this leads to strong constraints $m_{KK} \gtrsim 25 - 300$ TeV (with the range depending on whether FCNC processes violate CP) [23]. However in warped space for $c \gtrsim 1/2$, the Kaluza-Klein gauge boson coupling to fermions is universal even though fermions are split. This is because in warped space the Kaluza-Klein gauge boson wave functions are approximately flat near the UV brane. The corresponding bound for warped dimensions is $m_{KK} \gtrsim 2 - 30$ TeV. This fact that the Kaluza-Klein gauge bosons couple universally to spatially separated fermions is akin to a GIM-like mechanism in the 5D bulk [11]. Thus, warped dimensions ameliorate the bounds on the Kaluza-Klein scale.

4.3. Higgs as a pseudo Nambu-Goldstone boson

So far we have said very little about electroweak symmetry breaking and the Higgs mass. If the Higgs is confined to the IR (or TeV) brane then the tree-level Higgs mass parameter is naturally of order $\Lambda_{IR} = \Lambda_{UV} e^{-\pi kR}$. Since there are fermions and gauge bosons in the bulk the effects of their corresponding Kaluza-Klein modes must be sufficiently suppressed. This requires $ke^{-\pi kR} \sim \mathcal{O}(\text{TeV})$ and since in RS1 $\Lambda_{UV} \sim 10k$ we have $\Lambda_{IR} \sim \mathcal{O}(10 \text{ TeV})$. Consequently a modest amount of fine-tuning would be required to obtain a physical Higgs mass of order 100 GeV, as suggested by electroweak precision data [24]. Clearly, it is desirable to invoke a symmetry to keep the Higgs mass naturally lighter than the IR cutoff scale, such as the spontaneous breaking of a global symmetry[2].

Motivated by the fact that the dimensional reduction of a five-dimensional gauge boson $A_M = (A_\mu, A_5)$ contains a scalar field A_5, one can suppose that the Higgs scalar field is part of a higher-dimensional gauge field [25]. In a slice of AdS$_5$ the A_5 terms in the gauge boson kinetic term of the bulk Lagrangian (3.6) are

$$-\frac{1}{2} \int d^4x \, dy \, e^{-2ky}((\partial A_5)^2 - 2\eta^{\mu\nu}\partial_\mu A_5 \partial_5 A_\nu) + \ldots . \tag{4.7}$$

In particular notice that the higher-dimensional gauge symmetry prevents a tree-level mass for A_5. However the zero mode of the A_5 scalar field must be localised near the IR brane for the hierarchy problem to be solved. The solution for A_5 can be obtained by adding a gauge fixing term that cancels the mixed $A_5 A_\nu$ term [26, 27]. This gives the zero mode solution $A_5^{(0)}$ with y dependence $\propto e^{+2ky}$ and substituting back into the action leads to

$$-\frac{1}{2} \int d^4x \, dy \, e^{+2ky}(\partial_\mu A_5^{(0)}(x))^2 + \ldots . \tag{4.8}$$

Hence with respect to the flat 5D metric the massless scalar mode $A_5^{(0)}$ is indeed localised towards the IR brane and therefore can play the role of the Higgs boson.

To obtain a realistic model one assumes an SO(5)×U(1)$_{B-L}$ gauge symmetry in the bulk for the electroweak sector [27]. On the IR brane this symmetry is spontaneously broken by boundary conditions to SO(4)×U(1)$_{B-L}$. This leads to four Nambu-Goldstone bosons that can be identified with the Standard Model Higgs doublet. A Higgs mass is then generated because SO(5) gauge symmetry is explicitly broken in the fermion sector, in particular by the top quark. At one loop this generates an effective potential and electroweak symmetry is broken

---

[2]But this is not the unique possibility. As we will see later supersymmetry (instead of a global symmetry) can also be used to obtain a light Higgs mass.

dynamically via top-quark loop corrections [28]. This effect is finite and arises from the Hosotani mechanism with nonlocal operators in the bulk [29]. An unbroken SO(3) custodial symmetry guarantees that the Peskin-Takeuichi parameter $T = 0$. The important point however is that radiative corrections to the Higgs mass depend on $ke^{-\pi kR}$ and not on $\Lambda_{UV} e^{-\pi kR}$. Together with the accompanying one-loop factor $\frac{1}{16\pi^2}$ this guarantees a light Higgs mass of order $m_{Higgs} \lesssim 140$ GeV. Furthermore this model can be shown to pass stringent electroweak precision tests without a significant amount of fine-tuning [30].

In summary, the hierarchy problem can be solved by placing the Standard Model in the warped bulk together with identifying the Higgs scalar field as a pseudo Nambu-Goldstone boson. This leads to a very predictive scenario for the electroweak symmetry breaking sector. Moreover the setup is radiatively stable and valid up to the Planck scale with (as we will see later) grand unification incorporated in an interesting way! This makes the 5D warped bulk a compelling alternative framework to address the hierarchy problem in a complete scenario compared to the usual 4D scenarios. But even more compelling is that this framework can be given a purely 4D holographic description in terms of a strongly coupled gauge theory as will be shown in the next section.

5. AdS/CFT and holography

Remarkably 5D warped models in a slice of AdS can be given a purely 4D description. This holographic correspondence between the 5D theory and the 4D theory originates from the AdS/CFT correspondence in string theory. In 1997 Maldacena conjectured that [31]

$$
\begin{array}{ccc}
\text{type IIB string theory} & \overset{\text{DUAL}}{\Longleftrightarrow} & \mathcal{N} = 4 \text{ SU(N) 4D gauge theory} \quad (5.1) \\
\text{on AdS}_5 \times S^5 & &
\end{array}
$$

where $\mathcal{N}$ is the number of supersymmetry generators and

$$
\frac{R_{AdS}^4}{l_s^4} = 4\pi g_{YM}^2 N, \tag{5.2}
$$

with $R_{AdS} \equiv 1/k$, l_s is the string length and g_{YM} is the SU(N) Yang-Mills gauge coupling. Qualitatively we can see that the isometry of the five-dimensional sphere S^5 is the rotation group $SO(6) \cong SU(4)$, which is the R-symmetry group of the supersymmetric gauge theory. Moreover the $\mathcal{N} = 4$ gauge theory is a conformal field theory because the isometry group of AdS_5 is precisely the conformal group in four dimensions. In particular this means that the gauge couplings do not receive quantum corrections and therefore do not run with energy.

In the warped bulk we have only considered gravity. This represents the effective low energy theory of the full string theory. In order to neglect the string corrections, so that the bulk gravity description is valid, we require that $R_{AdS} \gg l_s$. This leads to the condition that $g_{YM}^2 N \gg 1$, which means that the 4D dual CFT is strongly coupled! Thus for our purposes the correspondence takes the form of a duality in which the weakly coupled 5D gravity description is dual to a strongly coupled 4D CFT. This remarkable duality means that any geometric configuration of fields in the bulk can be given a purely 4D description in terms of a strongly coupled gauge theory. Therefore warped models provide a new way to study strongly coupled gauge theories.

While there is no rigorous mathematical proof of the AdS/CFT conjecture, it has passed many nontrivial tests and an AdS/CFT dictionary to relate the two dual descriptions can be established. For every bulk field Φ there is an associated operator $\mathcal{O}$ of the CFT. In the AdS$_5$ metric (3.1) the boundary of AdS space is located at $y = -\infty$. The boundary value of the bulk field $\Phi(x^\mu, y = -\infty) \equiv \phi_0(x^\mu)$ acts as a source field for the CFT operator $\mathcal{O}$. The AdS/CFT correspondence can then be quantified in the following way [32, 33]

$$\int \mathcal{D}\phi_{CFT} \, e^{-S_{CFT}[\phi_{CFT}] - \int d^4x \, \phi_0 \mathcal{O}} = \int_{\phi_0} \mathcal{D}\phi \, e^{-S_{bulk}[\phi]} \equiv e^{i S_{eff}[\phi_0]}, \qquad (5.3)$$

where S_{CFT} is the CFT action with ϕ_{CFT} generically denoting the CFT fields and S_{bulk} is the bulk 5D action. The on-shell gravity action, S_{eff} is obtained by integrating out the bulk degrees of freedom for suitably chosen IR boundary conditions. In general n-point functions can be calculated via

$$\langle \mathcal{O} \ldots \mathcal{O} \rangle = \frac{\delta^n S_{eff}}{\delta \phi_0 \ldots \delta \phi_0}. \qquad (5.4)$$

In this way we see that the on-shell bulk action is the generating functional for connected Greens functions in the CFT.

So far the correspondence has been formulated purely in AdS$_5$ without the presence of the UV and IR branes. In particular notice from (5.3) that the source field ϕ_0 is a nondynamical field with no kinetic term. However since we are interested in the 4D dual of a slice of AdS$_5$ (and not the full AdS space) we will need the corresponding dual description in the presence of two branes.

Suppose that a UV brane is placed at $y = 0$. The $-\infty < y < 0$ part of AdS space is chopped off and the remaining $0 < y < \infty$ part is reflected about $y = 0$ with a Z_2 symmetry. The presence of the UV brane with an associated UV scale Λ_{UV} thus corresponds to an explicit breaking of the conformal invariance in the CFT at the UV scale (but only by nonrenormalisable terms) [34, 35, 36]. The fact that the CFT now has a finite UV cutoff means that the source field ϕ_0

becomes dynamical. A kinetic term for the source field will always be induced by the CFT but one can directly add an explicit kinetic term for the source field at the UV scale. Thus in the presence of a UV brane the AdS/CFT correspondence is modified to the form

$$\int \mathcal{D}\phi_0 \, e^{-S_{UV}[\phi_0]} \int_{\Lambda_{UV}} \mathcal{D}\phi_{CFT} \, e^{-S_{CFT}[\phi_{CFT}] - \int d^4x \, \phi_0 \mathcal{O}}$$

$$= \int \mathcal{D}\phi_0 \, e^{-S_{UV}[\phi_0]} \int_{\phi_0} \mathcal{D}\phi \, e^{-S_{bulk}[\phi]}, \tag{5.5}$$

where S_{UV} is the UV Lagrangian for the source field ϕ_0. It is understood that now the source field $\phi_0 = \Phi(x, y = 0)$. Moving away from the UV brane at $y = 0$ in the bulk corresponds in the 4D dual to running down from the UV scale to lower energy scales. Since the bulk is AdS the 4D dual gauge theory quickly becomes conformal at energies below the UV scale.

The presence of the IR brane at $y = \pi R$ corresponds to a spontaneous breaking of the conformal invariance in the CFT at the IR scale $\Lambda_{IR} = \Lambda_{UV} e^{-\pi k R}$ [34, 35]. The conformal symmetry is nonlinearly realised and particle bound states of the CFT can now appear. Formally this can be understood by noting that the (massless) radion field in RS1 is localised on the IR brane, since by sending the UV brane to the AdS boundary (at $y \to -\infty$), while keeping the IR scale fixed, formally decouples the source field and keeps the radion in the spectrum. This means that at the IR scale the CFT must contain a massless particle which is interpreted as the Nambu-Goldstone boson of spontaneously broken conformal symmetry. This so called dilaton is therefore the dual interpretation of the radion. A similar phenomenon also occurs in QCD where massless pions are the Nambu-Goldstone bosons of the spontaneously broken chiral symmetry at Λ_{QCD}. Indeed the AdS/CFT correspondence suggests that QCD may be the holographic description of a bulk (string) theory.

Thus, the dual interpretation of a slice of AdS not only contains a 4D dual CFT with a UV cutoff, but also a dynamical source field ϕ_0 with UV Lagrangian $S_{UV}[\phi_0]$. In particular note that the source field is an elementary (point-like) state up to the UV scale, while particles in the CFT sector are only effectively point-like below the IR scale but are composite above the IR scale. The situation is analogous to the (elementary) photon of QED interacting with the (composite) spin-1 mesons of QCD with cutoff scale Λ_{QCD}.

5.1. Holography of scalar fields

As a simple application of the AdS/CFT correspondence in a slice of AdS_5 we shall investigate in more detail the dual theory corresponding to a bulk scalar field Φ with boundary mass terms. The qualitative features will be very similar for other spin fields. In order to obtain the correlation functions of the dual theory we first need to compute the on-shell bulk action S_{eff}. According to (3.30) the bulk scalar solution is given by

$$\Phi(p, z) = \Phi(p)A^{-2}(z)\left[J_\alpha(iq) - \frac{J_{\alpha\pm1}(iq_1)}{Y_{\alpha\pm1}(iq_1)}Y_\alpha(iq)\right], \qquad (5.6)$$

where $z = (e^{ky} - 1)/k$, $A(z) = (1 + kz)^{-1}$, $q = p/(kA(z))$ and $\Phi(p, z)$ is the 4D Fourier transform of $\Phi(x, z)$. The $\pm$ refers to the two branches associated with $b = b_\pm = 2 \pm \alpha$. Substituting this solution into the bulk scalar action and imposing the IR boundary condition (3.12) leads to the on-shell action

$$S_{eff} = \frac{1}{2}\int \frac{d^4p}{(2\pi)^4}\left[A^3(z)\Phi(p, z)\left(\Phi'(-p, z) - b\,k\,A(z)\Phi(-p, z)\right)\right]\bigg|_{z=z_0}$$

$$= \frac{k}{2}\int \frac{d^4p}{(2\pi)^4}F(q_0, q_1)\Phi(p)\Phi(-p), \qquad (5.7)$$

where

$$F(q_0, q_1) = \mp iq_0\left[J_{\nu\mp1}(iq_0) - Y_{\nu\mp1}(iq_0)\frac{J_\nu(iq_1)}{Y_\nu(iq_1)}\right]$$

$$\times\left[J_\nu(iq_0) - Y_\nu(iq_0)\frac{J_\nu(iq_1)}{Y_\nu(iq_1)}\right], \qquad (5.8)$$

and $\nu \equiv \nu_\pm = \alpha \pm 1$.

• EXERCISE: *Verify (5.7) is obtained by substituting the bulk scalar solution (5.6) into the scalar part of the action (3.6).*

The dual theory two-point function of the operator $\mathcal{O}$ sourced by the bulk field Φ is contained in the self-energy $\Sigma(p)$ obtained by

$$\Sigma(p) = \int d^4x\, e^{-ip\cdot x}\frac{\delta^2 S_{eff}}{\delta(A^2(z_0)\Phi(x, z_0))\delta(A^2(z_0)\Phi(0, z_0))}$$

$$= \frac{k}{g_\phi^2}\frac{q_0(I_\nu(q_0)K_\nu(q_1) - I_\nu(q_1)K_\nu(q_0))}{I_{\nu\mp1}(q_0)K_\nu(q_1) + I_\nu(q_1)K_{\nu\mp1}(q_0)}, \qquad (5.9)$$

where a coefficient $1/g_\phi^2$ has been factored out in front of the scalar kinetic term in (3.6), so that g_ϕ is a 5D expansion parameter with $\dim[1/g_\phi^2] = 1$. The behaviour of $\Sigma(p)$ can now be studied for various momentum limits in order to obtain information about the dual 4D theory. When $A_1 \equiv A(z_1) \to 0$ the effects of the conformal symmetry breaking (from the IR brane) are completely negligible.

The leading nonanalytic piece in $\Sigma(p)$ is then interpreted as the pure CFT correlator $\langle \mathcal{O}\mathcal{O} \rangle$ that would be obtained in the string AdS/CFT correspondence with $A_0 \equiv A(z_0) \to \infty$. However in a slice of AdS the poles of $\langle \mathcal{O}\mathcal{O} \rangle$ determines the pure CFT mass spectrum with a nondynamical source field ϕ_0. These poles are identical to the poles of $\Sigma(p)$ since $\Sigma(p)$ and $\langle \mathcal{O}\mathcal{O} \rangle$ only differ by analytic terms. Hence the poles of the correlator $\Sigma(p)$ correspond to the Kaluza-Klein spectrum of the bulk scalar fields with Dirichlet boundary conditions on the UV brane.

There are also analytic terms in $\Sigma(p)$. In the string version of the AdS/CFT correspondence these terms are subtracted away by adding appropriate counterterms. However with a finite UV cutoff (corresponding to the scale of the UV brane) these terms are now interpreted as kinetic (and higher derivative terms) of the source field ϕ_0, so that the source becomes dynamical in the holographic dual theory. The source field can now mix with the CFT bound states and therefore the self-energy $\Sigma(p)$ must be resummed and the modified mass spectrum is obtained by inverting the whole quadratic term $S_{UV} + S_{eff}$. In the case with no UV boundary action S_{UV} this means that the zeroes of (5.9) are identical with the Kaluza-Klein mass spectrum (3.19) corresponding to (modified) Neumann conditions for the source field. In both cases (either Dirichlet or Neumann) these results are consistent with the fact that the Kaluza-Klein states are identified with the CFT bound states.

• EXERCISE: *Check that the zeroes of (5.9) agree with the Kaluza-Klein spectrum (3.19).*

At first sight it is not apparent that there are an infinite number of bound states in the 4D dual theory required to match the infinite number of Kaluza-Klein modes in the 5D theory. How is this possible in the 4D gauge theory? It has been known since the early 1970's that the two-point function in large-N QCD can be written as [37, 38]

$$\langle \mathcal{O}(p)\mathcal{O}(-p) \rangle = \sum_{n=1}^{\infty} \frac{F_n^2}{p^2 + m_n^2}, \tag{5.10}$$

where the matrix element for $\mathcal{O}$ to create the nth meson with mass m_n from the vacuum is $F_n = \langle 0|\mathcal{O}|n \rangle \propto \sqrt{N}/(4\pi)$. In the large N limit the intermediate states are one-meson states and the sum must be infinite because we know that the two-point function behaves logarithmically for large p^2. Since the 4D dual theory is a strongly-coupled SU(N) gauge theory that is conformal at large scales, it will have this same behaviour. This clearly has the same qualitative features as a Kaluza-Klein tower and therefore a dual 5D interpretation could have been posited in the 1970's!

To obtain the holographic interpretation of the bulk scalar field, recall that the scalar zero mode can be localised anywhere in the bulk with $-\infty < b < \infty$ where $b \equiv b_\pm = 2 \pm \alpha$ and $-\infty < b_- < 2$ and $2 < b_+ < \infty$. Since $b_\pm = 1 \pm \nu_\pm$ we have $-1 < \nu_- < \infty$ and $1 < \nu_+ < \infty$. The ν_- branch corresponds to $b_- < 2$, while the ν_+ branch corresponds to $b_+ > 2$. Hence the $\nu_-(\nu_+)$ branch contains zero modes which are localised on the UV (IR) brane.

5.1.1. ν_- branch holography

We begin first with the ν_- branch. In the limit $A_0 \to \infty$ and $A_1 \to 0$ one obtains

$$\Sigma(p) \simeq -\frac{2k}{g_\phi^2}\left[\frac{1}{\nu}\left(\frac{q_0}{2}\right)^2 + \left(\frac{q_0}{2}\right)^{2\nu+2}\frac{\Gamma(-\nu)}{\Gamma(\nu+1)} + \dots\right], \qquad (5.11)$$

where the expansion is valid for noninteger ν. The expansion for integer ν will contain logarithmns. Only the leading analytic term has been written in (5.11). The nonanalytic term is the pure CFT contribution to the correlator $\langle \mathcal{O}\mathcal{O}\rangle$. Formally it is obtained by rescaling the fields by an amount $A_0^{\nu+1}$ and taking the limit

$$\langle \mathcal{O}\mathcal{O}\rangle = \lim_{A_0 \to \infty}(\Sigma(p) + \text{counterterms}) = \frac{1}{g_\phi^2}\frac{\Gamma(-\nu)}{\Gamma(\nu+1)}\frac{p^{2(\nu+1)}}{(2k)^{2\nu+1}}. \qquad (5.12)$$

Since

$$\langle \mathcal{O}(x)\mathcal{O}(0)\rangle = \int \frac{d^4 p}{(2\pi)^4}e^{ipx}\langle \mathcal{O}\mathcal{O}\rangle, \qquad (5.13)$$

the scaling dimension of the operator $\mathcal{O}$ is

$$\dim \mathcal{O} = 3 + \nu_- = 4 - b_- = 2 + \sqrt{4 + a}, \qquad (5.14)$$

as shown in Table 1. If A_0 is finite then the analytic term in (5.11) becomes the kinetic term for the source field ϕ_0. Placing the UV brane at $z_0 = 0$ with $A_0 = 1$ leads to the dual Lagrangian below the cutoff scale $\Lambda \sim k$

$$\mathcal{L}_{4D} = -Z_0(\partial\phi_0)^2 + \frac{\omega}{\Lambda^{\nu_-}}\phi_0\mathcal{O} + \mathcal{L}_{CFT}, \qquad (5.15)$$

where Z_0, ω are dimensionless couplings. This Lagrangian describes a massless dynamical source field ϕ_0 interacting with the CFT via the mixing term $\phi_0\mathcal{O}$. This means that the mass eigenstate in the dual theory will be a mixture of the source field and CFT particle states. The coupling of the mixing term is irrelevant for $v_- > 0$ ($b_- < 1$), marginal if $v_- = 0$ ($b_- = 1$) and relevant for $v_- < 0$ ($b_- > 1$). This suggests the following dual interpretation of the massless bulk zero mode. When the coupling is irrelevant ($v_- > 0$), corresponding to a UV brane localised bulk zero mode, the mixing can be neglected at low energies, and hence to a very good approximation the bulk zero mode is dual to the massless 4D source field ϕ_0. However for relevant ($-1 < v_- < 0$) or marginal couplings ($v_- = 0$) the mixing can no longer be neglected. In this case the bulk zero mode is no longer UV-brane localised, and the dual interpretation of the bulk zero mode is a part elementary, part composite mixture of the source field with massive CFT particle states.

The first analytic term in (5.11) can be matched to the wavefunction constant giving $Z_0 = 1/(2vg_\phi^2 k)$. However at low energies the couplings in $\mathcal{L}$ will change. The low energy limit $q_1 \ll 1$ for $\Sigma(p)$ (and noninteger v) leads to

$$\Sigma(p)_{IR} \simeq -\frac{2k}{g_\phi^2}\left[(1 - A_1^{2v-})\left(\frac{q_0}{2}\right)^2\frac{1}{v} + \dots\right], \tag{5.16}$$

where $A_1 = e^{-\pi k R}$. Notice that there is no nonanalytic term because the massive CFT modes have decoupled. The analytic term has now also received a contribution from integrating out the massive CFT states. Note that when $v_- > 0$ the A_1 contribution to Z_0 is negligible and the kinetic term has the correct sign. On the other hand for relevant couplings the A_1 term now dominates the Z_0 term. The kinetic term still has the correct sign because $v_- < 0$.

● EXERCISE: *Show that for a marginal coupling ($v_- = 0$) the coefficient of the kinetic term is logarithmic and has the correct sign.*

The features of the couplings in (5.15) at low energies can be neatly encoded into a renormalisation group equation. If we define a dimensionless running coupling $\xi(\mu) = \omega/\sqrt{Z(\mu)}(\mu/\Lambda)^v$, which represents the mixing between the CFT and source sector with a canonically normalised kinetic term, then it will satisfy the renormalisation group equation [39]

$$\mu\frac{d\xi}{d\mu} = \gamma\,\xi + \eta\frac{N}{16\pi^2}\xi^3 + \dots, \tag{5.17}$$

where η is a constant and we have replaced $1/(g_\phi^2 k) = N/(16\pi^2)$. The first term arises from the scaling of the coupling of the mixing term $\phi_0\mathcal{O}$ (i.e. $\gamma = v_-$), and the second term arises from the CFT contribution to the wavefunction constant Z_0 (i.e. the second term in (5.11)). The solution of the renormalisation group equation for an initial condition $\xi(M)$ at the scale $M \sim \Lambda$ is

$$\xi(\mu) = \left(\frac{\mu}{M}\right)^\gamma \left\{\frac{1}{\xi^2(M)} + \eta\frac{N}{16\pi^2\gamma}\left[1 - \left(\frac{\mu}{M}\right)^{2\gamma}\right]\right\}^{-1/2}. \quad (5.18)$$

When $\gamma < 0$, the constant $\eta > 0$ and the renormalisation group equation (5.17) has a fixed point at $\xi_* \sim 4\pi\sqrt{-\gamma/(\eta N)}$, which does not depend on the initial value $\xi(M)$. This occurs when $-1 < \nu_- < 0$ and therefore since ξ_* is nonnegligible the mixing between the source and the CFT cannot be neglected.

In the opposite limit, $\gamma > 0$, the solution (5.18) for $M \sim \Lambda$ becomes $\xi(\mu) \sim 4\pi\sqrt{\gamma/N}(\mu/M)^\gamma$, where the solution (5.18) has been matched to the low energy value $Z(ke^{-\pi kR}) = 1/(2\gamma g_\phi^2 k)(1 - e^{-2\gamma\pi kR})$ arising from (5.16) (with $\gamma = \nu_-$). Thus when $\nu_- > 0$ the mixing between the source and CFT sector quickly becomes irrelevant at low energies.

5.1.2. ν_+ branch holography

Consider the case $\nu = \nu_+ > 1$. In the limit $A_0 \to \infty$ and $A_1 \to 0$ we obtain for noninteger ν

$$\Sigma(p) \simeq -\frac{2k}{g_\phi^2}\left[(\nu-1) + \left(\frac{q_0}{2}\right)^2\frac{1}{(\nu-2)} + \left(\frac{q_0}{2}\right)^{2\nu-2}\frac{\Gamma(2-\nu)}{\Gamma(\nu-1)}\right], \quad (5.19)$$

where only the leading analytic terms have been written. The nonanalytic term is again the pure CFT contribution to the correlator $\langle\mathcal{OO}\rangle$ and gives rise to the scaling dimension

$$\dim\mathcal{O} = 1 + \nu_+ = b_+ = 2 + \sqrt{4+a}. \quad (5.20)$$

This agrees with the result for the ν_- branch. At low energies $q_1 \ll 1$ one obtains

$$\Sigma(p)_{IR} \simeq -\frac{2k}{g_\phi^2}\left[(\nu-1) + \left(\frac{q_0}{2}\right)^2\frac{1}{(\nu-2)} - \nu(\nu-1)^2\frac{A_1^{2\nu}}{A_0^{2\nu}}\left(\frac{2}{q_0}\right)^2\right], \quad (5.21)$$

where the large-A_0 limit was taken first. We now see that at low energies the nonanalytic term has a pole at $p^2 = 0$ with the correlator

$$\langle\mathcal{OO}\rangle = \frac{8k^3}{g_\phi^2}\nu_+(\nu_+ - 1)^2 e^{-2\nu_+\pi kR}\frac{1}{p^2}, \quad (5.22)$$

where $A_0 = 1$ and $A_1 = e^{-\pi kR}$. This pole indicates that the CFT has a massless scalar mode at low energies! What about the massless source field? As can be seen from (5.19) and (5.21) the leading analytic piece is a constant term which corresponds to a mass term for the source field [36]. This leads to the dual Lagrangian below the cutoff scale $\Lambda \sim k$

$$\mathcal{L}_{4D} = -\widetilde{Z}_0(\partial\phi_0)^2 + m_0^2\phi_0^2 + \frac{\chi}{\Lambda^{\nu_+ - 2}}\phi_0\mathcal{O} + \mathcal{L}_{CFT}, \tag{5.23}$$

where $\widetilde{Z}_0$, χ are dimensionless parameters and m_0 is a mass parameter of order the curvature scale k. The bare parameters $\widetilde{Z}_0$ and m_0 can be determined from (5.19). Thus, the holographic interpretation is perfectly consistent. There is a massless bound state in the CFT and the source field ϕ_0 receives a mass of order the curvature scale and decouples. In the bulk the zero mode is always localised towards the IR brane. Indeed for $\nu_+ > 2$ the coupling between the source field and the CFT is irrelevant and therefore the mixing from the source sector is negligible. Hence to a good approximation the mass eigenstate is predominantly the massless CFT bound state. When $1 \le \nu_+ \le 2$ the mixing can no longer be neglected and the mass eigenstate is again part elementary and part composite.

5.2. Dual 4D description of the Standard Model in the bulk

The qualitative features of the scalar field holographic picture can now be used to give the 4D dual description of the Standard Model gauge fields and matter in the 5D bulk. In general for every bulk zero mode field there is a corresponding massless eigenstate in the dual 4D theory that is a mixture of (elementary) source and (composite) CFT fields. If the bulk zero mode is localised towards the UV brane, then in the dual theory the massless eigenstate is predominantly the source field. For example, as in RS1 the graviton zero mode is localised towards the UV brane so that in the dual theory the massless eigenstate is mostly composed of the graviton source field.

On the other hand the dual interpretation of a bulk zero mode localised towards the TeV brane is a state that is predominantly a CFT bound state. In this instance the source field obtains a mass of order the curvature scale and decouples from the low energy theory. Depending on the degree of localisation the bound state mixes with the massive source field. Only in the limit where the mode is completely localised on the TeV brane is the dual eigenmode a pure CFT bound state. Since the Higgs is confined to the IR brane, the Higgs field is interpreted as a pure bound state of the CFT in the dual theory. In this way we see that the RS1 solution to the hierarchy problem is holographically identical to the way 4D composite models solve the problem with a low-scale cutoff. The Higgs mass is quadratically divergent but only sensitive to the strong coupling scale Λ_{IR} which is hierarchically smaller than Λ_{UV}. To obtain a large top Yukawa coupling the top quark was localised near the IR brane, so in the dual theory the top quark is also (predominantly) a composite of the CFT. The rest of the light fermions are localised to varying degrees towards the UV brane, and therefore these states will be mostly elementary particle states in the dual theory. The detailed holographic picture of bulk fermions can be found in Ref. [39].

If zero modes are not localised in the bulk then the corresponding 4D dual massless eigenstate is partly composed of the elementary source field and the composite CFT state. In particular for bulk gauge fields whose zero modes are not localised in the bulk, the 4D dual massless eigenstate will be part composite and part elementary. Finally note that local symmetries in the bulk, such as gauge symmetries or general coordinate invariance, also have a 4D dual interpretation. The holographic dual of a local symmetry group G in the bulk is a CFT in which a subgroup G of the global symmetry group of the CFT is weakly gauged by the source gauge field [34, 40].

Thus, in summary the Standard Model in the warped 5D bulk is dual to a 4D hybrid theory with a mixture of elementary and composite states.

5.2.1. Yukawa couplings

The Yukawa coupling hierarchies can also be understood from the dual 4D theory. Consider first an electron (or light fermion) with $c > 1/2$. In the dual 4D theory the electron is predominantly an elementary field. The dual 4D Lagrangian is obtained from analysing $\Sigma(p)$ for fermions, where the CFT induces a kinetic term for the source field $\psi_L^{(0)}$. It is given by [39]

$$\mathcal{L}_{4D} = \mathcal{L}_{CFT} + Z_0 \bar{\psi}_L^{(0)} i\gamma^\mu \partial_\mu \psi_L^{(0)} + \frac{\omega}{\Lambda^{|c+\frac{1}{2}|-1}} (\bar{\psi}_L^{(0)} \mathcal{O}_R + h.c.), \qquad (5.24)$$

where Z_0, ω are dimensionless couplings and dim $\mathcal{O}_R = 3/2 + |c + 1/2|$. The source field $\psi_L^{(0)}$ pertains to the left-handed electron e_L and a similar Lagrangian is written for the right-handed electron e_R. At energy scales $\mu < k$ we have a renormalisation group equation like (5.17) for the mixing parameter ξ but with $\gamma = |c + 1/2| - 1$. Since $c > 1/2$ the first term in (5.17) dominates and the coupling ξ decreases in the IR. In particular at the TeV scale ($ke^{-\pi kR}$) the solution (5.18) gives

$$\xi(\text{TeV}) \sim \sqrt{c - \frac{1}{2}} \frac{4\pi}{\sqrt{N}} \left(\frac{ke^{-\pi kR}}{k} \right)^{c-\frac{1}{2}} = \sqrt{c - \frac{1}{2}} \frac{4\pi}{\sqrt{N}} e^{-(c-\frac{1}{2})\pi kR}. \qquad (5.25)$$

The actual physical Yukawa coupling λ follows from the three-point vertex between the physical states. Since both e_L and e_R are predominantly elementary they can only couple to the composite Higgs via the mixing term in (5.24). This is depicted in Fig. 3. In a large-N gauge theory the matrix element $\langle 0|\mathcal{O}_{L,R}|\Psi_{L,R}\rangle \sim \sqrt{N}/(4\pi)$, and the vertex between three composite states $\Gamma_3 \sim 4\pi/\sqrt{N}$ [38].

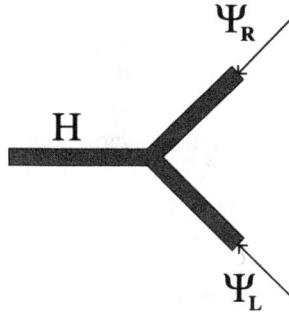

Fig. 3. The three-point Yukawa coupling vertex in the dual theory when the fermions are predominantly elementary source fields.

Thus if each of the elementary fields e_L and e_R mixes in the same way with the CFT so that $c_{eL} = c_{eR} \equiv c$ then

$$\lambda \propto \langle 0|\mathcal{O}_{L,R}|\Psi_{L,R}\rangle^2 \, \Gamma_3 \, \xi^2(\text{TeV}) = \frac{4\pi}{\sqrt{N}}(c - 1/2)e^{-2(c-\frac{1}{2})\pi k R}. \qquad (5.26)$$

This agrees precisely with the bulk calculation (4.3) where $\lambda_{ij}^{(5)}k \sim 4\pi/\sqrt{N}$.

• EXERCISE: *Show that when $|c| < 1/2$ the coupling of the mixing term in (5.24) is relevant and the renormalisation group equation (5.17) has a fixed point. Compare this with the bulk calculation following (4.3).*

Similarly we can also obtain the Yukawa coupling for the top quark with $c \lesssim -1/2$ in the dual theory. With this value of c the top quark is mostly a CFT bound state in the dual theory and we can neglect the mixing coupling with the CFT. As in the scalar field example this follows from the fact that the two point function $\langle \mathcal{O}_R \bar{\mathcal{O}}_R \rangle$ now has a massless pole. The CFT will again generate a mass term for the massless source field, so that the only massless state in the dual theory is the CFT bound state. The dual Lagrangian is given by [39]

$$\mathcal{L}_{4D} = \mathcal{L}_{CFT} + Z_0 \, \bar{\psi}_L^{(0)} i\gamma^\mu \partial_\mu \psi_L^{(0)} + \tilde{Z}_0 \, \bar{\chi}_R i\gamma^\mu \partial_\mu \chi_R$$
$$+ d \, k \, (\bar{\chi}_R \psi_L^{(0)} + h.c.) + \frac{\omega}{\Lambda^{|c+\frac{1}{2}|-1}} \, (\bar{\psi}_L^{(0)} \mathcal{O}_R + h.c.), \qquad (5.27)$$

where $Z_0, \tilde{Z}_0, d, \omega$ are dimensionless constants. The fermion $\psi_L^{(0)}$ pertains to t_L and a similar Lagrangian is written for t_R. Just as in the scalar case this dual Lagrangian is inferred from the behaviour of $\Sigma(p)$ for fermions. The CFT again induces a kinetic term for the source field $\psi_L^{(0)}$ but also generates a Dirac mass term of order the curvature scale k with a new elementary degree of freedom χ_R. Hence the elementary source field decouples from the low energy spectrum and the mixing term is no longer relevant for the Yukawa coupling. Instead the physical Yukawa coupling will arise from a vertex amongst three composite states so that $\lambda_t \sim \Gamma_3 \sim 4\pi/\sqrt{N} \sim \lambda^{(5)} k$, and consequently there is no exponential suppression in the Yukawa coupling. This is again consistent with the bulk calculation.

5.2.2. Minimal composite Higgs model

Similarly the Higgs as a pseudo Nambu-Goldstone boson scenario has a 4D dual interpretation [28]. Since A_5 is localised near the TeV-brane the Higgs boson is a composite state in the dual theory. The bulk SO(5) gauge symmetry is interpreted as an SO(5) *global* symmetry of the CFT. This global symmetry is then spontaneously broken down to SO(4) at the IR scale by the (unknown) strong dynamics of the CFT. This leads to a Nambu-Goldstone boson transforming as a **4** of SO(4), which is a real bidoublet of $SU(2)_L \times SU(2)_R$.

To break electroweak symmetry, an effective Higgs potential is generated at one loop by explicitly breaking the SO(5) symmetry in the elementary (fermion) sector and transmitting it to the CFT. The top quark plays the major role in breaking this symmetry. However the top quark must also be localised near the IR brane to obtain a large overlap with the Higgs field and therefore a large top mass. Moreover to prevent large deviations from the Standard Model prediction for $Z \rightarrow \bar{b}_L b_L$ the left-handed top quark must be localised towards the UV brane ($c_{tL} \lesssim 1/2$) [41, 28]. Thus to obtain a large top mass the right-handed top quark must be localised near the IR brane. This specific localisation is compatible with electroweak symmetry breaking where the Higgs field develops a vacuum expectation value and breaks SO(4) down to the custodial group SO(3). Hence in the dual theory the physical right-handed (left-handed) top quark is mostly composite (elementary) and the custodial symmetry prevents large contributions to the T parameter.

In summary this 4D composite Higgs model is very predictive, with minimal particle content and is consistent with electroweak precision tests [30]. This hybrid 4D theory with elementary and composite states successfully addresses the hierarchy problem, fermion masses and flavour problems in a complete framework.

6. Supersymmetric models in warped space

Supersymmetry elegantly solves the hierarchy problem because quadratic divergences to the Higgs mass are automatically cancelled thereby stabilising the electroweak sector. However this success must be tempered with the fact that super-symmetry has to be broken in nature. In order to avoid reintroducing a fine-tuning in the Higgs mass, the soft mass scale cannot be much larger than the TeV scale. Hence one needs an explanation for why the supersymmetry breaking scale is low. Since in warped space hierarchies are easily generated, the warp factor can be used to explain the scale of supersymmetry breaking, instead of the scale of electroweak breaking. This is one motivation for studying supersymmetric models in warped space. Thus, new possibilities open up for supersymmetric model building, and in particular for the supersymmetry-breaking sector. Moreover by the AdS/CFT correspondence these new scenarios have an interesting blend of supersymmetry and compositeness that lead to phenomenological consequences at the LHC.

A second motivation arises from the fact that electroweak precision data fa-vours a light (compared to the TeV scale) Higgs boson mass [24]. As noted earlier the Higgs boson mass in a generic warped model without any symme-try is near the IR cutoff (or from the 4D dual perspective the Higgs mass is near the compositeness scale). Introducing supersymmetry provides a simple reason for why the Higgs boson mass is light and below the IR cutoff of the theory.

6.1. Supersymmetry in a slice of AdS

It is straightforward to incorporate supersymmetry in a slice of AdS [11, 42]. The amount of supersymmetry allowed in five dimensions is determined by the dimension of the spinor representations. In five dimensions only Dirac fermions are allowed by the Lorentz algebra, so that there are eight supercharges which corresponds from the 4D point of view to an $\mathcal{N} = 2$ supersymmetry. This means that all bulk fields are in $\mathcal{N} = 2$ representations. At the massless level only half of the supercharges remain and the orbifold breaks the bulk supersymmetry to an $\mathcal{N} = 1$ supersymmetry.

Consider an $\mathcal{N} = 1$ (massless) chiral multiplet $(\phi^{(0)}, \psi^{(0)})$ in the bulk. We have seen that the zero mode bulk profiles of $\phi^{(0)}$ and $\psi^{(0)}$ are parametrised by their bulk mass parameters a and c, respectively. Since supersymmetry treats the scalar and fermion components equally, the bulk profiles of the component fields must be the same. It is clear that in general this is not the case except when $1 \pm \sqrt{4 + a} = 1/2 - c$, as follows from the exponent of the zero mode profiles in Table 1. This leads to the condition that

$$a = c^2 + c - 15/4, \tag{6.1}$$

and the one remaining mass parameter c determines the profile of the chiral mul-tiplet to be

$$\begin{pmatrix} \phi^{(0)} \\ \psi^{(0)} \end{pmatrix} \propto e^{(\frac{1}{2}-c)ky}. \tag{6.2}$$

Thus for $c > 1/2$ ($c < 1/2$) the chiral supermultiplet is localised towards the UV (IR) brane. It can be shown that the scalar boundary mass, that was tuned to be $b = 2 \pm \alpha$, follows from the invariance under a supersymmetry transformation [11] when (6.1) is satisfied.

Similarly a gauge boson with bulk profile $A_\mu^{(0)}(y) \propto 1$ and a gaugino with bulk profile $\lambda^{(0)}(y) \propto e^{(\frac{1}{2}-c_\lambda)ky}$ can be combined into an $\mathcal{N} = 1$ vector multiplet only for $c_\lambda = 1/2$. Of course this means that the gaugino zero-mode profile is flat like the gauge boson. At the massive level the on-shell field content of an $\mathcal{N} = 2$ vector multiplet is (A_M, λ_i, Σ) where λ_i is a symplectic-Majorana spinor (with $i = 1, 2$) and Σ is a real scalar in the adjoint representation of the gauge group. Invariance under supersymmetry transformations requires that Σ have bulk and boundary mass terms with $a = -4$ and $b = 2$, respectively. So, if Σ is even under the orbifold symmetry then these values will ensure a scalar zero mode.

Finally a graviton with bulk profile $h_{\mu\nu}^{(0)}(y) \propto e^{-ky}$ and a gravitino with bulk profile $\psi_\mu^{(0)}(y) \propto e^{(\frac{1}{2}-c_\psi)ky}$ can be combined into an $\mathcal{N} = 1$ gravity multiplet only for $c_\psi = 3/2$. In this case the gravitino zero-mode profile is localised on the UV brane.

6.2. The warped MSSM: a model of dynamical supersymmetry breaking

In the warped MSSM the warp factor is used to naturally generate TeV scale soft masses [43]. The UV (IR) scale is identified with the Planck (TeV) scale. The IR brane is associated with the scale of supersymmetry breaking, while the bulk and UV brane are supersymmetric. At the massless level the particle content is identical to the MSSM. The matter and Higgs superfields are assumed to be confined on the UV brane. This naturally ensures that all higher-dimension operators associated with proton decay and FCNC processes are sufficiently suppressed. In the bulk there is only gravity and the Standard Model gauge fields. These are contained in an $\mathcal{N} = 1$ gravity multiplet and vector multiplet, respectively.

Supersymmetry is broken by choosing different IR brane boundary conditions between the bosonic and fermionic components of the bulk superfields. On the boundaries the condition (3.20) defines a chirality since $(1 \mp \gamma_5)\Psi(y^*) = 0$ where $y^* = 0$ or πR. If opposite chiralities are chosen on the two boundaries then this leads to antiperiodic conditions for the fermions, namely

$$\Psi(y + 2\pi R) = -\Psi(y). \tag{6.3}$$

If the gauginos in the bulk are assumed to have opposite chiralities on the two boundaries then supersymmetry will be broken because the gauge bosons obey periodic boundary conditions. The gaugino zero mode is no longer massless and receives a mass

$$m_\lambda \simeq \sqrt{\frac{2}{\pi k R}}\, k\, e^{-\pi k R}. \tag{6.4}$$

Since the theory has a $U(1)_R$ symmetry this is actually a Dirac mass where the gaugino zero mode pairs up with a Kaluza-Klein mode [43]. The Kaluza-Klein mass spectrum of the gauginos also shifts relative to that of the gauge bosons by an amount $-\frac{1}{4}\pi k e^{-\pi k R}$. Similarly for the gravity multiplet the gravitino is assumed to have antiperiodic boundary conditions, while the graviton has periodic boundary conditions. The gravitino zero mode then receives a mass

$$m_{3/2} \simeq \sqrt{8}\, k\, e^{-2\pi k R}, \tag{6.5}$$

while the Kaluza-Klein modes are again shifted by an amount similar to that of the vector multiplet.

• EXERCISE: *Using the expressions (3.30) and (3.31), check that imposing opposite chiralities on the two boundaries for the gaugino and gravitino leads to the zero mode masses (6.4) and (6.5).*

If $ke^{-\pi k R} = \text{TeV}$ then the gaugino mass (6.4) is $m_\lambda \simeq 0.24$ TeV while the gravitino mass (6.5) is $m_{3/2} \simeq 3 \times 10^{-3}$ eV. Even though both the gaugino and gravitino are bulk fields the difference in their supersymmetry breaking masses follows from their coupling to the IR brane, which is where supersymmetry is broken. The gaugino is not localised in the bulk and couples to the IR brane with an $\mathcal{O}(1)$ coupling. Hence it receives a TeV scale mass. On the other hand the gravitino is localised on the UV brane and its coupling to the TeV brane is exponentially suppressed. This explains why the gravitino mass is much smaller than the gaugino mass.

The scalars on the UV brane will obtain a supersymmetry breaking mass at one loop via gauge interactions with the bulk vector multiplets. The gravity interactions with the gravity multiplet are negligible. A one loop calculation leads to the soft mass spectrum

$$\widetilde{m}_j^2 \propto \frac{\alpha_i}{4\pi}(\text{TeV})^2, \tag{6.6}$$

where $\alpha_i = g_i^2/(4\pi)$ are individual gauge contributions corresponding to the particular gauge quantum numbers of the particle state. The exact expressions are given in Ref. [43]. Unlike loop corrections to the usual 4D supersymmetric soft masses, the masses in (6.6) are finite. Normally UV divergences in a two-point

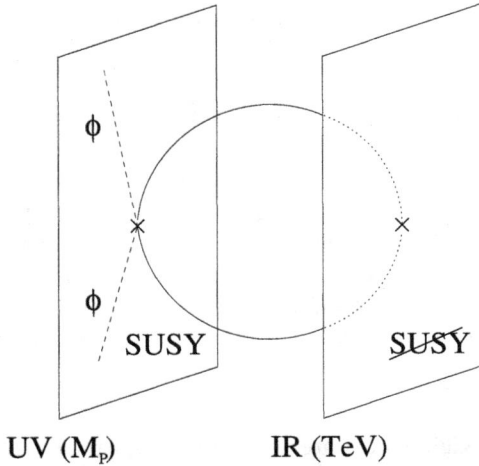

Fig. 4. The transmission of supersymmetry breaking in the warped MSSM to UV-brane localised matter fields via bulk gauge interactions which couple directly to the IR brane.

function arise when the two spacetime points coincide. But the spacetime points in the 5D loop diagram can never coincide, because the two branes are assumed to be a fixed distance apart, and therefore the 5D one-loop calculation leads to a finite result (see Figure 4). This is similar to the cancellation of divergences in the Casimir effect [44]. Since the contribution to the scalar masses is due to gauge interactions the scalar masses are naturally flavour diagonal. This means that the right-handed slepton is the lightest scalar particle since it has the smallest gauge coupling dependence. The lightest supersymmetric particle will be the superlight gravitino.

6.2.1. The dual 4D interpretation
We can use the AdS/CFT dictionary to obtain the dual 4D interpretation of the warped MSSM. Clearly the matter and Higgs fields confined to the UV brane are elementary fields external to the CFT. This is also true for the zero modes of the gravity multiplet since it is localised towards the UV brane. However, the bulk gauge field zero modes are partly composite since they are not localised. The Kaluza-Klein states, which are bound states of the CFT and localised near the IR brane, do not respect supersymmetry. Therefore at the TeV scale not only is conformal symmetry broken by the CFT but also supersymmetry. This requires some (unknown) nontrivial IR dynamics of the CFT, but the point is that supersymmetry is dynamically broken. Since the CFT is charged under the Standard Model gauge group, the gauginos (and gravitinos) will receive a tree-

level supersymmetry breaking mass, while the squarks and sleptons will receive their soft mass at one loop. In some sense this model is very similar to 4D gauge-mediated models except that there is no messenger sector since the CFT, responsible for supersymmetry breaking, is charged under the Standard Model gauge group.

In particular the bulk gaugino mass formula (6.4) can be understood in the dual theory. Since the gaugino mass is of the Dirac type the gaugino (source) field must marry a fermion bound state to become massive. This occurs from the mixing term $\mathcal{L} = \omega \lambda \mathcal{O}_\psi$. Since $c_\lambda = 1/2$, we have from Table 1 that $\dim \mathcal{O}_\psi = 5/2$ and therefore ω is dimensionless. This means that the mixing term coupling runs logarithmically so that at low energies the solution of (5.17) is

$$\xi^2(\mu) \sim \frac{16\pi^2}{N \log \frac{k}{\mu}}. \tag{6.7}$$

Thus at $\mu = k e^{-\pi k R}$ we obtain the correct factor in (6.4) since the Dirac mass $m_\lambda \propto \xi \langle 0 | \mathcal{O}_\psi | \Psi \rangle$, where in the large-$N$ limit the matrix element for $\mathcal{O}_\psi$ to create a bound state fermion is $\langle 0 | \mathcal{O}_\psi | \Psi \rangle \sim \sqrt{N}/(4\pi)$ [38].

Thus, in summary we have the dual picture

$$\begin{array}{ccc} \text{5D warped} & \text{DUAL} & \text{4D MSSM} \oplus \text{gravity} \\ \text{MSSM} & \Longleftrightarrow & \oplus \text{ strongly coupled 4D CFT} \end{array} \tag{6.8}$$

The warped MSSM is a very economical model of dynamical supersymmetry breaking in which the soft mass spectrum is calculable and finite, and unlike the usual 4D gauge-mediated models does not require a messenger sector. The soft mass TeV scale is naturally explained and the scalar masses are flavour diagonal. In addition, as we will show later, gauge coupling unification occurs with logarithmic running [45] arising primarily from the elementary (supersymmetric) sector as in the usual 4D MSSM [46].

6.3. The partly supersymmetric Standard Model: a natural model of high-scale supersymmetry breaking

Besides solving the hierarchy problem the supersymmetric standard model has two added bonuses. First, it successfully predicts gauge coupling unification and second, it provides a suitable dark matter candidate. Generically, however, there are FCNC and CP violation problems arising from the soft mass Lagrangian, as well as the gravitino and moduli problems in cosmology [47]. These problems stem from the fact that the soft masses are of order the TeV scale, as required for a natural solution to the hierarchy problem. Of course clever mechanisms exist that avoid these problems but perhaps the simplest solution would be to have all scalar masses at the Planck scale while still naturally solving the hierarchy problem. In the partly supersymmetric standard model [48] this is precisely what happens while still preserving the successes of the MSSM.

In 5D warped space the setup of the model is as follows. Supersymmetry is assumed to be broken on the UV brane while it is preserved in the bulk and the IR brane. The vector, matter, and gravity superfields are in the bulk while the Higgs superfield is confined to the IR brane. On the UV brane the supersymmetry breaking can be parametrised by a spurion field $\eta = \theta^2 F$, where $F \sim M_P^2$. In the gauge sector we can add the following UV brane term

$$\int d^2\theta \, \frac{\eta}{M_P^2} \frac{1}{g_5^2} W^\alpha W_\alpha \delta(y) + h.c. \tag{6.9}$$

This term leads to a gaugino mass for the zero mode $m_\lambda \sim M_P$, so that the gaugino decouples from the low energy spectrum. The gravitino also receives a Planck scale mass via a UV brane coupling and decouples from the low energy spectrum [49]. Similarly a supersymmetry breaking mass term for the squarks and sleptons can be added to the UV brane

$$\int d^4\theta \, \frac{\eta^\dagger \eta}{M_P^4} \, k \, S^\dagger S \, \delta(y), \tag{6.10}$$

where S denotes a squark or slepton superfield. This leads to a soft scalar mass $\tilde{m} \sim M_P$, so that the squark and slepton zero modes also decouple from the low energy spectrum.

The Higgs sector is different because the Higgs lives on the IR brane and there is no direct coupling to the UV brane. Hence, at tree-level the Higgs mass is zero, but a (finite) soft Higgs mass will be induced at one loop via the gauge interactions in the bulk of order

$$m_H^2 \sim \frac{\alpha}{4\pi} (ke^{-\pi kR})^2 \ll M_P^2. \tag{6.11}$$

As noted earlier the finiteness is due to the fact that the two 5D spacetime points on the UV and IR branes can never coincide (see Figure 5). Thus, we see that because of the warp factor the induced Higgs soft mass is much smaller than the scale of supersymmetry breaking at the Planck scale. So while at the massless level the gauginos, squarks and sleptons have received Planck scale masses, the Higgs sector remains (approximately) supersymmetric. In summary at the massless level the particle spectrum consists of the Standard Model gauge fields and matter (quarks and leptons) plus a Higgs scalar and Higgsino. This is why the model is referred to as *partly* supersymmetric.

At the massive level the Kaluza-Klein modes are also approximately supersymmetric. This is because they are localised towards the IR brane and have a

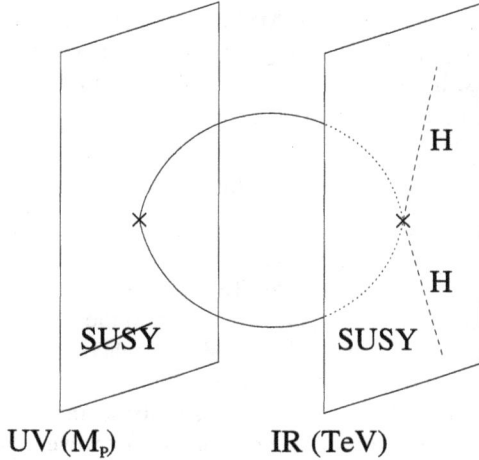

Fig. 5. The transmission of supersymmetry breaking in the partly supersymmetric standard model to the supersymmetric Higgs sector via bulk gauge interactions which couple directly to the UV brane.

small coupling to the UV brane. So the Planck scale supersymmetry breaking translates into an order TeV scale splitting between the fermionic and bosonic components of the Kaluza-Klein superfields.

Given that there are no gauginos, squarks or sleptons in the low energy spectrum it may seem puzzling how the quadratic divergences cancel in this model. Normally in the supersymmetric standard model the quadratic divergences in the Higgs mass are cancelled by a superpartner contribution of the opposite sign. However in the partly supersymmetric standard model there are no superpartners at the massless level. Instead what happens is that the difference between the Kaluza-Klein fermions and bosons sums up to cancel the zero mode quadratic divergence. Thus the Kaluza-Klein tower is responsible for keeping the Higgs mass natural even though supersymmetry is broken at the Planck scale.

6.3.1. Higgs sector possibilities

The motivation for making the Higgs sector supersymmetric is that the Higgs mass is induced at loop level and therefore the Higgs mass is naturally suppressed below the IR cutoff. In addition the supersymmetric partner of the Higgs, the Higgsino, provides a suitable dark matter candidate [48, 50]. However since the Higgsino is a fermion, gauge anomalies could be generated and these must be cancelled. This leads us to consider the following three possibilities:

(i) Two Higgs doublets: As in the MSSM we can introduce two Higgs doublet superfields H_1 and H_2, so that the gauge anomaly from the two Higgsinos cancel amongst themselves. In this scenario we can add the following superpotential on the IR brane

$$\int d^2\theta \, (y_d H_1 Q d + y_u H_2 Q u + y_e H_1 L e + \mu H_1 H_2) \,. \tag{6.12}$$

Thus the quarks and leptons receive their masses in the usual way. In addition the μ term in (6.12) is naturally of order the TeV scale so that there is no μ problem. The IR brane is approximately supersymmetric and the supersymmetric mass μ has a natural TeV value. This is unlike the MSSM where the natural scale of μ is M_P and consequently a problem for phenomenology.

(ii) One Higgs doublet: At first this possibility seems to be ruled since one massless Higgsino gives rise to a gauge anomaly. However starting with a bulk Higgs $\mathcal{N} = 2$ hypermultiplet $H = (H_1, H_2)$ with bulk mass parameter $c_H = 1/2$ that consists of two $\mathcal{N} = 1$ chiral multiplets $H_{1,2}$ we can generate a Higgsino Dirac mass and only one Higgs scalar doublet in the low energy spectrum. The trick is to use mixed boundary conditions where H_1 has Neumann (Dirichlet) boundary conditions on the UV (IR) brane and vice versa for H_2. This leads to a μ term

$$\mu \simeq \sqrt{\frac{2}{\pi k R}} \, k \, e^{-\pi k R}, \tag{6.13}$$

which is similar to the gaugino mass term (6.4) obtained in the warped MSSM. In this case the μ-term is naturally suppressed below TeV scale by the factor $1/\sqrt{\pi k R}$. Only one Higgs scalar remains in the low energy spectrum because the twisted boundary conditions localise one Higgs scalar doublet towards the UV brane where it obtains a Planck scale mass, and the other Higgs scalar is localised towards the IR brane where it obtains a mass squared μ^2.

(iii) Zero Higgs doublet–Higgs as a Slepton: No anomalies will occur if the Higgs is considered to be the superpartner of the tau (or other lepton). This idea is not new and dates back to the early days of supersymmetry [51]. The major obstacle in implementing this possibility in the MSSM is that the gauginos induce an effective operator $\frac{g^2}{m_\lambda} \nu \nu h h$ that leads to neutrino masses of order 10 GeV which are experimentally ruled out. However in the partly supersymmetric model $m_\lambda \sim M_P$ and neutrinos masses are typically of order 10^{-5} eV. This is phenomenologically acceptable and at least makes this a viable possibility. However the stumbling block is to generate a realistic spectrum of fermion masses without introducing abnormally large coefficients [48].

6.3.2. *Electroweak symmetry breaking*

In this model electroweak symmetry breaking can be studied and calculated using the 5D bulk propagators. Consider, for simplicity, a one Higgs doublet version of the model. The scalar potential is

$$V(h) = \mu^2 |h|^2 + \frac{1}{8}(g^2 + g'^2)|h|^4 + V_{gauge}(h) + V_{top}(h), \qquad (6.14)$$

where $V_{gauge}(h)$ and $V_{top}(h)$) are one-loop contributions to the effective potential arising from gauge boson and top quark loops, respectively. The first two terms in (6.14), which arise at tree-level, are monotonically increasing giving rise to a minimum at $\langle |h| \rangle = 0$ and therefore do not break electroweak symmetry. This is why we need to calculate the one-loop contributions. The one-loop gauge contribution is given by

$$V_{gauge}(h) = 6 \int_0^\infty \frac{dp}{8\pi^2} p^3 \log \left[\frac{1 + g^2 |h|^2 G_B(p)}{1 + g^2 |h|^2 G_F(p)} \right], \qquad (6.15)$$

where $G_{B,F}(p)$ are the boson (fermion) gauge propagators in the bulk whose expressions can be found in Ref. [48]. The contribution to the effective potential from V_{gauge} is again monotonically increasing. However there is also a sizeable contribution from top quark loops (due to the large top Yukawa coupling) given by

$$V_{top}(h) = 6 \int_0^\infty \frac{dp}{8\pi^2} p^3 \log \left[\frac{1 + p^2 y_t^2 |h|^2 G_B^2(p)}{1 + p^2 y_t^2 |h|^2 G_F(p)} \right]. \qquad (6.16)$$

This contribution generates a potential that monotonically decreases with $|h|$, destabilising the vacuum and thus triggering electroweak symmetry breaking. In order for this to occur the top quark needs to be localised near the IR brane with a bulk mass parameter $c_t \simeq -0.5$. Since the top quark $\mathcal{N} = 1$ chiral multiplet is localised near the IR brane, the top squark will only receive a TeV scale soft mass and consequently will remain in the low energy spectrum. In fact this radiative breaking of electroweak symmetry due to a large top Yukawa coupling is similar to that occuring in the usual MSSM. As in the MSSM the value of the Higgs mass is very model dependent but if no large tuning of parameters is imposed one obtains a light Higgs boson with mass $m_{Higgs} \lesssim 120$ GeV.

6.3.3. *Dual 4D interpretation*

The dual 4D interpretation of the partly supersymmetric model follows from applying the rules of the AdS/CFT dictionary. Supersymmetry is broken at the Planck scale in the dual 4D theory and is approximately supersymmetric at the IR scale. Thus from a 4D point of view supersymmetry is really just an accidental symmetry at low energies. At the massless level the Higgs is confined on the IR brane and the top quark is localised towards the TeV brane so both of these states are CFT composites and supersymmetric at tree level. The compositeness of the Higgs and stop explains why these states are not sensitive to the UV breaking of supersymmetry. These states are "fat" with a size of order TeV^{-1}, and are transparent to high momenta or short wavelength probes that transmit the breaking of supersymmetry. At one loop TeV-scale supersymmetry breaking effects arise from the small mixing with the elementary source fields, which directly feel the Planck scale supersymmetry breaking. The bulk gauge fields are partly composite and the light fermions which are localised to varying degrees near the UV brane are predominantly elementary fields. Since the light fermion superpartners are predominantly source fields they obtain Planck scale soft masses.

Thus the dual picture can be summarised as follows

$$
\begin{array}{ccc}
\begin{array}{c}\text{5D partly} \\ \text{supersymmetric SM}\end{array} & \overset{\text{DUAL}}{\Longleftrightarrow} & \begin{array}{l}\text{4D SM} \oplus \text{Higgsino} \oplus \text{stop} \\ \oplus \text{gravity} \\ \oplus \text{strongly coupled 4D CFT}\end{array}
\end{array}
\tag{6.17}
$$

The partly supersymmetric standard model is a natural model of high-scale supersymmetry breaking. Supersymmetry is realised in the most economical way. Only the Higgs sector and top quark are supersymmetric and composite, while all other squarks and sleptons have Planck scale masses. The Higgsino is the dark matter candidate and even gauge coupling unification is achieved as we show in the next section.

7. Grand unification

The unification of the gauge couplings strongly suggests that there is an underlying simplifying structure based on grand unified theories with gauge groups such as SU(5) and SO(10). We would like to preserve this structure in warped extra dimensions as well. At the quantum level gauge couplings are sensitive to particle states that can be excited in the vacuum and this leads to an energy dependence or running. At one loop the general expression for this energy dependence is

$$
\frac{1}{g_a^2(p)} = \frac{1}{g_U^2} + \frac{b_a}{8\pi^2} \log \frac{M_U}{p},
\tag{7.1}
$$

where g_U is the unified gauge coupling, M_U is the unification scale and $a = 1, 2, 3$ represents the gauge couplings of $U(1)_Y$, $SU(2)_L$ and $SU(3)$, respectively. The coefficients b_a depend on the charged particle states that can be excited in the quantum vacuum. At low energies the three gauge couplings are different and if these couplings are to unify at a high energy scale then b_a must be different. From the measurement of the three gauge couplings at the scale M_Z we can use the three equations (7.1) to obtain the experimental prediction for the ratio

$$B = \frac{b_3 - b_2}{b_2 - b_1} = 0.717 \pm 0.008, \tag{7.2}$$

where the error is due to the experimental values of the gauge couplings [52]. This ratio can also be calculated theoretically for the particle content of any model. Note that any constant or universal contribution to b_a does not affect the ratio B, since it is a ratio of differences.

In the Standard Model the quarks and leptons contribute universally to the running because they form complete SU(5) multiplets. Therefore they do not affect the relative running of the couplings and hence the value of B. Only the Higgs and Standard Model gauge bosons, which do not form complete SU(5) multiplets at low energies, affect the relative running. This leads to a Standard Model prediction $B = 0.528$, which does not agree with the experimental value (7.2), even allowing for a 10% theoretical uncertainty due to threshold corrections and higher-loop effects.

On the other hand the MSSM doubles the particle spectrum with the addition of gauginos, Higgsinos, squarks and sleptons. Again the squarks and sleptons form complete SU(5) multiplets and do not contribute to the relative running. However the extra contributions from the gauginos and Higgsinos combined with the gauge bosons and Higgs scalar field leads to the MSSM prediction $B = 0.714$. This is remarkably close to the experimental value (7.2), even if one accounts for theoretical uncertainties, and is one reason why supersymmetry is a leading solution to the hierarchy problem. The question we would like to now address is how does gauge coupling unification work for models in warped space?

7.1. Logarithmic running in 5D warped space

Given that models in 5D warped space can be given a 4D dual description the gauge couplings should run logarithmically. But how does this happen given that the 5D model has Kaluza-Klein modes? To answer this question let us consider the one-loop corrections to the U(1) gauge coupling of the zero mode generated by an even bulk 5D scalar ϕ with charge $+1$ [45]. The Kaluza-Klein spectrum of the scalar is given by the expressions (3.33) with $\alpha_\phi = 2$. To regulate the model we introduce a Pauli-Villars (PV) regulator scalar field Φ with bulk mass $\Lambda \lesssim k$ and no boundary mass terms. This means that the PV field has no massless mode and the zero mode obtains a mass $m_0 \simeq \Lambda/\sqrt{2}$. On the other hand the Kaluza-Klein spectrum remains relatively unaffected by the bulk mass and is given by

$$m_{\Phi_n} = (n + \tfrac{\alpha_\Phi}{2} - \tfrac{3}{4})\pi\, k\, e^{-\pi k R}, \text{ with } \alpha_\Phi = \sqrt{4 + \tfrac{\Lambda^2}{k^2}} \simeq 2 + \tfrac{\Lambda^2}{4k^2}. \text{ This can also}$$

be understood as follows: Adding a bulk mass term Λ has the effect of locally adding a mass term Λe^{-ky} at any point y in the bulk. Since the zero mode of a massless bulk scalar field is localised towards $y \simeq 0$ the affect of adding a bulk mass shifts the mass of the zero mode by $\sim \Lambda$. On the other hand the Kaluza-Klein modes are localised near $y = \pi R$, so adding the bulk mass term affects them by an amount $\sim \Lambda e^{-\pi kR}$. Thus the Kaluza-Klein spectrum of the ϕ and the PV field is approximately the same while zero modes are separated by a mass scale Λ.

The corrections to the photon self energy $\Pi_{\mu\nu}(q^2) = (q^2 \eta_{\mu\nu} - q_\mu q_\nu)\Pi(q^2)$ are given by [45]

$$
\begin{aligned}
\Pi(0) &\propto \sum_n \int d^4 p \left[\frac{1}{(p^2 + m_{\phi_n}^2)^2} - \frac{1}{(p^2 + m_{\Phi_n}^2)^2} \right] \\
&\simeq \frac{b_\phi}{8\pi^2} \log \frac{\mu}{\Lambda} - \frac{b_\phi}{64\pi^2} \frac{\Lambda^2}{k^2}(\pi kR),
\end{aligned} \tag{7.3}
$$

where we have introduced an infrared cutoff μ and $b_\phi = 1/3$ is the β-function coefficient. For $\mu \ll \Lambda < k$ the Kaluza-Klein contribution is negligible since the contributions from the Kaluza-Klein modes effectively cancels out. Instead the dominant contribution arises purely from the zero modes and is given by

$$
\Pi(0) \simeq \frac{b_\phi}{8\pi^2} \log \frac{\mu}{\Lambda}. \tag{7.4}
$$

This logarithmic dependence on the cutoff Λ is exactly what we would obtain in four dimensions and consistently agrees with the AdS/CFT correspondence.

This result can be contrasted with what one obtains in 5D flat space. If we add a bulk mass Λ to a bulk scalar field the whole Kaluza-Klein tower shifts by an amount $\sim \Lambda$. Only the Kaluza-Klein states above this scale cancel with the PV field, leaving behind the contribution from the Kaluza-Klein states below Λ. This leads to the contribution

$$
\Pi(0) \simeq \frac{b_\phi}{8\pi^2} \Lambda R, \tag{7.5}
$$

which are the power-law corrections [53].

It is also important to note that (7.4) can only be understood as the running of gauge couplings if matter is localised towards the UV brane. In this case the effective (field theory) cutoff is the Planck scale and the couplings can be interpreted as a running up to that scale. However for matter localised on the IR brane the effective (field theory) cutoff is the TeV scale. Above this scale the effective field theory description breaks down and there will be contributions from the fundamental (string) theory [45, 54].

7.2. *Partly supersymmetric grand unification*

Let us now consider gauge coupling unification in the partly supersymmetric standard model. The dominant contribution to the running will be logarithmic and it is just a question of determining the β-function coefficients. Consider a 5D SU(5) gauge theory that is broken on the Planck brane, but preserved on the IR brane. We can implement this setup by imposing boundary conditions on the bulk fields. These can be either Neumann $(+)$ or Dirichlet $(-)$ conditions corresponding to even or odd reflections, respectively, about the orbifold fixed points.

The SU(5) gauge bosons form an $\mathcal{N} = 2$ vector multiplet, $\mathcal{V} = (V, S)$, where $V(S)$ is an $\mathcal{N} = 1$ vector (chiral) multiplet. These fields are assumed to have the boundary conditions

$$V = \left[\begin{array}{c} V_\mu^a(+, +) \\ V_\mu^A(-, +) \end{array} \right], \qquad S = \left[\begin{array}{c} S^a(-, -) \\ S^A(+, -) \end{array} \right], \qquad (7.6)$$

where the indices $a(A)$ run over the unbroken (broken) generators, and the first (second) argument refers to the UV (IR) boundary condition. These boundary conditions break the SU(5) symmetry on the UV brane and the only zero modes are the Standard Model gauge bosons $V_\mu^a(+, +)$. At the nonzero mode level, the X, Y gauge bosons (contained in $V_\mu^A(-, +)$), and the SU(5) adjoint scalar states all obtain TeV-scale masses.

Similarly, the Higgs sector is supersymmetric and contains two Higgs doublets which are embedded into two $\mathcal{N} = 2$ bulk hypermultiplets, $\mathcal{H} = (H_5, H_5^c)$ and $\bar{\mathcal{H}} = (\bar{H}_5, \bar{H}_5^c)$, each transforming in the **5** of SU(5). The boundary conditions are

$$H_5 = \left[\begin{array}{c} H_2(+, +) \\ H_3(-, +) \end{array} \right], \qquad H_5^c = \left[\begin{array}{c} H_2^c(-, -) \\ H_3^c(+, -) \end{array} \right], \qquad (7.7)$$

and similarly for $\bar{\mathcal{H}}$. The only zero modes are the two Higgs doublets, $H_2(+, +)$ and $\bar{H}_2(+, +)$, as in the MSSM. This choice of boundary conditions neatly solves the doublet-triplet splitting problem [55].

The Standard Model matter fields are also embedded into bulk $\mathcal{N} = 2$ hypermultiplets. However, while it would seem natural to put all the fermions of one generation into a single **5** and **10**, parity assignments actually require the quarks and leptons to arise from different SU(5) bulk hypermultiplets [56, 57, 58]. Thus, for each generation we will suppose that there are bulk hypermultiplets $(\mathbf{5}_1, \mathbf{5}_1^c) + (\mathbf{5}_2, \mathbf{5}_2^c)$ and $(\mathbf{10}_1, \mathbf{10}_1^c) + (\mathbf{10}_2, \mathbf{10}_2^c)$ with boundary conditions

$$\mathbf{5}_1 = L_1(+, +) + d_1^c(-, +), \quad \mathbf{5}_1^c = L_1^c(-, -) + d_1(+, -), \qquad (7.8)$$
$$\mathbf{5}_2 = L_2(-, +) + d_2^c(+, +), \quad \mathbf{5}_2^c = L_2^c(+, -) + d_2(-, -), \qquad (7.9)$$

$$10_1 = Q_1(+,+) + u_1^c(-,+) + e_1^c(-,+), \qquad (7.10)$$

$$10_1^c = Q_1^c(-,-) + u_1(+,-) + e_1(+,-), \qquad (7.11)$$

$$10_2 = Q_2(-,+) + u_2^c(+,+) + e_2^c(+,+), \qquad (7.12)$$

$$10_2^c = Q_2^c(+,-) + u_2(-,-) + e_2(-,-), \qquad (7.13)$$

where the Standard Model fermions are identified with the zero modes of the fields with $(+,+)$ boundary conditions. Notice that this embedding elegantly explains why the fermions need not satisfy the SU(5) mass relations and although each Standard Model generation arises from different $\mathbf{5} + \mathbf{10}$ fields, the usual charge quantization and hypercharge assignments are still satisfied [56, 57]. This feature also explains why tree-level proton decay is not a problem in these models. There is simply no allowed coupling between $X, Y(-,+)$ gauge bosons and Standard Model fields $L_1(+,+)$ and $d_2^c(+,+)$ that is even under the orbifold symmetry. This is also true for couplings between Standard Model particles and the coloured Higgs triplets. However, a bulk U(1) symmetry must be introduced in order to prevent proton decay from higher-dimensional operators [59, 60].

The specific contributions to the gauge couplings are given by

$$\frac{1}{g_a^2(p)} = \frac{\pi R}{g_5^2} + \frac{1}{g_{Ba}^2(\Lambda)} + \frac{1}{8\pi^2}\Delta_a(p,\Lambda), \qquad (7.14)$$

where g_{Ba} are boundary couplings, and Δ_a are the one-loop corrections. The first term in (7.14) is the universal contribution from the tree-level gauge coupling g_5 in the bulk. The second term in (7.14) is an SU(5) violating term that follows from breaking the SU(5) symmetry on the Planck brane. It can be neglected because $g_{Ba}(\Lambda) \simeq 4\pi$, since the theory is effectively strongly coupled at the scale $\Lambda \gtrsim k$ [61, 46, 62]. Thus, we see that the dominant contributions to the gauge couplings will arise from the logarithmically enhanced terms of Δ_a. These terms cannot be obtained in the strongly coupled 4D dual theory, but instead can be calculated using the bulk zero-mode Green functions [63].

The exact Greens function expression can be expanded at low energies $p \lesssim$ TeV to obtain the dominant logarithmic contributions. For vector bosons the dominant term is

$$\Delta^a(V) = b_V^a \ln\frac{k}{p} + \dots, \qquad (7.15)$$

where $b_V^a = (0, -\frac{22}{3}, -11)$. Recall that $k \simeq M_P$ is the AdS curvature scale so that (7.15) is the usual logarithmic contribution from the Standard Model gauge bosons. There are no corresponding gaugino zero mode contributions because these modes have received a large supersymmetry breaking mass and decouple at low energy.

Instead for the Higgs sector, the leading contribution for the Higgs doublet is

$$\Delta^a(\mathcal{H}_{++}) = T_a(\mathcal{H}_{++}) \ln \frac{T}{p} + \dots, \tag{7.16}$$

where $T_a(\mathcal{H}_{++}) = \left(\frac{3}{10}, \frac{1}{2}, 0\right)$ and T is shorthand for the TeV scale. In this case the leading contribution is again a logarithmn but it is small. This can be interpreted as being due to the fact that the Higgs doublet is a composite particle. On the other hand the Higgs triplet contributions for $m_{-+} \lesssim p \lesssim T$, with m_{-+} the lowest lying massive state, are

$$\Delta^a(\mathcal{H}_{-+}) = T_a(\mathcal{H}_{-+}) \left[\frac{2}{3} \ln \frac{k}{p} + \ln \frac{T}{p} + \dots \right], \tag{7.17}$$

where $T_a(\mathcal{H}_{-+}) = \left(\frac{1}{5}, 0, \frac{1}{2}\right)$. There is now both a large and small logarithmic contribution. The small contribution is from the composite states, while the large contribution is due to elementary degrees of freedom which are required to form a Dirac state. These extra elementary states can also be inferred by directly studying the dual 4D theory [39]. Thus, the total Higgs contribution from both Higgs hypermultiplets $\mathcal{H}$ and $\bar{\mathcal{H}}$ is

$$\Delta^a(\mathcal{H} + \bar{\mathcal{H}}) = b^a_{\mathcal{H}+\bar{\mathcal{H}}} \ln \frac{k}{p} + \ln \frac{T}{p} + \dots, \tag{7.18}$$

where $b^a_{\mathcal{H}+\bar{\mathcal{H}}} = \left(\frac{4}{15}, 0, \frac{2}{3}\right)$.

Finally the first two matter generations are (predominantly) elementary and give rise to the contribution

$$\Delta^a(5_i^{(I)} + 10_i^{(I)}) = \frac{4}{3} \ln \frac{k}{p} + \dots. \tag{7.19}$$

This is the usual universal contribution from one generation of Standard Model fermions which form a complete SU(5) multiplet. On the other hand the third generation is partly composite with composite states t_R, b_R, τ_R and elementary states $t_L, b_L, \tau_L, \nu_{\tau L}$. This gives the nonuniversal contribution

$$\Delta^a(5_i^{(3)} + 10_i^{(3)}) = b^a_{(3)} \ln \frac{k}{p} + \frac{4}{3} \ln \frac{T}{p} + \dots, \tag{7.20}$$

where $b^a_{(3)} = \left(\frac{8}{15}, \frac{8}{3}, \frac{4}{3}\right)$. Clearly we see that the large logarithmic contribution that arises from the elementary states introduces a differential running in the gauge couplings.

If we now add up all the Δ^a contributions (7.15), (7.18), (7.19) and (7.20) arising from the elementary states in the model, then at the leading log level we obtain the total contribution [64]

$$\Delta^a = b^a_{\text{total}} \ln \frac{k}{p} + \dots, \tag{7.21}$$

where $b^a_{\text{total}} = \left(\frac{52}{15}, -2, -\frac{19}{3}\right)$. These β-function coefficient values give $B = 0.793$, which allowing for an approximately 10% theoretical uncertainty, agrees with the experimental value (7.2). Interestingly the partly composite third generation has restored the gauge coupling unification without the gauginos and Higgsinos.

• EXERCISE: *Use the one-loop gauge coupling expressions in Ref [63] for $p \lesssim$ TeV, to verify the individual logarithmic contributions in Δ^a.*

Note that even though the model is partly supersymmetric, supersymmetry plays no role in obtaining gauge coupling unification because all the differential running contributions come from the UV-brane localised elementary sector which is nonsupersymmetric. This means that a similar mechanism will also work for an inherently nonsupersymmetric model such as the minimal composite Higgs model [65].

8. Conclusion

Warped models in a slice of AdS$_5$ provide a new framework to study solutions of the hierarchy problem at the TeV scale. The warp factor naturally generates hierarchies and can be used to either stabilise the electroweak scale or explain why the scale of supersymmetry breaking is low. Remarkably by the AdS/CFT correspondence these 5D warped models are dual to strongly coupled 4D theories. The Higgs localised on the IR brane is dual to a composite Higgs. The corresponding Higgs boson mass can be light compared to the IR cutoff by using either a global symmetry and treating the Higgs as a pseudo Nambu-Goldstone boson or using supersymmetry to make only the Higgs sector supersymmetric. The good news is that these models are testable at the LHC (and an eventual linear collider), so it will be an exciting time to discover whether Nature makes use of the fifth dimension in this novel way. If not, there is no bad news, because the warped fifth dimension literally provides a new theoretical framework for studying the dynamics of strongly coupled 4D gauge theories and this will be an invaluable tool for many years to come.

Acknowledgements

I would like to thank the organisers for organising such an excellent summer school. I am especially grateful to Alex Pomarol for collaboration on the topics which formed the basis of these lectures. I also thank Kaustubh Agashe, Roberto Contino, Yasunori Nomura, Erich Poppitz and Raman Sundrum for discussions on warped matters, and Brian Batell for comments on the manuscript. This work was supported in part by a Department of Energy grant DE-FG02-94ER40823, a grant from the Office of the Dean of the Graduate School of the University of Minnesota, and an award from Research Corporation.

References

[1] V. A. Rubakov, Phys. Usp. **44** (2001) 871 [Usp. Fiz. Nauk **171** (2001) 913] [hep-ph/0104152]; lectures at the Les Houches Summer School: Particle Physics beyond the Standard Model, Aug. 2005 (this volume).

[2] G. Gabadadze, hep-ph/0308112.

[3] C. Csaki, hep-ph/0404096.

[4] A. Perez-Lorenzana, J. Phys. Conf. Ser. **18** (2005) 224 [hep-ph/0503177].

[5] R. Sundrum, hep-th/0508134.

[6] C. Csaki, J. Hubisz and P. Meade, hep-ph/0510275.

[7] C. Grojean, lectures at the Les Houches Summer School: Particle Physics beyond the Standard Model, Aug. 2005 (this volume).

[8] J. Hewett, lectures at the Les Houches Summer School: Particle Physics beyond the Standard Model, Aug. 2005 (this volume).

[9] L. Randall and R. Sundrum, Phys. Rev. Lett. **83** (1999) 3370 [hep-ph/9905221].

[10] S. Chang, J. Hisano, H. Nakano, N. Okada and M. Yamaguchi, Phys. Rev. D **62** (2000) 084025 [hep-ph/9912498].

[11] T. Gherghetta and A. Pomarol, Nucl. Phys. B **586** (2000) 141 [hep-ph/0003129].

[12] H. Davoudiasl, J. L. Hewett and T. G. Rizzo, Phys. Rev. D **63** (2001) 075004 [hep-ph/0006041].

[13] P. Breitenlohner and D. Z. Freedman, Phys. Lett. B **115** (1982) 197.

[14] Y. Grossman and M. Neubert, Phys. Lett. B **474** (2000) 361 [hep-ph/9912408].

[15] L. Randall and R. Sundrum, Phys. Rev. Lett. **83** (1999) 4690 [hep-th/9906064].

[16] H. Davoudiasl, J. L. Hewett and T. G. Rizzo, Phys. Lett. B **473** (2000) 43 [hep-ph/9911262].

[17] A. Pomarol, Phys. Lett. B **486** (2000) 153 [hep-ph/9911294].

[18] T. Gherghetta, M. Peloso and E. Poppitz, Phys. Rev. D **72** (2005) 104003 [hep-th/0507245].

[19] B. Batell and T. Gherghetta, hep-ph/0512356.

[20] S. J. Huber and Q. Shafi, Phys. Lett. B **498** (2001) 256 [hep-ph/0010195].

[21] T. Gherghetta, Phys. Rev. Lett. **92** (2004) 161601 [hep-ph/0312392].

[22] S. J. Huber, Nucl. Phys. B **666** (2003) 269 [hep-ph/0303183].

[23] A. Delgado, A. Pomarol and M. Quiros, JHEP **0001** (2000) 030 [hep-ph/9911252].

[24] LEP Electroweak Working Group, http://lepewwg.web.cern.ch/LEPEWWG/

[25] D. B. Fairlie, Phys. Lett. B **82** (1979) 97; N. S. Manton, Nucl. Phys. B **158** (1979) 141.

[26] L. Randall and M. D. Schwartz, JHEP **0111** (2001) 003 [hep-th/0108114].

[27] R. Contino, Y. Nomura and A. Pomarol, Nucl. Phys. B **671** (2003) 148 [hep-ph/0306259].

[28] K. Agashe, R. Contino and A. Pomarol, Nucl. Phys. B **719** (2005) 165 [hep-ph/0412089].

[29] Y. Hosotani, Phys. Lett. B **126** (1983) 309, Phys. Lett. B **129** (1983) 193.

[30] K. Agashe and R. Contino, hep-ph/0510164.

[31] J. M. Maldacena, Adv. Theor. Math. Phys. **2** (1998) 231 [Int. J. Theor. Phys. **38** (1999) 1113] [hep-th/9711200].

[32] S. S. Gubser, I. R. Klebanov and A. M. Polyakov, Phys. Lett. B **428** (1998) 105 [hep-th/9802109].

[33] E. Witten, Adv. Theor. Math. Phys. **2** (1998) 253 [hep-th/9802150].

[34] N. Arkani-Hamed, M. Porrati and L. Randall, JHEP **0108** (2001) 017 [hep-th/0012148].

[35] R. Rattazzi and A. Zaffaroni, JHEP **0104** (2001) 021 [hep-th/0012248].

[36] M. Perez-Victoria, JHEP **0105** (2001) 064 [hep-th/0105048].

[37] G. 't Hooft, Nucl. Phys. B **72** (1974) 461, Nucl. Phys. B **75** (1974) 461.

[38] E. Witten, Nucl. Phys. B **160** (1979) 57.

[39] R. Contino and A. Pomarol, JHEP **0411** (2004) 058 [hep-th/0406257].

[40] K. Agashe and A. Delgado, Phys. Rev. D **67** (2003) 046003 [hep-th/0209212].

[41] K. Agashe, A. Delgado, M. J. May and R. Sundrum, JHEP **0308** (2003) 050 [hep-ph/0308036].

[42] R. Altendorfer, J. Bagger and D. Nemeschansky, Phys. Rev. D **63** (2001) 125025 [hep-th/0003117].

[43] T. Gherghetta and A. Pomarol, Nucl. Phys. B **602** (2001) 3 [hep-ph/0012378].

[44] H. B. G. Casimir, Kon. Ned. Akad. Wetensch. Proc. **51** (1948) 793.

[45] A. Pomarol, Phys. Rev. Lett. **85** (2000) 4004 [hep-ph/0005293].

[46] W. D. Goldberger, Y. Nomura and D. R. Smith, Phys. Rev. D **67** (2003) 075021 [hep-ph/0209158].

[47] P. Binétruy, lectures at the Les Houches Summer School: Particle Physics beyond the Standard Model, Aug. 2005 (this volume).

[48] T. Gherghetta and A. Pomarol, Phys. Rev. D **67** (2003) 085018 [hep-ph/0302001].

[49] M. A. Luty, Phys. Rev. Lett. **89** (2002) 141801 [hep-th/0205077].

[50] M. Masip and I. Mastromatteo, hep-ph/0510311.

[51] P. Fayet, Phys. Lett. B **64** (1976) 159.

[52] M. E. Peskin, hep-ph/0212204.

[53] K. R. Dienes, E. Dudas and T. Gherghetta, Phys. Lett. B **436** (1998) 55 [hep-ph/9803466]; Nucl. Phys. B **537** (1999) 47 [hep-ph/9806292].

[54] W. D. Goldberger and I. Z. Rothstein, Phys. Rev. Lett. **89** (2002) 131601 [hep-th/0204160]; Phys. Rev. D **68** (2003) 125011 [hep-th/0208060].

[55] Y. Kawamura, Prog. Theor. Phys. **105** (2001) 999 [hep-ph/0012125].

[56] L. J. Hall and Y. Nomura, Phys. Rev. D **64** (2001) 055003 [hep-ph/0103125].

[57] A. Hebecker and J. March-Russell, Nucl. Phys. B **613** (2001) 3 [hep-ph/0106166].

[58] R. Barbieri, L. J. Hall and Y. Nomura, Phys. Rev. D **66** (2002) 045025 [hep-ph/0106190].

[59] W. D. Goldberger, Y. Nomura and D. R. Smith, Phys. Rev. D **67** (2003) 075021 [hep-ph/0209158].

[60] K. Agashe, A. Delgado and R. Sundrum, Annals Phys. **304** (2003) 145 [hep-ph/0212028].

[61] R. Contino, P. Creminelli and E. Trincherini, JHEP **0210** (2002) 029 [hep-th/0208002].

[62] W. D. Goldberger and I. Z. Rothstein, Phys. Rev. D **68** (2003) 125012 [hep-ph/0303158].

[63] K. w. Choi and I. W. Kim, Phys. Rev. D **67** (2003) 045005 [hep-th/0208071].

[64] T. Gherghetta, Phys. Rev. D **71** (2005) 065001 [hep-ph/0411090].

[65] K. Agashe, R. Contino and R. Sundrum, Phys. Rev. Lett. **95** (2005) 171804 [hep-ph/0502222].

Course 7

NEW APPROACHES TO
ELECTROWEAK SYMMETRY BREAKING

Christophe Grojean

Service de Physique Théorique, CEA Saclay, F91191 Gif–sur–Yvette, France

D. Kazakov, S. Lavignac and J. Dalibard, eds.
Les Houches, Session LXXXIV, 2005
Particle Physics Beyond the Standard Model

Contents

1. Preamble

We have lived a decade of great experimental successes [1]: discovery of the top, observation of solar and atmospherical neutrino oscillations, measure of direct CP violation in the K system, measure of CP violation in the B system, evidence of an accelerated phase in the expansion of the universe, determination of the dark energy and dark matter composition of the universe *etc...* These results have strengthened the Standard Model (SM) of particle physics as a successful description of Nature. Yet, these results also concluded that the SM matter only represents about 5% of the energy density of the Universe and therefore they called for a physics beyond the SM, albeit direct evidence for such a physics is desperately missing. The sector of electroweak symmetry breaking (EWSB) could well provide us with the first hints of this new physics in a detector. Indeed the usual Higgs mechanism jeopardizes our current understanding of the SM at the quantum level and electroweak precision measurements seriously contrive any extension beyond it. Better than a long introduction, the following tautology reveals that an understanding of the dynamics of EWSB is still missing

Why is EW symmetry broken?

because the Higgs potential is unstable at the origin

Why is the Higgs potential unstable at the origin?

because otherwise EW symmetry won't be broken

One should understand that the Higgs mechanism is only a description of EWSB and not an explanation of it since in particular there is no dynamics to explain the instability at the origin.

The aim of these lectures, after a quick review of the (SM) physics of EWSB, is to give a survey of recent approaches that go under the names of Little Higgs models, Gauge–Higgs unification models and Higgsless models[1].

---

[1] Some useful reviews/lecture notes on electroweak physics and the various bounds on the Higgs can be found in [2]. The reader interested to an introduction to technicolor models can have a look at [3]. Sections 4 and 5 will involve physics in extra dimensions. The material of these two sections is self-contained, still more advanced introductions to extra dimensions can be found in [4].

317

2. Electroweak symmetry breaking and new physics

The Standard Model (SM) of electroweak and strong interactions can be divided
into three sectors: the gauge sector, the flavor sector and the electroweak symmetry breaking sector. While the first two ones have been well tested in accelerator
experiments (like LEP, SLD, BABAR, BELLE, *etc...*), the third one is probably
offering the best clues for the physics beyond the SM since it involves two experimentally unknown parameters, namely the Higgs boson mass (M_h) and the
cutoff scale (Λ) of the SM itself. Yet, these two parameters are subject to theoretical consistency constraints — the well-known unitarity, triviality, stability
and naturality bounds — as well as indirect experimental constraints through the
electroweak precision data. While we know from flavor and CP physics that Λ
cannot be below few TeV, the fact that, within the SM, the ratio M_h^2/Λ^2 receives
some quantum corrections of order α_W/π, is telling us that the SM matter content is incomplete and that new physics at the TeV scale is required to stabilize
the weak scale. Three general questions are thus in order:
• Given the experimental results from LEP, SLD, Tevatron ..., what is the current bound on the scale of new physics in the EWSB sector?
• Is it possible to add new physics at the TeV scale that stabilizes the EW scale
and does not violate the above bound?
• What are the potentials to discover new physics in the EWSB sector at LHC?

Any heavy particle when integrated out will generate new non-renormalizable
interactions among the light SM particles. Would the SM make any sense as an
effective theory at low energy, these new interactions have to leave the SM gauge
symmetry unbroken. Yet, they can break some (accidental, approximate) SM
global symmetries and, depending on which global symmetry is actually broken,
these new interactions can manifest themselves at rather low energy. See Table 1.
The EWSB sector seems a good place to look for manifestation of new physics
in the energy range that will be explored at the LHC.

2.1. *SM Higgs physics*

2.1.1. *Higgs mechanism*
Experiments (see for instance Figure 1) show that above 100 GeV or so, electromagnetism and weak interaction are "unified", while at lower energy, the photon
is quite different than the Z and the $W^\pm$, in particular they have quite different
masses:

$$M_\gamma < 6 \times 10^{-17} \text{ eV},$$
$$M_{W^\pm} = 80.425 \pm .038 \text{ GeV}, \qquad (2.1)$$
$$M_{Z^0} = 91.1876 \pm .0021 \text{ GeV}.$$

Table 1

Examples of non-renormalizable interactions among SM particles obtained after integrating out some heavy degrees of freedom and bounds on the corresponding scales that suppress them. These interactions can be classified according to the global symmetries they break. Rather low scales can affect the EWSB sector. From [5].

broken symmetry	operators	scale Λ
B, L	$(QQQL)/\Lambda^2$	10^{13} TeV
flavor (1,2nd family), CP	$(\bar{d}s\bar{d}s)/\Lambda^2$	1000 TeV
flavor (2,3rd family)	$m_b(\bar{s}\sigma_{\mu\nu}F^{\mu\nu}b)/\Lambda^2$	50 TeV
custodial $SU(2)$	$(h^\dagger D_\mu h)^2/\Lambda^2$	5 TeV
none	$(h^\dagger h)^3/\Lambda^2$	–

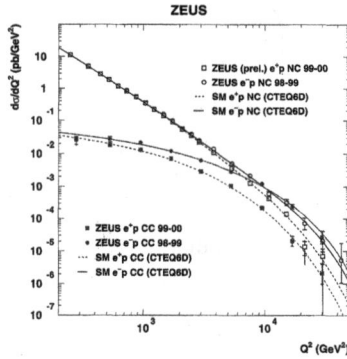

Fig. 1. Unification of electromagnetic and weak interactions as seen by the ZEUS collaboration [6].

The simplest way to explain the possible origin of these masses is to rely on a Higgs mechanism [7]: the initial Lagrangian is invariant under a full $SU(2)_L \times U(1)_Y$ but the vacuum only preserves a $U(1)_{em}$ subgroup. This *spontaneous* breaking can result for a scalar field taking a non zero vacuum expectation value (*vev*). More precisely, we assume that the theory involves a scalar field, H, of hypercharge $1/2$ and that transforms as a doublet under $SU(2)_L$. Using a $SU(2)_L \times U(1)_Y$ transformation, the *vev* of H can always be cast in the form:

Symmetry of the Lagrangian	Symmetry of the vaccum
$SU(2)_L \times U(1)_Y$	$U(1)_{em}$
Higgs doublet	Higgs *vev*
$H = \begin{pmatrix} H_+ \\ H_0 \end{pmatrix}$	$\langle H \rangle = \begin{pmatrix} 0 \\ \frac{v}{\sqrt{2}} \end{pmatrix}$

The non-zero *vev* of the Higgs is the consequence of the shape of the Higgs potential

$$V(H) = \lambda \left(H^\dagger H - \frac{v^2}{2} \right)^2, \qquad (2.2)$$

which has the shape of a Mexican hat and exhibits a local maximum at the origin.

The gauge boson masses are generated through the covariant derivative of the scalar field[2] ($W_\mu^\pm = \left(W_\mu^1 \mp W_\mu^2 \right)/\sqrt{2}$)

$$D_\mu H = \partial_\mu H - \frac{i}{2} \begin{pmatrix} gW_\mu^3 + g'B_\mu & \sqrt{2}gW_\mu^+ \\ \sqrt{2}gW_\mu^- & -gW_\mu^3 + g'B_\mu \end{pmatrix} H, \qquad (2.3)$$

and $|D_\mu H|^2$ contains the quadratic mass terms

$$\tfrac{1}{4} g^2 v^2 W_\mu^+ W^{-\mu} + \tfrac{1}{8} \left(W_\mu^3 \ B_\mu \right) \begin{pmatrix} g^2 v^2 & -gg'v^2 \\ -gg'v^2 & g'^2 v^2 \end{pmatrix} \begin{pmatrix} W^{3\mu} \\ B^\mu \end{pmatrix}. \qquad (2.4)$$

Thus the gauge boson spectrum contains
• a pair of electrically charged gauge bosons, the $W^\pm$'s, of mass

$$M_W^2 = \tfrac{1}{4} g^2 v^2; \qquad (2.5)$$

• a pair of electrically neutral gauge bosons, the photon γ and the Z which are the linear combinations of W^3 and B that diagonalize the mass matrix (2.4):

$$Z_\mu = c_W W_\mu^3 - s_W B_\mu \quad \text{and} \quad \gamma_\mu = s_W g' W_\mu^3 + c_W B_\mu, \qquad (2.6)$$

where the weak mixing angle is the ratio of the $SU(2)_L$ and $U(1)_Y$ gauge couplings

$$c_W = \cos\theta_W = \frac{g}{\sqrt{g^2 + g'^2}}, \qquad s_W = \sin\theta_W = \frac{g'}{\sqrt{g^2 + g'^2}}. \qquad (2.7)$$

[2]Throughout these lectures we will use a mostly minus signature metric $(+, -, \ldots, -)$. The greek indices, $\mu, \nu \ldots$, will denote our usual 4D spacetime coordinates, while coordinates along the extra dimensions will be denoted with lowercase latin indices. Uppercase indices will denote both 4D and extra dimensional coordinates.

As a consequence of the unbroken $U(1)_{em}$ symmetry, the photon remains massless while the mass of the Z is given by

$$M_Z^2 = \tfrac{1}{4}(g^2 + g'^2)v^2. \tag{2.8}$$

2.1.2. Counting the degrees of freedom

In the breaking of $SU2_L \times U(1)_Y$ down to $U(1)_{em}$, three gauge directions are broken, which require eating three Goldstone bosons. These Goldstone bosons are provided by the Higgs doublet which involves four real scalar degrees of freedom out of which three are the eaten Goldstone bosons which become the longitudinal polarizations of the massive gauge bosons. So in total there is one remaining physical real scalar degree of freedom: the famous "Higgs boson". It describes the fluctuations around the non-trivial vacuum.

This decomposition can be made explicit in the parametrization

$$H = e^{i\pi^a \sigma^a} \begin{pmatrix} 0 \\ \frac{v+h}{\sqrt{2}} \end{pmatrix}. \tag{2.9}$$

In the unitary gauge, the π^a are non-physical and correspond to the eaten Goldstone, while h is the physical Higgs boson. The physical Higgs has some non-trivial interactions with the gauge bosons and the quarks and leptons through the Yukawa couplings. It also has non-trivial self-interactions which are obtained by expanding the scalar potential around its minimum

$$V = \lambda \left(H^\dagger H - \frac{v^2}{2} \right)^2 = \lambda v^2 h^2 + \lambda v h^3 + \tfrac{1}{4}\lambda h^4. \tag{2.10}$$

In particular, the Higgs mass is

$$M_h^2 = 2\lambda v^2. \tag{2.11}$$

2.1.3. Custodial symmetry

From the spectrum of the gauge bosons given above, we easily obtain the value of the ρ parameter

$$\rho \equiv \frac{M_W^2}{M_Z^2 \cos^2 \theta_W} = \frac{\tfrac{1}{4}g^2 v^2}{\tfrac{1}{4}(g^2 + g'^2)v^2 \frac{g^2}{g^2+g'^2}} = 1. \tag{2.12}$$

There is an empirical theorem that says that every times at the end of a computation you find either 0 or 1, it means that either you made a mistake or there is a symmetry that can explain the result. In the present case, there is clearly not mistake so there must be a symmetry behind this particular value of the ρ parameter. Indeed, a $SU(2)$ custodial symmetry is present in the Higgs sector of the theory [8]. The Higgs doublet contains four real degrees of freedom and the point is that the Higgs potential is actually invariant under any rotation of these four components, hence a $SO(4)$ global symmetry, which mathematically is nothing else but $SU(2)_L \times SU(2)_R$, the $SU(2)_L$ being the gauge symmetry of the Standard Model. The origin of the $SU(2)_R$ symmetry can also be made more transparent as follows. A doublet of $SU(2)$ is a pseudo-real representation, which means that the complex conjugate of the doublet is equivalent to the doublet itself. In practise, if we consider the doublet to be a field with a lower $SU(2)$ index H_i, then the complex conjugate would automatically have an upper index. However, using the $SU(2$ epsilon $\epsilon_{ij} = i\sigma^2_{ij}$ this can be lowered again, and so H_i and $i\epsilon_{ij}(H^*)^j$ transform in the same way. This means that in addition to an $SU(2)_L$ acting on the usual $SU(2)$ index, there is another $SU(2)_R$ symmetry that mixes H with $i\sigma^2 H^*$. We can write a 2 by 2 matrix as

$$\mathcal{H} = \begin{pmatrix} i\sigma^2 H^* & H \end{pmatrix} = \begin{pmatrix} H_0^* & H_+ \\ -H_+^* & H_0 \end{pmatrix}, \tag{2.13}$$

on which the $SU(2)_L \times SU(2)_R$ symmetry acts as

$$\mathcal{H} \rightarrow U_L \mathcal{H} U_R^\dagger. \tag{2.14}$$

The $SU(2)_L \times SU(2)_R$ invariance of the Higgs potential is made explicit by noting that

$$\mathcal{H}^\dagger \mathcal{H} = H^\dagger H \begin{pmatrix} 1 & 0 \\ 0 & 1 \end{pmatrix}, \tag{2.15}$$

in such a way that the Higgs potential writes

$$V = \frac{\lambda}{4} \left(\text{tr } \mathcal{H}^\dagger \mathcal{H} - v^2 \right)^2. \tag{2.16}$$

The Higgs *vev*

$$\langle H \rangle = \begin{pmatrix} 0 \\ \frac{v}{\sqrt{2}} \end{pmatrix} \quad i.e. \quad \langle \mathcal{H} \rangle = \frac{v}{\sqrt{2}} \begin{pmatrix} 1 & \\ & 1 \end{pmatrix} \tag{2.17}$$

obviously breaks $SU(2)_L \times SU(2)_R$ down to the diagonal subgroup $SU(2)_D$ ($U_L = U_R$ in (2.14) leaves the vacuum invariant). Under this unbroken symmetry, the three gauge bosons $\left(W_\mu^1, W_\mu^2, W_\mu^3 \right)$ transforms as a triplet, which imposes in particular the same mass term for all W^i. The mass term for the W^3 gauge boson is read of from the mass matrix in the (γ, Z) basis:

$$
\begin{pmatrix} Z_\mu \ \gamma_\mu \end{pmatrix} \begin{pmatrix} M_Z^2 & 0 \\ 0 & 0 \end{pmatrix} \begin{pmatrix} Z^\mu \\ \gamma^\mu \end{pmatrix}
$$

$$
= \begin{pmatrix} W_\mu^3 \ B_\mu \end{pmatrix} \begin{pmatrix} c_W^2 M_Z^2 & -c_W s_W M_Z^2 \\ -c_W s_W M_Z^2 & s_W^2 M_Z^2 \end{pmatrix} \begin{pmatrix} W^{3\,\mu} \\ B^\mu \end{pmatrix}. \tag{2.18}
$$

So the $SU(2)_D$ symmetry finally imposes

$$
M_W^2 = c_W^2 M_Z^2, \tag{2.19}
$$

which is nothing else but $\rho = 1$. Actually, the custodial symmetry is explicitly broken by the $U(1)_Y$ interaction (H and $H^\star$ have opposite charges) and by the Yukawa interactions to quarks and leptons. As a consequence, at one-loop there will be corrections to $\rho = 1$. Actually, the screening theorem [9] states that these corrections only appear through some logarithms and thus the deviation to $\rho = 1$ remains rather small.

More generally, if $SU(2)_L \times U(1)_Y$ is broken not only through a doublet but also through a collection of scalar fields in the $2s_i + 1$ representation of $SU(2)_L$, carrying an hypercharge y_i and acquiring a *vev* v_i, we can in the same way obtain that the ρ parameter is now given by

$$
\rho = \frac{\sum_i (s_i(s_i + 1) - y_i^2) v_i^2}{\sum_i 2 y_i^2 v_i^2}. \tag{2.20}
$$

In particular, any non-doublet *vev* will contribute to a deviation from $\rho = 1$.

2.1.4. Unitarity bound

There is a fundamental motivation to generate the W and Z masses through a Higgs mechanism since these masses are actually inconsistent with the content of know particles and extra degrees of freedom, like the Higgs boson, are needed to soften the UV behavior of massive gauge bosons [10, 11]. One polarization of a massive spin-1 particle indeed grows with the energy of the particle.

Three polarizations of a massive spin-1 particle

(ϵ_μ is the polarization vector and k_μ is the momentum of the gauge boson)

$$
A_\mu = \epsilon_\mu \, e^{i k_\mu x^\mu}
$$

$$
\epsilon_\mu \epsilon^\mu = -1 \quad \text{and} \quad k^\mu \epsilon_\mu = 0
$$

$$
k^\mu = (E, 0, 0, k) \quad \text{with} \quad k_\mu k^\mu = E^2 - k^2 = M^2
$$

two transverse polarizations	one longitudinal polarization
$\epsilon_1^\mu = (0, 1, 0, 0)$ $\epsilon_1^\mu = (0, 0, 1, 0)$	$\epsilon_\perp^\mu = \left(\frac{k}{M}, 0, 0, \frac{E}{M}\right) \approx \frac{k^\mu}{M} + \mathcal{O}\left(\frac{E}{M}\right)$

Note that in the R_ξ gauge, the time-like polarization, $\epsilon^\mu \epsilon_\mu = 1, k^\mu \epsilon_\mu = M$, is arbitrarily heavy and decouples. The longitudinal polarization being proportional to the energy, when we look at the scattering of these longitudinal polarizations, we generically end up[3] with a tree-level amplitude that grows like E^4 (E being the energy in the center of mass frame)

$$\mathcal{A} = \mathcal{A}^{(4)} \left(\frac{E}{M}\right)^4 + \mathcal{A}^{(2)} \left(\frac{E}{M}\right)^2 + \dots, \tag{2.21}$$

and that becomes bigger than one around an energy scale of the order of the mass of the gauge boson. Thus in absence of any additional fundamental degrees of freedom, the theory will enter a strongly coupled regime (with a restoration of unitarity by higher loop effects and/or by the exchange of bound states) and we will lose perturbative control on the theory (one says that there is *perturbative unitarity* violation). This is precisely where the Higgs boson is useful for: it gives an additional contribution to the scattering amplitude that exactly cancels the previously energy-growing amplitude. Thus we are left with an amplitude, $\mathcal{A} = g^2 M_H^2/(4M_W^2)$, that remains finite at arbitrarily high energy. This finite amplitude still has to be small enough to maintain perturbative unitarity: that is the Higgs boson should be light enough since, if it is too heavy, the scattering amplitude will already be too big when the Higgs becomes efficient in the cancellation of the growing amplitude. To give a quantitative estimate of the unitarity bound on the Higgs mass, we need to go through a decomposition of the scattering amplitude into partial waves [12]:

$$\mathcal{A} = 16\pi \sum_{l=0}^{\infty} (2l + 1) P_l(\cos\theta) a_l, \tag{2.22}$$

where P_l are the Legendre polynomials ($P_0(x) = 1$, $P_1(x) = x$, $P_2(x) = 3x^2/2 - 1/2, \dots$). Using the orthonormality relation satisfied by the Legendre polynomials, we get the partial wave amplitudes, a_l

$$a_l = \frac{1}{32\pi} \int_{-1}^{1} d(\cos\theta) P_l(\cos\theta) \mathcal{A}. \tag{2.23}$$

The differential cross section is related to the scattering amplitude $\mathcal{A}$ by

$$\frac{d\sigma}{d\Omega} = \frac{1}{64\pi^2 s} |\mathcal{A}|^2, \tag{2.24}$$

---

[3] Actually, if the self-interactions of the gauge bosons are generated by the gauge invariant kinetic term $F_{\mu\nu} F^{\mu\nu}$, the quartic and cubic vertices are related to each other in such a way that the E^4 terms in the scattering amplitude cancel out and we are left with a scattering amplitude that only grows like E^2. This residual E^2 amplitude can only be cancelled by the exchange of a scalar degree of freedom associated to a Higgs mechanism [11].

which finally leads to

$$\sigma = \frac{16}{s} \sum_{l=0}^{\infty} (2l + 1)|a_l|^2. \tag{2.25}$$

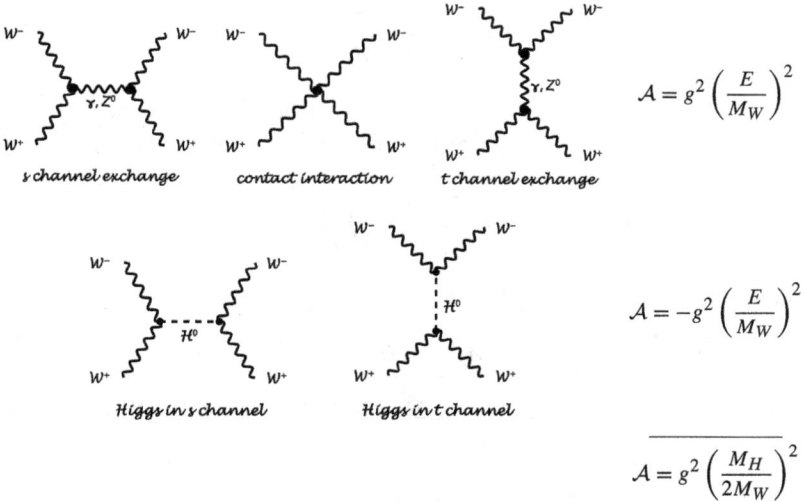

$$\mathcal{A} = g^2 \left(\frac{E}{M_W}\right)^2$$

$$\mathcal{A} = -g^2 \left(\frac{E}{M_W}\right)^2$$

$$\mathcal{A} = g^2 \left(\frac{M_H}{2M_W}\right)^2$$

Fig. 2. Scattering of the longitudinal components of the W in the Standard Model. The contact interaction and the exchange of other spin-1 particles give a contribution to the scattering amplitude that grows like E^2 (the expected E^4 term cancels out due the relation between the quartic and cubic gauge boson self-interactions). The exchange of the physical Higgs boson also gives a E^2 contribution to the W scattering amplitude. In a Higgs mechanism, the masses of the gauge bosons are related to their couplings to the Higgs, which as a consequence ensures that the E^2 contributions exactly cancel. The residual amplitude remains finite at arbitrarily large energy. The Higgs boson unitarizes the W scattering, as long as its mass is light enough. The conventions for the Feynman diagrams are the following: a wiggled line denotes a spin-1 field, a plain line denotes a spin-1/2 field and a dashed line denotes a spin-0 field.

The optical theorem

$$\sigma = \frac{1}{s} \text{Im} \left(\mathcal{A}(\theta = 0)\right), \tag{2.26}$$

simply amounts to

$$\text{Im} \left(a_l\right) = |a_l|^2, \tag{2.27}$$

which can be rewritten as

$$(\text{Im} \left(a_l\right) - 1/2)^2 + (\text{Re} \left(a_l\right))^2 = 1/4. \tag{2.28}$$

In the complex plane, a_l is thus located on the circle of radius $1/2$ centered at $(0, 1/2)$. Therefore

$$|\text{Re}\,(a_l)| \leq 1/2. \tag{2.29}$$

For the SM without a Higgs boson,

$$a_0 = \frac{g^2 E^2}{16\pi\, M_W^2}, \tag{2.30}$$

which is telling us that perturbative unitarity cannot be maintained without a Higgs above ~ 620 GeV (*i.e.* $\sqrt{s} \sim 1.2$ TeV). With a Higgs, we get

$$a_0 = \frac{g^2 M_H^2}{64\pi\, M_W^2}, \tag{2.31}$$

which leads to the upper bound for the Higgs mass:

$$M_H \leq 1.2 \text{ TeV}. \tag{2.32}$$

Actually, by considering the channel $2W^+W^- + ZZ \rightarrow 2W^+W^- + ZZ$, one would get the more stringent bound [12]

$$M_H \leq 780 \text{ GeV}. \tag{2.33}$$

In any case, these bounds give an order of magnitude estimate and they shouldn't be considered as tight bounds since non-perturbative effects will not turn on abruptly anyway. One should rather understand that the numerical values obtained here are the *raison d'être* of the LHC: we know that around one TeV, the dynamics of EWSB will be revealed.

2.1.5. Triviality bound

The other standard bounds on the Higgs mass are coming from the study of radiative corrections to the Higgs potential. At the quantum level, the coefficients of the Higgs potential

$$V(h) = -\tfrac{1}{2}\mu^2 h^2 + \tfrac{1}{4}\lambda h^4 \tag{2.34}$$

are running with the energy. At one loop level, the renormalization group equation for the Higgs quartic coupling is given by (see Fig. 3)

$$16\pi^2 \frac{d\lambda}{d\ln Q} = 24\lambda^2 - (3g'^2 + 9g^2 - 12y_t^2)\lambda$$
$$+ \tfrac{3}{8}g'^4 + \tfrac{3}{4}g'^2 g^2 + \tfrac{9}{8}g^4 - 6y_t^4 + \dots \text{(smaller Yukawa)}. \tag{2.35}$$

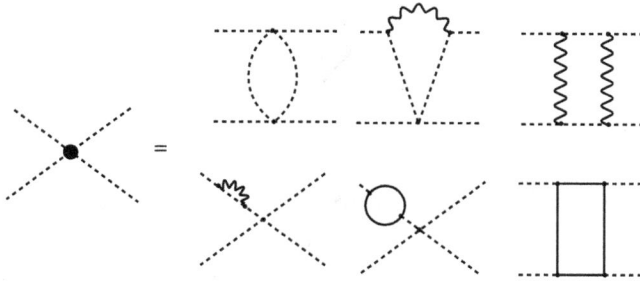

Fig. 3. One loop corrections to the Higgs quartic coupling.

In the large Higgs mass limit, the first term on the RHS dominates and makes the Higgs mass increasing with Q finally leading to an instability that can be seen as follows. The solution of the λ dominated renormalization groupe equation (RGE) is (M_h and v are the Higgs mass and the Higgs *vev* at the weak scale)

$$\lambda(Q) = \frac{M_h^2}{2v^2 - \frac{3}{2\pi^2} M_h^2 \ln(Q/v)}, \qquad (2.36)$$

which exhibits a Landau pole at

$$Q = v \, e^{4\pi^2 v^2 / 3 M_h^2}. \qquad (2.37)$$

New physics should appear before that point to prevent the instability to develop. We thus obtain an upper bound for the cutoff scale of the SM [13]

$$\Lambda \leq v \, e^{4\pi^2 v^2 / 3 M_h^2}. \qquad (2.38)$$

For a fixed value of the SM cutoff, this relation gives an upper bound on the Higgs mass (see Fig. 5). In particular, one cannot take $\lambda \to \infty$, since in this microscopic limit we necessarily have $\lambda = 0$ (trivial theory) and therefore no EWSB can occur.

2.1.6. Stability bound

In the previous section, we have looked at the large Higgs mass limit. Let us now consider the small Higgs mass limit. In that limit, the quartic coupling RGE is dominated by the top Yukawa coupling which makes the Higgs mass decreasing with Q and another instability develops. To obtain the energy dependence of λ, we further need the RG evolution of the top Yukawa coupling. At one loop (see Fig. 4), we have

$$16\pi^2 \frac{dy_t}{d \ln Q} = \frac{9}{2} y_t^3 + \dots \text{ (smaller Yukawa)}. \qquad (2.39)$$

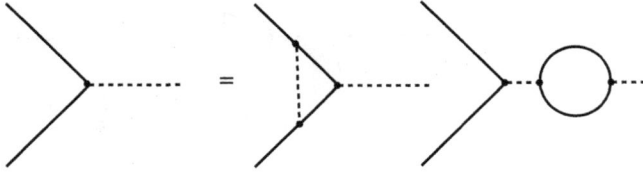

Fig. 4. One loop corrections to the top Yukawa coupling.

We now get

$$y^2(Q) = \frac{y_0^2}{1 - \frac{9}{16\pi^2} y_0^2 \ln \frac{Q}{Q_0}}, \qquad (2.40)$$

$$\lambda(Q) = \lambda_0 - \frac{\frac{3}{8\pi^2} y_0^4 \ln \frac{Q}{Q_0}}{1 - \frac{9}{16\pi^2} y_0^2 \ln \frac{Q}{Q_0}}. \qquad (2.41)$$

For large energy, the Higgs quartic coupling is driven to negative value and the Higgs potential is unbounded from below. Again, new physics should show up before the energy where $\lambda = 0$: (M_h and y_t are the Higgs mass and the top Yukawa coupling at the weak scale)

$$\Lambda \leq v \, e^{4\pi^2 M_h^2 / 3 y_t^4 v^2}. \qquad (2.42)$$

For a fixed value of the SM cutoff, this relation gives a lower bound on the Higgs mass (see Fig. 5). This is the vacuum stability bound [14].

The two instabilities of the Higgs quartic coupling that we just mentioned can be cured for instance if we could find a symmetry such that λ is related to the gauge coupling, e.g. $\lambda = g^2$, then λ would automatically inherit the good UV asymptotically free behavior of the gauge coupling. Such a relation holds in supersymmetric theories and also in 6D gauge–Higgs unification models where the Higgs is identified as a component of the gauge field along some extra dimensions (see Section 4).

2.1.7. Quadratic divergence and hierarchy problem
So far we have looked only at the running of the Higgs quartic coupling, a dimensionless parameter. The radiative corrections are actually more severe for the (tachyonic) mass term of the Higgs potential since it reveals to be highly dependent on the UV physics, which leads to the so called hierarchy problem [16].

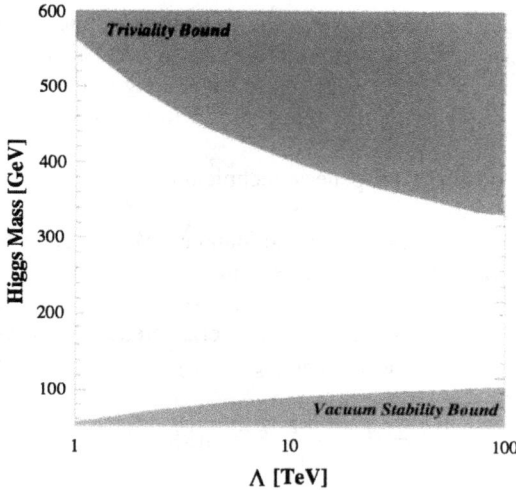

Fig. 5. Triviality and stability bounds on the physical Higgs mass as a function of the SM cutoff Λ. From [15].

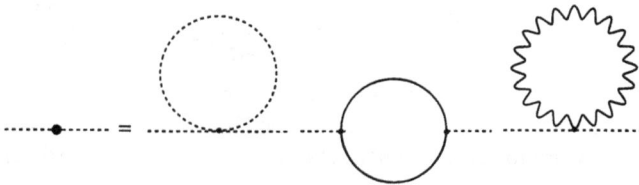

Fig. 6. One loop corrections to the Higgs mass. The three diagrams are quadratically divergent and make the Higgs mass highly UV sensitive.

The one-loop corrections to the M_h are depicted on Fig. 6 and we get

$$\delta M_h^2 = \left(\tfrac{1}{4}(9g^2 + 3g'^2) - 6y_t^2 + 6\lambda \right) \frac{\Lambda^2}{32\pi^2}. \tag{2.43}$$

As an example, for a 10 TeV cutoff, the gauge, top and Higgs contributions to the Higgs mass square corrections are respectively of the order of $(600 \text{ GeV})^2$, $-(1.5 \text{ TeV})^2$ and $(800 \text{ GeV})^2$, all quite far from what the Higgs mass should be. The SM particles give unnaturally large corrections to the Higgs mass: they destabilize the Higgs *vev* and tend to push it towards the UV cutoff of the SM. Some precise adjustment (fine-tuning) between the bare mass and the one-loop correction is needed to maintain the *vev* of the Higgs around the weak scale: take two large numbers, naturally their sum/difference will be of the same order unless these numbers are almost equal up to several significant digits (see [17] for a recent estimate of the amount of fine-tuning within the SM and various models BSM).

The hierarchy problem is a generic technical problem in any theory involving some light scalar fields.

In the study of any theory beyond the Standard Model, one needs to be able to quickly estimate the quadratically divergent corrections to the scalar potentials. One can calculate explicitly some Feynman diagrams or more conveniently rely on the computation of the Coleman–Weinberg potential [18]. At one-loop this effective potential for a scalar field ϕ is given by

$$V(\phi) = \int \frac{d^4 k_E}{2(2\pi)^4} \mathrm{STr} \ln(k_E^2 + M^2(\phi)), \qquad (2.44)$$

where the supertrace, *i.e.* the trace with an extra minus sign for the fermionic degrees of freedom, is over all the particles that acquire a mass when ϕ is away from the origin. After integrating over $d^4 k_E$, we get

$$V = -\frac{\Lambda^4}{128\pi^2} \mathrm{STr}\, 1 + \frac{\Lambda^2}{64\pi^2} \mathrm{STr}\, M^2(\phi) + \frac{1}{64\pi^2} \mathrm{STr}\, M^4(\phi) \ln \frac{M^2(\phi)}{\Lambda^2},$$

where we easily read off the quadratically divergent corrections to the scalar potential. Let us look explicitly at the case of the Higgs in the SM. The only things we need to know are the H-dependent masses of the different particles

H-dependent masses		
particles	number of polarizations	off-shell mass
$W^\pm$	3×2	$M_W^2 = \frac{1}{4} g^2 H^2$
Z^0	3	$M_Z^2 = \frac{1}{4}(g^2 + g'^2) H^2$
top	4×3	$M_t^2 = \frac{1}{2} y_t^2 H^2$
Higgs	1	$M_h^2 = \lambda(3h^2 - v^2)$
Goldstone	3	$M_G^2 = \lambda(H^2 - v^2)$

Notice that these masses are computed for a generic value of h: in particular, away from the true vacuum, $H = v$, the Goldstone are not massless and they contribute to the Coleman–Weinberg potential [19]. Summing over all the particles, we obtain the quadratically divergent correction we were after

$$V_{\Lambda^2} = \tfrac{1}{2} \left(\tfrac{1}{4}(9g^2 + 3g'^2) - 6y_t^2 + 6\lambda \right) \frac{\Lambda^2}{32\pi^2} H^2, \qquad (2.45)$$

in agreement with the diagrammatic computation.

2.2. *Stabilization of the Higgs potential by symmetries*

What we have learnt in the previous sections is that the description of EWSB with a Higgs suffers from several instabilities at the quantum level. Extra structures (particles and/or symmetries) are needed to stabilize the Higgs potential. To keep radiative corrections under control, a theorist can make use of two tools:

• the **spin trick** [20]: in general, a particle of spin s has $2s + 1$ degrees of polarization with the only exception of a particle moving at the speed of light in which case fewer polarizations may be physical. And conversely if a symmetry decouples some polarization states then the particle will necessarily propagate at the speed of light and thus will remain massless. For instance, gauge invariance ensures that the longitudinal polarization of a vector field is non-physical and chiral symmetry keeps only one fermion chirality: both spin-1 and spin-1/2 particles are protected from dangerous radiative corrections. Unfortunately, this spin trick cannot be used for spin-0 particle like the SM Higgs scalar boson.

• the **Goldstone theorem** [21]: when a global symmetry is spontaneously broken, the spectrum contains a *massless* spin-0 particle. However, here again, it seems difficult to invoke this trick to protect the SM Higgs boson from radiative corrections since a Goldstone boson can only have some derivative couplings unlike the Higgs field.

In the late 60's, the Coleman–Mandula's and Haag–Lopuszanski–Sohnius' theorems [22] taught us how to apply the previous tricks to spin-0 particles: the four-dimensional Poincaré symmetry has to be enlarged. The first construction of this type consists in embedding the 4D Poincaré algebra into a superalgebra. Then the supersymmetry between fermion and boson extends the spin trick to scalar particles. Actually there exists an even simpler way to enlarge the Poincaré symmetry which is going into extra dimensions: the 5D Poincaré algebra obviously contains the 4D Poincaré algebra as a subalgebra. After compactification of the extra dimensions, from a 4D dimensional point of view, the higher dimensional gauge field decomposes into a 4D gauge field (the components along our 4D world) and 4D scalar fields (the components along the extra dimensions). The symmetry between vectors and scalars allows to extend the spin trick to spin-0 particles.

Neither supersymmetry nor higher dimensional Poincaré symmetry are exact symmetry of Nature. Therefore, if they ever have a role to play, they have to be broken. In order not to lose any of their benefits, this breaking has to proceed without reintroducing any strong UV dependence into the renormalized scalar mass square: we need a *soft breaking*. This question has been well studied in supersymmetric theories and we would like, in Section 4, to discuss a soft breaking of higher dimensional gauge theories In Section 5, we will briefly report on Higgsless EWSB models [23, 24]. Finally, in Section 3, we will explain how to implement the idea of the Higgs as the pseudo-Goldstone boson (PGB).

2.3. EW precision tests

We have seen that we need new particles to stabilize the weak scale. They have to be massive to evade direct searches. They still influence SM physics and they can "detected" through precision measurements.

2.3.1. An example of EW corrections induced by a heavy particle
As an example, let us take an extra heavy B' gauge boson. The full Lagrangian is

$$\mathcal{L} = -\tfrac{1}{2} W_3 (p^2 - M_W^2) W_3 - t_0 M_W^2 W_3 B - \tfrac{1}{2} B (p^2 - t_0^2 M_W^2) B$$
$$+ g J_3 W_3 + g' J_y B - \tfrac{1}{2} B' (p^2 - M^2) B' + g' J_y B', \qquad (2.46)$$

where $t_0 = g'/g$ (later on, we will also use $c_0 = g/\sqrt{g^2 + g'^2}$ and $s_0 = g'/\sqrt{g^2 + g'^2}$). Let us now integrate out the heavy particle, which means that we freeze its dynamics and replace B' by its equation of motion

$$\frac{\partial \mathcal{L}}{\partial B'} = 0 \Leftrightarrow B' = \frac{g' J_y}{p^2 - M^2}. \qquad (2.47)$$

Plugging back this expression into the original Lagrangian and after an expansion for $M \gg p$, we obtain the effective Lagrangian

$$\mathcal{L} = -\tfrac{1}{2} W_3 (p^2 - M_W^2) W_3 - t_0 M_W^2 W_3 B - \tfrac{1}{2} B (p^2 - t_0^2 M_W^2) B$$
$$+ g J_3 W_3 + g' J_y B - \frac{(g' J_y)^2}{2M^2}. \qquad (2.48)$$

Using the equation of motion for B, $g' J_y = t_0 M_W^2 W_3 + (p^2 - t_0^2 M_W^2) B$, we can actually write the four-fermi interaction as a modification of the propagator of the gauge bosons

$$\mathcal{L} = -\tfrac{1}{2} W_3 \left(p^2 - M_W^2 \left(1 - \tfrac{t_0^2 M_W^2}{M^2} \right) \right) W_3$$

$$-t_0 M_W^2 \left(1 + \tfrac{p^2 - t_0^2 M_W^2}{M^2} \right) W_3 B$$

$$-\tfrac{1}{2} B \left(p^2 \left(1 - 2 \tfrac{t_0^2 M_W^2}{M^2} \right) - t_0^2 M_W^2 \left(1 - \tfrac{t_0^2 M_W^2}{M^2} \right) + \tfrac{p^4}{M^2} \right) B$$

$$+ g J_3 W_3 + g' J_y B + \mathcal{O}(p^6) + \mathcal{O}(1/M^4). \tag{2.49}$$

The mass matrix in the (W_3, B_{in}) basis thus reads

$$\left(1 - \tfrac{t_0^2 M_W^2}{M^2} \right) \begin{pmatrix} M_W^2 & -t_0 M_W^2 \\ -t_0 M_W^2 & t_0^2 M_W^2 \end{pmatrix}. \tag{2.50}$$

Note that the determinant of this mass matrix is vanishing as it should be to maintain the masslessness of the photon. Furthermore, we also note that the weak mixing angle is unaffected

$$Z = c_0 W_3 - s_0 B_{in} \quad \text{and} \quad \gamma = s_0 W_3 + c_0 B. \tag{2.51}$$

This is essential here to ensure that the photon actually couples to the electric charge $T_{3L} + Y$. The photon remains massless but the mass of the Z gets modified by the existence of the heavy B'

$$M_Z^2 = \tfrac{1}{c_0^2} M_W^2 \left(1 - \tfrac{t_0^2 M_W^2}{M^2} \right) \quad \text{and} \quad M_\gamma^2 = 0. \tag{2.52}$$

So at low energy, we obtain a deviation to $\rho = 1$ since we now have

$$\rho \equiv \frac{M_W^2}{M_Z^2 c^2} \approx 1 + \frac{t_0^2 M_W^2}{M^2}. \tag{2.53}$$

The deviation to $\rho = 1$ is usually called the T parameter

$$\rho \equiv 1 + \alpha_{em} T. \tag{2.54}$$

So in our example, we get

$$T = \frac{t_0^2 M_W^2}{\alpha_{em} M^2}. \tag{2.55}$$

The upper bound [1] on the T parameter, $T \leq .2$, gives a lower bound of the mass of the heavy B', $M \geq 3.7$ TeV. That is a rather generic result: the bound on new physics needed to stabilize the weak scale is at least one order of magnitude above the weak scale. This has been dubbed the *little hierarchy problem* or *LEP paradox* [25].

2.3.2. General structure of the EW corrections

Under mild assumptions (universality and heaviness), we can obtain the general form of the corrections induced by new physics [26–28] (see [29] for recent reviews). The most general $U(1)_{em}$ invariant quadratic Lagrangian for the SM gauge bosons

$$\mathcal{L} = -\tfrac{1}{2} W_3^\mu \, \Pi_{33}(p^2) \, W_{3\,\mu} - W_3^\mu \, \Pi_{3B}(p^2) \, B_\mu - \tfrac{1}{2} B^\mu \, \Pi_{BB}(p^2) \, B_\mu$$
$$- W_+^\mu \, \Pi_{+-}(p^2) \, W_{-\,\mu} \tag{2.56}$$

involves four vacuum polarizations that we expand in powers of momentum

$$\Pi_V(p^2) = \Pi_V(0) + p^2 \Pi_V'(0) + \tfrac{1}{2}(p^2)^2 \Pi_V''(0) + \mathcal{O}(p^6). \tag{2.57}$$

So 12 coefficients should describe the most general low energy effective Lagrangian. But 3 of them can actually be removed by normalizing the gauge bosons (which corresponds to the identification of the three SM parameters g, g' and v)

$$\Pi_{+-}'(0) = \Pi_{BB}'(0) = 1, \quad \Pi_{+-}(0) = -M_W^2 = -(80.425 \text{ GeV})^2. \tag{2.58}$$

The remaining 9 parameters are not yet fully independent since we need to impose the masslessness of the photon and its coupling to $Q = T_{3L} + Y$. So we are left with a total of 7 arbitrary coefficients [28]. They are given in Table 2 along with dimension six operators that generate them.

For instance, in the example discussed in the previous section, we obtain

$$t_0^2 \hat{S} = \hat{T} = t_0^2 Y = \frac{t_0^2 M_W^2}{M^2} \quad \text{and} \quad \hat{U} = V = X = W = 0. \tag{2.59}$$

The electroweak precise measurements can be analyzed using the parametrization just described and the results of the fits (assuming the existence of a light/heavy Higgs) are [28] (see also [30] for another recent analysis of EW precision measurements)

Fit	$10^3 \widehat{S}$	$10^3 \widehat{T}$	$10^3 Y$	$10^3 W$
115 GeV Higgs	0.0 ± 1.3	0.1 ± 0.9	0.1 ± 1.2	-0.4 ± 0.8
800 GeV Higgs	-0.9 ± 1.3	2.0 ± 1.0	0.0 ± 1.2	-0.2 ± 0.8

Table 2

Seven coefficients paramatrize the most general low energy Lagrangian beyond the SM. $SU(2)_L \times U(1)_Y$ invariant higher dimensional operators can give rise to these corrections. Notice that these corrections have definite symmetry properties under the gauge $SU(2)_L$ and the $SU(2)_c$ custodial symmetry. The more usual S, T and U coefficients are obtained by $S = 4s_w^2 \widehat{S}/\alpha_{em} \approx 119\,\widehat{S}$, $T = \widehat{T}/\alpha_{em} \approx 129\,\widehat{T}$ and $U = -4s_w^2 \widehat{U}/\alpha_{em} \approx -119\,\widehat{U}$. Note that since U, V and X are not generated by dimension six operators, we generically expect them to be further suppressed compare to the four other coefficients: $\widehat{U} \sim \frac{M_W^2}{\Lambda^2}\widehat{T}$, $\widehat{V} \sim \frac{M_W^4}{\Lambda^4}\widehat{T}$ and $X \sim \frac{M_W^2}{\Lambda^2}\widehat{S}$. From [28] ([28] uses a non-canonical normalization of the gauge bosons, hence the different factors of g and g' appearing in the definition of $\widehat{S}$, $\widehat{T}$ ...... W).

Coefficients	Dim. 6 Operators	$SU(2)_c$	$SU(2)_L$
$\widehat{S} = \frac{g}{g'}\Pi_{3B}'(0)$	$(H^\dagger \tau^a H)W_{\mu\nu}^a B_{\mu\nu}/gg'$	$+$	$-$
$\widehat{T} = \frac{1}{M_W^2}\left(\Pi_{33}(0) - \Pi_{+-}(0)\right)$	$|H^\dagger D_\mu H|^2$	$-$	$-$
$\widehat{U} = \Pi_{+-}'(0) - \Pi_{33}'(0)$	Dim. 8	$-$	$-$
$V = \frac{M_W^2}{2}\left(\Pi_{33}''(0) - \Pi_{+-}''(0)\right)$	Dim. 10	$-$	$-$
$X = \frac{M_W^2}{2}\Pi_{3B}''(0)$	Dim. 8	$+$	$-$
$Y = \frac{M_W^2}{2}\Pi_{BB}''(0)$	$(\partial_\rho B_{\mu\nu})^2/2g'^2$	$+$	$+$
$W = \frac{M_W^2}{2}\Pi_{33}''(0)$	$(D_\rho W_{\mu\nu}^a)^2/2g^2$	$+$	$+$

2.3.3. An example of EW corrections induced by a higher dimensional operator

As a concrete example of the analysis above, we would like to find out the consequence of the higher dimensional operator [27]

$$\mathcal{L}_T = \frac{a}{\Lambda^2}\left|H^\dagger D_\mu H\right|^2 , \qquad (2.60)$$

where a is a dimensionless coefficient. After EWSB, $\langle H \rangle = (0, v/\sqrt{2})$, the operator $\mathcal{L}_T$ simply appears as an additional mass term for the Z gauge boson

$$\mathcal{L}_T = \frac{av^4}{8\Lambda^2}\left(g'B_\mu - gW_\mu^3\right)^2 . \qquad (2.61)$$

Since the W mass and the weak mixing angle remained untouched, we easily obtain the correction to the ρ parameter induced by $\mathcal{L}_T$

$$\rho = \frac{M_W^2}{M_Z^2 c_W^2} = 1 + \frac{av^2}{\Lambda^2} , \qquad (2.62)$$

which corresponds to a non-vanishing T parameter

$$T = \frac{\rho - 1}{\alpha_{em}} = \frac{av^2}{\alpha_{em}\Lambda^2}. \tag{2.63}$$

This is also what we could derive using the formalism of the Table 2 since

$$\Pi_{+-} = -\tfrac{1}{4}g^2 v^2 \quad \text{and} \quad \Pi_{33} = -\tfrac{1}{4}g^2 v^2 + \frac{ag^2 v^4}{4\Lambda^2}. \tag{2.64}$$

3. Little Higgs models

Due to space constraints and since some very good reviews are already available [5, 31], I will not report on the construction of Little Higgs models in these lecture notes.

4. Gauge–Higgs unification models

The components of the gauge fields along some extra dimensions are seen from the 4D point of view as some 4D scalar fields (we will call them gauge-scalars). It is only above the compactification scale, when the extra dimensions open up, that the higher dimensional gauge structure reveals itself. We will now describe models of gauge–Higgs unification where the Higgs is identified as some gauge-scalars[4]. This approach is actually quite old [33,34] but it is only recently within the context of orbifolds [35] that it has been implemented in realistic models [36–43]. A series of questions immediately pops up

- which gauge group will contain the Higgs?
- how many extra dimensions do we need? how are they compactified?
- what are the radiative corrections?
- how can matter be incorporated? how are the Yukawa couplings generated?

It is interesting to note that the deconstruction versions of these gauge–Higgs unification models led to the idea of Little Higgs models [44]. The symmetry protecting the Higgs mass is there a discrete shift symmetry and the construction is much less constrained by the absence of 5D Lorentz invariance.

4.1. Orbifold breaking. 5D SU(3) model

Both 4D vectors and 4D scalars originating from higher dimensional gauge fields belong to an adjoint representation of the gauge group, while the SM Higgs boson is a fundamental representation of the weak symmetry. In order to identify the Higgs as components of a gauge field in extra dimensions we thus need to en-

---

[4]See [32] for recent pedagogical introductions to this approach.

large the $SU(2)_L \times U(1)_Y$ gauge symmetry into a bigger group G. This bigger group can be broken in different ways: (*i*) by introducing a higher dimensional Higgs field; (*ii*) by a Green–Schwarz mechanism; (*iii*) by compactification on a non-trivial background manifold; (*iv*) by compactification on an orbifold. This last method is not only well motivated in a stringy context but it also offers the advantage of easily accommodating the presence of 4D chiral matter[5].

The simplest example of an orbifold is $S^1/\mathbb{Z}_2$, *i.e.*, a circle $(-\pi R \leq y < \pi R)$ with a parity identification $(y \sim -y)$. The identification of the points y and $-y$ means that the values of any field evaluated at these points have to be physically equivalent, *i.e.*, equal up to a global symmetry transformation: $\phi(x, -y) = U\phi(x, y)$. For consistency, U has to be a $\mathbb{Z}_2$ symmetry, $U^2 = 1$. Note that there are two special points of the circle, 0 and πR, which are identified with themselves: they are *fixed points* of the orbifold. The invariance of the kinetic term dictates the transformation of the various components of the gauge field:

$$A_\mu(x, -y) = U A_\mu(x, y) U^{-1} \quad \text{and} \quad A_5(x, -y) = -U A_5(x, y) U^{-1}. \quad (4.1)$$

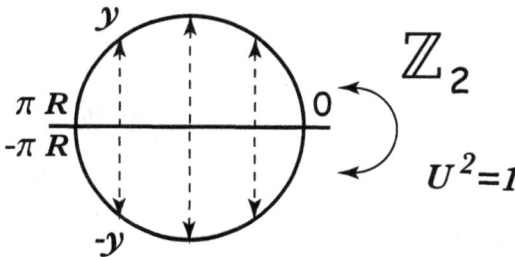

Fig. 7.   The simplest example of an orbifold: the points $-y$ and y on the circle are identified. The fields at the identified points have to be equal up to a global $\mathbb{Z}_2$ symmetry of the theory: for instance the components of a gauge field satisfy $A_\mu(x, -y) = U A_\mu(x, y) U^{-1}$ and $A_5(x, -y) = -U A_5(x, y) U^{-1}$, where U, a global symmetry of the theory, is the orbifold projection. The zero modes of A_μ correspond to the gauge directions that commute with the orbifold projection, $[A_\mu, U] = 0$, while the zero modes of A_5 anti-commute with U, $\{A_5, U\} = 0$.

In a Kaluza–Klein (KK) decomposition, the 4D mass is related to the derivative of the field along the extra dimension, thus a massless mode should be independent of y. From the orbifold boundary conditions (4.1), we obtain that the 4D massless vectors correspond to the generators of the gauge group that *commute* with the orbifold matrix U, while the 4D massless gauge-scalars correspond to the generators that *anticommute* with U. Let us consider the example of an $SU(3)$ gauge group broken by the orbifold projection $U = \text{diag}(-1, -1, 1)$

---

[5]See Section 5.5.1 for an explanation on how chiral matter is obtained in orbifold models.

down to $SU(2) \times U(1)$: from the eight gauge components of A_5, only a $SU(2)$ scalar doublet remains massless.

$$SU(3) \to SU(2) \times U(1)$$
$$U = \mathrm{diag}(-1, -1, 1)$$

$[A_\mu, U] = 0 \qquad A_\mu = \frac{1}{2} \begin{pmatrix} A_\mu^3 + A_\mu^8/\sqrt{3} & A_\mu^1 - iA_\mu^2 & \\ A_\mu^1 + iA_\mu^2 & -A_\mu^3 + A_\mu^8/\sqrt{3} & \\ & & -2A_\mu^8/\sqrt{3} \end{pmatrix} \qquad SU(2) \times U(1)$

$\{A_5, U\} = 0 \qquad A_5 = \frac{1}{2} \begin{pmatrix} & & A_5^4 - iA_5^5 \\ & & A_5^6 - iA_5^7 \\ A_5^4 + iA_5^5 & A_5^6 + iA_5^7 & \end{pmatrix} \qquad \dfrac{SU(3)}{SU(2) \times U(1)}$

It is tempting to identify the massless $SU(2)$ doublet contained in A_5 as the Higgs doublet: $H_0 = (A_5^6 - iA_5^7)/2$ and $H_+ = (A_5^4 - iA_5^5)/2$. For that, we need to know its $U(1)$ charge. Under any transformation of $SU(3)$, A_5 transforms as $\delta_T A_5 = g[T, A_5]$. In particular, under the $U(1)$ of $SU(2) \times U(1)$, $T = T_8 = \mathrm{diag}(1, 1, -2)/(2\sqrt{3})$, we obtain

$$\delta_T \begin{pmatrix} & H_+ \\ & H_0 \\ H_+^\star & H_0^\star \end{pmatrix} = g \frac{3}{2\sqrt{3}} \begin{pmatrix} & H_+ \\ & H_0 \\ -H_+^\star & -H_0^\star \end{pmatrix}. \qquad (4.2)$$

So the $U(1)$ charge of the doublet is equal to $\sqrt{3}/2$. We need to change the normalization of the $U(1)$ for the charge of the doublet to be $1/2$, this is achieved by picking up $U(1)_Y = T_8/\sqrt{3}$. The gauge coupling of $U(1)_Y$ is thus $g' = \sqrt{3}g$. Since we embedded $SU(2)_L \times U(1)_Y$ in a simple group, we get a prediction for the weak mixing angle

$$\sin^2 \theta_W = \frac{g'^2}{g^2 + g'^2} = \frac{3g^2}{g^2 + 3g^2} = \frac{3}{4}. \qquad (4.3)$$

This value is quite far from the experimental one ($\sin^2 \theta_W \approx 0.23$) which certainly invalidates this simple $SU(3)$ gauge–Higgs unification model. Furthermore, with this embedding of $SU(2)_L \times U(1)_Y$ into $SU(3)$, there is no way to get the quarks and leptons from $SU(3)$ irreducible representations.

At this point, we can envision at least two ways to proceed: *(i)* add an additional $U(1)$ factor to $SU(3)$; *(ii)* examine other embedding of $SU(2)_L \times U(1)_Y$ into simple groups. Though the former gives up one nice aspect of the gauge–Higgs unification models, namely the prediction of the weak mixing angle, recent developments seems to indicate that it is the right direction to go while, as we are going to see it, a radiative instability sullies the most promising models of the second class. Finally a third way to go is to modify the geometry of the extra dimensional space.

Before going on with the construction of gauge–Higgs unification models, we would like to mention that the orbifold projection can be reinterpreted as simple boundary conditions on an interval.

G → H orbifold breaking

	H subgroup	
$A_\mu^H(-y) = A_\mu^H(y)$		$\partial_5 A_\mu^H \mid_{y=0,\pi R} = 0$
	equivalent to	
$A_5^H(-y) = -A_5^H(y)$		$A_5^H \mid_{y=0,\pi R} = 0$
	G/H coset	
$A_\mu^{G/H}(-y) = -A_\mu^{G/H}(y)$		$A_\mu^{G/H} \mid_{y=0,\pi R} = 0$
	equivalent to	
$A_5^{G/H}(-y) = A_5^{G/H}(y)$		$\partial_5 A_5^{G/H} \mid_{y=0,\pi R} = 0$

It is also possible to accommodate a Scherk–Schwarz twist, *i.e.* $\phi(y + 2\pi R) = T\phi(y)$. The twist will manifest itself by different boundary conditions on both ends of the interval.

Let us also mention that orbifold breaking has been applied to break Grand Unified symmetries (see [45] for a review). In that latter case, the compactification is close to the GUT scale while in gauge–Higgs models it is of the order of the weak scale.

4.2. 6D G_2 model

We will restrict our analysis to abelian orbifold using inner automorphism (the orbifold matrix U is an element of the group itself). It is well-known that in that case the rank of the gauge group is preserved. Since $SU(2) \times U(1)$ is of rank two, we need to look for a simple group of rank two. There are only four possibilities: $SO(4)$, $SO(5)$, $SU(3)$ and the exceptional group G_2 (for an explicit matrix realization of G_2 that exhibits its $SU(3)$ subgroup, see [38]). The first three cases either do not provide a $SU(2)$ scalar doublet or lead to a prediction for the weak mixing angle too far from the experimental value. The most interesting possibility relies on G_2 [33, 38] which can be broken down to $SU(2) \times U(1)$ by compactification on a two dimensional orbifold, $T^2/\mathbb{Z}_4$, T^2 being a square torus and $\mathbb{Z}_4$, a rotation by ninety degrees (there is no way to break G_2 to $SU(2) \times U(1)$ just using a $\mathbb{Z}_2$ projection: a projection of order at least four is required, which can only be implemented in presence of at least two extra dimensions). There are again two *fixed points* on the torus left invariant by the rotation. The action of the orbifold on the fundamental representation of G_2 is defined by the matrix

$U = \text{diag}(i, i, -1, -i, -i, -1, 1)$. The low energy gauge group is found to be $SU(2) \times U(1)$ and the gauge couplings are such that the weak mixing angle satisfies $\sin^2 \theta_W = 1/4$. The massless spectrum also contains two $SU(2)$ scalar doublets, h and H, carrying respectively an hypercharge $1/2$ and $3/2$: h is thus a perfect candidate to be identified to the SM Higgs doublet.

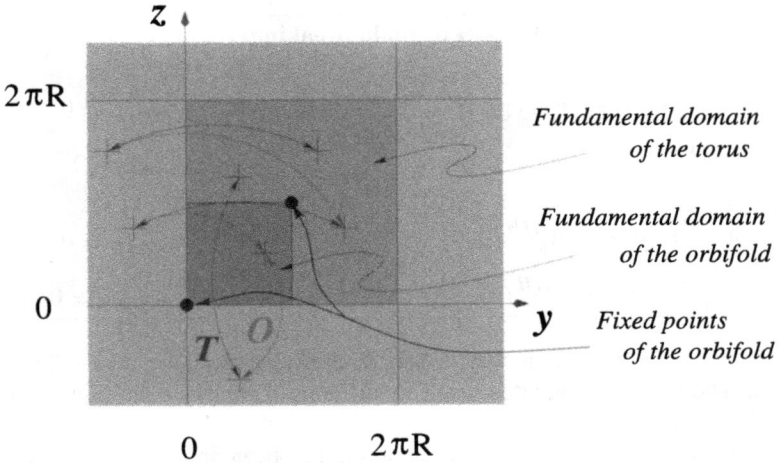

Fig. 8. $T^2/\mathbb{Z}_4$ orbifold: beside the torus identification T, $y \sim y + 2\pi R$ and $z \sim z + 2\pi R$, the orbifold projection, O, identifies the points $(y, z) \sim (-z, y)$ up to a global $\mathbb{Z}_4$ symmetry U: $A_\mu(-z, y) = U A_\mu(y, z) U^\dagger$, $A_y(-z, y) = -U A_z(y, z) U^\dagger$ and $A_z(-z, y) = U A_y(y, z) U^\dagger$. Using, $U = \text{diag}(i, i, -1, -i, -i, -1, 1)$, G_2 is broken dwon to $SU(2) \times U(1)$.

$G_2 \to SU(2) \times U(1)$
G_2 gauge group
adj $= 14 \qquad$ fundamental $= 7$
$SU(3)$ decomposition
$14 = 8 + 3 + \bar{3} \qquad 7 = 3 + \bar{3} + 1$
$SU(2) \times U(1)$ decomposition
T_8 normalization
$14 = \left(3_0 + (2 + \bar{2})_{\sqrt{3}/2} + 1_0\right) + \left(2 + \bar{2}\right)_{1/2\sqrt{3}} + \left(1 + \bar{1}\right)_{-1/\sqrt{3}}$
$7 = \left(2 + \bar{2}\right)_{1/2\sqrt{3}} + \left(1 + \bar{1}\right)_{-1/\sqrt{3}} + 1_0$
wek mixing angle
$U(1)_Y = \sqrt{3} T_8 \qquad \Leftrightarrow \qquad \sin^2 \theta_W = 1/4$

Another benefit of having two extra dimensions is that the non-abelian interaction piece contained in the gauge kinetic term will automatically generate a quartic coupling. In our model and after compactification, the potential for the canonically normalized scalars writes:

$$V = \frac{1}{6} g^2 \left(|h|^4 + 3|H|^4 + 3(H^\dagger \sigma^a h)(h^\dagger \sigma^a H) - 6|h|^2|H|^2 \right), \qquad (4.4)$$

where σ^a, $a = 1 \ldots 3$, are the Pauli matrices and g is the gauge coupling of the low energy $SU(2)$ gauge group in 4D. As in supersymmetry, we find that the Higgs quartic coupling is related to the square of the gauge coupling, though the details of the potential are different (in particular, in the present model, there is a single doublet of hypercharge 1/2).

4.3. Radiative corrections

There are two types of operators involving the 4D gauge-scalar fields that can be generated radiatively [46]:

(*i*) some bulk operators
(*ii*) some operators localized at the *fixed points* of the orbifold.

The higher dimensional gauge invariance acts on the gauge bosons and the gauge-scalars as

$$\delta A_M^A = \partial_M \epsilon^A + g f^{ABC} A_M^B \epsilon^C, \qquad (4.5)$$

where f^{ABC} are the structure constant. For consistency the gauge parameters, the gauge transformation parameter, ϵ^C, obey the same boundary conditions as A_μ

$$G \rightarrow H: \qquad \partial_5 \epsilon^H \big|_{\text{fixed point}} = 0 \text{ and } \epsilon^{G/H} \big|_{\text{fixed point}} = 0. \qquad (4.6)$$

So this is really gauge invariance that protects the first kind of operators: indeed above the compactification scale, the gauge-scalar fields really appear as some components of the higher dimensional gauge field and the Slavnov–Taylor identities forbids for instance the appearance of any mass term. Below the compactification scale, however, we have to deal with ordinary scalars which pick up some radiative but finite —since cut off at the compactification scale— mass. All the bulk operators are thus generated by IR effects and are finite. Another way to see it is that the only gauge invariant operator that can give rise to a Higgs potential must be non-local in the extra dimensions and expressed in term of the Wilson line $\mathcal{P} e^{i \int dx^i A_i}$. Being a non-local operator, the Higgs potential is finite to all orders in perturbation theory, it is UV insensitive and calculable once the degrees of freedom around the weak scale are known.

As far as the brane-localized operators are concerned, the situation is more complicated. At the fixed point, the bulk gauge group is partially broken: there is only a $SU(2) \times U(1)$ subgroup left unbroken. And acting on the gauge-scalar doublet, the unbroken gauge group acts linearly and certainly does not forbid any quadratically divergent localized mass term to be generated[6]

$$\delta_H A_\mu^H = \partial_\mu \epsilon^H + g f^{HHH} A_\mu^H \epsilon^H, \qquad \delta_H A_\mu^{G/H} = 0, \qquad (4.7)$$

$$\delta_H A_5^H = 0, \qquad \delta_H A_5^{G/H} = g f^{G/H \, G/H \, H} A_5^{G/H} \epsilon^H. \qquad (4.8)$$

Even though G/H is the only unbroken gauge symmetry at the fixed points, there is still some residual symmetry left over from the full G gauge symmetry in the bulk: indeed the broken generators of the higher dimensional gauge invariance act on the gauge-scalars as a shift symmetry proportional to the derivative of the gauge parameters [46]:

$$\delta_{G/H} A_\mu^H = 0, \qquad \delta_{G/H} A_\mu^{G/H} = 0, \qquad (4.9)$$

$$\delta_{G/H} A_5^H = 0, \qquad \delta_{G/H} A_5^{G/H} = \partial_5 \epsilon^{G/H}. \qquad (4.10)$$

This Peccei–Quinn like symmetry is enough to prevent the appearance of local mass counter-term. To construct an invariant, we need to use an object that transforms homogeneously under the gauge symmetry, like the gauge field strength tensor F_{MN}. In 5D orbifolds, there is no possible local counter-term involving the gauge field strength since it is an antisymmetric object while we have only one index at our disposal. In 6D orbifolds, however, the brane localized operators [38, 47]

$$\mathrm{Tr}(U^k F_{56}) \quad k = 1, 2, 3 \qquad (4.11)$$

are perfectly allowed and are invariant under the local gauge transformations:

$$\mathrm{Tr}(U F_{56}) \to \mathrm{Tr}(U g(0) F_{56} g^{-1}(0)) = \mathrm{Tr}(U F_{56} g^{-1}(0) g(0)) = \mathrm{Tr}(U F_{56}),$$

where in the first equality we used the fact that *at the fixed points* U and g commute (for concreteness, we considered that the fixed point is at the origin). These operators are potentially quite dangerous since they would correspond to a tadpole for some massive KK gauge-scalars along the unbroken $U(1)$'s directions and, through the non-abelian part of F_{56}, to a mass for the massless gauge-scalars. And by power-counting, these operators are *quadratically divergent*.

---

[6]Since $[H, H] \subset H$ and $[H, G/H] \subset G/H$, the only non-vanishing structure constants are f^{HHH} and $f^{G/H \, G/H \, H}$ and cyclic permutations.

In the case of a $T^2/\mathbb{Z}_2$ orbifold, a parity invariance, $(y, z) \rightarrow (-y, z)$, can be defined consistently with the orbifold projection

$$\mathbb{Z}_2$$

$$\phi(x, -y, -z) = U\phi(x, y, z)$$

$$P \nearrow \qquad\qquad \searrow P$$

$$\phi(x, y, -z) = U\phi(x, -y, z) \qquad U\phi(x, -y, z)$$

and this parity invariance forbids the appearance of the operators (4.11). On a $T^2/\mathbb{Z}_4$ orbifold, the parity symmetry is not consistent with the orbifold projection anymore

$$\mathbb{Z}_4$$

$$\phi(x, -z, y) = U\phi(x, y, z)$$

$$P \nearrow \qquad\qquad \searrow P$$

$$\phi(x, z, y) = U^{-1}\phi(x, -y, z) \qquad U\phi(x, -y, z)$$

and no discrete symmetry protects the operators (4.11). A direct computation (see Fig. 9) of the tadpole piece contained in (4.11) reveals that at one loop level the contributions of the ghost fields and of the gauge fields add up to indeed generate the operator $\mathrm{Tr}(U F_{56})$ at the fixed points[7]. Thus a quadratic divergence is reintroduced and it will destabilize the Higgs potential at one-loop. A possible way to (partially) cure this instability is to engineer a setup such that the sum of the quadratically divergent tadpoles at the different fixed points is vanishing (global cancellation) [40].

gauge contribution   *ghost contribution*

Fig. 9. One-loop diagrams contributing to the tadpole contained in the operator $\mathrm{Tr}(U F_{56})$ localized at the fixed point. These diagrams are quadratically divergent and they reintroduce a radiative instability in the Higgs potential.

In conclusion, though the gauge symmetry breaking on 5D orbifolds as well as on some 6D orbifolds is soft in the sense that the radiative corrections to the

---

[7]In the original computation [38] of the tadpole, we found that the two contributions cancel each other. But since no symmetry protects this tadpole operator, by virtue of the theorem mentioned in Section 2.1.3, our vanishing result was a sign of a mistake in our computation. More careful computations [40, 48] indeed concluded in a non-vanishing quadratically divergent tadpole.

Higgs potential are finite, in the case of the potentially most interesting orbifold, $T^2/\mathbb{Z}_4$, the gauge symmetry breaking is not soft.

4.4. Introducing matter and Yukawa interactions

The introduction of quarks and leptons and the generation of their masses are not straightforward [38,40]. If the matter was in the bulk, the Higgs *vev* would generate fermion masses that are controlled by the higher dimensional gauge coupling and we could only play with group theory factors associated to different representations to generate different masses, which certainly cannot account for the diversity of the matter spectrum. The other possibility is to resort to fermions localized at the orbifold fixed points. As far as the Yukawa interactions are concerned, one could try to directly linearly couple the SM fermions to the Higgs field at the fixed points. However introducing Yukawa couplings this way explicitly breaks the residual Peccei–Quinn shift symmetry G/H. One would like to look for operators that do not break this shift symmetry: this can be achieved by using operators that involve Wilson lines between the fixed points (the two fixed points may also coincide) $W = P e^{i \int A_i dx^i}$. Actually, it is natural to expect the appearance of these operators once some massive bulk fermions which could mix with fields at the fixed points are integrated out in a Froggatt–Nielsen like way. The masses of the light fermions can be obtained either by small mixing with the bulk fermions or by large bulk mass that will exponentially suppress the effective Yukawa and it is possible to easily reproduce the hierarchy of the matter spectrum [40]. Some interesting flavor structure in the first two generations has been revealed in the study of a particular $SU(3)$ model in 5D [41].

4.5. Experimental signatures

The collider signals have not been studied in details yet (may be due to the lack of a fully realistic model). We can still pin down some predictions of a generic gauge–Higgs unification model
- we should observe KK excitations of the W and Z around 500 GeV – 1 TeV;
- we should observe spin-1 KK excitations of the G/H coset, in particular, there will be some gauge bosons with the EW quantum numbers of the Higgs doublet;
- we should observe extra scalar fields;
- we should observe some bulk fermions that mix with the SM fermions to generate their masses.

4.6. Recent developments and open issues

In view of the quadratic divergence for the localized tadpole in the most promising 6D model, recent studies of gauge–Higgs unification models have focussed mainly on 5D. The main issue of 5D models is to accommodate the heaviness of the top quark and of the Higgs.

Regarding the top mass, since the Yukawa are generated via gauge coupling, it is hard to engineer a setup with a top heavier than the W mass. A possible way out is to embed the top in a large representation such that the effective Yukawa is enhanced by a group factor. For instance in the $SU(3)$ model of [42], the prediction $M_t = 2M_W$ has been reached at tree-level. The main drawback of this possibility is that the large representation will lower the scale where the extra dimensional theory becomes strongly coupled. Moreover, some rather large deviations in the coupling of the left bottom quark to the Z, $Zb_l\bar{b}_l$, will be introduced at tree level. Another possibility pursued in [43] is to explicitly give up Lorentz invariance along the extra dimension. In this case, each fermion will effectively feel an extra dimension of different length, alleviating the relation between the top and the W masses. The strong coupling scale is also lowered in that case and the Lorentz breaking reintroduce a UV sensitivity of the Higgs potential at higher loop (like in Little Higgs models). And again, corrections to $Zb_l\bar{b}_l$ and four fermion operators generated by the KK gauge bosons pose a bound on the scale of the fifth deimension of the order of few TeV.

Regarding the Higgs mass, it generically turns out to be too small, below the value currently excluded by LEP, because the quartic interaction is now generated at one-loop (contrary to the G_2 6D model where it was present at tree-level). Since the entire potential (mass and quartic) is loop generated, the potential will also generically prefer large values of the Higgs *vev* relative to the compactification scale so that the scale of new physics stays dangerously low. It was shown in [42] that the Higgs mass can be raised by the presence of several (twisted) bulk fermions. In the Lorentz violating model of [43], the Higgs mass is set by the scale of the top.

Another direction that has been explored [49, 50] is to embed the idea of gauge–Higgs unification in a warped extra dimension. The nice thing is that the warping enhances both the Higgs and the top mass. However, the non trivial background will also induce corrections to EW precision observables. Via the *AdS/CFT* correspondence (see Section 5.4.1 for a short introduction to it), these models will now be reinterpreted as weakly coupled duals of the old composite Higgs models of Georgi–Kaplan [51]. One highly valuable benefit of warped extra dimension is the ability to postpone the scale of new physics to very high energy with in particular the possibility to accommodate unification.

One final comment concerns the dynamics of the EW phase transition in these gauge–Higgs unification models. As in Little Higgs theories, the structure of the radiatively generated Higgs potential is more rich than just the ϕ^4 Mexican hat potential with the presence of a series of non-renormalizable interactions. It was shown that we could then obtained a moderately first order EW phase transition even for reasonably large values of the Higgs mass [52] (see also [53]). This revives the possibility of EW baryogenesis to generate the asymmetry between matter and anti-matter.

5. 5D Higgsless models

The idea behind Higgsless theories [23, 24] is that a momentum along an extra dimension is equivalent to a mass in 4D, so we can generate a mass by giving a momentum to a particule in an extra dimension. And like in quantum mechanics, a non-zero momentum along a compact direction can result from non-trivial boundary conditions (BC's). Therefore the work consists in identifying the appropriate boundary conditions and the geometry of the extra dimension to reproduce the spectrum and the couplings of the SM. Short presentations of Higgsless models can be found in [54–56]. A more comprehensive review, with significant overlap with the present lectures, is [57].

In the previous lectures, we have discussed 5D models where part of the gauge group is broken by orbifold compactification, *i.e.* by some particular boundary conditions[8]. This raises the hope to achieve a breaking of the EW symmetry directly by BC's. Before facing the details of an explicit construction, let us appreciate what the principle obstacles to the construction of a realistic model are [45]:

- **rank reduction:** in usual (abelian) orbifold compactifications, the rank of the gauge group cannot be reduced unless the orbifold projection corresponds to an outer automorphism of the gauge symmetry. For a given algebra, the number of automorphism is limited and in particular, it is not possible to break $SU(2) \times U(1)$ down to $U(1)$. So one has to consider more general BC's than the ones obtained from simple orbifold projection.

- **non rational mass ratio:** in usual Kaluza–Klein (KK) compactifications, the spectrum is dictated by the geometry of the extra dimensions and the mass gap between two KK states is given by an integer times the inverse size of the extra dimension. So it seems non-trivial to get a mass ratio of the W to the Z that is related to the gauge couplings.

---

[8]It might be useful to stress the difference between gauge–Higgs unification models and Higgsless models: in gauge-Higggs models, a bigger gauge group G is broken to $SU(2) \times U(1)$ by orbifold/boundary condition while the actual EW breaking is achieved via a Higgs mechanism, the Higgs being identified as some massless components of the bigger gauge group along the extra dimensions, the orbifold projection has indeed been carefully chosen such that a massless mode of A_5 remained massless. In Higgsless models, on the contrary, the boundary conditions are chosen such that no component of A_5 remained in the physical spectrum (the massless ones are thrown away by the boundary conditions while the massive ones are eaten up to give the longitudinal polarizations of the massive gauge bosons).

• **unitarity restoration:** we have shown in the first lecture that the Higgs boson was essential in restoring perturbative unitarity in the longitudinal massive gauge boson scattering. Thus the question that the 5D theories we want to consider raise is whether such a breaking of the gauge symmetries via BC's yields a consistent theory or not, or in other words, whether a momentum along a fifth dimension is UV safer than a regular 4D gauge boson mass. In order to verify that such a breaking is indeed soft, we need to investigate the issue of unitarity of scattering amplitudes in such 5D gauge theories compactified on an interval, with non-trivial BC's. We derive the general expression for the amplitude for elastic scattering of longitudinal gauge bosons, and wrote down the necessary conditions for the cancellation of the terms that grow with energy. We will find that all the consistent BC's are unitary in the sense that all terms proportional to E^4 and E^2 vanish. In fact, any theory with only Dirichlet or Neumann BC's is unitary. Surprisingly, this would also include theories where the boundary conditions can be thought of as coming from a very large expectation value of a brane localized Higgs field, in the limit when the expectation value diverges. In this limit, there is no scalar degrees of freedom at low energy, thus the name of Higgsless theories.

There are several aspects of Higgsless theories that I will not cover in these lectures. For instance the construction of 4D models from deconstruction, the construction of 6D models or some applications to GUT breaking. The reader is referred to the existing literature [58, 59]. Higgsless models only address the issue of EWSB. Ultimately, they should be embedded into a GUT and the $L - R$ structure seems to indicate that a $SO(10)$ or a Pati–Salam embedding is the good direction to go [60].

5.1. Gauge symmetry breaking by boundary conditions

5.1.1. Boundary conditions for a scalar field
We start with a bulk action for the scalar field

$$S_{bulk} = \int d^4x \int_0^{\pi R} dy \left(\frac{1}{2} \partial^M \phi \partial_M \phi - V(\phi) \right), \tag{5.1}$$

where we have assumed that the interval runs between 0 and πR. We will for simplicity first assume that there is no term added on the boundary of the interval. Let us apply the variational principle to this theory:

$$\delta S = \int d^4x \int_0^{\pi R} dy \left(\partial^M \phi \partial_M \delta \phi - \frac{\partial V}{\partial \phi} \delta \phi \right). \tag{5.2}$$

Separating out the ordinary 4D coordinates from the fifth coordinate (and integrating by parts in the ordinary 4D coordinates, where we apply the usual requirements that the fields vanish for large distances) we get

$$\delta S = \int d^4x \int_0^{\pi R} dy \left[-\partial_\mu \partial^\mu \phi \delta\phi - \frac{\partial V}{\partial \phi} \delta\phi - \partial_y \phi \partial_y \delta\phi \right]. \quad (5.3)$$

Since we have not yet decided what boundary conditions one wants to impose we will have to keep the boundary terms when integrating by parts in the fifth coordinate y:

$$\delta S = \int d^4x \int_0^{\pi R} dy \left[-\partial_M \partial^M \phi - \frac{\partial V}{\partial \phi} \right] \delta\phi - \int d^4x \left[\partial_y \phi \delta\phi \right]_0^{\pi R}. \quad (5.4)$$

To ensure that the variational principle is obeyed, we need $\delta S = 0$, but since this consists of a bulk and a boundary piece we require:
• The bulk equation of motion $\partial_M \partial^M \phi = -\partial V/\partial \phi$ as usual
• The boundary variation needs to also vanish. This implies that one needs to choose the BC such that

$$\partial_y \phi \, \delta\phi|_{bd} = 0. \quad (5.5)$$

    We will be calling a boundary condition *natural*, if it is obtained by letting the boundary variation of the field $\delta\phi|_{bd}$ to be arbitrary. In this case the natural BC would be $\partial_y \phi = 0$: a flat or Neumann BC. But at this stage this is not the only possibility: one could also satisfy (5.5) by imposing $\delta\phi|_{bd} = 0$ which would follow from the $\phi|_{bd} = 0$ Dirichlet BC. Thus we get two possible BC's for a scalar field on an interval with no boundary terms:
• Neumann BC $\partial_y \phi| = 0$
• Dirichlet BC $\phi| = 0$
However, we would only like to allow the natural boundary conditions in the theory since these are the ones that will not lead to explicit (hard) symmetry breaking once more complicated fields like gauge fields are allowed. Thus in order to still allow the Dirichlet BC one needs to reinterpret that as the natural BC for a theory with additional terms in the Lagrangian added on the boundary. The simplest possibility is to add a *mass term* to modify the Lagrangian as

$$S = S_{bulk} - \int d^4x \frac{1}{2} M_1^2 \phi^2|_{y=0} - \int d^4x \frac{1}{2} M_2^2 \phi^2|_{y=\pi R}. \quad (5.6)$$

These will give an additional contribution to the boundary variation of the action, which will now be given by:

$$\delta S_{bd} = -\int d^4x \, \delta\phi (\partial\phi + M_2^2 \phi)|_{y=\pi R} + \int d^4x \, \delta\phi (\partial_y \phi - M_1^2 \phi)|_{y=0}. \quad (5.7)$$

Thus the natural BC's will be given by

$$\partial_y \phi + M_2^2 \phi = 0 \text{ at } y = \pi R,$$
$$\partial_y \phi - M_1^2 \phi = 0 \text{ at } y = 0. \tag{5.8}$$

Clearly, for $M_i \rightarrow \infty$ we will recover the Dirichlet BC's in the limit. This is the way we will always understand the Dirichlet BC's: we will interpret them as the case with infinitely large boundary induced mass terms for the fields.

Let us now consider what happens when we add a *kinetic term* on the boundary for ϕ. For simplicity let us set the mass parameters on the branes to zero, and take as the action

$$S = S_{bulk} + \int d^4x \, \frac{1}{2M} \partial_\mu \phi \partial^\mu \phi|_{y=0}. \tag{5.9}$$

Note that the boundary term had to be added with a definite sign, that is we assume that the arbitrary mass parameter M is positive. This is in accordance with our expectations that kinetic terms have to have positive signs if one wants to avoid ghostlike states. For simplicity we have only added a kinetic term on one of the branes, but of course we could easily repeat the following analysis for the second brane as well. The boundary variation at $y = 0$ will be modified to

$$\delta S|_{y=0} = \int d^4x \, \delta\phi \left(\partial_y \phi - \frac{1}{M} \Box_4 \phi \right)_{|y=0}. \tag{5.10}$$

Thus the natural BC will be given by:

$$\partial_y \phi = \frac{1}{M} \Box_4 \phi. \tag{5.11}$$

Using the bulk equation of motion (in the presence of no bulk potential) $\Box_5 \phi = \Box_4 \phi - \phi'' = 0$ we could also write this BC as $M\phi' = \phi''$. The final form of the BC is obtained by using the Kaluza–Klein decomposition of the field ϕ where one usually assumes that the 4D modes ϕ_n have the x dependence $\phi_n(y)e^{ip_n \cdot x}$, where $p_n^2 = m_n^2$ is the nth KK mass eigenvalue. Using this form the BC will be given by:

$$\partial_y \phi = \frac{1}{M} \Box_4 \phi = -\frac{p_n^2}{M} \phi = -\frac{m_n^2}{M} \phi. \tag{5.12}$$

In either form this BC is quite peculiar: it depends on the actual mass eigenvalue in the final form, or involves second derivatives in the first form. This could be dangerous, since from the theory of differential equations we know that usually BC's that only involve first derivatives are the ones that will automatically lead to a hermitian differential operator on an interval. The usual reason is that the second derivative operator d^2/dy^2 is hermitian if the scalar product

$$\langle f, g \rangle = \int_0^{\pi R} dy \, f^*(y) g(y) \tag{5.13}$$

obeys the relation

$$\left\langle f, \frac{d^2}{dy^2} g \right\rangle = \left\langle \frac{d^2}{dy^2} f, g \right\rangle, \tag{5.14}$$

which is indeed satisfied if the functions f and g have boundary conditions of the form

$$f'_{|0,\pi R} = \alpha f_{|0,\pi R}. \tag{5.15}$$

From the hermiticity of the scalar product follows the usual properties of reality of the eigenvalues and completeness of the eigenfunctions. For the case of boundary conditions of the type (5.12), the scalar product has to be supplemented with a boundary term to remain hermitian:

$$\langle f, g \rangle = \int_0^{\pi R} dy \, f(y) g(y) + \frac{1}{M} f g_{|0}. \tag{5.16}$$

In particular, the completeness property of the eigenfunctions now reads:

$$\sum_n g_n(x) g_n(y) = \delta(x - y) - \frac{1}{M} \delta(x) \sum_n g_n(0) g_n(y). \tag{5.17}$$

5.1.2. Boundary conditions for a gauge field

The same exercise can now be repeated [45] for spin-1 particle[9]. One just has to be a bit more careful and fix a gauge to deal with gauge degrees of freedom in an appropriate way. A gauge field in 5D A_M contains a 4D gauge field A_μ and a 4D scalar A_5. The 4D vector will contain a whole KK tower of massive gauge bosons, however as we will see below the KK tower of the A_5 will be eaten by the massive gauge fields and (except for a possible zero mode) will be non-physical. That this is what happens can be guessed from the fact that the Lagrangian contains a mixing term between the gauge fields and the scalar, reminiscent of the usual 4D Higgs mechanism. The Lagrangian is given by the usual form

$$S = \int d^5x \; -\frac{1}{4} F^a_{MN} F^{MN\,a} = \int d^5x \; -\frac{1}{4} F^a_{\mu\nu} F^{\mu\nu\,a} - \frac{1}{2} F^a_{\mu5} F^{\mu5\,a}, \tag{5.18}$$

---

[9]The reader interested in boundary conditions for a spin-2 field can have a look at [61]. In Section 5.5.1, we will explain how to obtain boundary conditions for a spin-1/2.

where the field strength is given by the usual expression $F^a_{MN} = \partial_M A^a_N - \partial_N A^a_M + g_5 f^{abc} A^b_M A^c_N$ and g_5 is the 5D gauge coupling, which has mass dimension $-1/2$. The theory is non-renormalizable, so it has to be considered as a low-energy effective theory valid below a cutoff scale, that we will be calculating later on.

To determine the gauge fixing term, let us consider the mixing term between the 4D scalar and the 4D gauge fields:

$$\int d^5x \; -\frac{1}{2} F^a_{\mu 5} F^{\mu 5 a} = \int d^5x \; \partial_5 A^a_\mu \, \partial^\mu A^{5a} + \dots. \tag{5.19}$$

Integrated by parts, the mixing term becomes

$$[\partial_\mu A^{\mu a} A^a_5]_0^{\pi R} - \int_0^{\pi R} dy \, \partial^\mu A^a_\mu \, \partial_5 A^a_5. \tag{5.20}$$

The *bulk* mixing can be cancelled by adding a gauge fixing term of the form

$$S^{bulk}_{GF} = \int d^5x \; -\frac{1}{2\xi} (\partial_\mu A^{\mu a} - \xi \partial_5 A^a_5)^2. \tag{5.21}$$

This term is chosen such that the A_5 independent piece agrees with the usual Lorentz gauge fixing term, and such that the cross term exactly cancels the mixing term from (5.20). Thus within R_ξ gauge, which is what we have defined, the propagator for the 4D gauge fields will be the usual ones. Varying the full action we then obtain the bulk equations of motion and the possible BC's. After integrating by parts we find that $\delta S_{bulk} + \delta S^{bulk}_{GF}$ is given at the quadratic level by:

$$\int d^5x \left[\left(\partial_\mu(\partial^\mu A^{\nu a} - \partial^\nu A^{\mu a}) - \partial_5^2 A^{\nu a} + \frac{1}{\xi}\partial^\nu \partial_\sigma A^{\sigma a} \right) \delta A^a_\nu \right.$$
$$\left. - \left(\partial_\sigma \partial^\sigma A^a_5 - \xi \partial_5^2 A^a_5 \right) \delta A^a_5 \right]. \tag{5.22}$$

The bulk equations of motion will be that the coefficients of δA^a_ν and δA^a_5 in the above equation vanish. We can see, that the A^a_5 field has a term $\xi \partial_5^2 A^a_5$ in its equation. This will imply that if the wave function is not flat (e.g. the KK mode is not massless), then the field is not physical (since in the unitary gauge $\xi \to \infty$ this field will have an infinite effective 4D mass and decouples). This shows that as mentioned above, the scalar KK tower of A^a_5 will be completely unphysical due to the 5D Higgs mechanism, except perhaps for a zero mode for A^a_5. Whether or not there is a zero mode depends on the BC for the A_5 field. In Higgsless models, there won't be any A_5 zero mode.

In order to eliminate the *boundary* mixing term in (5.20), we also need to add a boundary gauge fixing term with an a priori unrelated boundary gauge fixing coefficient ξ_{bd}:

$$S^{bd}_{GF} - \frac{1}{2\xi_{bd}} \int d^4x (\partial_\mu A^{\mu a} \pm \xi_{bd} A^a_5)^2_{|0,\pi R},$$
(5.23)

where the $-$ sign is for $y = 0$ and the $+$ for $y = \pi R$. The boundary variations are then given by:

$$\left(\pm \partial_5 A^{\mu a} + \frac{1}{\xi_{bd}} \partial^\mu \partial_\nu A^{\nu a} \right) \delta A^a_{\mu|0,\pi R} - \left(\pm \xi \partial_5 A^a_5 + \xi_{bd} A^a_5 \right) \delta A^a_{5|0,\pi R}.$$

The natural boundary conditions in an arbitrary gauge ξ, ξ_{bd} are given by

$$\partial_5 A^{\mu a} \pm \frac{1}{\xi_{bd}} \partial_\mu \partial^\mu A^{\mu a} = 0, \quad \xi \partial_5 A^a_5 \pm \xi_{bd} A^a_5 = 0.$$
(5.24)

This simplifies quite a bit if we go to the unitary gauge on the boundary given by $\xi_{bd} \to \infty$. In this case we are left with the simple set of boundary conditions

$$\partial_5 A^{\mu a} = 0, \quad A^a_5 = 0.$$
(5.25)

This is the boundary condition that one usually imposed for gauge fields in the absence of any boundary terms. Note, that again we could have chosen some non-natural boundary conditions, where instead of requiring that the boundary variation be arbitrary we would require the boundary variation itself (and thus some of the fields on the boundary) to be vanishing. It turns out that these boundary conditions would lead to a hard (explicit) breaking of gauge invariance, and thus we will not consider them in the following discussion any further. We will see below how these simple BC's will be modified if one adds scalar fields on the branes.

5.1.3. Higgs mechanism localized on a boundary: scalar decoupling limit

Let us now consider the case when scalar fields that develop vacuum expectation value's (vev's) are added on the boundary [23, 45, 62, 63]. Instead of repeating a full and general analysis (that can be found in [63]), we will present a concrete example. We consider (see Fig. 10) a $SU(2)$ gauge group with Newmann BC's for the A_μ components at both ends of the interval. We then assume that at $y = \pi R$, $SU(2)$ is fully broken by the *vev* of a Higgs doublet. As for the scalar case, the boundary mass generated by the Higgs *vev* induces a mixed BC of the form

$$\partial_5 A^a_\mu (\pi R) = -\tfrac{1}{4} g^2_{5D} v^2 A^a_\mu (\pi R).$$
(5.26)

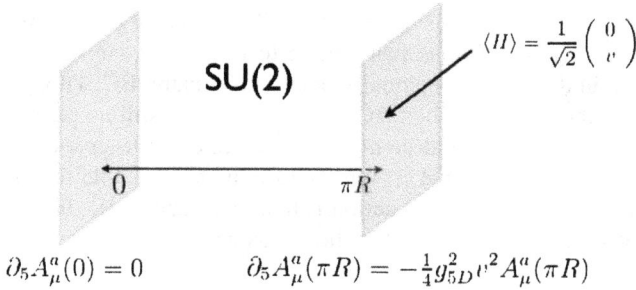

$$\partial_5 A_\mu^a(0) = 0 \qquad \partial_5 A_\mu^a(\pi R) = -\tfrac{1}{4}g_{5D}^2 v^2 A_\mu^a(\pi R)$$

Fig. 10. Example of a Higgs mechanism localized on a boundary. For a finite Higgs *vev*, we obtain a mixed BC which, in the infinite *vev* limit, simply becomes a Dirichlet BC: all the gauge bosons that couple to the Higgs have a wavefunction that vanishes at the point where the Higgs is localized. In that limit, there is no scalar degree of freedom in the low energy effective action and the gauge symmetry is entirely broken by the BC's, the mass of the lightest KK state is simply inversely proportional to the size of the extra dimension.

The canonically normalized ($\int_0^{\pi R} f_k^2(y) = 1$) KK modes are given by

$$A_\mu^a(x, y) = \sum_{k=1}^{\infty} f_k(y) A_\mu^{(k)}(x), \qquad (5.27)$$

with

$$f_k(y) = \frac{\sqrt{2}}{\sqrt{\pi R(1 + 16M_k^2/(g_{5D}^4 v^4)) + 4/(g_{5D}^2 v^2)}} \frac{\cos(M_k y)}{\sin(M_k \pi R)}. \qquad (5.28)$$

The BC at the origin, $y = 0$, is trivially satisfied while the condition at $y = \pi R$ determines the mass spectrum through the equation:

$$M_k \tan(M_k \pi R) = \frac{1}{4}g_{5D}^2 v^2. \qquad (5.29)$$

In the large *vev* limit, we obtain that the wavefunctions at the $y = \pi R$ boundary vanish like $1/v^2$

$$f_k(\pi R) \sim 2\sqrt{\frac{2}{\pi R}\frac{2k+1}{g_{5D}^2 R v^2}}, \qquad (5.30)$$

while

$$M_k \sim \frac{2k+1}{2R}\left(1 - \frac{4}{g_{5D}^2 \pi R v^2}\right). \qquad (5.31)$$

That limit exactly corresponds to a Dirichlet BC: in the large *vev* limit, the wave-functions of the gauge bosons that couple to the Higgs vanish. It can also be checked that, in that limit, A_5 actually obeys a Neumann BC. Though, in our example, due to the other Dirichlet BC at $y = 0$, there is still no physical massless mode for A_5, while the would be massive one are eaten to give the longitudinal polarizations of the massive A_μ. What allows us to decouple the Higgs degree of freedom from the low energy action is that, contrary to 4D, the masses of the gauge bosons are not proportional to the Higgs *vev*.

5.2. Unitarity restoration by KK modes. Sum Rules of Higgsless theories

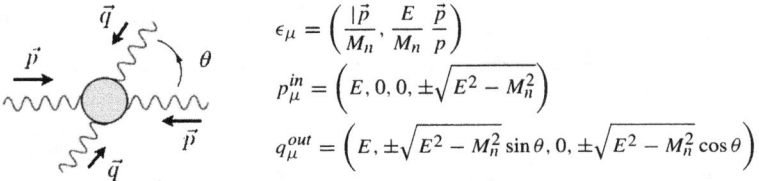

$$\epsilon_\mu = \left(\frac{|\vec{p}|}{M_n}, \frac{E}{M_n} \frac{\vec{p}}{p} \right)$$

$$p_\mu^{in} = \left(E, 0, 0, \pm \sqrt{E^2 - M_n^2} \right)$$

$$q_\mu^{out} = \left(E, \pm \sqrt{E^2 - M_n^2} \sin\theta, 0, \pm \sqrt{E^2 - M_n^2} \cos\theta \right)$$

Fig. 11. Elastic scattering of longitudinal modes of KK gauge bosons, $n + n \rightarrow n + n$, with the gauge index structure $a + b \rightarrow c + d$. The E-dependence can be estimated from $\epsilon \sim E$, $p_\mu \sim E$ and a propagator $\sim E^{-2}$.

Our aim is to build a Higgsless model of electroweak symmetry breaking using BC breaking in extra dimensions. However, there is a problem in theories with massive gauge bosons without a Higgs scalar: the scattering amplitude of longitudinal gauge bosons will grow with the energy and violate unitarity at a low scale [10–12] (see Section 2.1.4). What we would like to first understand is what happens to this unitarity bound [10] in a theory with extra dimensions [23, 65–67]. For simplicity we will be focusing on the elastic scattering of the longitudinal modes of the nth KK mode (see Fig. 11). The E-dependence can be estimated from $\epsilon \sim E$, $p_\mu \sim E$ and a propagator $\sim E^{-2}$. This way we find that the amplitude could grow as quickly as E^4, and then for $E \gg M_W$ can expand the amplitude in decreasing powers of E as

$$\mathcal{A} = \mathcal{A}^{(4)} \frac{E^4}{M_n^4} + \mathcal{A}^{(2)} \frac{E^2}{M_n^2} + \mathcal{A}^{(0)} + \mathcal{O}\left(\frac{M_n^2}{E^2} \right). \tag{5.32}$$

In the SM (and any theory where the gauge kinetic terms form the gauge invariant combination $F_{\mu\nu}^2$) the $\mathcal{A}^{(4)}$ term automatically vanishes, while $\mathcal{A}^{(2)}$ is only cancelled after taking the Higgs exchange diagrams into account.

---

[10]There are also some unitarity issues associated with the masses of the fermions. See [64].

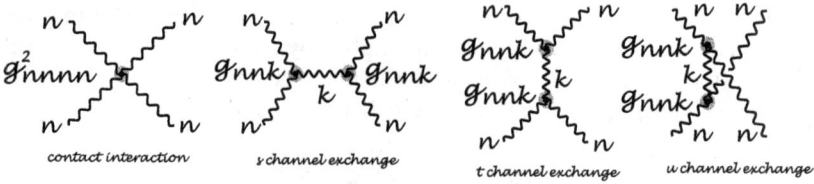

Fig. 12. The four diagrams contributing at tree level to the elastic scattering amplitude of the nth KK mode.

In the case of a theory with an extra dimension with BC breaking of the gauge symmetry there are no Higgs exchange diagrams, however one needs to sum up the exchanges of all KK modes, as in Fig. 12. As a result we will find the following expression for the terms in the amplitudes that grow with energy:

$$\mathcal{A}^{(4)} = i \left(g_{nnnn}^2 - \sum_k g_{nnk}^2 \right) a^{(4)}(\theta), \tag{5.33}$$

with

$$a^{(4)}(\theta) = \left(f^{abe} f^{cde} (3 + 6\cos\theta - \cos^2\theta) + 2(3 - \cos^2\theta) f^{ace} f^{bde} \right), \tag{5.34}$$

In order for the term $\mathcal{A}^{(4)}$ to vanish it is enough to ensure that the following sum rule among the coupling of the various KK modes is satisfied [23]:

$$E^4 \text{ sum rule:} \quad g_{nnnn}^2 = \sum_k g_{nnk}^2. \tag{5.35}$$

Assuming $\mathcal{A}^{(4)} = 0$ we get

$$\mathcal{A}^{(2)} = \frac{i}{M_n^2} \left(4g_{nnnn} M_n^2 - 3 \sum_k g_{nnk}^2 M_k^2 \right) a^{(2)}(\theta), \tag{5.36}$$

with

$$a^{(2)}(\theta) = \left(f^{ace} f^{bde} - \sin^2 \tfrac{\theta}{2} f^{abe} f^{cde} \right). \tag{5.37}$$

Here g_{nnnn}^2 is the quartic self-coupling of the nth massive gauge field, while g_{nnk} is the cubic coupling between the KK modes. In theories with extra dimensions these are of course related to the extra dimensional wave functions $f_n(y)$ of the various modes as

$$g_{mnk} = g_5 \int dy f_m(y) f_n(y) f_k(y), \tag{5.38}$$

$$g_{mnkl}^2 = g_5^2 \int dy f_m(y) f_n(y) f_k(y) f_l(y). \tag{5.39}$$

The most important point about the amplitudes in (5.33)–(5.36) is that they only depend on an overall kinematic factor multiplied by an overall expression of the couplings (the dynamics factorizes from the kinematics). Assuming that the relation (5.35) holds, we can find a sum rule that ensures the vanishing of the $\mathcal{A}^{(2)}$ term [23]:

$$E^2 \text{ sum rule:} \quad g_{nnnn} M_n^2 = \frac{3}{4} \sum_k g_{nnk}^2 M_k^2.\tag{5.40}$$

Amazingly, higher dimensional gauge invariance will ensure that both of these sum rules are satisfied as long as the breaking of the gauge symmetry is spontaneous. For example, it is easy to show the first sum rule via the completeness of the wave functions $f_n(y)$:

$$\int_0^{\pi R} dy\, f_n^4(y) = \sum_k \int_0^{\pi R} dy \int_0^{\pi R} dz\, f_n^2(y) f_n^2(z) f_k(y) f_k(z),\tag{5.41}$$

and using the completeness relation

$$\sum_k f_k(y) f_k(z) = \delta(y - z),\tag{5.42}$$

we can see that the two sides will indeed agree. One can similarly show [23] that the second sum rule will also be satisfied if the boundary conditions are natural ones and all terms in the Lagrangian (including boundary terms) are gauge invariant. Let us insist on the particular case of a Higgs mechanism localized at the boundary: for finite Higgs *vev*, the cancelation of the E^2 term requires the exchange of the brane Higgs scalar degree of freedom, however, in the infinite *vev* limit, the contribution of the Higgs exchange to the scattering amplitude actually cancels out and we are left with simple Dirichlet BC's for which the scattering amplitude is unitarized by the sole exchange of spin-1 KK excitations.

At this point, it should be noted that the two sum rules cannot be satisfied with a finite number of KK modes. This is full agreement with the old theorem by Cornwall et al. [11] who established that the only way to restore perturbative unitarity in the scattering of massive spin-1 particles is through the exchange of a scalar Higgs boson. Our 5D theory is non-renormalizable anyway, so it is valid up to a finite cutoff. What our result really shows is that, through the exchange of the KK gauge bosons, the perturbative unitarity breakdown is postponed from an energy scale of the order of the mass of the lightest KK state to the true 5D cutoff of the order of the mass of the heaviest KK state, see Fig. 13.

What we see from the above analysis is that in any gauge invariant extra dimensional theory the terms in the amplitude that grow with the energy will cancel. However, this will not automatically mean that the theory itself is unitary.

New Physics

(Higgs/strongly coupled theory?)   not directly set by the weak scale

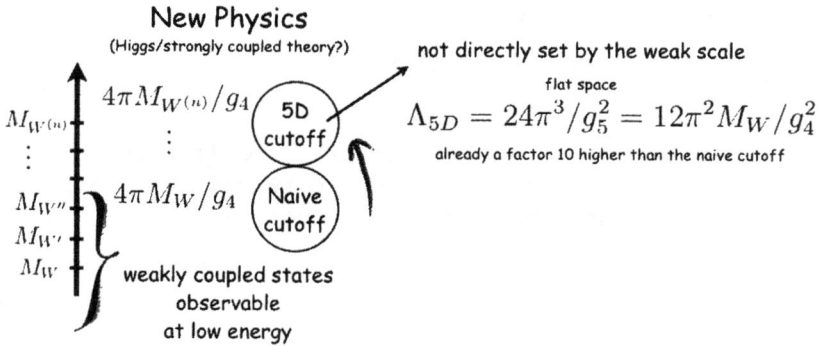

Fig. 13. The scattering amplitude of the longitudinal components of the lightest massive KK gauge boson would naively become non-perturbative at an energy scale $4\pi M_W/g_4$ However, before reaching that scale, the exchange of the KK excitations starts canceling the scattering amplitude. The story repeats itself until reaching the heaviest KK mode below the 5D cutoff for which no heavier excitations can intervene to smoothen its scattering amplitude. Thus the perturbative unitarity breakdown has be delayed and pushed to a scale which is not directly related to the mass of the lightest massive gauge bosons. A detailed analysis [66] of inelastic channels confirms the loss of perturbative unitarity at an energy scale related to the 5D cutoff.

The reason is that there are two additional worries: even if $\mathcal{A}^{(4)}$ and $\mathcal{A}^{(2)}$ vanish $\mathcal{A}^{(0)}$ could be too large and spoil unitarity. This is what happens in the SM if the Higgs mass is too large. In the extra dimensional case what this would mean is that the extra KK modes would make the scattering amplitude flatten out to a constant value. However if the KK modes themselves are too heavy then this flattening out will happen too late when the amplitude already violates unitarity. The other issue is that in a theory with extra dimensions there are infinitely many KK modes and thus as the scattering energy grows one should not only worry about the elastic channel, but the ever growing number of possible inelastic final states. The full analysis taking into account both effects has been performed in [66], where it was shown that after taking into account the opening up of the inelastic channels the scattering amplitude will grow linearly with energy, and will always violate unitarity at some energy scale. This is a consequence of the intrinsic non-renormalizability of the higher dimensional gauge theory. It was found in [66] that the unitarity violation scale due to the linear growth of the scattering amplitude is equal (up to a small numerical factor of order $2-4$) to the cutoff scale of the 5D theory obtained from naive dimensional analysis (NDA). This cutoff scale can be estimated in the following way. The one-loop amplitude in 5D is proportional to the 5D loop factor

$$\frac{g_5^2}{24\pi^3}. \tag{5.43}$$

The dimensionless quantity obtained from this loop factor is

$$\frac{g_5^2 E}{24\pi^3},$$
(5.44)

where E is the scattering energy. The cutoff scale can be obtained by calculating the energy scale at which this loop factor will become order one (that is the scale at which the loop and tree-level contributions become comparable). From this we get

$$\Lambda_{NDA} = \frac{24\pi^3}{g_5^2}.$$
(5.45)

We can express this scale using the matching of the higher dimensional and the lower dimensional gauge couplings. In the simplest theories this is usually given by

$$g_5^2 = \pi R g_4^2,$$
(5.46)

where πR is the length of the interval, and g_4 is the effective 4D gauge coupling. So the final expression of the cutoff scale can be given as

$$\Lambda_{NDA} = \frac{24\pi^2}{g_4^2 R}.$$
(5.47)

We will see that in the Higgsless models $1/R$ will be replaced by M_W^2/M_{KK}, where M_W is the physical W mass, and M_{KK} is the mass of the first KK mode beyond the W. Thus the cutoff scale will indeed be lower if the mass of the KK mode used for unitarization is higher. However, this Λ_{NDA} could be significantly higher than the cutoff scale in the SM without a Higgs, which is around 1.2 TeV. We will come back to a more detailed discussion of Λ_{NDA} in Higgsless models at the end of this section.

5.3. Toys models

Now that we have convinced ourselves that one can use KK gauge bosons to delay the unitarity violation scale basically up to the cutoff scale of the higher dimensional gauge theory, we should start looking for a model that actually has these properties and resembles the SM. It should have a massless photon, a massive charged gauge boson to be identified with the W and a somewhat heavier neutral gauge boson to be identified with the Z. Most importantly, we need to have the correct SM mass ratio (at tree-level)

$$\frac{M_W^2}{M_Z^2} = \cos^2\theta_W = \frac{g^2}{g^2 + g'^2},$$
(5.48)

where g is the $SU(2)_L$ gauge coupling and g' the $U(1)_Y$ gauge coupling of the SM. We would like to use BC's to achieve this. As stressed in the introduction, this seems to be very hard at first sight, since we need to somehow get a theory where the masses of the KK modes are related to the gauge couplings. Usually the KK masses are simply integer or half-integer multiples of $1/R$.

5.3.1. En route vers a Higgsless model

For example, if we look at a very naive toy model with an $SU(2)$ gauge group in the bulk (see Fig. 14), we could consider the following BC's for the various gauge directions:

$$\partial_y A_\mu^3 = 0 \text{ at } y = 0, \pi R, \tag{5.49}$$

$$\partial_y A_\mu^{1,2} = 0 \text{ at } y = 0, \qquad A_\mu^{1,2} = 0 \text{ at } y = \pi R. \tag{5.50}$$

Solving the bulk equations of motion and enforcing the BC's, one obtains the following KK decomposition

$$A_\mu^1(x, y) = \sum_{k=0}^{\infty} \frac{1}{\sqrt{\pi R}} \sin \frac{(2k+1)y}{2R} \left(W_\mu^{+\,(k)}(x) + W_\mu^{-\,(k)}(x) \right), \tag{5.51}$$

$$A_\mu^2(x, y) = \sum_{k=0}^{\infty} \frac{1}{\sqrt{\pi R}} \sin \frac{(2k+1)y}{2R} \left(W_\mu^{+\,(k)}(x) - W_\mu^{-\,(k)}(x) \right), \tag{5.52}$$

$$A_\mu^3(x, y) = \sum_{k=0}^{\infty} \sqrt{\frac{2}{2^{\delta_{k0}}\pi R}} \cos \frac{ky}{R} \gamma_\mu^{(k)}(x). \tag{5.53}$$

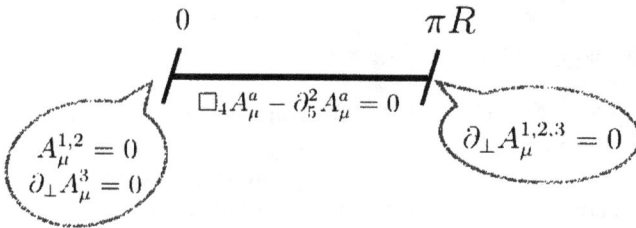

Fig. 14. Example of breaking of $SU(2)$ down to $U(1)$ achieved by boundary conditions. The spectrum consists of a massless gauge boson and all its KK excitations, a pair of electrically charged massive gauge bosons and all their KK excitations. With a lot of imagination one could see something starting to resemble the SM with a massless photon, a pair of massive $W^\pm$ and the first KK excitation of the photon that can be seen as a Z. This model is still quite far from reality (bad W/Z mass ration, too light resonances) but it illustrates the basic idea that the masses of the W and the Z are generated by the boundary conditions and not through a usual Higgs mechanism.

This spectrum somewhat resembles that of the SM in the sense that there is a massless gauge boson that can be identified with the γ, a pair of charged massive gauge bosons that can be identified with the $W^\pm$, and a massive neutral gauge boson that can be identified with the Z. However, we can see that the mass ratio of the W and Z is given by

$$\frac{M_Z}{M_W} = 2, \tag{5.54}$$

and another problem is that the first KK modes of the W, Z are given by

$$\frac{M_{Z'}}{M_Z} = 2, \quad \frac{M_{W'}}{M_W} = 3. \tag{5.55}$$

Thus, besides getting the totally wrong W/Z mass ratio there would also be additional KK states at masses of order 250 GeV, which is phenomenologically unacceptable. We will see that both of these problems can be resolved by going to a warped Higgsless model with custodial $SU(2)$.

5.3.2. Flat Higgsless model

From the above discussion it is clear that in order to find a Higgsless model with the correct W/Z mass ratio one needs to find an extra dimensional model that has the custodial $SU(2)$ symmetry incorporated [68]. Once such a construction is found, the gauge boson mass ratio will automatically be the right one. Therefore we need to somehow involve $SU(2)_R$ in the construction. The simplest possibility is to put an entire $SU(2)_L \times SU(2)_R \times U(1)_{B-L}$ gauge group in the bulk of an extra dimension [23]. In order to mimic the symmetry breaking pattern in the SM most closely, we assume that on one of the branes the symmetry breaking is $SU(2)_L \times SU(2)_R \to SU(2)_D$, with $U(1)_{B-L}$ unbroken. On the other boundary one needs to reduce the bulk gauge symmetry to that of the SM, and thus have a symmetry breaking pattern $SU(2)_R \times U(1)_{B-L} \to U(1)_Y$, which is illustrated in Fig. 15.

We denote by A_M^{Ra}, A_M^{La} and B_M the gauge bosons of $SU(2)_R$, $SU(2)_L$ and $U(1)_{B-L}$ respectively; g_{5L} and g_{5R} are the gauge couplings of the two $SU(2)$'s and $\tilde{g}_5$, the gauge coupling of the $U(1)_{B-L}$. In order to obtain the desired BC's as discussed above we need to follow the procedure laid out in section 5.1. We assume that there is a boundary Higgs on the left brane in the representation $(1, 2)_{\frac{1}{2}}$ under $SU(2)_L \times SU(2)_R \times U(1)_{B-L}$, which will break $SU(2)_R \times U(1)_{B-L}$ to $U(1)_Y$. We could also use the more conventional triplet representation under $SU(2)_R$ which will allow us to get neutrino masses later on. On the right brane we assume that there is a bi-doublet Higgs in the representation $(2, 2)_0$ which breaks the electroweak symmetry as in the SM: $SU(2)_L \times SU(2)_R \to SU(2)_D$. We will then take all the Higgs vev's to infinity in order to decouple the boundary

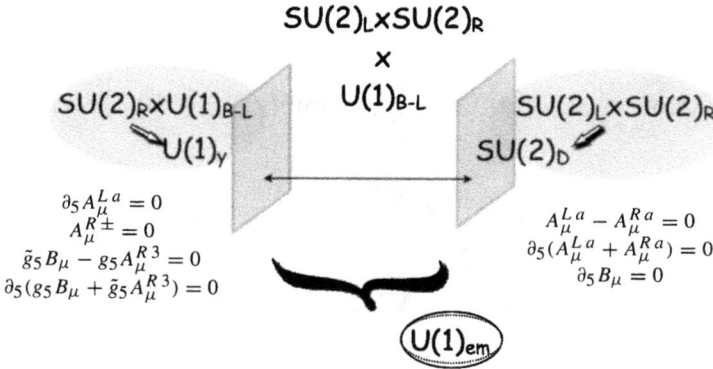

Fig. 15. The symmetry breaking structure of the flat space Higgsless toy model [23].

scalars from the theory, and impose the natural boundary conditions as described earlier. The BC's we will arrive at then are:

at $y = 0$: (5.56)

$$\partial_z (g_{5R} B_\mu + \tilde{g}_5 A_\mu^{R3}) = 0, \ \partial_z A_\mu^{La} = 0, \ A_\mu^{R1,2} = 0,$$

$$\tilde{g}_5 B_\mu - g_{5R} A_\mu^{R3} = 0,$$

at $y = \pi R$: (5.57)

$$\partial_z (g_{5R} A_\mu^{La} + g_{5L} A_\mu^{Ra}) = 0, \ \partial_z B_\mu = 0, \ g_{5L} A_\mu^{La} - g_{5R} A_\mu^{Ra} = 0.$$

The BC's for the A_5 and B_5 components will be the opposite of the 4D gauge fields as usual, *i.e.* all Dirichlet conditions should be replaced by Neumann and *vice versa*. The next step to determine the mass spectrum is to find the right KK decomposition of this model. First of all, none of the A_5 and B_5 components have a flat BC on both ends. This means that there will be no zero mode in these fields, and as we have seen all the massive scalars are unphysical, since they are just gauge artifacts (supplying the longitudinal components of the massive KK towers). The main point to observe about the KK decomposition of the gauge fields is that the BC's will mix up the states in the various components. This will imply that a single 4D mode will live in several different 5D fields. Since in the bulk there is no mixing, and we are discussing at the moment a flat 5D background, the wave functions will be of the form $f_k(y) \propto a \cos M_k y + b \sin M_k y$. If we make the simplifying assumption that $g_{5L} = g_{5R} = g_5$, then the KK decomposition will be somewhat simpler than the most generic one, and given by (we denote by $A_\mu^{L,R\pm}$ the linear combinations $(A^{L,R1} \mp iA^{L,R2})/\sqrt{2}$):

$$B_\mu(x,y) = g_5\, a_0\gamma_\mu(x) + \tilde{g}_5 \sum_{k=1}^{\infty} b_k \cos(M_k^Z(y-\pi R))\, Z_\mu^{(k)}(x), \qquad (5.58)$$

$$A_\mu^{L3}(x,y) = \tilde{g}_5\, a_0\gamma_\mu(x) - g_5 \sum_{k=1}^{\infty} b_k \frac{\cos(M_k^Z y)}{2\cos(M_k^Z \pi R)}\, Z_\mu^{(k)}(x), \qquad (5.59)$$

$$A_\mu^{R3}(x,y) = \tilde{g}_5 a_0\gamma_\mu(x) - g_5 \sum_{k=1}^{\infty} b_k \frac{\cos(M_k^Z(y-2\pi R))}{2\cos(M_k^Z \pi R)}\, Z_\mu^{(k)}(x) \qquad (5.60)$$

$$A_\mu^{L\pm}(x,y) = \sum_{k=1}^{\infty} c_k \cos(M_k^W y)\, W_\mu^{(k)\pm}(x), \qquad (5.61)$$

$$A_\mu^{R\pm}(x,y) = \sum_{k=1}^{\infty} c_k \sin(M_k^W y)\, W_\mu^{(k)\pm}(x). \qquad (5.62)$$

The BC's further impose the mass spectrum that is made up of a massless photon, the gauge boson associated with the unbroken $U(1)_Q$ symmetry, and some towers of massive charged and neutral gauge bosons, $W^{(k)}$ and $Z^{(k)}$ respectively. The $W^\pm$ masses are solutions of the quantization equation:

$$\cos(2M_W \pi R) = 0, \qquad (5.63)$$

thus

$$M_k^W = \frac{2k-1}{4R}, \quad k=1,2\ldots. \qquad (5.64)$$

The quantization equation giving the masses of the neutral gauge boson is a little bit more complicated due to the mixing of the various $U(1)$ factors of the bulk gauge group:

$$\tan^2(M_Z \pi R) = 1 + \frac{\tilde{g}_5^2}{g_5^2}. \qquad (5.65)$$

The KK Z's masses are thus given by

$$M_k^Z = \left(M_0 + \frac{k-1}{R}\right)_{k=1,2\ldots} \quad \text{and} \quad M_k^{Z'} = \left(-M_0 + \frac{k}{R}\right)_{k=1,2\ldots} \qquad (5.66)$$

where $M_0 = \frac{1}{\pi R}\arctan\sqrt{1+2\tilde{g}_5^2/g_5^2}$. Note that $1/(4R) < M_0 < 1/(2R)$ and thus the Z''s are heavier than the Z's ($M_k^{Z'} > M_k^Z$). We also get that the lightest Z is heavier than the lightest W ($M_1^Z > M_1^W$), in agreement with the SM spectrum. However, the mass ratio of W/Z is given by

$$\frac{M_W^2}{M_Z^2} = \frac{\pi^2}{16} \arctan^{-2} \sqrt{1 + \frac{\tilde{g}_{4D}^{'2}}{g_{4D}^2}} \sim 0.85, \tag{5.67}$$

and hence the ρ parameter is

$$\rho = \frac{M_W^2}{M_Z^2 \cos^2 \theta_W} \sim 1.10. \tag{5.68}$$

To arrive at these expression, we have assumed that the SM quarks and leptons are localized on the $SU(2)_L \times U(1)_Y$ boundary, leading to the following relations between the 4D and 5D gauge couplings:

$$g_4 = \frac{g_5}{\sqrt{\pi R}} \quad g_4' = \frac{\sqrt{2}\tilde{g}_5}{\sqrt{\pi R}}. \tag{5.69}$$

The W/Z mass ratio is close to the SM value, however the ten percent deviation is still huge compared to the experimental precision. The reason for this deviation is that while the bulk and the right $SU(2)_D \times U(1)_{B-L}$ brane are symmetric under custodial $SU(2)$, the left $SU(2)_L \times U(1)_Y$ brane is not, and the KK wave functions do have a significant component around the left brane, which will give rise to the large deviation from $\rho = 1$. Thus one needs to find a way of making sure that the KK modes of the gauge fields do not very much feel that left brane, but are repelled from there, and only the lightest (almost zero modes) γ, Z, $W^\pm$ will have a large overlap with the left brane.

In summary, the flat Higgsless model suffers from two serious drawbacks: *(i)* there are too light KK excitations of the W and Z, *(ii)* the deviation of the ρ parameter from its custodial value is too important. We are now going to see that embedding the model in a warped space actually cure these two problems at once.

5.4. Warped Higgsless model with custodial symmetry

5.4.1. A bit of AdS/CFT

In order to ensure an unbroken custodial symmetry, we need to devise a set-up such that the KK modes are localized away from the point where the custodial symmetry is broken. One can think to add large kinetic terms localized on the $SU(2)_L \times U(1)_Y$ boundary, which will indeed repeal the wavefunctions of the massive KK gauge bosons [69]. Actually, there is an even simpler way to lo-

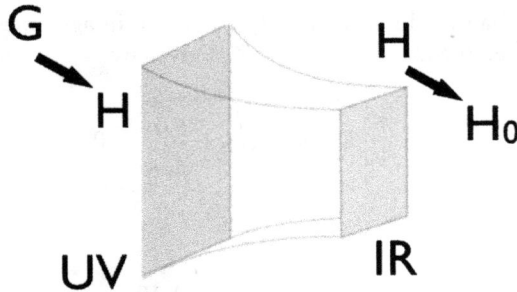

Fig. 16. Correspondence between a 5D gauge theory in an Anti-de-Sitter (*AdS*) space and a 4D conformal field theory (*CFT*). The UV brane is interpreted as a UV cutoff of the *CFT* while the IR brane mimics the spontaneous breaking of the conformal symmetry by the *CFT*. The subgroup H unbroken on the UV brane corresponds to a gauge symmetry of the *CFT* while the coset G/H is a global symmetry of the *CFT*. The strong dynamics of the *CFT* spontaneously breaks H to H_0 at the IR scale. 5D fields localized close to the IR boundary correspond to composites states of the *CFT*, while UV matter are elementary fields coupled to the *CFT*.

calize the wavefunctions: to warp the space. Indeed in an Anti-de-Sitter (*AdS*) space, the KK wavefunctions are Bessel functions of order one that are generically exponentially peaked at one end of the interval. In views of the *AdS/CFT* correspondence [70, 71], this localization property allows to infer the global and local symmetries left out by particular BC's. As illustrated in Fig. 16, let us consider a gauge symmetry G in the bulk of AdS_5. The BC's on the UV brane breaks G to a subgroup H which is further broken to H_0 by the BC's on the IR brane. The corresponding 4D *CFT* will posses a H gauge invariance and a G/H global symmetry. The interpretation of the IR BC's is that H is actually spontaneously broken down to H_0. If matter is added in the 5D theory, it will be interpreted as composite states of the *CFT* if it localized close the IR brane or as elementary fields coupled to the *CFT* if it is peaked on the UV brane. For instance a 5D Higgs localized on the IR brane can be seen as the dual version of composite Higgs models [49, 68]. When the Higgs vacuum expectation value is sent to infinity, the 4D theory is more a technicolor-like model. In that sense, the 5D warped Higgsless model can be seen as a weakly coupled dual version of walking technicolor [24].

5.4.2. Towards a realistic Higgsless model
From the correspondence discussed above, we can now relatively easily find the theory that we are after [24]. We want a theory that has an $SU(2)_L \times SU(2)_R \times$

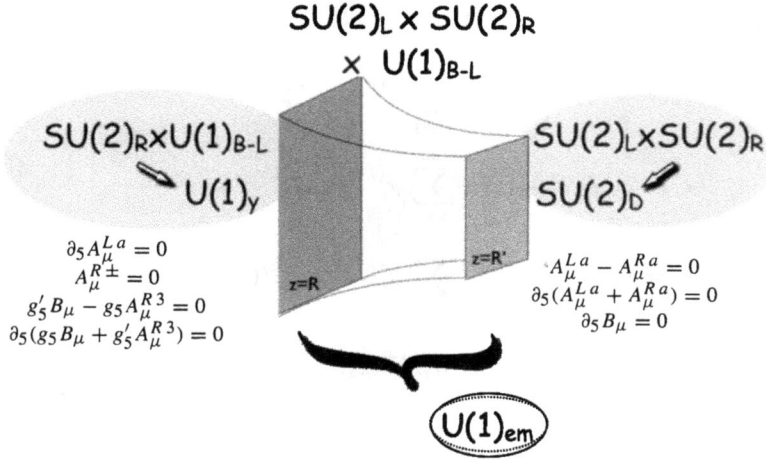

SU(2)$_L$ × SU(2)$_R$
× U(1)$_{B-L}$

SU(2)$_R$×U(1)$_{B-L}$
U(1)$_Y$

SU(2)$_L$×SU(2)$_R$
SU(2)$_D$

$$\partial_5 A_\mu^{L\,a} = 0$$
$$A_\mu^{R\,\pm} = 0$$
$$g_5' B_\mu - g_5 A_\mu^{R\,3} = 0$$
$$\partial_5 (g_5 B_\mu + g_5' A_\mu^{R\,3}) = 0$$

$z=R$

$z=R'$

$$A_\mu^{L\,a} - A_\mu^{R\,a} = 0$$
$$\partial_5 (A_\mu^{L\,a} + A_\mu^{R\,a}) = 0$$
$$\partial_5 B_\mu = 0$$

U(1)$_{em}$

Fig. 17. The symmetry breaking structure of the warped Higgsless model [24]. We will be considering a 5D gauge theory in the fixed gravitational Anti-de-Sitter (*AdS*) background. The UV brane is
located at $z = R$ and the IR brane is located at $z = R'$. R is the *AdS* curvature scale. In conformal
coordinates, the *AdS* metric is given by $ds^2 = (R/z)^2 \left(\eta_{\mu\nu} dx^\mu dx^\nu - dz^2\right)$.

$U(1)_{B-L}$ global symmetry, with the $SU(2)_L \times U(1)_Y$ subgroup weakly gauged,
and broken by BC's on the IR brane. To have the full global symmetry, we need
to take $SU(2)_L \times SU(2)_R \times U(1)_{B-L}$ in the bulk of AdS_5. To make sure that we
do not get unwanted gauge fields at low energies, we need to break $SU(2)_R \times
U(1)_{B-L}$ to $U(1)_Y$ on the UV brane, which we will do by BC's as in the flat case.
Finally, the boundary conditions on the TeV brane break $SU(2)_L \times SU(2)_R$ to
$SU(2)_D$. This setup is illustrated in Fig. 17. Note, that it is practically identical
to the flat space toy model considered before, except that the theory is in *AdS*
space.

The only difference between the flat and the warped Higgsless models is the
shape of the wavefunctions. Indeed the solution of the bulk equations of motion
in *AdS* space involves some Bessel fuctions of order one:

$$\psi_k^{(A)}(z) = z \left(a_k^{(A)} J_1(q_k z) + b_k^{(A)} Y_1(q_k z) \right), \tag{5.70}$$

where A labels the corresponding gauge boson.

Due to the mixing of the various gauge groups, the KK decomposition is
slightly complicated but it is obtained by simply enforcing the BC's with wave-

functions of the form (5.70):

$$B_\mu(x, z) = g_5 \, a_0 \gamma_\mu(x) + \sum_{k=1}^{\infty} \psi_k^{(B)}(z) \, Z_\mu^{(k)}(x), \tag{5.71}$$

$$A_\mu^{L\,3}(x, z) = \tilde{g}_5 \, a_0 \gamma_\mu(x) + \sum_{k=1}^{\infty} \psi_k^{(L3)}(z) \, Z_\mu^{(k)}(x), \tag{5.72}$$

$$A_\mu^{R\,3}(x, z) = \tilde{g}_5 \, a_0 \gamma_\mu(x) + \sum_{k=1}^{\infty} \psi_k^{(R3)}(z) \, Z_\mu^{(k)}(x), \tag{5.73}$$

$$A_\mu^{L\,\pm}(x, z) = \sum_{k=1}^{\infty} \psi_k^{(L\pm)}(z) \, W_\mu^{(k)\,\pm}(x), \tag{5.74}$$

$$A_\mu^{R\,\pm}(x, z) = \sum_{k=1}^{\infty} \psi_k^{(R\pm)}(z) \, W_\mu^{(k)\,\pm}(x). \tag{5.75}$$

Here $\gamma(x)$ is the 4D photon, which has a flat wavefunction due to the unbroken $U(1)_Q$ symmetry, and $W_\mu^{(k)\,\pm}(x)$ and $Z_\mu^{(k)}(x)$ are the KK towers of the massive W and Z gauge bosons, the lowest of which are supposed to correspond to the observed W and Z. Enforcing the BC's with these wavefunctions lead to the quantization equations from which the spectrum is read of. For the W's we get

$$(R_0 - \tilde{R}_0)(R_1 - \tilde{R}_1) + (\tilde{R}_1 - R_0)(\tilde{R}_0 - R_1) = 0, \tag{5.76}$$

where the ratios $R_{0,1}$ and $\tilde{R}_{0,1}$ are given by

$$R_i \equiv \frac{Y_i(MR)}{J_i(MR)}, \qquad \tilde{R}_i \equiv \frac{Y_i(MR')}{J_i(MR')}. \tag{5.77}$$

To leading order in $1/R$ and for $\log(R'/R) \gg 1$, the lightest solution for this equation for the mass of the $W^\pm$'s is

$$M_W^2 \approx \frac{1}{R'^2 \log\left(\frac{R'}{R}\right)}. \tag{5.78}$$

Note, that this result does not depend on the 5D gauge coupling, but only on the scales R, R'. Taking $R = 10^{-19}$ GeV^{-1} will fix $R' = 2 \cdot 10^{-3}$ GeV^{-1}. The equation determining the masses of the KK tower for the Z (the states that are mostly A^{L3} or A^{R3}) is given by

$$2\tilde{g}_5^2 (R_0 - \tilde{R}_1)(\tilde{R}_0 - R_1) =$$
$$g_5^2 \big[(R_0 - \tilde{R}_0)(R_1 - \tilde{R}_1) + (\tilde{R}_1 - R_0)(\tilde{R}_0 - R_1) \big]. \tag{5.79}$$

The lowest mass of the Z tower is approximately given by

$$M_Z^2 = \frac{g_5^2 + 2\tilde{g}_5^2}{g_5^2 + \tilde{g}_5^2} \frac{1}{R'^2 \log\left(\frac{R'}{R}\right)}. \tag{5.80}$$

Finally, there is a third tower of states, corresponding to the excited modes of the photon (the particles that are mostly B-type), whose masses are given by

$$R_0 = \tilde{R}_0. \tag{5.81}$$

This does not have a light mode (the zero mode corresponding to the massless photon has been separated out explicitly in (5.71)–(5.73)).

If the SM fermions are localized on the Planck brane then the leading order expression for the effective 4D couplings will be given by (see sections 5.5 and 5.6 for more details)

$$\frac{1}{g^2} = \frac{R \log\left(\frac{R'}{R}\right)}{g_5^2},$$

$$\frac{1}{g'^2} = R \log\left(\frac{R'}{R}\right)\left(\frac{1}{g_5^2} + \frac{1}{\tilde{g}_5^2}\right), \tag{5.82}$$

thus the 4D Weinberg angle will be given by

$$\sin\theta_W = \frac{\tilde{g}_5}{\sqrt{g_5^2 + 2\tilde{g}_5^2}} = \frac{g'}{\sqrt{g^2 + g'^2}}. \tag{5.83}$$

We can see that to leading order the SM expression for the W/Z mass ratio is reproduced in this theory as expected. In fact the full structure of the SM coupling is reproduced at leading order in $1/\log(R'/R)$, which implies that at the leading log level there is no S-parameter either. An S-parameter in this language would have manifested itself in an overall shift of the coupling of the Z compared to its SM value evaluated from the W and γ couplings, which are absent at this order of approximation. The corrections to the SM relations will appear in the next order of the log expansion. Since $\log\left(R'/R\right) \sim \mathcal{O}(10)$, this correction could still be too large to match the precision electroweak data. We will be discussing the issue of electroweak precision observables in section 5.6

The KK masses of the W (and the Z bosons as well due to custodial $SU(2)$ symmetry) will be given approximately by

$$M_{W_n} = \frac{\pi(n + 1/2)}{2R'}, \quad n = 1, 2, \ldots. \tag{5.84}$$

We can see that the ratio between the physical W mass and the first KK mode is given by

$$\frac{M_W}{M_W'} = \frac{4}{3\pi} \frac{1}{\sqrt{\log\left(\frac{R'}{R}\right)}}. \tag{5.85}$$

We can see that warping achieves two desirable properties: it enforces custodial $SU(2)$ and thus automatically generates the correct W/Z mass ratio, but it also pushes up the masses of the KK resonances of W and Z. This implies that we can get a theory where the W', Z' bosons are not so light that they would already be excluded by the LEP or the Tevatron experiments. In the flat model, the mass spectrum involved only one scale and the mass gap between the W, Z and their next KK excitations was too small. In the warped models, the presence of two scales (the size of the extra dimension as well as the curvature scale of the space) allows to increase the mass gap. The $\log R'/R$ supression of the mass of the W compare to the mass of the W' can actually be easily understood from naive dimensional analysis and the AdS/CFT correspondence depicted in section 5.4.1. Indeed, the W mass originates from the IR spontaneous breaking of gauge symmetry weakled coupled to the CFT, so

$$M_W^2 = \frac{g_4^2}{R_{IR}^2}. \tag{5.86}$$

Furthermore, the 4D gauge coupling is obtained from the 5D one (from NDA, $g_5^2 = R_{UV}$) and the normalization condition of the (flat) wavefunction of the massless gauge boson:

$$\frac{1}{g_4^2} = \frac{\int_{R_{UV}}^{R_{IR}} dz \, R_{UV}/z}{g_5^2}. \tag{5.87}$$

Hence

$$M_W^2 = \frac{1}{R_{IR}^2 \log R_{IR}/R_{UV}}. \tag{5.88}$$

Finally, we can return to the issue of perturbative unitarity in these models. In the flat space case we have seen that the scale of unitarity violation is basically given by the NDA cutoff scale (5.45). However, in a warped extra dimension all scales will be dependent on the location along the extra dimension, so the lowest cutoff scale that one has is at the IR brane given by

$$\Lambda_{NDA} \sim \frac{24\pi^3}{g_5^2} \frac{R}{R'}. \tag{5.89}$$

Using our expressions for the 4D couplings and the W and W' masses above we can see that [66,72]

$$\Lambda_{NDA} \sim \frac{12\pi^4 M_W^2}{g^2 M_{W^{(1)}}}. \qquad (5.90)$$

From the formula above, it is clear that the heavier the resonance, the lower the scale where perturbative unitarity is violated. This also gives a rough estimate, valid up to a numerical coefficient, of the actual scale of non–perturbative physics. An explicit calculation of the scattering amplitude, including inelastic channels, shows that this is indeed the case and the numerical factor is found to be roughly $1/4$ [66].

Since the ratio of the W to the first KK mode mass squared is of order

$$\frac{M_W^2}{M_{W^{(1)}}^2} = \mathcal{O}\left(1/\log\left(R'/R\right)\right), \qquad (5.91)$$

raising the value of R (corresponding to lowering the 5D UV scale) will significantly increase the NDA cutoff. With R chosen to be the inverse Planck scale, the first KK resonance appears around 1.2 TeV, but for larger values of R this scale can be safely reduced down below a TeV (see Fig. 21).

5.5. Fermion masses

In the SM, quarks and leptons acquire a mass, after EWSB, through their Yukawa couplings to the Higgs. In absence of a Higgs, one cannot write any Yukawa coupling and one should expect the fermions to remain massless. However, as for the gauge fields, appropriate boundary conditions will force the fermions to acquire a momentum along the extra dimension and this how they will become massive from the 4D point of view. We are now going to review this construction [24, 69, 73, 74] (see also [75] for earlier works on fermions in a RS warped background). A general discussion on spin-1/2 boundary condition can be found in [73].

The SM fermions cannot be completely localized on the UV boundary: since the unbroken gauge group on that boundary coincides with the SM $SU(2)_L \times U(1)_Y$ symmetry, the theory on that brane would be chiral and there is no way for the chiral zero mode fermions to acquire a mass. The SM fermions cannot live on the IR brane either since the unbroken $SU(2)_D$ gauge symmetry will impose an isospin invariant spectrum and the up-type and down-type quarks will be degenerate, as the electron and the electron neutrino will be. The only possibility is thus to embed the SM fermions into 5D fields living in the bulk and feeling the gauge symmetry breakings on both boundaries. Since the irreducible spin-1/2 representations of the 5D Lorentz group correspond to 4-component Dirac spinor, extra fermionic degrees of freedom are needed to complete the SM chiral spinors to 5D Dirac spinors and we are back again to a vector-like spectrum. However, as it is well known, orbifold like projections can get rid of half of the spectrum at the lowest KK level to actually provide a 4D effective chiral theory.

5.5.1. *Chiral fermions from orbifold projection/boundary condition*
A 5D Dirac spinor decomposes under the 4D Lorentz subgroup into two two-component spinors (technically speaking, we say that the simplest 5D irreducible spin-1/2 representation breaks up as $(0, 1/2) + (1/2, 0)$ under the 4D Lorentz subgroup)

$$\Psi = \begin{pmatrix} \chi_\alpha \\ \bar{\psi}^{\dot{\alpha}} \end{pmatrix} \tag{5.92}$$

in the 5D chiral representation of the Dirac matrices:

$$\Gamma^\mu = \begin{pmatrix} 0 & \sigma^\mu \\ \bar{\sigma}^\mu & 0 \end{pmatrix}_{\mu=0,1,2,3} \quad \text{and} \quad \Gamma^5 = i \begin{pmatrix} 1 & 0 \\ 0 & -1 \end{pmatrix}, \tag{5.93}$$

where $\sigma^i = -\bar{\sigma}^i$ are the usual Pauli spin matrices[11], while $\sigma^0 = \bar{\sigma}^0 = -1$.

Under the $\mathbb{Z}_2$ orbifold projection, $y \sim -y$, discussed in Section 4.1, and in order to leave the 5D Dirac equation invariant, Ψ has to satisfy

$$\Psi(-y) = -i\Gamma^5\Psi(y), \tag{5.94}$$

i.e.

$$\chi(-y) = \chi(y) \quad \text{and} \quad \psi(-y) = -\psi(y). \tag{5.95}$$

Therefore, only χ will have a KK zero mode

$$\chi(x, y) = \sum_{n=0}^{\infty} \cos \frac{ny}{R} \chi^{(n)}(x) \quad \text{and} \quad \psi(x, y) = \sum_{n=1}^{\infty} \sin \frac{ny}{R} \psi^{(n)}(x). \tag{5.96}$$

Let us now quickly explain how we can recover this result on the interval approach through boundary conditions. With a mass in the bulk, the 5D Lagrangian for Ψ reads

$$S = \int d^5x \left(\frac{i}{2}(\bar{\Psi} \Gamma^M \partial_M \Psi - \partial_M \bar{\Psi} \Gamma^M \Psi) - m\bar{\Psi}\Psi \right). \tag{5.97}$$

---

[11]Explicitly, $\sigma^1 = \begin{pmatrix} 0 & 1 \\ 1 & 0 \end{pmatrix}$, $\sigma^2 = \begin{pmatrix} 0 & -i \\ i & 0 \end{pmatrix}$, $\sigma^3 = \begin{pmatrix} 1 & 0 \\ 0 & -1 \end{pmatrix}$.

which in 4D components writes[12]

$$S = \int d^5x \left(-i\bar{\chi}\bar{\sigma}^\mu\partial_\mu\chi - i\psi\sigma^\mu\partial_\mu\bar{\psi} + (\psi \overset{\leftrightarrow}{\partial_5} \chi - \bar{\chi} \overset{\leftrightarrow}{\partial_5} \bar{\psi}) \right.$$
$$\left. + m(\psi\chi + \bar{\chi}\bar{\psi}) \right), \tag{5.98}$$

where $\overset{\leftrightarrow}{\partial_5} = \frac{1}{2}(\overset{\rightarrow}{\partial_5} - \overset{\leftarrow}{\partial_5})$. The variation of this 5D Lagrangian leads to the bulk equations of motion:

$$-i\bar{\sigma}^\mu\partial_\mu\chi - \partial_5\bar{\psi} + m\bar{\psi} = 0 \quad \text{and} \quad -i\sigma^\mu\partial_\mu\bar{\psi} + \partial_5\chi + m\chi = 0. \tag{5.99}$$

On top of that, requiring that the variation of the Lagrangian at the boundary is also vanishing gives

$$-\delta\psi\chi + \psi\delta\chi + \delta\bar{\chi}\bar{\psi} - \bar{\chi}\delta\bar{\psi} = 0. \tag{5.100}$$

Naively, one might think that, since there are two independent spinors, χ and $\bar{\psi}$, one would require two independent boundary conditions for each spinor. However, because the bulk equations of motion are only first order, there is only one integration constant. So for the Dirac pair, $(\chi, \bar{\psi})$, there is only one boundary condition $f(\chi, \psi) = 0$ at each boundary, where f is some function of the spinors and their conjugates. The form of f together with the bulk equations of motion in (5.99) then determines all of the arbitrary coefficients in the complete solution to the spinor equation of motion on the interval. We can, for instance, require that the spinor ψ vanishes on both boundaries. Then the bulk equations of motion will enforce

$$(\partial_5 + m)\,\chi_{|0,\pi R} = 0. \tag{5.101}$$

Solving the equations of motion with these boundary conditions results in a zero mode for χ, but not for ψ. That is, the low energy theory is a chiral theory.

As with gauge and scalar fields, there will be, in the 4D effective theory, a tower of massive Dirac fields. The spectrum is obtained by solving the bulk equations, that will dictate the general form of the wavefunctions, and by enforcing the boundary conditions. Let us perform this KK decomposition explicitly. The 5D spinors χ and ψ can be written as a sum of products of 4D KK Dirac fermions with 5D wavefunctions:

$$\chi = \sum_n g_n(y)\,\chi_n(x) \quad \text{and} \quad \bar{\psi} = \sum_n f_n(y)\,\bar{\psi}_n(x). \tag{5.102}$$

---

[12]Usually, the terms with the left acting derivatives are generally integrated by parts, so that all derivatives act to the right. However, since we are working here in a compact space with boundaries, the integration by parts produces boundary terms which can not be neglected. Note that both (5.97) and (5.98) are hermitian.

The KK fermions obey the 4D Dirac equation:

$$-i\bar{\sigma}^\mu \partial_\mu \chi^{(n)} + m_n \bar{\psi}^{(n)} = 0 \quad \text{and} \quad -i\sigma^\mu \partial_\mu \bar{\psi}^{(n)} + m_n \chi^{(n)} = 0. \quad (5.103)$$

Substituting this decomposition into the 5D bulk equations of motions gives the following equations

$$g'_n + m\, g_n - m_n\, f_n = 0 \quad \text{and} \quad f'_n - m\, f_n + m_n\, g_n = 0. \quad (5.104)$$

The standard approach to solving this system of equations is to combine the two first order coupled equations into two second order uncoupled wave equations:

$$g''_n + (m_n^2 - m^2)g_n = 0 \quad \text{and} \quad f''_n + (m_n^2 - m^2)f_n = 0. \quad (5.105)$$

The solution is simply a sum of sines and cosines, with coefficients that are determined by reimposing the first order equations, and imposing the boundary conditions. For instance, if we impose that $\psi = 0$ both at $y = 0$ and $y = \pi R$, we obtain

$$m_n = \sqrt{m^2 + \tfrac{n^2}{R^2}}, \quad n = 1, 2 \dots, \quad (5.106)$$

$$f_n(y) = a_n \sin \tfrac{ny}{R}, \quad (5.107)$$

$$g_n(y) = \tfrac{a_n}{m_n} \left(\tfrac{n}{R} \cos \tfrac{ny}{R} - m \sin \tfrac{ny}{R} \right), \quad (5.108)$$

and the remaining coefficient a_n is fixed by the normalization condition[13]

$$\int_0^{\pi R} dy f_n^2(y) = 1. \quad (5.109)$$

On top of the massive spectrum, the boundary conditions also allow for a zero mode for χ:

$$g_0(y) = \sqrt{\frac{2m}{1 - e^{-2m\pi R}}}\, e^{-my}. \quad (5.110)$$

Note that the 5D mass doesn't contribute to the mass of the lightest fermion (that remains massless because of chirality) but it dictates the shape of its wavefunction.

---

[13]The normalization of a 4D Dirac fermion actually imposes *two* equations:

$$\int_0^{\pi R} dy f_n^2(y) = 1 \quad \text{and} \quad \int_0^{\pi R} dy g_n^2(y) = 1.$$

However, thanks to the quantization equation, the second equation is redundant. This can be easily checked on the simple example presented here. In more complicated cases, the redundancy of the two normalization conditions is a good consistency check that the right KK decomposition has been obtained.

Table 3

Embedding of the SM fermions into 5D Dirac spinors. We have indicated the quantum numbers of the different components under the bulk $SU(2)_L \times SU(2)_R \times U(1)_{B-L}$ symmetry, the subgroup $SU(2)_L \times U(1)_Y$ that remains unbroken on the UV boundary, the subgroup $SU(2)_D \times U(1)_{B-L}$ unbroken on the IR brane and finally the electric charge. The shaded spinors are the fields with the right quantum numbers to be identified as the massless SM fermions while the other spinors correspond to partners needed to complete 5D Dirac spinors. The latter become massive by the orbifold projection/boundary conditions. Through the Dirac mass added on the IR boundary, there will be a mixing between the would be zero modes and some partners and at the end the guy that would be identified as the SM u_L is a mix of χ_{u_L} and a small amount of χ_{u_R}. Since this last field has wrong SM quantum numbers, we would end up with deviations in the couplings of the fermions to the gauge bosons. These deviations will be particularly sizable for the third generation due to the heaviness of the top.

Particle	bulk $L \times R \times (B-L)$	UV $L \times Y$	IR $D \times (B-L)$	Q_{em}
$\begin{pmatrix} \chi_u \\ \chi_d \end{pmatrix}_L$	$(\square, 1, 1/6)$	$(\square, 1/6)$	$(\square, 1/6)$	$\begin{matrix} 2/3 \\ -1/3 \end{matrix}$
$\begin{pmatrix} \psi_u \\ \psi_d \end{pmatrix}_L$	$(\bar\square, 1, -1/6)$	$(\bar\square, -1/6)$	$(\bar\square, -1/6)$	$\begin{matrix} -2/3 \\ 1/3 \end{matrix}$
$\begin{pmatrix} \chi_u \\ \chi_d \end{pmatrix}_R$	$(1, \square, 1/6)$	$\begin{matrix} (1, 2/3) \\ (1, -1/3) \end{matrix}$	$(\square, 1/6)$	$\begin{matrix} 2/3 \\ -1/3 \end{matrix}$
$\begin{pmatrix} \psi_u \\ \psi_d \end{pmatrix}_R$	$(1, \bar\square, -1/6)$	$\begin{matrix} (1, -2/3) \\ (1, 1/3) \end{matrix}$	$(\bar\square, -1/6)$	$\begin{matrix} -2/3 \\ 1/3 \end{matrix}$

$$Q_{em} = Y + T_{3L} \qquad Y = (B-L) + T_{3R}$$

In conclusion, an orbifold projection or equivalently appropriate boundary conditions allow to obtain a 4D chiral spectrum from a 5D theory. This way we can embed the SM quarks and leptons into 5D Dirac spinors following Table 3.

5.5.2. Fermions in AdS background

In principle when one is dealing with fermions in a non-trivial background, one needs to work with the "square-root" of the metric also known as *vielbeins* and to introduce the spin connection to covariantize derivatives. Fortunately, in an *AdS* background[14], the spin connection drops out from the spin-1/2 action that simply

---

[14]We recall that, in conformal coordinates, the *AdS* metric is given by (R is the *AdS* curvature scale)

$$ds^2 = (R/z)^2 \left(\eta_{\mu\nu} dx^\mu dx^\nu - dz^2 \right).$$

reads

$$S = \int d^5x \frac{R^4}{z^4} \left(-i\bar{\chi}\bar{\sigma}^\mu \partial_\mu \chi - i\psi\sigma^\mu \partial_\mu \bar{\psi} + (\psi \overleftrightarrow{\partial_5} \chi - \bar{\chi} \overleftrightarrow{\partial_5} \bar{\psi}) \right. $$
$$\left. + \frac{c}{z} (\psi\chi + \bar{\chi}\bar{\psi}) \right)$$

where the coefficient $c = mR$ is the bulk Dirac mass in units of the *AdS* curvature (and as before $\overleftrightarrow{\partial_5} = (\overrightarrow{\partial_5} - \overleftarrow{\partial_5})/2$). The bulk equations of motion are:

$$-i\bar{\sigma}^\mu \partial_\mu \chi - \partial_5 \bar{\psi} + \frac{c+2}{z}\bar{\psi} = 0, \qquad -i\sigma^\mu \partial_\mu \bar{\psi} + \partial_5 \chi + \frac{c-2}{z}\chi = 0.$$

The KK decomposition is performed as in the flat case (5.102) with the wave-functions now obeying the coupled first order differential equations

$$f_n' + m_n g_n - \frac{c+2}{z} f_n = 0, \qquad g_n' - m_n f_n + \frac{c-2}{z} g_n = 0, \qquad (5.111)$$

which can be combine into uncoupled second order differential equations

$$f_n'' - \frac{4}{z} f_n' + \left(m_n^2 - \frac{c^2-c-6}{z^2} \right) f_n = 0, \qquad (5.112)$$

$$g_n'' - \frac{4}{z} g_n' + \left(m_n^2 - \frac{c^2+c-6}{z^2} \right) g_n = 0. \qquad (5.113)$$

The solutions are now linear combinations of Bessel functions, as opposed to sin and cos functions:

$$g_n(z) = z^{\frac{5}{2}} \left(A_n J_{c+\frac{1}{2}}(m_n z) + B_n Y_{c+\frac{1}{2}}(m_n z) \right), \qquad (5.114)$$

$$f_n(z) = z^{\frac{5}{2}} \left(C_n J_{c-\frac{1}{2}}(m_n z) + D_n Y_{c-\frac{1}{2}}(m_n z) \right). \qquad (5.115)$$

The first order bulk equations of motion (5.111) further impose that

$$A_n = C_n \quad \text{and} \quad B_n = D_n. \qquad (5.116)$$

The remaining undetermined coefficients are determined by the boundary conditions, and the wave function normalization.

Finally, when the boundary conditions permit, there can also be a zero mode. For instance, if $\psi_{|R,R'} = 0$, the zero mode is given by

$$g_0(y) = A_0 \left(\frac{z}{R} \right)^{2-c}, \quad f = 0. \qquad (5.117)$$

The coefficient A_0 is determined by the normalization condition

$$\int_R^{R'} dz \left(\frac{R}{z}\right)^5 \frac{z}{R} A_0^2 \left(\frac{z}{R}\right)^{4-2c} = A_0^2 \int_R^{R'} \left(\frac{z}{R}\right)^{-2c} dz = 1. \quad (5.118)$$

To understand from these equations where the fermions are localized, we study the behavior of this integral as we vary the limits of integration. If we send R' to infinity, we see that the integral remains convergent if $c > 1/2$, and the fermion is then localized on the UV brane. If we send R to zero, the integral is convergent if $c < 1/2$, and the fermion is localized on the IR brane. The value of the Dirac mass determines whether the fermion is localized towards the UV or IR branes. We note that the opposite choice of boundary conditions that yields a zero mode ($\chi|_{R,R'} = 0$) results in a zero mode solution for ψ with localization at the UV brane when $c < -1/2$, and at the IR brane when $c > -1/2$. The interesting feature in the warped case is that the localization transition occurs not when the bulk mass passes through zero, but at points where $|c| = 1/2$. This is due to the curvature effects of the extra dimension. The *CFT* interpretation of the c parameter is an anomalous dimension that controls the amount of compositeness of the fermion [71].

5.5.3. Higgsless fermions masses
We have already explained how to embed SM fermions into 5D Dirac spinors. To get the zero modes we desire, the following boundary conditions have to be imposed

$$\begin{pmatrix} \chi_{u_L} \\ \bar{\psi}_{u_L} \\ \chi_{d_L} \\ \bar{\psi}_{d_L} \end{pmatrix} \quad \begin{matrix} + & + \\ - & - \\ + & + \\ - & - \end{matrix} \qquad \begin{pmatrix} \chi_{u_R} \\ \bar{\psi}_{u_R} \\ \chi_{d_R} \\ \bar{\psi}_{d_R} \end{pmatrix} \quad \begin{matrix} - & - \\ + & + \\ - & - \\ + & + \end{matrix} \qquad (5.119)$$

Where the $+$ and $-$ refer to Neumann and Dirichlet boundary conditions, the first/second sign denoting the BC on the UV/IR brane respectively. These boundary conditions give massless chiral modes that match the fermion content of the standard model. However, the u_L, d_L, u_R, and d_R are all massless at this stage, and we need to lift the zero modes to achieve the standard model mass spectrum. While simply giving certain boundary conditions for the fermions will enable us to lift these zero modes, in the following discussion, we talk about boundary operators, and the boundary conditions that these operators induce. There are some subtleties in dealing with boundary operators for fermions. These arise from the fact that the fields themselves are not always continuous in the presence of a boundary operator. This is due to the fact that the equations of motion for fermions are first order. The most straightforward approach is to enforce the boundary conditions that give the zero modes as shown in Eq. (5.119) on the real boundary at $z = R, R'$ while the boundary operators are added on a fictitious brane a distance ϵ away from it. The distance between the fictitious brane and the physical one is taken to be ϵ. The new boundary condition is then obtained by taking the distance ϵ to be small. This physical picture is quite helpful in understanding what the different boundary conditions will do. The details can be found in [73].

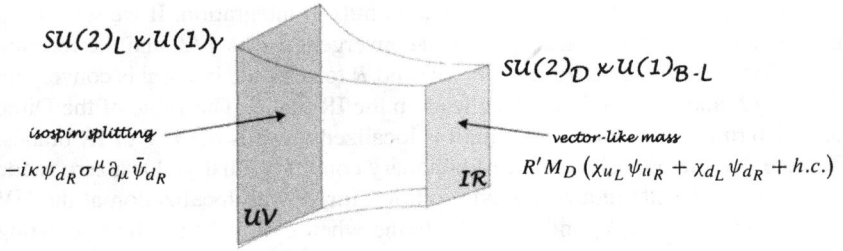

Fig. 18. Brane localized operators needed to lift up the masses of the SM fermions.

The IR brane being vector-like, we can now form an $SU(2)_D$ mass term that will mix the L and R SM helicities. However, this Dirac mass term has to be the same for the up and the down quarks (the mass term is isospin invariant). Fortunately, the $SU(2)_R$ invariance is broken on the UV brane and there we can introduce operators that will distinguish between u_R and d_R. Technically, the effects of the brane localized operators is to modified the BCs. Explicitly, the IR Dirac mass affect the BCs as follows

$$
\begin{array}{ccc}
\chi_L & + & \\
\psi_L & - & \psi_{L|_{IR}} = 0 \\
\chi_R & - & \chi_{R|_{IR}} = 0 \\
\psi_R & +
\end{array}
\quad M_D \quad
\begin{array}{c}
\text{discontinuities} \\
\text{in} \\
\chi_L \ \& \ \psi_R
\end{array}
\Longrightarrow
\quad
\begin{array}{l}
\psi_{L|_{IR}} = -M_D R' \, \psi_{R|_{IR}} \\
\chi_{R|_{IR}} = M_D R' \, \chi_{L|_{IR}}
\end{array}
$$

In the same way, the UV brane operator will modify the BCs as follows

$$
\begin{array}{cc}
\chi_{u_R} & - \\
\psi_{u_R} & +
\end{array}
\quad \chi_{u_R|_{UV}} = 0 \quad
\begin{array}{c}
\kappa \\
\Longrightarrow
\end{array}
\quad
\begin{array}{c}
\text{discontinuity} \\
\text{in} \\
\psi_{u_R}
\end{array}
\quad \chi_{u_R|_{UV}} = \kappa m \, \psi_{u_R|_{UV}}
$$

It is now easy to enforce these modified boundary conditions using the general form of the wavefunctions (5.114)–(5.115) that satisfy the bulk equations. of motion. For fermions localized towards the UV brane ($c_L > 1/2$ and $c_R < -1/2$), we obtain the approximate expression

$$
m \approx \frac{\sqrt{2c_L - 1}}{\sqrt{\kappa^2 - 1/(2c_R + 1)}} M_D \left(\frac{R_{UV}}{R_{IR}} \right)^{c_L - c_R - 1}. \tag{5.120}
$$

The spectrum of the light generations of quarks can be easily reproduced along these lines. The top mass poses a difficulty, however. Indeed, increasing M_D won't arbitrarily increase the fermion mass which will saturate: the situation is similar to what happens with a large Higgs *vev* localized on the boundary, the gauge boson masses remain finite even when the *vev* is sent to infinity. The maximum value of the fermion mass can be inferred by noticing that in the infinite M_D limit, there is a *chirality flip* in the BCs that become

$$
\begin{array}{c}
\chi_L \quad + \\
\psi_L \quad - \\
\chi_R \quad - \\
\psi_R \quad +
\end{array}
\qquad
\begin{array}{l}
\psi_{L|\mathrm{IR}} = -M_D R' \, \psi_{R|\mathrm{IR}} \\
\chi_{R|\mathrm{IR}} = M_D R' \, \chi_{L|\mathrm{IR}}
\end{array}
\qquad
\begin{array}{c}
M_D \to \infty \\
\Longrightarrow
\end{array}
\qquad
\begin{array}{l}
\psi_{R|\mathrm{IR}} = 0 \\
\chi_{L|\mathrm{IR}} = 0
\end{array}
\qquad
\begin{array}{c}
\chi_L \quad - \\
\psi_L \quad + \\
\chi_R \quad + \\
\psi_R \quad -
\end{array}
$$

And the corresponding mass is

$$
m^2 = \frac{2}{R'^2 \log R'/R} = 2 M_W^2. \tag{5.121}
$$

Where in the last equality, we used the expression of the W mass in terms of R and R' and we have assumed $g_{5R} = g_{5L}$. If we want to go above this saturated mass, one needs to localize the fermions towards the IR brane. However, even in this case a sizable Dirac mass term on the TeV brane is needed to obtain a heavy enough top quark. The consequence of this mass term is the boundary condition for the bottom quarks

$$
\chi_{bR} = M_D R' \, \chi_{bL}. \tag{5.122}
$$

This implies that if $M_D R' \sim 1$ then the left handed bottom quark has a sizable component also living in an $SU(2)_R$ multiplet, which however has a coupling to the Z that is different from the SM value. Thus there will be a large deviation in the $Zb_L\bar{b}_L$. Note, that the same deviation will not appear in the $Zb_R\bar{b}_R$ coupling, since the extra kinetic term introduced on the Planck brane to split top and bottom will imply that the right handed b lives mostly in the induced fermion on the Planck brane which has the correct coupling to the Z.

The only way of getting around this problem would be to raise the value of $1/R'$, and thus lower the necessary mixing on the TeV brane needed to obtain a heavy top quark. One way of raising the value of $1/R'$ is by increasing the ratio g_{5R}/g_{5L} (at the price of also making the gauge KK modes heavier and thus the theory more strongly coupled). Another possibility for rasing the value of $1/R'$ is to separate the physics responsible for electroweak symmetry breaking from that responsible for the generation of the top mass. In technicolor models this is usually achieved by introducing a new strong interaction called topcolor. In the extra dimensional setup this would correspond to adding two separate AdS_5

bulks, which meet at the Planck brane [63]. One bulk would then be mostly responsible for electroweak symmetry breaking, the other for generating the top mass. The details of such models have been worked out in [63]. The main consequences of such models would be the necessary appearance of an isotriplet pseudo-Goldstone boson called the top-pion, and depending on the detailed implementation of the model there could also be a scalar particle (called the top-Higgs) appearing. This top-Higgs would however not be playing a major role in the unitarization of the gauge boson scattering amplitudes, but rather serve as the source for the top-mass only.

5.6. Electroweak precision tests

In order to compare Higgsless models to precision electroweak measurements, we need to compute the Peskin–Takeuchi parameters S, T and U [69,72,76–79]. We use such parameters to fit the Z-pole observables at LEP1. In [28], Barbieri et al. proposed an enlarged set of parameters, to take into account also differential cross section measurements at LEP2. However, the only new information contained by the new parameters is the bound on four-fermi operators generated by the exchange of KK bosons, that we take into account to bound the lighter resonances at LEP2 and Tevatron. Effectively, our S, T and U are linear combinations of the parameters in [28].

In [78] we computed the oblique corrections in the standard way, in terms of mass eigenstates, in the limit where the light fermions are localized on the Planck brane. The only relevant technical point in the calculation, is the matching of the 4D gauge couplings. Indeed, if one writes down the couplings of the fermions, only two quantities does not depend on the overall Z and W normalizations and are completely fixed by the boundary condition. Namely, the electric charge and the ratio between the hypercharge and T_3 couplings to the Z. Matching such quantities with the SM predictions, it is possible to cast all the corrections in the oblique parameters.

In the basic model, with $g_{5L} = g_{5R} = g_5$ and vanishing localized kinetic terms, the leading contribution to S in the $1/\log \frac{R'}{R} \approx .3$ expansion is:

$$S \approx \frac{6\pi}{g^2 \log \frac{R'}{R}} \approx 1.15, \tag{5.123}$$

while $T \approx U \approx 0$. This value of S is clearly too large to be compared with the experimental result[15].

---

[15]Actually, this number should not be compared with the usual SM fit, but we should disentangle the contribution of the Higgs. Namely, it is enough to do the fit assuming a large Higgs mass, equal to the cut-off of the theory [28]. We are also neglecting loop corrections from the gauge KK modes.

However, the theory has more parameters: for instance, kinetic operators can be localized on the boundaries for the locally unbroken gauge symmetries:

$$\mathcal{L} = -\left[\frac{r}{4}W_{\mu\nu}^{L}{}^{2} + \frac{r'}{4}B_{\mu\nu}^{Y}{}^{2}\right]\delta(z-R) - \frac{R'}{R}\left[\frac{\tau'}{4}B_{\mu\nu}{}^{2} + \frac{\tau}{4}W_{\mu\nu}^{D}{}^{2}\right]\delta(z-R').$$

Let us first study the effect of *asymmetric bulk gauge couplings* and *Planck brane kinetic terms*. The leading contribution to S is:

$$S \approx \frac{6\pi}{g^2 \log\frac{R'}{R}} \frac{2}{1 + \frac{g_{5R}^2}{g_{5L}^2}} \frac{1}{1 + \frac{r}{R\log R'/R}}, \tag{5.124}$$

where, again, $T \approx U \approx 0$. Now, in case of large g_{5R}/g_{5L} ratio (or large $SU(2)_L$ kinetic term) S is suppressed. However, the W mass squared is also parametrically multiplied by the same factor. This means that the smaller S the larger the scale of the KK resonances, $1/R'$. So, in order to have small corrections we possibly enter a strong coupling regime, where the above calculation becomes meaningless.

Another set of parameters are the *TeV kinetic terms*. Their contribution is more complicated, so we will show some results at leading order for $\tau, \tau' \ll R\log\frac{R'}{R}$. The $SU(2)_D$ kinetic term appears at linear order, and effectively multiplies eq. 5.124 by a factor $1 + \frac{\tau}{R}$. On the other hand, the $U(1)_{B-L}$ kinetic term contributes at quadratic order. If only τ' is turned on,

$$S \approx \frac{6\pi}{g^2 \log\frac{R'}{R}} - \frac{8\pi}{g^2}\left(1 - \left(\frac{g'}{g}\right)^2\right)\frac{\tau'^2}{(R\log R'/R)^2}, \tag{5.125}$$

$$T \approx -\frac{2\pi}{g^2}\left(1 - \left(\frac{g'}{g}\right)^4\right)\frac{\tau'^2}{(R\log R'/R)^2}, \tag{5.126}$$

while $U \approx 0$. So, S vanishes for $\tau' \approx 0.15 \, R\log\frac{R'}{R}$. However, another effect is to make one of the Z' lighter, namely the one that couples with the hypercharge.

We also numerically scanned the parameter space to seek for a region where the model is not ruled out. For different values of g_{5R}/g_{5L}[16], we scanned the $\tau - \tau'$ space (see Fig. 19). Requiring both $|S|$ and $|T|$ to be smaller that 0.3, there is an allowed region only for large ratio, $g_{5R}/g_{5L} > 2.5$, where the theory is most likely strongly coupled. These results are in agreement with similar studies in [80] and [28].

---

[16]Using the Planck kinetic terms instead would only result in slightly different Z' couplings, and so different exclusion plots.

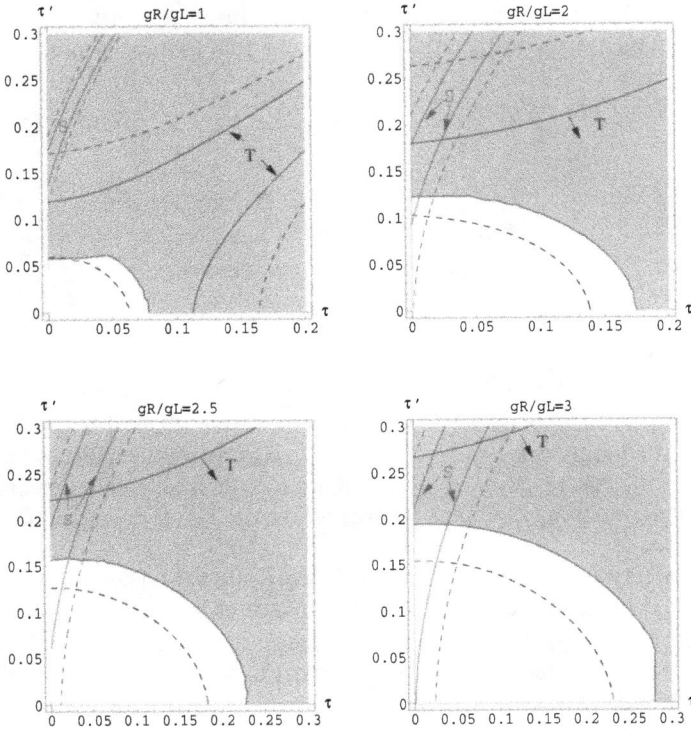

Fig. 19. Combined plots of the experimental constraints on Higgsless models for different values of the g_{5R}/g_{5L} ratio, in the parameter space $\tau-\tau'$ (normalized by $R \log R'/R$). The solid contours for S (red) and T (blue) are at 0.25; the dashed contours at 0.5. The black solid (dashed) line corresponds to a deviation in the differential cross section of 3% (2%) at LEP2. The shaded region is excluded by a deviation larger that 3% at LEP and/or direct search at Run1 at Tevatron.

As originally proposed, the model seems to be disfavoured by the experiments, if one wants strong coupling to arise above 3 TeV. However, this isn't the end of the story since there is a solution [72, 81] to the S problem which has additional beneficial side-effects. It has been known for a long time in Randall-Sundrum models with a Higgs that the effective S parameter is large and negative [82] if the fermions are localized on the TeV brane as originally proposed. When the fermions are localized on the Planck brane the contribution to S is positive, and so for some intermediate localization the S parameter vanishes, as first pointed out for RS models by Agashe et al. [68]. The reason for this is fairly simple. Since the W and Z wavefunctions are approximately flat, and the gauge KK mode wavefunctions are orthogonal to them, when the fermion wavefunctions are also approximately flat the overlap of a gauge KK mode with two fermions will approximately vanish. Since it is the coupling of the gauge KK modes to the fermions that induces a shift in the S parameter, for approximately flat fermion wavefunctions the S parameter must be small. Note that not only does reducing the coupling to gauge KK modes reduce the S parameter, it also weakens the experimental constraints on the existence of light KK modes. This case of delocalized bulk fermions is not covered by the no-go theorem of [28], since there it was assumed that the fermions are localized on the Planck brane.

In order to quantify these statements, it is sufficient to consider a toy model where all the three families of fermions are massless and have a universal delocalized profile in the bulk. Before showing some numerical results, it is useful to understand the analytical behavior of S in interesting limits. For fermions almost localized on the Planck brane, it is possible to expand the result for the S-parameter in powers of $(R/R')^{2c_L-1} \ll 1$. The leading terms, also expanding in powers of $1/\log$, are:

$$S = \frac{6\pi}{g^2 \log \frac{R'}{R}} \left(1 - \frac{4}{3} \frac{2c_L - 1}{3 - 2c_L} \left(\frac{R}{R'} \right)^{2c_L-1} \log \frac{R'}{R} \right), \qquad (5.127)$$

and $U \approx T \approx 0$. The above formula is actually valid for $1/2 < c_L < 3/2$. For $c_L > 3/2$ the corrections are of order $(R'/R)^2$ and numerically negligible. As we can see, as soon as the fermion wave function starts leaking into the bulk, S decreases.

Another interesting limit is when the profile is almost flat, $c_L \approx 1/2$. In this case, the leading contributions to S are:

$$S = \frac{2\pi}{g^2 \log \frac{R'}{R}} \left(1 + (2c_L - 1) \log \frac{R'}{R} + \mathcal{O}\left((2c_L - 1)^2 \right) \right). \qquad (5.128)$$

In the flat limit $c_L = 1/2$, S is already suppressed by a factor of 3 with respect to the Planck brane localization case. Moreover, the leading terms cancel out for:

$$c_L = \frac{1}{2} - \frac{1}{2 \log \frac{R'}{R}} \approx 0.487. \qquad (5.129)$$

For $c_L < 1/2$, S becomes large and negative and, in the limit of TeV brane localized fermions ($c_L \ll 1/2$):

$$S = -\frac{16\pi}{g^2} \frac{1 - 2c_L}{5 - 2c_L}, \qquad (5.130)$$

while, in the limit $c_L \to -\infty$:

$$T \quad \to \quad \frac{2\pi}{g^2 \log \frac{R'}{R}} (1 + \tan^2 \theta_W) \approx 0.5, \tag{5.131}$$

$$U \quad \to \quad -\frac{8\pi}{g^2 \log \frac{R'}{R}} \frac{\tan^2 \theta_W}{2 + \tan^2 \theta_W} \frac{1}{c_L} \approx 0. \tag{5.132}$$

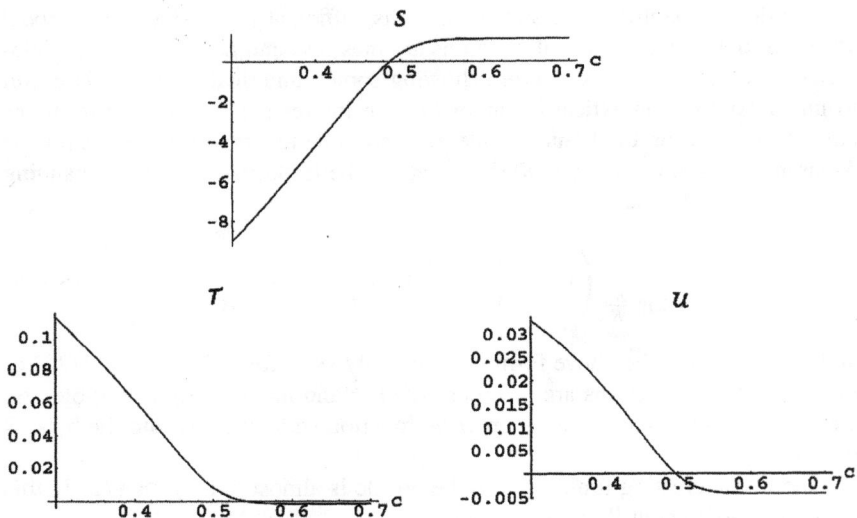

Fig. 20. Plots of the oblique parameters as function of the bulk mass of the reference fermion. The values on the right correspond to localization on the Planck brane. S vanishes for $c = 0.487$.

In Fig. 20 we show the numerical results for the oblique parameters as function of c_L. We can see that, after vanishing for $c_L \approx 1/2$, S becomes negative and large, while T and U remain smaller. With R chosen to be the inverse Planck scale, the first KK resonance appears around 1.2 TeV, but for larger values of R this scale can be safely reduced down below a TeV. Such resonances will be weakly coupled to almost flat fermions and can easily avoid the strong bounds from direct searches at LEP or Tevatron. If we are imagining that the AdS space is a dual description of an approximate conformal field theory (CFT), then $1/R$ is the scale where the CFT is no longer approximately conformal and perhaps becomes asymptotically free. Thus it is quite reasonable that the scale $1/R$ would be much smaller than the Planck scale.

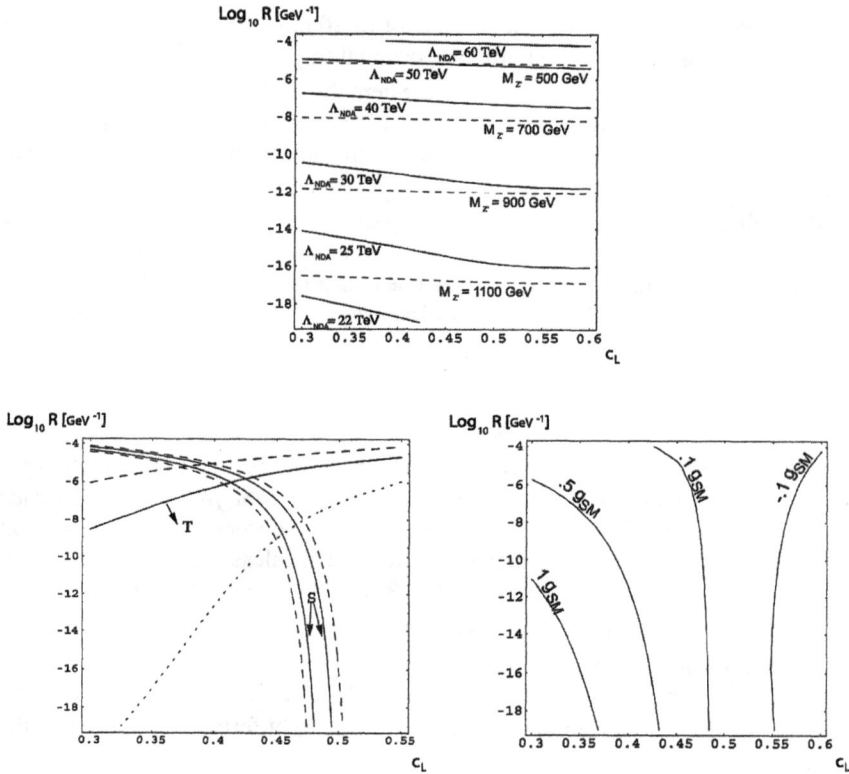

Fig. 21. In the left plot we show the contour plots of Λ_{NDA} (solid blue lines) and $M_{Z(1)}$ (dashed red lines) in the parameter space c_L–R. The shaded region is excluded by direct searches of light Z' at LEP. In the center, the contours of S (red), for $|S| = 0.25$ (solid) and 0.5 (dashed) and T (blue), for $|T| = 0.1$ (dotted), 0.3 (solid) and 0.5 (dashed), as function of c_L and R are shown. On the right, contours for the generic suppression of fermion couplings to the first resonance with respect to the SM value can be seen. The region for c_L, allowed by S, is between 0.43 ± 0.5, where the couplings are suppressed at least by a factor of 10.

In Fig. 21 we have plotted the value of the NDA scale (5.89) as well as the mass of the first resonance in the $(c_L - R)$ plane. Increasing R also affects the oblique corrections. However, while it is always possible to reduce S by delocalizing the fermions, T increases and puts a limit on how far R can be raised. One can also see from Fig. 21 that in the region where $|S| < 0.25$, the coupling of the first resonance with the light fermions is generically suppressed to less than 10% of the SM value. This means that the LEP bound of 2 TeV for SM–like Z' is also decreased by a factor of 10 at least (the correction to the differential cross section is roughly proportional to $g^2/M_{Z'}^2$). In the end, values of R as large as 10^{-7} GeV^{-1} are allowed, where the resonance masses are around 600 GeV. So, even if, following the analysis of [66], we take into account a factor of roughly $1/4$ in the NDA scale, we see that the appearance of strong coupling regime can be delayed up to 10 TeV.

It is fair to say that, to date, the major challenge facing Higgsless models is really the incorporation of the third family of quarks while the oblique corrections can be kept under control, at a price of some conspiracy in the localization of the SM quarks and leptons along the extra dimension.

5.7. Collider signatures

The non-observation of a physical scalar Higgs would be the first indication for a Higgsless scenario. Yet, *the absence of proof is not the proof of the absence* and some other models exist in which the Higgs is unobservable at the LHC and we need to look for other distinctive features of Higgsless models. This section follows closely the original works of [80, 83] and the reviews [55, 56].

The main predictions of Higgsless scenarios are

• the absence of a Higgs;
• the presence of spin-1 KK resonances with the W, Z quantum numbers;
• some slight deviations in the universality of the light fermion couplings to the SM gauge bosons;
• some deviations in the gauge boson self-interactions compared to the SM.

5D Higgsless models are non-renormalizable and will become strong coupled at some cutoff scale Λ. But Higgsless models have been devised to push Λ high enough and to avoid any trouble with EW precision test, which also means that the strong coupling sector won't be observable at the LHC. Still, beyond the SM spectrum, additional weakly coupled states are required to raise Λ and these ones should be observable.

Many different realizations of Higgsless models have been proposed, differing in the way the SM fermions are introduced or even in the number of extra dimensions. All these models will have different particular signatures. However, the fundamental mechanism by which Λ is raised is a common feature to all these models: new massive spin-1 particles, with the same quantum numbers as the SM gauge bosons, appear at the TeV scale and their couplings to the W, Z and γ obey unitarity sum rules like (5.35)–(5.40) that enforce the cancellation of the energy-growing contributions to the scattering amplitudes of the longitudinal W, Z. Vector boson fusion processes will thus provide a model-independent test of the Higgsless scenario.

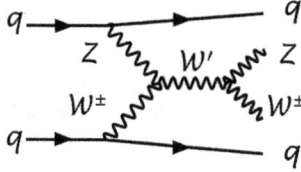

Generically, the sum rules are saturated by the inclusion of the first/few resonance(s). Moreover, as required by the smallness of the oblique corrections, the couplings of the KK gauge bosons to the light SM fermions will also be small. Therefore model-independently, we can predict the existence narrow and light resonances in the scattering of W and Z and at least one of these resonances has to appear below 1 TeV or so otherwise it will be inefficient to restore unitarity. For instance, the authors of [83] focussed on the $W_L^{\pm} Z_L \to W_L^{\pm} Z_L$ elastic scattering (charged resonances like W' will be present and the final state is easily disentangled from the background).

In Fig. 22, we show the WZ elastic cross-section and the number of events per 100 GeV bin in the $2j + 3l + \nu$ channel at the LHC with an integrated luminosity of 300 fb^{-1} and appropriate cuts. It is found that with 10 fb^{-1} of data, corresponding to one year of running at low luminosity, the LHC will probe a Higgsless W' up to 550 GeV, while covering the whole preferred range up to 1 TeV will require 60 fb^{-1}.

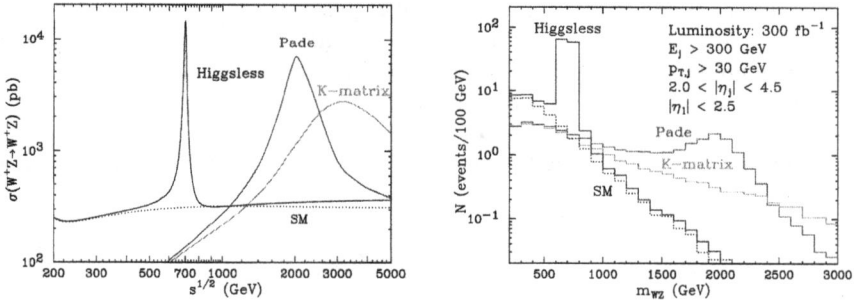

Fig. 22. Left: $W_L^{\pm} Z_L \to W_L^{\pm} Z_L$ elastic cross-sections in the SM (dotted), the Higgsless model with a 700 GeV W' (blue) and two unitarization models: Padé (red) and K-matrix (green). Right: the number of events per 100 GeV bin in the $2j + 3l + \nu$ channel at the LHC with an integrated luminosity of 300 fb^{-1} and cuts as indicated in the figure (the different curves are the same as on the left). The couplings and the partial width of W' can be predicted in a model-indepent way from the unitarity sum rules: $g_{WZW'} < g_{WWZ} M_Z^2/(\sqrt{3} M_{W'} M_W)$, $\Gamma(W' \to WZ) \sim \alpha M_{W'}^3/(144 s_W^2 M_W^2)$. From [83].

While a Higgsless W' could hardly escape detection at the LHC, we will have
to wait for a linear collider (ILC) to precisely measure its couplings and thus to
experimentally check the saturation of the unitarity sum rules.

Another way to look for the W', Z' KK resonances is through Drell–Yan
processes [80]. However, these analyses are more model-dependent since they
rely on the couplings of the way SM fermions are embedded in the model.

Let us also mention that KK gluon could easily show up as resonances in dijet
spectrum [80]. See Fig. 23. Again, the analysis depends on the localization of
the fermions in the bulk.

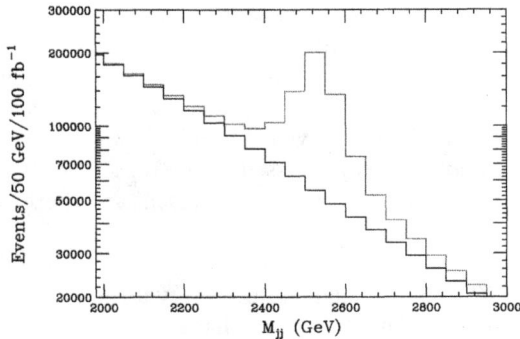

Fig. 23. Dijet invariant mass spectrum at the LHC showing a prominent resonance due to the first
gluon KK state. The black histogram corresponds to the SM background. From [80].

Finally another interesting prediction of Higgsless models is the presence of
anomalous 4- and 3-boson couplings. Indeed, in the SM the sum rules cancelling
the terms growing with the fourth power of the energy are already satisfied by
gauge invariance. In order to accomodate the contribution od the new states,
the couplings between SM gauge bosons have to be corrected. Assuming that
the sum rules are satisfied by the first resonance only it is easy to evaluate such
deviations:

$$\delta = \frac{\delta g_{WWZ}}{g_{WWZ}} \sim -\frac{1}{3} \frac{M_Z^2}{{M_Z'}^2}. \tag{5.133}$$

Here δ is an overall shift of the coupling, and the deviation has been evaluated
from the W elastic scattering sum rules. In Figure 24 we plotted the deviations
in the $WWWW$ and WWZ gauge couplings in the Higgsless model: the red
lines encircle the preferred region by EWPTs. A deviation of order 1% to 3% is
expected in the trilinear gauge couplings. This deviation is close to the present
experimental bound from LEP and might be probed by LHC. A linear collider
(ILC) will surely be able to measure such deviations. Here we stress again that

such deviations are a solid prediction of the Higgsless mechanism and are independent on the details of the specific Higgsless model we are interested in.

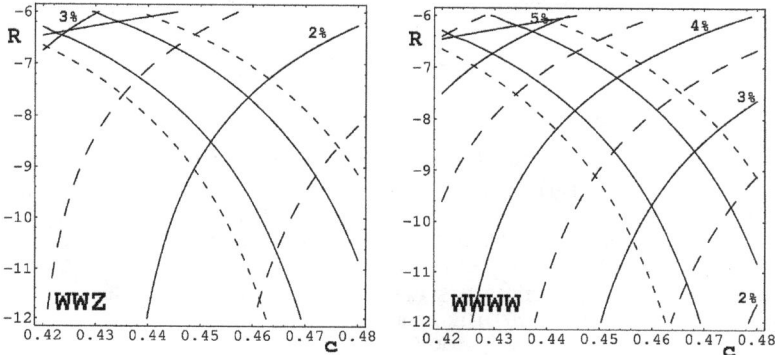

Fig. 24. Deviations in the WWZ and $WWWW$ gauge boson couplings in the Higgsless model as a function of c, the fermion localization parameter, and R, the energy scale on the UV brane. The red and blue lines are the regions preferred by S and T. From [55].

6. Recent developments and conclusion

I have tried to give an overview of recent models of EWSB. This overview inevitably suffers from a bias and it certainly cannot claim for any exhaustivity. I have chosen few models, mostly those I have worked on. To repair any injustice, if ever it is possible, I must at least mention some other approaches that appear recently and that would definitively deserve other lectures on their own:

- fat Higgs models [84];
- gauge extensions of the minimal supersymmetric standard model [85];
- bosonic seesaw [86];
- supersymmetric Little Higgs models [87];
- twin Higgs models [88];
- 6D RG induced EWSB [89]

Within few years, the LHC will tell us for sure how EW symmetry is broken. We might well see a simple Higgs. But I hope that I convinced you that there is still room for more interesting/exciting possibilities with a complete dynamics at the origin of the symmetry breaking. Only one model, most likely not even within the ones that I described in these lectures, will survive. May be by wandering along our own road, you will be led to it... Meanwhile, let us enjoy life with or without a Higgs.

Acknowledgements

I would like to thank the organizers and the students of the Les Houches summer school, as well as my collaborators on the works I presented here.

References

[1] S. Eidelman et al. [Particle Data Group], Phys. Lett. B **592**, 1 (2004).

[2] R. S. Chivukula, hep-ph/9803219. S. Dawson, hep-ph/9901280. C. F. Kolda and H. Murayama, JHEP **0007**, 035 (2000) [hep-ph/0003170]. M. Carena and H. E. Haber, Prog. Part. Nucl. Phys. **50**, 63 (2003) [hep-ph/0208209]. M. Quirós, hep-ph/0302189. A. Djouadi, hep-ph/0503172. S. Dawson, hep-ph/0510385. L. Reina, hep-ph/0512377.

[3] R. S. Chivukula, hep-ph/9803219. R. S. Chivukula, hep-ph/0011264. C. T. Hill and E. H. Simmons, Phys. Rept. **381**, 235 (2003) [Erratum-ibid. **390**, 553 (2004)] [hep-ph/0203079]. K. Lane, hep-ph/0202255. W. Kilian, hep-ph/0303015.

[4] V. A. Rubakov, Phys. Usp. **44**, 871 (2001) [Usp. Fiz. Nauk **171**, 913 (2001)] [hep-ph/0104152]. G. Gabadadze, hep-ph/0308112. C. Csáki, hep-ph/0404096. T. G. Rizzo, eConf **C040802**, L013 (2004) [hep-ph/0409309]. G. Burdman, AIP Conf. Proc. **753**, 390 (2005) [hep-ph/0409322]. A. Perez-Lorenzana, J. Phys. Conf. Ser. **18**, 224 (2005) [hep-ph/0503177]. R. Sundrum, hep-th/0508134.

[5] M. Schmaltz and D. Tucker-Smith, hep-ph/0502182.

[6] http://giroz.desy.de/physics/sfew/PUBLIC/sfew_results/preliminary/dis04/dis04.php

[7] P. W. Anderson, Phys. Rev. **130**, 439 (1963). P. W. Higgs, Phys. Lett. **12**, 132 (1964). P. W. Higgs, Phys. Rev. Lett. **13**, 508 (1964). P. W. Higgs, Phys. Rev. **145**, 1156 (1966). G. S. Guralnik, C. R. Hagen and T. W. B. Kibble, Phys. Rev. Lett. **13**, 585 (1964). F. Englert and R. Brout, Phys. Rev. Lett. **13**, 321 (1964). T. W. B. Kibble, Phys. Rev. **155**, 1554 (1967).

[8] S. Weinberg, Phys. Rev. D **19**, 1277 (1979). L. Susskind, Phys. Rev. D **20**, 2619 (1979). P. Sikivie, L. Susskind, M. B. Voloshin and V. I. Zakharov, Nucl. Phys. B **173**, 189 (1980).

[9] M. J. G. Veltman, Acta Phys. Polon. B **8**, 475 (1977).

[10] C. H. Llewellyn Smith, Phys. Lett. B **46**, 233 (1973). D. A. Dicus and V. S. Mathur, Phys. Rev. D **7**, 3111 (1973).

[11] J. M. Cornwall, D. N. Levin and G. Tiktopoulos, Phys. Rev. Lett. **30**, 1268 (1973) [Erratum-ibid. **31**, 572 (1973)]. J. M. Cornwall, D. N. Levin and G. Tiktopoulos, Phys. Rev. D **10**, 1145 (1974) [Erratum-ibid. D **11**, 972 (1975)].

[12] B. W. Lee, C. Quigg and H. B. Thacker, Phys. Rev. Lett. **38**, 883 (1977). B. W. Lee, C. Quigg and H. B. Thacker, Phys. Rev. D **16**, 1519 (1977). M. S. Chanowitz and M. K. Gaillard, Nucl. Phys. B **261**, 379 (1985).

[13] K. G. Wilson, Phys. Rev. B **4**, 3184 (1971). K. G. Wilson and J. B. Kogut, Phys. Rept. **12**, 75 (1974).

[14] A. D. Linde, JETP Lett. **23**, 64 (1976) [Pisma Zh. Eksp. Teor. Fiz. **23**, 73 (1976)]. A. D. Linde, Phys. Lett. B **70**, 306 (1977). S. Weinberg, Phys. Rev. Lett. **36**, 294 (1976).

[15] T. Hambye and K. Riesselmann, hep-ph/9708416. C. F. Kolda and H. Murayama, JHEP **0007**, 035 (2000) [hep-ph/0003170].

[16] V.F. Weisskopf Phys. Rev. **56**, 72 (1939). G. 't Hooft, in: *Recent Developments In Gauge Theories*, Proc. of the Nato Advanced Study Institute, Cargese, France, August 26 - September 8, 1979.

[17] J. A. Casas, J. R. Espinosa and I. Hidalgo, JHEP **0411**, 057 (2004) [hep-ph/0410298]. J. A. Casas, J. R. Espinosa and I. Hidalgo, JHEP **0503**, 038 (2005) [hep-ph/0502066].

[18] S. R. Coleman and E. Weinberg, Phys. Rev. D **7**, 1888 (1973).

[19] M. B. Einhorn and D. R. T. Jones, Phys. Rev. D **46**, 5206 (1992).

[20] C. S. Lam, hep-ph/0211230. C. Grojean, in: Proc. of the SUGRA20 Conference, Boston, Massachusetts, 17-20 March 2003.

[21] J. Goldstone, Nuovo Cim. **19**, 154 (1961). J. Goldstone, A. Salam and S. Weinberg, Phys. Rev. **127**, 965 (1962).

[22] S. R. Coleman and J. Mandula, Phys. Rev. **159**, 1251 (1967). R. Haag, J. T. Lopuszanski and M. Sohnius, Nucl. Phys. B **88**, 257 (1975).

[23] C. Csáki, C. Grojean, H. Murayama, L. Pilo and J. Terning, Phys. Rev. D **69**, 055006 (2004) [hep-ph/0305237].

[24] C. Csáki, C. Grojean, L. Pilo and J. Terning, Phys. Rev. Lett. **92**, 101802 (2004) [hep-ph/0308038].

[25] R. Barbieri and A. Strumia, hep-ph/0007265.

[26] M. Golden and L. Randall, Nucl. Phys. B **361**, 3 (1991). B. Holdom and J. Terning, Phys. Lett. B **247**, 88 (1990). G. Altarelli and R. Barbieri, Phys. Lett. B **253**, 161 (1991). M. E. Peskin and T. Takeuchi, Phys. Rev. Lett. **65**, 964 (1990). M. E. Peskin and T. Takeuchi, Phys. Rev. D **46**, 381 (1992). G. Altarelli, R. Barbieri and S. Jadach, Nucl. Phys. B **369**, 3 (1992) [Erratum-ibid. B **376**, 444 (1992)]. I. Maksymyk, C. P. Burgess and D. London, Phys. Rev. D **50**, 529 (1994) [hep-ph/9306267]. C. P. Burgess, S. Godfrey, H. Konig, D. London and I. Maksymyk, Phys. Rev. D **49**, 6115 (1994) [hep-ph/9312291].

[27] W. Buchmuller and D. Wyler, Nucl. Phys. B **268**, 621 (1986). B. Grinstein and M. B. Wise, Phys. Lett. B **265**, 326 (1991).

[28] R. Barbieri, A. Pomarol, R. Rattazzi and A. Strumia, Nucl. Phys. B **703**, 127 (2004) [hep-ph/0405040].

[29] K. Matchev, hep-ph/0402031. J. D. Wells, hep-ph/0512342.

[30] Z. Han and W. Skiba, Phys. Rev. D **71**, 075009 (2005) [hep-ph/0412166].

[31] M. Schmaltz, Nucl. Phys. Proc. Suppl. **117**, 40 (2003) [hep-ph/0210415]. T. Han, H. E. Logan and L. T. Wang, hep-ph/0506313. M. Perelstein, hep-ph/0512128.

[32] M. Serone, AIP Conf. Proc. **794**, 139 (2005) [hep-ph/0508019]. G. Cacciapaglia, in: Proc. of the workshop "Les Houches 2005: Physics at TeV Colliders 2005".

[33] N. S. Manton, Nucl. Phys. B **158**, 141 (1979).

[34] D. B. Fairlie, J. Phys. G **5**, L55 (1979). D. B. Fairlie, Phys. Lett. B **82**, 97 (1979). P. Forgács and N. S. Manton, Commun. Math. Phys. **72**, 15 (1980). S. Randjbar-Daemi, A. Salam and J. Strathdee, Nucl. Phys. B **214**, 491 (1983). I. Antoniadis and K. Benakli, Phys. Lett. B **326**, 69 (1994) [hep-th/9310151]. D. Kapetanakis and G. Zoupanos, Phys. Rept. **219**, 1 (1992). G. R. Dvali, S. Randjbar-Daemi and R. Tabbash, Phys. Rev. D **65**, 064021 (2002) [hep-ph/0102307].

[35] P. Candelas, G. T. Horowitz, A. Strominger and E. Witten, Nucl. Phys. B **258**, 46 (1985). L. J. Dixon, J. A. Harvey, C. Vafa and E. Witten, Nucl. Phys. B **261**, 678 (1985). L. J. Dixon, J. A. Harvey, C. Vafa and E. Witten, Nucl. Phys. B **274**, 285 (1986).

[36] P. Candelas, G. T. Horowitz, A. Strominger and E. Witten, Nucl. Phys. B **258**, 46 (1985). E. Witten, Nucl. Phys. B **258**, 75 (1985). L. E. Ibáñez, H. P. Nilles and F. Quevedo, Phys. Lett.

B **187**, 25 (1987). L. E. Ibáñez, J. Mas, H. P. Nilles and F. Quevedo, Nucl. Phys. B **301**, 157 (1988).

[37] Y. Hosotani, Phys. Lett. B **126**, 309 (1983), Phys. Lett. B **129**, 193 (1983), Annals Phys. **190**, 233 (1989). H. Hatanaka, T. Inami and C. S. Lim, Mod. Phys. Lett. A **13**, 2601 (1998) [hep-th/9805067]. H. Hatanaka, Prog. Theor. Phys. **102**, 407 (1999) [hep-th/9905100]. M. Kubo, C. S. Lim and H. Yamashita, hep-ph/0111327. L. J. Hall, H. Murayama and Y. Nomura, Nucl. Phys. B **645**, 85 (2002) [hep-th/0107245]. Nucl. Phys. B **639**, 307 (2002) [hep-ph/0107331]. I. Antoniadis, K. Benakli and M. Quirós, New J. Phys. **3**, 20 (2001) [hep-th/0108005].

[38] C. Csáki, C. Grojean and H. Murayama, Phys. Rev. D **67**, 085012 (2003) [hep-ph/0210133].

[39] I. Gogoladze, Y. Mimura and S. Nandi, Phys. Lett. B **562**, 307 (2003) [hep-ph/0302176]. G. Burdman and Y. Nomura, Nucl. Phys. B **656**, 3 (2003) [hep-ph/0210257]. N. Haba, M. Harada, Y. Hosotani and Y. Kawamura, Nucl. Phys. B **657**, 169 (2003) [Erratum-ibid. B **669**, 381 (2003)] [hep-ph/0212035]. N. Haba and Y. Shimizu, Phys. Rev. D **67**, 095001 (2003) [Erratum-ibid. D **69**, 059902 (2004)] [hep-ph/0212166]. I. Gogoladze, Y. Mimura and S. Nandi, Phys. Rev. Lett. **91**, 141801 (2003) [hep-ph/0304118]. N. Haba, Y. Hosotani and Y. Kawamura, Prog. Theor. Phys. **111**, 265 (2004) [hep-ph/0309088]. K. W. Choi, N. Y. Haba, K. S. Jeong, K. I. Okumura, Y. Shimizu and M. Yamaguchi, JHEP **0402**, 037 (2004) [hep-ph/0312178]. N. Haba, Y. Hosotani, Y. Kawamura and T. Yamashita, Phys. Rev. D **70**, 015010 (2004) [hep-ph/0401183]. N. Haba and T. Yamashita, JHEP **0402**, 059 (2004) [hep-ph/0401185]. Y. Hosotani, S. Noda and K. Takenaga, Phys. Rev. D **69**, 125014 (2004) [hep-ph/0403106]. Y. Hosotani, hep-ph/0408012. K. Hasegawa, C. S. Lim and N. Maru, Phys. Lett. B **604**, 133 (2004) [hep-ph/0408028]. J. L. Diaz-Cruz, Mod. Phys. Lett. A **20**, 2397 (2005) [hep-ph/0409216]. Y. Hosotani, S. Noda and K. Takenaga, Phys. Lett. B **607**, 276 (2005) [hep-ph/0410193]. Y. Hosotani and M. Mabe, Phys. Lett. B **615**, 257 (2005) [hep-ph/0503020]. I. Gogoladze, T. Li, Y. Mimura and S. Nandi, Phys. Rev. D **72**, 055006 (2005) [hep-ph/0504082]. A. Aranda, J. L. Diaz-Cruz and A. Rosado, hep-ph/0507230. A. Aranda and J. L. Diaz-Cruz, hep-ph/0510138. N. Haba, S. Matsumoto, N. Okada and T. Yamashita, hep-ph/0511046.

[40] C. A. Scrucca, M. Serone and L. Silvestrini, Nucl. Phys. B **669**, 128 (2003) [hep-ph/0304220]. C. A. Scrucca, M. Serone, L. Silvestrini and A. Wulzer, JHEP **0402**, 049 (2004) [hep-th/0312267].

[41] G. Martinelli, M. Salvatori, C. A. Scrucca and L. Silvestrini, JHEP **0510**, 037 (2005) [hep-ph/0503179].

[42] G. Cacciapaglia, C. Csáki and S. C. Park, hep-ph/0510366.

[43] G. Panico, M. Serone and A. Wulzer, hep-ph/0510373.

[44] N. Arkani-Hamed, A. G. Cohen and H. Georgi, Phys. Lett. B **513**, 232 (2001) [hep-ph/0105239].

[45] A. Hebecker and J. March-Russell, Nucl. Phys. B **625**, 128 (2002) [hep-ph/0107039].

[46] G. von Gersdorff, N. Irges and M. Quirós, Nucl. Phys. B **635**, 127 (2002) [hep-th/0204223], hep-ph/0206029. R. Sundrum, unpublished.

[47] G. von Gersdorff, N. Irges and M. Quirós, Phys. Lett. B **551**, 351 (2003) [hep-ph/0210134].

[48] C. Biggio and M. Quirós, Nucl. Phys. B **703**, 199 (2004) [hep-ph/0407348].

[49] R. Contino, Y. Nomura and A. Pomarol, Nucl. Phys. B **671**, 148 (2003) [hep-ph/0306259]. K. Agashe, R. Contino and A. Pomarol, Nucl. Phys. B **719**, 165 (2005) [hep-ph/0412089]. K. Agashe and R. Contino, hep-ph/0510164.

[50] K. y. Oda and A. Weiler, Phys. Lett. B **606**, 408 (2005) [hep-ph/0410061]. Y. Hosotani and M. Mabe, Phys. Lett. B **615**, 257 (2005) [hep-ph/0503020]. Y. Hosotani, S. Noda, Y. Sakamura and S. Shimasaki, hep-ph/0601241.

[51] D. B. Kaplan and H. Georgi, Phys. Lett. B **136**, 183 (1984). D. B. Kaplan, H. Georgi and S. Dimopoulos, Phys. Lett. B **136**, 187 (1984). H. Georgi, D. B. Kaplan and P. Galison, Phys. Lett. B **143**, 152 (1984). H. Georgi and D. B. Kaplan, Phys. Lett. B **145**, 216 (1984). M. J. Dugan, H. Georgi and D. B. Kaplan, Nucl. Phys. B **254**, 299 (1985).

[52] G. Panico and M. Serone, JHEP **0505**, 024 (2005) [hep-ph/0502255]. N. Maru and K. Takenaga, Phys. Rev. D **72**, 046003 (2005) [hep-th/0505066].

[53] J. R. Espinosa, M. Losada and A. Riotto, Phys. Rev. D **72**, 043520 (2005) [hep-ph/0409070]. C. Grojean, G. Servant and J. D. Wells, Phys. Rev. D **71**, 036001 (2005) [hep-ph/0407019].

[54] G. Cacciapaglia, C. Csáki, C. Grojean and J. Terning, eConf **C040802**, FRT004 (2004). C. Csáki, hep-ph/0412339.

[55] G. Cacciapaglia, in: Proc. of the workshop "Les Houches 2005: Physics at TeV Colliders 2005".

[56] B. Lillie and J. Tening, in: Report of the working group on CP Studies and Non-Standard Higgs Physics (2006).

[57] C. Csáki, J. Hubisz and P. Meade, hep-ph/0510275.

[58] R. Foadi, S. Gopalakrishna and C. Schmidt, JHEP **0403**, 042 (2004) [hep-ph/0312324]. R. Casalbuoni, S. De Curtis and D. Dominici, Phys. Rev. D **70**, 055010 (2004) [hep-ph/0405188]. N. Evans and P. Membry, hep-ph/0406285. H. Georgi, Phys. Rev. D **71**, 015016 (2005) [hep-ph/0408067]. M. Perelstein, JHEP **0410**, 010 (2004) [hep-ph/0408072]. R. Sekhar Chivukula, E. H. Simmons, H. J. He, M. Kurachi and M. Tanabashi, Phys. Rev. D **71**, 035007 (2005) [hep-ph/0410154].

[59] J. Hirn and J. Stern, Eur. Phys. J. C **34**, 447 (2004) [hep-ph/0401032]. J. Hirn and J. Stern, JHEP **0409**, 058 (2004) [hep-ph/0403017]. J. Hirn and J. Stern, hep-ph/0504277. S. Gabriel, S. Nandi and G. Seidl, Phys. Lett. B **603**, 74 (2004) [hep-ph/0406020]. T. Nagasawa and M. Sakamoto, Prog. Theor. Phys. **112**, 629 (2004) [hep-ph/0406024]. C. D. Carone and J. M. Conroy, Phys. Rev. D **70**, 075013 (2004) [hep-ph/0407116]. S. Chang, S. C. Park and J. Song, Phys. Rev. D **71**, 106004 (2005) [hep-ph/0502029]. N. K. Tran, hep-th/0502205.

[60] K. Agashe and G. Servant, Phys. Rev. Lett. **93**, 231805 (2004) [hep-ph/0403143]. K. Agashe and G. Servant, JCAP **0502**, 002 (2005) [hep-ph/0411254].

[61] Z. Chacko, M. Graesser, C. Grojean and L. Pilo, Phys. Rev. D **70**, 084028 (2004) [hep-th/0312117]. M. Carena, J. D. Lykken and M. Park, Phys. Rev. D **72**, 084017 (2005) [hep-ph/0506305]. R. Bao, M. Carena, J. Lykken, M. Park and J. Santiago, hep-th/0511266.

[62] Y. Nomura, D. R. Smith and N. Weiner, Nucl. Phys. B **613**, 147 (2001) [hep-ph/0104041].

[63] G. Cacciapaglia, C. Csáki, C. Grojean, M. Reece and J. Terning, Phys. Rev. D **72**, 095018 (2005) [hep-ph/0505001].

[64] C. Schwinn, Phys. Rev. D **69**, 116005 (2004) [hep-ph/0402118]. D. A. Dicus and H. J. He, Phys. Rev. D **71**, 093009 (2005) [hep-ph/0409131]. C. Schwinn, Phys. Rev. D **71**, 113005 (2005) [hep-ph/0504240].

[65] R. S. Chivukula, D. A. Dicus and H. J. He, Phys. Lett. B **525**, 175 (2002) hep-ph/0111016. R. S. Chivukula and H. J. He, Phys. Lett. B **532**, 121 (2002) hep-ph/0201164. R. S. Chivukula, D. A. Dicus, H. J. He and S. Nandi, Phys. Lett. B **562**, 109 (2003) hep-ph/0302263. S. De Curtis, D. Dominici and J. R. Pelaez, Phys. Lett. B **554**, 164 (2003) hep-ph/0211353. Phys. Rev. D **67**, 076010 (2003) hep-ph/0301059. Y. Abe, N. Haba, Y. Higashide, K. Kobayashi and M. Matsunaga, Prog. Theor. Phys. **109**, 831 (2003) hep-th/0302115. H. J. He, hep-ph/0412113.

[66] M. Papucci, hep-ph/0408058.

[67] T. Ohl and C. Schwinn, Phys. Rev. D **70**, 045019 (2004) [hep-ph/0312263]. Y. Abe, N. Haba, K. Hayakawa, Y. Matsumoto, M. Matsunaga and K. Miyachi, Prog. Theor. Phys. **113**, 199 (2005) [hep-th/0402146]. A. Muck, L. Nilse, A. Pilaftsis and R. Ruckl, Phys. Rev. D **71**, 066004 (2005) [hep-ph/0411258]. L. Nilse, hep-ph/0512216.

[68] K. Agashe, A. Delgado, M. J. May and R. Sundrum, JHEP **0308**, 050 (2003) [hep-ph/0308036].

[69] R. Barbieri, A. Pomarol and R. Rattazzi, Phys. Lett. B **591**, 141 (2004) [hep-ph/0310285].

[70] J. M. Maldacena, Adv. Theor. Math. Phys. **2**, 231 (1998) [Int. J. Theor. Phys. **38**, 1113 (1999)] [hep-th/9711200]. N. Arkani-Hamed, M. Porrati and L. Randall, JHEP **0108**, 017 (2001) [hep-th/0012148]. R. Rattazzi and A. Zaffaroni, JHEP **0104**, 021 (2001) [hep-th/0012248]. M. Perez-Victoria, JHEP **0105**, 064 (2001) [hep-th/0105048]. K. Agashe and A. Delgado, Phys. Rev. D **67**, 046003 (2003) [hep-th/0209212].

[71] R. Contino and A. Pomarol, JHEP **0411**, 058 (2004) [hep-th/0406257].

[72] G. Cacciapaglia, C. Csáki, C. Grojean and J. Terning, Phys. Rev. D **71**, 035015 (2005) [hep-ph/0409126].

[73] C. Csáki, C. Grojean, J. Hubisz, Y. Shirman and J. Terning, Phys. Rev. D **70**, 015012 (2004) [hep-ph/0310355].

[74] Y. Nomura, JHEP **0311**, 050 (2003) [hep-ph/0309189].

[75] Y. Grossman and M. Neubert, Phys. Lett. B **474**, 361 (2000) [hep-ph/9912408]. T. Gherghetta and A. Pomarol, Nucl. Phys. B **586**, 141 (2000) [hep-ph/0003129]. S. J. Huber and Q. Shafi, Phys. Lett. B **498**, 256 (2001) [hep-ph/0010195]. J. A. Bagger, F. Feruglio and F. Zwirner, Phys. Rev. Lett. **88**, 101601 (2002) [hep-th/0107128]. J. Bagger, F. Feruglio and F. Zwirner, JHEP **0202**, 010 (2002) [hep-th/0108010].

[76] G. Burdman and Y. Nomura, Phys. Rev. D **69**, 115013 (2004) [hep-ph/0312247].

[77] H. Davoudiasl, J. L. Hewett, B. Lillie and T. G. Rizzo, Phys. Rev. D **70**, 015006 (2004) [hep-ph/0312193].

[78] G. Cacciapaglia, C. Csáki, C. Grojean and J. Terning, Phys. Rev. D **70**, 075014 (2004) [hep-ph/0401160].

[79] R. S. Chivukula, E. H. Simmons, H. J. He, M. Kurachi and M. Tanabashi, Phys. Rev. D **70**, 075008 (2004) [hep-ph/0406077]. R. S. Chivukula, E. H. Simmons, H. J. He, M. Kurachi and M. Tanabashi, Phys. Lett. B **603**, 210 (2004) [hep-ph/0408262].

[80] H. Davoudiasl, J. L. Hewett, B. Lillie and T. G. Rizzo, JHEP **0405**, 015 (2004) [hep-ph/0403300]. J. L. Hewett, B. Lillie and T. G. Rizzo, JHEP **0410**, 014 (2004) [hep-ph/0407059]. B. Lillie, hep-ph/0505074.

[81] R. Foadi, S. Gopalakrishna and C. Schmidt, Phys. Lett. B **606**, 157 (2005) [hep-ph/0409266]. R. S. Chivukula, E. H. Simmons, H. J. He, M. Kurachi and M. Tanabashi, Phys. Rev. D **71**, 115001 (2005) [hep-ph/0502162]. R. Casalbuoni, S. De Curtis, D. Dolce and D. Dominici, Phys. Rev. D **71**, 075015 (2005) [hep-ph/0502209]. R. Sekhar Chivukula, E. H. Simmons, H. J. He, M. Kurachi and M. Tanabashi, Phys. Rev. D **72**, 015008 (2005) [hep-ph/0504114]. H. Georgi, hep-ph/0508014. R. S. Chivukula, E. H. Simmons, H. J. He, M. Kurachi and M. Tanabashi, Phys. Rev. D **72**, 075012 (2005) [hep-ph/0508147]. R. Foadi and C. Schmidt, hep-ph/0509071. R. Sekhar Chivukula, E. H. Simmons, H. J. He, M. Kurachi and M. Tanabashi, Phys. Rev. D **72**, 095013 (2005) [hep-ph/0509110].

[82] C. Csáki, J. Erlich and J. Terning, Phys. Rev. D **66**, 064021 (2002) [hep-ph/0203034].

[83] A. Birkedal, K. Matchev and M. Perelstein, Phys. Rev. Lett. **94**, 191803 (2005) [hep-ph/0412278]. A. Birkedal, K. T. Matchev and M. Perelstein, hep-ph/0508185.

[84] R. Harnik, G. D. Kribs, D. T. Larson and H. Murayama, Phys. Rev. D **70**, 015002 (2004) [hep-ph/0311349]. R. Kitano, G. D. Kribs and H. Murayama, Phys. Rev. D **70**, 035001 (2004) [hep-ph/0402215]. S. Chang, C. Kilic and R. Mahbubani, Phys. Rev. D **71**, 015003 (2005) [hep-ph/0405267]. A. Delgado and T. M. P. Tait, JHEP **0507**, 023 (2005) [hep-ph/0504224].

[85] M. A. Luty, J. Terning and A. K. Grant, Phys. Rev. D **63**, 075001 (2001) [hep-ph/0006224]. P. Batra, A. Delgado, D. E. Kaplan and T. M. P. Tait, JHEP **0402**, 043 (2004) [hep-ph/0309149]. J. A. Casas, J. R. Espinosa and I. Hidalgo, JHEP **0401**, 008 (2004) [hep-ph/0310137].

[86] H. D. Kim, Phys. Rev. D **72**, 055015 (2005) [hep-ph/0501059].

[87] A. Birkedal, Z. Chacko and M. K. Gaillard, JHEP **0410**, 036 (2004) [hep-ph/0404197]. P. H. Chankowski, A. Falkowski, S. Pokorski and J. Wagner, Phys. Lett. B **598**, 252 (2004) [hep-ph/0407242]. Z. Berezhiani, P. H. Chankowski, A. Falkowski and S. Pokorski, hep-ph/0509311. C. Csaki, G. Marandella, Y. Shirman and A. Strumia, hep-ph/0510294.

[88] Z. Chacko, H. S. Goh and R. Harnik, hep-ph/0506256. R. Barbieri, T. Gregoire and L. J. Hall, hep-ph/0509242. Z. Chacko, Y. Nomura, M. Papucci and G. Perez, hep-ph/0510273.

[89] E. Dudas, C. Papineau and V. A. Rubakov, hep-th/0512276.

Course 8

ASPECTS OF STRING PHENOMENOLOGY

Emilian Dudas

CPhT, Ecole Polytechnique 91128 Palaiseau Cedex, France
and
LPT, Bat. 210, Univ. Paris-Sud, 91405 Orsay Cedex, France

D. Kazakov, S. Lavignac and J. Dalibard, eds.
Les Houches, Session LXXXIV, 2005
Particle Physics Beyond the Standard Model
© 2006 Elsevier B.V. All rights reserved

Contents

1. Preamble

These lectures contain an introduction in string theory with emphasize towards phenomenological applications. We present in particular the basis of orientifold constructions, the compactification from ten to four dimensions, intersecting brane world models, mechanisms for supersymmetry breaking, moduli stabilisation and the large extra dimensions models.

2. Introduction

The search for a microscopic quantum theory containing the four fundamental interactions: electromagnetic, weak, strong and gravitational haunted the greatest physicists of the twenties century, starting with the unified field theory program of Einstein. Since the discovery of the anomaly cancelation for superstrings in ten dimensions (10d) [1], the construction of the heterotic strings [2] and the seminal papers on the compactification to four dimensions [3], [4], string theory has become the best candidate for a fundamental quantum theory of all interactions including Einstein gravity. The theory contains only one free dimensionfull parameter, the string scale M_s, while the four-dimensional (4d) gauge group and the matter content are manifestations of the geometric properties of the internal space that, however, we are unable to select in a unique fashion. The Standard Model hopefully would correspond to a particular internal space or vacuum configuration chosen by nature by some still unknown mechanism. The 4d low-energy couplings depend only on the string scale and on various vacuum expectation values (vev's) of fields describing the string coupling constant, the size and the shape of the internal manifold. There is therefore, in principle, the hope to understand the empirically observed pattern of the parameters in the Standard Model. A long activity in heterotic strings [5], [6] was devoted to this program [7], in the hope that these rather tight constraints would determine in some way the right vacuum describing our world. Despite serious insights into the structure and the phenomenological properties of 4d models [5, 6], no unique candidate having as low-energy limit the Standard Model emerged[1]. Moreover, there were (and

---

[1]This embarrassing for the string model builders was reconsidered recently and converted into a new picture, *the landscape* [8], which takes advantage of the huge number of vacua of string com-

E. Dudas

there still are) conceptual problems to be solved, as for example the large degeneracy of the string vacua and the related problems of spacetime supersymmetry breaking and moduli stabilization. Most of these problems asked for a better understanding of the strong coupling limit of string theory, of which very little was known for a long time. The other string theories were, for a long time, discarded as inappropriate for phenomenological purposes. Type II strings were considered unable to produce a realistic gauge group, while Type I strings, despite serious advances made over the years [11–14], that revealed striking differences with respect to heterotic strings, were less studied and their consistency rules not widely known as for heterotic vacua.

By the middle of the last decade, it became clear that all known string theories are actually related by various dualities to each other and to a mysterious eleven dimensional theory, the so-called M-theory [15]. It therefore became possible to obtain some nonperturbative string results, at least for theories with enough supersymmetry. Moreover, the discovery and the study of D-branes [16] put the duality predictions on a firmer quantitative basis and, on the other hand, was an important step in unraveling the geometric structure underlying the consistency conditions of orientifold vacua, stimulating a new activity in this field [17]. The first chiral 4d Type I model was proposed [18], and efforts on 4d model building allowed a better understanding of supersymmetric 4d vacua [19] and of their gauge and gravitational anomaly cancelation mechanisms [20], similar to the 6d generalized Green-Schwarz mechanism [21]. The presence of D-branes in orientifold models led to new mechanisms for breaking supersymmetry by compactification [22], by internal magnetic fields [23, 24] and their T-dual version, the intersecting brane worlds [28] or directly on some (anti)branes [30–32], providing perturbatively stable non-BPS analogs of Type IIB configurations [33]. More recently, various NS-NS and RR fluxes in type II string theories [37] brought a new perspective on the moduli stabilization problem, of crucial importance for phenomenological applications [38].

On the phenomenological side, the M-theory compactification of Horava and Witten [36], with a fundamental scale $M_{11} \sim 2 \times 10^{16}$ GeV, provided a framework [39] for the perturbative MSSM unification of gauge couplings [40], and stimulated studies of 4d compactifications [41, 42] and of supersymmetry breaking along the new (eleventh) dimension [41,43,70]. Moreover, it was noticed [44] that in Type I strings the string scale can be lowered all the way down to the TeV range. Similar ideas appeared for lowering the fundamental Planck scale in theories with (sub)millimeter gravitational dimensions [45], as an alternative solution to the gauge hierarchy problem, and, simultaneously, a new way for lowering the

---

pactifications in order to analyze in a statistical manner the low energy physics, by invoking for the cosmological problem [9] and eventually the hierarchy problem [10] anthropic type arguments.

GUT scale in theories with large (TeV) dimensions [46] was proposed. The new emerging picture found a simple realization in a perturbative Type I setting [47] with low string scale. On the other hand, a new perspective on the dynamical (technicolor-like) mechanisms for electroweak symmetry breaking appeared in strongly curved (or warped) compactifications [48], in which the strongly coupled theory has a weakly coupled, dual gravitational description via the AdS/CFT correspondence [49]. In a noncompact version of this proposal, gravity is localized and a standard 4d cosmology arises without the standard problems of compactified higher-dimensional theories.

The goal of the present lectures is to present a comprehensive review on the basic necessary tools for an phenomenologically oriented string theorist and to introduce the new ideas underlying the modern string phenomenology. The convention for the metric signature throughout the paper is $(-, +, \cdots +)$, ten-dimensional (10d) indices are denoted by $A, B, \cdots$, eleven dimensional indices by $I, J, \cdots$, five-dimensional (5d) indices by $M, N, \cdots$ and four-dimensional indices (4d) by $\mu, \nu, \cdots$.

3. String theories and D-branes: spectra and dualities

To date, the (super)strings are the only known consistent quantum theories including Einstein gravity. They are therefore promising candidates for a unifying picture of elementary particles and fundamental interactions.

It has been known for a long time that there are five consistent (anomaly-free) superstring theories in 10d, namely:

– The heterotic closed string theories, with gauge groups $SO(32)$ and $E_8 \times E_8$ and $\mathcal{N} = 1$ spacetime supersymmetry, that after a toroidal compactification corresponds to $\mathcal{N} = 4$ supersymmetry in four dimensions. There are also nonsupersymmetric heterotic vacua, in particular a non-tachyonic one based on the gauge group $SO(16) \times SO(16)$ [50, 51].

– The (non-chiral) Type IIA and (chiral) Type IIB closed string theories, with 10d $\mathcal{N} = 2$ spacetime supersymmetry, that after a toroidal compactification corresponds to $\mathcal{N} = 8$ supersymmetry in four dimensions. Of particular importance in type II theories are the presence in the massless spectrum of Ramond-Ramond forms of odd (even) rank for type IIB (type IIA). As shown by Polchinski [16], their presence signals the presence of localized dynamical objects, *the BPS D-branes*, of worldvolume dimension appropriate to couple to the Ramond-Ramond (RR) forms, more precisely:

IIB: D1, D3, D5, D7 and D9
IIA: D0, D2, D4, D6 and D8

For type IIB, there is also an instanton-like BPS object, the D(-1) brane. Moreover, since the D9 brane is a space-filling object, only non-BPS $D9$-$\overline{D}9$ pairs are
actually possible. D-branes are dynamical, since they contain gauge and matter
fields living on their vorldvolume. In a toroidal compactification, M coincident
Dp-branes have an $\mathcal{N} = 4$ 4d supersymmetric spectrum and generate a gauge
group $U(M)$, with M^2 gauge bosons coming from open strings $|i, \bar{j}\rangle$ with the two
ends on any D-brane, $i = 1 \cdots M$. They are described by Chan-Paton hermitian
matrices $\lambda_{i\bar{j}}$. Splitting the M D-branes into two groups of coincident M_1 and
M_2 branes, with $M_1 + M_2 = M$ breaks the gauge group into $U(M_1) \otimes U(M_2)$.
As shown in a series of papers by A.Sen, the complementary branes also exist but they are non-BPS with a non-supersymmetric open string spectrum and
are unstable [33]. In particular, there is a complex open-string tachyon, whose
condensation could play a role in cosmology [34].

– The Type I open string theory, with gauge group $SO(32)$ and $\mathcal{N} = 1$
10d supersymmetry. The theory contains D1, D5 and D9 BPS branes and also
non-dynamical mirror-like hyperplanes of spacetime dimension two, six and ten
called O1, O5 and O9 planes. If D branes are non-coincident with the O-planes
(more precisely, no massless states in the mixed D-brane / O-plane Mobius amplitude), the gauge group is unitary, but physical states have to be invariant under
a mirror symmetry defined by the O-planes, called *orientifold projection*. If the
D-branes are coincident with the O-planes, (more precisely, there are massless
states in the mixed D-brane / O-plane Mobius amplitude), the gauge group is
broken by the orientifold projection to to orthogonal $O(M)$ or $USp(M)$ gauge
groups [53], depending on the nature of the O-planes. This theory can be defined
more precisely as a projection (orientifold) type I = type IIB/Ω of the Type IIB
string [11], as we will see in more detail later on.

The massless modes of the above superstring theories and their interactions are
described by the known effective 10d supergravity theories, namely:

– The low energy limit of the the two heterotic theories *and* of the Type I
open string are described by the ten dimensional $\mathcal{N} = 1$ (or $(1, 0)$) supergravity,
coupled to the super Yang-Mills system based on the gauge groups $E_8 \times E_8$ and
$SO(32)$, respectively.

– The low energy limit of the Type II strings is described by the $\mathcal{N} = 2$ Type
IIA supergravity (with $(1, 1)$ supersymmetry) and Type IIB supergravity (with
$(2, 0)$ supersymmetry).

The above picture evolved considerably in the last ten years, in the sense of
connecting seemingly different string theories. First of all, the T-duality of type
IIB and IIA strings, which in particular maps the even RR forms of type IIB
into the odd RR forms of type IIA, maps also appropriately the corresponding
D-branes. Indeed, a T-duality in a coordinate parallel to a Dp brane generate a
$D(p - 1)$ brane perpendicular to the T-dual coordinate, whereas a T-duality in a

coordinate perpendicular to the Dp brane generates a $D(p+1)$ brane parallel to the T-dual coordinate.

Secondly, from the very beginning of their construction, it was a puzzle that the heterotic $SO(32)$ and the Type I ten dimensional strings share the same low-energy theory. Indeed, the two low-energy actions coincide if the following identifications are made

$$\lambda_I = \frac{1}{\lambda_H}, \quad M_I = \frac{M_H}{\sqrt{\lambda_H}}, \tag{3.1}$$

where M_I, M_H are the heterotic and Type I string scales and λ_I, λ_H are the corresponding string couplings. A natural conjecture was made, that the two string theories are dual (in the sense of the weak-strong mapping of the gauge coupling) to each other [55, 56]. New arguments in favor of this duality came soon:

– The heterotic $SO(32)$ string can be obtained as a soliton solution of the Type I string [57].

– There is a precise mapping of BPS states (and their masses) betwwen the two theories. If we compactify, for example, both theories to nine dimensions on a circle of radius $R_I(R_H)$ in Type I (heterotic) units, we can relate states with the masses

$$\mathcal{M}_I^2 = l^2 R_I^2 M_I^4 + \frac{m^2 R_I^2 M_I^4}{\lambda_I^2} + \frac{n^2}{R_I^2} \quad \leftrightarrow$$
$$\mathcal{M}_H^2 = m^2 R_H^2 M_H^4 + \frac{l^2 R_H^2 M_H^4}{\lambda_H^2} + \frac{n^2}{R_H^2}, \tag{3.2}$$

where n, l (n, m) are Kaluza-Klein and winding numbers on Type I (heterotic) side. It is interesting to notice in this formula how perturbative heterotic winding states (m) become non perturbative on the Type I side. An important role in checking dualities in various dimensions is played by the D-branes [16], which correspond on the heterotic side to non perturbative states.

A second, even more surprising conclusion, was reached in studying the strong coupling limit of the ten dimensional Type IIA string. It was already known that a simple truncation of eleven dimensional supergravity [54] on a circle of radius R_{11} gives the Type IIA supergravity in 10d, and that the Type IIA string coupling λ is related to the radius by [55]

$$M_{11} R_{11} = \lambda^{2/3}. \tag{3.3}$$

On the other hand, if we consider Kaluza-Klein masses of the compactified 11d supergravity and map them in Type IIA string units, we find

$$
m_n = \frac{n}{R_{11}} \quad \leftrightarrow \quad m_n = \frac{n}{\lambda} M_{IIA}. \tag{3.4}
$$

Therefore, on Type IIA side, they can be interpreted as non perturbative, and, with a bit more effort, as BPS D0 brane states. The natural conclusion is that in the strong coupling limit $\lambda \to \infty$ of the Type IIA string, a new dimension emerges ($R_{11} \to \infty$ using (3.3)) and the low energy theory becomes the uncompactified 11d supergravity [58], [55]! As there is no known quantum theory whose low energy limit describes the 11d supergravity, a new name was invented for this underlying structure, the M-theory [15].

Soon after, Horava and Witten gave convincing arguments that the 11d supergravity compactified on a line segment S^1/Z_2 (or, equivalently, on a circle with opposite points identified) should describe the strong coupling limit of the $E_8 \times E_8$ heterotic string [36]. They argued that the two gauge factors sit at the ends of the interval, very much like the gauge quantum numbers of open strings are sitting at their ends. The basic argument is that only half (one Majorana-Weyl) of the original (Majorana) 11d gravitino lives on the boundary. This would produce gravitational anomalies unless 248 new Majorana-Weyl fermions appear at each end. This is exactly the dimension of the gauge group E_8.

Finally, let us notice that in ten dimensions the Type IIB string is conjectured to be self-dual in the sense of an $SL(2, Z)$ strong-weak coupling S-duality. Moreover, the $SO(32)$ heterotic string compactified on a circle of radius R is T-dual to $E_8 \times E_8$ heterotic string compactified on a circle of radius $1/R$. By combining all the above information one can build a whole web of dualities, which becomes richer and richer when new space dimensions are compactified [59].

4. Orientifolds and the Type I string

The Type I string describes the dynamics of open and closed superstrings. Denoting by $0 \leq \sigma \leq \pi$ the coordinate describing the open string at a given time, the two ends $\sigma = 0, \pi$ contain the gauge group (Chan-Paton) degrees of freedom and the corresponding charged matter fields. The open string quantum states can be conveniently described by matrices

$$
|k; a \rangle = \sum_{i,j=1}^{M} \lambda_{i,j}^{a} |k; i, j \rangle, \tag{4.1}
$$

where $i, j = 1 \cdots M$ denote Chan-Paton indices and k other internal quantum numbers. At the ends of open strings, we must add boundary conditions, which for string coordinates can be of two types

$$\frac{\partial X^{\mu}}{\partial \sigma}\bigg|_{\sigma=0,\pi} = 0 \quad (N), \quad X^{\mu}|_{\sigma=0,\pi} = \text{cst} \quad (D), \tag{4.2}$$

where the two different possibilities denote the Neumann (N) and Dirichlet (D) strings. As will be explained later on, even if we start with a theory containing only Neumann strings, the Dirichlet strings can arise after performing various T-duality operations or in orbifolds containing Z_2-type elements. Since by joining two open strings one can create a closed string, propagation of closed strings must be added for consistency. The corresponding quantum fluctuations produce the closed (gravitational-type) spectrum of the theory, neutral under the Chan-Paton gauge group, that always contains the gravitational (super)multiplet. The string oscillators are defined by the Fourier modes of the string coordinates. For closed string coordinates, the oscillator expansion reads

$$\begin{aligned}
X_c^{\mu} &= x^{\mu} + \alpha' p^{\mu} \tau + \frac{i}{2}\sqrt{2\alpha'} \sum_{n \neq 0} \frac{1}{n} \left[\alpha_n^{\mu} e^{-2in(\tau-\sigma)} + \tilde{\alpha}_n^{\mu} e^{-2in(\tau+\sigma)} \right] \\
&\equiv X(\tau - \sigma) + \tilde{X}(\tau + \sigma).
\end{aligned} \tag{4.3}$$

The usual canonical quantization gives the following commutators for the left movers: $[\alpha_m^{\mu}, \alpha_n^{\mu}] = m\delta_{m+n}\eta^{\mu\nu}$, and similarly for the right movers. For open strings with Neumann boundary conditions, for example, the oscillator expansion reads

$$X_o^{\mu} = x^{\mu} + 2\alpha' p^{\mu} \tau + i\sqrt{2\alpha'} \sum_{n \neq 0} \frac{1}{n} \left[\alpha_n^{\mu} e^{-in\tau} \cos n\sigma \right]. \tag{4.4}$$

Type I can be seen as a projection (or *orientifold*) of Type IIB theory, obtained by projecting the Type IIB spectrum by the involution Ω, exchanging the left and right closed oscillators $\alpha_m^{\mu}, \tilde{\alpha}_m^{\mu}$ and acting on the open-strings ones by phases

$$\begin{aligned}
\text{closed:} \quad & \Omega X(\tau, \sigma) \Omega^{-1} = X(\tau, -\sigma), \quad \alpha_m^{\mu} \leftrightarrow \tilde{\alpha}_m^{\mu}, \\
\text{open:} \quad & \Omega X(\tau, \sigma) \Omega^{-1} = X(\tau, \pi - \sigma), \quad \alpha_m^{\mu} \to \pm(-1)^m \alpha_m^{\mu}.
\end{aligned} \tag{4.5}$$

Ω is by definition an involution, *i.e.* $\Omega^2 = 1$. The action of Ω on the zero-modes (compactification lattice) of closed strings is by interchanging left and right momenta $\mathbf{p}_L \leftrightarrow \mathbf{p}_R$. Ω acts as a truncation in the closed string spectrum, keeping in particular one linear combination of the two gravitini (and dilatini) of type IIB superstring and projecting out the orthogonal linear combination.

The truncation produces therefore a gravitational anomaly, which asks for its cancellation the presence of additional fermions in the spectrum from the open string sector.

As will be seen in the next subsection, the orientifold projection introduces, from the world-sheet point of view, new surfaces in the Polyakov topological expansion of the string perturbation theory. From the spacetime viewpoint, they correspond to non-dynamical, mirror-like objects called *orientifold planes*, defined by T-duality as fixed points of the orientifold projection. More precisely, let us T-dualize a compact coordinate $X(\tau, \sigma) = X(\tau - \sigma) + \tilde{X}(\tau + \sigma)$, of radius R, on which Ω acts according to (4.5). After T-duality, the new closed string coordinate and the corresponding orientifold projection are defined as

$$X'(\tau, \sigma) = X(\tau - \sigma) - \tilde{X}(\tau + \sigma),$$
$$\Omega\, X'(\tau, \sigma)\, \Omega^{-1} = -X'(\tau, -\sigma). \tag{4.6}$$

The orientifold projection acts therefore on X' as $\Omega' = \Omega\, \Pi_{X'}$, where Π_X is a spacetime parity in X'. The new orientifold operation Ω' has two *fixed hypersurfaces* of spacetime dimension nine

$$X' = -X' \quad \to \quad X' = 0,$$
$$X' = -X' + 2\pi R \quad \to \quad X' = \pi R, \tag{4.7}$$

called O8 planes. By T-dualizing again we find $\Omega' = \Omega\, \Pi_1 \Pi_2$ whose fixed surfaces (on a square torus of radii R_1 and R_2, for simplicity) are the four O7 planes of coordinates

$$(0, 0), \quad (\pi R_1, 0), \quad (0, \pi R_2), \quad (\pi R_1, \pi R_2). \tag{4.8}$$

One subtlety arises however, namely a parity in two (or more generally an even number of) coordinates is a π rotation in a two-plane, whose square is the spacetime fermion number $(-1)^F$. Summarizing, n T-dualities generates 2^n nondynamical $O(9 - n)$ planes and the orientifold projection acts on the various closed string states according to

$$\Omega' = \Omega\, \Pi_1 \Pi_2 \cdots \Pi_n, \quad \text{odd } n,$$
$$\Omega' = \Omega\, \Pi_1 \Pi_2 \cdots \Pi_n (-1)^{F_L}, \quad \text{even } n, \tag{4.9}$$

where $(-1)^{F_L}$ is the spacetime left fermion number. The peculiarity of the O-planes is that they have *negative tension*, which is consistent since they are nondynamical objects, with no degrees of freedom localised on their world-volume. Their tension can be computed from the string propagation between the O-planes, described by the Klein bottle (see the next section for its definition) and the result for an Op plane of standard type is

$$T_{Op} = -2^{p-5}T_{Dp}, \tag{4.10}$$

where T_{Dp} is the tension of one Dp brane. The type I superstring is only a particular example of an orientifold construction, which can be defined more generally as IIB/Ω' or IIA/Ω', where $\Omega' = \Omega \times g$, and g is a spacetime and/or worldsheet involution $g^2 = 1$. The geometry and the nature of the O-planes depend on the details of the operation g, as we will see in some specific examples later on. The orientifold projection has generically a nontrivial action on the Chan-Paton factors

$$\Omega' : \lambda \to \gamma_{\Omega'} \lambda^T \gamma_{\Omega'}^{-1}, \quad \text{where} \quad \gamma_{\Omega'}^T \gamma_{\Omega'}^{-1} = \pm 1, \tag{4.11}$$

as a consequence of $\Omega'^2 = 1$.

4.1. Perturbative expansion, one-loop amplitudes

The parent Type IIB string contains D(-1),D1,D3,D5,D7 (and D9) BPS branes, coupling electrically or magnetically to the various RR forms present in the massless spectrum. Out of them, the D1, D5 and D9 branes are invariant under Ω and therefore are present in the Type I theory, as (sub)spaces on which open string ends can terminate. In some sense, open strings can be considered as twisted states of the Ω involution [11], in analogy with twisted states in orbifold compactifications of closed strings.

The perturbative, topological expansion in orientifolds involves two-dimensional surfaces with holes h, boundaries b and crosscaps c. Each surface has an associated factor $\lambda^{-\chi}$, where

$$\chi = 2 - 2h - b - c \tag{4.12}$$

is the Euler genus of the corresponding surface. Tree-level diagrams include, in addition to the sphere with genus $\chi = 2$, the disk with one boundary $\chi = 1$, where open string vertex operators can be attached, and the projective plane RP^2 with one crosscap ($\chi = 1$). One-loop diagrams include, in addition to the usual torus $\mathcal{T}$ with one handle, the Klein bottle $\mathcal{K}$ with two crosscaps, the annulus $\mathcal{A}$ with two boundaries and the Möbius $\mathcal{M}$ with one boundary and one crosscap, all of them having $\chi = 0$. The last two diagrams allow the propagation of open strings with Chan-Paton charges $|k; ij >$ in the annulus and $|k; ii >$ in the Möbius, containing the gauge group and the charged matter degrees of freedom. On the other hand, the torus and the Klein bottle describe the propagation of closed string degrees of freedom.

One-loop string diagrams may be constructed as generalizations of the one-loop vacuum energy in field-theory. In d noncompact dimensions, the contribution of a real boson of mass m to the vacuum energy in field theory is

$$
\Gamma = \frac{1}{2} \int \frac{d^d p}{(2\pi)^d} \ln(p^2 + m^2) = -\frac{1}{2} \int_0^\infty \frac{dt}{t} \int \frac{d^d p}{(2\pi)^d} e^{-(p^2+m^2)t}
$$
$$
= -\frac{1}{2(4\pi)^{d/2}} \int_0^\infty \frac{dt}{t^{1+d/2}} e^{-tm^2}, \tag{4.13}
$$

where we introduced a Schwinger proper-time parameter through the identity

$$
\ln\left(\frac{A}{B}\right) = -\int_0^\infty \frac{dt}{t} (e^{-tA} - e^{-tB}), \tag{4.14}
$$

and where we also neglected in (4.13) an (infinite) irrelevant mass-independent term. The result (4.13) readily generalizes to the case of more particles in the loop with mass operator m and different spin, as

$$
\Gamma = -\frac{1}{2(4\pi)^{d/2}} \, \mathrm{Str} \int_0^\infty \frac{dt}{t^{1+d/2}} e^{-tm^2}, \tag{4.15}
$$

where Str takes into account the multiplicities of particles and their spin and reduces in 4d to the usual definition $\mathrm{Str}\, m^{2k} = \sum_J (-1)^{2J}(2J+1)\,\mathrm{tr}\, m_J^{2k}$, where m_J denotes the mass matrix of particles of spin J.

The generalization of (4.15) to the Type IIB torus partition function in d noncompact (and $10-d$ compact) dimensions is (keeping only internal metric moduli here for simplicity)

$$
\mathcal{T} = \mathrm{Tr} \frac{1+(-1)^G}{2} \frac{1+(-1)^{\tilde G}}{2} \mathcal{P} q^{L_0} \bar q^{\bar L_0} = \frac{1}{(4\pi^2\alpha')^{\frac{d}{2}}}
$$
$$
\times \sum_{rs} X_{rs} \int_F \frac{d^2\tau}{(\mathrm{Im}\,\tau)^{1+\frac{d}{2}}} \chi_r(\tau)\,\chi_s(\bar\tau)\,\Gamma_{rs}^{(10-d,10-d)}(\tau,\bar\tau,g_{i\bar j}), \tag{4.16}
$$

where $q = exp(2\pi i\tau)$ and τ is the modular parameter of the torus, L_0, $\bar L_0$ are Virasoro operators for the left and the right movers, $(-1)^G$ $((-1)^{\tilde G})$ is the worldsheet left (right) fermion number implementing the GSO projection and $\mathcal{P}$ an operator needed in orbifold compactifications (see Section 4) in order to project onto physical states. In (4.16), the χ's are a set of modular functions of the underlying conformal field theory, $\Gamma_{rs}^{(10-d,10-d)}$ is the contribution from the compactification lattice depending on the compact metric components $g_{i\bar j}$ and X is a matrix of integers. The integral in (4.16) is performed over the fundamental region

$$F : \text{Im}\,\tau \geq 0, \quad -\frac{1}{2} \leq \tau_1 \leq \frac{1}{2}, \quad |\tau| \geq 1, \tag{4.17}$$

and the $\text{Im}\,\tau$ factors come from integrating over noncompact momenta. The typical form of the characters is

$$\chi_r = q^{h_r - \frac{c}{24}} \sum_{n=0}^{\infty} d_n^r q^n, \tag{4.18}$$

where h_r is the conformal weight, c is the central charge of the conformal field theory and the d_n^r are positive integers.

Let us start with a brief review of the algorithm used in the following. This was introduced in [12, 13], and developed further in [62]. The starting point consists in adding to the (halved) one-loop torus amplitude the Klein-bottle $\mathcal{K}$. This completes the projection induced by Ω, and is a linear combination of the diagonal contributions to the torus amplitude, with argument $q\bar{q}$. Then one obtains[2]

$$\begin{aligned}
\mathcal{K} &= \text{Tr}\,\frac{\Omega}{2}\frac{1 + (-1)^G}{2}\mathcal{P}\,e^{-4\pi\tau_2 L_0} \\
&= \frac{1}{2(4\pi^2\alpha')^{\frac{d}{2}}} \int_0^{\infty} \frac{d\tau_2}{\tau_2^{1+\frac{d}{2}}} \sum_r X_{rr}\,\chi_r(2i\tau_2)\,\Gamma_{\mathcal{K},r}^{(10-d)}(i\tau_2, g_{ij}),
\end{aligned} \tag{4.19}$$

with τ_2 the proper time for the closed string and $\Gamma_{\mathcal{K},r}^{(10-d)}(i\tau_2, g_{i\bar{j}})$ the contribution of the compactification lattice. In order to identify the corresponding open sector, it is useful to perform the S modular transformation induced by

$$\mathcal{K}: \quad 2\tau_2 \quad \xrightarrow[S]{} \quad \frac{1}{2\tau_2} \equiv l, \tag{4.20}$$

thus turning the loop-channel Klein-bottle amplitude $\mathcal{K}$ into the tree-level channel amplitude. The latter describes the propagation of the closed spectrum on a

---

[2] As discussed in [62], in general one has the option of modifying eq. (4.19), altering X_{ii} by signs ϵ_i. These turn sectors symmetrized under left-right interchange into antisymmetrized ones, and vice-versa, and are in general constrained by compatibility with the fusion rules. This freedom, which has the spacetime interpretation of flipping the RR charge of some orientifold planes, will turn out to be important later on since it can induce by consistency constraints supersymmetry breaking.

cylinder of length l terminating at two crosscaps[3], and has the generic form

$$\mathcal{K} = \frac{1}{2(4\pi^2\alpha')^{\frac{d}{2}}} \int_0^\infty dl \sum_r \Gamma_r^2 \, \chi_r(il) \, \tilde{\Gamma}_{\mathcal{K},r}^{(10-d)}(il, g_{i\bar{j}})$$

$$\equiv \frac{1}{2(4\pi^2)^{\frac{d}{2}}\alpha'^5} \int_0^\infty dl \, \tilde{\mathcal{K}}, \qquad (4.21)$$

where $\tilde{\Gamma}_{\mathcal{K},r}^{(10-d)}(il, g_{i\bar{j}})$ is the Poisson transform of $\Gamma_{\mathcal{K},r}$ and the coefficients Γ_r can be related to the one-point functions of the closed-string fields in the presence of a crosscap. Alternatively, in a spacetime language, the Ω involution has fixed (hyper)surfaces called orientifold (O) planes, carrying RR charge. The Klein bottle amplitude is then interpreted as describing the tree-level closed string (e.g. graviton, dilaton and RR fields) propagation starting and ending on orientifold (O) planes. Since the modulus of the Klein loop amplitude is $0 \le \tau_2 < \infty$, the τ_2 integral is not cut in the ultraviolet (UV) and is generically UV divergent. Physically, this divergence is related to the presence of an uncanceled RR charge from the O planes, which asks for the introduction of D branes and corresponding open strings. It will be important later on to distinguish between several types of O-planes. First of all, in supersymmetric models there are O_+ planes carrying negative tension and RR charge and O_- planes carrying positive tension and RR charge, in order to preserve supersymmetry. In nonsupersymmetric models, there can exist $\overline{O}_+$ planes with flipped RR charge compared to their supersymmetric O_+ cousins, but with the same NS-NS couplings, therefore breaking supersymmetry. Analogously, we can define $\overline{O}_-$ planes, starting from O_- planes and flipping only the RR charge. We will exemplify later on in detail the couplings of these four different types of O-planes to supergravity fields in different models.

The open strings spectrum may be deduced from the closed-string spectrum in a similar fashion. A very important property of one-loop open string amplitudes is that they all have a dual interpretation as tree-level closed string propagation between D-branes and/or O-planes. First, the loop-channel annulus amplitude may be deduced from the tree-level channel boundary-to-boundary amplitude. This has the general form [12] (see also [14])

$$\mathcal{A} = \frac{1}{(8\pi^2\alpha')^{\frac{d}{2}}} \int_0^\infty dl \sum_r B_r^2 \, \chi_r(il) \, \tilde{\Gamma}_{\mathcal{A},r}^{(10-d)}(il, g_{i\bar{j}})$$

$$\equiv \frac{1}{(8\pi^2)^{\frac{d}{2}}\alpha'^5} \int_0^\infty dl \, \tilde{\mathcal{A}}, \qquad (4.22)$$

---

[3]The crosscap, or real projective plane, is a non-orientable surface that may be defined starting from a 2-sphere and identifying antipodal points.

where the coefficients B_r can be related to the one-point functions of closed-string fields on the disk and on the RP^2 crosscap. In a spacetime interpretation, the annulus amplitudes describe open strings with ends stuck on D branes. The relevant S modular transformation now maps the closed string proper time l on the tube to the open-string proper time t on the annulus, according to

$$\mathcal{A}: \quad l \xrightarrow{s} \frac{1}{l} \equiv \frac{t}{2}. \tag{4.23}$$

The direct-channel annulus amplitude then takes the form

$$
\begin{aligned}
\mathcal{A} &= \frac{1}{2} Tr \frac{1 + (-1)^G}{2} \mathcal{P} e^{-\pi t L_0} = \frac{1}{2(8\pi^2 \alpha')^{\frac{d}{2}}} \\
&\times \int_0^\infty \frac{dt}{t^{1+\frac{d}{2}}} \sum_{r,a,b} A_{ab}^r n_a n_b \chi_r \left(\frac{it}{2} \right) \Gamma_{\mathcal{A},r}^{(10-d)} \left(\frac{it}{2}, g_{i\bar{j}} \right),
\end{aligned}
\tag{4.24}
$$

where L_0 in (4.24) is the Virasoro operator in the open sector, the n's are integers that have the interpretation of Chan-Paton multiplicities for the boundaries (D branes) and the A^r are a set of matrices with integer elements. These matrices are obtained solving diophantine equations determined by the condition that the modular transform of eq. (4.24) involves only integer coefficients, while the Chan-Paton multiplicities arise as free parameters of the solution. Supersymmetric models contain only D-branes, i.e. objects carrying positive RR charges. Nonsupersymmetric models ask generically also for *antibranes*, objects carrying negative RR charges but with NS-NS couplings identical to those of branes.

Finally, the tree-level channel Möbius amplitude $\mathcal{M}$ describes the propagation of closed strings between D branes and O planes (or boundaries and crosscaps, in worldsheet language), and is determined by factorization from $\tilde{\mathcal{K}}$ and $\tilde{\mathcal{A}}$. It contains the fields (characters) common to the two amplitudes, with coefficients that are geometric means of those present in $\tilde{\mathcal{K}}$ and $\tilde{\mathcal{A}}$ [12], [13]. Thus

$$
\begin{aligned}
\mathcal{M} &= -2 \frac{1}{(8\pi^2 \alpha')^{\frac{d}{2}}} \int_0^\infty dl \sum_r B_r \Gamma_r \hat{\chi}_r \left(il + \frac{1}{2} \right) \tilde{\Gamma}_{\mathcal{M},r}^{(10-d)} (il, g_{i\bar{j}}) \\
&\equiv \frac{1}{(8\pi^2)^{\frac{d}{2\alpha'^5}}} \int_0^\infty dl \, \tilde{\mathcal{M}},
\end{aligned}
\tag{4.25}
$$

where the hatted characters form a real basis and are obtained by the redefinitions

$$\hat{\chi}_r \left(il + \frac{1}{2} \right) = e^{-i\pi h_r} \chi_r \left(il + \frac{1}{2} \right). \tag{4.26}$$

The loop-channel Möbius amplitude can then be related to $\tilde{\mathcal{M}}$ by a modular P transformation and by the redefinition (4.26)

$$\mathcal{M}: \qquad \frac{it}{2} + \frac{1}{2} \quad \underset{P}{\longrightarrow} \quad \frac{i}{2t} + \frac{1}{2} \equiv il + \frac{1}{2}. \qquad (4.27)$$

This is realized on the hatted characters by the sequence $P = T^{1/2}ST^2ST^{1/2}$, with S the matrix that implements the transformation $\tau \to -1/\tau$ and T the diagonal matrix that implements the transformation $\tau \to \tau + 1$. The direct-channel Möbius amplitude then takes the form

$$\mathcal{M} = \mathrm{Tr} \frac{\Omega}{2} \frac{1 + (-1)^G}{2} \mathcal{P}(-e^{-\pi t})^{L_0} = -\frac{1}{2(8\pi^2\alpha')^{\frac{d}{2}}}$$

$$\times \int_0^\infty \frac{dt}{t^{1+\frac{d}{2}}} \sum_{r,a} M_a^r \, n_a \, \hat{\chi}_r \left(\frac{it}{2} + \frac{1}{2}\right) \Gamma_{\mathcal{M},r}^{(10-d)} \left(\frac{it}{2}, g_{i\bar{j}}\right), \qquad (4.28)$$

where by consistency the integer coefficients M_a^r satisfy constraints [63] that make $\mathcal{M}$ to be the orientifold Ω projection of $\mathcal{A}$. The full one-loop vacuum amplitude is

$$\int \left[\frac{1}{2}\mathcal{T}(\tau, \bar{\tau}) + \mathcal{K}(2i\tau_2) + \mathcal{A}\left(\frac{it}{2}\right) + \mathcal{M}\left(\frac{it}{2} + \frac{1}{2}\right)\right], \qquad (4.29)$$

where the different measures of integration are left implicit. In the remainder of this paper, we shall often omit the dependence on world-sheet modular parameters.

The absence of UV divergences ($l \to \infty$ limit) in the above amplitudes asks for constraints on the Chan-Paton factors, called *tadpole consistency conditions* [61]. They are equivalent to the absence of tree-level one-point functions for some closed string fields and ensure that the total RR charge in the theory is zero. In the notations used here, they read

$$B_r = \Gamma_r, \qquad (4.30)$$

and generically determine the Chan-Paton multiplicities, that in ten dimensions equals $M = 32$. The tadpoles for RR fields can be related [61] to inconsistencies in the field equations of RR forms (often reflected in the presence of gauge and gravitational anomalies). Indeed, D branes and O planes are electric and magnetic sources for RR forms. The Bianchi identities and field equations for a form of order n then read (in the language of differential forms)

$$dH_{n+1} = *J_{8-n}, \qquad d * H_{n+1} = *J_n, \qquad (4.31)$$

where the subscript on the electric and magnetic sources denotes their rank. The field equations of an n-form A_n are globally consistent if

$$\int_{C_{9-n}} *J_{n+1} = 0, \tag{4.32}$$

for all closed (sub)manifolds C_{9-n} transverse to the D-brane worldvolume. In particular, in a compact space the RR flux must be zero, and this gives nontrivial constraints on the spectrum of D-branes in the theory. The sources generated by D-brane /O-planes in the field eqs. of the RR field A_n are generically of the form

$$J_{n+1} = \sum_i q_i^{(n)} \, \delta(\mathbf{y} - \mathbf{y}_i) \, dV_{n+1}, \tag{4.33}$$

where $q_i^{(n)}$ are the RR charges of the D-branes / O-planes with respect to the RR form A_n and dV_{n+1} is the volume element $n+1$ form on the branes/O-planes worldvolume. The RR tadpole conditions is then the global neutrality condition

$$\sum_{Dp} q_{Dp}^{(n)} + \sum_{Op} q_{Op}^{(n)} = 0, \tag{4.34}$$

and clearly asks for objects with positive *and* negative charges in order to be satisfied. The presence of negative charges in a consistent string compactification is the main reason to introduce orientifolds.

The situation is different for NS-NS tadpoles. Indeed, suppose that there is a dilaton tadpole, of the type $\exp(-\Phi)$ in the string frame, generated by the presence of (anti)brane-(anti)orientifold Dp-Oq systems. The dilaton classical field equation reads

$$\partial_A(\sqrt{g} \, g^{AB} \partial_B \Phi) = \sum_i \alpha_i \sqrt{g} \, e^{\frac{(p-3)\Phi}{4}} \, \delta^{(9-p)}(y - y_i), \tag{4.35}$$

where $A, B = 1 \cdots 10$ and y_i denote the position of the brane-orientifold planes in the space transverse to the brane. The uncancelled dilaton tadpole is expressed as

$$\sum_i \alpha_i \neq 0, \quad \text{while} \quad \sum_i \alpha_i \int_C \sqrt{g} \, e^{\frac{(p-3)\Phi}{4}} \, \delta^{(9-p)}(y - y_i) = 0. \tag{4.36}$$

The first inequality means that, around the *flat vacuum*, the r.h.s. source in (4.35) does not integrate to zero and violates the integrability condition coming from the l.h.s. of (4.35). As stressed in [64], however, this simply means that the real background is *not* the flat background, but a curved one. This explains the second equality in (4.36), where C is any closed curve or (hyper)surface in the internal space. Explicit examples of such Type I backgrounds were given in [66, 67].

4.2. Various D-branes and O-planes: BPS and non-BPS configurations

Like particles and antiparticles in quantum field theory, there are D-branes and anti D-branes in type II and orientifold spectra, differing in the sign of their RR charge. In perturbative constructions there are four types of orientifold planes. The analog of the particles p-objects (antiparticles, anti p-objects) satisfy BPS-type conditions. In type II strings with the two supercharges denoted Q, $\tilde{Q}$, they preserve 1/2 supersymmetry defined by

$$\text{p-objects:} \quad T_p = q_p, \quad Q + \prod_{i=p+1}^{9-p} (\Gamma^i \Gamma^{11}) \tilde{Q},$$

$$\text{anti p-objects:} \quad T_p = -q_p, \quad Q - \prod_{i=p+1}^{9-p} (\Gamma^i \Gamma^{11}) \tilde{Q}, \quad (4.37)$$

where Γ^i in (4.37) denote gamma matrices along the directions transverse to the worldvolume of the objects, whereas $\Gamma^{11} = \Gamma_0 \cdots \Gamma_9$ denotes the fermion chirality in 10d. A model containing simultaneously objects and anti-objects clearly breaks completely supersymmetry. We display below the various (anti)objects appearing in orientifold compactifications, along with their tensions and RR charges, constrained in a specific model to satisfy the RR tadpole conditions (4.34).

Table 1
The RR tensions, NS–NS charges and typical gauge groups in perturbative orientifold constructions.

	Tension	RR charge	Typical gauge group for m Dp-branes (first 2 lines) for m Dp /Op (last 4 lines)
Dp	$+1$	$+1$	$U(m)$
$\overline{D}p$	$+1$	-1	$U(m)$
$Op+$	-2^{p-5}	-2^{p-5}	$SO(m)$
$Op-$	$+2^{p-5}$	$+2^{p-5}$	$USp(m)$
$\overline{O}p+$	-2^{p-5}	$+2^{p-5}$	$SO(m)$
$\overline{O}p-$	$+2^{p-5}$	-2^{p-5}	$USp(m)$

5. Compactification to four-dimensions

The 4d theories are defined after a compactification à la Kaluza-Klein. Typically, the ten dimensional spacetime is decomposed as $M_{10} = M_4 \times K_6$, where M_4 is our four dimensional Minkowski spacetime and K_6 is a compact manifold whose volume V traditionally defines the compactification scale M_c

$$V = M_c^{-6} \equiv M_{GUT}^{-6}, \tag{5.1}$$

which defines the scale of the Kaluza-Klein mass excitations in the internal space. The compactification scale was also identified above with the grand unified scale M_{GUT} in a naive string unification picture, because the field theory description breaks down above M_c. However, we will see later that (5.1) can be substantially altered in string models with large extra dimensions.

The massless fields in a toroidal compactification are the zero modes of the 10d fields, that in more general settings depend on the topology of the compact space K_6. If we denote by i, j six dimensional internal indices, then we have, for example, the following decompositions:

$$g_{AB} : g_{\mu\nu}, \ g_{mn}, \ g_{\mu m},$$
$$B_{AC} : B_{\mu\nu}, \ B_{mn}, \ B_{\mu m},$$

where in 4d $g_{\mu\nu}$ is the graviton, $g_{\mu m}$, $B_{\mu m}$ are gauge fields and g_{mn} are scalars describing the shape of the compact space. On the other hand, $B_{\mu\nu}$ and B_{mn} are pseudoscalar, axion-type fields.

The toroidal compactification of superstring theories to four dimensions gives rise to spectra with $\mathcal{N} = 4$ supersymmetry. This can easily be seen by decomposing massless 10d states, which fill representations of $SO(8)$, into representations of $SO(2) \otimes SO(6)$, where $SO(2)$ refers to the four noncompact dimensions and $SO(6)$ to the 6d toroidal compact space. Under this decomposition, a Majorana-Weyl 10d spinor (e.g. a supercharge) decomposes as $\mathbf{8}_s = (\mathbf{2}, \mathbf{4})$ and corresponds to four supercharges in 4d. The number of unbroken 4d supersymmetries is more generally given by the number of covariantly constant spinors ϵ^a satisfying

$$\nabla_\mu \epsilon^a = 0. \tag{5.2}$$

Their number $\mathcal{N}$ is governed by the *holonomy group* of the compact space. Imposing the $SU(3)$ subgroup of $SO(6)$ to be the holonomy group reduces to one the number of covariantly constant spinors, according to the decomposition

$$(\mathbf{2}, \mathbf{4}) = (\mathbf{2}, \mathbf{1}) + (\mathbf{2}, \mathbf{3}), \tag{5.3}$$

since only $(\mathbf{2}, \mathbf{1})$ is $SU(3)$ invariant. A well-known example of this type are the *Calabi-Yau* spaces [3]. Another particularly simple way of reducing the number of supersymmetries and of producing fermion chirality is to compactify on *orbifolds* [4]. A d-dimensional orbifold O^d can be constructed starting with the d-dimensional euclidean space R^d or the d-dimensional torus T^d and identifying points as

$$O^d = R^d/S = T^d/P, \qquad (5.4)$$

where the *space group* S contains rotations θ and translations v and the *point group* P is the discrete group of rotations obtained from the space group ignoring the translations. A typical element of S acts on coordinates as $X \to \theta X + v$ and is usually denoted (θ, v). The subgroup of S formed by pure translations $(1, v)$ is called the *lattice* Γ of S. The identification of points of R^d under Γ defines the torus T^d. Points of T^d can then be further identified under P to form the orbifold O^d. This is clearly consistent only if P consists of rotations which are automorphisms of the lattice Γ.

In most of the following sections we will be interested in 4d $\mathcal{N} = 1$ orientifold models, obtained by orbifolding the six real (three complex) internal coordinates

$$z_1 = \frac{1}{\sqrt{2}}(x_4 + ix_5), \quad z_2 = \frac{1}{\sqrt{2}}(x_6 + ix_7), \quad z_3 = \frac{1}{\sqrt{2}}(x_8 + ix_9), \qquad (5.5)$$

by the twist $\theta(z_1, z_2, z_3) = (e^{2i\pi v_1} z_1, e^{2i\pi v_2} z_2, e^{2i\pi v_3} z_3)$, where $\mathbf{v} \equiv (v_1, v_2, v_3)$ is called the twist vector and where for a Z_N orbifold $\theta^N = 1$. The action of the orbifold on a 10d Majorana-Weyl spinor denoted as $|s_1 s_2 s_3 s_4 >$, where $s_i = \pm 1/2$ are the helicities in the spacetime and the three compact torii, is given by

$$\begin{aligned}\theta|s_1 s_2 s_3 s_4 > &= e^{2\pi i (v_1 s_2 + v_2 s_3 + v_3 s_4)})\, |s_1 s_2 s_3 s_4 > \\ &= e^{\pi i (\pm v_1 \pm v_2 \pm v_3)})\, |s_1 s_2 s_3 s_4 > . \end{aligned} \qquad (5.6)$$

If $v_1 \pm v_2 \pm v_3 = 0$ with some fixed sign choice, with all $v_i \neq 0$, the holonomy group is $SU(3)$ and the orientifold has generically $\mathcal{N} = 1$ supersymmetry (the $\mathcal{N} = 2$ of the parent Type IIB model broken to half of it by the orientifold projection Ω) while if, for example, $v_3 = 0$ and $v_1 + v_2 = 0$, the corresponding orientifold has $\mathcal{N} = 2$ supersymmetry. The group structure of the orientifolds we use in the following[4] is $(1, \Omega, \theta^k, \Omega\theta^k \equiv \Omega_k)$. The independent possible orbifolds were classified long time ago [4] and in 4d they are Z_3, Z_4, Z_6, Z_6', Z_7, Z_8, Z_8', Z_{12}, Z_{12}' and $Z_N \times Z_M$ for some integers N and M. All of the simplest corresponding orientifold constructions contain a set of 32 D9 branes.

---

[4]The group structure is however not unique, see for example [71].

In addition, the ones containing Z_2-type elements (Z_4, Z_6, Z'_6, Z_8, Z_{12}, Z'_{12}) have sets of 32 D5 branes, needed here for the perturbative consistency of the compactified theory. The presence of the D5 branes can be understood as follows. The orientifold group element $\Omega \theta^{N/2}$ (and sometimes other elements, too) has fixed hyperplanes called O5$_+$ planes, negatively charged under the (twisted) RR fields. By flux conservation, they ask for a corresponding set of D5 branes with opposite RR charge. The actual position of the D5 branes is not completely fixed. They can naturally sit at the orbifold fixed points or they can live "in the bulk" in sets of 2N branes in a Z_N orbifold. This brane displacement [17,71] can be understood as a Higgs phenomenon breaking the open string gauge group and the sets of 2N bulk branes can be regarded as one brane and its various images through the orbifold and orientifold operations.

The projection operator introduced in eqs. (4.19), (4.24) and (4.28) is, for a Z_N orbifold, expressed as

$$\mathcal{P} = \frac{1}{N}(1 + \theta + \cdots + \theta^{N-1}), \qquad (5.7)$$

and therefore projects into orbifold invariant states $\mathcal{P}|phys> = |phys>$. The untwisted massless closed spectrum, defined by

$$Z^i(\tau, \sigma + 2\pi) = Z^i(\tau, \sigma), \qquad (5.8)$$

where Z^i are the string coordinates of the three compact torii, is found by first displaying the right (and left) massless states

Sector	State	θ^k	helicity	
NS	$\Psi^\mu_{-1/2}|0>$	1	± 1	
NS	$\Psi^{j,\pm}_{-1/2}|0>$	$e^{\pm 2\pi i k v_j}$	6×0	(5.9)
R	$|s_0 s_1 s_2 s_3>$	$e^{2\pi i k(s_1 v_1 + s_2 v_2 + s_3 v_3)}$	$4 \times \left(\pm\frac{1}{2}\right)$	

where $s_i = \pm 1/2$, $s_0 + s_1 + s_2 + s_3 = 0$ (mod 2) is the GSO projection in the R sector and $j = 1, 2, 3$ denote (complex) compact indices.

The physical closed string spectrum is obtained taking left-right tensor products $|L> \otimes |R>$ invariant under the orbifold and orientifold involution. The compact space is flat, up a finite number of singularities, defined as the fixed points of the orbifold operation, whose number is given by

$$N_f = \det(1 - \theta) = 64 \prod_{i=1}^{3} \sin^2\left(\frac{v_i}{2}\right). \qquad (5.10)$$

There are fields living only in the fixed points, the *twisted states*, coming from strings which are closed up to an orbifold operation

$$Z^i(\tau, \sigma + 2\pi) = e^{2\pi i v_i} Z^i(\tau, \sigma). \tag{5.11}$$

Each fixed point gives a physically distinct spectrum. If some fixed points are equivalent (they have the same massless spectrum), it results in a multiplicity of states in four dimensions. In the heterotic string constructions, which contain only closed strings, this degeneracy could account for the number of generations of quarks and leptons of the Standard Model. In the type II strings, orbifolds generate twisted RR forms and therefore the D-branes and the O-planes carry generically untwisted and twisted charges. In addition to the untwisted RR tadpole conditions (4.34), there will then be additional, twisted RR tadpole conditions, which severely constrain the gauge group and the matter content of the orientifold models.

The action of a twist element θ^k in the open (N) and (D) sectors can be described by 32×32 matrices $\gamma_{\theta^k} \equiv \gamma^k = (\gamma_\theta)^k$ acting on the Chan-Paton degrees of freedom denoted by $\lambda^{(0)}$ for gauge bosons and by $\lambda^{(i)}$ ($i = 1, 2, 3$) for matter scalars. Imposing that vertex operators for the corresponding physical states be invariant under the orbifold projection defines this action to be

$$\theta^k : \lambda^{(0)} \to \gamma^k \lambda (\gamma^k)^{-1}, \quad \lambda^{(i)} \to e^{2\pi i k v_i} \gamma^k \lambda (\gamma^k)^{-1}. \tag{5.12}$$

Since $\theta^N = 1$, it follows from (5.12) that $\gamma^N = \pm 1$. For $\gamma^N = 1$ the gauge groups in the D9 and D5 brane sectors are subgroups of $SO(32)$, while for $\gamma^N = -1$ the D9 and D5 gauge groups are subgroups of $U(16)$. The two choices correspond, in the notation of the previous section, to "real" charges n and to pairs of complex charges $(m, \bar{m})$, respectively. For every element $\Omega \theta^k \equiv \Omega_k$ there is an associated matrix acting on the CP indices γ_{Ω_k} as

$$\Omega_k : \lambda^{(0)} \to -\gamma_{\Omega_k} (\lambda^{(0)})^T (\gamma_{\Omega_k})^{-1},$$
$$\lambda^{(i)} \to -\gamma_{\Omega_k} (\lambda^{(i)})^T (\gamma_{\Omega_k})^{-1}. \tag{5.13}$$

Since $\Omega^2 = 1$ it follows also that $\gamma_\Omega = \pm\gamma_\Omega^T$. The corresponding Möbius amplitudes are multiplied by the multiplicity factor $Tr(\gamma_{\Omega_k}^{-1} \gamma_{\Omega_k}^T)$. Without loss of generality the matrices γ^k can be chosen to be diagonal. The tadpole consistency conditions fix the size of the Chan-Paton matrices γ analogously to (4.30), whereas correct spin-statistics for bosons and fermions in the loop channel and correct couplings of closed states to branes and O-planes in the tree-level channel determine the nature of gauge group and the gauge group representation of the charged matter content. A generic supersymmetric model contains in the closed and in the open spectrum states having a 10d origin, which contain a compactification lattice depending on all six compact coordinates; they form the so-called $\mathcal{N} = 4$ sector. There could also exist states having a 6d origin, with a compactification lattice depending on two compact coordinates, the so-called $\mathcal{N} = 2$ sectors. Finally, there are states without any excitations in the compact coordinates, forming the $\mathcal{N} = 1$ sectors.

Any string compactification has a large number of *moduli fields*, related to the sizes and the shape of the compact space and including the original 10d dilaton. A simple example is obtained by considering a compact two torus and the geometric fields: the metric (symmetric) g_{ij} and antisymmetric tensor $B_{ij} = \epsilon_{ij} B$ with indices in the torus $i, j = 1, 2$. Then from the 4d point of view, the four fields organize into two complex fields

$$T = \sqrt{\det g} + i B, \quad U = \frac{\sqrt{\det g} + i g_{12}}{g_{22}}, \tag{5.14}$$

where T describes basically (its real part) the size of the torus and is generically known as a *Kähler modulus*, whereas U describes mostly the shape of the torus and is generically known as a *complex structure* modulus.

In a flat space (no fluxes or warping) supersymmetric compactifications, the moduli fields have no scalar potential and are therefore *flat directions* of the 4d theory, associated to massless 4d fields, which generate unacceptable modifications of the gravitational force by inducing new macroscopic forces. Lifting the flat directions and stabilizing the moduli fields is one of the most important problems of string phenomenology. We will come back to this important issue in section 8 of these lectures.

6. Branes at angles: intersecting brane worlds

One of the simplest ways of partially or totally breaking supersymmetry in orientifold models is by rotating the branes in the compact space. For definiteness we discuss in some detail the case of D6 branes in type IIA orientifolds. In this case, there are three angles $\theta_1, \theta_2, \theta_3$ that one (or a stack of parallel and coincident) D6 brane(s) can make with the horizontal axis x_4, x_6, x_8 of the three torii of the compact space. The supercharge preserved by a given stack of D6 branes is

$$Q + P \tilde{Q}, \tag{6.1}$$

where P is the parity in the space transverse to the D6 brane(s). For example for "unrotated" D6 branes with worldvolume along $x_0 \cdots x_3 x_4 x_6 x_8$, the parity is equal to $P = P_1 P_2 P_3$, where $P_1 = \Gamma^4 \Gamma^{11}$, etc. In the absence of a second stack or orientifold planes, the number of supersymmetries of a stack of D-branes in a toroidal compactification is of course independent of the rotation in the compact space. For two distinct stacks of D-branes/O-planes, generically denoted for brevity as $D^{(1)}$ and $D^{(2)}$, the relevant quantities are the relative angles $\theta_i^{(12)} = \theta_i^{(1)} - \theta_i^{(2)}$. The supercharges preserved by each stack are $Q + P^{(1)} \tilde{Q}$ and $Q + P^{(2)} \tilde{Q}$. The number of unbroken supersymmetries (supercharges) is then given by the intersection of the two supercharges, which is equal to the number of $+1$ eigenvalues of the matrix $[P^{(1)}]^{-1} P^{(2)}$. Denoting by R the rotation needed to go from the brane $D^{(1)}$ to the brane $D^{(2)}$, we find $P^{(2)} = R^{-1} P^{(1)} R$ and moreover

$$[P^{(1)}]^{-1} R^{-1} P^{(1)} R = R^2, \tag{6.2}$$

where the action of the double rotation on a 10d spinor is given by

$$\begin{aligned} R^2 |s_1 s_2 s_3 s_4 > &= e^{2i(s_2 \theta_1^{(12)} + s_3 \theta_2^{(12)} + s_4 \theta_3^{(12)})} |s_1 s_2 s_3 s_4 > \\ &= e^{i(\pm \theta_1^{(12)} \pm \theta_2^{(12)} \pm \theta_3^{(12)})} |s_1 s_2 s_3 s_4 > . \end{aligned} \tag{6.3}$$

The outcome can be summarized as follows [25]:

$$\begin{aligned} \theta_3^{(12)} = 0, \theta_1^{(12)} \pm \theta_2^{(12)} = 0 \ (\text{mod } 2\pi) &\rightarrow \mathcal{N} = 2 \text{ unbroken SUSY}, \\ \theta_1^{(12)} \pm \theta_2^{(12)} \pm \theta_3^{(12)} = 0 \ (\text{mod } 2\pi) &\rightarrow \mathcal{N} = 1 \text{ unbroken SUSY}, \\ \theta_1^{(12)} \pm \theta_2^{(12)} \pm \theta_3^{(12)} \neq 0 \ (\text{mod } 2\pi) &\rightarrow \mathcal{N} = 0 \text{ SUSY}. \end{aligned} \tag{6.4}$$

In the compact space case, there are two important additional ingredients. First of all, the rotation of branes in the compact space is quantized, according to

$$\tan \theta_i^{(a)} = \frac{m_i^{(a)} R_{i2}}{n_i^{(a)} R_{i1}}, \tag{6.5}$$

where $(m_i^{(a)}, n_i^{(a)})$ are positive or negative integers called the *wrapping numbers* of the brane(s) $D^{(a)}$ along the two compact directions of the compact torus T_i^2 [26]. This quantization reflects the fact that in each of the three two dimensional compactification lattices, the worldvolume of the brane should cross points identified by the lattice, such that the total internal volume of the branes stays finite. In the opposite case, the internal volume of the brane would be infinite which is inconsistent with the propagation of D6 branes in a compact space. The total internal volume of the brane $D^{(a)}$ is then

$$V^{(a)} = (2\pi)^3 \prod_{i=1}^{3} \sqrt{m_i^{(a),2} R_{i2}^2 + n_i^{(a),2} R_{i1}^2}. \tag{6.6}$$

For two stacks of branes $D^{(a)}$ and $D^{(b)}$, it can easily be shown geometrically that the number of times they intersect in the compact torus T_i^2 is given by the *intersection number*

$$I_i^{(ab)} = m_i^{(a)} n_i^{(b)} - n_i^{(a)} m_i^{(b)}. \tag{6.7}$$

The remarkably simple property of the branes at angles constructions is that they easily generate 4d chirality. In the simplest (but unrealistic) example of the type IIA string with two sets of M_a coincident $D^{(a)}$ and M_b coincident $D^{(b)}$ D6 intersecting branes, the gauge group is $U(M_a) \otimes U(M_b)$, the $D^{(a)} - D^{(a)}$ and $D^{(b)} - D^{(b)}$ open spectra are non-chiral, whereas the strings stretched between the two sets of D-branes have a chiral fermionic spectrum in the representation $(M_a, \bar{M}_b)$ of the gauge group, of multiplicity equal to the total number of times the branes $D^{(a)}$ and $D^{(b)}$ intersect in the compact space

$$D^{(a)} - D^{(b)} : I^{(ab)} = \prod_{i=1}^{3} I_i^{(ab)} = \prod_{i=1}^{3} \left(m_i^{(a)} n_i^{(b)} - n_i^{(a)} m_i^{(b)} \right). \tag{6.8}$$

The second important additional ingredient in the compact space case is that, due to the positive RR charge of the D-branes, as already stressed, the RR tadpole consistency conditions in supersymmetric compactifications can be satisfied only by including the negative charge O-planes in orientifolds of type II strings. Furthermore, in the presence of non-trivial rotations, D-branes acquire additional RR charges which make impossible to satisfy tadpole conditions in a toroidal compactification, whereas supersymmetric models can be found in orbifold compactifications [27]. This effect is easier to describe in the T-dual formulation of the type I string with unrotated but "magnetized" D9 branes $D^{(a)}$, i.e. branes with internal magnetic fields $H_i^{(a)}$ at the ends of the open strings. The effect of the magnetic field depends on the charges $q_{L,R} = \pm 1/2, 0$ carried by the left and right string endpoints. The relation between the angles and the internal magnetic fields is

$$\theta_i^{(a)} = \arctan\left(\pi q_L H_i^{(a)}\right) + \arctan\left(\pi q_R H_i^{(a)}\right), \tag{6.9}$$

whereas the Dirac quantization condition is actually the T-dual version of the angle quantization (6.5) after the T-dualities $R_{i2}' = \alpha'/R_{i2}$, with the ratio $q_i^{(a)} \equiv m_i^{(a)}/n_i^{(a)}$ being a generalized Dirac quantum. In this T-dual formulation, the induced charges on a D9 brane can be easily worked out starting from the Wess-Zumino couplings of D-branes to the RR fields in the presence of the magnetic fields. By expresing the result in the intersecting branes language, we find for a general configuration of D6-branes $D^{(a)}$ the RR tadpole conditions

$$\sum_a M_a n_1^{(a)} n_2^{(a)} n_3^{(a)} = 16, \quad \sum_a M_a n_1^{(a)} m_2^{(a)} m_3^{(a)} = -16\epsilon_1,$$

$$\sum_a M_a m_1^{(a)} n_2^{(a)} m_3^{(a)} = -16\epsilon_2, \quad \sum_a M_a m_1^{(a)} m_2^{(a)} n_3^{(a)} = -16\epsilon_3, \quad (6.10)$$

where

$(\epsilon_1, \epsilon_2, \epsilon_3) = (0, 0, 0)$ in toroidal comp.,

$(\epsilon_1, \epsilon_2, \epsilon_3) = (\pm 1, 0, 0)$ in Z_2 comp.,

$(\epsilon_1, \epsilon_2, \epsilon_3) = (\pm 1, \pm 1, \pm 1)$ in $Z_2 \times Z_2$ comp.. $\qquad (6.11)$

The first tadpole condition in (6.10) comes from the D9 type couplings to the ten form A_{10}, whereas the other three ones come from the three different $D5_i$ type couplings to the six forms $A_{6,i}$. These last couplings are always present in orbifold compactifications containing Z_2 factors acting in four internal coordinates, which generate O5 planes as fixed points of the operation ΩZ_2, which in turn ask for the presence of either D5 branes or appropriately magnetized D9 branes. In the IIA language with D6 branes at angles, the type I O9 plane becomes an O6 plane which can be characterized, in analogy with the D6 branes, by wrapping numbers

$$O6 : (m_i, n_i) = (0, 1), \ (0, 1), \ (0, 1), \qquad (6.12)$$

whereas the three different type of $O5_i$ planes, $i = 1, 2, 3$ of type I orbifold compactifications become $O6_i$ planes, characterized by the wrapping numbers

$$O6_1 : (m_i, n_i) = (0, -\epsilon_1), \ (1, 0), \ (1, 0),$$

$$O6_2 : (m_i, n_i) = (1, 0), \ (0, -\epsilon_2), \ (1, 0),$$

$$O6_3 : (m_i, n_i) = (1, 0), \ (1, 0), \ (0, -\epsilon_3). \qquad (6.13)$$

On the other hand, the condition that each stack of D-branes preserve the same $\mathcal{N} = 1$ supersymmetry can be written as

$$m_1^{(a)} n_2^{(a)} n_3^{(a)} v_2 v_3 + n_1^{(a)} m_2^{(a)} n_3^{(a)} v_1 v_3 + n_1^{(a)} n_2^{(a)} m_3^{(a)} v_1 v_2 = \prod_{i=1}^{3} m_i^{(a)}, \quad (6.14)$$

for any a, where v_i are the volumes of the three compact torii. The conditions (6.10), (6.13) are the crucial constraints to be imposed on any $\mathcal{N} = 1$ intersecting brane orientifold model.

In IIA orientifolds with D6 branes at angles, each stack $D^{(a)}$ has a mirror $D^{(a')}$ with respect to the O6 planes, of wrapping numbers $(-m_i^{(a)}, n_i^{(a)})$. The chiral spectrum for toroidal compactification contains chiral fermions in

sector	representation	multiplicity of states
$D^{(a)} - D^{(b)}$	$(\bar{M}_a, M_b)$	I_{ab}
$D^{(a')} - D^{(b)}$	(M_a, M_b)	I_{ab}
$D^{(a')} - D^{(a)}$	$\dfrac{M_a(M_a - 1)}{2}$	$\dfrac{1}{2}(I_{a'a} + I_{0a})$
$D^{(a')} - D^{(a)}$	$\dfrac{M_a(M_a + 1)}{2}$	$\dfrac{1}{2}(I_{a'a} - I_{0a})$

$$ \tag{6.15} $$

Various quasi-realistic models along these lines were constructed in the last couple of years [28]. The generic Standard Model type construction contains four (or more) stacks, containing D-branes with a minimal gauge group $U(3) \times U(2) \times U(1)^2 = SU(3) \times SU(2) \times U(1)^4$. Out of the four abelian gauge factors, three are generically anomalous by mixing a la Stueckelberg with string axions and get masses of the order the string scale. One linear combination is massless and is to be identified with the hypercharge. The quarks and leptons come typically from the byfundamental states of the open strings stretched between the various D-brane stacks. Right-handed neutrinos are usually part of the massless spectrum, whereas the number of Higgs scalars is typically large, but it can be reduced in particular constructions.

7. Mechanisms for supersymmetry breaking

7.1. The Scherk-Schwarz mechanism

The Scherk-Schwarz mechanism for breaking supersymmetry takes advantage of the presence of compact spaces in compactifications of higher-dimensional supersymmetric field theories or of superstrings. The main idea is to use symmetries S of the higher-dimensional theory which do not commute with supersymmetry, typically R-symmetries or the fermion number $(-1)^F$. Then, after being transported around the compact space (a circle of radius R, for concreteness and coordinate $0 \le y \le 2\pi R$), bosonic and fermionic fields Φ_i return to the initial value (at $y = 0$) only up to a symmetry operation

$$ \Phi_i(2\pi R, x) = U_{ij}(\omega)\Phi_j(0, x), \tag{7.1} $$

where the matrix $U \in S$ is different for bosons and fermions and x are noncompact coordinates. At the field theory level, (7.1) implies that the Kaluza-Klein decomposition on the circle is changed so that zero modes acquire a nontrivial dependence on the y coordinate, according to

$$\Phi_i(y, x) = U_{ij}(\omega, y/R) \sum_m e^{\frac{imy}{R}} \Phi_j^{(m)}(x),$$ \hfill (7.2)

where ω is a number, quantized in String Theory. The matrix U satisfies some additional constraints in supergravity in order for the generated scalar potential to be positive definite. The ansatze considered by Scherk and Schwarz is $U = \exp(My)$, where M is an antihermitian matrix. Then kinetic terms in the y direction generate mass terms and break supersymmetry, the resulting fermion-boson splittings being equal to ω/R. This twisting procedure is very similar to the breaking of supersymmetry at *finite temperature* and, because of this, the terms breaking supersymmetry are UV finite, even at the field theory level.

The mechanism can be applied in globally supersymmetric models, in supergravity models and in superstrings. The breaking, being induced by the different boundary conditions for bosons and fermions, is therefore an explicit breaking. However, at the supergravity level, it appears to be *spontaneous*.

In heterotic strings the only available perturbative method of breaking supersymmetry is the Scherk-Schwarz mechanism[5]. In this case, soft masses M_{SUSY} $\sim R^{-1}$ are generated at tree-level for the gauginos, so that phenomenologically interesting values require $R^{-1} \sim$ TeV. The reason for this is that the gauge fields live in the full (bulk) 10d space and directly feel [82] supersymmetry breaking. Phenomenological reasons ask therefore for radii of the TeV size, a rather unnatural possibility in heterotic models [90], since it asks for a string coupling of the order of 10^{32}. The most popular mechanism invoked in this case for breaking supersymmetry is gaugino condensation in a hidden sector [98] $< \lambda\lambda >= \Lambda^3$, while the transmission to the observable sector is mediated by gravitational interactions, and thus

$$M_{SUSY} \sim \frac{\Lambda^3}{M_P^2}, \quad \text{where} \quad \Lambda \sim M_P \, e^{-1/[2b_0 g^2(M_P)]}.$$ \hfill (7.3)

This mechanism singles out intermediate scales $\Lambda \sim 10^{12} - 10^{13}$ GeV, naturally realized by the one-loop running of the hidden sector gauge coupling and could also be useful for purposes like neutrino masses or PQ axions. Gaugino condensation, however, is a nonperturbative field theory phenomenon and there is little hope to discover a string theory description of it. A third possibility is to start directly with nonsupersymmetric strings, possibly interpreted as models

---

[5] Supersymmetry is also broken by orbifolding the internal space. However, the resulting breaking is not soft, in the sense that there is typically no trace of the original supersymmetry in the resulting spectrum.

with supersymmetry broken at the string scale $M_{SUSY} \sim M_H$. As M_H is very large, however, this possibility was completely ignored since there was no clear way to solve the hierarchy problem in this case.

In models with D-branes there are two different ways in which supersymmetry can be broken by compactification in a phenomenologically interesting way. This is due to the two main new features of these theories:

– The Standard Model can be confined to a subspace (D-brane) of the full ten or eleven dimensional space.

– The string scale in these models can be lowered all the way down to the TeV range.

There are essentially two distinct subclasses of constructions. In the first, the D brane under consideration is parallel to the direction of breaking and the massless D brane spectrum feels at tree-level supersymmetry breaking. This situation was called "Scherk-Schwarz" breaking in [22], since it is the analog of the heterotic constructions [80] and the spectrum is a discrete deformation of a supersymmetric model. The corresponding spectra have heterotic duals. In the second class, the D brane under consideration is perpendicular to the direction of the breaking and the massless D brane spectrum is supersymmetric at tree-level. This was called "M-theory breaking" in [22], since it describes in particular supersymmetry breaking in M-theory along the eleventh dimension [41,43]. These models ask also for the presence of antibranes (and antiorientifold planes) in the spectrum, interacting with the branes. Supersymmetry breaking is transmitted by radiative corrections from the brane massive states or from the gravitational sector to the massless modes.

All RR and NS-NS tadpoles can be set to zero in both subclasses of these models. Moreover, in these models the closed (gravitational) sector has a softly broken supersymmetry.

7.2. Models based on non-BPS systems: brane supersymmetry breaking

In these constructions [30–32], the closed (bulk) sector is exactly supersymmetric to lowest order, whereas supersymmetry is broken at the string scale on some stack of (anti)branes.

Ex: the 10d USp(32) orientifold

As already explained, there is an important difference between tadpoles of RR closed fields and tadpoles of NS-NS closed fields. While the first signal an internal inconsistency of the theory and must therefore always be cancelled, the latter ask for a background redefinition and remove flat directions, producing potentials for the corresponding fields and leading actually to consistent models [64]. The difference between RR and NS-NS tadpoles turns out to play an important role in (some) models with broken supersymmetry. Indeed, there is a consistent model in

10d described by the same closed spectrum as the $SO(32)$ type I superstring, but with a nonsupersymmetric open spectrum. In this case the orientifold projection is $\Omega' = \Omega(-1)^{F_L}$ and breaks spacetime supersymmetry. Indeed, the massless gauge bosons are in this case projected by $\gamma_\Omega = -\gamma_\Omega^T$, implying $\lambda = \lambda^T$ and the gauge group $USp(N)$, with $N = 32$ whereas the fermions are in the symmetric representation. The $USp(32)$ model is interpreted as an "exotic" $O9_-$ plane of positive RR charge (instead of the negative charged $O9_+$ of the supersymmetric case), asking for 32 $D\bar{9}$ (anti)branes in the open sector. The $D\bar{9}$- $O9_-$ system is our first example of a non-BPS configuration, breaking supersymmetry at the string scale, with no parameter to tune in order to recover supersymmetry. Notice that in contrast to the well-known brane-antibrane pair, the antibrane-exotic O-plane configuration has no tachyon excitation in the open spectrum, fact reflecting the impossibility of this configuration to annihilate and disappear into the vacuum. The dynamics of this and related systems [67] is, to our knowledge, still an open question. In this class of models we are forced to live with a dilaton tadpole. There is a singlet in the open string fermionic spectrum which can be correctly identified with the goldstino realising a nonlinear supersymmetry on the antibranes. The spectrum is free of gauge and gravitational anomalies. The effective action contains the bosonic terms

$$S = \int d^{10}x\{\sqrt{g}\,\mathcal{L}_{SUGRA} - (N + 32)T_9\sqrt{g}e^{-\Phi} + (N - 32)q_9 A_{10}\} + \cdots \quad (7.4)$$

The coupling to the dilaton reflects that antibranes couple to NS-NS fields in the same way as branes, while O_- planes couple with a flipped sign compared to O_+ planes. The NS-NS tadpoles generate scalar potentials for the corresponding (closed-string) fields, in our case the (10d) dilaton. The dilaton potential reads

$$V \sim (N + 32)\,e^{-\Phi}, \qquad\qquad\qquad\qquad (7.5)$$

and in the Einstein frame is proportional to $(N + 32)\exp(3\Phi/2)$. It has therefore the (usual) runaway behaviour towards zero string coupling, a feature which is common to all perturbative constructions. Due to the dilaton tadpole, the background of this model is not the 10d Minkowski space. However, it was shown in [66] that a background with $SO(9)$ Poincare symmetry can be explicitly found. In this background, the tenth dimension is spontaneously compactified and the geometry is $R^9 \times S^1/Z_2$, with localized gravity. Notice that there is no obvious candidate for the endpoint of the decay to a supersymmetric ground state in this model, contrary to the popular folklore that non-supersymmetric string vacua decay one way or another into supersymmetric ones.

7.3. *Internal magnetic fields / intersecting branes*

Internal background magnetic fields H_i in a compact torus T^i (of radii $R_1^{(i)}$, $R_2^{(i)}$) can couple to the open string endpoints [23], carrying charges $q_L^{(i)}, q_R^{(i)}$ under H_i. Particles of different spin couple differently to the magnetic field and acquire different masses, breaking supersymmetry [24]. Defining $\pi \epsilon_i = \arctan(\pi q_L^{(i)} H_i) + \arctan(\pi q_R^{(i)} H_i)$, the mass splittings of all string states can be summarized by the formula

$$\delta m^2 = (2n+1)|\epsilon_i| + 2\Sigma_i \epsilon_i, \tag{7.6}$$

where n are the Landau levels of the charged particles in the magnetic field and Σ_i are internal helicities. Possible values of the magnetic fields satisfy a Dirac quantization condition $H_i \sim k/(R_1^{(i)} R_2^{(i)})$. For weak fields, $\epsilon_i \simeq (q_L^{(i)} + q_R^{(i)}) H_i$ and the resulting mass splittings are inversely proportional to the area of the magnetized torus $m_{SUSY}^2 \sim k/(R_1^{(i)} R_2^{(i)})$ [24]. The spectrum generically contains charged tachyons coming from scalars having internal helicities $\Sigma_i = -1$ ($\Sigma_i = 1$) for positive (negative) magnetic field, which can however be avoided in special models. The mechanism can easily accomodate several magnetic fields pointing out in several compact torii and can also be implemented in orbifold models.

After an odd number of T-dualities, as discussed in section 5, internal magnetic fields on open string ends are mapped into rotation of D-branes in the internal space. The internal magnetic field picture is however simpler to use in order to argue, through the magnetic field - spin coupling, for partial or complete supersymmetry breakdown.

Models with non-BPS configurations or internal magnetic fields are characterized by the fact that all RR tadpoles cancel, while some NS-NS tadpoles are left uncanceled. As discussed in Section 3, the proper interpretation of the NS-NS tadpoles is that scalar potentials are generated for appropriate NS-NS moduli fields.

8. Anomalies and generalized four dimensional Green-Schwarz mechanism

A consistent quantum field theory should have no gauge anomalies. In four-dimensions, this implies

$$Tr(Q_i Q_j Q_k) = 0, \tag{8.1}$$

where Q_i is the generator of a local (abelian or non-abelian) gauge symmetry and the trace is over the whole chiral fermionic spectrum of the theory. A consistent string theory is also anomaly-free, however anomaly cancelation can be achieved in a non-trivial way. This is mainly due to axionic type fields with nonlinear gauge transformations, whose couplings to gauge fields can produce local gauge variations compensating the Adler–Bardeen-Jackiw triangle anomalies. The simplest example of this type is the so-called "anomalous $U(1)_X$" factor, often present in heterotic string compactifications to four dimensions. In this case, by denoting the gauge group as $G = \prod_a G_a \otimes U(1)_X$, where G_a are simple gauge group factors, an explicit computation by using the massless spectrum shows that there can be non-vanishing mixed gauge anomalies

$$U(1)_X G_a^2 : C_a = \frac{1}{4\pi^2} Tr(Q_a^2 X),$$

$$U(1)_X^3 : C_X = \frac{1}{4\pi^2} Tr(X^3),$$

$$U(1)_X SO(1,3)^2 : C_{\text{grav}} = \frac{1}{192\pi^2} Tr\, X, \tag{8.2}$$

where the last anomaly is the $U(1)_X$-graviton-graviton mixed gauge-gravitational anomaly. All the other gauge anomalies have to vanish. However, the values of the mixed anomalies are not independent, but they are related through the relation

$$\delta_{GS} = \frac{C_a}{k_a} = \frac{C_X}{k_X} = \frac{1}{192\pi^2} Tr X, \tag{8.3}$$

where the rational numbers k_a, called Kac-Moody levels, define the gauge couplings. In superspace notation, the gauge kinetic function is

$$\int d^2\theta \frac{1}{4} k_a S W^{\alpha,a} W_{\alpha,a} + \text{h.c.}, \tag{8.4}$$

where S is the universal dilaton-axion superfield and $W_{\alpha,a}$ denotes the G_a chiral gauge superfield. The reason why the anomalies (8.3) define a consistent theory, as explained in [35], is the fact that the Kahler potential of S is of the form

$$K(S, \bar{S}) = -\ln(S + \bar{S} - \delta_{GS} V_X), \tag{8.5}$$

and contains a Stueckelberg type mixing $\delta_{GS} A_X^\mu \partial_\mu \text{Im}\, S$ between the axion $\text{Im}\, S$ and the gauge field. The supergauge transformations which leave the Kähler potential invariant are

$$V_X \to V_X + \Lambda + \bar{\Lambda}, \quad S \to S + \delta_{GS}\Lambda. \tag{8.6}$$

The gauge variation of the whole effective action is then

$$\delta S = -\frac{1}{4} \int d^2\theta \, \Lambda \sum_{A=a,x} \left[(C_A - \delta_{GS} k_A) W^{\alpha,A} W_{\alpha,A}\right] + \text{h.c.}$$
$$+ \text{grav. contribution}, \tag{8.7}$$

which vanishes precisely when (8.3) holds.

In orientifold models the anomaly cancelation works slightly differently. Let us start by defining the Type I compactification moduli, obtained by a straightforward reduction of the 10d action (10.4). By defining complex coordinates $i = 1, 2, 3$ and the associated components of the metric, $g_i^{\alpha\beta}$, $\alpha, \beta = 1, 2$ (with the dimension of a squared radius), the dilaton S and the geometric moduli T_i, U_i for the three complex planes are [72]

$$S = a^{RR} + i \frac{\sqrt{g_1 g_2 g_3} M_I^6}{\lambda_I}, \quad U_i = \frac{g_i^{12} + i\sqrt{g_i}}{g_i^{22}}, \quad T_i = b_i^{RR} + i \frac{\sqrt{g_i} M_I^2}{\lambda_I}, \quad (8.8)$$

where $g_i \equiv \det g_i^{\alpha\beta}$. We expect here surprises compared to the heterotic models where at tree-level the gauge couplings are universal, $1/g_a^2 = Re\,f_a = k_a Re\,S$. For example, in the first chiral orientifold model base on the Z_3 orbifold [18], the abelian $U(1)_X$ factor is anomalous and the mixed $U(1)_X G_a^2$ anomalies $(C_{SU(12)}, C_{SO(8)}, C_{U(1)}) = (1/4\pi^2)(-18, 36, -432)$ are incompatible with the standard 4d Green-Schwarz mechanism, involving the S field. The solution to this puzzle was proposed in [20], in analogy with the generalized Green-Schwarz mechanism found in 6d by Sagnotti [21]. It was conjectured in [20] that the gauge fields in Z_3 do couple at tree-level to a linear symmetric combination M of the 27 closed twisted moduli

$$f_a = S + s_a M. \quad (8.9)$$

Under a $U(1)_X$ gauge transformation with (superfield) parameter Λ, there are cubic gauge anomalies. The generalized Green-Schwarz mechanism requires a shift of the twisted moduli field combination M [73, 76]

$$V_X \to V_X + \Lambda + \bar{\Lambda}, \qquad M \to M + \epsilon \Lambda, \quad (8.10)$$

such that the gauge-invariant combination appearing in the Kähler potential is $M + \bar{M} - \epsilon V_X$. The mixed anomalies in this case cancel provided the following conditions holds

$$\epsilon = \frac{C_{SU(12)}}{s_{SU(12)}} = \frac{C_{SO(8)}}{s_{SO(8)}} = \frac{C_{U(1)_X}}{s_{U(1)_X}}. \quad (8.11)$$

It was shown in [20] that actually the mixed anomalies C_a are proportional to $tr(Q_X\gamma)tr(Q_a^2\gamma)$, where Q_X, Q_a are gauge group generators of $U(1)_X$ and the gauge group factor G_a, respectively. By an explicit check they showed that indeed this proportionality is valid and therefore the fields playing a role in canceling gauge anomalies are the twisted (linear combination of) fields M. Surprisingly, the dilaton S plays no role in anomaly cancelation, since, as $tr\,Q_X = 0$, it does not mixes with the gauge fields. The actual computation of the coefficients

s_a and ϵ (and therefore the check of the overall normalization in (8.11)) was performed in [73], by coupling the theory to a background spacetime magnetic field B. The tree-level gauge couplings were computed to be equal to

$$\frac{4\pi^2}{g_{a,0}^2} = \frac{1}{\ell} + \sum_{k=1}^{[\frac{N-1}{2}]} s_{ak} m_k \tag{8.12}$$

$$= \frac{1}{\ell} + \sum_{k=1}^{[\frac{N-1}{2}]} \frac{8\pi^2}{\sqrt{2\pi N}} (\mathrm{tr} Q_a^2 \gamma^k) \left| \prod_{i=1}^{3} \sin \pi k v_i \right|^{1/2} m_k, \tag{8.13}$$

where l is the Hodge dual of the axion $Re\, S$ in (8.8). Analogously, the coefficient ϵ in (8.11) can be found from the mixing between gauge fields and the twisted RR antisymmetric tensors, which can be computed by introducing a background magnetic field B' for the abelian gauge factor $U(1)_X$. The $U(1)_X$ gauge boson becomes massive, breaking spontaneously the symmetry, even for zero VEV's of the twisted fields m_k. However, the corresponding global symmetry $U(1)_X$ remains unbroken, since the Fayet-Iliopoulos terms vanish in the orbifold limit $m_k = 0$ [75]. The result is

$$\epsilon = \sqrt{\frac{2}{N\pi^3}} \sum_k \prod_{i=1}^{3} |\sin \pi k v_i|^{\frac{1}{2}} (-i\mathrm{tr} Q_X \gamma^k). \tag{8.14}$$

The above discussion generalizes easily for other models, in the case of more anomalous $U(1)_\alpha$ ($\alpha = 1 \cdots N_X$) and more linear combinations of twisted moduli fields M_k coupling to gauge fields. In this case the gauge kinetic function becomes

$$f_a = S + \sum_k s_{ak} M_k, \tag{8.15}$$

and (8.10) generalizes to

$$V_\alpha \to V_\alpha + \Lambda_\alpha + \bar{\Lambda}_\alpha, \qquad M_k \to M_k + \epsilon_{k\alpha} \Lambda_\alpha, \tag{8.16}$$

in an obvious notation. Cancelation of gauge anomalies $\mathrm{tr}(X_\alpha Q_a^2)$ described by the coefficients $C_{\alpha a}$ asks for the Green-Schwarz conditions

$$C_{\alpha a} = \sum_k s_{ak} \epsilon_{k\alpha}, \tag{8.17}$$

valid for each α, a. The gauge-invariant field combination appearing in the Kähler potential is $M_k + \bar{M}_k - \sum_\alpha \epsilon_{k\alpha} V_\alpha$ and generates, by supersymmetry, the D-terms

$$V_D = \sum_\alpha \frac{g_\alpha^2}{2} \left(\sum_A X_A^\alpha G_A \Phi^A - \frac{1}{2} \sum_k \epsilon_{k\alpha} \frac{\partial G}{\partial M_k} \right)^2, \tag{8.18}$$

in Planck units $M_P = 1$, where Φ^A denotes the set of charged chiral fields of $U(1)_\alpha$ charge X_A^α, $G_A = \partial G / \partial \Phi^A$ and

$$G = K + \ln |W|^2, \tag{8.19}$$

where here W is the superpotential. Notice that for standard gauge symmetries, commuting with supersymmetry, gauge invariance of the superpotential implies

$$V_D = \sum_\alpha \frac{g_\alpha^2}{2} \left(\sum_A X_A^\alpha K_A \Phi^A - \frac{1}{2} \sum_k \epsilon_{k\alpha} \frac{\partial K}{\partial M_k} \right)^2. \tag{8.20}$$

There is a further subtlety with the anomaly cancellation in the (generic) case of several abelian factors A_α which mix between themselves and with string axions a_α, the relevant terms in the effective action being

$$S = -\frac{1}{4} s_{\alpha\beta k} a_k F_\alpha \wedge F_\beta + \frac{1}{2} (\partial_\mu a_k - \epsilon_{k\alpha} A_\mu^\alpha)^2. \tag{8.21}$$

In this case, working for example in a scheme where the anomaly is distributed symmetrically among the abelian currents, it is clear from (8.17) that anomaly cancellation cannot be achieved if $s_{\alpha\beta k} \epsilon_{k\gamma}$ has an antisymmetric part in the indices β and γ, since the triangle anomaly coefficients $C_{\alpha\beta\gamma}$ are symmetric in all indices. In this case, it turns out that consistency conditions ask for the presence of generalized Chern-Simons terms [29] of the form

$$\frac{1}{4} E_{\alpha\beta,\gamma} A^\alpha A^\beta F^\gamma, \tag{8.22}$$

where the coefficients $E_{\alpha\beta,\gamma}$ satisfy the conditions

$$E_{\alpha\beta,\gamma} = -E_{\beta\alpha,\gamma},$$
$$E_{\alpha\beta,\gamma} + E_{\gamma\alpha,\beta} + E_{\beta\gamma,\alpha} = 0. \tag{8.23}$$

Anomaly cancellation is then satisfied provided that

$$C_{\alpha\beta\gamma} - s_{\alpha\beta k} \epsilon_{k\gamma} + 2 E_{\alpha\beta,\gamma} = 0. \tag{8.24}$$

As well known, the fermionic matter content in the Standard Model is completely anomaly-free. If in the near future the LHC collider discovers new $U(1)$ gauge bosons with masses in the TeV range and new fermions with an anomalous spectrum, this would be a signal that axionic like fields, coupled to the gauge fields and with nonlinear gauge transformations, should exist and cancel (at least partially, up to generalized Chern-Simons terms) the gauge anomalies. Even if this would not be a proof (indeed, very heavy chiral fermions could produce at low energies a similar effect)), their discovery would be a serious hint for the existence of an underlying string theory with a Green-Schwarz anomaly cancelation mechanism.

9. Moduli stabilization

Moduli stabilization is one of the major problems in string phenomenology. Indeed, in supersymmetric compactifications moduli fields are massless and correspond to flat directions in the scalar potential. Very often, supersymmetry breaking generate runaway moduli potentials which are generically unacceptable due to the induced time dependence and rolling of the moduli towards uninteresting configurations. Whereas part of the moduli fields can be stabilized in a perturbative manner, some of them and most notably the 10d dilaton asks clearly for going beyond perturbation theory. We will discuss some related mechanisms in the context of type IIB strings and the Horava-Witten M-theory and also comment on applications to finding de Sitter vacua in string theory.

9.1. Flux moduli stabilization in orientifolds of the type IIB string

The Type IIB string has NS-NS fields and RR fields from the closed string sector, of particular relevance for us being the NS-NS dilaton ϕ, antisymmetric tensor B_{AB}, the RR zero form A, two-form A_{MN} and the four-form A_{MNPR} with self-dual field strength. We consider orientifolds of the type IIB/Ω', where the definition and action of the orientifold projection on the relevant fields is

$$\Omega' = \Omega\,\Pi_1 \cdots \Pi_6 (-1)^{F_L}, \quad \text{where } \mathbf{y} \to -\mathbf{y},$$

$$\text{even fields:} \quad \phi, B_{\mu m}, A, A_{\mu m}, A_{\mu_1 \cdots \mu_4},$$

$$\text{odd fields:} \quad B_{\mu\nu}, B_{mn}, A_{\mu\nu}, A_{mn}, A_{\mu_1\mu_2\mu_3 m}, A_{m_1 m_2 m_3 \mu}, \tag{9.1}$$

where μ, ν, etc are 4d spacetime indices and m, n, etc are 6d internal indices. The orientifold projection Ω' has fixed 4d surfaces corresponding to O3 planes, which by the RR tadpole conditions asks for the addition of a certain number (16 in the absence of fluxes) of D3 branes. Supersymmetric orbifold compactifications contain generically also D7 branes, which can in the most general case be of three different types, depending on which four internal coordinates their worldvolume spans.

Even if the fields B_{mn} and A_{mn} are odd under the orientifold projection, their fields strength $H_3 = dB$ and $F_3 = dA_2$ with all indices in the internal space are even and can therefore support background fluxes. We will define for future purposes

$$S = e^{-\phi} - iA, \quad G_3 = F_3 - iSH_3. \tag{9.2}$$

The three-form fluxes satisfy a Dirac quantization condition in the compact space

$$\frac{1}{2\pi\alpha'} \int_{A_a} F_3 = 2\pi M_a, \quad \frac{1}{2\pi\alpha'} \int_{B^a} H_3 = 2\pi N^a, \tag{9.3}$$

where (A_a, B^a) are dual three-cycles of the compact Calabi-Yau type space and (M_a, N^a) are integers describing the various fluxes passing through the available cycles[6]. Giddings, Kachru and Polchinski [83] showed that with the three-form fluxes plus five-form flux F_5 and an appropriate dilaton-axion background, the 10d supergravity field eqs. have a supersymmetric solution provided that the fluxes satisfy a self-duality condition

$$*_6 G_3 = i G_3, \tag{9.4}$$

where $*_6 G_3$ denote the Hodge dual in the 6d internal space of the complex three-form flux. The self-duality condition (9.4) can be satisfied only for discrete choices of the dilaton S and also the complex moduli fields U_a, which are therefore *stabilized* by the presence of fluxes [83]. The 10d type IIB supergravity effective action, in the string frame, contains in particular the bosonic terms

$$S_{IIB} = \frac{1}{2k_{10}^2} \int d^{10}x \left\{ e^{-2\phi} [R + 4(\nabla\phi)^2] - \frac{1}{2\times 3!} G_3 \bar{G}_3 - \frac{1}{4\times 5!} F_5^2 \right\}$$

$$+ S_{D3,O3} - \frac{i}{8k_{10}^2} \int e^{\phi} C_4 \wedge G_3 \wedge \bar{G}_3 + \cdots, \tag{9.5}$$

where the last term in (9.5) is written in form language and we omitted other bosonic terms in the effective action irrelevant for the present discussion. The

---

[6]Since there are two type of cycles, there are really two different type of fluxes for each field strength H_3, F_3. We consider here for simplicity only one type in order to keep the discussion as simplest as possible.

last term in (9.5) shows that in the presence of fluxes, the RR form, which couples to the D3/O3 localized objects has an additional source which changes the RR tadpole conditions (4.34). Indeed, from the five-form field eqs. integrated over the compact space and by using the quantization conditions (9.3), we get the modified Gauss law

$$\sum_{D3} q_{D3} + \sum_{O3} q_{O3} + \sum_{a} M_a N^a = 0. \tag{9.6}$$

For self-dual fluxes, the flux contribution is always positive and therefore forces the total number of D3 brane to be less than the value without the flux $N_{D3} < 16$. This reduces therefore the rank of the gauge group, which is a welcome consequence of the fluxes.

The Calabi-Yau spaces have a complex $(3, 0)$ (meaning that it has purely holomorphic indices) form Ω whose integral over the cycles (A_a, B^a) are functions of the complex structure moduli fields z_α (one example being the field U we defined in section 4)

$$X^a(z_\alpha) = \int_{A_a} \Omega, \quad F_a(z_\alpha) = \int_{B^a} \Omega, \tag{9.7}$$

where X^a, F_a can be explicitly computed for a given Calabi-Yau space. It was shown [84] that the effect of the three-form fluxes can be nicely encoded in 4d in a contribution to the superpotential

$$W_{\text{flux}}(S, z_\alpha) = \int G_3 \wedge \Omega = (2\pi)^2 \alpha' \left(M_a X^a(z_\alpha) - i S N^a F_a(z_\alpha) \right), \tag{9.8}$$

which depends on the dilaton S and the complex moduli fields (z_α) but *not* on the Kahler moduli fields. The Kahler potential of type IIB orientifold vacua in the lowest approximation is independent of fluxes and is of the form

$$K = K(T_r + \bar{T}_r) - \ln(S + S^\dagger) - \ln\left(-i \int \Omega \wedge \bar{\Omega}\right), \tag{9.9}$$

where the last term defines the Kahler potential of complex structure moduli fields z_α. The Kahler potential of Kahler moduli is of no-scale type, satisfying

$$K^{r\bar{s}} K_r K_{\bar{s}} = 3, \tag{9.10}$$

the simplest and best known example being the case of only one (overall) Kahler modulus described by

$$K(T + \bar{T}) = -3 \ln(T + \bar{T}). \tag{9.11}$$

Generically the flux superpotential (9.8) stabilizes the dilaton and most or even all complex structure moduli fields. The Kahler moduli remain flat directions and additional ingredients are needed for their stabilisation.

9.2. Supersymmetry breaking and moduli stabilization in Horava-Witten M-theory

The example we discuss now is the compactified version of the supersymmetry breaking in Horava-Witten M-theory [36]. As it will become transparent, this mechanism stabilizes the same fields as the type IIB flux stabilization discussed above and is probably a dual version of it. At strong coupling, the ten-dimensional $E_8 \times E_8$ heterotic string becomes M-theory on $R^{10} \times S^1/Z_2$ [36]. The gravitational field propagates in the bulk of the eleventh dimension, while the $E_8 \times E_8$ gauge fields live at the Z_2 fixed points 0 and πR_{11}. We write M^{11} for $R^{10} \times S^1$ and M_i^{10}, $i = 1, 2$ for the two fixed (hyper)planes. The gauge and gravitational kinetic energies take the form

$$L = \frac{1}{2\kappa_{11}^2} \int_{M^{11}} d^{11}x \sqrt{g} R - \sum_i \frac{3^{1/3}}{4\pi (2\pi\kappa_{11}^2)^{2/3}} \int_{M_i^{10}} d^{10}x \sqrt{g} \, \text{tr} \, F_i^2, \quad (9.12)$$

where κ_{11} is here the eleven-dimensional gravitational coupling and F_i, for $i = 1, 2$, is the field strength of the ith E_8, which propagates on the fixed plane M_i^{10}.

The compactification pattern of this theory down to 4d is different according to the relative value of the eleventh radius compared to the other radii, that are denoted collectively R in the following. Assuming for simplicity an isotropic compact space, there are two distinct compactification patterns

$$R_{11} < R : 11d \rightarrow 10d \rightarrow 4d,$$
$$R_{11} > R : 11d \rightarrow 5d \rightarrow 4d. \quad (9.13)$$

In the strong coupling limit $R_{11} > R$, there is therefore an energy range where the spacetime is effectively five dimensional. Consider the simplest truncation of 11d supergravity down to 5d, keeping only the breathing mode of the compact space $g_{i\bar{j}} = \delta_{i\bar{j}} \exp(\sigma)$, and concentrate for simplicity on zero modes only. In this case, the only matter multiplet in 5d (in addition to the 5d gravitational multiplet with bosonic fields (g_{MN}, C_M), where $C_{Mi\bar{j}} = (1/6) A_M \delta_{i\bar{j}}$ is a vector field originating from the 3-from of 11d supergravity) is the universal hypermultiplet of bosonic fields (σ, C_{MNP}, a), with $C_{ijk} = (1/6)\epsilon_{ijk}a$, whose scalar fields parametrize the coset $SU(2, 1)/SU(2) \times U(1)$ [106]. The relevant part of the bosonic

5d supergravity lagrangian is

$$
\begin{aligned}
S_5 = \frac{1}{2k_5^2} \int d^5x \sqrt{g} \Big\{ & R - \frac{9}{2}(\partial_M \sigma)^2 - \frac{1}{24} e^{6\sigma} G_{MNPQ} G^{MNPQ} \\
& - \frac{3}{2} F_{MN} F^{MN} - 2 e^{-6\sigma} |\partial_M a|^2 \Big\} \\
& - \frac{1}{k_5^2} \int d^5 x \epsilon^{MNPQR} \Big\{ \frac{i}{\sqrt{2}} C_{MNP} \partial_Q a \partial_R a^\dagger + \frac{1}{2\sqrt{2}} A_M F_{NP} F_{QR} \Big\},
\end{aligned}
$$

(9.14)

where $F_{MN} = \partial_M A_N - \partial_N A_M$. The compactification from 5d to 4d is on the orbifold S^1/Z_2^{HW}, with an orbifold action Z_2^{HW} whose action on our relevant fields will be defined below. The lagrangian of the universal hypermultiplet can be derived from the 4d Kähler potential [106]

$$
\mathcal{K} = -\ln(S + S^\dagger - 2a^\dagger a),
$$

(9.15)

lifted back to 5d, where $S = \exp(3\sigma) + a^\dagger a + i a_1$ and the axion a_1 is defined by the Hodge duality $\sqrt{2} \exp(6\sigma) G_{MNPQ} = \epsilon_{MNPQR}(\partial^R a_1 + i a^\dagger \partial^R a)$. The lagrangian (9.14) has a global $SU(2)_R$ symmetry, acting linearly on the redefined hypermultiplet fields

$$
z_1 = \frac{1-S}{1+S}, \quad z_2 = \frac{2a}{1+S},
$$

(9.16)

which form a doublet (z_1, z_2). In the gravitational multiplet, the 5d Dirac gravitino is equivalent to two 4d Majorana gravitinos, transforming as an $SU(2)_R$ doublet. One of the gravitini is even under Z_2^{HW} and has a zero mode (before the Scherk-Schwarz twisting), while the other is odd and has only massive KK excitations. The Z_2^{HW} projection acts on the hypermultiplet as $Z_2^{HW} S = S$, $Z_2^{HW} a = -a$, which translates on the $SU(2)$ doublet in the obvious way

$$
Z_2^{HW} \begin{pmatrix} z_1 \\ z_2 \end{pmatrix} = \begin{pmatrix} 1 & 0 \\ 0 & -1 \end{pmatrix} \begin{pmatrix} z_1 \\ z_2 \end{pmatrix}.
$$

(9.17)

The Horava-Witten projection then asks for using the $U(1)_R$ subgroup of $SU(2)_R$ and the corresponding Scherk-Schwarz decomposition reads [41]

$$
\begin{pmatrix} \hat{z}_1 \\ \hat{z}_2 \end{pmatrix} = \begin{pmatrix} \cos \omega x_5 & \sin \omega x_5 \\ -\sin \omega x_5 & \cos \omega x_5 \end{pmatrix} \begin{pmatrix} z_1 \\ z_2 \end{pmatrix},
$$

(9.18)

corresponding to the Scherk-Schwarz matrix $M = i\omega \sigma_2$. Notice that, thanks to the anticommutation relation $\{Z_2^{HW}, M\} = 0$, the fields $\hat{z}_i$ have the same Z_2^{HW} parities as the fields z_i. The 4d complex superfields of the model are S (with the zero mode $a = 0$) and T, where $T = g_{55} + iC_5$ and the axion C_5 is the fifth component of the vector field in the 5d gravitational multiplet. The resulting scalar potential in 4d in the Einstein frame is computed from the kinetic terms of the $(\hat{z}_1, \hat{z}_2)$ fields derived form (9.15). After putting $z_2 = 0$ at the zero mode level, the result is

$$V = \int dx_5 \sqrt{g_{55}} \, g^{55} K^{a\bar{b}} \, \partial_5 z_a \partial_5 z_{\bar{b}} = \frac{4\omega^2}{(T + T^\dagger)^3} \frac{|1 - S|^2}{S + S^\dagger}, \qquad (9.19)$$

where $a, b = 1, 2$ and $K^{a\bar{b}}$ is the inverse of the Kähler metric $K_{a\bar{b}} = \partial_a \partial_{\bar{b}} K$. This result is interpreted as a superpotential generated for S. The 4d theory is completely described by

$$K = -\ln(S + S^\dagger) - 3\ln(T + T^\dagger), \qquad W = 2\omega(1 + S). \qquad (9.20)$$

Notice that the superpotential corresponds to a non-perturbative effect from the heterotic viewpoint. The minimum of the scalar potential is $S = 1$ and corresponds to a spontaneously broken supergravity with zero cosmological constant. The order parameter for supersymmetry breaking is the gravitino mass $m_{3/2}^2 = e^K |W|^2 = 2\omega^2 / (T + T^\dagger)^3$. In 4d supergravity units, by using the relation between the 5d Planck mass $M_5^3 = 1/k_5^2$ and the 4d one $t M_5^3 = M_P^2$, we get $m_{3/2} = \omega/R_5$, where R_5 is the radius of the fifth coordinate $R_5 = t M_5^{-1}$. The Goldstone fermion is the fifth component of the $(Z_2^{HW}$ even) 5d gravitino Ψ_5. The resulting models are of no-scale type [107].

A general Calabi-Yau compactification of Horava-Witten M-theory, on a CY space of Hodge numbers $h_{(1,1)}$, $h_{(2,1)}$ contains, from a 5d viewpoint, in addition to the gravity and the universal hypermultiplet, a number $h_{(1,1)}$ of vector multiplets whose 5d bosonic components are $(A_{Mi\bar{j}}, g_{i\bar{j}})$ with det $g_{i\bar{j}} = 1$ and $h_{(2,1)}$ hypermultiplets $(g_{ij} C_{ij\bar{k}})$. The $SU(2)_R$ symmetry acts non trivially on the fermions of the vector multiplets and on the scalars of the hypermultiplets. As a result, the Scherk-Schwarz along the 5th dimension generate a scalar potential for the hypermultiplets, which are the dilaton and the complex structure moduli. Some explicit examples were worked out in the second paper of [41]. For example, for a CY space of Hodge numbers $(h_{(1,1)}, h_{(2,1)}) = (3, 3)$, the resulting effective action is described by

$$K = -\ln(S + S^\dagger) - \sum_{\alpha=1}^{3} \ln(U_\alpha + \bar{U}_\alpha^i) - \sum_{i=1}^{3} \ln(T_i + \bar{T}_i),$$

$$W = \frac{\omega}{\sqrt{2}}(1 + S)\prod_\alpha (1 + U_\alpha). \qquad (9.21)$$

Due to the generalized dimensional reduction, there are background values for $G_{\mu\nu\rho\sigma}$, G_{5ijk}, $G_{5ij\bar{k}}$. The qualitative relation with the type IIB flux compactifications goes as follows. The heterotic M-theory compactified to nine dimensions is M-theory on $S^1 \times S^1/Z_2$, which can be also interpreted, by reversing the two circles, as type IIA compactified on S^1/Z_2, the so-called type I' theory. From the type IIA viewpoint, the Scherk-Schwarz background fluxes are the RR 4-form field strength and the 3-form NS-NS one. By a sequence of five T-dualities, the RR fluxes are mapped into type IIB fluxes of various ranks, including 3-form flux, whereas the NS-NS 3-form flux is still present. This argument is only of qualitative value, since T-duality in the presence of fluxes are involved.

9.3. Moduli stabilization and de Sitter vacua

Stabilisation of all moduli is mandatory for any application of string theory to low energy particle physics phenomenology and cosmology. However, it turns out that, even if this is realized, generically the resulting vacuum energy is negative, whereas recent observations bring more and more evidence that the vacuum energy is very tiny but positive. Recently there were attempts based on toy models, like the KKLT scenario [85] and variants of them [86], to obtain de Sitter vacua, which we describe in the following. These toy models starts with Calabi-Yau flux compactifications of type IIB orientifolds with D3 and D7 branes, with one Kahler modulus. The generic effect of the fluxes is to stabilize the dilaton and the complex structure fields. If the fluxes break supersymmetry, under some assumptions these moduli fields can be considered to be heavy and integrated out, procedure which, by using (9.8) can generate a residual constant W_0 in the superpotential. The gauge couplings on D7 branes at lowest order are given by the Kahler (volume) modulus T. Nonperturbative gaugino condensation on D7 branes or D3 brane instantons can generate a *nonperturbative* superpotential for T. The Kahler and superpotential of the resulting model is of the simple form [85]

$$K = -3\ln(T + \bar{T}), \quad W = W_0 + a\,e^{-bT}. \tag{9.22}$$

The resulting scalar potential is no longer of the no-scale type and has a stable AdS minimum for the volume type field T. Notice however a generic feature of this scenario. The effective theory analysis is valid in the large volume limit $t = Re\,T >> 1$, in which case it can be shown that the flux induced masses are much lighter than the KK masses, which are in turn much lighter than the fundamental scale

$$M_{\text{flux}} \sim 1/R^3 << M_{KK} \sim 1/R << M_{string} \sim M_s. \tag{9.23}$$

Then the nonperturbative effects generate a small superpotential which by minimization, asks for small values of W_0. A similar problem does occurs in the heterotic strings, where W_0 is provided by the quantization condition of the antisymmetric tensor and is (up to some 2π factors) an integer. In the type II strings case, the novelty is that there are many different possible fluxes available and W_0 receives many different contributions. Models with many fluxes have a large number of vacua, generating the so-called *landscape* of string theory [8], whereas W_0 varies from one vacuum to another. A small number of these vacua can have $W_0 << 1$ and for some (anthropic?) reasons our universe could choose one of these vacua. This combination of nonperturbative effects stabilizing moduli with the landscape anthropic type arguments generated recently a slight shift in the perspective on various problems like the cosmological constant, the hierarchy problem and the scale of supersymmetry breaking.

The next step undertaken in KKLT is the lifting of the AdS minimum to a dS metastable one by adding by hand D3 antibranes. This uplifting does not admit a spontaneous supersymmetric realization and can at best be described as a nonlinear realization of supersymmetry. Another proposal using D-terms generated by internal magnetic fields on D7 branes was subsequently put forward in [86], where however the effective lagrangian was inconsistent with gauge invariance. It is also simple to show that there is not possible to uplift a supersymmetric anti de Sitter vacuum to a de Sitter one by D-term contributions [87]. Indeed, we already saw previously that the general expression for D-terms in supergravity is

$$D^\alpha = \sum_A X_A^\alpha G_A \Phi^A - \frac{1}{2} \sum_k \epsilon_{k\alpha} \frac{\partial G}{\partial M_k}, \qquad (9.24)$$

where M_k are moduli fields with nonlinear gauge transformations. On the other hand, on a flat 4d background, G_A and G_k are a measure of the F contributions to supersymmetry breaking $F_k = \exp(K/2)G_k$. So $\langle D^\alpha \rangle \neq 0$ only if $\langle F_k \rangle \neq 0$.

More recently, another simple proposal was made in [89] using a D-term coming from an R-symmetry, with a corresponding FI term composed of a constant and a T field-dependent pieces. In this last proposal the minimization forces the nonperturbatively generated superpotential to have values close to the Planck scale, which is rather unnatural. To our knowledge, a satisfactory uplifting to a dS metastable vacuum still lacks in the literature. We attempt here to provide one possible consistent framework to achieve this goal, by using the D-terms of an anomalous abelian $U(1)_X$ gauge group with a Green-Schwarz mechanism, by following the effective lagrangian analysis in [88]. The minimal example has an anomalous $U(1)$ factor which generate a T-dependent Fayet-Iliopoulos term and a charged (of charge -1 for simplicity) field ϕ. There is a hidden sector with gauge group $SU(N_c)$ and $N_f < N_c$ flavors Q_a^i, $\bar{Q}_{\bar{j}}^{\bar{a}}$, with i, $\bar{j}$ denoting the flavor index and $a, \bar{a}$ the color index, of $U(1)_X$ charges $q, \bar{q}$ on D7 branes, generating nonperturbative effects by gaugino and matter condensates. The nonperturbative degrees of freedom in the effective theory are the mesons

$$M_{\bar{j}}^i = Q_a^i \bar{Q}_{\bar{j}}^{\bar{a}}. \tag{9.25}$$

The dynamical scale of the theory is

$$\Lambda = e^{-\frac{8\pi^2 kT}{3N_c - N_f}} M_P, \tag{9.26}$$

whereas the effective lagrangian of the model is

$$K = -3\ln(T + \bar{T} - |\phi|^2 - Tr(\bar{M}M)^{1/2} - \delta_{GS}V_X),$$
$$W = W_0 + (N_c - N_f)\left(\frac{e^{-8\pi^2 kT}}{\det M}\right)^{\frac{1}{N_c - N_f}} + \phi^{q+\bar{q}}\lambda_i^{\bar{j}} M_{\bar{j}}^i. \tag{9.27}$$

Under gauge transformations, T transforms non-linearly

$$V_X \to V_X + \Lambda + \bar{\Lambda}, \qquad T \to T + \delta_{GS}\Lambda, \tag{9.28}$$

and fixes the gauge invariant combination $T + \bar{T} - \delta_{GS}V_X$ in the Kahler potential. Notice that the nonperturbative superpotential cannot come from a pure super-Yang-Mills sectors without matter, since the exponential $a \exp(-bT)$ is not $U(1)_X$ gauge invariant. It can be checked however [88] that the meson charge is precisely such that the nonperturbative meson superpotential *is* gauge invariant. The scalar potential contains a D-term

$$V_D = \frac{1}{2(T + \bar{T})}\left(-\phi\frac{\partial K}{\partial \phi} + (q + \bar{q})Tr\left(M\frac{\partial K}{\partial M}\right) + \frac{3\delta_{GS}}{2(T + \bar{T})}\right)^2. \tag{9.29}$$

It is easy to see the supersymmetry is necessarily broken in the model. The resulting vaccum energy cannot be computed exactly analytically, but under some approximations it can be checked that in a certain region of the parameter space there are consistent de-Sitter metastable vacua solutions.

10. Small and large extra dimensions

10.1. Compactification, mass scales and couplings

In string theory, the string coupling constant is a dynamical variable $\lambda = \exp(\phi)$, and the only free parameter is the string length $\alpha' = 1/M_s^2$, where M_s is the string mass scale.

Four dimensional string couplings and scales are predicted in terms of M_s and of various dynamical fields: dilaton, volume of compact space, etc. In contrast to the usual GUT models, which do not incorporate gravity and thus make no predictions for Newton's constant, the perturbative string models do make a definite prediction for the gravitational coupling strength. Since the length scale of string theory $\sqrt{\alpha'}$, the volume V of the internal manifold and the expectation value of the dilaton field ϕ are not directly known from experiment, one might naively think that by adjusting α', V, and $\langle \phi \rangle$ one could fit to any desired values the Newton's constant, the GUT scale M_{GUT}, and the GUT coupling constant α_{GUT}. However, this is not true for the weakly coupled heterotic strings. In 10d, the low energy supergravity effective action looks like

$$S_{eff} = \int d^{10}x \sqrt{g}\, e^{-2\phi} \left[\frac{4}{(\alpha')^4} R - \frac{1}{(\alpha')^3} \mathrm{tr} F^2 + \cdots \right], \tag{10.1}$$

where R is the scalar curvature and $\mathrm{tr} F^2$ is the Yang-Mills kinetic term. After compactification on an internal manifold of volume V (in the string metric), one gets a four-dimensional effective action that looks like

$$S_{eff} = \int d^4x \sqrt{g}\, e^{-2\phi} V \left[\frac{4}{(\alpha')^4} R - \frac{1}{(\alpha')^3} \mathrm{tr} F^2 + \cdots \right]. \tag{10.2}$$

Notice that the same function $Ve^{-2\phi}$ multiplies both R and $\mathrm{tr} F^2$. From (10.2), defining the heterotic scale $M_H = \alpha'^{-1/2}$, one thus gets

$$M_H = \left(\frac{\alpha_{GUT}}{8} \right)^{1/2} M_P, \qquad \lambda_H = 2(\alpha_{GUT} V)^{1/2} M_H^3, \tag{10.3}$$

where $M_P = G_N^{-1/2}$ is the Planck mass. Then $M_H \sim 5 \times 10^{17}$ GeV, and therefore there is some (slight) discrepancy between the GUT scale M_{GUT} and the string scale M_H. Indeed, from (5.1) and (10.3) we find $M_{GUT}/M_H = (4\alpha_{GUT}/\lambda_H^2)^{1/6}$ which asks, in order to find $M_{GUT} \sim 2 - 3 \times 10^{16}$ GeV, for a very large string coupling λ_H. The problem might be alleviated by considering an anisotropic Calabi-Yau space and a lot of effort in this direction was made over the years [7].

In the light of the new string duality picture arising in the last ten years, let us see what changes in the strong coupling regime and let us investigate whether, for a string scale of the order of the GUT scale, the gauge unification problem has a natural solution in a region of *large* string coupling constant. The behavior is completely different depending on whether one considers the $SO(32)$ or the $E_8 \times E_8$ heterotic string.

Let us first consider the strongly-coupled $SO(32)$ heterotic string, equivalent to the weakly-coupled Type I string. We repeat the above discussion, using the Type I dilaton ϕ_I, metric g_I, and scalar curvature R_I. The analog of (10.1) is

$$L_{eff} = \int d^{10}x \sqrt{g_I} \left[e^{-2\phi_I} \frac{4}{(\alpha')^4} R_I - e^{-\phi_I} \frac{1}{(\alpha')^3} \mathrm{tr} F^2 + \ldots \right]. \tag{10.4}$$

Contrary to the heterotic string case, the gravitational and gauge actions multiply different functions of ϕ_I, $e^{-2\phi_I}$ and $e^{-\phi_I}$, since the first is generated by a world-sheet path integral on the sphere, while the second arises from the disk. The analog of (10.2) is then

$$L_{eff} = \int d^4x \sqrt{g_I}\, V \left[\frac{4e^{-2\phi_I}}{(\alpha')^4} R - \frac{e^{-\phi_I}}{(\alpha')^3} \mathrm{tr} F^2 + \ldots \right]. \tag{10.5}$$

The 4d quantities can be expressed as

$$M_I = \left(\frac{2}{\alpha_{GUT}^2 M_P^2} \right)^{1/4} V^{-1/4}, \qquad \lambda_I = 4\alpha_{GUT} M_I^6 V. \tag{10.6}$$

Hence

$$M_I = \left(\frac{\alpha_{GUT}\lambda_I}{8} \right)^{1/2} M_P, \tag{10.7}$$

showing that after taking α_{GUT} from experiment one can make M_I as small as one wishes simply by taking e^{ϕ_I} to be small, that is, by taking the Type I superstring to be weakly coupled[7]. In particular, as mentioned in the Introduction, M_I can be lowered down to the weak scale [44]. In this case the unification picture is completely different [46, 60], as we will see later on.

We will now argue that the $E_8 \times E_8$ heterotic string has a similar strong coupling behavior: one retains the standard GUT relations among the gauge couplings, but loosing the prediction for Newton's constant, which can thus be considerably below the weak coupling bound.

At strong coupling, the ten-dimensional $E_8 \times E_8$ heterotic string becomes M-theory on $R^{10} \times S^1/Z_2$ [36]. Let us compactify the effective action (9.12) to four/five dimensions on a compact manifold whose volume (in the eleven-dimensional metric, from now on) is V. Let S^1 have a radius R_{11} and define the

---

[7]In this case, however, $M_I^6 V \ll 1$ and a better physical picture is obtained by performing T-dualities, thus generating lower-dimensional branes.

eleven dimensional scale $M_{11} = 2\pi(4\pi\kappa_{11}^2)^{-1/9}$. Upon reducing (9.12) down to 4d, one can express M_{11} and R_{11} in terms of four-dimensional parameters

$$M_{11} = (2\alpha_{GUT}V)^{-1/6}, \quad R_{11}^{-1} = \left(\frac{2}{\alpha_{GUT}}\right)^{3/2} M_P^{-2} V^{-1/2}. \tag{10.8}$$

From the first relation we find that $M_{11} \sim M_{GUT}$, and therefore the Horava-Witten theory can accomodate a traditional MSSM unification-type scenario with fundamental scale $M_{11} \sim 10^{16}$ GeV. The second one, for $V = M_{GUT}^{-6}$, gives $R_{11}^{-1} \sim 10^{13} - 10^{15}$ GeV. This is again a sensible result, since R_{11} has to be large compared to the eleven-dimensional Planck scale in order to have a reliable field-theory description of the theory.

10.2. Large extra dimensions

The presence of branes in strings and M-theory opens new perspectives for particle physics phenomenology. Indeed, we already saw in (10.7) that in Type I strings the string scale is not necessarily tied to the Planck scale. In view of the new D-brane picture, let us take a closer look at the simplest example of compactified Type I string, with only D9 branes present. We found in (10.7) that the string scale can be in the TeV range if the string coupling is extremely small, $\lambda_I \sim 10^{-32}$. Then, from the second relation (10.6) one can see that in this case the compact volume is very small $VM_I^6 \sim 10^{-32}$. Let us split the compact volume into two parts, $V = V^{(1)}V^{(2)}$, where $V^{(1)}$, of dimension $6 - n$, is of order one in string units and $V^{(2)}$, of dimension n, is very small. The Kaluza-Klein states of the brane fields along $V^{(2)}$ are much heavier than the string scale and therefore are difficult to excite. The physics is then better captured in this case by performing T-dualities along $V^{(2)}$, which read

$$\lambda_I' = \frac{\lambda_I}{V^{(2)}M_I^n}, \qquad V_\perp = \frac{1}{V^{(2)}M_I^{2n}}. \tag{10.9}$$

In the T-dual picture, neglecting numerical factors, the relations (10.9) become

$$M_P^2 \sim \frac{1}{\alpha_{GUT}\lambda_I'}V_\perp M_I^{2+n}, \qquad \frac{1}{\alpha_{GUT}} \sim \frac{V_\parallel M_I^{6-n}}{\lambda_I'}, \tag{10.10}$$

where for transparency of notation we redefined $V^{(1)} \equiv V_\parallel$. After the n T-dualities, the D9 brane becomes a D(9-n) brane, since the T-dual winding modes of the bulk (orthogonal) compact space are very heavy and therefore the brane fields cannot propagate in the bulk. As seen from (10.10), for a very large bulk volume the string scale can be very low $M_I \ll M_P$. The geometric picture here is that we have a D-brane with some compact radii parallel to it, of order M_I^{-1}, and some very large, orthogonal compact radii. In particular, if the full compact space is orthogonal to the brane ($n = 6$), from (10.10) the T-dual string coupling is fixed by the unified coupling $\lambda'_I \sim \alpha_{GUT}$, and therefore we find [47]

$$M_P^2 \sim \frac{1}{\alpha_{GUT}^2} V_\perp M_I^{2+n}, \tag{10.11}$$

a relation similar to that proposed in the field-theoretical scenario of [45].

Let us now imagine a "brane-world" picture[8], in which the Standard Model gauge group and charged fields are confined to the D-brane under consideration. We can then ask a very important question: what are the present experimental limits on parallel $R_\parallel$ and perpendicular $R_\perp$ type radii? The Standard Model fields have light KK states in the parallel directions $R_\parallel$. Their possible effects in accelerators were studied in detail [90], and the present limits are $R_\parallel^{-1} \geq 4 - 5$ TeV. On the other hand, Standard Model excitations related to $R_\perp$ are very heavy and are thus irrelevant at low energy. The main constraints on $R_\perp$ come from the presence of very light winding (KK after T-dualities) gravitational excitations, which can therefore generate deviations from the Newton law of gravitational attraction. The actual experimental limits on such deviations are limited to the cm range and experiments in the near future are planned to improve them [96]. There are strong collider constraints also, due to the fact that gravity become strong at $M_*^{2+n} = \frac{1}{\alpha_{GUT}^2} M_I^{2+n}$ [99]. For $M_I \sim$ TeV in (10.11), the case of only one extra dimension is clearly excluded, since it asks for $R_\perp^{-1} \sim 10^8$ Km. However, for two extra dimensions, we find $R_\perp^{-1} \sim$ 1mm, not yet excluded by the present experimental data. On the other hand, if all compact dimensions are perpendicular and large, one finds $R_\perp^{-1} \sim$ fm, distance scale completely inaccessible for Newton law measurements. Such a physical picture with $M_I \sim$ TeV provides in principle a new solution to the gauge hierarchy problem, i.e. of why the Higgs mass M_h is much lower than the 4d Planck mass M_P, provided the physical cutoff M_I has similar values $M_I \simeq M_h$.

In Type I strings, the brane we considered can be a D9 or a D5 brane, up to T-dualities. Our brane world can live on any of the branes; let us choose for concreteness that our Standard Model gauge group be on a D9 brane. Notice that, while D9 branes fill (before T-dualities) the full 10d space, D5 branes fill only six dimensions. The D5 degrees of freedom can of course propagate in what we called previously bulk space, and can change slightly our previous picture. The relation between the corresponding D9 and D5 gauge couplings is

$$g_9^2/g_5^2 = V_\perp, \tag{10.12}$$

---

[8] For earlier proposals of such a "brane-world" picture, see [95].

where $V_\perp$ denotes here (before T-dualities) the compact volume perpendicular to the D5 brane. If $V_\perp \gg 1$ in string units, then D5 branes live in (at least part of) the bulk and, by (10.12) their gauge coupling is very suppressed compared to our (D9) gauge coupling. In particular, if $V_\perp$ in (10.12) is as in (10.11), the D5 gauge couplings are of gravitational strength. The fields in mixed 95 representations are charged under both gauge groups. Then, due to their very small gauge couplings, the D5 gauge groups manifest themselves as global symmetries on our D-brane, and could be used for protecting baryon and lepton number nonconservation processes. Indeed, global symmetries are presumably violated by nonrenormalizable operators suppressed by the fundamental scale M_I and, since M_I can be very low, we need suppression of many higher-dimensional operators.

There are clearly many challenging questions that such a scenario must answer [97] in order to be seriously considered as an alternative to the conventional "desert picture" of supersymmetric unification at energies of the order of 10^{16} GeV. The gauge hierarchy problem still has a counterpart here, understanding the possible mm size of the compact dimensions (perpendicular to our brane) in a theory with a fundamental length (energy) in the 10^{-16} mm (TeV) range. There are several ideas concerning this issue in the literature which however need further studies in realistic models in order to prove their viability [105].

10.3. Accelerated unification

Models with gauge-coupling unification at low energy triggered by Kaluza-Klein states were independently proposed in [46], at the same time as brane-world models with a low-string scale. It is transparent, however, that low-scale string models are the natural framework for this fast-driven unification. In this chapter we separate the discussion into two steps: we begin with the field-theoretic picture originally proposed in [46], and then move to the Type I string approach developed in [60, 73] which brings some new, interesting features.

In a field theory approach, the KK excitations of the Standard Model gauge bosons and matter multiplets contribute to the energy evolution of the physical gauge couplings above the compactification scale. The early paper [100] pointed out that the KK excitations give large threshold corrections. As shown in [46], if the energy is higher than the KK compactification scale $1/R$, these corrections should be interpreted as power-law accelerated evolution of gauge couplings.

The one-loop evolution of gauge couplings in 4d between energy scales μ_0 and μ can be computed with standard methods, and the final result can be cast into the form

$$\frac{1}{\alpha_a(\mu)} = \frac{1}{\alpha_a(\mu_0)} + \frac{1}{2\pi} \sum_r \mathrm{Str} \int_{1/\mu^2}^{1/\mu_0^2} \frac{dt}{t} Q_{a,r}^2 \left(\frac{1}{12} - \chi_r^2 \right) e^{-tm_r^2}, \quad (10.13)$$

where $Q_{a,r}$ is the gauge group generator in the representation r of the gauge group, m_r^2 is the mass operator and χ_r is the helicity of various charged particles contributing in the loop.

Let us start, for reasons to be explained later on, with the MSSM in 4d and try to extend it to 5d, where the fifth dimension is a circle of radius $R_{||}$, in the notation introduced in the previous section. In this case, the second term in the RHS of (10.13) is replaced by

$$\frac{1}{2\pi} \sum_r \text{Str} \int_{1/\mu^2}^{1/\mu_0^2} \frac{dt}{t} Q_{a,r}^2 \left(\frac{1}{12} - \chi_r^2\right) \left(\sum_n e^{-tm_{n,r}^2(R_{||})} + e^{-tm_r^2}\right), \quad (10.14)$$

where we separated the mass operator into a part containing fields with KK modes and a part containing fields without KK modes. Evaluating (10.14) with $\mu_0 = M_Z$, one finds

$$\frac{1}{\alpha_a(\mu)} = \frac{1}{\alpha_a(M_Z)} - \frac{b_a}{2\pi} \ln \frac{\mu}{M_Z} - \frac{\tilde{b}_a}{2\pi} \int_{1/\mu^2}^{1/M_Z^2} \frac{dt}{t} \theta_3^\delta \left(\frac{it}{\pi R_{||}^2}\right) \quad (10.15)$$

$$\simeq \frac{1}{\alpha_a(M_Z)} - \frac{b_a}{2\pi} \ln \frac{\mu}{M_Z} + \frac{\tilde{b}_a}{2\pi} \ln(\mu R_{||}) - \frac{\tilde{b}_a}{2\pi} [(\mu R_{||})^\delta - 1].$$

The coefficients $\tilde{b}_a$ in (10.15) denote one-loop beta-function coefficients of the massive KK modes, to be computed in each specific model. The important term contained in (10.15) is the power-like term $(\mu R_{||})^\delta \gg 1$, which overtakes the logarithmic terms in the higher-dimensional regime and governs the eventual unification pattern.

The power-like term is proportional to the coefficients $\tilde{b}_a$, that *are not* the usual 4d MSSM ones which successfully predict unification. Let us however go on and find the *minimal* possible embedding of the MSSM in a 5d spacetime. Before doing it, notice that compactifying on a circle a supersymmetric theory in 5d gives a 4d theory with at least $\mathcal{N} = 2$ supersymmetries. The simplest way to avoid this is to compactify on an *orbifold*. We consider as example the case of a Z_2 orbifold which breaks supersymmetry down to $\mathcal{N} = 1$. 5d fields can be even or odd under this operation, in particular 5d Dirac fermions in 4d truncate into one even Weyl fermion containing a zero mode and its KK tower and one odd Weyl fermion, with no associated zero mode, and its KK tower. It is easy to realize that a 4d chiral multiplet (ψ_1, ϕ_1) can arise from a 5d hypermultiplet containing KK modes $(\psi_1^{(n)}, \psi_2^{(n)}, \phi_1^{(n)}, \phi_2^{(n)})$ or from a 5d vector multiplet. Similarly, a 4d massless vector multiplet (λ, A_μ) arises from a 5d vector multiplet containing the KK modes $(\lambda^{(n)}, \psi_3^{(n)}, A_\mu^{(n)}, a^{(n)})$, where $\psi_i^{(n)}, i = 1, 2, 3$ are 4d Weyl fermions and $\phi_i^{(n)}$, $a^{(n)}$ are complex scalars. The massive KK representations are clearly nonchiral, while chirality is generated at the level of zero modes.

The simplest embedding of the MSSM in 5d is the following [46]. The gauge bosons and the two Higgs multiplets of MSSM are already in real representations of the gauge group and naturally extend to KK representations ($\lambda^{(n)}$, $\psi_3^{(n)}$, $A_\mu^{(n)}$, $a^{(n)}$) and ($\psi_1^{(n)}$, $\psi_2^{(n)}$, $H_1^{(n)}$, $H_2^{(n)}$), respectively[9]. The matter fermions of MSSM, being chiral, can either contain only zero modes or, alternatively, can have associated mirror fermions and KK excitations for $\eta = 0, 1, 2, 3$ families. The unification pattern does not depend on η (the value of the unified coupling, on the other hand, does), since each family forms a complete $SU(5)$ representation. The massive beta-function coefficients for this simple 5d extension of the MSSM are

$$(\tilde{b}_1, \tilde{b}_2, \tilde{b}_3) = \left(\frac{3}{5}, -3, -6\right) + \eta(4, 4, 4), \tag{10.16}$$

where, as usual, we use the $SU(5)$ embedding $\tilde{b}_1 \equiv (3/5)\tilde{b}_Y$. Interestingly enough, as seen from Figure 1, the couplings unify with a surprisingly good precision, for any compact radius 10^3 GeV $\leq R_{\parallel}^{-1} \leq 10^{15}$ GeV, at a energy scale roughly a factor of 20 above the compactification scale $R_{\parallel}^{-1}$. The algebraic reason for this is that, in order to have MSSM unification, the conditions that must be fulfilled are $B_{12}/B_{13} = B_{13}/B_{23} = 1$, where $B_{ac} \equiv \frac{\tilde{b}_a - \tilde{b}_c}{b_a - b_c}$. Although these relations are not satisfied exactly in our case, they are nonetheless approximately satisfied $B_{12}/B_{13} = 72/77 \simeq 0.94$, $B_{13}/B_{23} = 11/12 \simeq 0.92$.

This fast unification with KK states is another numerical miracle (coincidence?), similar to the MSSM unification. There are clearly a lot of questions that this scenario can raise, which were discussed in detail in the literature [46, 101, 102].

In order to have a physical interpretation of the unification scale discussed above, we now turn to the string approach. In a superstring model, the threshold corrections to gauge couplings can be generically written as

$$\mathcal{B}_a = \mathcal{B}_a^{(\mathcal{N}=4)} + \mathcal{B}_a^{(\mathcal{N}=2)} + \mathcal{B}_a^{(\mathcal{N}=1)}, \tag{10.17}$$

where the different terms in the rhs of (10.17) denote contributions from $\mathcal{N} = 4$, $\mathcal{N} = 2$ and $\mathcal{N} = 1$ sectors, respectively. The $\mathcal{N} = 4$ sectors have a 10d origin and give no contribution to threshold corrections. The $\mathcal{N} = 2$ sectors contain the lattice of one compact torus $\Gamma^{(2)}$. In these sectors only BPS KK states contribute to threshold corrections and string oscillators decouple [60,77]. Therefore, their contribution to the evolution of gauge couplings does not stop at

---

[9]As one of the two Higgses in a hypermultiplet is odd under Z_2, the simplest extension actually has one KK Higgs hypermultiplet and one Higgs without KK excitations.

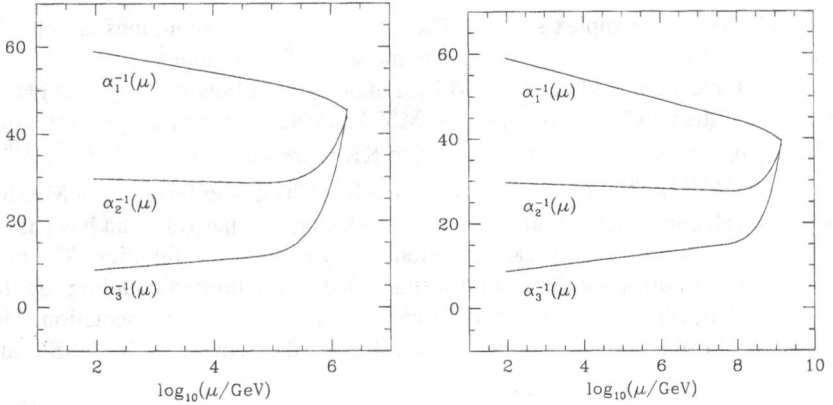

Fig. 1. Unification of gauge couplings in the presence of extra spacetime dimensions. We consider two representative cases: $R^{-1} = 10^5$ GeV (left), $R^{-1} = 10^8$ GeV (right). In both cases we have taken $\delta = 1$ and $\eta = 0$.

the string scale M_I, but rather, as we will see, at a heavy KK scale. The $\mathcal{N} = 1$ sectors have no KK excitations and give a moduli-independent contribution to threshold-corrections, interpreted as the $\mathcal{N} = 1$ contribution to gauge couplings, running up to M_I.

The string one-loop threshold corrections coming from $\mathcal{N} = 2$ sectors were computed in [73, 77]. In addition to the dilaton $\text{Im } S \sim 1/l$, there are tree-level (disk) contributions from couplings of gauge fields to the twisted moduli denoted m_k in what follows. Putting all the terms together, we find the complete one-loop gauge couplings

$$\frac{4\pi^2}{g_a^2(\mu)} = \frac{1}{l} + \sum_k s_{ak} m_k + \frac{1}{4} b_a^{(\mathcal{N}=1)} \ln \frac{M_I^2}{\mu^2}$$

$$- \frac{1}{4} \sum_{i=1}^{3} b_{ai}^{(\mathcal{N}=2)} \ln(\sqrt{G_i}\mu^2 |\eta(U_i)|^4 \text{Im } U_i), \qquad (10.18)$$

where for a rectangular torus of radii R_1, R_2, we have $\sqrt{G_i} = R_1 R_2$ and $\text{Im } U = R_1/R_2$. In (10.18), $b_{ai}^{(\mathcal{N}=2)}$ denote beta function coefficients from $\mathcal{N} = 2$ sectors having KK excitations in the compact torus T^i. The total beta function of the model has several contributions

$$b_a = b_a^{(\mathcal{N}=1)} + \sum_{i=1}^{3} b_{ai}^{(\mathcal{N}=2)}. \qquad (10.19)$$

Let us now consider the field-theory limit of the corrections given by an $\mathcal{N} = 2$ sector, depending on a torus of radii $R_{1,2}$. In the limit $R_1 \to \infty$ and R_2 fixed, the corrections are linearly divergent as $\Lambda_2 \sim R_1/R_2$. These power-law corrections can be used to lower the unification scale [46] in models with a low value of the string scale M_I. In this case, to get unification one needs 10^3 GeV $\leq R_1^{-1} \leq 10^{15}$ GeV. On the other hand, it is at first sight surprising that in the opposite limit of very heavy KK states (windings after T-duality) $R_1 \to 0$, there is a divergent contribution $\Lambda_2 \sim R_2/R_1$. In particular, this applies to mm perpendicular dimensions and can therefore spoil the solution to the hierarchy problem [60, 103].

Another interesting and unexpected feature is that in the limit $R_1, R_2 \to \infty$ with R_1/R_2 fixed, $\Lambda_2 \sim \ln(R_1 R_2 \mu^2)$, instead of the quadratic divergence ($\delta = 2$ in (10.15)) expected in the field theory approach. The same result holds in the $R_1, R_2 \to 0$ limit. This result can be understood by the following argument [103]. After T-duality, the two directions are very large and perpendicular to the brane under consideration. One-loop threshold corrections can also be understood as tree-level coupling of gauge fields to closed sector fields, which have a bulk variation reproducing the threshold dependence on the compact space. The bulk variation can be computed in a supergravity approximation, solving classical field equations for closed fields coupled to various sources subject to global neutrality (or global tadpole cancelation) in the compact space. As the Green function in two dimensions has a logarithmic behavior, this explains the logarithmic term $\ln(R_1 R_2 \mu^2)$. The same argument in one compact dimension explains the linearly divergent term previously discussed.

The logarithmic evolution $\Lambda_2 \sim \ln(R_1 R_2 \mu^2)$ can also be used to achieve unification at a high energy scale, even if the fundamental string scale has much lower values [60], by "running" beyond the string scale.

Notice that both the power-law and the logarithmic evolution of gauge couplings use $\mathcal{N} = 2$ beta-functions. As shown in the field-theory approach, a simple higher-dimensional extension succeeds in producing unification with $\mathcal{N} = 2$ sectors. In this case however, MSSM unification would be just a numerical accident.

11. Conclusions

We are two years ahead of the most exciting period in the recent decades for the theoretical and experimental particle physics, the beginning of the Large Hadron Collider physics at CERN in 2007. Supersymmetry, a crucial and universal ingredient of most string constructions, in its extreme low energy version, will finally be either experimentally confirmed or ruled out, at least as a solution to the hierarchy problem. String constructions, supersymmetric or not, can be tested at LHC only in the extreme limit of string scale in the couple of TeV range or if there are large extra dimensions. A possible hint (but certainly not a proof) for an underlying string theory would be the discovery of new abelian (Z') gauge bosons in the TeV range with anomalous fermionic spectra, asking for a (generalized or not) 4d Green-Schwarz mechanism with axionic couplings and/or eventually generalized Chern-Simons terms. Another possible hint would be the discovery of very light neutral states interacting through non-renormalizable interactions with the Standard Model fields, whose little role from a particle physics perspective can only be justified from first principles by their identification with the modes describing the quantum fluctuations of the internal space of a ten or eleven dimensional string theory.

For other beyond the standard physics models like e.g. ones based on dynamical electroweak symmetry breaking, if experimentally confirmed, string theory, through its less perturbative aspects like the AdS/CFT correspondence [49], can still be instrumental in order to provide a calculable framework for nonperturbative gauge dynamics.

Acknowledgments

I would like to thank the organizers of the 2005 Les Houches school "Physics Beyond the Standard Model" for providing a very nice and stimulating environement. This work is supported in part by the CNRS PICS no. 2530 and 3059, INTAS grant 03-51-6346, the RTN grants MRTN-CT-2004-503369, MRTN-CT-2004-005104 and by a European Union Excellence Grant, MEXT-CT-2003-509661.

References

[1] M.B. Green and J.H. Schwarz, Phys. Lett. B **149** (1984) 117.

[2] D. Gross, J. Harvey, E. Martinec and R. Rohm, Nucl. Phys. B **256** (1985) 468.

[3] P. Candelas, G. Horowitz, A. Strominger and E. Witten, Nucl. Phys. B **258** (1985) 46.

[4] L. Dixon, J. Harvey, C. Vafa and E. Witten, Nucl. Phys. B **256** (1985) 620 and Nucl. Phys. B **274** (1986) 285

[5] D. Gepner and E. Witten, Nucl. Phys. B **278** (1986) 493; I. Antoniadis, C. Bachas and C. Kounnas, Nucl. Phys. B **289** (1987) 87; K.S. Narain, M.H. Sarmadi and C. Vafa, Nucl. Phys. B **288** (1987) 551; H. Kawai, D.C. Lewellen and S.-H.H. Tye, Nucl. Phys. B **288** (1987) 1; W. Lerche, D. Lüst and A.N. Schellekens, Nucl. Phys. B **287** (1987) 477.

[6] L.E. Ibáněz, H.P. Nilles and F. Quevedo, Phys. Lett. B **192** (1987) 332.

[7] For review papers, see L.E. Ibanez, hep-th/9505098; K.R. Dienes, hep-th/9602045; F. Quevedo, hep-th/9603074.

[8] L. Susskind, hep-th/0302219; M. R. Douglas, JHEP **0305** (2003) 046; S. Ashok and M. R. Douglas, JHEP **0401** (2004) 060; F. Denef and M. R. Douglas, JHEP **0405** (2004) 072.

[9] R. Bousso and J. Polchinski, JHEP **0006** (2000) 006.

[10] N. Arkani-Hamed and S. Dimopoulos, JHEP **0506** (2005) 073; N. Arkani-Hamed, S. Dimopoulos, G. F. Giudice and A. Romanino, Nucl. Phys. B **709** (2005) 3.

[11] A. Sagnotti, in: Cargese '87, Non-Perturbative Quantum Field Theory, eds. G. Mack et al. (Pergamon Press, Oxford, 1988) p. 521.

[12] G. Pradisi and A. Sagnotti, Phys. Lett. B **216** (1989) 59; P. Horava, Nucl. Phys. B **327** (1989) 461, Phys. Lett. B **231** (1989) 251; M. Bianchi, G. Pradisi and A. Sagnotti, Nucl. Phys. B **376** (1992) 365.

[13] M. Bianchi and A. Sagnotti, Phys. Lett. B **247** (1990) 517 and Nucl. Phys. B **361** (1991) 519.

[14] J.A. Harvey and J.A. Minahan, Phys. Lett. B **188** (1987) 44; N. Ishibashi and T. Onogi, Nucl. Phys. B **318** (1989) 239; P. Horava, Nucl. Phys. B **327** (1989) 461, Phys. Lett. B **231** (1989) 251.

[15] J.H. Schwarz, hep-th/9607201; M.J. Duff, hep-th/9608117; P.K. Townsend, hep-th/9612121.

[16] J. Polchinski, Phys. Rev. Lett. **75** (1995) 4724.

[17] E. Gimon and J. Polchinski, Phys. Rev. D **54** (1996) 1667.

[18] C. Angelantonj, M. Bianchi, G. Pradisi, A. Sagnotti and Y. S. Stanev, Phys. Lett. B **385** (1996) 96.

[19] M. Berkooz and R. Leigh, Nucl. Phys. B **483** (1997) 187; Z. Kakushadze, Nucl. Phys. B **512** (1998) 221; Z. Kakushadze and G. Shiu, Phys. Rev. D **56** (1997) 3686 and Nucl. Phys. B **520** (1998) 75.

[20] L.E. Ibáñez, R. Rabadan and A. Uranga, Nucl. Phys. B **542** (1999) 112.

[21] A. Sagnotti, Phys. Lett. B **294** (1992) 196; S. Ferrara, F. Riccioni and A. Sagnotti, Nucl. Phys. B **519** (1998) 115.

[22] I. Antoniadis, E. Dudas and A. Sagnotti, Nucl. Phys. B **544** (1999) 469; I. Antoniadis, G. D'Appollonio, E. Dudas and A. Sagnotti, Nucl. Phys. B **553** (1999) 133; I. Antoniadis, G. D'Appollonio, E. Dudas and A. Sagnotti, Nucl. Phys. B **565** (2000) 123; A. L. Cotrone, Mod. Phys. Lett. A **14** (1999) 2487.

[23] E.S. Fradkin and A.A. Tseytlin, Phys. Lett. B **163** (1985) 123; A. Abouelsaood, C.G. Callan, C.R. Nappi and S.A. Yost, Nucl. Phys. B **280** (1987) 599.

[24] C. Bachas, hep-th/9503030; M. Bianchi and Ya. Stanev, Nucl. Phys. B **253** (1998) 193.

[25] M. Berkooz, M. R. Douglas and R. G. Leigh, Nucl. Phys. B **480** (1996) 265; N. Ohta and P. K. Townsend, Phys. Lett. B **418** (1998) 77.

[26] R. Blumenhagen, L. Goerlich, B. Kors and D. Lust, JHEP **0010** (2000) 006; R. Blumenhagen, B. Kors and D. Lust, JHEP **0102** (2001) 030.

[27] C. Angelantonj, I. Antoniadis, E. Dudas and A. Sagnotti, Phys. Lett. B **489** (2000) 223; C. Angelantonj and A. Sagnotti, hep-th/0010279.

[28] G. Aldazabal, S. Franco, L. E. Ibanez, R. Rabadan and A. M. Uranga, JHEP **0102** (2001) 047; M. Cvetic, G. Shiu and A. M. Uranga, Nucl. Phys. B **615** (2001) 3 and Phys. Rev. Lett. **87** (2001) 201801; C. Kokorelis, JHEP **0209** (2002) 029; D. Bailin, G. V. Kraniotis and A. Love, Phys. Lett. B **553** (2003) 79; G. Honecker, Nucl. Phys. B **666** (2003) 175; M. Larosa and G. Pradisi, Nucl. Phys. B **667** (2003) 261; C. M. Chen, G. V. Kraniotis, V. E. Mayes, D. V. Nanopoulos and J. W. Walker, Phys. Lett. B **611** (2005) 156; E. Dudas and C. Timirgaziu, Nucl. Phys. B **716** (2005) 65; R. Blumenhagen, M. Cvetic, F. Marchesano and G. Shiu, JHEP **0503** (2005) 050.

[29] I. Antoniadis, E. Kiritsis and T. N. Tomaras, Phys. Lett. B **486** (2000) 186; E. Dudas, A. Falkowski and S. Pokorski, Phys. Lett. B **568** (2003) 281; L. Andrianopoli, S. Ferrara and M. A. Lledo, JHEP **0404** (2004) 005.

[30] I. Antoniadis, E. Dudas and A. Sagnotti, Phys. Lett. B **464** (1999) 38; for phenomenology of the model, see I. Antoniadis, K. Benakli and M. Quiros, Nucl. Phys. B **583** (2000) 35.

[31] G. Aldazabal and A. M. Uranga, JHEP **9910** (1999) 024, JHEP **0001** (2000) 031, JHEP **0002** (2000) 015; G. Aldazabal, L. E. Ibanez, F. Quevedo and A. M. Uranga, JHEP **0008** (2000) 002.

[32] C. Angelantonj, I. Antoniadis, G. D'Appollonio, E. Dudas and A. Sagnotti, Nucl. Phys. B **572** (2000) 36; C. Angelantonj, R. Blumenhagen and M. R. Gaberdiel, Nucl. Phys. B **589** (2000) 545.

[33] A. Sen, JHEP **9806** (1998) 007, JHEP **9808** (1998) 010, JHEP **9809** (1998) 023, JHEP **9812** (1998) 021; For a review, see e.g. M. R. Gaberdiel, Class. Quant. Grav. **17** (2000) 3483.

[34] A. Sen, JHEP **0207** (2002) 065; G. W. Gibbons, Phys. Lett. B **537** (2002) 1; M. Fairbairn and M. H. G. Tytgat, Phys. Lett. B **546** (2002) 1 and Mod. Phys. Lett. A **17** (2002) 1797.

[35] M. Dine, N. Seiberg and E. Witten, Nucl. Phys. B **289** (1987) 589.

[36] P. Horava and E. Witten, Nucl. Phys. B **460** (1996) 506 and Nucl. Phys. B **475** (1996) 94.

[37] S. B. Giddings, S. Kachru and J. Polchinski, Phys. Rev. D **66** (2002) 106006; S. Kachru, M. B. Schulz and S. Trivedi, JHEP **0310** (2003) 007; S. Kachru, R. Kallosh, A. Linde and S. P. Trivedi, Phys. Rev. D **68** (2003) 046005; M. Grana, T. W. Grimm, H. Jockers and J. Louis, Nucl. Phys. B **690** (2004) 21.

[38] F. Marchesano and G. Shiu, Phys. Rev. D **71** (2005) 011701; P. G. Camara, L. E. Ibanez and A. M. Uranga, Nucl. Phys. B **708** (2005) 268; A. Font and L. E. Ibanez, JHEP **0503** (2005) 040; M. Cvetic, T. Li and T. Liu, Phys. Rev. D **71** (2005) 106008; I. Antoniadis, A. Kumar and T. Maillard, hep-th/0505260.

[39] E. Witten, Nucl. Phys. B **471** (1996) 135.

[40] H. Georgi, H. Quinn and S. Weinberg, Phys. Rev. Lett. **33** (1974) 451; S. Dimopoulos, S. Raby and F. Wilczek, Phys. Rev. D **24** (1981) 1681.

[41] E. Dudas and C. Grojean, Nucl. Phys. B **507** (1997) 553; E. Dudas, Phys. Lett. B **416** (1998) 309.

[42] A. Lukas, B. A. Ovrut, K. S. Stelle and D. Waldram, Phys. Rev. D **59** (1999) 086001; Nucl. Phys. B **552** (1999) 246; J. R. Ellis, Z. Lalak and W. Pokorski, Nucl. Phys. B **559** (1999) 71; J. P. Derendinger and R. Sauser, Nucl. Phys. B **582** (2000) 231.

[43] I. Antoniadis and M. Quiros, Phys. Lett. B **392** (1997) 61, Nucl. Phys. B **505** (1997) 109, Phys. Lett. B **416** (1998) 327.

[44] J.D. Lykken, Phys. Rev. D **54** (1996) 3693.

[45] N. Arkani-Hamed, S. Dimopoulos and G. Dvali, Phys. Lett. B **429** (1998) 263 and Phys. Rev. D **59** (1999) 086004.

[46] K.R. Dienes, E. Dudas and T. Gherghetta, Phys. Lett. B **436** (1998) 55, Nucl. Phys. B **537** (1999) 47, hep-ph/9807522.

[47] I. Antoniadis, N. Arkani-Hamed, S. Dimopoulos and G. Dvali, Phys. Lett. B **436** (1998) 257.

[48] L. Randall and R. Sundrum, Phys. Rev. Lett. **83** (1999) 3370 and Phys. Rev. Lett. **83** (1999) 4690; M. Gogberashvili, Int. J. Mod. Phys. D **11** (2002) 1635 and Europhys. Lett. **49** (2000) 396.

[49] J. Maldacena, Adv. Theor. Phys. **2** (1998) 231; S.S. Gubser, I.R. Klebanov and A.M. Polyakov, Phys. Lett. B **428** (1998) 105; E. Witten, Adv. Theor. Phys. **2** (1998) 253.

[50] L. Dixon and J. Harvey, Nucl. Phys. B **274** (1986) 93.

[51] L. Alvarez-Gaume, P. Ginsparg, G. Moore and C. Vafa, Phys. Lett. B **171** (1986) 155.

[52] N. Seiberg and E. Witten, Nucl. Phys. B **276** (1986) 272.

[53] J. Paton and H.M. Chan, Nucl. Phys. B **10** (1969) 519; J.H. Schwarz, Phys. Rep. **89** (1982) 223; N. Marcus and A. Sagnotti, Phys. Lett. B **119** (1982) 97.

[54] E. Cremmer, B. Julia and J. Scherk, Phys. Lett. B **76** (1978) 409.

[55] E. Witten, Nucl. Phys. B **443** (1995) 85.

[56] J. Polchinski and E. Witten, Nucl. Phys. B **460** (1996) 525.

[57] A. Dabholkar, Phys. Lett. B **357** (1995) 307; C.M. Hull, Phys. Lett. B **357** (1995) 545.

[58] C.M. Hull and P. Townsend, Nucl. Phys. B **438** (1995) 409.

[59] See, for example, A. Sen, Nucl. Phys. Proc. Suppl. **58** (1997) 5.

[60] C. Bachas, JHEP **9811** (1998) 023.

[61] J. Polchinski and Y. Cai, Nucl. Phys. B **296** (1988) 91; J. Polchinski, Phys. Rev. Lett. **75** (1995) 4724; M. Bianchi and J. F. Morales, JHEP **0003** (2000) 030.

[62] D. Fioravanti, G. Pradisi and A. Sagnotti, Phys. Lett. B **321** (1994) 349; G. Pradisi, A. Sagnotti and Ya.S. Stanev, Phys. Lett. B **354** (1995) 279, Phys. Lett. B **356** (1995) 230, Phys. Lett. B **381** (1996) 97; L. R. Huiszoon and A. N. Schellekens, Nucl. Phys. B **584** (2000) 705.

[63] For a review, see A. Sagnotti and Y. S. Stanev, Fortsch. Phys. **44** (1996) 585 [Nucl. Phys. Proc. Suppl. **55B** (1997) 200].

[64] W. Fischler and L. Susskind, Phys. Lett. B **171** (1986) 383 and Phys. Lett. B **173** (1986) 262.

[65] S. Sugimoto, Prog. Theor. Phys. **102** (1999) 685.

[66] E. Dudas and J. Mourad, Phys. Lett. B **486** (2000) 172; R. Blumenhagen and A. Font, Nucl. Phys. B **599** (2001) 241.

[67] E. Dudas, J. Mourad and C. Timirgaziu, Nucl. Phys. B **660** (2003) 3; E. Dudas and C. Timirgaziu, JHEP **0403** (2004) 060 and Nucl. Phys. B **716** (2005) 65.

[68] A. Sen, Mod. Phys. Lett. **A11** (1996) 1339; R. Gopakumar and S. Muhki, Nucl. Phys. B **479** (1996) 260.

[69] E. Dudas and J. Mourad, Phys. Lett. B **423** (1998) 281.

[70] M. Fabinger and P. Horava, Nucl. Phys. B **580** (2000) 243.

[71] E. Gimon and C. Johnson, Nucl. Phys. B **477** (1996) 715; A. Dabholkar and J. Park, Nucl. Phys. B **477** (1996) 710; For reviews on orientifolds, see A. Dabholkar, hep-th/9804208; E. Dudas, Class. Quant. Grav. **17** (2000) R41; C. Angelantonj and A. Sagnotti, Phys. Rept. **371** (2002) 1 [Erratum-ibid. **376** (2003) 339].

[72] G. Aldazabal, A. Font, L.E. Ibáñez and G. Violero, Nucl. Phys. B **536** (1998) 29; Z. Kakushadze, G. Shiu and S.H.H. Tye, Nucl. Phys. B **533** (1998) 25.

[73] I. Antoniadis, C. Bachas and E. Dudas, Nucl. Phys. B **560** (1999) 93; P. Bain and M. Berg, JHEP **0004** (2000) 013.

[74] C. Scrucca and M. Serone, JHEP **9912** (1999) 024 and Nucl. Phys. B **564** (2000) 555.

[75] E. Poppitz, Nucl. Phys. B **542** (1999) 31.

[76] Z. Lalak, S. Lavignac and H. P. Nilles, Nucl. Phys. B **559** (1999) 48; M. Cvetic, L. Everett, P. Langacker and J. Wang JHEP **9904** (1999) 020.

[77] C. Bachas and C. Fabre, Nucl. Phys. B **476** (1996) 418.

[78] L. Dixon, V. Kaplunovsky and J. Louis, Nucl. Phys. B **355** (1991) 649; I. Antoniadis, K. Narain and T. Taylor, Phys. Lett. B **267** (1991) 37; I. Antoniadis, E. Gava and K. Narain, Nucl. Phys. B **383** (1992) 93; I. Antoniadis, E. Gava, K.S. Narain and T.R. Taylor, Nucl. Phys. B **407** (1993) 706; P. Mayr and S. Stieberger, Nucl. Phys. B **407** (1993) 725 and Phys. Lett. B **355** (1995) 107; E. Kiritsis and C. Kounnas, hep-th/9410212 and Nucl. Phys. B **442** (1995) 472; H.P. Nilles and S. Stieberger, Nucl. Phys. B **499** (1997) 3.

[79] M. Cvetic, M. Plumacher and J. Wang, JHEP **0004** (2000) 004; M. Cvetic and P. Langacker, Nucl. Phys. B **586** (2000) 287.

[80] R. Rohm, Nucl. Phys. B **237** (1984) 553; C. Kounnas and M. Porrati, Nucl. Phys. B **310** (1988) 355; S. Ferrara, C. Kounnas, M. Porrati and F. Zwirner, Nucl. Phys. B **318** (1989) 75; C. Kounnas and B. Rostand, Nucl. Phys. B **341** (1990) 641; E. Kiritsis and C. Kounnas, Nucl. Phys. B **503** (1997) 117.

[81] J. Scherk and J.H. Schwarz, Nucl. Phys. B **153** (1979) 61 and Phys. Lett. B **82** (1979) 60; E. Cremmer, J. Scherk and J.H. Schwarz, Phys. Lett. B **84** (1979) 83; P. Fayet, Phys. Lett. B **159** (1985) 121 and Nucl. Phys. B **263** (1986) 649; for some phenomenological studies, see A. Pomarol and M. Quiros, Phys. Lett. B **438** (1998) 255; I. Antoniadis, S. Dimopoulos, A. Pomarol and M. Quiros, Nucl. Phys. B **544** (1999) 503.

[82] I. Antoniadis, C. Bachas, D. Lewellen and T. Tomaras, Phys. Lett. B **207** (1988) 441.

[83] S. B. Giddings, S. Kachru and J. Polchinski, Phys. Rev. D **66** (2002) 106006.

[84] S. Gukov, C. Vafa and E. Witten, Nucl. Phys. B **584** (2000) 69 [Erratum-ibid. B **608** (2001) 477]; T. R. Taylor and C. Vafa, Phys. Lett. B **474** (2000) 130.

[85] S. Kachru, R. Kallosh, A. Linde and S. P. Trivedi, Phys. Rev. D **68** (2003) 046005.

[86] C. P. Burgess, R. Kallosh and F. Quevedo, JHEP **0310** (2003) 056.

[87] K. Choi, A. Falkowski, H. P. Nilles and M. Olechowski, Nucl. Phys. B **718** (2005) 113.

[88] E. Dudas and S. K. Vempati, Nucl. Phys. B **727** (2005) 139.

[89] G. Villadoro and F. Zwirner, Phys. Rev. Lett. **95** (2005) 231602.

[90] I. Antoniadis, Phys. Lett. B **246** (1990) 377; I. Antoniadis and K. Benakli, Phys. Lett. B **326** (1994) 69; I. Antoniadis, K. Benakli and M. Quirós, Phys. Lett. B **331** (1994) 313; A. Delgado, A. Pomarol and M. Quiros, Phys. Rev. D **60** (1999) 095008.

[91] J. Blum and K.R. Dienes, Phys. Lett. B **414** (1997) 260 and Nucl. Phys. B **516** (1998) 93.

[92] E. Witten, Nucl. Phys. B **460** (1996) 541.

[93] T. Banks and L. Susskind, hep-th/9511194.

[94] For reviews, see S. Weinberg. Rev. Mod. Phys. **61** (1989) 1; E. Witten, hep-ph/0002297.

[95] V.A. Rubakov and M.E. Shaposhnikov, Phys. Lett. B **125** (1983) 139 and Phys. Lett. B **125** (1985) 372; K. Akama, Lect. Notes Phys. **176** (1982) 267 and Prog. Theor. Phys. **78** (1987) 184; M. Visser, Phys. Lett. B **159** (1985) 22.

[96] J.C. Long, H.W. Chan and J.C. Price, Nucl. Phys. B **539** (1999) 23.

[97] N. Arkani-Hamed and S. Dimopoulos, Phys. Rev. D **65** (2002) 052003; N. Arkani-Hamed and M. Schmaltz, Phys. Rev. D **61** (2000) 033005; N. Arkani-Hamed, L. J. Hall, D. R. Smith and N. Weiner, Phys. Rev. D **61** (2000) 116003; A. Delgado, A. Pomarol and M. Quiros, JHEP **0001** (2000) 030.

[98] S. Ferrara, L. Girardello and H.P. Nilles, Phys. Lett. B **125** (1983) 457; J.P. Derendinger, L.E. Ibáñez and H.P. Nilles, Phys. Lett. B **155** (1985) 467; M. Dine, R. Rohm, N. Seiberg and E. Witten, Phys. Lett. B **156** (1985) 55.

[99] G.F. Giudice, R. Rattazzi and J.D. Wells, Nucl. Phys. B **544** (1999) 3; E.A. Mirabelli, M. Perelstein and M.E. Peskin, Phys. Rev. Lett. **82** (1999) 2236; T. Han, J.D. Lykken and R.J. Zhang, Phys. Rev. D **59** (1999) 105006; J.L. Hewett, Phys. Rev. Lett. **82** (1999) 4765; M. Besancon, hep-ph/9909364; E. Dudas and J. Mourad, Nucl. Phys. B **575** (2000) 3; E. Accomando, I. Antoniadis and K. Benakli, Nucl. Phys. B **579** (2000) 3; S. Cullen, M. Perelstein and M. E. Peskin, Phys. Rev. D **62** (2000) 055012.

[100] T.R. Taylor and G. Veneziano, Phys. Lett. B **212** (1988) 147.

[101] D. Ghilencea and G.G. Ross, Phys. Lett. B **442** (1998) 165 and Nucl. Phys. B **569** (2000) 391; Z. Kakushadze, Nucl. Phys. B **548** (1999) 205; S. Abel and S. King, Phys. Rev. D **59** (1999) 095010; T. Kobayashi, J. Kubo, M. Mondragon and G. Zoupanos, Nucl. Phys. B **550** (1999) 99; C.D. Carone, Phys. Lett. B **454** (1998) 70; A. Delgado and M. Quiros, Nucl. Phys. B **559** (1999) 235; G. K. Leontaris and N. D. Tracas, Phys. Lett. B **454** (1999) 53 and Phys. Lett. B **470** (1999) 84.

[102] Z. Kakushadze and T. R. Taylor, Nucl. Phys. B **562** (1999) 78; D. M. Ghilencea and S. Groot Nibbelink, Nucl. Phys. B **641** (2002) 35.

[103] I. Antoniadis and C. Bachas, Phys. Lett. B **450** (1999) 83.

[104] L. E. Ibanez, C. Munoz and S. Rigolin, Nucl. Phys. B **553** (1999) 43; I. Antoniadis, E. Kiritsis and T. N. Tomaras, Phys. Lett. B **486** (2000) 186.

[105] N. Arkani-Hamed, S. Dimopoulos and J. March-Russell, hep-th/9908146.

[106] S. Ferrara and S. Sabharwal, Class. Quantum Grav. **6** (1989) L77 and Nucl. Phys. B **332** (1990) 317.

[107] E. Cremmer, S. Ferrara, C. Kounnas and D.V. Nanopoulos, Phys. Lett. B **133** (1983) 61; J. Ellis, C. Kounnas and D.V. Nanopoulos, Nucl. Phys. B **241** (1984) 406 and Nucl. Phys. B **247** (1984) 373.

[108] E. Cremmer and J. Scherk, Nucl. Phys. B **50** (1972) 222; L. Clavelli and J.A. Shapiro, Nucl. Phys. B **57** (1973) 490; C.G. Callan, C. Lovelace, C.R. Nappi and S.A. Yost, Nucl. Phys. B **293** (1987) 83 and Nucl. Phys. B **308** (1988) 221; J. Polchinski and Y. Cai, Nucl. Phys. B **296** (1988) 91.

Course 9

PARTICLE ASTROPHYSICS AND COSMOLOGY

Pierre Binétruy

AstroParticule et Cosmologie, Université Paris 7,
Collège de France, 11 place Marcellin Berthelot,
F-75231 Paris Cedex 05, France

D. Kazakov, S. Lavignac and J. Dalibard, eds.
Les Houches, Session LXXXIV, 2005
Particle Physics Beyond the Standard Model

457

Contents

1. What is particle astrophysics or astroparticle?

Astroparticle, or equivalently particle astrophysics, is the field at the interface between astrophysics and particle physics. For a long time, astronomy has relied on observations in the visible range of electromagnetic waves. In the XXth century, this has been extended to the whole electromagnetic spectrum. However, as one goes to higher frequencies, one is moving from a classical wave description to the concept of a (increasingly energetic) particle: one reaches the realm of astroparticles.

Thus, particle astrophysics typically deals with sources of energetic particles in the Universe (Active Galactic Nuclei [AGN], supernovae, Gamma Ray Bursts [GRB], . . .). It uses a multi-wavelength approach to study the extreme sources of the Universe. For example, a classical source such as Markarian 421 has been shown to cover a spectrum that ranges from radio wavelengths to high energy gamma rays, thus spanning some 18 orders of magnitude in frequency.

All this uses the photon as a messenger. One of the goals of particle astrophysicists is to use other messengers, thus opening new windows on the sky. An unprecedented experimental effort aims at obtaining new information on sources, and eventually a whole map of the sky using:

- high energy gamma rays (HIREs, AGASA, AUGER),
- high energy neutrinos (AMANDA, ANTARES, NESTOR),
- gravitational waves (LIGO, VIRGO, LISA).

We will describe in what follows some of this effort. The goal is to use a multi-messenger approach to study a given source and understand its dynamics:

- gravitational waves provide information on the dynamics of the bulk mass,
- high energy photons trace populations of accelerated particles, as well as dark matter annihilation,
- high energy cosmic rays allow to identify the nature of the acceleration process that takes place,
- neutrinos are probing the innermost parts of the source, opaque to photons.

The more energetic are the particles, the more compact is the source. This is why the most compact objects, namely neutron stars and black holes, play a central role in particle astrophysics. On may note *en passant* the analogy between black holes and fundamental particles. Black holes are indeed very simple

461

objects, characterized solely by their mass, charge and angular momentum. However, just as understanding the dynamics of fundamental particles is not enough to understand physics at larger scales (high T superconductivity or quantum Hall effect),there is much more to astrophysics than understanding the inner structure of a few building blocks such as black holes or neutron stars.

The other connection between astrophysics and particle physics is found in the early universe. Indeed, because of the expansion of the universe, the primordial plasma is hot and dense, and is made of fundamental particles. It thus provides the physical conditions to study these particles. In parallel, a more precise understanding of the fundamental theory is likely to provide the boundary conditions necessary to fully describe the evolution of the universe. Thus, nowadays, the study of the early universe is becoming increasingly intricate with the understanding of the dynamics of fundamental particles at high energies.

It is often instructive to list the main open questions that one addresses in a given field. The way the questions are phrased evolves with time and is thus a good tracker of our current knowledge. For particle astrophysics and cosmology, this gives:

- Did the Universe start with a big bang?
- If there was inflation, what was its dynamics?
- What is the origin of matter-antimatter asymmetry?
- What were the first luminous objects, and when did they appear?
- How to detect and identify dark matter?
- What is the content and geometry of the universe at large scale?
- How to confirm or infirm the present acceleration of the expansion?
- Is there a dark energy? Why is the vacuum energy so small?
- What are the limits of validity of general relativity?
- How do cosmic accelerators function?
- What can the study of energetic sources teach us on the laws of physics?
- How are black holes formed?
- Are there new states of matter at extreme densities or energies?
- How do supernovae explode? How are heavy elements formed?

We see that this covers some of the most fondamental questions of modern science.

2. The (not so) quiet universe

2.1. *Gravity rules the evolution of the universe*

The evolution of the universe at large is governed by gravity, and thus described by Einstein's equations. We recall that, in the context of general relativity, the metric is a dynamical field, i.e. is spacetime dependent: its fluctuating part is the gravitational field. Einstein's equations are highly non-linear second order differential equations of motion for this field. They read[1]:

$$G_{\mu\nu} \equiv R_{\mu\nu} - \frac{1}{2}g_{\mu\nu}R = 8\pi G_N T_{\mu\nu} + \lambda g_{\mu\nu}. \tag{2.1}$$

where $R_{\mu\nu}$ is the Ricci tensor, R the associated curvature scalar[2] and $T_{\mu\nu}$ the energy-momentum tensor; finally λ the cosmological constant.

Thus Einstein's equations relate the geometry of spacetime (the left-hand side of (2.1)) with its matter field content (the right-hand side). As we will see, it is still an open question whether the cosmological term belongs to the left or the right-hand side.

One may try to apply Einstein's equations to describe not a given gravitational system like a planet or a star but the evolution of the whole universe. In Einstein's days, this is a bold move: it should be remembered how little of the universe was known at the time these equations were written. "In 1917, the world was supposed to consist of our galaxy and presumably a void beyond. The Andromeda nebula had not yet been certified to lie beyond the Milky Way."[Pais [1] p. 286] Indeed, it is in this context that Einstein introduced the cosmological constant in order to have a static solution (until it was observed by Hubble that the universe is expanding) for the universe.

Under the assumption that the Universe is homogeneous and isotropic on scales of order 100 Mpc (1 pc = 3.262 light-year = 3.086×10^{16} m) and larger, one may first try to find a homogeneous and isotropic metric as a solution of Einstein's equations. The most general ansatz is, up to coordinate redefinitions, the

---

[1]The metric signature we adopt is Einstein's choice: $(+, -, -, -)$.

[2]In the context of general relativity, one defines the Christoffel symbol or spin connection $\Gamma^\rho{}_{\mu\nu}$ which is the analogue of the gauge field (it appears in covariant derivatives). It is defined in terms of the metric as:

$$\Gamma^\rho{}_{\mu\nu} = \frac{1}{2}g^{\rho\sigma}[\partial_\mu g_{\nu\sigma} + \partial_\nu g_{\mu\sigma} - \partial_\sigma g_{\mu\nu}], \tag{2.2}$$

where $g^{\rho\sigma}$ is the inverse metric tensor: $g^{\rho\sigma}g_{\sigma\tau} = \delta^\rho_\tau$.

In the same way that one defines the field strength by differentiating the gauge field, one introduces the Riemann curvature tensor:

$$R^\mu{}_{\nu\alpha\beta} = \partial_\alpha\Gamma^\mu{}_{\nu\beta} - \partial_\beta\Gamma^\mu{}_{\nu\alpha} + \Gamma^\mu{}_{\alpha\sigma}\Gamma^\sigma{}_{\nu\beta} - \Gamma^\mu{}_{\beta\sigma}\Gamma^\sigma{}_{\nu\alpha}. \tag{2.3}$$

By contracting indices, one then defines the Ricci tensor $R_{\mu\nu}$ and the curvature scalar R

$$R_{\mu\nu} \equiv R^\alpha{}_{\mu\alpha\nu}, \qquad R \equiv g^{\mu\nu}R_{\mu\nu}. \tag{2.4}$$

Robertson-Walker metric:

$$ds^2 = c^2 dt^2 - a^2(t)\,\gamma_{ij} dx^i dx^j, \tag{2.5}$$

$$\gamma_{ij} dx^i dx^j = \frac{dr^2}{1 - kr^2} + r^2\left(d\theta^2 + \sin^2\theta\, d\phi^2\right), \tag{2.6}$$

where $a(t)$ is the cosmic scale factor, which is time-dependent in an expanding or contracting universe. Such a universe is called a Friedmann-Lemaître universe. The constant k which appears in the spatial metric γ_{ij} can take the values ± 1 or 0: the value 0 corresponds to flat space, i.e. usual Minkowski spacetime; the value $+1$ to closed space ($r^2 < 1$) and the value -1 to open space. Note that r is dimensionless whereas a has the dimension of a length. From now on, we set $c = 1$, except otherwise stated.

The components of the Einstein tensor now read (see Exercise 2-1):

$$G_{tt} = 3\left(\frac{\dot{a}^2}{a^2} + \frac{k}{a^2}\right), \tag{2.7}$$

$$G_{ij} = -\gamma_{ij}\left(\dot{a}^2 + 2a\ddot{a} + k\right), \tag{2.8}$$

where we use standard notations: $\dot{a}$ is the first time derivative of the cosmic scale factor, $\ddot{a}$ the second time derivative.

For the energy-momentum tensor, we follow our assumption of homogeneity and isotropy and assimilate the content of the Universe to a perfect fluid:

$$T_{\mu\nu} = -pg_{\mu\nu} + (p + \rho)U_\mu U_\nu, \tag{2.9}$$

where U^μ is the velocity 4-vector ($U^t = 1$, $U^i = 0$). It follows from (2.9) that $T_{tt} = \rho$ and $T_{ij} = a^2 p\gamma_{ij}$. The pressure p and energy density ρ usually satisfy the equation of state:

$$p = w\rho. \tag{2.10}$$

The constant w takes the value $w \sim 0$ for non-relativistic matter (negligible pressure) and $w = 1/3$ for relativistic matter (radiation). In all generality, the perfect fluid consists of several components with different values of w.

One now obtains from the $(0, 0)$ and (i, j) components of the Einstein equations (2.1):

$$3\left(\frac{\dot{a}^2}{a^2} + \frac{k}{a^2}\right) = 8\pi G_N \rho + \lambda, \tag{2.11}$$

$$\dot{a}^2 + 2a\ddot{a} + k = -8\pi G_N a^2 p + a^2 \lambda, \tag{2.12}$$

The first of the preceding equations can be written as the Friedmann equation, which gives an expression for the Hubble parameter $H \equiv \dot{a}/a$ measuring the rate of the expansion of the Universe:

$$H^2 \equiv \frac{\dot{a}^2}{a^2} = \frac{1}{3}(\lambda + 8\pi G_N \rho) - \frac{k}{a^2}. \tag{2.13}$$

Note that the cosmological constant appears as a constant contribution to the Hubble parameter. For the time being, we will set it to zero and return to it in subsequent chapters.

Friedmann equation can be understood on very simple grounds: since the universe at large scale is homogeneous and isotropic, there is no specific location and motion in the universe should not allow to identify any such location. This implies that the most general motion has the form $\mathbf{v}(t) = H(t)\mathbf{x}$ where $\mathbf{x}$ and $\mathbf{v}$ denote the position and the velocity and $H(t)$ is an arbitrary function of time. Since $\mathbf{v} = \dot{\mathbf{x}}$, one obtains $\mathbf{x} = a(t)\mathbf{r}$, where $\mathbf{r}$ is a constant for a given body (called the comoving coordinate) and $a(t)$ is related to $H(t)$ through $H = \dot{a}/a$. Now, consider a particle of mass m located at position $\mathbf{x}$: the sum of its kinetic and gravitational potential energy is constant. Denoting by ρ the energy density of the (homogeneous) universe, we have

$$\frac{1}{2}m\mathbf{v}^2 - \frac{4\pi}{3}G_N m\rho \, |\mathbf{x}|^2 = \text{cst.} \tag{2.14}$$

Writing this constant $-km\mathbf{r}^2/2$, we obtain from (2.14)

$$\frac{\dot{a}^2}{a^2} = \frac{8\pi}{3}G_N \rho - \frac{k}{a^2}, \tag{2.15}$$

which is nothing but Friedmann equation (2.13) (with vanishing cosmological constant).

Friedmann equation should be supplemented by the conservation of the energy-momentum tensor which simply yields:

$$\dot{\rho} = -3H(p + \rho). \tag{2.16}$$

Hence a component with equation of state (2.10) has its energy density scaling as $\rho \sim a(t)^{-3(1+w)}$. Thus non-relativistic matter (often referred to as matter) energy density scales as a^{-3}. In other words, the energy density of matter evolves in such a way that ρa^3 remains constant. Radiation scales as a^{-4} and a component with equation of state $p = -\rho$ ($w = -1$) has constant energy density[3].

---

[3] The latter case corresponds to a cosmological constant as can be seen from (2.11–2.12) where the cosmological constant can be replaced by a component with $\rho_\Lambda = -p_\Lambda = \lambda/(8\pi G_N)$.

We note for future use that, if a component with equation of state (2.10) dominates the energy density of the universe (as well as the curvature term $-k/a^2$), then (2.15) has a scaling solution $a(t) \sim t^n$ with $n = 2/[3(1+w)]$. For example, in a matter-dominated universe, $a(t) \sim t^{2/3}$.

Differentiating the Friedmann equation with respect to time, and using the energy-momentum conservation (2.16), one easily obtains

$$\ddot{a} = -\frac{4\pi G_N}{3} a(3p + \rho) + a\frac{\lambda}{3}. \tag{2.17}$$

This allows to recover (2.12) from Friedmann equation and energy-momentum conservation.

In an expanding or contracting universe, the light emitted by a distant source undergoes a frequency shift which gives a direct information on the time dependence of the cosmic scale factor $a(t)$. To obtain the explicit relation, we consider a photon propagating in a fixed direction (θ and ϕ fixed). Its equation of motion is given as in special relativity by setting $ds^2 = 0$ in (2.5):

$$c^2 dt^2 = a^2(t)\frac{dr^2}{1 - kr^2}. \tag{2.18}$$

Thus, if a photon (an electromagnetic wave) leaves at time t a galaxy located at distance r from us, it will reach us at time t_0 such that

$$\int_t^{t_0} \frac{cdt}{a(t)} = \int_0^r \frac{dr}{\sqrt{1 - kr^2}}. \tag{2.19}$$

The electromagnetic wave is emitted with the same amplitude at a time $t + T$ where the period T is related to the wavelength of the emitted wave λ by the relation $\lambda = cT$. It is thus received with the same amplitude at the time $t_0 + T_0$ given by

$$\int_{t+T}^{t_0+T_0} \frac{cdt}{a(t)} = \int_0^r \frac{dr}{\sqrt{1 - kr^2}}, \tag{2.20}$$

the wavelength of the received wave being simply $\lambda_0 = cT_0$. Since $T_0, T \ll t_0, t$, we obtain from comparing (2.19) and (2.20)

$$\frac{cT_0}{a_0} = \frac{cT}{a(t)}, \quad \text{i.e.} \quad \frac{\lambda_0}{\lambda} = \frac{a_0}{a(t)}, \tag{2.21}$$

where a_0 is the present value of the cosmic scale factor.

Defining the redshift parameter z as the fractional increase in wavelength $z = (\lambda_0 - \lambda)/\lambda$, we have

$$1 + z = \frac{a_0}{a(t)}. \tag{2.22}$$

One may thus replace time by redshift since time decreases monotonically as redshift increases.

Exercise 2-1: In the case of the Robertson-Walker metric (2.5),
a) compute the non-vanishing Christoffel symbols (2.2),
b) using the fact that the Ricci tensor associated with the 3-dimensional metric γ_{ij} is simply $R_{ij}(\gamma) = 2k\gamma_{ij}$, compute the components of the Ricci tensor and the scalar curvature (2.4),
c) deduce the components of the Einstein tensor (2.7) and (2.8).

Hints: a) $\Gamma^i{}_{jt} = \delta^i_j \dot{a}/a$, $\Gamma^t{}_{ij} = a\dot{a}\gamma_{ij}$, $\Gamma^i{}_{jk} = \Gamma^i{}_{jk}(\gamma)$.
b) $R_{tt} = -3\ddot{a}/a$, $R_{ij} = \left(2k + \ddot{a}a + 2\dot{a}^2\right)\gamma_{ij}$, $R = 6\left(k + \ddot{a}a + \dot{a}^2\right)/a^2$.

2.2. A first try at solving Einstein's equations: the Schwarzschild solution

Because Einstein's equations are non-linear, there are few solutions known. The first exact non-trivial solution was found in late 1915 by Schwarzschild, who was then fighting in the German army, within a month of the publication of Einstein's theory and presented on his behalf by Einstein at the Prussian Academy in the first days of 1916 [2], just before Schwarzschild death from a illness contracted at the front. It describes static isotropic regions of empty spacetime, such as the ones encountered in the exterior of static stars.

If we look for static isotropic solutions, we may always write the spacetime metric as[4]:

$$ds^2 = e^{2\nu(r)}dt^2 - e^{2\lambda(r)}dr^2 - r^2\left(d\theta^2 + \sin^2\theta d\phi^2\right). \tag{2.23}$$

In other words, the only non-vanishing elements of the metric are:

$$g_{tt} = e^{2\nu(r)}, \quad g_{rr} = -e^{2\lambda(r)}, \quad g_{\theta\theta} = -r^2, \quad g_{\phi\phi} = -r^2\sin^2\theta. \tag{2.24}$$

It is then lengthy but straightforward to work out the Christoffel symbols from (2.2):

$$\Gamma^r{}_{tt} = \nu' e^{2(\nu-\lambda)}, \quad \Gamma^t{}_{rt} = \nu',$$
$$\Gamma^r{}_{rr} = \lambda', \quad \Gamma^\theta{}_{r\theta} = \Gamma^\phi{}_{r\phi} = 1/r,$$

---

[4]We have absorbed a general function $e^{2\mu(r)}$ in front of the last term by redefining the variable r.

$$\Gamma^r{}_{\theta\theta} = -re^{-2\lambda}, \qquad \Gamma^\phi{}_{\theta\phi} = \cot\theta,$$
$$\Gamma^r{}_{\phi\phi} = -r\sin^2\theta e^{-2\lambda}, \qquad \Gamma^\theta{}_{\phi\phi} = -\sin\theta\cos\theta, \tag{2.25}$$

and the Ricci tensor components from (2.4)

$$R_{tt} = \left(\nu'' + \nu'^2 - \lambda'\nu' + \frac{2\nu'}{r}\right)e^{2(\nu-\lambda)},$$

$$R_{rr} = -\nu'' - \nu'^2 + \lambda'\nu' + \frac{2\lambda'}{r},$$

$$R_{\theta\theta} = 1 - \left(1 + r\nu' - r\lambda'\right)e^{-2\lambda},$$

$$R_{\phi\phi} = R_{\theta\theta}\sin^2\theta. \tag{2.26}$$

We are now ready to solve the Einstein's equations in the vacuum under the assumption of isotropy and staticity. From (2.1), we have simply $R_{\mu\nu} - \frac{1}{2}g_{\mu\nu}R = 0$. Contracting with $g^{\mu\nu}$, we obtain $R = 0$. Hence Einstein's equations amount to a condition of vanishing Ricci tensor:

$$R_{\mu\nu} = 0. \tag{2.27}$$

From the vanishing of R_{tt} and R_{rr}, we obtain $\lambda' + \nu' = 0$. Since, at large distance from the star, space should be flat, both λ and ν should vanish at spatial infinity. Hence

$$\lambda = -\nu. \tag{2.28}$$

It then follows from the vanishing of $R_{\theta\theta}$ that

$$\left(1 + 2r\nu'^2\right)e^{2\nu} = 1, \tag{2.29}$$

which is solved by

$$g_{tt} = e^{2\nu} = 1 - \frac{2G_N M}{r}. \tag{2.30}$$

The constant of integration is identified with the mass M because, in the Newtonian limit, $g_{tt} = 1 + 2\phi$ where ϕ is the Newtonian potential.

The Schwarzschild solution thus reads, for $r > R$,

$$ds^2 = \left(1 - \frac{2G_N M}{r}\right)dt^2 - \left(1 - \frac{2G_N M}{r}\right)^{-1}dr^2 - r^2 d\theta^2 - r^2\sin^2\theta d\phi^2. \tag{2.31}$$

The requirement of staticity might seem to narrow the field of application of the Schwarzschild metric. It turns out that this solution is very general because of

a powerful theorem. *Birkhoff theorem* states that a spherically symmetric gravitational field in empty space *is* static and thus that the metric is given by the
Schwarzschild solution (see [3] section 11.7). One consequence is that a pul-
sating spherically symmetric body cannot radiate gravitational waves into empty
space. We will return to this when we discuss gravitational waves.

The Schwarschild solution is singular at $r = R_S \equiv 2G_N M$, a distance known
as the Schwarzschild radius. This is not a problem as long as $R_S < R$ since this
solution describes the exterior region of the star. A different metric describes the
interior. On the other hand, we will see in section 2.6 that, in the case where
$R < R_S$, i.e. $2G_N M/R > 1$, the system undergoes gravitational collapse and
turns into a black hole.

<u>Exercise 2-2</u>: What is the Schwarzschild radius of the sun? of an object of
mass $3 \times 10^6 \, M_\odot$?

Hints: Do not forget that we have set $c = 1$. Otherwise, $R_S = 2G_N M/c^2$, that
is 2.95 km for the sun, 8.85×10^9 m $= 0.06$ au for an object of mass $3 \times 10^6 \, M_\odot$.

2.3. *Gravitational instability I: galaxies*

We have assumed in section 2.1 a homogeneous universe. If some inhomogeneity
appears, gravitational instability will develop it. In order to see how this may
happen, let us consider a matter dominated universe ($\rho \propto a^{-3}$) and assume that
some inhomogeity $\delta\rho$ appears locally[5]: it is associated with a variation δa of the
cosmic scale factor a such that $\delta\rho/\rho = -3\delta a/a$. Developing (2.17) to first
order, one obtains

$$\delta\ddot{a} = \frac{8\pi}{3}G_N \rho \delta a = H^2 \delta a. \tag{2.32}$$

Since $a(t) \sim t^{2/3}$, $H = 2/(3t)$ and we find scaling solutions $\delta a \sim t^{\frac{1}{2}\pm\frac{5}{6}}$. The
solution $\delta a \sim t^{4/3}$ corresponds to $\delta\rho/\rho \sim t^{2/3}$ and thus to a gravitational instability.

There is one ingredient that we have not taken into account until now: the early
universe is hot. This is why the gravitational instability that we have just found
does not necessarily lead to gravitational collapse: gravitational energy may be
counterbalanced by thermal energy. To illustrate, let us consider a system of size
R and mass M with temperature $k_B T \sim G_N M m_N/R$ (i.e. the thermal energy
of a nucleon in the system is of the order of its gravitational energy). A typical

---

[5]We should, in principle, allow for a spatial dependence which introduces extra terms. It turns out
that, under suitable hypotheses, such terms are negligible (see [3] chapter 15, sect. 9).

time scale for gravitational collapse is obtained by dimensional analysis (see later (2.62)):

$$t_g \sim \left(\frac{R^3}{G_N M} \right)^{1/2}. \tag{2.33}$$

Concurrently, the plasma cools down, with a time constant t_c. If $t_c < t_g$, then gravity wins over thermal energy and the system collapses. If the cooling of the primordial plasma is dominantly through thermal bremsstrahlung, then ([4], section 1.5.1)

$$t_c \sim \frac{1}{c} \left(\frac{\hbar}{m_e c} \right)^{-2} \frac{1}{n\alpha^3} \left(\frac{k_B T}{m_e c^2} \right)^{1/2}, \tag{2.34}$$

where n is the number density. The condition $t_c < t_g$ turns into a condition on R:

$$R < \frac{\alpha^3}{\alpha_G} \frac{\hbar}{m_e c} \left(\frac{m_N}{m_e} \right)^{1/2} \sim 75 \text{ kpc}, \tag{2.35}$$

where $\alpha_G \equiv (G_N m_N^2/\hbar c) \sim 6 \times 10^{-39}$ measures the strength of gravity at low energy. This length scale is in the ball park of galaxy sizes, typically 10 to 20 kpc. Typical masses may also be inferred (by making explicit the assumption that the thermal energy $k_B T$ is higher than the ionisation potential $\alpha^2 m_e c^2$; see section 1.5.1 of [4]):

$$M < \frac{\alpha^5}{\alpha_G} m_N \left(\frac{m_N}{m_e} \right)^{1/2} \sim 3 \times 10^{41} \text{ kg}. \tag{2.36}$$

One may infer from these numbers that galaxies were formed at a redshift smaller than 9 (see Exercise 2-3).

Exercise 2-3: Consider a galaxy of typical size $R_{gal} \sim 20$ kpc and mass $M_{gal} \sim 3 \times 10^{44}$ g. Assuming that regions of density ρ 100 times larger than the average density of the universe collapse, deduce at which approximate redshift took place galaxy formation. One gives the present mean density of the universe: $\rho_0 \sim 10^{-30}$ g.cm^{-3}.

Hints: infer from $\rho_{gal} \sim 10^{-25}$ g.cm^{-3}, that $z \sim 10$.

2.4. Gravitational instability II: compact objects, from dwarfs to black holes

In extreme cases, gravitational pressure may be counterbalanced by the quantum degeneracy pressure[6]. Since matter is made of fermions of spin 1/2, Pauli principle applies: two fermions cannot be in the same state. Since there are $4\pi p^2 dp/(2\pi\hbar)^3$ levels per unit volume with momentum between p and $p + dp$ and two spin states per level, the number of fermions per unit volume is given in terms of the maximal momentum by

$$n = \frac{2}{(2\pi\hbar)^3} \int_0^{p_F} 4\pi k^2 dk = \frac{p_F^3}{3\pi^2\hbar^3}. \tag{2.37}$$

The energy of the highest level, or Fermi energy ϵ_F, is therefore given in terms of the number density n. If the particles are non-relativistic, then

$$\epsilon_F = \frac{p_F^2}{2m} = \frac{1}{2}\left(3\pi^2\right)^{2/3} \hbar^2 \frac{n^{2/3}}{m}. \tag{2.38}$$

On the other hand, the gravitational energy per nucleon of a system of size R and mass M (with $N = 4\pi R^3 n_N/3 = M/m_N$ nucleons) is

$$\epsilon_g = \frac{G_N M^2}{NR} = G_N m_N^2 \frac{N}{R} = \left(\frac{4\pi}{3}\right)^{1/3} G_N m_N^2 N^{2/3} n_N^{1/3}. \tag{2.39}$$

The Fermi energy starts to dominate over the gravitational energy for

$$n_N^{1/3} > \frac{2}{(3\pi^2)^{2/3}} \left(\frac{G_N m_N^2 m}{\hbar^2}\right) N^{2/3} \nu^{2/3}, \tag{2.40}$$

where $\nu = n_N/n$ (ν depends on the species of the fermions that are degenerate; see below), or

$$RM^{1/3} \sim \frac{1}{\alpha_G} \frac{\hbar}{mc} m_N^{1/3} \nu^{-2/3}, \tag{2.41}$$

where, as above, $\alpha_G \equiv (G_N m_N^2/\hbar c) \sim 6 \times 10^{-39}$.

We see from (2.40) that gravitational collapse is first stopped by the quantum degeneracy of electrons: the corresponding astrophysical objects are known as *white dwarfs*. Writing thus $m = m_e$ and $\nu = 2$ (two nucleons per electron), we find that $RM^{1/3} \sim 10^{-2} R_\odot M_\odot^{1/3}$. A white dwarf with $M = M_\odot$ has radius $R \sim 10^{-2} R_\odot$ and density $\rho \sim 10^6 \rho_\odot$. It is more compact than a star.

---

[6]Since, in the cases considered here, pressure is proportional to internal energy, and since we deal here with orders of magnitude (neglecting factors of order one), our arguments will be based on a balance between internal energies.

If density continues to increase, the value of the Fermi energy is such that the fermions are relativistic: it follows from (2.37) that $p_F > mc$ reads $\left(3\pi^2\right)^{1/3} \hbar n^{1/3} > mc$ or, using $n = 3N/\left(4\pi v R^3\right)$,

$$R < \left(\frac{9\pi}{4}\right)^{1/3} \frac{\hbar c}{mc^2} N^{1/3} v^{-1/3}. \tag{2.42}$$

But, since $\epsilon_F \sim p_F c = \left(3\pi^2\right)^{1/3} \hbar c n^{1/3}$, both ϵ_F and ϵ_g scale like $n^{1/3}$. Quantum degeneracy pressure can overcome gravitational collapse only for $N < 3\sqrt{\pi}\alpha_G^{-3/2}/(2v^2)$, or

$$M < 3\sqrt{\pi}\alpha_G^{-3/2} m_N/(2v^2) \sim 1 \, M_\odot/v^2. \tag{2.43}$$

This bound is the well-known Chandrasekhar limit for white dwarf masses (a more careful computation gives a numerical factor of 5.87 [3]). The condition (2.42) then gives

$$R < \frac{3\sqrt{\pi}}{2}\alpha_G^{-1/2} \frac{\hbar c}{mc^2} \frac{1}{v}. \tag{2.44}$$

Setting $m = m_e$ gives a limit value of some 10^4 km.

For even higher densities, most electrons and protons are converted into neutrons through inverse beta decay ($p + e^- \rightarrow n + v$). A new object called *neutron star* forms when the neutron Fermi energy balances the gravitational energy. Writing $m = m_n$ instead of m_e in (2.41), we now have $RM^{1/3} \sim 10^{-5} R_\odot M_\odot^{1/3}$: a neutron star with $M = M_\odot$ has radius $R \sim 10^{-5} R_\odot$ and density $\rho \sim 10^{15} \rho_\odot$.

The bound (2.43) obtained above in the case of relativistic fermions (neutrons in this case) is called the Oppenheimer-Volkoff bound: more precisely, the maximal mass of a neutron star is $M = 0.7 \, M_\odot$, with a corresponding radius $R = 9.6$ km (cf. (2.44) with $m = m_n$). If the mass is larger, the star undergoes gravitational collapse and forms a black hole. We will return below to the study of black holes.

2.5. Stars

The evolution of a generic gravitational system is governed by the two processes that we have discussed in the preceding sections: gravitational pressure is balanced by thermal pressure and quantum degeneracy pressure of electrons. The dynamics of the evolution is controlled by the relative magnitudes of the latter two contributions. To illustrate this, let us simply write[7] for the non-gravitational energy $\epsilon = k_B T + \epsilon_F$. Then using (2.38) and (2.39), we have

$$k_B T \sim \left(\frac{4\pi}{3}\right)^{1/3} G_N m_N^2 N^{2/3} v^{1/3} n^{1/3} - \frac{1}{2} \left(3\pi^2\right)^{2/3} \hbar^2 \frac{n^{2/3}}{m_e}. \tag{2.45}$$

As n increases (for example because the radius decreases), the temperature increases, following the first term on the right-hand side, before the second term takes over. At a number density

$$n_c^{1/3} \sim \frac{4^{1/3}}{3\pi} \alpha_G \frac{m_e c}{\hbar} N^{2/3} v^{1/3}, \tag{2.46}$$

the temperature reaches a maximum

$$k_B T_{\max} = \frac{1}{2} \left(\frac{4}{9\pi}\right)^{2/3} \alpha_G^2 m_e c^2 N^{4/3} v^{2/3}. \tag{2.47}$$

If this temperature is sufficient to trigger nuclear reactions, then the system will turn into a gravitationally bound self-sustained nuclear reactor, i.e. a star.

A typical nuclear energy scale is[8] $\epsilon \sim \alpha^2 m_N c^2$. The condition for the gravitationally bound system to be a star powered by nuclear reactions is thus $k_B T_{\max} > \epsilon$ or in terms of the mass:

$$M = N m_N > M_* \equiv 3(\pi^2/2)^{1/4} \left(\frac{\alpha}{\alpha_G}\right)^{3/2} \left(\frac{m_N}{m_e}\right)^{3/4} m_N v^{-1/2}$$
$$\sim 0.2 \, M_\odot. \tag{2.48}$$

The associated radius R_* is of order

$$R_* \sim \frac{G_N M_* m_N}{k_B T_{\max}} \sim 0.4 \, R_\odot. \tag{2.49}$$

Masses of stars in the universe range from 0.1 to 50 $M_\odot$. Systems of lower mass end up as planets or brown dwarfs. Systems of higher mass are unstable.

We may derive from the previous considerations the luminosity of a star, that is the energy emitted per unit time. If we assimilate the plasma at the core of

---

[7] I follow in this section [4] section 1.5.3, to which I refer for more details.

[8] If you are surprised to see here the fine structure constant, remember that two protons have to overcome Coulomb repulsion before strongly interacting. See eq. (B.3) of Appendix Appendix B and the corresponding footnote.

the star to a black-body of temperature T, then the energy density is given by the standard expression:

$$\rho = aT^4, \quad a = \frac{\pi^2}{15} \frac{k_B^4}{\hbar^3 c^3} = 7.56 \times 10^{-16} \text{J.m}^{-3}.\text{K}^{-4}. \tag{2.50}$$

The available energy E thus scales as $T^4 R^3$. The time it takes to radiate depends on the time t_e that it takes photons to escape from the core through random walk. As is well-known this time scales as R^2. More precisely, the mean free path λ of photons among nucleons is $\lambda = (n_N \sigma_T)^{-1}$ where $\sigma_T = (8\pi/3)/(e^2/m_e c^2)^2$ is the Thomson scattering cross-section. A photon undergoes[9] $(R/\lambda)^2$ collisions on its way from the core to the surface and thus goes a distance $R^2/\lambda \equiv ct_e$. The luminosity of the star scales like

$$L = \frac{E}{t_e} \sim T^4 R^3 \frac{\lambda}{R^2} \sim \frac{RT^4}{n_N \sigma_T} \sim \frac{R^4 T^4}{M \sigma_T}, \tag{2.52}$$

since $n \sim N/R^3 \sim M/R^3$. Using $k_B T \sim G_N M m_N / R$, we deduce that the luminosity scales M^3.

If the central plasma is only partially ionized, one finds that L scales like M^5 (see [5], section 1.3).

In order to discuss the various types of stars, it has proven useful to compare their absolute luminosity L with their surface temperature T_s. The famous Hertzsprung-Russell or H-R diagram plots $\log T_s$ vs $\log L$. If we identify the surface of the star with the surface of a black body of temperature T_s, we expect the standard Planck distribution:

$$\frac{dE}{dSdtd\Omega d\nu} = \frac{2h\nu^3/c^2}{e^{h\nu/k_B T_s} - 1}. \tag{2.53}$$

Fitting the spectrum received from the star to this spectrum gives T_s.

Integration gives the famous Stefan-Boltzmann law: $L = 4\pi R^2 \sigma T_s^4$ with $\sigma = (\pi^2 k^4 / 60\hbar^3 c^2)$. Hence T_s scales like $L^{1/4} R^{-1/2} = L^{1/4} M^{-1/2}$. In the case where $L \sim M^3$, we have $T_s \sim L^{1/12}$ (in the alternate case of partial ionisation, $T_s \sim L^{3/20}$). Thus, in the H-R diagram, these stars should be along a line of approximate slope between 0.1 and 0.15. This is known as the *main sequence band* in the H-R diagram.

---

[9]The ratio R/λ is called the optical depth τ of the system of size R

$$\tau \equiv n\sigma R. \tag{2.51}$$

2.6. *Gravitational collapse: black holes*

We have seen that the fate of sufficiently massive star is not to be a white dwarf or a neutron star. We have also identified a potential problem for describing the metric outside a star of mass M and radius smaller than the Schwarzschild radius $R_S = 2G_N M$. Moving away from the quiet universe, we will see in this section that such systems gravitationally collapse to form a black hole.

Let us take this opportunity to present a classical interpretation of the Schwarz-schild radius. Remember that the existence of black holes was conceived by Michell [6] and Laplace [7] centuries earlier than general relativity. Indeed, the classical condition for escape a body of mass m and velocity v from a spherical star of mass M and radius R is

$$\frac{1}{2}mv^2 > \frac{G_N mM}{R}. \tag{2.54}$$

Thus, not even light ($v = c$) can escape the attraction of the star if $R < 2G_N M/c^2$, the Schwarzschild radius.

The collapse into a black hole was first discussed in the framework of general relativity by Oppenheimer and Snyder [8]. We will follow their analysis. We consider a fluid of negligible pressure, thus described by the energy-momentum tensor (see (2.9)) $T_{\mu\nu} = \rho U_\mu U_\nu$, and study its spherically symmetric collapse.

It turns out that we have already studied this system when we discussed the evolution of a homogeneous and isotropic universe in section 2.1. The metric is given by

$$ds^2 = d\hat{t}^2 - a^2(\hat{t}) \left(\frac{d\hat{r}^2}{1 - k\hat{r}^2} + \hat{r}^2 d\hat{\theta}^2 + \hat{r}^2 \sin^2 \hat{\theta} d\hat{\phi}^2 \right), \tag{2.55}$$

as in $(2.5)^{10}$ and the Einstein tensor components are the same as in (2.7,2.8). We normalize the coordinate $\hat{r}$ so that $a(0) = 1$. Thus

$$\rho(\hat{t}) = \rho(0)/a^3(\hat{t}), \tag{2.56}$$

and Einstein's equations simply read:

$$\dot{a}^2 + k = \frac{8\pi G_N}{3} \frac{\rho(0)}{a}, \tag{2.57}$$

$$\dot{a}^2 + 2a\ddot{a} + k = 0. \tag{2.58}$$

---

[10] except that we do not normalize k to ± 1 or 0 because we are looking at a different system. We will see just below that it is fixed by initial conditions. We add a hat to this system of coordinates to distinguish it from the Robertson-Walker coordinates, as well as from the Schwarzschild coordinates that we will use later.

Assuming that the fluid is initially at rest ($\dot{a} = 0$), we obtain from (2.57)

$$k = \frac{8\pi G_N}{3}\rho(0). \tag{2.59}$$

Thus, (2.57) simply reads

$$\dot{a}^2(\hat{t}) = k\left[a^{-1}(\hat{t}) - 1\right]. \tag{2.60}$$

The solution is given by the parametric equation of a cycloid:

$$\hat{t} = \frac{\psi + \sin\psi}{2\sqrt{k}},$$
$$a = \frac{1 + \cos\psi}{2}. \tag{2.61}$$

We see that a vanishes for $\psi = \pi$, that is after a time

$$\tau = \frac{\pi}{2\sqrt{k}} = \frac{\pi}{2}\left(\frac{3}{8\pi G_N\rho(0)}\right)^{1/2}. \tag{2.62}$$

Thus a sphere initially at rest with energy density $\rho(0)$ and negligible pressure collapses to a state of infinite energy density in a finite time τ.

In the case of a star of radius R and mass M, this solution for the interior of the star should be matched with the Schwarzschild solution describing the exterior:

$$ds^2 = \left(1 - \frac{2G_N M}{r}\right)dt^2 - \left(1 - \frac{2G_N M}{r}\right)^{-1}dr^2 - r^2 d\theta^2 - r^2\sin^2\theta d\phi^2. \tag{2.63}$$

The correspondence between the interior and exterior coordinates is simply $r = Ra(\hat{t})$, $\theta = \hat{\theta}$ and $\phi = \hat{\phi}$, with a more complicate relation between t and $\hat{t}$ (see [3] section 11.9 and below). The first relation ensures that

$$k = \frac{2MG_N}{R^3} = \frac{R_S}{R^3}, \tag{2.64}$$

in agreement with (2.59) and $M = (4\pi/3)\rho(0)R^3$.

We note that the singularity at $r = R_S$ is not a harmful singularity. One can for example check that the curvature computed from the metric (2.63) has no singularity at $r = R_S$; hence there is no singularity in the gravitational field. The singularity is only an artifact of the system of coordinates that we have used (see Exercise 2-4). It remains that the sphere $r = R_S$, which defines what one calls the horizon of the black hole, plays an important role in discussing the physics of black holes, as we will now see. On the other hand, the singularity that we have found at $\psi = \pi$ in (2.61) (where $a = 0$) is a true singularity!

Let us consider sending a light signal radially from some point r_1 to $r_2 > r_1$ where it is received a time Δt later. Since $ds^2 = 0$ (as well as $d\theta = d\phi = 0$), we have simply

$$\Delta t = \int_{r_1}^{r_2} \frac{dr}{(1 - R_S/r)}. \tag{2.65}$$

If $r_1 < R_S$, this is finite only for $r_2 < R_S$, in which case it is simply $r_2 - r_1 + R_S \ln\left[(R_S - r_2)/(R_S - r_1)\right]$. In other words, signals emitted from within the Schwarschild radius never reach the outside. There is really a breach of communication. This is why the surface $r = R_S$ is called an *event horizon*.

We also note that, at large distance, the black hole is only caracterized by its mass M. Black holes are indeed very simple objects, somewhat similar to particles: Schwarzschild black holes are caracterized by their mass. We will see in what follows that one can only add to this spin (rotating black holes) and charge (charged black holes) but no other independent characteristics: in the picturesque language used by black hole aficionados, it is said that black holes can have no hair. Indeed very similar to fundamental particles, which are caracterized by a finite set of numbers (including mass, spin, and electric charge).

Falling into a black hole

Let us investigate the radial motion of an object in free fall into the black hole, using the Schwarzschild coordinates. According to general relativity, a particle (or an astronaut) in free fall has an equation of motion:

$$\frac{dU^\mu}{d\tau} + \Gamma^\mu{}_{\nu\lambda} U^\nu U^\lambda = 0, \quad U^\mu = \frac{dx^\mu}{d\tau}, \tag{2.66}$$

where τ is the proper time (i.e. the time in the rest frame: $ds^2 = c^2 d\tau^2$). The velocity U^μ obviously satisfies the condition

$$g_{\mu\nu} U^\mu U^\nu = 1. \tag{2.67}$$

Since we are studying radial motion, $U^\theta = U^\phi = 0$. As for the time component, we have

$$\frac{dU^t}{d\tau} = -2\Gamma^t{}_{rt} U^r U^t = -g^{tt} \partial_r g_{tt} U^r U^t. \tag{2.68}$$

Now, since

$$\frac{dg_{tt}}{d\tau} = \frac{dg_{tt}}{dr}\frac{dr}{d\tau} = \partial_r g_{tt} U^r, \tag{2.69}$$

we conclude that $d\left(g_{tt}U^t\right)/dr = 0$. Hence

$$g_{tt}U^t = C, \tag{2.70}$$

where C is a constant fixed by initial conditions. If at $t = 0$, we drop the object at rest ($U^i = 0$) from a distance $r = R > R_S = 2G_N M$, then one deduces from (2.67) that $U^t(R) = 1/\sqrt{g_{tt}(R)}$. Thus (2.70) gives

$$C = \sqrt{g_{tt}(R)} = \sqrt{1 - R_S/R}. \tag{2.71}$$

We have now fully determined U^t (from (2.70)):

$$U^t = C\left(1 - \frac{R_S}{r}\right)^{-1}. \tag{2.72}$$

We obtain U^r from (2.67) which simply reads $1 = g_{tt}\left(U^t\right)^2 + g_{rr}\left(U^r\right)^2$. Multiplying with g_{tt} and using (2.70) and $g_{tt}g_{rr} = -1$, we have

$$U^r = -\left(C^2 - 1 + \frac{R_S}{r}\right)^{1/2}, \tag{2.73}$$

where we have chosen the minus sign for the infall. Since $U^t/U^r = dt/dr$, we finally obtain the trajectory of the infalling astronaut:

$$t = \int_r^R dr\, C\left(1 - \frac{R_S}{r}\right)^{-1}\left(C^2 - 1 + \frac{R_S}{r}\right)^{-1/2}. \tag{2.74}$$

This can easily be integrated in the limit $R \to \infty$ i.e. $C \to 1$ (see Exercise 2-5). For completeness, the integration of (2.74) in the general case gives

$$r = R\frac{1 + \cos\psi}{2},$$

$$t = \left[\left(\frac{R}{2} + R_S\right)\left(\frac{R}{R_S} - 1\right)^{1/2}\right]\psi + \frac{R}{2}\left(\frac{R}{R_S} - 1\right)^{1/2}\sin\psi \tag{2.75}$$

$$+ R_S \ln\left|\frac{(R/R_S - 1)^{1/2} + \tan(\psi/2)}{(R/R_S - 1)^{1/2} - \tan(\psi/2)}\right|,$$

where ψ is the cycloid variable already encountered in (2.61).

We are interested in the motion near the Schwarzschild radius. We can thus make an expansion $r = R_S + \rho$ to the first order:

$$\frac{dt}{dr} = -C\left(1 - \frac{R_S}{R_S + \rho}\right)^{-1}\left(C^2 - 1 + \frac{R_S}{R_S + \rho}\right)^{-1/2}$$

$$\sim -\frac{R_S}{\rho} = -\frac{R_S}{r - R_S}. \tag{2.76}$$

Integration gives

$$t = -R_S \ln\left(\frac{r}{R_S} - 1\right) + \text{cst.}, \quad \text{or} \quad r = R_S\left(1 + \text{cst.}e^{-t/R_S}\right). \tag{2.77}$$

Thus, for an distant observer, it takes the astronaut an infinite time to reach the Schwarzschild radius.

On the other hand, if we count in proper time τ i.e. the time that shows on the astronaut watch, we have from (2.73)

$$\frac{d\tau}{dr} = \frac{1}{U^r} = -\left(c^2 - 1 + \frac{R_S}{r}\right)^{-1/2}. \tag{2.78}$$

For $r \to R_S$, $d\tau/dr \to -1/C$; hence the infall of the astronaut to the horizon takes a finite amount of *proper* time. More precisely, integration gives

$$\tau = \int_r^R \frac{dr}{[R_S/r - R_S/R]^{1/2}} = \frac{1}{2}\left(\frac{R^3}{R_S}\right)^{1/2}(\psi + \sin\psi). \tag{2.79}$$

You may compare this with (2.61) and (2.62) to realize that the time of collapse that we had computed earlier was the proper time.

Rotating black hole and Kerr metric

We said earlier that the Schwarzschild solution was found in the early days of general relativity. This is a static solution whereas one expects that some of the systems that gravitationally collapse should be in rotation. One had to wait till 1963 to obtain a solution that would apply to a rotating black hole. This is the Kerr metric [9] later completed by Newman et al. [10] to allow for a non-vanishing electric charge. The Kerr-Newman metric describing a black hole of mass M, charge Q and angular momentum $J \equiv aM$ has the following form in the so-called Boyer-Lindquist coordinates (for simplicity we set here $G_N = 1$):

$$
\begin{aligned}
ds^2 = {}& \left(1 - \frac{2Mr - Q^2}{\rho^2}\right)dt^2 + \frac{2a(2Mr - Q^2)}{\rho^2}\sin^2\theta dt d\phi \\
& - \frac{\rho^2}{r^2 + a^2 - 2Mr + Q^2}dr^2 - \rho^2 d\theta^2 \\
& - \left[r^2 + a^2 + \frac{a^2\sin^2\theta}{\rho^2}(2Mr - Q^2)\right]\sin^2\theta d\phi^2,
\end{aligned} \tag{2.80}
$$

where

$$\rho^2 \equiv r^2 + a^2\cos^2\theta. \tag{2.81}$$

For simplicity we will only consider here the Kerr metric which describes a spinning black hole of vanishing charge[11]: we thus set $Q = 0$.

The event horizon is again determined by the location where the coefficient of dr^2 blows up[12]:

$$r_\pm = M \pm \sqrt{M^2 - a^2}. \tag{2.82}$$

We see that the event horizon exists only in the case where $a \leq M$ i.e. $J \leq G_N M^2/c$ (reinstating for a moment G_N and c that we have set to 1). The "cosmic censorship" hypothesis states that holes always form with $J \leq M^2$: all black hole singularities are protected by an horizon. A black hole spinning at the maximum rate ($J = M^2$) is called an extreme Kerr black hole. We also note that there are two horizons:

• the outer horizon $r_H \equiv r_+$ corresponds to the Schwarzschild horizon in the limit of vanishing a; we will refer to it simply as the horizon;

• the inner horizon r_- is called the Cauchy horizon; near the Cauchy horizon, spacetime becomes unstable and is no longer described by the Kerr metric [11]. Another distance plays an important role: it is the place where the metric coefficient g_{tt} vanishes:

$$r = M + \sqrt{M^2 - a^2 \cos^2\theta} \equiv r_{stat}. \tag{2.83}$$

It is called static limit for reasons that I now explain.

We consider an observer at fixed r and θ rotating around the spinning black hole with constant angular velocity

$$\Omega = \frac{d\phi}{dt} = \frac{U^\phi}{U^t}. \tag{2.84}$$

The condition (2.67) then reads

$$1 = \left(U^t\right)^2 \left[g_{tt} + 2\Omega g_{t\phi} + \Omega^2 g_{\phi\phi}\right]. \tag{2.85}$$

Since the last factor has to be positive, Ω has to lie between the two roots of the corresponding quadratic equation:

$$-\frac{g_{t\phi} + \sqrt{g_{t\phi}^2 - g_{tt}g_{\phi\phi}}}{g_{\phi\phi}} < \Omega < \frac{-g_{t\phi} + \sqrt{g_{t\phi}^2 - g_{tt}g_{\phi\phi}}}{g_{\phi\phi}}. \tag{2.86}$$

---

[11] Alternatively, the form of the metric for a non-spinning charged black hole (i.e. setting $a = 0$ in (2.80) is called Reissner-Nordstrom.

[12] It can be checked indeed that wordlines run through it only in the inward direction, not outward, which makes it a horizon.

As one approaches the black hole, one reaches the distance r_{stat} where $g_{tt} = 0$ and beyond which the value of Ω_{min} is strictly positive. Thus, in the region $r_{stat} < r < r_H$, there is no static observer: all observers are inexorably dragged along in the direction of rotation of the black hole. This region is known as the *ergosphere*. Figure 1 represents the ergosphere for an extreme Kerr black hole.

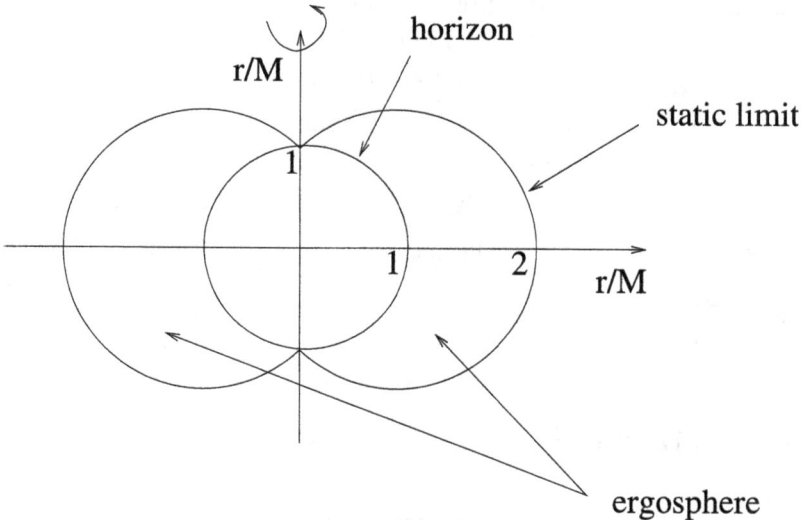

Fig. 1. The ergosphere for an extreme Kerr black hole.

The presence of the ergosphere is important because it allows to extract rotational energy from the spinning black hole. Let us illustrate this with the Penrose process [12]. If a particle enters the ergosphere and splits into two, in such a way that one fragment goes into the hole, the remaining fragment may leave the ergosphere with more energy than the original particle. This possible because rotational energy is extracted from the hole.

A Kerr hole thus has two kinds of energy: a rotational one which can be extracted, and an irreducible one. The fraction of rotational energy is

$$1 - \left(\frac{1 + \left[1 - (a/M)^2 \right]^{1/2}}{2} \right)^{1/2}, \tag{2.87}$$

which is 29% for an extreme Kerr black hole ($a = M$). Once it has lost all its spinning energy, it will turn into a Schwarzschild black hole.

This illustrates a general theorem of black hole physics, often referred to as the second law: the area of the event horizon cannot decrease. Here, a Kerr hole horizon has an area $4\pi \left(M + \sqrt{M^2 - a^2}\right)^2$ which is smaller than for a Schwarzschild black hole ($4\pi \times 4M^2$).

<u>Exercise 2-4</u>: Define the Kruskal coordinates (v, u, θ, ϕ) related to the Schwarzschild coordinates (t, r, θ, ϕ) through [13]:

$$\text{for } r > R_S, \quad u = (r/R_S - 1)^{1/2}\, e^{r/2R_S} \cosh(t/2R_S),$$
$$v = (r/R_S - 1)^{1/2}\, e^{r/2R_S} \sinh(t/2R_S),$$
$$\text{for } r < R_S, \quad u = (1 - r/R_S)^{1/2}\, e^{r/2R_S} \sinh(t/2R_S),$$
$$v = (1 - r/R_S)^{1/2}\, e^{r/2R_S} \cosh(t/2R_S). \tag{2.88}$$

Deduce from (2.31) the form of the metric in Kruskal coordinates:

$$ds^2 = \frac{4R_S^3}{r} e^{-r/R_S} \left(dv^2 - du^2\right) - r^2 \left(d\theta^2 + \sin^2\theta d\phi^2\right), \tag{2.89}$$

where r is given as an implicit function of u and v:

$$\left(\frac{r}{R_S} - 1\right) e^{r/R_S} = u^2 - v^2. \tag{2.90}$$

<u>Exercise 2-5</u>: Consider the case where an object is dropped at rest from infinity into a black hole (i.e. $R \to \infty$ in (2.74)).

a) Determine explicitly the trajectory in terms of the Schwarzschild time t (it proves to be convenient to change the origin of time to the instant of the final catastrophe).

b) Show that reaching Schwarzschild radius requires an infinite amount of time t.

Solution: a) Integrate (2.74) explicitly with $C = 1$ (the change in the origin of time changes the integral upper limit to 0):

$$\frac{t}{R_S} = -\frac{2}{3}\left(\frac{r}{R_S}\right)^{3/2} - 2\left(\frac{r}{R_S}\right)^{1/2} + \ln\left|\frac{(r/R_S)^{1/2} + 1}{(r/R_S)^{1/2} - 1}\right|. \tag{2.91}$$

b) One recovers $r/R_S = 1 + \text{cst} \times \exp(-t/R_S)$ in the limit as $t \to \infty$ as in (2.77).

An entertaining introduction to black holes for the educated reader (not necessarily a professional physicist) may be found in the little book by Taylor and Wheeler [14].

2.7. *Gravitational waves*

In the following, gravitational waves will play an important role as a possible probe of some of the violent phenomena that take place in the universe. Crudely speaking, gravitational waves are ripples propagating across spacetime.

In order to understand their dynamics, it is easiest to linearize Einstein equations i.e. to consider them in the weak gravitational field approximation and to show that they amount ot a wave equation. We thus consider a small departure from flat spacetime:

$$g_{\mu\nu} = \eta_{\mu\nu} + h_{\mu\nu}, \tag{2.92}$$

where $\eta_{\mu\nu}$ is the flat metric $(+1, -1, -1, -1)$ and the coefficients $|h_{\mu\nu}|$ are small compared to 1.

The Riemann tensor is then easily computed to first order in $h_{\mu\nu}$ to be

$$R_{\mu\nu\alpha\beta} = \frac{1}{2} \left(\partial_\nu \partial_\alpha h_{\mu\beta} + \partial_\mu \partial_\beta h_{\nu\alpha} - \partial_\nu \partial_\beta h_{\mu\alpha} - \partial_\mu \partial_\alpha h_{\nu\beta} \right), \tag{2.93}$$

from which one can easily deduce the Einstein tensor $G_{\mu\nu}$. In order to write this tensor, we introduce the so-called trace reverse[13] of $h_{\mu\nu}$:

$$\bar{h}_{\mu\nu} \equiv h_{\mu\nu} - \frac{1}{2}\eta_{\mu\nu} h, \tag{2.94}$$

where $h \equiv h^\mu_{\ \mu} = \eta^{\mu\nu} h_{\mu\nu}$ is the trace of $h_{\mu\nu}$ (note that all indices are raised or lowered here and in what follows with the metric $\eta_{\mu\nu}$). One then finds, to lowest order in $h_{\mu\nu}$,

$$G_{\mu\nu} = -\frac{1}{2} \left[\partial^\alpha \partial_\alpha \bar{h}_{\mu\nu} + \eta_{\mu\nu} \partial^\alpha \partial^\beta \bar{h}_{\alpha\beta} - \partial^\alpha \partial_\nu \bar{h}_{\mu\alpha} - \partial^\alpha \partial_\mu \bar{h}_{\nu\alpha} \right]. \tag{2.95}$$

At this point, it is important to note that different $h_{\mu\nu}$ correspond to the same physical situation. Indeed, if we consider an infinitesimal coordinate transformation $x'^\mu = x^\mu + \xi^\mu(x)$ under which, as usual, the metric transforms as

$$g'^{\mu\nu} = \frac{\partial x'^\mu}{\partial x^\lambda} \frac{\partial x'^\nu}{\partial x^\rho} g^{\lambda\rho}, \tag{2.96}$$

we have correspondingly from (2.92)

$$h'_{\mu\nu} = h_{\mu\nu} - \partial_\mu \xi_\nu - \partial_\nu \xi_\mu. \tag{2.97}$$

Hence, the two fields $h_{\mu\nu}$ and $h'_{\mu\nu}$ are within the same class of physically equivalent configurations. This is reminiscent of two vector fields that differ by a

---

[13]One has $\bar{h} \equiv \bar{h}^\mu_{\ \mu} = -h$ and thus $h_{\mu\nu} = \bar{h}_{\mu\nu} - \frac{1}{2}\eta_{\mu\nu}\bar{h}$.

gauge transformation. This is why (2.97) is also referred to as a gauge transformation on the linearized metrics. Just as in the vector case, one may choose a specific configuration within the class of gauge equivalents by fixing a gauge condition.

The expression (2.95) simplifies drastically if we go to the Lorentz gauge[14]:

$$\partial^\nu \bar{h}_{\mu\nu} = 0, \tag{2.98}$$

since then $G_{\mu\nu} = -\frac{1}{2}\Box \bar{h}_{\mu\nu}$. The Einstein equations simply read

$$\Box \bar{h}_{\mu\nu} = -16\pi G_N T_{\mu\nu}, \text{ or } \Box h_{\mu\nu} = -16\pi G_N \left(T_{\mu\nu} - \frac{1}{2}\eta_{\mu\nu} T \right), \tag{2.99}$$

which is the standard equation in flat space for a wave propagating at the speed of light, with the energy-momentum tensor as a source.

Plane wave solutions are of the form

$$\bar{h}_{\mu\nu} = A_{\mu\nu} \exp\left(-ik_\alpha x^\alpha\right), \tag{2.100}$$

where $A_{\mu\nu}$ is some constant symmetric tensor. The Lorentz gauge condition then reads

$$A_{\mu\nu} k^\nu = 0. \tag{2.101}$$

One may use the remaining gauge freedom (see the last footnote) to further constrain $A_{\mu\nu}$ with conditions of tracelessness and transversality:

$$A^\mu{}_\mu = 0, \qquad A_{\mu\nu} U^\nu = 0, \tag{2.102}$$

where U^μ is some fixed velocity (timelike unit vector). The equations (2.101) and (2.102) fix what is known as the traceless transverse gauge[15].

If we choose $U^\mu = \delta_0^\mu$ and **k** along the z axis, then $A_{\mu 0} = 0 = A_{\mu z}$ and we are left with two independent components: $A_{xx} = -A_{yy}$ (using tracelessness) and $A_{xy} = A_{yx}$. We conclude that a gravitational wave has two independent polarizations.

In order to get a more physical understanding, we may study how the distance ξ^μ between two free particles evolves with time in presence of a gravitational

---

[14] From (2.97), one obtains $\partial^\nu \bar{h}'_{\mu\nu} = \partial^\nu \bar{h}_{\mu\nu} - \Box \xi_\mu$. One thus chooses the coordinate transformation in such a way that $\Box \xi_\mu = \partial^\nu \bar{h}_{\mu\nu}$. Note that ξ_μ is defined up to a vector field η_μ which satisfies $\Box \eta_\mu = 0$.

[15] Note that in this gauge $\bar{h}_{\mu\nu} = h_{\mu\nu}$.

wave [see [15] p. 218]. It obeys the equation

$$\frac{d^2}{d\tau^2}\xi^\mu = R^\mu{}_{\alpha\beta\nu}U^\alpha U^\beta \xi^\nu,\qquad(2.103)$$

where τ is proper time and U^μ is the velocity of the particles. In the traceless transverse gauge, one finds two types of motions summarized in Figure 2: the + polarization corresponds to $A_{xy} = 0$ whereas the $\times$ polarisation corresponds to $A_{xx} = 0$.

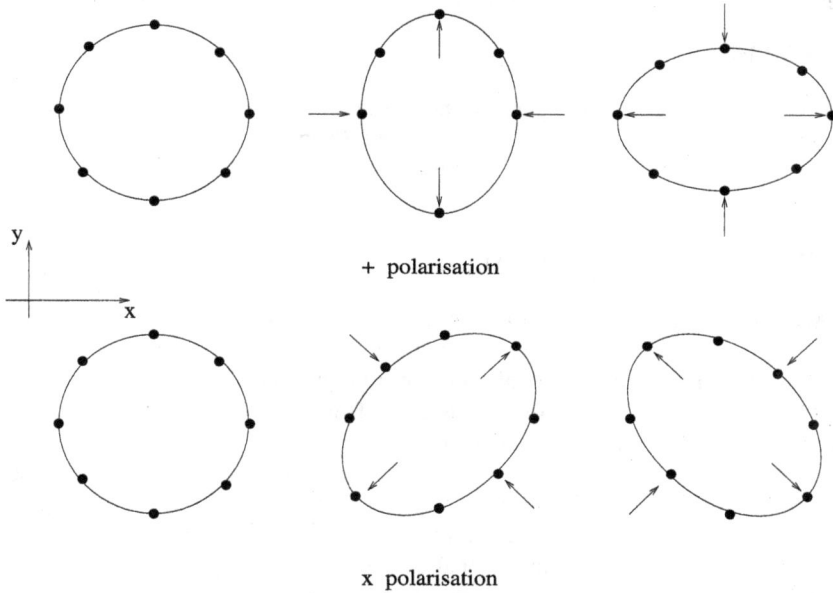

Fig. 2. Distortion of a circle of free particles by a gravitational wave travelling in the z direction.

3. The violent universe

Astrophysics is primarily an observational science: it describes observed facts and puts them into categories before getting to the heart of the phenomena. The latter task is often extremely difficult because of the complexity of the processes involved. It turns out however that, at the core of some of the most violent events that take place in the universe (the subject matter of particle astrophysics), there are generic and somewhat simpler processes in action. It is the purpose of this Chapter to unravel them and to show how they can be studied in order to eventually understand their structure and dynamics.

3.1. Stellar evolution

We have seen in the preceding Chapter that stars are gravitationally sustained nuclear reactors and that, once the nuclear fuel is exhausted, they end up as white dwarfs, neutron stars or black holes. We will now discuss rather sketchily the corresponding dynamical processes.

It turns out that the key parameter here is the mass of the star. We have seen earlier that stars have masses between 0.1 and around 50 solar masses. Let us review the different regimes as one increases the mass M of the star[16].

- $0.1 M_\odot < M < 1.3 M_\odot$

Nuclear reactions taking place at the centre burn hydrogen to produce helium following the pp chain. This uses a fraction 0.007 of the total rest mass energy. Thus this phase of the evolution lasts for a time

$$t_H \sim 0.007 M/L \sim 10^9 \text{ yr} \left(\frac{M}{M_\odot} \right)^{-2}, \tag{3.1}$$

if we can use equation (2.52) of Chapter 2. This is the longest period of time in the evolution of a star. We have seen that such stars populate the main sequence band in the H-R diagram. It follows that a large fraction of the observed stars will be found in this band. We also conclude from (3.1) that the lightest stars are the longer-lived (a factor 10^5 between $0.1 M_\odot$ and $30 M_\odot$).

Once hydrogen is exhausted, the core collapses which increases the temperature. Hydrogen begins to burn in the outer regions and the envelope expands slowly to become a red giant. Eventually, the core becomes hot enough to start helium burning. But the gravitational pressure is already balanced by electron degenacy pressure and helium is immediately released. The ultimate fate of the core is to become a white dwarf. On the other hand, outer regions may undergo violent instabilities and most of the star be shaken off to form a planetary nebula. The sun is expected to live up to 12×10^9 years (to be compared to its present age of 4.5×10^9 years) before collapsing to a white dwarf.

- $1.3 M_\odot < M < 8 M_\odot$

In this range of mass, the CNO cycle is preferred to burn hydrogen. We stress in Appendix Appendix B that such burning (as well as heavier elements) is more sensitive to temperature. This results in a convective core (contrary to the previous case where the core was radiative). This time, helium burning takes place

---

[16]I will make heavy use here of the various nuclear reactions at work, as described in Appendix Appendix B.

before the core becomes degenerate. The red giant phase eventually leads to a white dwarf.

- $8M_\odot < M < 50M_\odot$

In the case of more massive stars, the nuclear burning process reaches heavier elements. In fact, since each time the core has run out of a lighter species, it collapses until it reaches a temperature where the burning of the next element starts, the star has an onion-like structure: the heavier elements have started burning in the inner layers while lighter elements are still burning in the outer layers.

Because of the high temperatures reached, high energy photons photodisintegrate nuclei. This together with inverse beta decay which converts protons and electrons into neutrons, reduces the pressure and leads to a collapse which is eventually sustained only by the neutron degeneracy pressure. The stellar remnant is then a protoneutron star.

Particular attention has to be paid to the neutrinos during the evolution. It is well-known that neutrinos are copiously produced in nuclear reactions. In the early stages, neutrinos escape freely from the star. For example, the reaction $\bar{\nu} + p \rightarrow n + e^+$ has a cross-section $\sigma_\nu = 10^{-47}$ m^2 $(E_\nu/1 \text{ MeV})^2$ and thus a mean free path

$$\lambda_\nu = (n_N \sigma_\nu)^{-1} \sim 10^4 R_\odot \left(\frac{E_\nu}{1 \text{ MeV}} \right)^{-2} \left(\frac{M}{10 M_\odot} \right)^{-2} \nu^{-2}, \qquad (3.2)$$

where we have used (2.46) of Chapter 2. Thus, neutrinos escape freely from the star. However, once the collapse of the iron core has started, the neutrinos may be trapped inside. This is believed to play a central role in the stiffening of the core and to lead to an explosion, known as supernova explosion. We will describe it in more details in section 3.4.

For a fraction of such stars, the explosion fails or is insufficient to expel enough mass. The process then continues till the gravitational collapse into a black hole, since one is beyond the Oppenheimer-Volkoff mass limit for a neutron star. The formation of black holes of mass between $8M_\odot$ and $50M_\odot$ is believed to occut under such circumstances. Delayed collapse but also accretion from a companion in a binary system is believed to lead to lower mass black hole. Regarding binary systems, which are very common, we will first have to learn a little more about accretion in section 3.3.

Supermassive black holes are the subject of the next section.

3.2. Active Galactic Nuclei and supermassive black holes

Active galaxies are galaxies where the dominant energy output is not due to stars. In the case of Active Galactic Nuclei (AGN), the non-thermal radiation comes from a central region of a few parsecs around the centre of the galaxy. The most famous example of such AGNs is provided by quasi-stellar objects (QSOs) or quasars: these starlike objects turn out to be associated with the point-like optical emission from the nucleus of an active galaxy.

The typology of active extragalactic objects is very complex: radio loud and radio quiet quasars, Seyfert galaxies, BL Lacs or blazars... There has been an effort to build a unified picture [16]: the apparent diversity in the observations would then result from the diversity of perspectives from which we observers see these highly non-isotropic objects. Typically, the model for radio-loud AGNs includes: a central engine, a pair of oppositely directed relativistic jets (cones of semi-angle around 1°) an accretion disk (of size of the order of 1 parsec), and a torus of material (of size of the order of 100 parsec) which obscures the central engine when one observes it sideways. Depending on the relative angle between the line of sight and the jet axis, observation may vary in important ways.

It is now believed that the power ouput of active galaxies is derived from gravitational energy associated with a supermassive black hole which thus acts as the central engine. The presence of these massive black holes seems in fact to go beyond active galaxies: the centres of most nearby galaxies, which show no or little activity, seem to host a massive black hole. This is usually seen by observing the motion of gas or stars around the centre.

For example, observation [17] in the centre of the spiral galaxy NGC4258 of a disk of rotating gas has indicated the presence of a mass of $3.6 \times 10^7 M_\odot$ in a region less than 0.13 parsec in radius.

In our own galaxy, the center is a moderately active region which contains a dense and luminous star cluster. The last two decades have seen mounting evidence that, at the center of this cluster, there is a massive black hole associated with the compact radio source Sagittarius A*. Recent observations of the motions of nearby stars by the imager/spectrometer NAOS/CONICA working in the infrared [18] have confirmed the presence of a black hole of mass $(2.6 \pm 0.2).10^6$ solar mass (see Figure 3).

There is thus an obvious interest to point all available instruments towards the galactic center to get the most complete information about its emission. The satellite INTEGRAL, launched in 2002, has provided images of the galactic centre for a gamma energy in the range 20 − 40 keV [19]. Two independent observations have identified a source which is less than a arc-minute from Sgr A*.

More recently, the H.E.S.S. collaboration has announced a detection in the direction of the galactic center. H.E.S.S. consists of 4 Cherenkov detectors in Namibia, which detect gamma rays in the 100 GeV to 10 TeV range. The Galactic Center data [20] is 2003 data based on two telescopes. The position of the source is compatible with Sgr A* ($14 \pm 30''$). It is consistent with a point-like source.

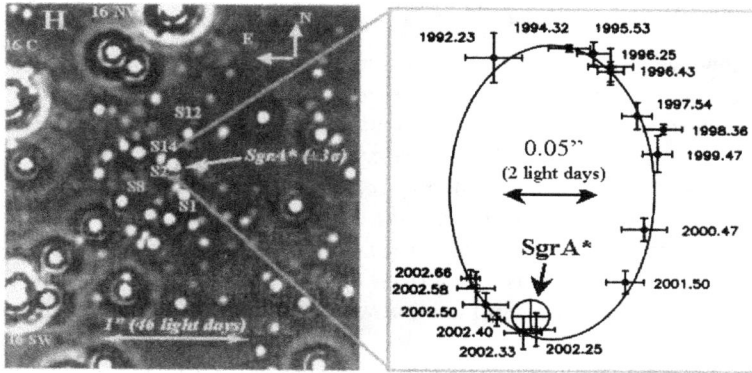

Fig. 3. Left panel: NAOS/CONICA image of the central galactic 2" (around Sgr A*). Right panel: orbit of the star S2 around Sgr A*.

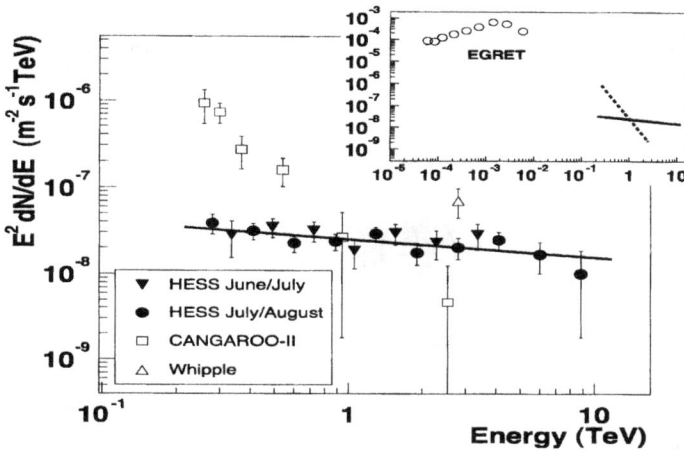

Fig. 4. Energy spectrum of γ-rays from the Galactic Center as measured by H.E.S.S.

The formation of these supermassive black holes is still a source of debate. One may envisage:

• some of the gas in a newly-formed galaxy undergoes gravitational collapse to form a superstar, which eventually collapses into a black hole;

• stellar mass or intermediate (a few hundred solar masses) black holes form and merge into a single supermassive black hole;

• a stellar mass black hole undergoes runaway growth via accretion.

3.3. Accretion to a compact object

Until now, we have discussed isolated astrophysical objects. Most of the astro-physical sources of high energy particles and most of the brightest objects are the consequence of the interaction of a compact object (white dwarf, neutron star or black hole) with its surroundings (a companion star, a galaxy, ...). It is there-fore important to understand how this interaction proceeds, i.e. to unravel the mysteries of accretion and ejection.

Let us consider a mass M falling from far away into the gravitational field of a compact object of mass M_c. If the kinetic energy that it gains i.e. $E = G_N M_c M/R$ is converted into radiation with efficiency ε, the luminosity of the accreting system will be

$$L = \varepsilon \frac{dE}{dt} = \varepsilon \frac{GM_c}{R} \frac{dM}{dt}. \tag{3.3}$$

To illustrate our discussion, we will use two distinct examples in what follows:
- (a) a stellar remnant of mass $M = M_\odot$ and radius $R = 1$ km;
- (b) a supermassive black hole of mass $M = 10^8 M_\odot$ and of gravitational radius

$$R = R_g \equiv \frac{G_N M_c}{c^2} = 1.5 \times 10^{11} \text{ m} \left(\frac{M_c}{10^8 M_\odot} \right). \tag{3.4}$$

Thus (3.3) reads

$$(a) \quad L = 7.9 \times 10^{27} \text{ W}\varepsilon \left(\frac{M_c}{M_\odot} \right) \left(\frac{R}{1 \text{ km}} \right)^{-1} \left(\frac{\dot{M}}{10^{-12} M_\odot \text{yr}^{-1}} \right), \tag{3.5}$$

$$(b) \quad L = 10^{40} \text{ W}\varepsilon \left(\frac{\dot{M}}{1.5 M_\odot \text{yr}^{-1}} \right). \tag{3.6}$$

There is however a limiting process to accretion: photons emitted by the ac-creting star exert a radiation pressure on the accreting matter which limits the rate of accretion. To be more quantitative, we have to compute the corresponding electromagnetic force. The number dn of photons which cross a sphere of radius r in a time dt is simply per unit surface

$$dn = \frac{Ldt}{4\pi r^2} \frac{1}{\hbar\omega}, \tag{3.7}$$

with ω some average frequency. These photons interact with the electrons of ion-ized matter through Thompson scattering, transfering them a momentum $\hbar\omega/c$:

$$\frac{dp}{dt} = \frac{L\sigma_T}{4\pi cr^2}. \tag{3.8}$$

This force is transferred to the protons because electrons and ions are strongly coupled in the plasma. Accretion is stopped when this force is counterbalanced by the gravitational force exerted at distance r on the protons i.e. $G_N M_c m_p / r^2$. This gives a limit value for the luminosity called the *Eddington luminosity*:

$$L_E = \frac{4\pi c}{\sigma_T} G_N m_p M_c. \tag{3.9}$$

This gives for our two examples:

$$(a) \quad L_E = 1.3 \times 10^{31} \text{ W} \left(\frac{M_c}{M_\odot}\right), \tag{3.10}$$

$$(b) \quad L_E = 1.3 \times 10^{39} \text{ W} \left(\frac{M_c}{10^8 M_\odot}\right). \tag{3.11}$$

The black body temperature T_E if Eddington luminosity is radiated from a sphere of radius R is given by $L_E = 4\pi R^2 \sigma T_E^4$, i.e.

$$(a) \quad T_E = 1.8 \times 10^8 \text{ K} \left(\frac{M_c}{M_\odot}\right)^{1/4} \left(\frac{R}{1 \text{ km}}\right)^{-1/2}, \tag{3.12}$$

$$(b) \quad T_E = 1.5 \times 10^6 \text{ K} \left(\frac{M_c}{10^8 M_\odot}\right)^{-1/4}. \tag{3.13}$$

The characteristic time t_E it takes an object to radiate its entire mass if its luminosity were L_E is given by $L_E t_E = M_c c^2$, i.e.

$$t_E = \frac{\sigma_T c}{4\pi G_N m_p} \sim 4 \times 10^8 \text{ yr}. \tag{3.14}$$

The maximum accretion rate $\dot{M}_E$ is given by imposing $L \sim L_E$ (and $\varepsilon = 1$) in (3.3): $\dot{M}_E = (4\pi m_p / \sigma_T) R$. In other words,

$$(a) \quad \dot{M}_E = 1.5 \times 10^{-9} M_\odot \text{ yr}^{-1} \left(\frac{R}{1 \text{ km}}\right), \tag{3.15}$$

$$(b) \quad \dot{M}_E = 2 M_\odot \text{ yr}^{-1} \left(\frac{M_c}{10^8 M_\odot}\right). \tag{3.16}$$

3.4. Supernovae

Supernovae explosions provide some of the brightest events in the sky, some of them being visible to the naked eye. Supernovae explosions were thus recorded in 1006, 1054, 1181, 1572 and 1604. The Sn 1987 A explosion allowed the detection of neutrinos and gamma emission.

The modern theory of supernovae was initiated in the 30s by Baade and Zwicky [21].

Supernovae follow a classification according to spectroscopy. In type I supernovae, hydrogen lines are absent whereas they are present in type II. Moreover, type I has the following subclasses:
- type Ia: intermediate mass elements (Si),
- type Ib: helium lines present,
- type Ic: helium lines weak or absent.

Each type corresponds to a different mechanism for the explosion. In particular, type II and type Ia have a completely different interpretation. We will focus in what follows on these two.

Supernovae of type II

We have seen in section 3.1 that presupernova stars ($M > 8M_\odot$) have an onion-like structure. From the outer to the inner layers, one finds increasingly heavy elements: H, He, C, O, Ne, Si and Fe. For example, for a star of mass $M = 15M_\odot$ just before the explosion:

Composition	Fe	$CONeSi$	He	H
Mass (in $M_\odot$)	1.4	1	1.5	11
Radius (in km)	10^3	4×10^4	3×10^5	4 to 60×10^7

Note that, whereas the tiny iron core is the engine of the explosion, it is the envelope that will reveal it.

As Si is burned, the mass of the Fe core increases. The resulting density increase then turns the electrons relativisitic and makes electronic capture ($p + e \rightarrow n + \nu$) energetically favorable. This diminishes the degenerate electron pressure and leads to the collapse of the core. Since $\rho_{core} \sim 10^{12}$ kg.m^{-3}, the collapse time is typically $(G_N \rho)^{-1/2} \sim 0.1$ s.

This time, neutrinos produced as electrons are turned into neutrons are trapped in the imploding core. The critical density for which neutrinos are trapped, typically $\rho \sim 2 \times 10^{14}$ kg/m^3, is obtained by equating their diffusion velocity

$v_{diff} = c/\tau$ ($\tau = n\sigma R \sim \rho E_\nu^2 R$ is the optical depth defined in (2.51) of Chapter 2) with the collapse velocity $v_{coll} \sim (G_N M/R)^{1/2}$. Since $R \sim \rho^{-1/3}$, and $E_\nu \sim \epsilon_F \sim \rho^{1/3}$ (for relativistic fermions, see section 2.4 of chapter 2), we have $v_{diff} \propto \rho^{-4/3}$ and $v_{coll} \propto \rho^{1/6}$.

As the core is crushed to higher densities, the density approaches that of a neutron star ($\rho \sim 2 \times 10^{17}$ kg/m^3) and matter becomes almost incompressible. If the process was elastic, the kinetic energy would be enough to bring it back to the initial state. Typically

$$E \sim G_N M_{core} \left(\frac{1}{R_{NS}} - \frac{1}{R_{WD}} \right) \sim \frac{G_N M_{core}}{R_{NS}} \sim 3 \times 10^{46} \text{ J}. \qquad (3.17)$$

This is not completely so but there is a rebound of the core which sends a shock wave outward. Meanwhile, the stellar matter has started to free fall since it is no longer sustained by its core. The falling matter meets the outgoing shock wave and turns it into an accretion wave.

Neutrinos emitted from the core heat up and expand the bubble thus formed. Convection and neutrino heating thus convey a fraction of the order of one percent of the neutron star gravitational mass (3.17) to the accretion front. This is enough to make it explode.

One word of caution however: numerical models that try to reproduce supernovae explosions have been until now unable to explode the supernovae! One needs to start the explosion artificially. It therefore remains possible that one is still missing a key ingredient in the recipe.

The bulk of the star blown off by the explosion makes what is known as a supernova remnant. It sweeps the interstellar medium at great velocity (10000 km/s) and may remain visible for 10^5 to 10^6 years. A large fraction of the interstellar medium is thus swept by supernovae remnants (see Exercise 3-1). This is important since this is believed to be the way the heaviest nuclear elements are scattered in the universe (primordial nucleosynthesis produces no element heavier than ^{7}Li).

Supernovae of type Ia

SNIa events are thermonuclear explosions of white dwarfs. More precisely, a carbon-oxygen white dwarf accretes matter (from a companion star or by coalescence with another white dwarf) which causes its mass to exceed the Chandrasekhar limit. The central core collapses, making the carbon burn and causing a wave of combustion to propagate through the star, disrupting it completely. The total production of energy is thus almost constant. For a white dwarf of radius 1500 to 2000 km, about 2×10^{51} ergs is released in a few seconds during

P. Binétruy

which takes place the acceleration of the material. This is followed by a period of free expansion. Virtually all the energy of the explosion goes into the expansion. The luminosity of the supernova, on the other hand, finds its origin in the nuclear decay of the ^{56}Ni freshly synthesized (see Appendix Appendix B). The energy release in the nuclear decays ^{56}Ni $\rightarrow$ ^{56}Co $\rightarrow$ ^{56}Fe, with respective lifetimes of 8.8 and 111 days, represents a few percent of the initial energy release.

This model allows to understand the homogeneity of the observed type Ia supernovae explosions and why they have been used successfully as standard candles in cosmology (see section 4.5 of Chapter 4). The structure of a white dwarf is determined by degenerate electrons and thus independent of detailed chemical composition (see section 2.3 of Chapter 2). The rate of expansion is set by the total energy available since the complete white dwarf is disrupted. Finally, the absolute brightness is determined by the radioactive decay of ^{56}Ni produced during the explosion. Less Ni means a lower luminosity but also lower temperature in the gas and thus lower opacity and more rapide energy escape. Thus dimmer supernovae are quicker i.e. have narrower light curves.

There are discussions regarding the propagation of the combustion front: *deflagration* (subsonic speed for the burning front, see Figure 5) or *detonation* (the combustion front coincides with a shock front moving outward supersonically)? To agree with observations of intermediate mass elements at the outer layers, the white dwarf must be pre-expanded. Most likely, an initial delagration phase does the job. Thus favored models [22] are of the type deflagration-detonation: the production of ^{56}Ni is fixed by a parameter characterizing the transition between deflagration and detonation.

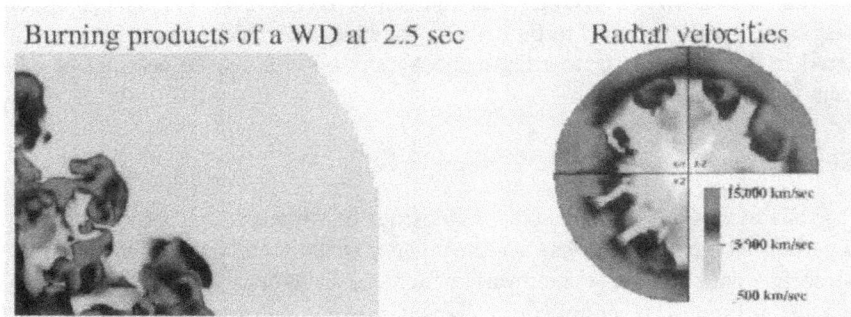

Fig. 5. Structure of the deflagration front in an exploding C/O white dwarf (left) and the velocity field at about 2 seconds after runaway based on 3-D calculations (see [22] for details and references).

Exercise 3-1: a) Assuming approximately one supernova explosion every 30 years in our galaxy (assimilated to a disk of radius 15 kpc and thickness 200 pc), compute the corresponding rate $\mathcal{R}$ of supernovae explosions per pc^3 and per year.

b) If every supernova leads to a remnant of radius $R = 100$ pc that lasts for $t \sim 10^6$ yrs, what fraction of the galaxy volume is filled by the supernova remnant?

Hints: a) $\mathcal{R} \sim 2.3 \times 10^{-13}$ pc^{-3}.yr^{-1}.
b) $1 - \exp\left[-(4\pi/3)R^3\mathcal{R}t\right] \sim 0.5$.

3.5. Gamma ray bursts

Gamma ray bursts (GRB) are the most luminous events observed in the universe. They were discovered accidentally by the American military satellites VEGA which were designed to monitor the nuclear test ban treaty of 1963. The first burst was found in 1969, buried in gamma-ray data from 1967: two Vela satellites had detected more or less identical signals, showing the source to be roughly the same distance from each satellite [23].

A GRB explosion can be as luminous as objects which are in our vicinity, such as the Crab nebula, although they are very distant. The initial flash is short (from a few seconds to a few hundred for a long GRB, a fraction of a second for a short one). From 1991 to 2000, BATSE (Burst and Transient Source Experiment) has allowed to detect some 2700 bursts and showed that their distribution is isotropic, a good argument in favor of their cosmological origin. In 1997 (February 28), the precise determination of the position of a GRB (hence named 970228) by the Beppo-SAX satellite allowed ground telescopes to discover a rapidly decreasing optical counterpart, called afterglow. Typically in the afterglow, the photon energy decreases with time as a power law (from X ray to optical, IR and radio) as well as the flux: it stops after a few days or weeks.

The study of afterglows gives precious information on the dynamics of GRBs.

Spectroscopic observations allow to measure the redshift, through the identification of absorption lines. For example, they allowed to confirm that GRB 050904 recently observed by the Swift satellite [24], launched in 2004, has a redshift $z = 6.295$ [25].

One may then infer the amount of energy released by the GRB. Typical values range from 10^{44} to 10^{47} J. One must however correct for a factor $\Omega/(4\pi)$ which takes into account the fact that emission is collimated in a solid angle Ω. Once this is taken into account, it seems that the energy radiated in gamma rays is always close to 10^{44} J.

Given the time scales involved (a few milliseconds for the rise time of the gamma signal), the size of the source must be very small: it cannot exceed the distance radiation can travel in the same time interval, i.e. at most a few hundred kilometers. Energy must have been ejected in an ultra-relativistic flow which converted its kinetic energy into radiation away from the source: the Lorentz factors involved are typically of the order of 100!

In the collapsar model of Woosley [26], long GRBs are associated with the explosion of a rapidly rotating massive star which collapses into a spinning black hole. The burst and its afterglow have been successfully explained by the interaction of a highly relativistic jet with itself and with the circumstellar medium. Typically, one expects per day 10^6 collapses of massive stars in the Universe; 10^3 give rise to a GRB and approximately 1 of these is pointing towards us its jet. Hence, one may observe from earth about one GRB per day.

One expects a supernova for every long GRB because the energetic jet explodes the star [27]. The first evidence came from the connection established between GRB980425 and the very energetic SN1998bw [28]. But the evidence was not conclusive because this GRB was atypical: the energy released was unusually small. Recent observations have made a much stronger case for such a connection. For example, Figure 6 [29] shows that the afterglow spectral emission of GRB030329 is replaced after some 30 days by an emission typical of a supernova.

3.6. Gravitational waves

The measurement of the energy loss of the binary pulsar PSR 1913+16 discovered by Hulse and Taylor [30] has provided an indirect evidence of gravitational waves. This system consists of two bodies of mass of order $1.4 M_\odot$ orbiting in some 8 hours on an elliptical orbit with a maximum separation of the order of $R_\odot$! One of the objects is a pulsar, a neutron star emtting radio pulses which sweep space with a period of 59 ms. It is the quasi-periodical fluctuations in the arrival time of these pulses that allowed to infer that this pulsar was in a binary system. Moreover, a speedup in the orbital frequency was observed whose magnitude was in perfect agreement with the energy loss due to gravitational radiation [31]. Hulse and Taylor were awarded the Nobel prize in physics in 1993 for this discovery.

Direct detection is however still ahead of us, although much progress has been made in this respect in the last 40 years. There are three types of gravitational detectors, which operate at various sensitivities and frequency ranges.

Resonant or acoustic detectors

The antenna is typically a cylindrical bar which is set into oscillation by the periodic stresses associated with the tidal forces induced by the gravitational wave. The mechanical vibration could be detectable if the frequency of the wave is close to the resonant frequency of the bar. Current resonant detectors are sensitive to

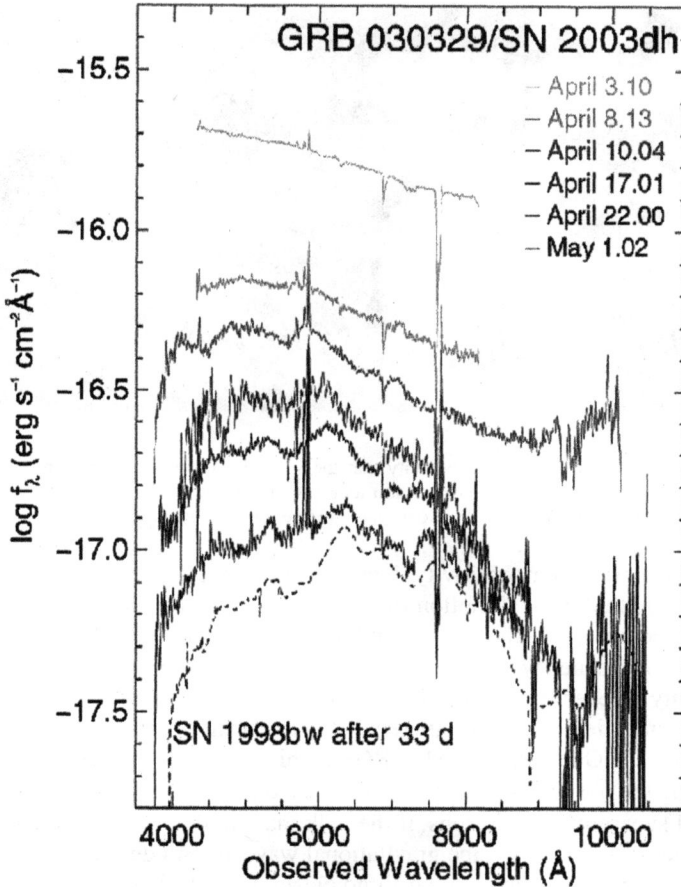

Fig. 6. Spectral evolution of the combined optical flux density of the afterglow of GRB030329, the associated SN2003dh, and its host galaxy. For comparison, the lowest curve (dashed) gives the same quantity for SN1998bw after 33 days [29].

a signal with a frequency in a narrow window around the kHz: typical sources are thus galactic supernovae or millisecond pulsars. But the expected event rate is quite small and a network analysis is required to reject fake events.

Ground based interferometers

Interferometers detect gravitational waves by measuring the distance between mirrors which follow free geodesics in spacetime (at least in the direction of the

P. Binétruy

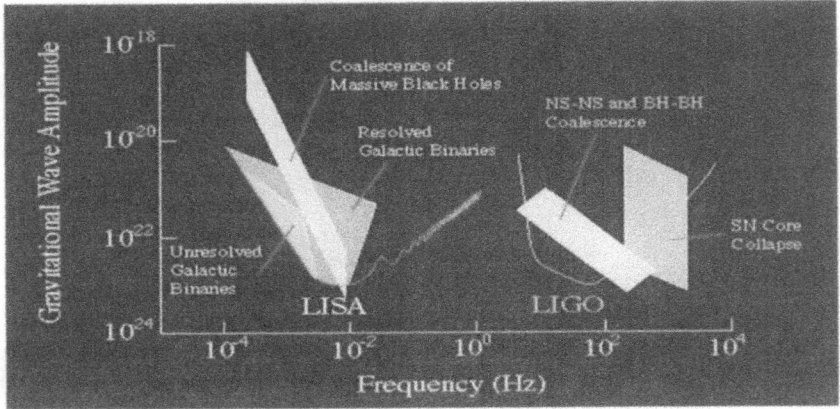

Fig. 7. The spectrum shows the sensitivity that the space interferometer LISA and an advanced version of the ground interferometers (a 4 km long LIGO) will obtain in their respective operating frequency bands, as well as spectral regions where various sources are predicted to be.

reflected beam). Their sensitivity depends on the interferometer arm length and is limited by the residual motion of the mirrors (in particular the seismic noise for a ground-based interferometer) and the photon shot noise. As can be seen on Figure 7, the frequency range extends from a few Hz to several kHz. Given the uncertainty on event rates, detection of gravitational waves is not guaranteed for the first generation of interferometers but is expected for the advanced detectors (Advanced LIGO and Advanced VIRGO) which are now considered.

Among the possible astrophysical sources, one may select as in Figure 7:
• type II supernovae explosions, if the bulk mass (collapsed or ejected) motions are asymetrical. Typically, the gravitational wave amplitude h is estimated to be 10^{-21} to 10^{-24} for a source in the Virgo cluster (i.e. at 15 Mpc).
• coalescence of two compact objects such as neutrons stars or black holes. This is preceded by an inspiraling phase which enables precision tests of general relativity at post-Newtonian orders. A typical amplitude is $h \sim 10^{-23}$ to 10^{-22} for a source of two objects of mass $1.4 M_\odot$ located at 10 Mpc. Advanced interferometers could observe of the order of 10 solar mass black hole binaries up to redshift of $z = 0.4$.

Space interferometers

Going to space allows to reach much larger arm lengths and probe a different frequency range, and thus different astrophysical sources. LISA, scheduled to be launched in 2015, is a set of three spacecrafts distant from one another by 5.10^6 km. Each spacecraft houses two masses in free fall: lasers are reflected on these masses and interferometry measures the relative variations in distances.

Figure 7 gives the sensitivity curve of LISA. Below 3 mHz it is dominated by spurious accelerations of the reference proof masses. The sensitivity from 3 mHz to 1 Hz reflects the shot noise limited performance of the interferometer. The sensitivity above 10 mHz has a complex behavior caused by the interferometer arm length being greater than the wavelength of the gravitational waves. In the region marked "Unresolved Galactic Binaries", gravity wave sources are so numerous that they will appear to LISA as an unresolved background.

There are a certain number of galactic binary sources, called verification binaries, where detection is guaranteed (through indirect means, see above). Numerous other galactic binaries should be resolved. But the most spectacular signal expected from LISA is the observation of the inspiral and final plunge of two supermassive black holes of mass in the range 10^6 to $10^8 M_\odot$.

4. The universe at large

Some of the bright sources that we have been discussing will allow us to shed some light on the distant universe (which, in some sense, is also an earlier universe). We are now ready to discuss the properties of our universe in its largest scales.

4.1. Energy budget

The Friedmann equation

$$H^2 \equiv \frac{\dot{a}^2}{a^2} = \frac{1}{3}\left(\lambda + 8\pi G_N \rho\right) - \frac{k}{a^2}. \tag{4.1}$$

allows to define the Hubble constant H_0, i.e. the present value of the Hubble parameter, which sets the scale of our Universe at present time. Because of the troubled history of the measurement of the Hubble constant, it has become customary to express it in units of $100 \text{ km.s}^{-1}.\text{Mpc}^{-1}$ which gives its order of magnitude. Present measurements give

$$h_0 \equiv \frac{H_0}{100 \text{ km.s}^{-1}.\text{Mpc}^{-1}} = 0.7 \pm 0.1.$$

The corresponding length and time scales are:

$$\ell_{H_0} \equiv \frac{c}{H_0} = 3000\, h_0^{-1} \text{ Mpc} = 9.25 \times 10^{25}\, h_0^{-1} \text{ m}, \tag{4.2}$$

$$t_{H_0} \equiv \frac{1}{H_0} = 3.1 \times 10^{17}\, h_0^{-1} \text{ s} = 9.8\, h_0^{-1} \text{ Gyr}. \tag{4.3}$$

A reference energy density at present time t_0 is obtained from the Friedmann equation for vanishing cosmological ($\lambda = 0$) and flat space ($k = 0$):

$$\rho_c \equiv \frac{3H_0^2}{8\pi G_N} = 1.9\ 10^{-26}\ h_0^2\ \text{kg/m}^3. \tag{4.4}$$

This corresponds to approximately one galaxy per Mpc^3 or 5 protons per m^3. In fundamental units where $\hbar = c = 1$, this is of the order of $\left(10^{-3}\text{eV}\right)^4$. In the case of a vanishing cosmological constant, it follows from (4.1) that, depending on whether the present energy density of the Universe ρ_0 is larger, equal or smaller than ρ_c, the present Universe is spatially open ($k > 0$), flat ($k = 0$) or closed ($k < 0$). Hence the name critical density for ρ_c.

It has become customary to normalize the different forms of energy density in the present Universe in terms of this critical density. Separating the energy density ρ_{M0} presently stored in non-relativistic matter (baryons, neutrinos, dark matter, ...) from the density ρ_{R0} presently stored in radiation (photons, relativistic neutrino if any), one defines:

$$\Omega_M \equiv \frac{\rho_{M0}}{\rho_c}, \quad \Omega_R \equiv \frac{\rho_{R0}}{\rho_c}, \quad \Omega_\Lambda \equiv \frac{\lambda}{3H_0^2}, \quad \Omega_k \equiv -\frac{k}{a_0^2 H_0^2}. \tag{4.5}$$

The last term comes from the spatial curvature and is not strictly speaking a contribution to the energy density.

Then the Friedmann equation taken at time t_0 simply reads

$$\Omega_M + \Omega_R + \Omega_\Lambda = 1 - \Omega_k = 1. \tag{4.6}$$

Since matter dominates over radiation in the present Universe, we may neglect Ω_R in the preceding equation. Using the dependence of the different components with the scale factor $a(t) = a_0/(1+z)$, one may then rewrite the Friedmann equation at any time as:

$$H^2(t) = H_0^2 \left[\Omega_\Lambda + \Omega_M \left(\frac{a_0}{a(t)}\right)^3 + \Omega_R \left(\frac{a_0}{a(t)}\right)^4 + \Omega_k \left(\frac{a_0}{a(t)}\right)^2 \right], \tag{4.7}$$

$$\text{or} \quad H^2(z) = H_0^2 \left[\Omega_M (1+z)^3 + \Omega_R (1+z)^4 + \Omega_k (1+z)^2 + \Omega_\Lambda \right]. \tag{4.8}$$

where a_0 is the present value of the cosmic scale factor and all time dependences (or alternatively redshift dependence) have been written explicitly. We note that, even if Ω_R is negligible in (4.6), this is not so in the early Universe because the radiation term increases faster than the matter term in (4.7) as one gets back in time (i.e. as $a(t)$ decreases). If we add an extra component X with equation of state $p_X = w_X \rho_X$, it contributes an extra term $\Omega_X (a_0/a(t))^{3(1+w_X)}$ where $\Omega_X = \rho_X/\rho_c$.

An important information about the evolution of the universe at a given time is whether its expansion is accelerating or decelerating. The acceleration of our universe is usually measured by the deceleration parameter q which is defined as:

$$q \equiv -\frac{\ddot{a}a}{\dot{a}^2}. \tag{4.9}$$

Using (2.17) of Chapter 2 and separating again matter and radiation, we may write it at present time t_0 as:

$$q_0 = -\frac{1}{H_0^2}\left(\frac{\ddot{a}}{a}\right)_{t=t_0} = \frac{1}{2}\Omega_M + \Omega_R - \Omega_\Lambda. \tag{4.10}$$

Once again, the radiation term Ω_R can be neglected in this relation. We see that in order to have an acceleration of the expansion ($q_0 < 0$), we need the cosmological constant to dominate over the other terms.

We can also write the deceleration parameter in (4.9) in terms of redshift as in (4.8)

$$q = \frac{1}{H(z)^2}\left[\frac{1}{2}\Omega_M(1+z)^3 + \Omega_R(1+z)^4 - \Omega_\Lambda\right]. \tag{4.11}$$

This shows that the universe starts accelerating at redshifts $1+z \sim (2\Omega_\Lambda/\Omega_M)^{1/3}$ (neglecting Ω_R), that is typically redshifts of order 1. If we introduce an extra component X as above, it contributes a term $\Omega_X(1+z)^{3(1+w_X)}(1+3w_X)/2$: only components with equation of state parameter $w_X < -1/3$ tend to accelerate the expansion of the universe.

The measurements of the Hubble constant and of the deceleration parameter at present time allow to obtain the behaviour of the cosmic scale factor in the last stages of the evolution of the universe:

$$a(t) = a_0\left[1 + \frac{t-t_0}{t_{H_0}} - \frac{q_0}{2}\frac{(t-t_0)^2}{t_{H_0}^2} + \cdots\right]. \tag{4.12}$$

4.2. Measure of distances

Measuring cosmological distances allows to study the geometry of spacetime. Depending on the type of observation, one may define several distances.

First consider a photon travelling in an expanding or contracting Friedmann universe. Its equation of motion is fixed by the condition $ds^2 = 0$ (as in (2.18) of Chapter 2). One then defines the proper distance as

$$d(t) \equiv a(t) \int_0^r \frac{dr}{\sqrt{1-kr^2}} = a(t) \int_t^{t_0} \frac{cdt'}{a(t')}. \tag{4.13}$$

Using

$$\int_t^{t_0} \frac{cdt}{a(t)} = \int_{a(t)}^{a_0} \frac{cda}{a\dot{a}} = \int_{a(t)}^{a_0} \frac{cda}{a^2 H} = \int_0^z \frac{cdz}{H(z)},$$

we may extract from (4.13) the proper distance at time t_0:

$$d(t_0) = a_0 \int_0^r \frac{dr}{\sqrt{1-kr^2}} = a_0 \begin{cases} \sin^{-1} r & k = +1 \\ r & k = 0 \\ \sinh^{-1} r & k = -1 \end{cases} \tag{4.14}$$

$$= \ell_{H_0} \int_0^z \frac{dz}{[\Omega_M(1+z)^3 + \Omega_R(1+z)^4 + \Omega_k(1+z)^2 + \Omega_\Lambda]^{1/2}},$$

where $\ell_{H_0} = cH_0^{-1}$.

If a photon source of luminosity L (energy per unit time) is placed at a distance r from the observer, then the energy flux ϕ (energy per unit time and unit area) received by the observer is given by

$$\phi = \frac{L}{4\pi a_0^2 r^2 (1+z)^2} \equiv \frac{L}{4\pi d_L^2}. \tag{4.15}$$

The two powers of $1 + z$ account for the photon energy redshift and the time dilatation between emission and observation. The quantity $d_L \equiv a_0 r(1 + z)$ is called luminosity distance.

If the source is at a redshift z of order one or smaller, the effect of spatial curvature is unimportant and we can approximate the integral $\int_0^r dr/\sqrt{1-kr^2}$ in (4.13) by simply r (i.e. the value for $k = 0$). This equation gives

$$a_0 r \sim \int_t^{t_0} \frac{a_0 c dt}{a(t)} = \int_a^{a_0} \frac{a_0 c da}{a\dot{a}} \sim \ell_{H_0} \int_a^{a_0} \frac{da}{a[1 - q_0 H_0(t - t_0)]}, \tag{4.16}$$

where we have used the development (4.12) with $t_{H_0} = \ell_{H_0}/c = H_0^{-1}$. Using $H_0(t - t_0) \sim (a - a_0)/a_0 \ll 1$ and $a = a_0/(1 + z)$, we obtain for $z \ll 1$

$$a_0 r = \ell_{H_0} z \left(1 - \frac{1+q_0}{2} z + \cdots \right). \tag{4.17}$$

Thus, the luminosity distance reads, for $z \ll 1$,

$$d_L = \ell_{H_0} z \left(1 - \frac{1+q_0}{2} z + \cdots \right) (1+z) = \ell_{H_0} z \left(1 + \frac{1-q_0}{2} z + \cdots \right). \quad (4.18)$$

Hence measurement of deviations to the Hubble law ($d_L = \ell_{H_0} z$) at moderate redshift allow to measure the combination $\Omega_M/2 - \Omega_\Lambda$ (see (4.10)).

Another distance is defined in cases where one measures the angular diameter δ of a source in the sky. If D is the diameter of the source, then D/δ would be the distance of the source in Euclidean geometry. In a universe with a Robertson-Walker metric, it turns out to be $a(t)r = a_0 r/(1+z)$. This defines the angular diameter distance d_A

$$d_A = \frac{d_L}{(1+z)^2}. \quad (4.19)$$

Several distance measurements tend to point towards an evolution of the present universe dominated by the cosmological constant contribution[17] and thus a late acceleration of its expansion, as we will now see.

4.3. Cosmic Microwave Background [CMB]

We will recall briefly in Section 5 the history of the Universe (see Table 1 below). We will see that, soon after matter-radiation equality, electrons recombine with the protons to form neutral atoms of hydrogen, which induces the decoupling of matter and photon. From this epoch on, the universe becomes transparent. The primordial gas of photons produced at this epoch cools down as the universe expands and forms nowadays the cosmic microwave background. It is primarily homogeneous and isotropic but includes fluctuations at a level of 10^{-5}, which are of much interest since they are imprints of the recombination and earlier epochs.

Before discussing the spectrum of CMB fluctuations, we introduce the important notion of a particle horizon in cosmology.

Because of the speed of light, a photon which is emitted at the big bang ($t = 0$) will have travelled a finite distance at time t. The proper distance (4.13) measured at time t is simply given by the integral:

$$d_h(t) = a(t) \int_0^t \frac{c\,dt'}{a(t')} \quad (4.20)$$

$$= \frac{\ell_{H_0}}{1+z} \int_z^\infty \frac{dz}{\left[\Omega_M(1+z)^3 + \Omega_R(1+z)^4 + \Omega_k(1+z)^2 + \Omega_\Lambda \right]^{1/2}},$$

---

[17]at least when analyzed in the framework of the model discussed in this section, i.e. including non-relativistic matter, radiation and a cosmological constant.

where, in the second line, we have used (4.14). This is the maximal distance that a photon (or any particle) could have travelled at time t since the big bang. In other words, it is possible to receive signals at a time t only from comoving particles within a sphere of radius $d_h(t)$. This distance is known as the particle horizon at time t.

A quantity of relevance for our discussion of CMB fluctuations is the horizon at the time of the recombination i.e. $z_{\rm rec} \sim 1100$. We note that the integral on the second line of (4.20) is dominated by the lowest values of z: $z \sim z_{\rm rec}$ where the universe is still matter dominated. Hence

$$d_h(t_{\rm rec}) \sim \frac{2\ell_{H_0}}{\Omega_M^{1/2} z_{\rm rec}^{3/2}} \sim 0.3 \text{ Mpc}. \tag{4.21}$$

We note that this is simply twice the Hubble radius at recombination $H^{-1}(z_{\rm rec})$, as can be checked from (4.8):

$$R_H(t_{\rm rec}) \sim \frac{\ell_{H_0}}{\Omega_M^{1/2} z_{\rm rec}^{3/2}}. \tag{4.22}$$

This radius is seen from an observer at present time under an angle

$$\theta_H(t_{\rm rec}) = \frac{R_H(t_{\rm rec})}{d_A(t_{\rm rec})}, \tag{4.23}$$

where the angular distance has been defined in (4.19). We can compute analytically this angular distance under the assumption that the universe is matter dominated (see Exercise 4-1). Using (4.34), we have

$$d_A(t_{\rm rec}) = \frac{a_0 r}{1 + z_{\rm rec}} \sim \frac{2\ell_{H_0}}{\Omega_M z_{\rm rec}}. \tag{4.24}$$

Thus, since, in our approximation, the total energy density Ω_T is given by Ω_M,

$$\theta_H(t_{\rm rec}) \sim \Omega_T^{1/2}/(2z_{\rm rec}^{1/2}) \sim 0.015 \text{ rad } \Omega_T^{1/2} \sim 1° \, \Omega_T^{1/2}. \tag{4.25}$$

We have written in the latter equation Ω_T instead of Ω_M because numerical computations show that, in case where Ω_Λ is non-negligible, the angle depends on $\Omega_M + \Omega_\Lambda = \Omega_T$.

We can now discuss the evolution of photon temperature fluctuations. For simplicity, we will assume a flat primordial spectrum of fluctuations: this leads to predictions in good agreement with experiment; moreover, as we will see in the next Section, it is naturally explained in the context of inflation scenarios.

Before decoupling, the photons are tightly coupled with the baryons through Thomson scattering. In a gravitational potential well, gravity tends to pull this

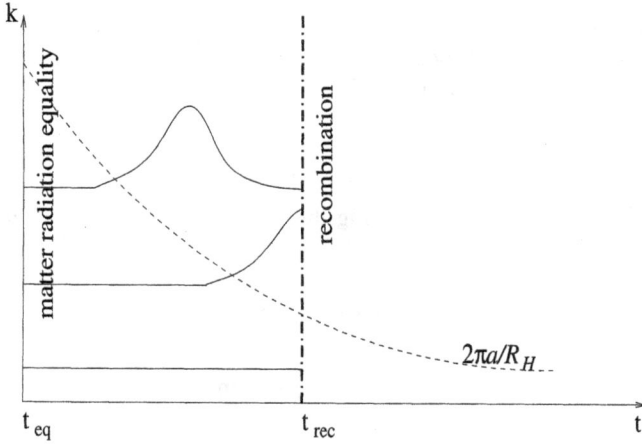

Fig. 8. Evolution of the photon temperature fluctuations before the recombination. This diagram illustrates that oscillations start once the corresponding Fourier mode enters the Hubble radius (these oscillations are fluctuations in temperature, along a vertical axis orthogonal to the two axes that are drawn on the figure).

baryon-photon fluid down the well whereas radiation pressure tends to push it out. Thus, the fluid undergoes a series of acoustic oscillations. These oscillations can obviously only proceed if they are compatible with causality i.e. if the corresponding wavelength is smaller than the horizon scale or the Hubble radius: $\lambda = 2\pi a(t)/k < R_H(t)$ or

$$k > 2\pi \frac{a(t)}{R_H(t)} \sim t^{-1/3}. \tag{4.26}$$

Starting with a flat primordial spectrum, we see that the first oscillation peak corresponds to $\lambda \sim R_H(t_{rec})$, followed by other compression peaks at $R_H(t_{rec})/n$ (see Figure 8). They correspond to an angular scale on the sky:

$$\theta_n \sim \frac{R_H(t_{rec})}{d_A(t_{rec})} \frac{1}{n} = \frac{\theta_H(t_{rec})}{n}. \tag{4.27}$$

Since photons decouple at t_{rec}, we observe the same spectrum presently (up to a redshift in the photon temperature)[18].

---

[18] A more careful analysis indicates the presence of Doppler effects besides the gravitational effects that we have taken into account here. Such Doppler effects turn out to be non-leading for odd values of n.

Experiments usually measure the temperature difference of photons received by two antennas separated by an angle θ, averaged over a large fraction of the sky. Defining the correlation function

$$C(\theta) = \left\langle \frac{\Delta T}{T_0}(\mathbf{n}_1) \frac{\Delta T}{T_0}(\mathbf{n}_2) \right\rangle, \tag{4.28}$$

averaged over all $\mathbf{n}_1$ and $\mathbf{n}_2$ satisfying the condition $\mathbf{n}_1 \cdot \mathbf{n}_2 = \cos\theta$, we have indeed

$$= \left\langle \left(\frac{T(\mathbf{n}_1) - T(\mathbf{n}_2)}{T_0} \right)^2 \right\rangle = 2\left(C(\theta) - C(0)\right). \tag{4.29}$$

We may decompose $C(\theta)$ over Legendre polynomials:

$$C(\theta) = \frac{1}{4\pi} \sum_{l}^{\infty} (2l+1) C_l P_l(\cos\theta). \tag{4.30}$$

The monopole ($l = 0$) related to the overall temperature T_0, and the dipole ($l = 1$) due to the Solar system peculiar velocity, bring no information on the primordial fluctuations. A given coefficient C_l characterizes the contribution of the multipole component l to the correlation function. If $\theta \ll 1$, the main contribution to C_l corresponds to an angular scale[19] $\theta \sim \pi/l \sim 200°/l$. The previous discussion (see (4.25) and (4.27)) implies that we expect the first acoustic peak at a value $l \sim 200\Omega_T^{-1/2}$.

The power spectrum obtained by the WMAP experiment is shown in Figure 9. One finds the first acoustic peak at $l \sim 200$, which constrains the ΛCDM model used to perform the fit to $\Omega_T = \Omega_M + \Omega_\Lambda \sim 1$. Many other constraints may be inferred from a detailed study of the power spectrum [32].

4.4. Baryon acoustic oscillations

We noted in the previous section that, before decoupling, baryons and photons were tightly coupled and the baryon-photon fluid underwent a series of acoustic oscillations. We have until now followed the fate of photons after decoupling. Similarly, once baryons decouple from the radiation, their oscillations freeze in, which leads to specific imprints in the galaxy power spectrum.

Recently, Einsenstein and collaborators [33], using data from the Sloan Digital Sky Survey, have identified a baryon acoustic peak in the matter power spectrum on scales of 100 Mpc.

---

[19]The C_l are related to the coefficients a_{lm} in the expansion of $\Delta T/T$ in terms of the harmonic spherics Y_{lm}: $C_l = \langle |a_{lm}|^2 \rangle_m$. The relation between the value of l and the angle comes from the observation that Y_{lm} has $(l - m)$ zeros for $-1 < \cos\theta < 1$ and $\mathrm{Re}(Y_{lm})$ m zeros for $0 < \phi < 2\pi$.

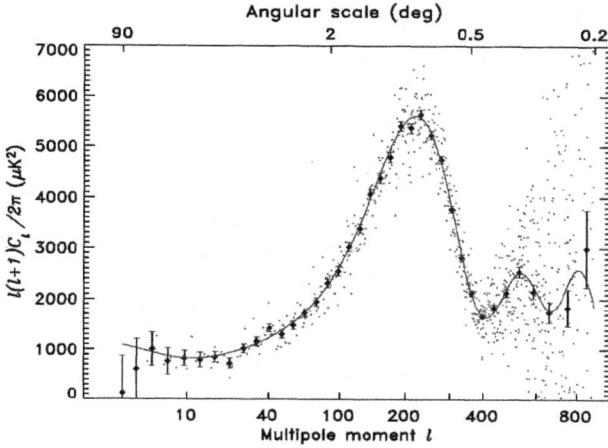

Fig. 9. This figure compares the best fit power law ΛCDM model to the temperature angular power spectrum observed by WMAP. The gray dots are the unbinned data [32].

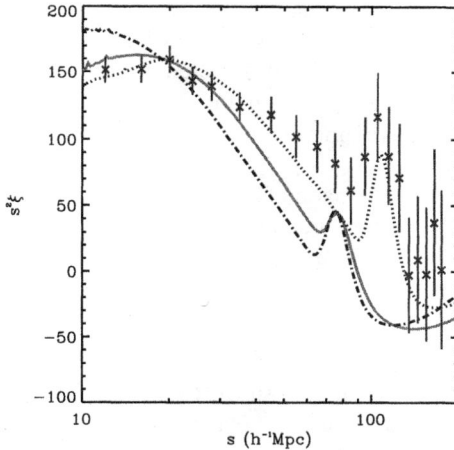

Fig. 10. Correlation function in redshift space for the best-fit power-law ΛCDM model (dotted line) and for the best-fit Einstein-de Sitter models ($\Omega_\Lambda = 0$) with a neutrino component $\Omega_\nu = 0.12$ (solid line) and a quintessence component $\Omega_Q = 0.12$ (dot-dashed line) [34].

This provides a clean test on cosmological models. Figure 10 shows how the ΛCDM model just described fares with respect to observations by comparison with a Einstein-de Sitter model with $\Omega_\Lambda = 0$.

4.5. Acceleration of the universe

The approach that has made the first strong case for a flat space ($\Omega_T = 1$) uses supernovae of type Ia as standard candles (see section 3.4 of Chapter 3 for a discussion of this point). Two groups, the Supernova Cosmology Project [35] and the High-z Supernova Search [36] have found that distant supernovae appear to be fainter than expected in a flat matter-dominated Universe. If this is to have a cosmological origin, this means that, at fixed redshift, they are at larger distances than expected in such a context and thus that the Universe is accelerating.

More precisely, one uses the relation (4.15) between the flux ϕ received on earth and the luminosity L of the supernova. Traditionally, flux and luminosity are expressed on a log scale as apparent magnitude m_B and absolute magnitude M (magnitude is $-2.5 \log_{10}$ luminosity + constant). The relation then reads

$$m_B = 5 \log(H_0 d_L) + M - 5 \log H_0 + 25. \tag{4.31}$$

The last terms are z-independent, *if one assumes that supernovae of type Ia are standard candles*; they are then measured by using low z supernovae. The first term, which involves the luminosity distance d_L, varies logarithmically with z up to corrections which depend on the geometry, more precisely on $q_0 = \Omega_M/2 - \Omega_\Lambda$ for small z as can be seen from (4.18). This allows to compare with data cosmological models with different components participating to the energy budget, as can be seen from Figure 11.

This can be turned into a limit in the $\Omega_M - \Omega_\Lambda$ plane for the model considered here (see Figure 12).

Let us note that the combination $\Omega_M/2 - \Omega_\Lambda$ is 'orthogonal' to the combination $1 - \Omega_k = \Omega_M + \Omega_\Lambda$ measured in CMB experiments (see next Chapter). The two measurements are therefore complementary: this is sometimes referred to as 'cosmic complementarity'.

4.6. Standard or standardizable candles?

As we noted in section 3.4, dimmer supernovae are quicker. In practice, one thus has to correct the light curves using a phenomenological stretch factor. It is thus more precise to state that supernovae of type Ia are standardizable. Moreover, the type of measurement discussed above is sensitive to many possible systematic effects (evolution besides the light-curve timescale correction, presence of dust, etc.), and this has fueled a healthy debate on the significance of supernova data as well as a thorough study of possible systematic effects by the observational groups concerned.

Other cosmic events may be used as standardizable candles. Two of them have received special attention recently: gamma ray bursts and coalescence of binary black holes.

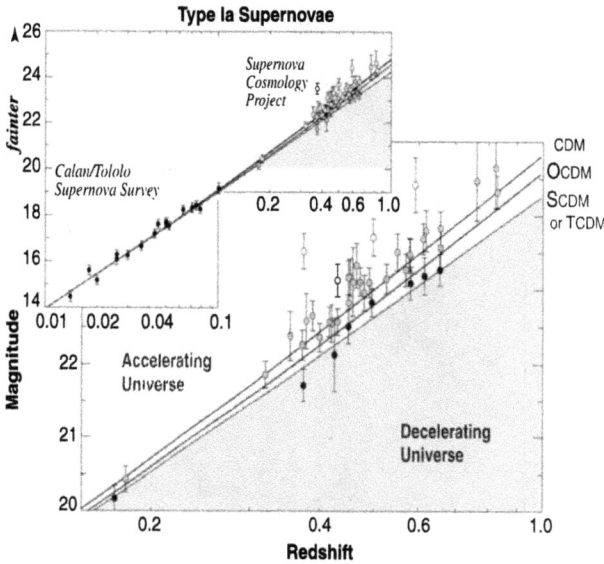

Fig. 11. Hubble plot (magnitude versus redshift) for Type Ia supernovae observed at low redshift by the Calan-Tololo Supernova Survey and at moderate redshift by the Supernova Cosmology Project.

Since GRBs are extremely bright, they can be very distant (up to redshifts larger than 10) and can be used for cosmology. One may discuss in what respect GRBs are normalizable candles [37–39]. The question, still debated [40], is whether one can determine the luminosity of a GRB through a relation between the collimation corrected energy E_γ and the peak E_{peak} in the rest frame prompt burst spectrum.

We have seen that black holes are very simple objects. Coalescence of black holes may thus provide the ultimate standard candle. The precision that can be reached is however limited by gravitational lensing (see below) [41].

Exercise 4-1: We compute exactly the luminosity distance $d_L = a_0 r(1+z)$ or angular distance $d_A = a_0 r/(1+z)$ in the case of a matter-dominated universe. Defining

$$\zeta_k(r) \equiv \begin{cases} \sin^{-1} r & k = +1 \\ r & k = 0 \\ \sinh^{-1} r & k = -1 \end{cases}, \tag{4.32}$$

P. Binétruy

Supernova Cosmology Project
Perlmutter *et al.* (1998)

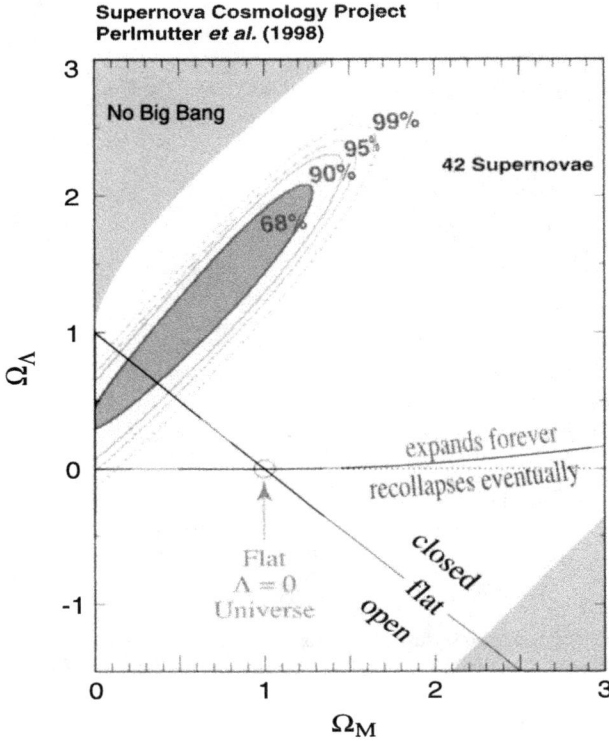

Fig. 12. Best-fit coincidence regions in the $\Omega_M - \Omega_\Lambda$ plane, based on the analysis of 42 type Ia supernovae discovered by the Supernova Cosmology Project [35].

use (4.14) which reads, in the case of a matter-dominated universe,

$$a_0 \zeta_k(r) = \ell_{H_0} \int_0^z \frac{dz}{\left[\Omega_M (1+z)^3 + (1-\Omega_M)(1+z)^2\right]^{1/2}}, \tag{4.33}$$

to prove Mattig's formula [42]:

$$a_0 r = 2\ell_{H_0} \frac{\Omega_M z + (\Omega_M - 2)\left[\sqrt{1 + \Omega_M z} - 1\right]}{\Omega_M^2 (1+z)}. \tag{4.34}$$

Hints: For $k \neq 0$, change to the coordinate $u^2 = k(\Omega - 1)/[\Omega(1+z)]$ in order to compute the integral (4.33). Using the last of equations (4.5), which reads $\ell_{H_0}^2/a_0^2 = k(\Omega - 1)$, one obtains

$$\zeta_k(r) = 2\left(\zeta_k\left[\sqrt{\frac{k(\Omega - 1)}{\Omega}}\right] - \zeta_k\left[\sqrt{\frac{k(\Omega - 1)}{(1+z)\Omega}}\right]\right),$$

from which (4.34) can be inferred.

5. The early universe

Table 1 summarizes the history of the universe in the context of the big bang model with an inflationary epoch.

As we have seen in Section 4, it follows from the Friedmann equation that, if the Universe is dominated by a component of equation of state $p = w\rho$, then the cosmic scale factor $a(t)$ varies with time as $t^{2/[3(1+w)]}$. We start at time t_0 with the energy budget obtained in the previous Chapter, say $\Omega_M = 0.3, \Omega_\Lambda = 0.7, \Omega_k \sim 0$. Radiation consists of photons and relativistic neutrinos:

$$\rho_R(t) = \rho_\gamma(t)\left[1 + \frac{7}{8}\left(\frac{4}{11}\right)^{4/3} N_\nu^{\text{rel}}(t)\right], \tag{5.1}$$

where $N_\nu^{\text{rel}}(t)$ is the number of relativistic neutrinos at time t. At present time t_0, we have $\Omega_\gamma = \rho_\gamma(t_0)/\rho_c = 2.48 \times 10^{-5} \, h^{-2}$ and the mass limits on neutrinos imply $N_\nu^{\text{rel}}(t_0) \leq 1$. In any case, $\Omega_R \ll \Omega_M$.

For redshifts larger than 1, the vacuum energy becomes subdominant and the universe is matter-dominated ($a(t) \sim t^{2/3}$). However radiation energy density increases more rapidly (as $a(t)^{-4}$) than matter ($a(t)^{-3}$) as one goes back in time (as $a(t)$ decreases). At time t_{eq}, there is equality. This corresponds to

$$\frac{1}{1+z_{\text{eq}}} = \frac{a(t_{\text{eq}})}{a_0} = \frac{1.68 \, \Omega_\gamma}{\Omega_M} = \frac{4.17 \times 10^{-5}}{\Omega_M h_0^2}, \tag{5.2}$$

where we have assumed 3 relativistic neutrinos at this time.

However, at $t_{\text{rec}} > t_{\text{eq}}$ i.e. still in the matter-dominated epoch, electrons recombine with the protons to form atoms of hydrogen and, because hydrogen is neutral, this induces the decoupling of matter and photon: as we have seen, from then on ($t_{\text{rec}} < t < t_0$), the universe becomes transparent[20]. This is the impor-

---

[20]It is believed that hydrogen is later reionized by the photons produced by the first stars or quasars but the universe is then sufficiently dilute to prevent recoupling.

Table 1

The different stages of the cosmological evolution in the standard scenario, given in terms of time t since the big bang singularity, the energy kT of the background photons and the redshift z. The double line following nucleosynthesis indicates the part of the evolution which has been tested through observation. The values ($h_0 = 0.7$, $\Omega_M = 0.3$, $\Omega_\Lambda = 0.7$) are adopted to compute explicit values.

t	kT_γ (eV)	z	
$t_0 \sim 15$ Gyr	2.35×10^{-4}	0	now
$\sim$ Gyr	$\sim 10^{-3}$	~ 10	formation of galaxies
$t_{rec} \sim 4 \times 10^5$ yr	0.26	1100	recombination
$t_{eq} \sim 4 \times 10^4$ yr	0.83	3500	matter-radiation equality
3 min	6×10^4	2×10^8	nucleosynthesis
1 s	10^6	3×10^9	e^+e^- annihilation
4×10^{-6} s	4×10^8	10^{12}	QCD phase transition
$< 4 \times 10^{-6}$ s	$> 10^9$		baryogenesis
			inflation
$t = 0$		∞	big-bang

tant recombination stage. After decoupling the energy density $\rho_\gamma \sim T^4$ of the primordial photons is redshifted according to the law

$$\frac{T(t)}{T_0} = \frac{a_0}{a(t)} = 1 + z. \tag{5.3}$$

One observes presently this cosmic microwave background (CMB) as a radiation with a black-body spectrum at temperature $T_0 = 2.725$ K or energy $kT_0 = 2.35 \times 10^{-4}$ eV.

Since the binding energy of the ground state of atomic hydrogen is $E_b = 13.6$ eV, one may expect that the energy kT_{rec} is of the same order. It is substantially smaller because of the smallness of the ratio of baryons to photons $\eta = n_b/n_\gamma \sim 5 \times 10^{-10}$. Indeed, according to the Saha equation, the fraction x of ionized atoms is given by

$$\frac{n_p n_e}{n_H n_\gamma} = \frac{x^2}{(1-x)} \eta = \frac{4.05 c^3}{\pi^2} \left(\frac{m_e}{2\pi kT} \right)^{3/2} e^{-E_b/kT} . \tag{5.4}$$

Hence, because $\eta \ll 1$, the ionized fraction x becomes negligible only for energies much smaller than E_b. A careful treatment gives $kT_{\text{rec}} \sim 0.26$ eV.

After recombination, the intergalactic medium remains neutral during a period often called the dark ages, until the first stars ignite and the first quasars are formed. The ultraviolet photons produced by these sources progressively re-ionize the universe. This period, called the re-ionization period, may be long since only small volumes around the first galaxies start to be ionized until these volumes coalesce to re-ionize the full intergalactic medium.

The measurement of the polarization of the CMB at large scale by WMAP [43] shows that the intergalactic medium is already significantly ionized at $z \sim 10 - 20$. On the other hand, the fact that it is only for $z \geq 6$ that the Gunn-Peterson effect[21] is observed, shows that one reaches the end of the re-ionization period at $z = 6$ [44].

5.1. Inflation

The inflation scenario has been proposed to solve a certain number of problems faced by the cosmology of the early universe [45]. Among these one may cite:

- the flatness problem

If the total energy density of the universe is presently close to the critical density, it should have been even more so in the primordial universe. Indeed, we can write (2.13) as

$$\frac{\rho_t(t)}{\rho_c(t)} - 1 = \frac{k}{\dot{a}^2} , \tag{5.5}$$

where $\rho_c(t) = 3H^2(t)/(8\pi G_N)$ and the total energy density ρ_t includes the vacuum energy. If we take for example the radiation-dominated era where $a(t) \sim t^{1/2}$, then (5.5) can be written as ($\dot{a} \sim t^{-1/2} \sim a^{-1}$)

$$\frac{\rho_t(t)}{\rho_c(t)} - 1 = \left[\frac{\rho_t(t_U)}{\rho_c(t_U)} - 1 \right] \left(\frac{a(t)}{a(t_U)} \right)^2 = \left[\frac{\rho_t(t_U)}{\rho_c(t_U)} - 1 \right] \left(\frac{kT_U}{kT} \right)^2 , \tag{5.6}$$

---

[21] a discontinuity in the level of the continuum observed in the spectrum of quasars, which results from Lyman α absorption from the *neutral* intergalactic medium.

where we have used the fact that $T(t) \propto a(t)^{-1}$ and we have taken as a reference point the epoch t_U of the grand unification phase transition. This means that, if the total energy density is close to the critical density at matter-radiation equality (as can be inferred from the present value), it must be even more so at the time of the grand unification phase transition: by a factor $(1\mathrm{eV}/10^{16}\mathrm{GeV})^2 \sim 10^{-50}$! This seems to put unbearable fine tuning on the behaviour of the very early universe. It tends to indicate that the evolution of the universe between nucleosynthesis and grand unification is somewhat different from what is expected in the standard big bang scenario.

• the horizon problem
Because of the speed of light, a photon which is emitted at the big bang ($t = 0$) will have travelled a finite distance at time t. The proper distance (4.13) measured at time t is

$$d_H(t) = a(t) \int_0^r \frac{dr'}{\sqrt{1 - kr'^2}} = a(t) \int_0^t \frac{cdt'}{a(t')} \tag{5.7}$$

$$= \frac{\ell_{H_0}}{1+z} \int_z^\infty \frac{dz}{\left[\Omega_M(1+z)^3 + \Omega_R(1+z)^4 + \Omega_k(1+z)^2 + \Omega_\Lambda\right]^{1/2}},$$

where, in the second line, we have used (4.14). This is the maximal distance that a photon (or any particle) could have travelled at time t since the big bang. It thus gives the maximum size of a causally connected patch of the universe and is known as the (particle) horizon at time t.

As an example we may compute the horizon at the time of the recombination i.e. $z_{\mathrm{rec}} \sim 1100$. We assume flat space ($k = 0$) and note that the integral on the second line of (5.7) is dominated by the lowest values of z: $z \sim z_{\mathrm{rec}}$ where the universe is still matter dominated. Hence

$$d_H(t_{\mathrm{rec}}) \sim \frac{2\ell_{H_0}}{\Omega_M^{1/2}(1+z_{\mathrm{rec}})^{3/2}} \sim 0.3 \text{ Mpc.} \tag{5.8}$$

We note that this is simply twice the Hubble radius at recombination $H^{-1}(z_{\mathrm{rec}})$, as can be checked from (4.8). This horizon distance is seen from an observer at present time under an angle

$$\theta_H(t_{\mathrm{rec}}) = \frac{d_H(t_{\mathrm{rec}})}{d_A(t_{\mathrm{rec}})}, \tag{5.9}$$

where the angular distance has been defined in (4.19):

$$d_A(t_{\mathrm{rec}}) \sim \frac{\ell_{H_0}}{1+z_{\mathrm{rec}}} \int_0^{z_{\mathrm{rec}}} \frac{dz}{\left[\Omega_M(1+z)^3 + \Omega_\Lambda\right]^{1/2}}. \tag{5.10}$$

This is easily computed for $\Omega_\Lambda \ll 1$ to be

$$\theta_H(t_{\text{rec}}) \sim (1 + z_{\text{rec}})^{-1/2} \sim 0.03 \text{ rad} \sim 2^o, \tag{5.11}$$

a result which gives the right order of magnitude for more general values of Ω_Λ as long as $\Omega_M + \Omega_\Lambda \sim 1$. This means that two points opposite on the sky are separated by about 100 horizons at the time of recombination. It is then extremely difficult to understand why the cosmic microwave background should be isotropic and homogeneous over the whole sky.

• the monopole problem
Monopoles occur whenever a simple gauge group is broken to a group with a $U(1)$ factor. This is precisely what happens in grand unified theories. In this case their mass is of order M_U/g^2 where g is the value of the coupling at grand unification. Because we are dealing with stable particles with a superheavy mass, there is a danger to overclose the universe, i.e. to have an energy density much larger than the critical density.. We then need some mechanism to dilute the relic density of monopoles.

Inflation provides a remarkably simple solution to these problems: it consists in a period of the evolution of the universe where the expansion is exponential. Indeed, if the energy density of the universe is dominated by the vacuum energy ρ_{vac}, then the Friedmann equation reads

$$H^2 = \frac{\dot{a}^2}{a^2} = \frac{\rho_{\text{vac}}}{3m_P^2}, \tag{5.12}$$

where $m_P \equiv (8\pi G_N)^{-1/2}$ is the reduced Planck mass. If $\rho_{\text{vac}} > 0$, this is readily solved as

$$a(t) = H_{\text{vac}}^{-1} e^{H_{\text{vac}}t} \quad \text{with} \quad H_{\text{vac}} \equiv \sqrt{\frac{\rho_{\text{vac}}}{3m_P^2}}. \tag{5.13}$$

Such a behaviour is in fact observed whenever the magnitude of the Hubble parameter changes slowly with time i.e. is such that $|\dot{H}| \ll H^2$.

Such a space was first proposed by de Sitter [46, 47] with very different motivations and is thus called de Sitter space.

Obviously a period of inflation will ease the horizon problem. Indeed, during inflation a photon may travel a proper distance (following (5.7))

$$d(t) = a(t) \int_0^t \frac{c\,dt'}{a(t')} = \frac{c}{H_{\text{vac}}} e^{H_{\text{vac}}t} \text{ for } H_{\text{vac}}t \gg 1. \tag{5.14}$$

It follows that a period of inflation extending from t_i to $t_f = t_i + \Delta_t$ contributes to the horizon size a value $c e^{H_{vac} \Delta_t} / H_{vac}$, which can be very large.

Similarly, a period of exponential expansion of the universe may solve the monopole problem by diluting the concentration of monopoles by a very large factor. It also dilutes any kind of matter. Indeed, a sufficiently long period of inflation "empties" the universe. However matter and radiation may be produced at the end of inflation by converting the energy stored in the vacuum. This conversion is known as reheating (because the temperature of the matter present in the initial stage of inflation behaves as $a^{-1}(t) \propto e^{-H_{vac}t}$, it is very cold at the end of inflation; the new matter produced is hotter). If the reheating temperature is lower than the scale of grand unification, monopoles are not thermally produced and remain very scarce.

Finally, it is not surprising that the universe comes out very flat after a period of exponential inflation. Indeed, the spatial curvature term in the Friedmann equation is then damped by a factor $a^{-2} \propto e^{-2H_{vac}\Delta_t}$. For example, a value $H_{vac} \Delta_t \sim 60$ (one refers to it as 60 e-foldings) would easily account for the huge factor 10^{50} of adjustment that we found earlier.

Most inflation models rely on the dynamics of a scalar field in its potential. Inflation occurs whenever the scalar field evolves slowly enough in a region where the potential energy is large. The set up necessary to realize this situation has evolved with time: from the initial proposition of Guth [45] where the field was trapped in a local minimum to "new inflation" with a plateau in the scalar potential, chaotic inflation [48,49] where the field is trapped at values much larger than the Planck scale and more recently hybrid inflation [50] with at least two scalar fields, one allowing an easy exit from the inflation period.

The equation of motion of a homogeneous scalar field $\phi(t)$ with potential $V(\phi)$ evolving in a Friedmann-Robertson-Walker universe is:

$$\ddot{\phi} + 3H\dot{\phi} = -V'(\phi). \tag{5.15}$$

where $V'(\phi) \equiv dV/d\phi$. The term $3H\dot{\phi}$ is a friction term due to the expansion. The corresponding energy density and pressure are:

$$\rho = \frac{1}{2}\dot{\phi}^2 + V(\phi), \tag{5.16}$$

$$p = \frac{1}{2}\dot{\phi}^2 - V(\phi). \tag{5.17}$$

We may note that the equation of conservation of energy $\dot{\rho} = -3H(p+\rho)$ takes here simply the form of the equation of motion (5.15). These equations should be complemented with the Friedmann equation (5.12).

When the field is slowly moving in its potential, the friction term dominates over the acceleration term in the equation of motion (5.15) which reads:

$$3H\dot{\phi} \simeq -V'(\phi). \tag{5.18}$$

The curvature term may then be neglected in the Friedmann equation (5.12) which gives

$$H^2 \simeq \frac{\rho}{3m_P^2} \simeq \frac{V}{3m_P^2}. \tag{5.19}$$

Then the equation of conservation $\dot{\rho} = -3H(p + \rho) = -3H\dot{\phi}^2$ simply gives

$$\dot{H} \simeq -\frac{\dot{\phi}^2}{2m_P^2}. \tag{5.20}$$

It is easy to see that the condition $|\dot{H}| \ll H^2$ amounts to $\dot{\phi}^2/2 \ll \rho/3 \sim V(\phi)/3$, i.e. a kinetic energy for the scalar field much smaller than its potential energy. Using (5.18) and (5.19), the latter condition then reads

$$\varepsilon \equiv \frac{1}{2}\left(\frac{m_P V'}{V}\right)^2 \ll 1. \tag{5.21}$$

The so-called slow roll regime is characterized by the two equations (5.18) and (5.19), as well as the condition (5.21). It is customary to introduce another small parameter:

$$\eta \equiv \frac{m_P^2 V''}{V} \ll 1, \tag{5.22}$$

which is easily seen to be a consequence of the previous equations[22].

An important quantity to be determined is the number of Hubble times elapsed during inflation. From some arbitrary time t to the time t_e marking the end of inflation (i.e. of the slow roll regime), this number is given by

$$N(t) = \int_t^{t_e} H(t)dt. \tag{5.23}$$

It gives the number of e-foldings undergone by the scale factor $a(t)$ during this period (see (5.13)). Since $dN = -H dt = -H d\phi/\dot{\phi}$, one obtains from (5.18) and (5.19)

$$N(\phi) = \int_{\phi_e}^{\phi} \frac{1}{m_P^2} \frac{V'}{V} d\phi. \tag{5.24}$$

---

[22]Differentiating (5.18), one obtains $\eta = \epsilon - \ddot{\phi}/(H\dot{\phi})$.

In the de Sitter phase i.e. in the phase of exponential growth of the cosmic scale factor, quantum fluctuations of the scalar field value are transmitted to the metric. Because the size of the horizon is fixed (to H^{-1}) in this phase, the comoving scale a/k associated with these fluctuations eventually outgrows the horizon, at which time the fluctuations become frozen. It is only much later when the universe has recovered a radiation or matter dominated regime that these scales reenter the horizon and evolve again. They have thus been protected from any type of evolution throughout most of the evolution of the universe (this is in particular the case for the fluctuations on a scale which reenters the horizon *now*). Fluctuations in the cosmic microwave background provide detailed information on the fluctuations of the metric. In particular, the observation by the COBE satellite of the largest scales puts a important constraint on inflationary models. Specifically, in terms of the scalar potential, this constraint known as COBE normalization, reads:

$$\frac{1}{m_P^3} \frac{V^{3/2}}{V'} = 5.3 \times 10^{-4}. \tag{5.25}$$

5.2. Inflation scenarios

Since the central prediction of inflation, namely that the total energy density of the universe is very close to the critical energy density ρ_c for which space is flat (i.e. $\Omega \equiv \rho/\rho_c \sim 1$), seems to be in good agreement with observation, any theory of fundamental interactions should provide an inflation scenario.

Such a scenario was first imagined in the context of the phase transition associated with grand unification [45]. There is thus an obvious connection with supersymmetry. Let us see indeed why supersymmetry provides a natural setting for inflation scenarios.

We recall that a standard scenario for inflation involves a scalar field ϕ evolving slowly in its potential $V(\phi)$, satisfying the conditions (5.21) and (5.22). Supersymmetry is a natural set up for discussing fundamental scalar fields. Moreover flat directions of the potential are a very general property of supersymmetric models.

Using the slowroll parameter introduced above, the COBE normalization condition can be written as

$$V^{1/4} \sim \varepsilon^{1/4} \, 6.7 \times 10^{16} \text{ GeV}. \tag{5.26}$$

In most of the models that we will be discussing, ε is very small. However as long as $\varepsilon \gg 10^{-52}$, we have $V^{1/4} \gg 1$ TeV. In other words, the typical scale associated with inflation is then much larger than the TeV, in which case it makes little sense to work outside a supersymmetric context.

As we just pointed out, one might expect that the presence of numerous flat directions may ease the search for an inflating potential. One possible difficulty arises from the condition (5.22) which may be written as a condition on the mass of the inflaton field

$$m^2 \ll H^2. \tag{5.27}$$

Any fundamental theory with a single dimensionful scale (such as string theory or just plain supergravity with a standard scenario of supersymmetry breaking) runs into the danger of having to fine tune parameters in order to satisfy this constraint. This is known as the η *problem*.

Since supersymmetric scalar potentials consist of F-terms and D-terms, the discussion of suitable potentials for inflation naturally follows this classification. As we will see in the following, they easily provide models for what is known as hybrid inflation which involves two directions in field space: one is slow-rolling whereas the other ensures the exit from inflation (and is fixed during slowroll).

F term inflation

Let us start with a simple illustrative model [51]. We consider two chiral supermultiplets of respective scalar components σ and χ with superpotential

$$W(\sigma, \chi) = \sigma \left(\lambda \psi^2 - \mu^2 \right). \tag{5.28}$$

Writing $|\sigma| \equiv \phi/\sqrt{2}$, one obtains for the scalar potential

$$V = 2\lambda^2 \phi^2 |\psi|^2 + \left| \lambda \psi^2 - \mu^2 \right|^2. \tag{5.29}$$

The global supersymmetric minimum is found for $\psi^2 = \mu^2$ and $\phi = 0$ but, for fixed ϕ, we may write the potential as ($\psi \equiv A + iB$)

$$V = \mu^4 + 2\lambda(\lambda\phi^2 - \mu^2)A^2 + 2\lambda(\lambda\phi^2 + \mu^2)B^2 + \lambda^2(A^2 + B^2)^2. \tag{5.30}$$

We conclude that, for $\phi^2 > \phi_c^2 \equiv \mu^2/\lambda$, there is a local minimum at $A = B = 0$ for which $V = \mu^4$. In other words, the ϕ direction is flat for $\phi > \phi_c$ with a non-vanishing potential energy. This may lead to inflation if one is trapped there. Since global supersymmetry is broken along this direction, one expects that loop corrections yield some slope which allows slowroll. Once ϕ reaches ϕ_c, ψ starts picking up a vacuum expectation value and one quickly falls into the global minimum.

This simple example of F-term hybrid inflation may easily be generalized. However F-term inflation suffers from a major drawback when one tries to consider it in the context of supergravity [51, 52]. We recall the form of the scalar potential in supergravity,

$$V = e^{K/m_P^2} \left[D_i W g^{i\bar{j}} D_{\bar{j}} \bar{W} - 3\frac{|W|^2}{m_P^2} \right] + \frac{g^2}{2} \mathrm{Re} f_{ab}^{-1} D^a D^b, \qquad (5.31)$$

where

$$D_i W = \frac{\partial W}{\partial \phi^i} + \frac{1}{m_P^2} \frac{\partial K}{\partial \phi^i} W, \qquad (5.32)$$

and

$$D^a = -\frac{\partial K}{\partial \phi^i} (t^a)^i{}_j \phi^j + \xi^a. \qquad (5.33)$$

Here ξ^a is a Fayet-Iliopoulos term which is present only in the case of a $U(1)$ symmetry.

In what follows the crucial role is played by the exponential factor e^{K/m_P^2} in front of the F-terms. Thus, inflation necessarily breaks supersymmetry. Let us assume for a moment that the inflation is dominated by some of the F-terms and that the D-terms are vanishing or negligible. Then the slow roll conditions (5.21) and (5.22) can be written as

$$\varepsilon = \frac{1}{2} \left(\frac{K_I}{m_P} + \cdots \right)^2 \ll 1, \quad \eta = K_{I\bar{I}} + \cdots \ll 1. \qquad (5.34)$$

Here the subscript I denotes a derivative with respect to the inflaton noted ϕ^I. The extra terms noted $\cdots$ in these expressions are typically of the same order as the ones written explicitly. Their precise value is model dependent. They might lead to cancellations but generically, this requires a fine tuning.

During inflation, unless a very special form is chosen, K_I is typically of the order of ϕ^I. Thus, in principle, one can satisfy the ε constraint if the inflation scenario is of the small field type ($\phi^I \ll 1$, see Section 5.1). Bu the η condition is more severe. The quantity $K_{I\bar{I}}$ stands in front of the kinetic term and therefore in the true vacuum it should be normalized to one. Then it is very unlikely to expect it to be much smaller during inflation. Indeed, this condition can be written as (5.27) since the mass of the inflaton m^2 receives a contribution $K_{I\bar{I}} V/m_P^2 \sim K_{I\bar{I}} H^2$.

These arguments indicate that it is not easy to implement F-type inflation in supergravity theories. All the solutions proposed involve specific non-minimal forms of the Kähler potential [52].

D term inflation

What is interesting about inflation supported by D-terms is that the problems discussed above can be automatically avoided because of the absence of a factor e^{K/m_P^2} in front of them. Indeed for inflation dominated by some of the D-terms the slow roll conditions can be easily satisfied.

Let us show how such a scenario can naturally emerge in a theory with a $U(1)$ gauge symmetry [53,54]. We first consider an example with global supersymmetry and a non-anomalous $U(1)$ symmetry. We introduce three chiral superfields ϕ_0, ϕ_+ and ϕ_- with charges equal to 0, +1 and −1 respectively. The superpotential has the form

$$W = \lambda \phi_0 \phi_+ \phi_-, \tag{5.35}$$

which can be justified by several choices of discrete or continuous symmetries and in particular by R-symmetry. The scalar potential in the global supersymmetry limit reads:

$$V = \lambda^2 |\phi_0|^2 \left(|\phi_-|^2 + |\phi_+|^2 \right) + \lambda^2 |\phi_+ \phi_-|^2 + \frac{g^2}{2} \left(|\phi_+|^2 - |\phi_-|^2 - \xi \right)^2, \tag{5.36}$$

where g is the gauge coupling and ξ is a Fayet-Iliopoulos D-term (which we choose to be positive). This system has a unique supersymmetric vacuum with broken gauge symmetry

$$\phi_0 = \phi_- = 0, \quad |\phi_+| = \sqrt{\xi}. \tag{5.37}$$

Minimizing the potential, for fixed values of ϕ_0, with respect to other fields, we find that for $|\phi_0| > \phi_c \equiv g\sqrt{\xi}/\lambda$, the minimum is at $\phi_+ = \phi_- = 0$. Thus, for $|\phi_0| > \phi_c$ and $\phi_+ = \phi_- = 0$ the tree level potential has a vanishing curvature in the ϕ_0 direction and large positive curvature in the remaining two directions ($m_\pm^2 = \lambda^2 |\phi_0|^2 \mp g^2 \xi$). Along the ϕ_0 direction ($|\phi_0| > \phi_c$, $\phi_+ = \phi_- = 0$), the tree level value of the potential remains constant: $V = g^2 \xi^2 / 2 \equiv V_0$. Thus ϕ_0 provides a natural candidate for the inflaton field.

Along the inflationary trajectory all the F-terms vanish and Universe is dominated by the D-term which splits the masses of the Fermi-Bose components in the ϕ_+ and ϕ_- superfields. Such splitting results in a one-loop effective potential. In the present case this potential can be easily evaluated and for large ϕ_0 it behaves as

$$V_{eff} = \frac{g^2}{2} \xi^2 \left(1 + \frac{g^2}{16\pi^2} \ln \frac{\lambda^2 |\phi_0|^2}{\Lambda^2} \right) \equiv V_0 \left(1 + \frac{C g^2}{8\pi^2} \ln \frac{\lambda \varphi}{\Lambda} \right), \tag{5.38}$$

where $\varphi \equiv |\phi_0|$ and $C \sim 1$.

Along this potential, the value of φ that leads to the right number $N \sim 50$ of e-foldings can be directly obtained from (5.24):

$$\frac{\varphi}{m_P} = \sqrt{\frac{NCg^2}{4\pi^2}}. \tag{5.39}$$

This is safely of order g in the model that we consider but might be dangerously close to 1 for models which yield a larger value for C (see below). The values of the slowroll parameters (5.21) and (5.22) are correspondingly

$$\varepsilon = \frac{Cg^2}{32N\pi^2}, \quad \eta = -\frac{1}{2N}, \tag{5.40}$$

which yields a spectral index

$$1 - n_S = \frac{1}{N}\left(1 + \frac{3Cg^2}{32\pi^2}\right). \tag{5.41}$$

Finally, the COBE normalisation (5.25) fixes the overall scale:

$$\xi^{1/2} \sim \left(\frac{C}{N}\right)^{1/4} \times 1.9 \; 10^{16} \text{GeV}. \tag{5.42}$$

Let us now consider the supergravity extension of our model. For definiteness we will assume canonical normalization for the gauge kinetic function f and the Kähler potential (i.e. $K = |\phi_-|^2 + |\phi_+|^2 + |\phi_0|^2$; note that this form maximizes the problems for the F-type inflation). The scalar potential reads

$$V = e^{(|\phi_-|^2 + |\phi_+|^2 + |\phi_0|^2)/m_P^2} \lambda^2 \left[|\phi_+\phi_-|^2 \left(1 + \frac{|\phi_0|^4}{m_P^4}\right) \right.$$

$$+ |\phi_+\phi_0|^2 \left(1 + \frac{|\phi_-|^4}{m_P^4}\right) + |\phi_-\phi_0|^2 \left(1 + \frac{|\phi_+|^4}{m_P^4}\right) - 3\frac{|\phi_+\phi_-\phi_0|^2}{M^2} \right]$$

$$+ \frac{g^2}{2}\left(|\phi_+|^2 - |\phi_-|^2 - \xi\right)^2. \tag{5.43}$$

Again for values of $|\phi_0| > \phi_c$, other fields than ϕ_0 vanish and the behaviour is much similar to the global supersymmetry case. The zero tree level curvature of the inflaton potential is not affected by the exponential factor in front of the first term since this term is vanishing during inflation. This solves the problems of the F-type inflation.

5.3. Dark energy

In this section, we assume that some unknown mechanism relaxes the vacuum energy to zero or to a very small value and will introduce some new dynamical component which accounts for the present observation regarding a late acceleration of the universe (see section 4.5 of Chapter 4. We thus try to identify a new component of the energy density with negative pressure:

$$p = w\rho, \quad w < 0. \tag{5.44}$$

For example, a network of light, nonintercommuting topological defects [55, 56] gives $w = -n/3$ where n is the dimension of the defect *i.e.* 1 for a string and 2 for a domain wall. The equation of state for a minimally coupled scalar field necessarily satisfies the condition $w \geq -1$.

Experimental data may constrain such a dynamical component, referred to in the literature as dark energy, just as it did with the cosmological constant. For example, in a spatially flat Universe with only matter and an unknown component X with equation of state $p_X = w_X \rho_X$, one obtains from (2.17) of Chapter 2 with $\rho = \rho_M + \rho_X$, $p = w_X \rho_X$ the following form for the deceleration parameter (compare with (4.10) of Chapter 4)

$$q_0 = \frac{\Omega_M}{2} + (1 + 3w_X)\frac{\Omega_X}{2}, \tag{5.45}$$

where $\Omega_X = \rho_X / \rho_c$. Supernovae results give a constraint on the parameter w_X.

Another important property of dark energy is that it does not appear to be clustered (just as a cosmological constant). Otherwise, its effects would have been detected locally, as for the case of dark matter.

A particularly interesting candidate in the context of fundamental theories is a scalar[23] field ϕ slowly evolving in its potential $V(\phi)$. Indeed, since the corresponding energy density and pressure are, for a minimally coupled scalar field,

$$\rho_\phi = \frac{1}{2}\dot{\phi}^2 + V(\phi), \tag{5.46}$$

$$p_\phi = \frac{1}{2}\dot{\phi}^2 - V(\phi), \tag{5.47}$$

we have

$$w_\phi = \frac{\frac{1}{2}\dot{\phi}^2 - V(\phi)}{\frac{1}{2}\dot{\phi}^2 + V(\phi)}. \tag{5.48}$$

---

[23] A vector field or any field which is not a Lorentz scalar must have settled down to a vanishing value. Otherwise, Lorentz invariance would be spontaneously broken.

If the kinetic energy is subdominant ($\dot{\phi}^2/2 \ll V(\phi)$), we clearly obtain $-1 \leq w_\phi \leq 0$.

A consequence is that the corresponding speed of sound $c_s = \delta p/\delta \rho$ is of the order of the speed of light: the scalar field pressure resists gravitational clustering.

We will see below that the scalar field must be extremely light. We therefore have two possible situations:

• a scalar potential slowly decreasing to zero as ϕ goes to infinity [57–59]. This is often referred to as *quintessence* or runaway qintessence.

• a very light field (pseudo-Goldstone boson) which is presently relaxing to its vacuum state [60].

In both cases one is relaxing to a position where the vacuum energy is zero. This is associated with our assumption that some unknown mechanism wipes the cosmological constant out. We discuss the two cases in turn.

Runaway quintessence

A runaway potential is frequently present in models where supersymmetry is dynamically broken. We have seen that supersymmetric theories are characterized by a scalar potential with many flat directions, *i.e.* directions ϕ in field space for which the potential vanishes. The corresponding degeneracy is lifted through dynamical supersymmetry breaking. In some instances (dilaton or compactification radius), the field expectation value $< \phi >$ actually provides the value of the strong interaction coupling. Then at infinite ϕ value, the coupling effectively goes to zero together with the supersymmetry breaking effects and the flat direction is restored: the potential decreases monotonically to zero as ϕ goes to infinity.

Let us take the example of an exponentially decreasing potential: it rather naturally arises in the context of string theories (through gaugino condensation in weakly coupled heterotic string theory or through fluxes in more modern version, moduli acquire this type of potential). More explicitly, we consider the following action

$$S = \int d^4x \sqrt{g} \left[-\frac{m_P^2}{2} R + \frac{1}{2} \partial^\mu \phi \partial_\mu \phi - V(\phi) \right], \tag{5.49}$$

which describes a real scalar field ϕ minimally coupled with gravity and the self-interactions of which are described by the potential:

$$V(\phi) = V_0 e^{-\lambda \phi/m_P}, \tag{5.50}$$

where V_0 is a positive constant.

The energy density and pressure stored in the scalar field are given by (5.46) and (5.47). We will assume that the background (matter and radiation) energy density ρ_B and pressure p_B obey a standard equation of state

$$p_B = w_B \rho_B. \tag{5.51}$$

If one neglects the spatial curvature ($k \sim 0$), the equation of motion for ϕ simply reads

$$\ddot{\phi} + 3H\dot{\phi} = -\frac{dV}{d\phi}, \tag{5.52}$$

with

$$H^2 = \frac{1}{3m_P^2}(\rho_B + \rho_\phi). \tag{5.53}$$

This can be rewritten as

$$\dot{\rho}_\phi = -3H\dot{\phi}^2. \tag{5.54}$$

We are looking for *scaling solutions i.e.* solutions where the ϕ energy density scales as a power of the cosmic scale factor: $\rho_\phi \propto a^{-n_\phi}$ or $\dot{\rho}_\phi/\rho_\phi = -n_\phi H$. In this case, one easily obtains from (5.46), (5.47) and (5.54) that the ϕ field obeys a standard equation of state

$$p_\phi = w_\phi \rho_\phi, \tag{5.55}$$

with

$$w_\phi = \frac{n_\phi}{3} - 1. \tag{5.56}$$

Hence

$$\rho_\phi \propto a^{-3(1+w_\phi)}. \tag{5.57}$$

If one can neglect the background energy ρ_B, then (5.53) yields a simple differential equation for $a(t)$ which is solved as:

$$a \propto t^{2/[3(1+w_\phi)]}. \tag{5.58}$$

Since $\dot{\phi}^2 = (1 + w_\phi)\rho_\phi \sim t^{-2}$, one deduces that ϕ varies logarithmically with time. One then easily obtains from (5.52,5.53) that

$$\phi = \phi_0 + \frac{2}{\lambda}m_P \ln(t/t_0). \tag{5.59}$$

and[24]

$$w_\phi = \frac{\lambda^2}{3} - 1, \tag{5.60}$$

It is clear from (5.60) that, for λ sufficiently small, the field ϕ can play the role of quintessence. We note that, even if we started with a small value ϕ_o, ϕ reaches a value of order m_p.

But the successes of the standard big-bang scenario indicate that clearly ρ_ϕ cannot have always dominated: it must have emerged from the background energy density ρ_B. Let us thus now consider the case where ρ_B dominates. It turns out that the solution just discussed with $\rho_\phi \gg \rho_B$ and (5.60) is a late time attractor [61] only if $\lambda^2 < 3(1 + w_B)$. If $\lambda^2 > 3(1 + w_B)$, the global attractor turns out to be a scaling solution [57, 62, 63] with the following properties:

$$\Omega_\phi \equiv \frac{\rho_\phi}{\rho_\phi + \rho_B} = \frac{3}{\lambda^2}(1 + w_B), \tag{5.61}$$

$$w_\phi = w_B. \tag{5.62}$$

The second equation (5.62) clearly indicates that this does not correspond to a dark energy solution (5.44). The semi-realistic models discussed earlier tend to give large values of λ and thus the latter scaling solution as an attractor. Moreover [64], on the observational side, the condition that ρ_ϕ should be subdominant during nucleosynthesis (in the radiation-dominated era) imposes to take rather large values of λ. Typically requiring $\rho_\phi/(\rho_\phi + \rho_B)$ to be then smaller than 0.2 imposes $\lambda^2 > 20$.

Ways to obtain a quintessence component have been proposed however. Let us sketch some of them in turn.

One is the notion of *tracker field* [65]. This idea also rests on the existence of scaling solutions of the equations of motion which play the role of late time attractors, as illustrated above. An example is provided by a scalar field described by the action (5.49) with a potential

$$V(\phi) = \lambda \frac{\Lambda^{4+\alpha}}{\phi^\alpha}, \tag{5.63}$$

with $\alpha > 0$. In the case where the background density dominates, one finds an attractor scaling solution [58, 66] $\phi \propto a^{3(1+w_B)/(2+\alpha)}$, $\rho_\phi \propto a^{-3\alpha(1+w_B)/(2+\alpha)}$. Thus ρ_ϕ decreases at a slower rate than the background density ($\rho_B \propto a^{-3(1+w_B)}$)

---

[24]under the condition $\lambda^2 \leq 6$ ($w_\phi \leq 1$ since $V(\phi) \geq 0$).

and tracks it until it becomes of the same order at a given value a_Q. We thus have:

$$\frac{\phi}{m_P} \sim \left(\frac{a}{a_Q}\right)^{3(1+w_B)/(2+\alpha)}, \tag{5.64}$$

$$\frac{\rho_\phi}{\rho_B} \sim \left(\frac{a}{a_Q}\right)^{6(1+w_B)/(2+\alpha)}. \tag{5.65}$$

One finds

$$w_\phi = -1 + \frac{\alpha(1+w_B)}{2+\alpha}. \tag{5.66}$$

Shortly after ϕ has reached for $a = a_Q$ a value of order m_P, it satisfies the standard slow roll conditions (Eqs. (5.21) and (5.22)) and therefore (5.66) provides a good approximation to the present value of w_ϕ. Thus, at the end of the matter-dominated era, this field may provide the quintessence component that we are looking for.

Two features are interesting in this respect. One is that this scaling solution is reached for rather general initial conditions, *i.e.* whether ρ_ϕ starts of the same order or much smaller than the background energy density [65]. Regarding the cosmic coincidence problem, it can be rephrased here as follows (since ϕ is of order m_P in this scenario): why is $V(m_P)$ of the order of the critical energy density ρ_c? It is thus the scale Λ which determines the time when the scalar field starts to emerge and the universe expansion reaccelerates. Indeed, using (5.65), the constraint reads:

$$\Lambda \sim \left(H_0^2 m_P^{2+\alpha}\right)^{1/(4+\alpha)}. \tag{5.67}$$

We may note that this gives for $\alpha = 2$, $\Lambda \sim 10$ MeV, not such an atypical scale for high energy physics.

Recently a model [67] has been proposed which goes one step further: the dynamical component, a scalar field, is called k-essence and the model is based on the property observed in string models that scalar kinetic terms may have a non-trivial structure. Tracking occurs only in the radiation-dominated era; a new attractor solution where quintessence acts as a cosmological constant is activated by the onset of matter domination.

Quintessential problems

However appealing, the quintessence idea is difficult to implement in the context of realistic models [68, 69]. The main problem lies in the fact that the quintessence field must be extremely weakly coupled to ordinary matter. This problem can take several forms:

• we have assumed until now that the quintessence potential monotonically decreases to zero at infinity. In realistic cases, this is difficult to achieve because the couplings of the field to ordinary matter generate higher order corrections that are increasing with larger field values, unless forbidden by a symmetry argument. For example, in the case of the potential (5.63), the generation of a correction term $\lambda_d\, m_p^{4-d}\phi^d$ puts in jeopardy the slowroll constraints on the quintessence field, unless very stringent constraints are imposed on the coupling λ_d. But one typically expects from supersymmetry breaking $\lambda_d \sim M_{SB}^4/m_p^4$ where M_{SB} is the supersymmetry breaking scale.

Similarly, because the *vev* of ϕ is of order m_p, one must take into account the full supergravity corrections. One may then argue [70] that this could put in jeopardy the positive definiteness of the scalar potential, a key property of the quintessence potential. This may point towards models where $\langle W \rangle = 0$ (but not its derivatives) or to no-scale type models: in the latter case, the presence of 3 moduli fields T^i with Kähler potential $K = -\sum_i \ln(T^i + \bar{T}^i)$ cancels the negative contribution $-3|W|^2$ in the supergravity potential.

• the quintessence field must be very light. If we return to our example in (5.63), $V''(m_P)$ provides an order of magnitude for the mass-squared of the quintessence component:

$$m_\phi \sim \Lambda \left(\frac{\Lambda}{m_P} \right)^{1+\alpha/2} \sim H_0 \sim 10^{-33} \text{ eV}. \qquad (5.68)$$

using (5.67). This might argue for a pseudo-Goldstone boson nature of the scalar field that plays the rôle of quintessence. This field must in any case be very weakly coupled to matter; otherwise its exchange would generate observable long range forces. Eötvös-type experiments put very severe constraints on such couplings.

• it is difficult to find a symmetry that would prevent any coupling of the form $\beta(\phi/m_P)^n F^{\mu\nu} F_{\mu\nu}$ to the gauge field kinetic term. Since the quintessence behavior is associated with time-dependent values of the field of order m_P, this would generate, in the absence of fine tuning, corrections of order one to the gauge coupling. But the time dependence of the fine structure constant for example is very strongly constrained: $|\dot{\alpha}/\alpha| < 5 \times 10^{-17} \text{yr}^{-1}$. This yields a limit [68]:

$$|\beta| \leq 10^{-6} \frac{m_P H_0}{<\dot{\phi}>}, \qquad (5.69)$$

where $<\dot{\phi}>$ is the average over the last 2×10^9 years.

Pseudo-Goldstone boson

There exists a class of models [60] very close in spirit to the case of runaway quintessence: they correspond to a situation where a scalar field has not yet reached its stable groundstate and is still evolving in its potential.

More specifically, let us consider a potential of the form:

$$V(\phi) = M^4 v\left(\frac{\phi}{f}\right),$$ (5.70)

where M is the overall scale, f is the vacuum expectation value $<\phi>$ and the function v is expected to have coefficients of order one. If we want the potential energy of the field (assumed to be close to its *vev* f) to give a substantial fraction of the energy density at present time, we must set

$$M^4 \sim \rho_c \sim H_0^2 m_P^2.$$ (5.71)

However, requiring that the evolution of the field ϕ around its minimum has been overdamped by the expansion of the Universe until recently imposes

$$m_\phi^2 = \frac{1}{2} V''(f) \sim \frac{M^4}{f^2} \leq H_0^2.$$ (5.72)

Let us note that this is again one of the slowroll conditions familiar to the inflation scenarios.

From (5.71) and (5.72), we conclude that f is of order m_P (as the value of the field ϕ in runaway quintessence) and that $M \sim 10^{-3}$ eV (not surprisingly, this is the scale Λ typical of the cosmological constant). As we have seen, the field ϕ must be very light: $m_\phi \sim h_0 \times 10^{-60} m_P \sim h_0 \times 10^{-33}$ eV. Such a small value is only natural in the context of an approximate symmetry: the field ϕ is then a pseudo-Goldstone boson. A typical example of such a field is provided by the string axion field. In this case, the potential simply reads:

$$V(\phi) = M^4 [1 + \cos(\phi/f)].$$ (5.73)

All the preceding shows that there is extreme fine tuning in the couplings of the quintessence field to matter, unless they are forbidden by some symmetry. This is somewhat reminiscent of the fine tuning associated with the cosmological constant. In fact, *the quintessence solution does not claim to solve the cosmological constant (vacuum energy) problem* described above. If we take the example of a supersymmetric theory, the dynamical cosmological constant provided by the quintessence component clearly does not provide enough amount of supersymmetry breaking to account for the mass difference between scalars (sfermions) and fermions (quarks and leptons): at least 100 GeV. There must be other sources of supersymmetry breaking and one must fine tune the parameters of the theory in order not to generate a vacuum energy that would completely drown ρ_ϕ.

However, the quintessence solution shows that, once this fundamental problem is solved, one can find explicit fundamental models that effectively provide the small amount of cosmological constant that seems required by experimental data.

Appendix A. Astrophysical constants and scales

Constants

The tradition in astrophysics is to use the CGS system. Whereas there are in some specific cases useful quantities to be defined (such as the parsec), centimeter and gram seem hardly relevant. We thus use here the international system. Note that $1\ kg.m^{-3} = 10^{-3}\ g.cm^{-3}$, $1\ J = 10^7\ erg$, $1\ W = 10^7\ erg.s^{-1}$.

Newtonian gravitational constant: $G_N = 6.6742 \times 10^{-11}\ \text{m}^3\ \text{kg}^{-1}\ \text{s}^{-2}$
$\alpha_G \equiv G_N m_p^2/(\hbar c) = 5.906 \times 10^{-39}$
Fine structure constant: $\alpha \equiv e^2/(4\pi\epsilon_0\hbar c) = 7.297 \times 10^{-3} = 1/137$
Classical electron radius: $r_e = e^2/(4\pi\epsilon_0 m_e c^2) = 2.817 \times 10^{-15}\ \text{m}$
Thomson cross section: $\sigma_T = 8\pi r_e^2/3 = 0.665\ \text{barn} = 0.665 \times 10^{-28}\ \text{m}^2$.
Boltzmann constant: $k_B = 1.380 \times 10^{-23}\ \text{J.K}^{-1} = 8.617 \times 10^{-5}\ \text{eV.K}^{-1}$.

Typical length scales

Astronomical unit (au) = Sun-Earth distance = 1.4960×10^{11} m
Parsec (au/arc sec): 1 pc = 3.262 light-year = 3.086×10^{16} m
Solar radius $R_\odot = 6.9598 \times 10^8$ m
Sun-galactic center distance: 10 kpc
Milky way galaxy disk radius (luminous matter): 15 kpc

Typical mass scales

Solar mass: $M_\odot = 1.989 \times 10^{30}$ kg
Milky Way galaxy mass: 4 to $10 \times 10^{11}\ M_\odot$

Typical luminosities

Solar luminosity: $L_\odot = 3.85 \times 10^{33}$ erg/s = 3.85×10^{26} W

Typical densities

Present mean density of the universe: $\rho_0 = 10^{-30}$ g.cm^{-3}
Interstellar medium: 10^{-25} g.cm^{-3}
Neutron star: 10^{15} g.cm^{-3}

Appendix B. Burning fuel: nuclear reactions in stars

Thermonuclear reactions

In order to obtain fusion of two particles of charge Z_1 and Z_2, one has to overcome the Coulomb barrier of height

$$E_{Coul} = \frac{Z_1 Z_2 e^2}{4\pi \epsilon_0 r_0} \sim Z_1 Z_2 \text{ MeV,} \tag{B.1}$$

where we have used

$$r_0 \sim A^{1/3} 1.44 \text{ fm} \quad \text{(nuclear radius).} \tag{B.2}$$

One expects that the thermal energy might help to overcome this barrier. If we take the example of the sun where $T \sim 10^7$ K at the centre, i.e. $k_B T \sim 10^3$ eV which is totally insufficient.

Soon after the tunnel effect was discovered by G. Gamow, R. Atkinson and F. Houtermans showed in 1929 that thermonuclear reactions provide the energy source for stars. Indeed, for energies $E < E_{Coul}$, there is the possibility of tunneling through the Coulomb barrier. The tunnelling probability reads approximately:

$$P_0 = p_0 E^{-1/2} e^{-2\pi \eta}, \quad \eta = \left(\frac{\mu}{2E}\right)^{1/2} Z_1 Z_2 \frac{e^2}{\hbar}, \tag{B.3}$$

where μ is the reduced mass[25]. This gives for $Z_1 Z_2 = 1$ and again $T = 10^7$ K, $P_0 \sim 10^{-20}$ for particles with average kinetic energy ($E \sim k_B T$). Since this decreases rapidly with $Z_1 Z_2$, this allows only the lightest nuclei to react.

Nuclear burning

One calls nuclear "burning" of a given element the thermonuclear fusion reactions by which this element is converted into heavier ones. Since we have seen

---

[25]Note that the probability is of order 1 for $\eta \sim 1$ i.e. $E \sim \alpha^2 \mu c^2$, if we disregard factors of order one. This is what we used for ϵ_ν in section 2.5 of Chapter 2.

that only light elements can be produced at first, this process is sequential leading to the conversion into heavier and heavier elements, until one reaches the element $A = 56$ i.e. ^{56}Fe, which corresponds to the maximum of the binding energy per nucleon, and thus to a particularly stable species. We thus follow the sequence of nuclear burnings with increasing A.

Hydrogen burning

Hydrogen burning produces helium ^{4}He, through the fusion of four ^{1}H. The gain in binding energy is 26.731 MeV, which represents a fraction 0.7% of the mass energy. This is approximately ten times more than any other fusion process, even though a fraction of this energy is taken away by neutrinos.

There are two main series of reactions which produce ^{4}He:

• the proton-proton (pp) chain was discovered by Bethe and Critchfield; it does not require the presence of heavier elements.

• the Carbon-Nitrogen-Oxygen (CNO) cycle was independently discovered by von Weizsäcker and Bethe; it requires the presence of heavier elements to catalyze the cycle.

The pp chain starts with the following reactions:

$$^1\text{H} + {}^1\text{H} \rightarrow {}^2\text{H} + e^+ + \nu$$
$$^2\text{H} + {}^1\text{H} \rightarrow {}^3\text{He} + \gamma, \qquad \text{(B.4)}$$

followed by

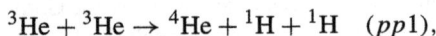

$$^3\text{He} + {}^3\text{He} \rightarrow {}^4\text{He} + {}^1\text{H} + {}^1\text{H} \quad (pp1),$$

or by

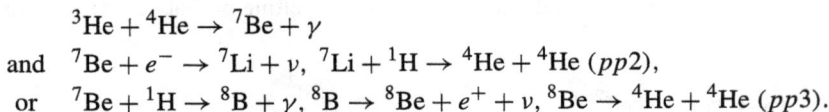

$$^3\text{He} + {}^4\text{He} \rightarrow {}^7\text{Be} + \gamma$$
$$\text{and} \quad ^7\text{Be} + e^- \rightarrow {}^7\text{Li} + \nu, \ {}^7\text{Li} + {}^1\text{H} \rightarrow {}^4\text{He} + {}^4\text{He} \ (pp2),$$
$$\text{or} \quad ^7\text{Be} + {}^1\text{H} \rightarrow {}^8\text{B} + \gamma, \ {}^8\text{B} \rightarrow {}^8\text{Be} + e^+ + \nu, \ {}^8\text{Be} \rightarrow {}^4\text{He} + {}^4\text{He} \ (pp3).$$

The energies released which are available for stellar matter are $Q = 26.20$ MeV (*pp1*), $Q = 25.67$ MeV (*pp2*), $Q = 19.20$ MeV (*pp3*). The rest of the energy is taken away by neutrinos[26].

---

[26]These neutrinos were detected with great care in the case of the Sun to try to understand the "enigma" of solar neutrinos. Note that the so-called *pp* neutrinos (associated with (B.4)) have a continuous spectrum as normal in beta decay: $0 < E_\nu < 0.42$ MeV; similarly for the ^{8}B neutrinos: $0.81 \text{ MeV} < E_\nu < 14.06$ MeV. On the other hand, the ^{7}Be neutrinos have a discrete spectrum: $E_\nu = 0.861$ MeV or $E_\nu = 0.383$ MeV. To be complete, one should add a second chain of reactions (*pep*) which occurs only in 0.04% of cases (and has thus been neglected here): it starts with ^{1}H $+ e^- + {}^1\text{H} \rightarrow {}^2\text{H} + \nu$ and the corresponding *pep* neutrino has $E_\nu = 1.44$ MeV. As has now been established, the deficit of solar neutrinos observed on earth has nothing to do with the solar model but is due to oscillations associated with the non-zero neutrino masses.

The sequence of reactions associated with the CNO cycle consists of the CN cycle:

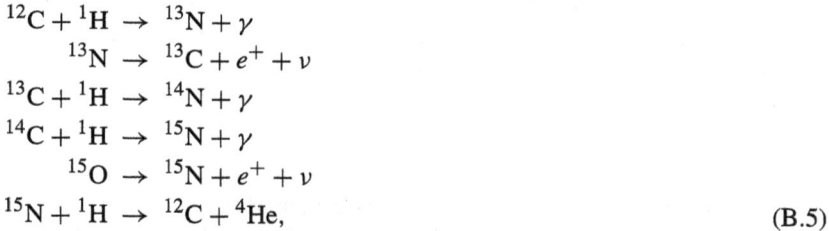

$$^{12}C + {}^1H \rightarrow {}^{13}N + \gamma$$
$$^{13}N \rightarrow {}^{13}C + e^+ + \nu$$
$$^{13}C + {}^1H \rightarrow {}^{14}N + \gamma$$
$$^{14}C + {}^1H \rightarrow {}^{15}N + \gamma$$
$$^{15}O \rightarrow {}^{15}N + e^+ + \nu$$
$$^{15}N + {}^1H \rightarrow {}^{12}C + {}^4He, \tag{B.5}$$

which is completed by the following sequence (10^4 times rarer) which involves ^{15}O:

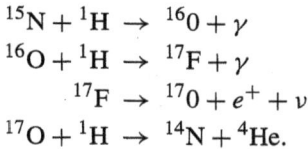

$$^{15}N + {}^1H \rightarrow {}^{16}O + \gamma$$
$$^{16}O + {}^1H \rightarrow {}^{17}F + \gamma$$
$$^{17}F \rightarrow {}^{17}O + e^+ + \nu$$
$$^{17}O + {}^1H \rightarrow {}^{14}N + {}^4He.$$

The whole cycle is completed by restoring the heavier elements C, N and O at the end. The energy gain of the whole cycle is 24.97 MeV. At around $T \sim 1.4 \times 10^7$ K, the CNO cycle starts to dominate over the pp chain.

Helium burning

Helium burning requires $T > 10^8$ K. The key reaction is the formation of ^{12}C through the triple alpha (3α) reaction:

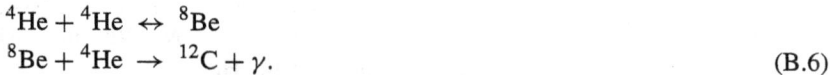

$$^4He + {}^4He \leftrightarrow {}^8Be$$
$$^8Be + {}^4He \rightarrow {}^{12}C + \gamma. \tag{B.6}$$

The net energy release per ^{12}C nucleus is $Q = 7.275$ MeV.

Once sufficient ^{12}C abundance has been achieved, further α capture leads to the formation of heavier elements:

$$^{12}C + {}^4He \rightarrow {}^{16}O + \gamma$$
$$^{16}O + {}^4He \rightarrow {}^{20}Ne + \gamma \tag{B.7}$$

. . .

Heavier elements

With enough ^{12}C and ^{16}O produced at the core of the star by helium burning, carbon burning starts when T reaches 5 to 10×10^8 K. Typically, two ^{12}C produce an excited ^{24}Mg nucleus which decays through various channels:

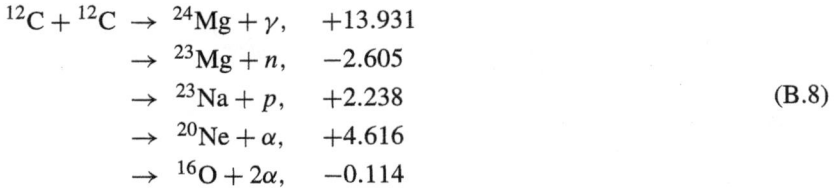

$$
\begin{aligned}
^{12}C + {}^{12}C &\rightarrow {}^{24}Mg + \gamma, & +13.931 \\
&\rightarrow {}^{23}Mg + n, & -2.605 \\
&\rightarrow {}^{23}Na + p, & +2.238 \\
&\rightarrow {}^{20}Ne + \alpha, & +4.616 \\
&\rightarrow {}^{16}O + 2\alpha, & -0.114
\end{aligned}
\tag{B.8}
$$

where the last column gives the energy release Q in MeV.

Oxygen burning starts at $T > 10^9$ K because of the higher Coulomb barrier:

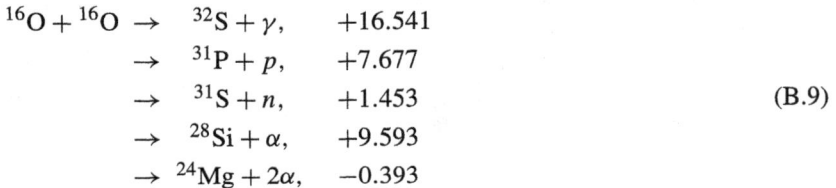

$$
\begin{aligned}
^{16}O + {}^{16}O &\rightarrow {}^{32}S + \gamma, & +16.541 \\
&\rightarrow {}^{31}P + p, & +7.677 \\
&\rightarrow {}^{31}S + n, & +1.453 \\
&\rightarrow {}^{28}Si + \alpha, & +9.593 \\
&\rightarrow {}^{24}Mg + 2\alpha, & -0.393
\end{aligned}
\tag{B.9}
$$

For temperatures higher than 3×10^9 K, heavier nuclei build up, such as ^{27}Al and ^{24}mg, up to ^{56}Fe, which corresponds, as we already stressed, to a maximum of the binding energy per nucleon.

Fe and beyond

Higher elements are thus not produced by burning, since the binding energy per nucleon starts decreasing after ^{56}Fe. However, in situations where $N = Z$ is imposed by initial conditions, the dominant nucleus is $^{56}_{28}$Ni, since it has the highest binding energy per nucleon of all nuclei with $Z = N$. With increasing temperature, this shifts to ^{54}Fe $+ 2p$ and finally to $14\,^4$He.

Neutron capture:

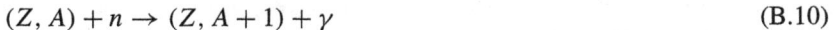

$$
(Z, A) + n \rightarrow (Z, A+1) + \gamma
\tag{B.10}
$$

may lead to a heavier element if $(Z, A+1)$ is unstable under beta decay:

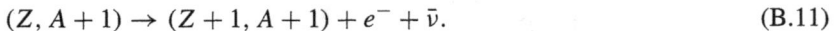

$$
(Z, A+1) \rightarrow (Z+1, A+1) + e^- + \bar{\nu}.
\tag{B.11}
$$

References

[1] A. Pais, *Subtle is the Lord... The science and life of Albert Einstein.* Oxford University Press, 1982.
[2] K. Schwarzschild, Sitzungsber. Preuss. Akad. Wiss. (1916) 189.

[3] S. Weinberg, *Gravitation and Cosmology*. J. Wiley and sons, 1972.

[4] T. Padmanabhan, *Theoretical astrophysics*, vol. I: Astrophysical processes. Cambridge University Press, 2000.

[5] T. Padmanabhan, *Theoretical astrophysics*, vol. II: Stars and stellar systems. Cambridge University Press, 2001.

[6] R. J. Michell, Phil. Trans. R. Soc. (London) **74** (1784) 35.

[7] P. S. Laplace, *Le système du monde*, vol. II. Paris, 1795.

[8] J. R. Oppenheimer and H. Snyder, Phys. Rev. **56** (1939) 455.

[9] R. P. Kerr, Phys. Rev. Lett. **11** (1963) 237.

[10] E. T. Newman, E. Couch, K. Chinnapared, A. Exton, A. Prakash, and R. Torrence, J. Math. Phys. **6** (1965) 918.

[11] E. Poisson and W. Israel, Phys. Rev. Lett. **63** (1989) 1663.

[12] R. Penrose, Revista del Nuovo Cimento **1** (1969) 252.

[13] M. D. Kruskal, Phys. Rev. **119** (1960) 1743.

[14] E. F. Taylor and J. A. Wheeler, *Exploring black holes: an introduction to general relativity*. Addison Wesley Longman, 2000.

[15] B. F. Schutz, *A first course in general relativity*. Cambridge University Press, 1985.

[16] M. C. Begelman, R. D. Blandford, and M. J. Rees, Rev. Mod. Phys. **56** (1984) 255.

[17] M. Miyoshi et al., Nature **373** (1995) 127.

[18] T. Ott et al., ESO Messenger **111** (2003) 1.

[19] G. Bélanger, ApJ **601** (2004) L163.

[20] F. Aharonian et al., [H.E.S.S. Collaboration], Astron. Astrophys. **425** (2004) L13 [astro-ph/0408145].

[21] W. Baade and F. Zwicky, Phys. Rev. **45** (1934) 138.

[22] P. Hoeflich, C. Gerardy, E. Linder, and H. Marion, astro-ph/0301334.

[23] R. W. Klebesadel, I. B. Strong, and R. A. Olson, ApJ **182** (1973) L85.

[24] G. Cusumano et al., astro-ph/0509737 (submitted to Nature).

[25] N. Kawai et al., astro-ph/0512052 (submitted to Nature).

[26] S. E. Woosley, ApJ **405** (1993) 273.

[27] A. I. MacFadyen and S. E. Woosley, ApJ **524** (1999) 262 [astro-ph/9810274].

[28] T. Galama, Nature **395** (1998) 670.

[29] J. Hjorth, Nature **423** (2003) 847 [astro-ph/0306347].

[30] R. A. Hulse and J. H. Taylor, ApJ **195** (1975) L51.

[31] J. H. Taylor and J. M. Weisberg, ApJ **345** (1989) 434.

[32] D. N. Spergel et al., Astrophys. J. Supp. **148** (2003) 175.

[33] D. J. Einsenstein et al., ApJ **633** (2005) 560.

[34] A. Blanchard, M. Douspis, M. Rowan-Robinson, and S. Sarkar, astro-ph/0512085.

[35] S. cosmology project, Astrophys. J. **517** (1999) 565.

[36] A. G. Riess et al., Astron. J. **116** (1998) 1009.

[37] Z. G. Dai, E. W. Liang, and D. Xu, ApJ **612** (2004) L101 [astro-ph/0407497].

[38] G. Ghirlanda, G. Ghisellini, and D. Lazzati, ApJ **616** (2004) 331 [astro-ph/0405602].

[39] G. Ghirlanda, G. Ghisellini, D. Lazzati, and C. Firmani, ApJ **613** (2004) L13 [astro-ph/0408350].

[40] A. S. Friedman and J. S. Bloom, ApJ **627** (2005) 1 [astro-ph/0407497].

[41] D. E. Holz and S. A. Hughes, astro-ph/0212218.

[42] S. Mattig, Astron. Nachr. **284** (1958) 109.

[43] A. Kogut et al., ApJS **148** (2003) 161.

[44] S. G. Djorgovski, S. Castro, D. Stern, and A. A. Mahabal, ApJ **560** (2001) L5.

[45] A. H. Guth, Phys. Rev. D **23** (1981) 347.

[46] W. de Sitter, Proc. Kon. Ned. Akad. Wer. **19** (1917) 217.

[47] W. de Sitter, Proc. Kon. Ned. Akad. Wer. **20** (1917) 229.

[48] A. D. Linde, Phys. Lett. B **129** (1983) 177.

[49] A. Albrecht and P. J. Steinhardt, Phys. Rev. Lett. **48** (1982) 1220.

[50] A. D. Linde, Phys. Lett. B **259** (1991) 38.

[51] E. J. Copeland, A. R. Liddle, D. Lyth, E. D. Stewart, and D. Wands, Phys. Rev. D **49** (1994) 6410.

[52] E. D. Stewart, Phys. Rev. D **51** (1995) 6847.

[53] P. Binétruy and G. Dvali, Phys. Lett. B **388** (1996) 241.

[54] E. Halyo, Phys. Lett. B **387** (1996) 43.

[55] A. Vilenkin, Phys. Rev. Lett. **53** (1984) 1016.

[56] D. Spergel and U. Pen, Astrophys. J. **491** (1997) L67.

[57] C. Wetterich, Nucl. Phys. B **302** (1988) 668.

[58] B. Ratra and P. J. E. Peebles, Phys. Rev. D **37** (1988) 3406.

[59] R. R. Caldwell, R. Dave, and P. J. Steinhardt, Phys. Rev. Lett. **75** (1995) 2077.

[60] J. Frieman, C. Hill, A. Stebbins, and I. Waga, Phys. Rev.Lett. **75** (1995) 2077.

[61] J. Halliwell, Phys. Lett. B **185** (1987) 341.

[62] E. J. Copeland, A. R. Liddle, and D. Wands, Phys. Rev. D **57** (1998) 4686.

[63] P. Ferreira and M. Joyce, Phys. Rev. Lett. **79** (1997) 4740.

[64] P. Ferreira and M. Joyce, Phys. Rev. D **58** (1998) 023503.

[65] I. Zlatev, L. Wang, and P. J. Steinhardt, Phys. Rev. Lett. **82** (1999) 896.

[66] P. J. E. Peebles and B. Ratra, Astrophys. J. **325** (1988) L17.

[67] C. Armendariz-Pico, V. Mukhanov, and P. J. Steinhardt, Phys. Rev. Lett. **85** (2000) 4438.

[68] S. M. Carroll, Phys. Rev. Lett. **81** (1998) 3067.

[69] C. Kolda and D. Lyth, Phys. Lett. B **458** (1999) 197.

[70] P. Brax and J. Martin, Phys. Lett. B **468** (1999) 40.

Course 10

ULTRA-HIGH ENERGY COSMIC RAYS

Peter Tinyakov

Service de Physique Théorique, Université Libre de Bruxelles, CP225, blv. du Triomphe,
B-1050 Bruxelles, Belgium

D. Kazakov, S. Lavignac and J. Dalibard, eds.
Les Houches, Session LXXXIV, 2005
Particle Physics Beyond the Standard Model
© *2006 Elsevier B.V. All rights reserved*

Contents

1. Introduction

The purpose of these lectures is to provide a introduction into the subject for those who, like the author, have a particle-physics rather than astrophysics or cosmic-ray-physics background. Accordingly, I will not assume any special knowledge and will try to provide the background information where necessary. Some parts of these lectures are oversimplified, both for the lack of time and expertise. I will try to compensate it with references to the specialized literature. A very useful information can also be found in the recent reviews [1–5].

This is perhaps a right time to get interested in cosmic rays, particularly in their ultra-high-energy part. One reason is that the existing data contain an apparent problem: the spectrum of cosmic rays seems to extend beyond the theoretically predicted cutoff. Second, the new experimental data are expected in a few years. There is one more reason: once again, as in the past, cosmic rays may become a major source of information for the physics of elementary particles, at least in the highest-energy domain.

There is a long history of relationship between these two areas of physics. Long before the accelerator era, many elementary particles were discovered in cosmic rays. Here are some of the key events. The very first (indirect) observation of cosmic rays goes back to 1912 when Victor Hess in his balloon experiment discovered the "penetrating radiation" coming from space by observing the discharge of an electrometer. The more rapid discharge at high latitudes was attributed to a "radiation" coming from space, the "cosmic rays". It took nearly 20 years to confirm this guess by a direct evidence. In 1929, Dmitry Skobelzyn observed first cosmic ray tracks using a newly invent cloud chamber. In 1933 in his cloud chamber Carl Anderson discovered a positively charged counterpart of the electron later called positron. In 1937 Seth Neddermeyer and Carl Anderson discover the muon. It followed in 1947 by the discoveries of the pion by Cecil Powell at Bristol and of "V-particles" (kaons) by George Rochester and Clifford Butler in Manchester. The experimental particle physics began.

One more event directly related to the development of the cosmic ray physics itself and to the subject of these lectures must be mentioned. In 1938 Pierre Auger discovered that detectors put far apart record simultaneous signals. He has attributed these events to "extensive air showers", showers of the secondary particles produced in collisions of the primary high-energy particles with the air atoms. He estimated the energy of primary particles to be $\sim 10^6$ GeV, much higher than any energy observed at that time. We know now that particles of energies up to 10^{11} GeV exist in cosmic rays. This is this last part of the spectrum starting at $E \gtrsim 10^{19}$ eV which will be the main focus of these lectures. We will refer to it as ultra-high energy cosmic rays, or UHECR.

2. Background information

2.1. The observed spectrum

The measured spectrum of cosmic rays ranges from 1 GeV to $\sim 10^{11}$ GeV. The highest-energy event observed has the energy of 3×10^{11} GeV ~ 50 J. Over this enormous range of energies the slope is nearly constant and equals roughly -3. Thus, for each decade up in energy the integral flux drops by 2 orders of magnitude. While the total flux of particles with energies higher than 100 GeV is about 1 particle per m^2 per second, at $E > 10^{10}$ GeV the overall flux is less than 1 particle per km^2 per year.

There are two well-established features in the cosmic ray spectrum, the "knee" between 10^6 GeV and 10^7 GeV where the spectrum softens, and the "ankle" between 10^9 GeV and 10^{10} GeV where the spectrum hardens. Their origin is debated.

The primary particles do not penetrate to the ground. Their direct observation is therefore only possible in space and, for a realistic size of the space detector of 1m^2, is limited to energies $E \lesssim 10^5$ GeV. The composition is directly measured only up to these energies. The flux consists of 85% protons, 2% electrons, 12% alpha particles and 1% heavier nuclei.

2.2. Detection techniques

At the high-energy end of the spectrum where the flux is low, the only way to detect cosmic ray particles is to use the atmosphere as the detector. After the first interaction with the air the cascade develops, as shown schematically in Fig. 1.

The penetration depth of the incident particle is determined by the probability distribution

$$P_{\text{pass}} = \exp\left\{-\sigma \int n(x)dx\right\},$$

where x is a distance along the trajectory, σ is the cross section and $n(x)$ is the air density. Note that the penetration depth depends on the quantity $\int n dx$ which, when multiplied by the mass of scattering centers, is the "column density". Its dimension is g/cm^2. This quantity is a convenient "coordinate" along the shower

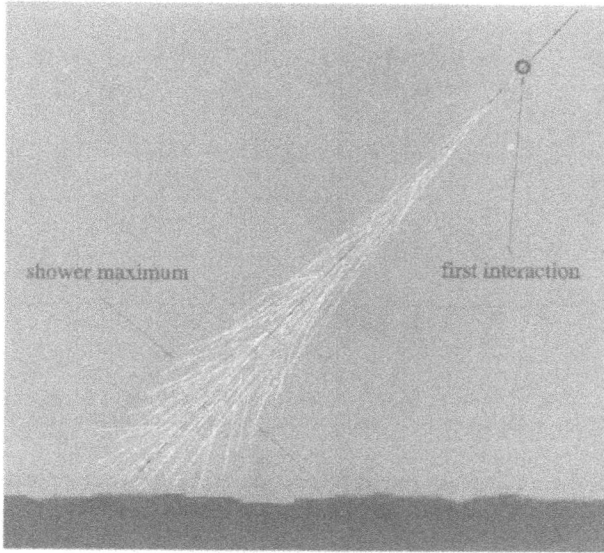

Fig. 1. Airshower produced by an ultra-high energy particle in the atmosphere.

as it takes into account the variable density of the atmosphere. When viewed vertically the atmosphere has a column density of $\sim 10^3$ g/cm^2.

Observationally, the first interaction of high-energy cosmic rays with the air happens at about 100 g/cm^2. One deduces from this fact the estimate for the interaction cross section,

$$\sigma \sim 0.04 \text{ barn} \sim (100 \text{ MeV})^{-2}.$$

This is roughly consistent with theoretical expectations for both primary hadrons and photons.

The full development of a shower produced by a particle with energy of the order of 10^{10} GeV or higher requires about 800 g/cm^2. An example of the shower profile is shown in Fig. 2 where the number of particles is plotted versus column density for the highest-energy Fly's Eye event. Note that the number of particles in the shower maximum is very large, of order 10^{11} or larger.

The particles in the shower are concentrated near the core. Their distribution, the lateral distribution function, is approximately given by the formula [7]

$$S(r) = S_0 \left(\frac{r}{r_0}\right)^{-\alpha} \left(1 + \frac{r}{r_0}\right)^{\alpha - \eta},$$

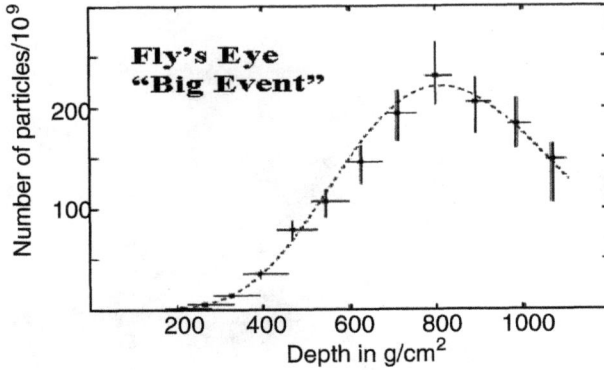

Fig. 2. Longitudinal development of the shower (the number of particles in the shower versus column density as counted from the top of the atmosphere) for the highest-energy Fly's Eye event [6].

where $S(r)$ is the particle density as a function of the distance r from the core, r_0 is the so-called Moliere unit determined by the characteristics of the atmosphere (typically $r_0 \sim 100$ m), while α and η are the exponents determined empirically. Their values vary from experiment to experiment; α is of order 1.2–1.3, while η depends on the zenith angle and varies roughly from 3.9 for vertical showers to 3 for the inclined ones.

There are two basic techniques to detect the airshowers: ground array and fluorescent telescope. In the first case the particle detectors are arranged in the array on the ground level. The typical spacing between detectors is of order 1 km. Scintillator (AGASA) or water Cherenkov (Pierre Auger experiment) detectors are used; they may be supplemented with the muon detectors as in the Yakutsk experiment. The energy is estimated from the particle density at a certain distance from the core of a shower. This distance is chosen as to minimize the fluctuations which are due to the fluctuations of the depth of the first interaction point. The particle density at 600 m from the core is typically used, although 1000 m is adopted in the Pierre Auger experiment.

The fluorescent method consists in observing, by a specially designed telescope, the fluorescent light emitted by air during the shower propagation. The active element of the telescope consists in the set of phototubes arranged in an array resembling the fly's eye. When the shower front propagates in the atmosphere, its image moves across the array of phototubes. The counts in phototubes and time information allow, in principle, to restore the shower direction and the total amount of light emitted, which in turn can be converted into the energy of the incident particle.

Despite totally different principle of operation, the two methods have similar accuracy. The typical error in energy determination is estimated to be $\lesssim 30\%$,

Table 1

The existing UHECR experiments. The number of events listed refers to published events only.

experiment	year	type	angular resolution	published events at $E > 10^{19}$ eV
Volcano Ranch	1959	scintillator	3°	27
Haverah Park	1968	water Cherenkov	3°	87
SUGAR	1968	μ-detectors	4°	175
Yakutsk	1974	scintillator + atmospheric Cherenkov + μ-detectors	3°	163
Fly's Eye	1981	fluorescent	asymmetric	1
AGASA	1990	scintillator	1.8°	775
HiRes mono	1994	fluorescent	asymmetric	–
HiRes stereo	1994	fluorescent	0.6°	271
Pierre Auger	being built	fluorescent + scintillator	0.6°	–

while the arrival direction is typically determined with the accuracy of 2–3 degrees. The combination of 2 fluorescent detectors (in the stereo mode) or hybrid detectors consisting of a fluorescent telescope and a ground array have better angular resolution which may be smaller than 1°. Table 1 lists main UHECR experiments and some of their characteristics.

2.3. The arrival direction distribution

The large-angle distribution of the arrival directions of UHECR is compatible with uniform [8]. One should keep in mind, however, that most of the existing experiments (except SUGAR) are situated in the Northern hemisphere. Their acceptance area thus covers only the Northern half of the sky. In particular, the Galactic center is not seen. Also, the number of events at highest energies is rather limited. Thus, the conclusion about isotropy of the cosmic ray flux should be taken with care.

At small angles (comparable with the experimental angular resolution) the clustering at energies $E > 4 \times 10^{19}$ eV has been found at about 4σ level [8–10]. If confirmed, this may indicate the existence of sufficiently bright point sources. We postpone the discussion of this issue till Section 5.

3. Propagation of UHECR

3.1. General remarks

One generally assumes that the highest-energy cosmic rays are of extragalactic origin. Firstly, their space distribution is isotropic which is difficult to arrange if they would be of the Galactic origin. Secondly, starting with energy of order 10^{18} eV protons are no longer confined in the Galactic magnetic field. Thus, both propagation in the Galaxy and in the extragalactic medium have to be considered.

The propagation of UHECR primary particles may be affected by their scattering off matter in the Universe and by deflections in the magnetic fields (if the particles are charged). Scattering off protons may be neglected regardless the nature of the primary particles. Indeed, we know from observations of airshowers that the cross section of the scattering of the primary particles on nucleons is of the order of $\sigma \sim (100 \text{MeV})^2$. For the average baryon density in the Universe $n \sim 10^{-6}$ cm^{-3} the mean free path is

$$\lambda_{\text{EG}} = \frac{1}{n\sigma} \sim 10^4 \text{ Gpc},$$

which is 3 orders of magnitude larger than the size of the Universe. Likewise in the Galaxy, taking the density of baryons to be 0.1cm^{-3}, we find

$$\lambda_{\text{G}} \sim 100 \text{ Mpc},$$

which is much larger than the Galactic size. Thus scattering on baryons can be neglected.

Scattering on photons does depend on both the nature of primary particles and on the photon background. The dominant and best known contribution to the latter is the CMB. The total photon background in the relevant range is shown in Fig. 3. The central peak represents the CMB. The parts other that CMB are model-dependent (for the more detailed discussion see [11]).

3.2. Pion photoproduction

In the case when the primary particles are protons (the most conservative assumption) their cross section on CMB photons is known. It is peaked near the threshold of pion production, which in the laboratory frame corresponds to the energy

$$E \sim \frac{m_\pi (m_p + m_\pi/2)}{2\omega_\gamma} \sim 10^{20} \text{ eV}.$$

It is important that the cross section of pion photoproduction in the relevant center-of-mass energy range is measured experimentally. Its dependence on energy is shown in Fig. 4. The corresponding interaction and energy attenuation lengths are represented in Fig. 5.

Pion photoproduction on CMB leads to the rapid energy loss by protons, as is shown in Fig. 6. For realistic assumptions about the particle energy at the source, the proton which is observed with the energy of 10^{20} eV had to be emitted relatively nearby, certainly within 100 Mpc.

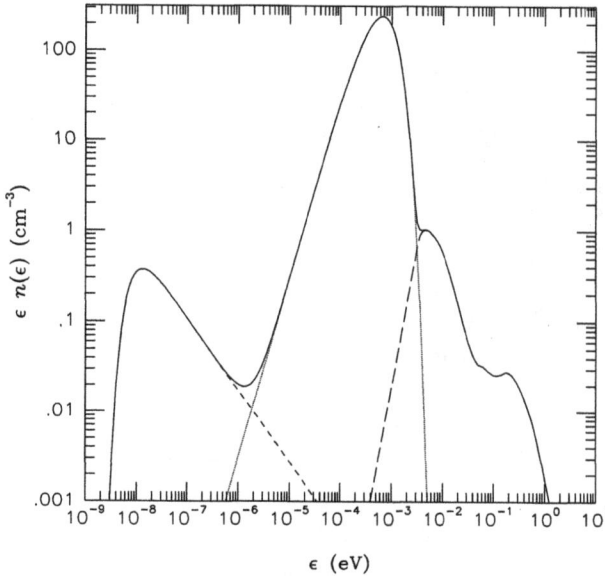

Fig. 3. Extragalactic photon background [11]. The central peak is the CMB.

3.3. The GZK cutoff

The rapid decrease of the proton attenuation length just below 10^{20} eV (see Fig. 5) leads to the sharp cutoff in the proton spectrum, the celebrated Greisen-Zatsepin-Kuzmin effect [12, 13]. Indeed, for the uniform distribution of sources the observed flux is proportional to the linear size R of the region from which it is collected,

$$\text{flux} \propto \int_0^R \frac{1}{r^2} r^2 dr \propto R.$$

At energies (a few) $\times 10^{19}$ eV this region is roughly the whole Universe, while at energies larger than 10^{20} eV it shrinks by a factor of ~ 100 to $R_{GZK} \sim 50$ Mpc. Thus, there should be a sharp drop in the flux by the same factor ~ 100. This is confirmed by detailed calculations, as is shown in Fig. 7.

The assumption of the uniform distribution of sources is essential for the argument. If there is a local over-density of sources, the cutoff is less pronounced [14], although for realistic distributions it cannot be eliminated [15]. The under-density leads to a sharper cutoff. Even when the source distribution

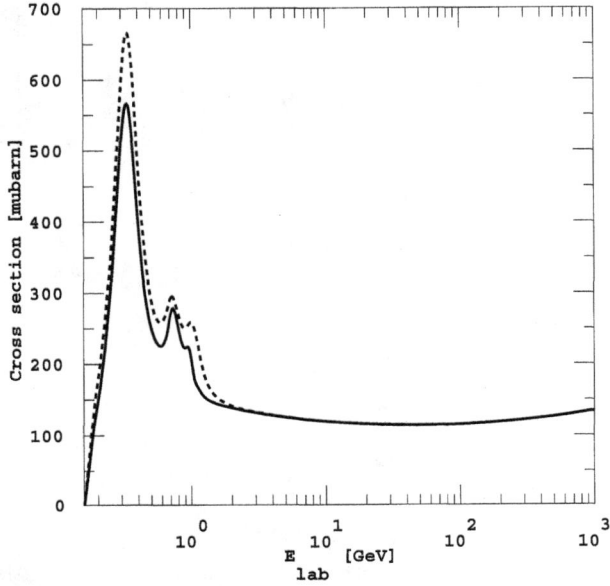

Fig. 4. Pion photoproduction cross section on proton (solid line) and neutron (dashed line).

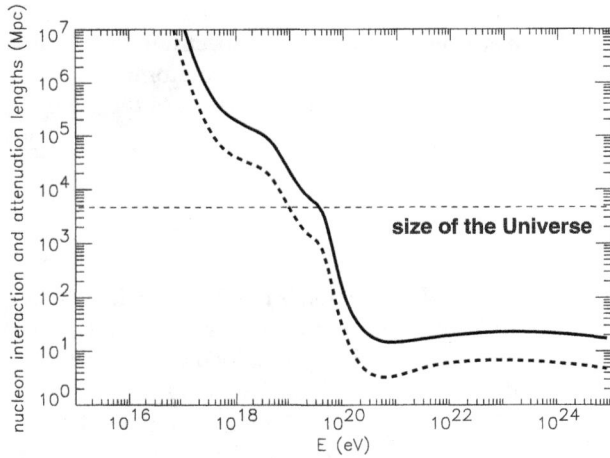

Fig. 5. Proton interaction (dashed line) and attenuation (solid line) lengths on the CMB photons [11].

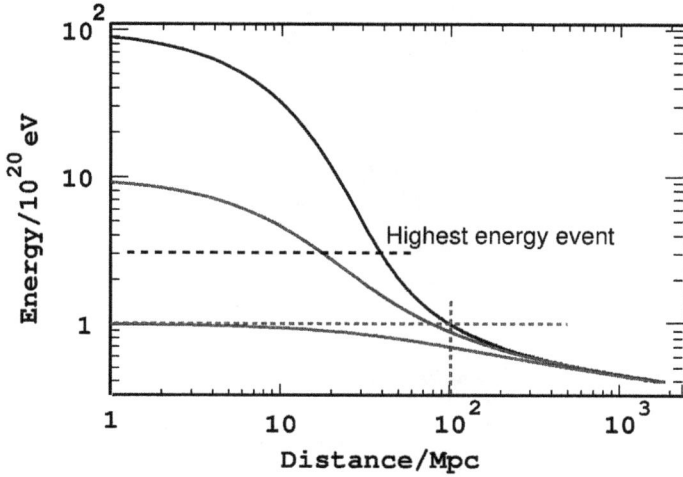

Fig. 6. Energy of a proton versus propagation distance. Dashed line shows the highest-energy event observed.

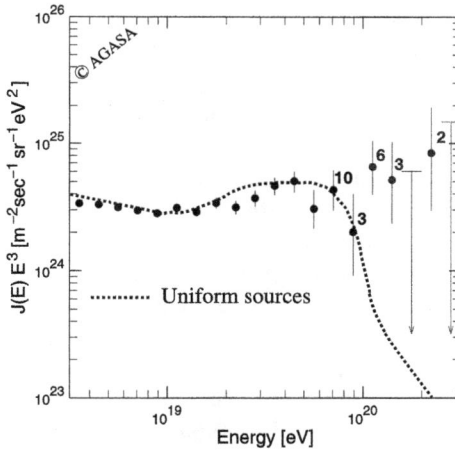

Fig. 7. The expected proton spectrum (dashed line) in the case of the uniform distribution of sources. Dots with error bars represent the actual measurement by AGASA.

P. Tinyakov

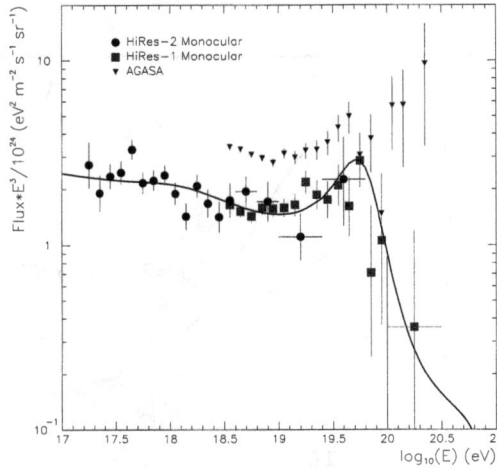

Fig. 8. The comparison between the AGASA and HiRes measurements of the UHECR spectrum as presented by the HiRes collaboration [16]. Circles – HiRes 2 mono; squares – HiRes 1 mono; triangles – AGASA.

is uniform *in average* the picture is different if the source-to-source distance is comparable to R_{GZK}. The cutoff is sharper at larger source separations.

The measurement of the spectrum at its highest-energy end is not conclusive at the moment. The AGASA claims the absence of the cutoff (see Fig. 7), while the HiRes collaboration insists on the spectrum which is compatible with the cutoff. Taken at face value, the difference between the two experiments is at the level of 4σ. The comparison between the results is shown in Fig. 8.

Note that the two experiments use different methods of energy estimation: AGASA is a ground array, while HiRes is a fluorescent telescope. The fact that they do not agree even in the range $10^{18} - 10^{19}$ eV where the number of observed events is large suggests that something is missing in our understanding of the energy estimation methods. Both techniques contain potentially subtle points. In the case of the ground array one has to rely heavily on the Monte-Carlo simulations of the showers and hadron interaction models which include the extrapolation of the measured cross sections into the high-energy domain (the center-of-mass energy in the UHECR-air collision is ~ 100 TeV). In the case of the fluorescent telescope technique the uncertainties are related to the light yield which is not directly measured, and to the atmospheric conditions on which it might depend.

The existing disagreement may be resolved by the Pierre Auger experiment which uses the hybrid technique. Observation of the same showers by the two

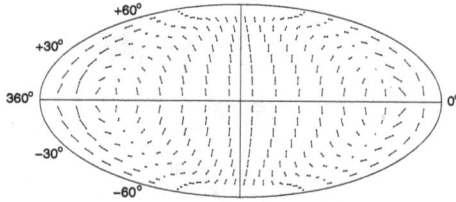

Fig. 9. The map of deflections of a proton with energy 4×10^{19} eV in the Galactic magnetic field of a spiral structure.

different methods will eliminate all the effects introduced by the selection procedures and related to strong fluctuations of parameters of individual showers. If these effects are indeed the reason of disagreement, this will allow to cross-calibrate the two methods. If the reason is different, even hybrid-detector measurements may not be conclusive.

3.4. Deflections in the magnetic fields

If UHECR primary particles are charged they will be deflected in the Galactic and extragalactic magnetic fields. One should distinguish the cases of regular and random deflections. The following expressions represent the estimates in these two cases:

$$\theta_{reg} \sim 0.52° \, q \left(\frac{E}{10^{20} \text{ eV}}\right)^{-1} \left(\frac{R}{1 \text{ kpc}}\right) \left(\frac{B_\perp}{10^{-6} \text{ G}}\right),$$

$$\theta_{rand} \sim 1.8° \, q \left(\frac{E}{10^{20} \text{ eV}}\right)^{-1} \left(\frac{l_c R}{50 \text{ Mpc}^2}\right)^{1/2} \left(\frac{B}{10^{-9} \text{ G}}\right).$$

Here q is the particle charge ($q = 1$ for protons), R is the propagation distance, $B_\perp$ is the projection of the regular magnetic field onto the plane perpendicular to the line of sight, l_c is the correlation length of the random magnetic field and B is its mean square value.

The Galactic magnetic field is thought to have both regular and random components of the comparable strength of order few μG. The deflections due to the random component are negligible in most of the directions [17]. The deflections due to the regular part may reach the value of $5 - 10°$ depending on energy and the direction on the sky [18–21]. The Galactic magnetic fields are not very well known; for a particular spiral model [19] the deflection map of a proton with energy 4×10^{19} eV is represented on Fig. 9.

The extragalactic magnetic field is known even worse. It is usually thought to be smaller than 10^{-9} G [22], however larger fields may not be excluded [23].

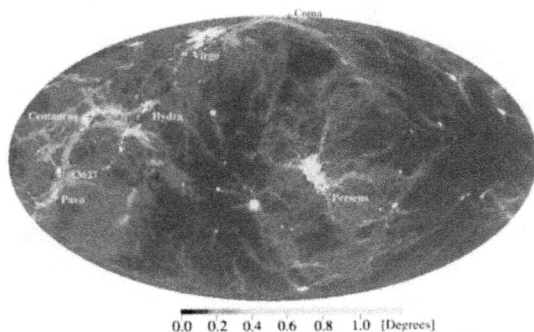

0.0 0.2 0.4 0.6 0.8 1.0 [Degrees]

Fig. 10. The map of deflections of a proton with energy 4×10^{19} eV in the extragalactic magnetic field obtained in the numerical simulation of the structure formation [25].

There have been attempts to clarify this issue theoretically by means of the simulation of the structure formation [24–26]. The idea is to introduce the seed magnetic field at the initial stage of the simulation and to track its evolution during structure formation. The seed field should be normalized so as to obtain the observed strength of the magnetic field in clusters of galaxies. With the map of the extragalactic magnetic field thus obtained one may then simulate the deflections of UHECR on their way to Earth. The results the simulation [25] are shown in Fig. 10. According to these results, the deflections of a 4×10^{19} eV proton in the extragalactic magnetic field is negligible everywhere except small regions in the direction of the nearby galaxy clusters. If correct this indicates that charged particle astronomy may be possible at highest energies. A different simulation [26] gives qualitatively larger deflections. This controversy is not resolved at present.

3.5. Summary: the UHECR puzzle

The spectrum of the highest-energy cosmic rays extends beyond 10^{20} eV. Their chemical composition is unknown, however they interact strongly (with hadronic cross section) in the atmosphere. The large-scale distribution of arrival directions is compatible with uniform, which suggests extragalactic origin. At highest energies the observed events should point back to their sources. This raises a number of questions (which constitute the UHECR puzzle):

• How particles of such a high energy can be produced?

• Why there is (apparently) no cut-off in the spectrum?

• Why there are no (obvious) candidates for sources behind highest-energy cosmic ray events?

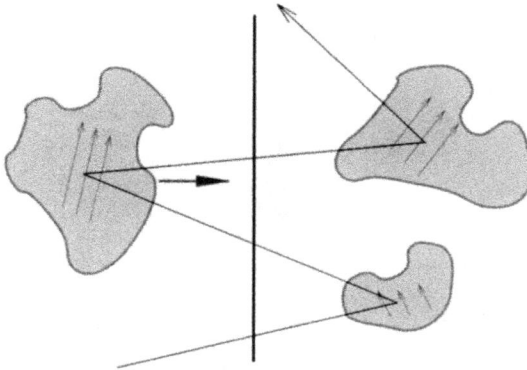

Fig. 11. Schematic picture of the Fermi acceleration mechanism. A particle bouncing off the magnetic field impurities frozen in two media which have a relative velocity toward each other gets accelerated like a tennis ball bouncing in a contracting space between two moving walls.

4. Ideas on the market

There exists a large number of models attempting to solve one or more problems related to UHECR — the acceleration to highest energies, propagation without attenuation, isotropic distribution over the sky. There is no, however, the "standard" or strongly preferable one. All the existing models are not completely. Rather, they typically attempt to address one of the issues at the expense of the others. The purpose of this section is not, however, to criticize them but to provide an overview of the ideas involved.

4.1. Particle acceleration

The key question concerning UHECR is their production. There are two ways to obtain high-energy particles: by the acceleration of the low-energy ones, and in decays of sufficiently massive progenitors. Correspondingly, there are two types of the mechanisms of UHECR production, the "bottom-up" and the "top-down". The "bottom-up" mechanisms can in turn be divided into stochastic and direct acceleration.

Stochastic acceleration. The general idea of the stochastic acceleration is illustrated in Fig. 11. In various astrophysical environments one generically has the situation when two media are moving toward each other creating a shock wave as an interface separating them. A particle flying across the shock may be reflected back from the impurities of the magnetic field which are frozen in the

P. Tinyakov

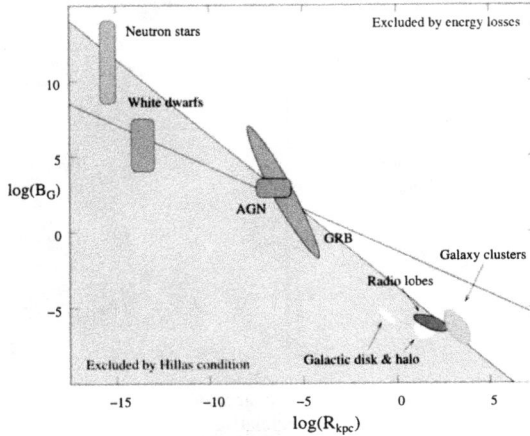

Fig. 12. The $B - R$ plot showing various astrophysical accelerators. The grey areas correspond to exclusion regions for $E > 10^{20}$ eV.

medium. The reflection may occur again, and so on until the particle leaves the shock region. Clearly, particles which bounce in this manner several times get accelerated, acquiring part of the kinetic energy of the bulk motion of the media. The mechanism is in fact the same as the one heating the gas molecules when the gas is compressed.

There are many variations of this mechanism (see Refs. [27–31] for more details). They all have two common features. Firstly, they generically predict the power-law spectrum. This is an important argument in favor of this type of mechanisms since, as we have seen above, the observed spectrum maintains the power-law shape over the energy range extending by more than 10 orders of magnitude.

Secondly, all these mechanisms require the possibility of *magnetic reflection* of the accelerated particles from the medium. Since the magnetic field impurities must be smaller in size than the acceleration site, and the Larmor radius of the accelerated particle in turn has to be smaller than the magnetic field impurities, there is a limit on the maximum energy of the accelerated particles which depends on the size of the acceleration site R and the value of the magnetic field B [32],

$$E_{\max} < eBR, \tag{4.1}$$

where e is the particle charge. This is known as the "Hillas condition". It can be visualized as the exclusion region on the $B - R$ plane, Fig. 12. An additional constraint arises if one takes into account the losses during the acceleration process.

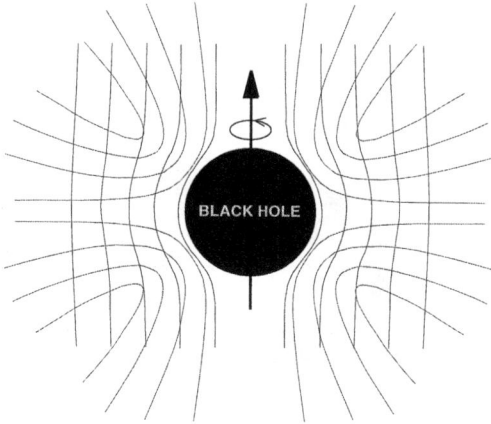

Fig. 13. The magnetic and electric field lines near the horizon of a rotating black hole embedded in the uniform magnetic field.

Note that there exist losses intrinsic to the acceleration process, namely the synchrotron radiation of particles at the time of their reflection off the magnetic field impurities. These losses also depend on the magnetic field and the size of the acceleration site and produce an additional exclusion region on the $B - R$ plane. For the acceleration of particles to energies higher than 10^{20} eV the resulting allowed region is rather small (white region on Fig. 12): there are very few objects potentially capable of accelerating to such a high energy.

Direct acceleration. The possibility of the "direct" particle acceleration is best illustrated by the following simple model. Consider the black hole embedded in a uniform magnetic field. The latter may be thought of, for instance, as produced by the accretion disc around the black hole (the accretion itself we will neglect and assume that there are no particles around the black hole). In the vicinity of the black hole, at the distance of several Schwarzshild radii, the magnetic field may be approximately considered as uniform. There exists an exact solution to the Einstein-Maxwell equations describing this system [33]. Near the horizon the magnetic field lines get "dragged" by the gravitational field of the rotating black hole; this produces the electric field capable of accelerating particles. The filed line configuration is shown schematically on Fig. 13.

One may naively think that the direct acceleration allows to attain much higher energies than the stochastic one. This is not the case. In fact, the direct acceleration is subject to severe constraints. One of them is given by the same eq. (4.1) as above, as it is clear from the fact that near the horizon of the black hole $E \lesssim B$.

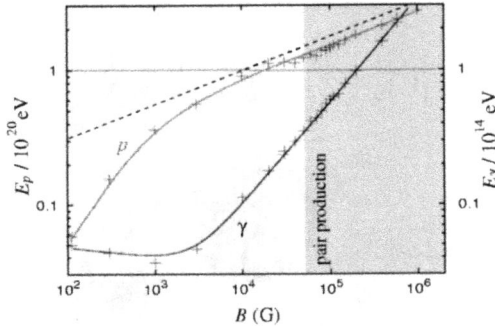

Fig. 14. The dependence of maximum energy of accelerated protons (left scale) and the average energy of accompanying curvature photons (right scale) on the magnetic field near the maximally rotating black hole of $10^{10} M_\odot$ [34]. Crosses represent the results of the numerical simulation.

There are also constraints on maximum energy which follow from the consideration of losses during the acceleration (curvature radiation). Finally, there are the consistency constraints which arise from the requirement that the acceleration region is not polluted by the $e^+ e^-$-pair production. The situation for the black hole of the mass $10^{10} M_\odot$ is summarized in Fig. 14. One can see that the acceleration to 10^{20} eV is only marginally possible. Interestingly, even in the most favorable conditions the direct acceleration is rather inefficient: only a tiny fraction of $< 0.1\%$ of the energy extracted from the black hole goes into kinetic energy of accelerated particles. The bulk of energy is emitted in the form of the electromagnetic radiation, predominantly in the TeV range. If such accelerators exist in Nature they should be very bright TeV sources.

4.2. AGN models

This is a most conventional and most thoroughly studied class of models of UHECR production [29, 30]. Active Galactic Nuclei (AGN) are the most powerful objects in the Universe. At the same time they are very common. Most of the galaxies are thought to contain a supermassive black hole in the center; some of these black holes produce jets which can be as long as 1 Mpc. The shocks in jets and at the points where the jets hit the surrounding medium and stop (so-called hot spots) are sites of stochastic acceleration. The very fact that the jets and hot spots are visible in radio and x-rays implies that particle acceleration takes place, although does not mean that the maximum energy is sufficiently high to explain UHECR. The vicinity of the supermassive black hole may provide conditions for the direct acceleration.

One of the important advantages of the AGN models (at least of those based on stochastic acceleration) is that they naturally lead to the observed power-law spectrum. Another one is that only known physics (known particles and interactions) is required for these models to work.

The disadvantage is that this is precisely the class of models which created the UHECR "paradox". The uniform distribution of AGNs in space implies the GZK cutoff. Thus, the AGN models in their conventional form cannot explain the apparent excess of events beyond the cutoff energy. The problem with particular sources is also quite hard: the AGNs are very well visible, and it is difficult to explain why there are no obvious AGN candidates behind the observed events.

4.3. *Gamma-ray bursts as sources of UHECR*

The gamma-ray burst (GRB) models [35–40] is an attempt to solve two problems of the AGN scenario: the acceleration to highest energies and the absence of the obvious source candidates. In the GRB models the acceleration of UHECR is attributed to GRBs. The advantage of the GRBs as acceleration sites of UHECR is that the former are associated with ultra-relativistic jets which have Lorentz factors as large as 100 or even 1000. Thus, the requirements on the parameters characterizing the particle-acceleration site are correspondingly weaker.

GRBs are transient events with the time scale of up to several minutes. If one assumes that UHECR particles are protons, then there must be a random delay in their arrival times due to the deflections in the intergalactic magnetic fields. We have discussed typical deflection angles above; let us take $1°$ at 50 Mpc for the estimate. The time delay (with respect to gamma rays) which corresponds to such a deflection is of the order of 10^6 yr. Thus the bulk of the cosmic rays produced in a given gamma-ray burst would come million years after the burst itself. This obviously solves the problem of sources. There is a prediction which follows immediately from this picture: since individual bursts are spread in time over millions of years, the flux corresponding to each one is tiny, and no burst should produce more than one event. In other words, all observed events must have come from different locations. If the clustering found in the UHECR events (see Section 5.1) is real, the GRB model is ruled out. There exists also a Galactic version of the GRB model [40].

4.4. *Young rapidly rotating neutron stars (magnetars)*

If extreme values of parameters are assumed, one may argue that young rapidly rotating neutron stars can accelerate iron nuclei to sufficiently high energies [41]. The heavy nuclei, in particular iron, are favored for acceleration: they have larger charge (the acceleration is faster) and mass (the synchrotron losses are suppressed). They are not viable candidates for extragalactic UHECR as they

Cen A

Fig. 15. The "teapot" model of UHECR: cosmic rays produced by one or a few nearby sources are isotropized by a strong Mpc-scale extragalactic magnetic field.

disintegrate easily into individual protons. However, if produced in the Galaxy they may well be primary UHECR particles.

The generic problem with the Galactic origin of UHECR, no matter what is the mechanism, is the anisotropy. Galaxy matter is distributed non-isotropically with respect to us, the proof is the Milky Way on the sky. Similarly, the UHECR flux tends to be anisotropic. In the case of heavy nuclei one may try to use the Galactic magnetic field to save the situation. Rather extreme values of the field are then required [41] to isotropize UHECR.

4.5. Strong magnetic fields

A somewhat similar idea is that the *extragalactic* magnetic field may be strong enough (at the level of several μGauss) to confine UHE protons within several Mps around the Galaxy [23, 42]. These protons may be produced by one or several powerful AGNs in the nearby region, see Fig. 15. As a possible candidate Cen A has been proposed. This may solve the source problem and smoothen the cutoff. Interestingly, the detectable flux of UHE neutrons from the source (Cen A) is predicted in this model [43].

4.6. Z-burst model

An interesting class of models called Z-burst models [44, 45] is based on the fact that neutrinos are massive. Neutrinos by themselves cannot be particles producing air showers for their cross section with the air is too small. So, they have to be converted to hadrons and photons. The mechanism proposed is the scattering on (anti)neutrino background at the Z resonance, as shown schematically in Fig. 16.

The neutrino and Z-boson masses determine unambiguously the incident neutrino energy,

$$E_{res} \sim \frac{4\text{eV}}{m_\nu} \times 10^{21} \text{ eV}.$$

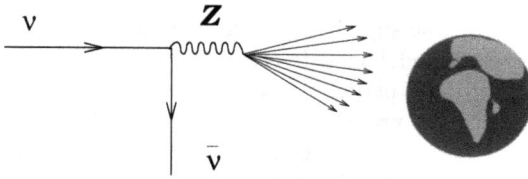

Fig. 16. The schematic picture of the Z-burst model.

This raises severe astrophysical problems, even though neutrinos are penetrating particles and hypothetical sources may be very far away. The origin of these problems may be explained as follows (see Ref. [46] and references therein). In order to produce neutrinos with energies exceeding 10^{21} eV one has to accelerate protons to at least 10^{22} eV. This is by itself very difficult if possible at all. What is even more problematic is to produce sufficient flux: the UHE neutrino flux required is more than 3 orders of magnitude larger than the observed flux of UHECR. Any source producing such neutrinos would also radiate in the electromagnetic channel. Regardless of the emission spectrum most part of this electromagnetic radiation will soften (with total energy approximately preserved) until it reaches the GeV-TeV region where the Universe becomes transparent to photona. There exists an experimental bound on the diffuse gamma-ray flux in this energy range set by the EGRET satellite [47] which the Z-burst models generically overshoot [46, 48–50].

4.7. Exotic primary particles

The existence of the GZK cutoff depends crucially on the Δ-resonance. Therefore, it is specific to protons as primary particles. There were several proposals to replace protons by hypothetical particles in order to eliminate the cutoff or shift it to higher energies [51–56]. The requirements on such particles are quite strict: in order to produce airshowers similar to the observed ones they have to interact strongly in the atmosphere, and their mass cannot be too large — should be lighter than roughly ~ 50 GeV. The candidates proposed include, for instance, light SUSY hadrons like uds-gluino bound state, glieballino, bottom squark containing hadrons, the axion-like particle. The generic problem with these scenarios is that the new particle has to escape observations at the accelerators, which is difficult in view of their large cross section and small mass. Only marginal window in parameter space still remains open, if any.

4.8. Superheavy dark matter

Let us move to the "top-down" scenarios. The most viable one is the decay of the superheavy dark matter [57–59]. One assumes that (a fraction of) the dark matter is composed of the superheavy relic particles which can (very slowly) decay into the standard-model ones. In order to produce UHECR of the observed energies the mass of these particles has to be in the range $10^{13} - 10^{15}$ GeV. These particles have to be primordial, and thus their lifetime has to lie within $10^{10} - 10^{22}$ yr (the lower limit is the age of the Universe, while the upper limit arises from the requirement that these particles do not overclose the Universe). It is an interesting theoretical problem to explain such a long but finite lifetime of heavy particles. This can be achieved, e.g., by the instanton-type mechanism [58], or by the quantum gravity effects [57].

Despite the fact that the dark matter distribution in the Universe is uniform, this model can explain the absence of the GZK cutoff. The point is that there is an overdensity of dark matter in the halo of our Galaxy. It is convenient to divide the observed flux of UHECR into Galactic and extragalactic components. Only the extragalactic contribution exhibits the GZK cutoff. The relative strength of these two components is comparable. Indeed, as discussed above, the flux is proportional to the linear size of the region from which it is collected. The relevant size is $\sim 10^5$ times smaller in the case of the galactic contribution compared to the extragalactic one. At the same time the dark matter overdensity in The Galaxy is about 10^5. The two factors roughly compensate each other. So, at the cutoff energy the extragalactic contribution dies away, and the flux drops by a factor of 2 instead of 100 for the truly uniform distribution. Although measurable in principle, such a drop is difficult to see because the spectrum is falling off very rapidly with energy.

The large Galactic contribution to the total UHECR flux leads to a distinct signature, the anisotropy toward the Galactic center [60, 61]. The value of the anisotropy depends on the halo model, as is shown on Fig. 17. For the harmonic analysis of the anisotropy see Refs. [61, 62]. Another important signature of the superheavy dark matter models is the domination of photons over baryons in the UHECR flux. It should be noted that the photon fraction in the UHECR flux is constrained experimentally [63].

A difficult problem of the superheavy dark matter scenario is the fine tuning of the flux which is required to explain the observed smooth continuation of the spectrum around $10^{19} - 10^{20}$ eV.

4.9. Superheavy dark matter + AGNs

The AGN model explains well the UHECR events at energies up to (a few) $\times$ 10^{19} eV, the problem being the apparent continuation of the spectrum beyond this point [64, 65]. One may try to combine this model with the superheavy dark matter scenario in order to explain the whole spectrum including the highest-energy events. The combined model produces quite a good fit to the AGASA

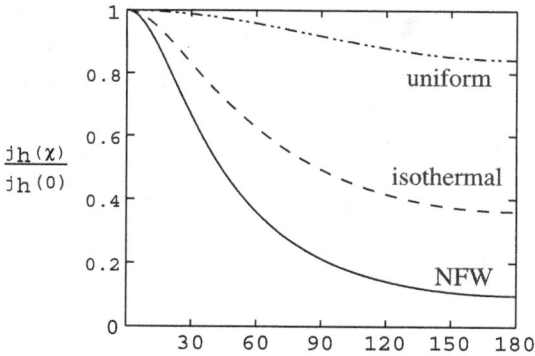

Fig. 17. The UHECR flux normalized to the flux in the direction of the Galactic center, as a function of the angle for the Navarro-Frenk-White (solid line), isothermal (dashed line) and uniform halo profile (dashed-dotted line) of the radius 100 kpc.

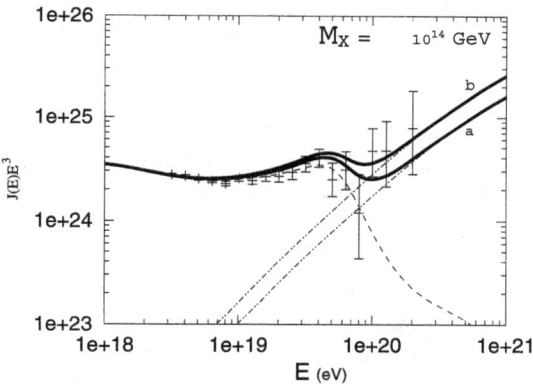

Fig. 18. The UHECR spectrum in the combined SHDM + AGN model [65] (solid lines). Two curves correspond to the uncertainty in the SHDM component.

data, including the highest-energy part [65]. The fitted spectrum is shown in Fig. 18.

Similarly to the pure SHDM scenario this combined model is subject to the anisotropy constraints. In order to analyze these constraints quantitatively it is best to use the data of the experiment which covered the region around the galactic center. The only such experiment is SUGAR. Although this experiment is considered unreliable as far as energy determination is concerned, it certainly has sufficient angular resolution for the wide-angle analysis. Fig. 19 shows the SUGAR data versus the prediction of the pure SHDM model [66] (see also

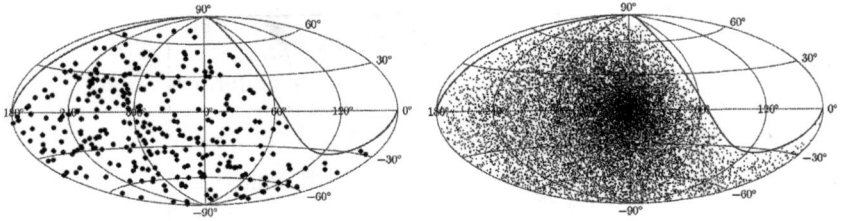

Fig. 19. The SUGAR data (left) vs. the prediction of the pure SHDM model (right) [66]. The Galactic center is at the center of each plot. Dotted line marks the boundary of the SUGAR acceptance region.

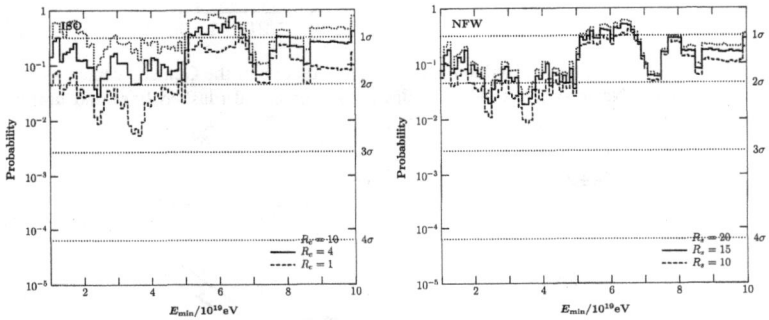

Fig. 20. The significance of the difference between the observed and predicted UHECR distributions in the case of isothermal (left panel) and NFW (right panel) halo profiles [66]. Different curves correspond to different core sizes R_C as indicated on the plots.

Ref. [67]). Clearly, the two distributions are very different. The statistical significance of the difference, however, becomes not very large when one mixes the SHDM flux with the uniformly distributed contribution from AGNs in the proportion dictated by Fig. 18. It is shown in Fig. 20 as a function of the energy for the isothermal and NFW halo profiles. One concludes from these plots that the combined SHDM + AGN model is disfavored at 2σ level by the SUGAR data.

4.10. Topological defects

The topological defect model of UHECR is a "top-down" scenario in which the decaying superheavy particles are not primordial but instead are permanently produced during the evolution of the topological defects, e.g. cosmic strings, monopoles and string-monopole networks. These models were popular at the time when topological defects were considered as a possible explanation of the structure formation. With the recent WMAP data such explanation is excluded, and the topological defect models look less attractive.

The major difficulty of the topological defect models is to produce sufficient flux of UHECR [68]. This can nevertheless be achieved in the combined string-monopole networks ("necklaces") which may arise in some grand-unified theories [69]. The main signature of these models is that UHECR flux should be dominated by the photons.

4.11. Violation of Lorentz invariance

UHECR particles are the most energetic particles ever observed. Their energies differ from the ones attainable at accelerators by 8 orders of magnitude. One may speculate [70] that special relativity is not valid at such high energies. If this were the case nearly all the problems related to UHECR (except the acceleration) would be solved. As has been already mentioned, the existence of the GZK cut-off and the rapid (in cosmological standards) energy attenuation of UHE protons is related to the Δ-resonance. If this resonance or even the threshold of pion production is eliminated or shifted to higher energies the protons would propagate freely through the Universe. The flux at all energies including the highest ones would then be dominated by the most distant sources which would therefore be numerous and weak. This would solve both the GZK and the source problems.

These ideas may be put on the quasi-quantitative ground in the framework of the phenomenological model where one assumes different maximum attainable speeds for different particle species [70]. The relative difference which is required is of the order of 10^{-23}. It is consistent at present with other experimental constraints. Interestingly, the relevant range of parameters can be probed experimentally in the near future.

This scenario is very viable phenomenologically. It is one of the few models where one can have neutral primary particles (the neutrons which may become stable at high energies) [70, 71]. The problem is that one has to accept and incorporate into theory the violation of the Lorentz invariance.

4.12. Strong neutrino interactions

To escape the GZK cutoff and solve the source problem one needs a penetrating particle. In the standard model such particle is the neutrino. One needs, however, to provide the strong interaction of primary particles with the atmosphere. Note that this interaction happens at a much higher center-of-mass energy than the interaction with the CMB photons during the propagation, namely at around 100 TeV. Thus, if one could construct a model in which the neutrino interaction cross section grows rapidly and becomes large at energies of order 100 TeV, neutrino itself could be the particles which initiates the airshowers. This would be a very attractive possibility, as no new particles would be required and all problems would be solved.

There were several attempts to construct such a model [72–75]. The most "conservative" idea involving only the standard model physics is to use the instanton effects. There has been a discussion in the literature [76–79] concerning the possibility that the effects of electroweak instantons may become unsuppressed at energies above ~ 10 TeV. One might then speculate that the neutrino cross section becomes large. The straightforward extrapolation of the energy dependence of the instanton cross section suggests that this might be the case [80]. However, the numerical calculations show the opposite [79]: the cross section always stays exponentially suppressed.

Another idea used to make the neutrino cross section large is to employ extra dimensions [72]. This is also unlikely to work [74]. So, at present there is no viable model based on the strong neutrino interactions at high energies.

5. Quest for sources

5.1. Clustering of UHECR events

It has been first noted by the AGASA collaboration that the small-scale distribution of the observed UHECR events deviates from uniform – there are clusters of events of the size of the experimental angular resolution [9, 81]. The mere existence of such clusters does not imply point sources as they (clusters) may be due to fluctuations over the uniform background. The real question is whether the clustering is statistically significant. As is standard in statistical methods, the way to answer this question is as follows. One makes a hypothesis about the distribution of events, then chooses an observable and tries to *reject* the hypothesis at some significance level comparing the predicted distribution of the observable with its distribution in the real data. Note that *inability to reject the hypothesis* (e.g., in the case of the bad choice of the observable) *is not an argument in its favor*. In the case of clustering, the hypothesis is obviously the uniform distribution of events. The existing approaches differ, thus, only by the choice of the observable.

A convenient observable characterizing clustering quantitatively is the angular correlation function evaluated at the angular resolution scale [10, 81]. This quantity is easy to calculate: one has to count the number of pairs of events falling within given angular distance from each other and compare to the same quantity evaluated for the uniform distributions. The result for AGASA data is presented in Fig. 21. One clearly sees the peak at small angles. Its angular size is compatible with the AGASA angular resolution. Similar results were obtained in Ref. [81]. The significance is of order 4σ.

The analysis of other experiments is not conclusive at present. The Yakutsk data exhibit clustering as well, with the significance larger than one would predict

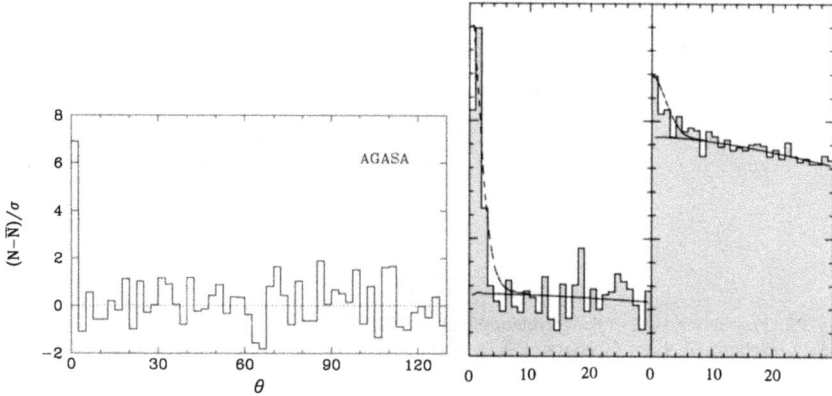

Fig. 21. Correlation function of AGASA events. Left panel: the analysis of Ref. [10]. N and $\bar{N}$ are the number of pairs separated by the given angle in the real AGASA data with $E > 4 \times 10^{19}$ eV and Monte-Carlo samples, respectively, σ is the standard deviation in the Monte-Carlo samples in the corresponding angular bin. Right panel: the analysis of Ref. [81]. Histogram represents the number of pairs in the real data, solid lines show Monte-Carlo expectation. The energy cuts are $E > 4 \times 10^{19}$ eV and $E > 10^{19}$ eV on left and right parts, respectively.

from the AGASA data given the different number of events and different angular resolution of Yakutsk experiment. The other ground array experiments have too few events. A real competitor to AGASA could be HiRes. However, the set of events observed in the monocular mode is difficult to analyze because of their essentially asymmetric angular resolution. The set of stereo events is currently rather small; there is no clustering observed in this set. This is compatible with the AGASA result.

5.2. Correlations with BL Lacs

The apparent clustering of UHECR events suggests that there exist localized sources which may be identifiable with the already existing data and whose number can be estimated from statistical arguments [82–84]. In the standard analysis (see, e.g., Ref. [47]) one identifies sources as a statistically significant excess of events from a given direction with respect to the background. Given the set of such "sources" (excesses) one then tries to find their astrophysical counterparts. In the case of cosmic rays such an analysis is not conclusive. First, the largest cluster observed is a triplet, whose statistical significance in case of AGASA is of order 1% [9]. Second, the angular resolution of UHECR experiments is too poor to identify convincingly astrophysical counterparts of UHECR clusters on a source-by-source basis. Thus, the statistical approach has to be used.

P. Tinyakov

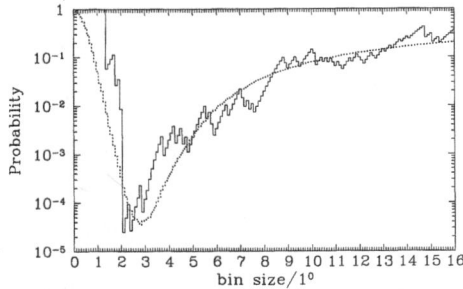

Fig. 22. The significance of the correlation between the combined AGASA and Yakutsk dataset and the 22 brightest BL Lacs (histogram) [87]. The dotted curve is the Monte-Carlo simulation of the correlation signal.

Given the set of candidate sources, one may apply the correlation method. Although one may not be able to name individual sources, the fact of correlations may be established reliably. Similar to autocorrelation analysis, one tries to reject the hypothesis that the given set of sources and UHECR events are uncorrelated. Just as above, the failure to reject the hypothesis does not prove it.

Several candidates for sources have been considered which include AGNs (in particlar, radio-loud quasars [85, 86] and BL Lacs [87]), colliding galaxies [88, 89], dead quasars [90], pulsars, seyfert galaxies [91, 92] and others. Of AGNs, the promising candidates are the BL Lacs, the AGNs with jets directed along the line of sight.

Correlations of the UHECR events with BL Lacs were first claimed in Ref. [87] where the subsample of the 22 brightest BL Lacs was selected. The primary particles were assumed to be neutral, i.e. no correction for the galactic magnetic field was performed. The significance of the correlations was estimated to be better than 4σ. The significance of the correlation as a function of the angle is shown in Fig. 22. This estimate was later criticized [93] and defended [94]. The case of protons was considered in Ref. [19] where the correction for the Galactic magnetic field was performed. Note that the two approaches do not contradict to each other since the UHECR flux may contain both neutral and charged components. A different set of BL Lacs selected on the basis of their gamma-ray emission was considered in Ref. [95].

There recently appeared an independent set of data from the HiRes experiment in the stereo mode. This set is relatively small (271 events with energies $E > 10^{19}$ eV), but has unprecedented angular resolution of $0.6°$. With this set, correlations of the neutral component of UHECR with the sources can be tested. (The charged correlations cannot be studied because the energies of the events are not published.)

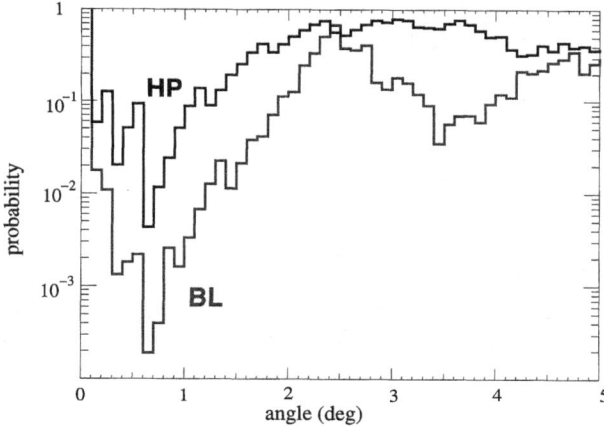

Fig. 23. The significance of the correlations between the HiRes stereo dataset and the set of "confirmed" (BL) and "high polarization" (HP) BL Lacs from the Veron catalog [97] as a function of the angle.

In Ref. [96] three previously considered sets of BL Lacs were correlated with the HiRes events: the set of 22 most powerful BL Lacs, the set of 14 potentially gamma-ray loud BL Lacs of Ref. [95] and the set of all BL Lacs with the visible magnitude $m < 18$ considered in Ref. [19]. The first two sets show no significant correlations. The third set exhibits significant correlations at the angles compatible with the HiRes angular resolution. The significance of these correlations as a function of the angle is shown in Fig. 23.

5.3. Predictions for future data

The first correlation analyses had one important drawback: they involved the adjustment of the BL Lac subsets. In the calculation of the significance this adjustment may be compensated by the penalty factor [94]. However, the UHECR luminosity of sources cannot be estimated directly. Correspondingly, the number of coincidences in the future experiments cannot be predicted directly as is necessary for the independent test of correlations.

In the case of the HiRes data the statistical analysis is straightforward, so one can estimate quantitatively the fraction of the neutral particles in the total UHECR flux which contribute to these correlations. The estimate is shown in Fig. 24. One can see from the picture that the fraction of $1.5 - 3\%$ is the most probable.

On the basis of this estimate one can make predictions for the future experiments and future datasets of the existing experiments. These predictions are

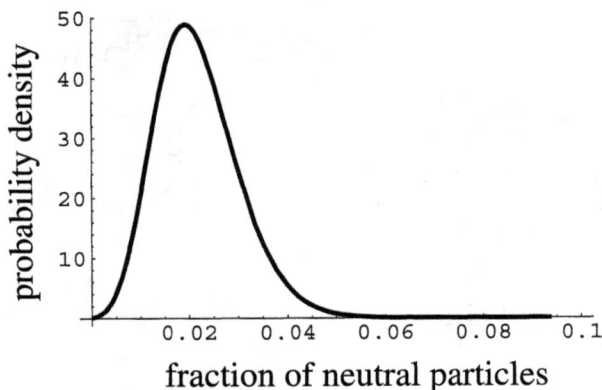

Fig. 24. The estimate of the fraction of the neutral particles in the total UHECR flux contributing to correlations of the HiRes stereo events with BL Lacs.

Table 2

The 95% confidence interval $\Delta n_{95\%}$ for the predicted number of coincidences between confirmed BL Lacs from the Veron catalog [97] and UHECR events within the angle $\sqrt{2}\sigma$ where σ is the angular resolution of the experiment. N is the total number of UHECR events, S is the mean number of events from BL Lacs and B is the mean number of the random coincidences within the angle $\sqrt{2}\sigma$. PAO = Pierre Auger Observatory; TA = Telescope Array. N_0 is the number of the events which the experiment has to accumulate to reach the current sensitivity of the HiRes to these correlations.

Experiment	σ	N	S	B	$\Delta n_{95\%}$	N_0
HiRes, original	0.6°	271	7.46	3.54		
HiRes (stereo)	0.6°	190	5.23	2.48	1–17	271
AGASA	2.4°	1500	43.4	310	308–417	3870
PAO (surface)	1.4°	8000	87.9	216	239–413	3517
PAO (hybrid)	0.6°	2000	24.8	9.10	15–66	467
TA (surface)	1.55°	8000	239	710	785–1235	1560
TA (hybrid)	0.62°	2000	58.6	30.3	47–161	277

summarized in Table 2. We see from the table that the angular resolution plays the crucial role. This is understandable since the background (the number of random coincidences) grows like σ^2. For this reason the surface array of Pierre Auger experiment alone is much less sensitive to correlations than the hybrid detector. Another important factor is the location of the experiment, since the BL Lacs of the Veron catalog are more numerous in the Northern hemisphere. This explains why the Telescope Array needs nearly twice less events than the Pierre Auger to reach the same sensitivity.

6. Conclusions and outlook

It seems clear from what we have discussed that there is little known for certain concerning ultra-high energy cosmic rays. These are the facts which are established, in my view, without doubt: i) The cosmic rays exist which have energies up to 10^{20} eV. ii) The primary particles have hadronic cross sections with the air at the center-of-mass energy of order 100 TeV. iii) The large-scale distribution of arrival directions is nearly uniform. One should add to this some theoretical input: if Lorentz symmetry is exact, protons of energy 10^{20} eV must have come from distances less than 50 Mpc. This is about all.

These are the major questions which are still open, in the order of importance for particle physics.

• What are the primary particles? Because of the very indirect observation of the *first* interaction we may never know the answer. For the moment the possibilities range from exotic hadrons to photon and strongly interacting neutrino. It is extremely important to narrow this range.

• Is there a cutoff in the spectrum? Present data are rather contradictory. One may argue that the discrepancy is not so big: one may imagine a spectrum which is roughly 2σ away from both HiRes and AGASA. This would, however, mean the *absence* of the cutoff. Notably, the discrepancy exists between different methods of the energy determination. The cross-calibration is therefore necessary. One may expect progress in this direction with the full operation of the Pierre Auger experiment.

• What are the sources of UHECR? The key issue here is the analysis of the arrival directions. Present data can give no new information. With the new datasets coming soon, the hypotheses formulated so far can be tested. This includes clustering of UHECR, large-scale isotropy toward the Galactic center, correlations with BL Lacs.

Clearly, these questions are interrelated, and answering one would trigger the resolution of the others. The feeling is that we are one step away from resolving the puzzle.

References

[1] M. Nagano and A. A. Watson, Rev. Mod. Phys. **72** (2000) 689.

[2] P. Bhattacharjee and G. Sigl, Phys. Rept. **327** (2000) 109 [astro-ph/9811011].

[3] D. F. Torres and L. A. Anchordoqui, Rept. Prog. Phys. **67** (2004) 1663 [astro-ph/0402371].

[4] A. V. Olinto, Phys. Rept. **333** (2000) 329 [astro-ph/0002006].

[5] X. Bertou, M. Boratav and A. Letessier-Selvon, Int. J. Mod. Phys. A **15** (2000) 2181 [astro-ph/0001516].

[6] D. J. Bird et al., Astrophys. J. **441** (1995) 144.

[7] J. Linsley, L. Scarsi and B. Rossi, Phys. Rev. Lett. **6** (1961) 485.

[8] M. Takeda et al., Astrophys. J. **522**, 225 (1999) [astro-ph/9902239].

[9] Y. Uchihori, M. Nagano, M. Takeda, M. Teshima, J. Lloyd-Evans and A. A. Watson, Astropart. Phys. **13** (2000) 151 [astro-ph/9908193].

[10] P. G. Tinyakov and I. I. Tkachev, JETP Lett. **74** (2001) 1 [Pisma Zh. Eksp. Teor. Fiz. **74** (2001) 3] [astro-ph/0102101].

[11] S. Lee, Phys. Rev. D **58** (1998) 043004 [astro-ph/9604098].

[12] K. Greisen, Phys. Rev. Lett. **16** (1966) 748.

[13] G. T. Zatsepin and V. A. Kuzmin, JETP Lett. **4** (1966) 78 [Pisma Zh. Eksp. Teor. Fiz. **4** (1966) 114].

[14] G. A. Medina-Tanco, astro-ph/9905239.

[15] M. Blanton, P. Blasi and A. V. Olinto, Astropart. Phys. **15**, 275 (2001) [astro-ph/0009466].

[16] T. Abu-Zayyad et al. [High Resolution Fly's Eye Collaboration], Astropart. Phys. **23** (2005) 157 [astro-ph/0208301].

[17] P. G. Tinyakov and I. I. Tkachev, Astropart. Phys. **24**, 32 (2005) [astro-ph/0411669].

[18] T. Stanev, Astrophys. J. **479** (1997) 290 [astro-ph/9607086].

[19] P. G. Tinyakov and I. I. Tkachev, Astropart. Phys. **18** (2002) 165 [astro-ph/0111305].

[20] P. Tinyakov and I. Tkachev, astro-ph/0305363.

[21] M. Prouza and R. Smida, Astron. Astrophys. **410** (2003) 1 [astro-ph/0307165].

[22] P. P. Kronberg, Rept. Prog. Phys. **57** (1994) 325.

[23] G. R. Farrar and T. Piran, Phys. Rev. Lett. **84** (2000) 3527 [astro-ph/9906431].

[24] K. Dolag, D. Grasso, V. Springel and I. Tkachev, JETP Lett. **79** (2004) 583 [Pisma Zh. Eksp. Teor. Fiz. **79** (2004) 719] [astro-ph/0310902].

[25] K. Dolag, D. Grasso, V. Springel and I. Tkachev, JCAP **0501** (2005) 009 [astro-ph/0410419].

[26] G. Sigl, F. Miniati and T. A. Ensslin, Phys. Rev. D **68** (2003) 043002 [astro-ph/0302388].

[27] P. L. Biermann and P. A. Strittmatter, Astrophys. J. **322** (1987) 643.

[28] R. D. Blandford, Phys. Scripta **T85** (2000) 191 [astro-ph/9906026].

[29] P. L. Biermann, J. Phys. G **23** (1997) 1.

[30] J. P. Rachen and P. L. Biermann, Astron. Astrophys. **272** (1993) 161 [astro-ph/9301010].

[31] A. P. Szabo and R. J. Protheroe, Astropart. Phys. **2** (1994) 375 [astro-ph/9405020].

[32] A. M. Hillas, Ann. Rev. Astron. Astrophys. **22** (1984) 425.

[33] R. M. Wald, Phys. Rev. D **10** (1974) 1680; J. Bičak and L. Dvořak, Gen. Rel. Grav. **7** (1976) 959; J. Bičak and V. Janis, MNRAS **212** (1985) 899.

[34] A. Neronov, P. Tinyakov and I. Tkachev, J. Exp. Theor. Phys. **100** (2005) 656 [Zh. Eksp. Teor. Fiz. **100** (2005) 744] [astro-ph/0402132].

[35] M. Milgrom and V. Usov, Astrophys. J. **449** (1995) L37 [astro-ph/9505009].

[36] M. Milgrom and V. Usov, Astropart. Phys. **4** (1996) 365 [astro-ph/9506099].

[37] M. Vietri, Astrophys. J. **453** (1995) 883 [astro-ph/9506081].

[38] E. Waxman, Phys. Rev. Lett. **75** (1995) 386 [astro-ph/9505082].

[39] A. Dar, Phys. Rev. Lett. **76** (1996) 3878 [astro-ph/9811196].

[40] A. Dar and R. Plaga, Astron. Astrophys. **349** (1999) 259 [astro-ph/9902138].

[41] P. Blasi, R. I. Epstein and A. V. Olinto, Astrophys. J. **533** (2000) L123 [astro-ph/9912240].

[42] E. J. Ahn, G. A. Medina-Tanco, P. L. Biermann and T. Stanev, astro-ph/9911123.

[43] L. A. Anchordoqui, H. Goldberg and T. J. Weiler, Phys. Rev. Lett. **87** (2001) 081101 [astro-ph/0103043].

[44] D. Fargion, B. Mele and A. Salis, Astrophys. J. **517** (1999) 725 [astro-ph/9710029].

[45] T. J. Weiler, Astropart. Phys. **11** (1999) 303 [hep-ph/9710431].

[46] D. S. Gorbunov, P. G. Tinyakov and S. V. Troitsky, Astropart. Phys. **18** (2003) 463 [astro-ph/0206385].

[47] P. Sreekumar et al. [EGRET Collaboration], Astrophys. J. **494** (1998) 523 [astro-ph/9709257].

[48] V. S. Berezinsky and A. Yu. Smirnov, Astrophys. Space Sci. **32** (1975) 461.

[49] O. E. Kalashev et al., Phys. Rev. D **65** (2002) 103003.

[50] Z. Fodor, S. D. Katz and A. Ringwald, JHEP **0206** (2002) 046 [hep-ph/0203198].

[51] T. W. Kephart and T. J. Weiler, Astropart. Phys. **4** (1996) 271 [astro-ph/9505134].

[52] D. J. H. Chung, G. R. Farrar and E. W. Kolb, Phys. Rev. D **57** (1998) 4606 [astro-ph/9707036].

[53] I. F. M. Albuquerque, G. R. Farrar and E. W. Kolb, Phys. Rev. D **59** (1999) 015021 [hep-ph/9805288].

[54] D. S. Gorbunov, G. G. Raffelt and D. V. Semikoz, Phys. Rev. D **64** (2001) 096005 [hep-ph/0103175].

[55] J. Madsen and J. M. Larsen, Phys. Rev. Lett. **90** (2003) 121102 [astro-ph/0211597].

[56] M. Kachelriess, D. V. Semikoz and M. A. Tortola, Phys. Rev. D **68** (2003) 043005 [hep-ph/0302161].

[57] V. Berezinsky, M. Kachelriess and A. Vilenkin, Phys. Rev. Lett. **79** (1997) 4302 [astro-ph/9708217].

[58] V. A. Kuzmin and V. A. Rubakov, Phys. Atom. Nucl. **61** (1998) 1028 [Yad. Fiz. **61** (1998) 1122] [astro-ph/9709187].

[59] M. Birkel and S. Sarkar, Astropart. Phys. **9** (1998) 297 [hep-ph/9804285].

[60] S. L. Dubovsky and P. G. Tinyakov, JETP Lett. **68** (1998) 107 [hep-ph/9802382].

[61] V. Berezinsky and A. A. Mikhailov, Phys. Lett. B **449** (1999) 237 [astro-ph/9810277].

[62] G. A. Medina-Tanco and A. A. Watson, Astropart. Phys. **12** (1999) 25 [astro-ph/9903182].

[63] M. Ave, J. A. Hinton, R. A. Vazquez, A. A. Watson and E. Zas, Phys. Rev. D **65** (2002) 063007 [astro-ph/0110613].

[64] V. Berezinsky, A. Z. Gazizov and S. I. Grigorieva, hep-ph/0204357.

[65] V. Berezinsky, A. Z. Gazizov and S. I. Grigorieva, astro-ph/0210095.

[66] H. B. Kim and P. Tinyakov, Astropart. Phys. **21** (2004) 535 [astro-ph/0306413].

[67] M. Kachelriess and D. V. Semikoz, Phys. Lett. B **577** (2003) 1 [astro-ph/0306282].

[68] V. Berezinsky, P. Blasi and A. Vilenkin, Phys. Rev. D **58** (1998) 103515 [astro-ph/9803271].

[69] V. Berezinsky and A. Vilenkin, Phys. Rev. Lett. **79** (1997) 5202 [astro-ph/9704257].

[70] S. R. Coleman and S. L. Glashow, Phys. Rev. D **59** (1999) 116008 [hep-ph/9812418].

[71] S. L. Dubovsky and P. G. Tinyakov, Astropart. Phys. **18** (2002) 89 [astro-ph/0106472].

[72] P. Jain, D. W. McKay, S. Panda and J. P. Ralston, Phys. Lett. B **484** (2000) 267 [hep-ph/0001031].

[73] L. Anchordoqui, H. Goldberg, T. McCauley, T. Paul, S. Reucroft and J. Swain, Phys. Rev. D **63** (2001) 124009 [hep-ph/0011097].

[74] R. Emparan, M. Masip and R. Rattazzi, Phys. Rev. D **65** (2002) 064023 [hep-ph/0109287].

[75] Z. Fodor, S. D. Katz, A. Ringwald and H. Tu, Phys. Lett. B **561** (2003) 191 [hep-ph/0303080].

[76] A. Ringwald, Nucl. Phys. B **330** (1990) 1.

[77] S. Y. Khlebnikov, V. A. Rubakov and P. G. Tinyakov, Nucl. Phys. B **350** (1991) 441.

[78] P. G. Tinyakov, Int. J. Mod. Phys. A **8** (1993) 1823.

[79] F. Bezrukov, D. Levkov, C. Rebbi, V. A. Rubakov and P. Tinyakov, Phys. Lett. B **574** (2003) 75 [hep-ph/0305300].

[80] A. Ringwald, JHEP **0310** (2003) 008 [hep-ph/0307034].

[81] M. Takeda et al., talk prepared for 27th International Cosmic Ray Conference (ICRC 2001), Hamburg, Germany, 7-15 Aug 2001.

[82] S. L. Dubovsky, P. G. Tinyakov and I. I. Tkachev, Phys. Rev. Lett. **85** (2000) 1154 [astro-ph/0001317].

[83] Z. Fodor and S. D. Katz, Phys. Rev. D **63** (2001) 023002 [hep-ph/0007158].

[84] M. Kachelriess and D. Semikoz, Astropart. Phys. **23** (2005) 486 [astro-ph/0405258].

[85] G. R. Farrar and P. L. Biermann, Phys. Rev. Lett. **81** (1998) 3579 [astro-ph/9806242].

[86] A. Virmani, S. Bhattacharya, P. Jain, S. Razzaque, J. P. Ralston and D. W. McKay, Astropart. Phys. **17** (2002) 489 [astro-ph/0010235].

[87] P. G. Tinyakov and I. I. Tkachev, JETP Lett. **74** (2001) 445 [Pisma Zh. Eksp. Teor. Fiz. **74** (2001) 499] [astro-ph/0102476].

[88] N. Hayashida et al., Astron. J. **120** (2000) 2190 [astro-ph/0008102].

[89] S. S. Al-Dargazelli, A. W. Wolfendale, A. Smialkowski and J. Wdowczyk, J. Phys. G **22** (1996) 1825.

[90] E. Boldt and P. Ghosh, astro-ph/9902342.

[91] A. V. Uryson, J. Exp. Theor. Phys. **89** (1999) 597 [Zh. Eksp. Teor. Fiz. **116** (1999) 1121].

[92] A. V. Uryson, astro-ph/0303347.

[93] N. W. Evans, F. Ferrer and S. Sarkar, Phys. Rev. D **67** (2003) 103005 [astro-ph/0212533].

[94] P. G. Tinyakov and I. I. Tkachev, Phys. Rev. D **69** (2004) 128301.

[95] D. S. Gorbunov, P. G. Tinyakov, I. I. Tkachev and S. V. Troitsky, Astrophys. J. **577** (2002) L93 [astro-ph/0204360].

[96] D. S. Gorbunov, P. G. Tinyakov, I. I. Tkachev and S. V. Troitsky, astro-ph/0508329.

[97] M. P. Véron-Cetty and P. Véron, ESO scientific report (2000); M. P. Véron-Cetty and P. Véron, Astron. Astrophys. **374** (2001) 92.

Course 11

NEUTRINO MASS AND MIXING: TOWARD THE UNDERLYING PHYSICS

Alexei Yu. Smirnov

International Centre for Theoretical Physics,
Strada Costiera 11, Trieste, Italy
Institute for Nuclear Research, RAS, Moscow, Russia

D. Kazakov, S. Lavignac and J. Dalibard, eds.
Les Houches, Session LXXXIV, 2005
Particle Physics Beyond the Standard Model
© *2006 Elsevier B.V. All rights reserved*

Contents

1. Preamble

The central issue of these lectures is the neutrino mass and lepton mixing, non-standard neutrino interactions.

Neutrino mass is considered as the first manifestation of physics beyond the standard model, as a window to new physics. What is this New physics? What do we see in the window? How far beyond?

The statement requires some clarification. The quark and lepton mass hierarchies as well as the structure of CKM mixing have no explanation in the Standard Model either. And in this sense they are also manifestations of physics beyond SM.

Quark mass and mixing as well as neutrino mass and mixing are new physics. Some part of this new physics may be in common. At the same time, neutrinos may require something more.

The bottom line is that new physics behind the neutrino masses and mixing has not been identified yet. It is difficult to say with confidence what is correct context or domain of new physics involved. There are plenty of models, scenarios and approaches and only few simplest possibilities have been excluded so far. Typically models accommodate but not really explain the results. And we can argue only what is the most plausible possibility.

2. Introduction

Structure of the course: reflects two existing approaches to the problem or lines of research.

1. Bottom-up: toward the underlying physics. It includes the following issues:
– analysis of experimental results;
– description of physics effects involved;
– determination of the mass and mixing
– reconstruction of the mass matrix;
– identification (searches for) possible symmetries, scales, dynamics.

Shortly, this is an attempt to uncover the underlying physics starting from observations.

2. Top-down approach: from certain theoretical context to data. Here theoretical models motivated by some other, in general, unrelated to neutrinos reasons are used as the starting point. This line includes

– formulation of the theoretical context;

– study of mechanisms of neutrino mass generation;

– predictions of neutrino mass and mixing and other observables.

At some point these two lines should meet and apparently both lines are necessary.

Speaking on the style of lectures, the main emphasis is on

1. motivations of various developments;

2. physical picture of the effects;

3. fun and then formalism: where it is possible, I will derive precise results on the basis of physical picture alone without numerical calculations.

Two zeros. The main salient feature of neutrinos is the neutrality:

$$Q_\gamma = Q_c = 0. \tag{2.1}$$

So it would be natural to explain all unusual properties of neutrinos using this feature. The neutrality opens the following possibilities:

• neutrinos can be Majorana particles, and therefore have the Majorana mass terms;

• they can mix with singlets of the SM symmetry group;

• the right-handed components (RH), if exist, are singlets of $SU(3) \times SU(2) \times U(1)$. So, their masses are unprotected by the symmetry and therefore can be large: $M_R \gg v_{EW}$, where v_{EW} is the electroweak scale.

In turn, properties of the RH components allow them to propagate in extra dimensions, or be located on the "hidden" (not ours) brane in contrast to other fermions, *etc.*

Introduction of the RH neutrino has a number of attractive features [1], in particular, it allows one to extend the electroweak symmetry to the gauged $SU(2)_L \times SU(2)_R \times U(1)_{B-L}$.

Is this enough to explain the properties of the mass spectrum and mixings?

Window to hidden world Neutrinos are the only known fermions which can mix with particles from the hidden sector (related *e.g.* to SUSY or strings or mirror world). Properties of the neutrino mass and mixing can be associated to this mixing.

In the first part of the course I will explain how neutrino parameters (masses and mixing) have been determined. I will argue why we are confident that the interpretation in terms of vacuum masses and mixing is correct.

In the second part I will review analysis of these results and their possible implications.

3. Notions and notations

3.1. Flavors, masses and mixing

The *flavor* neutrino states: $\nu_f \equiv (\nu_e, \nu_\mu, \nu_\tau)$ are defined as the states which correspond to certain charge leptons: e, μ and τ. The correspondence is established by weak interactions: ν_l and l ($l = e, \mu, \tau$) form the charged currents. It is not excluded that additional neutrino states, the sterile neutrinos, ν_s, exist. The neutrino *mass states*, ν_1, ν_2, and ν_3, with masses m_1, m_2, m_3 are the eigenstates of mass matrix as well as the eigenstates of the total Hamiltonian in vacuum.

The *vacuum mixing* means that the flavor states do not coincide with the mass eigenstates. The flavor states are combinations of the mass eigenstates:

$$\nu_l = U_{li}\nu_i, \quad l = e, \mu, \tau, \quad i = 1, 2, 3, \tag{3.1}$$

where the mixing parameters U_{li} form the PMNS mixing matrix U_{PMNS} [2,3]. The mixing matrix can be conveniently parameterized as

$$U_{PMNS} = V_{23}(\theta_{23})V_{13}(\theta_{13})I_\delta V_{12}(\theta_{12}), \tag{3.2}$$

where V_{ij} is the rotation matrix in the ij-plane, θ_{ij} is the corresponding angle and $I_\delta \equiv diag(1, 1, e^{i\delta})$ is the matrix of CP violating phase.

3.2. Two aspects of mixing

Many conceptual points can be clarified using just 2ν mixing. Also at the present level of accuracy of measurements the 2ν dynamics is enough to describe the data. For two neutrino mixing, *e.g.* $\nu_e - \nu_a$, we can write:

$$\nu_e = \cos\theta\, \nu_1 + \sin\theta\, \nu_2, \quad \nu_a = \cos\theta\, \nu_2 - \sin\theta\, \nu_1, \tag{3.3}$$

where ν_a is the non-electron neutrino state, and θ is the vacuum mixing angle.

There are two important physical aspects of mixing. According to (3.3) the flavor neutrino states are combinations of the mass eigenstates. One can think in terms of wave packets. Then propagation of ν_e (ν_a) is described by a system of two wave packets which correspond to ν_1 and ν_2.

In fig. 1a). we show representation of ν_e and ν_a as the combination of mass states. The lengths of the boxes, $\cos^2\theta$ and $\sin^2\theta$, give the *admixtures* of ν_1 and ν_2 in ν_e and ν_a.

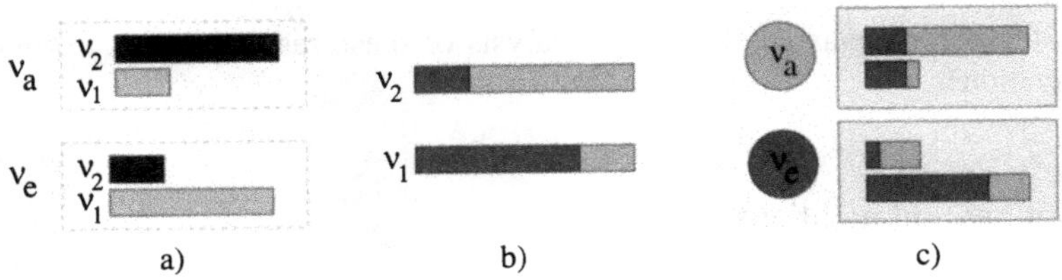

Fig. 1. a). Representation of the flavor neutrino states as the combination of the mass eigenstates. The length of the box gives the admixture of (or probability to find) corresponding mass state in a given flavor state. (The sum of the lengths of the boxes is normalized to 1. b). Flavor composition of the mass eigenstates. The electron flavor is shown in dark and the non-electron flavor in grey. The sizes of the dark and grey parts give the probability to find the electron and non-electron neutrino in a given mass state. c). Portraits of the electron and non-electron neutrinos: shown are representations of the electron and non-electron neutrino states as combinations of the eigenstates for which, in turn, we show the flavor composition.

The key point is that the flavor states are *coherent* mixtures (combinations) of the mass eigenstates. The *relative phase* or phase difference of ν_1 and ν_2 in ν_e as well as ν_a is fixed: according to (3.3) it is zero in ν_e and π in ν_a. Consequently, there are certain *interference* effects between ν_1 and ν_2 which depend on the relative phase.

Second aspect: the relations (3.3) can be inverted:

$$\nu_1 = \cos\theta\, \nu_e - \sin\theta\, \nu_a, \quad \nu_2 = \cos\theta\, \nu_a + \sin\theta\, \nu_e. \tag{3.4}$$

In this form they determine the *flavor composition* of the mass states (eigenstates of the Hamiltonian), or shortly, the flavors of eigenstates. According to (3.4) the probability to find the electron flavor in ν_1 is given by $\cos^2\theta$, whereas the probability that ν_1 appears as ν_a equals $\sin^2\theta$. This flavor decomposition is shown in fig. 1b). by colors (different shadowing).

Inserting the flavor decomposition of mass states in the representation of the flavors states, we get the "portraits" of the electron and non-electron neutrinos fig. 1c). According to this figure, ν_e is a system of two mass eigenstates which in turn have a composite flavor. On the first sight the portrait has a paradoxical feature: there is the non-electron (muon and tau) flavor in the electron neutrino! The paradox has the following resolution: in the ν_e-state the ν_a-components of ν_1 and ν_2 are equal and have opposite phases. Therefore they cancel each other and the electron neutrino has pure electron flavor as it should be. The key point is interference: the interference of the non-electron parts is destructive in ν_e. The electron neutrino has a "latent" non-electron component which can not be seen due to particular phase arrangement. However during propagation the phase difference changes and the cancellation disappears. This leads to an appearance of the non-electron component in propagating neutrino state which was originally produced as the electron neutrino. This is the mechanism of neutrino oscillations. Similar consideration holds for the ν_a state.

3.3. Who mixes neutrinos?

How mixed neutrino states (that is, the coherent mixtures on the mass eigenstates) are created? Why neutrinos and not charged leptons? In fact, these are non trivial questions.

Creation (preparation – in quantum mechanics terms) of the mixed neutrino states is a result of interplay of the charged current weak interactions and kinematic features of specific reactions. Differences of masses of the charged leptons play crucial role.

Let us consider three neutrino species separately.

1) Electron neutrinos: The combination of mass eigenstates which we call the electron neutrino is produced, *e.g.*, in the beta decay (together with electron). The reason is the energy conservation: no other combination can be produced because the energy release is about few MeV, so that neither muon nor tau lepton can appear.

2) Muon neutrino. Almost pure ν_μ state is produced together with muons in the charged pion decay: $\pi^+ \rightarrow \mu^+ \nu_\mu$. Here the reason is "chirality suppression" - essentially the angular momentum conservation and V-A character of the charged current weak interactions. The amplitude is proportional to the mass of the charged lepton squared. Therefore the channel with the electron neutrino: $\pi^+ \rightarrow e^+ \nu_e$ is suppressed as $\propto m_e^2/m_\mu^2$. Also coherence between ν_μ and small admixture of ν_e is lost almost immediately due to difference of kinematics.

3) Tau neutrino. Enriched ν_τ - flux can be obtained in the beam-dump experiments at high energies: In the thick target all light mesons (π, K which are sources of usual neutrinos) are absorbed before decay, and only heavy short living particles, like D mesons, have enough time to decay. The D mesons have also modes of decay with emission of ν_e and ν_μ which are chirality suppressed in comparison with $D \rightarrow \tau \nu_\tau$. Furthermore, coherence of ν_e and ν_μ with ν_τ is lost due to strongly different energies and momenta.

What about neutral currents? Clearly here the flavor blind state is produced.

4. Physical effects

4.1. To determination of oscillation parameters

In the Table 1 we show parameters to be determined, sources of information for their determination and physical effects involved. In the first approxima-

Table 1
Parameters and effects.

Parameters	Source of information	Physics effect
$\Delta m_{12}^2, \theta_{12}$	Solar neutrinos	Adiabatic conversion
		Averaged vacuum oscillations
	KamLAND	Non-averaged vacuum oscillations
$\Delta m_{23}^2, \theta_{23}$	Atmospheric neutrinos	Vacuum oscillations
	K2K	Vacuum oscillations
θ_{13}	CHOOZ	Vacuum oscillations
	Atmospheric neutrinos	oscillations in matter

tion (when 1-3 mixing is neglected) the three neutrino problem splits into two neutrino problems and parameters of the 1-2 and 2-3 sectors can be determined independently.

Essentially two effects are relevant for interpretation of the present data in the lowest approximation:

- vacuum oscillations (both averaged and non-averaged) [2–4];
- adiabatic conversion in medium [5,6].

Furthermore the 2ν mixing effects are enough. Notice that in the next order, when sub-leading effects are included, the problem becomes much more difficult and degeneracy of parameters appear. We will comment on this later.

4.2. Neutrino oscillation in vacuum

In vacuum the neutrino mass states are the eigenstates of the Hamiltonian. Therefore dynamics of propagation has the following features:

- Admixtures of the eigenstates (mass states) in a given neutrino state do not change. In other words, there is no $\nu_1 \leftrightarrow \nu_2$ transitions. ν_1 and ν_2 propagate independently. The admixtures are determined by mixing in a production point (by θ, if pure flavor state is produced).
- Flavors of the eigenstates do not change. They are also determined by θ. Therefore the picture of neutrino state (fig. 1 c) does not change during propagation.
- Relative phase (phase difference) of the eigenstates monotonously increases.

The phase is the only operating degree of freedom and we will consider it here in details.

Phase difference. Due to difference of masses, the states ν_1 and ν_2 have different phase velocities $v_{phase} = E_i/p_i \approx 1 + m_i^2/2E^2$ (for ultrarelativistic neutrinos), so that

$$\Delta v_{phase} \approx \frac{\Delta m^2}{2E}, \quad \Delta m^2 \equiv m_2^2 - m_1^2. \tag{4.1}$$

The phase difference changes as

$$\Delta \phi = \Delta v_{phase} t. \tag{4.2}$$

Explicitly, in the plane wave approximation we have the phases of two mass states as $\phi_i = E_i t - p_i x$. Apparently, to find the phase difference which determines the interference effect one should take the phases of mass states in the same space-time point:

$$\phi \equiv \phi_1 - \phi_2 = \Delta E t - \Delta p x. \tag{4.3}$$

Since $p = \sqrt{E^2 - m^2}$, we have

$$\Delta p = \frac{dp}{dE} \Delta E + \frac{dp}{dm^2} \Delta m^2 = \frac{1}{v_g} \Delta E - \frac{\Delta m^2}{2p}, \tag{4.4}$$

where $v_g = dE/dp$ is the group velocity. Plugging (4.4) into (4.3) we obtain

$$\phi = \Delta E \left(t - \frac{x}{v_g} \right) + \frac{\Delta m^2}{2p} x. \tag{4.5}$$

Depending on physical conditions either $\Delta E \approx 0$ or/and $(t - x/v_g)$ is small which imposes the bound on size of the wave packet. As a consequence, the first term is small and we reproduce the result (4.2).

Increase of the phase leads to oscillations. Indeed, the change of phase modifies the interference: in particular, cancellation of the non-electron parts in the state produced as v_e disappears and the non-electron component becomes observable. The process is periodic: when $\Delta \phi = \pi$, the interference of non-electron parts is constructive and at this point the probability to find v_a is maximal. Later, when $\Delta \phi = 2\pi$, the system returns to its original state: $v(t) = v_e$. The oscillation length is the distance at which this return occurs:

$$l_v = \frac{2\pi}{v_{phase}} = \frac{4\pi E}{\Delta m^2}. \tag{4.6}$$

The depth of oscillations A_P is determined by the mixing angle. It is given by maximal probability to observe the "wrong" flavor v_a. From the fig. 1c. one finds immediately (summing up the parts with the non-electron flavor in the amplitude)

$$A_P = (2 \sin \theta \cos \theta)^2 = \sin^2 2\theta. \tag{4.7}$$

Putting things together we obtain expression for the transition probability

$$P = A_P \left(1 - \cos \frac{2\pi L}{l_\nu} \right) = \sin^2 2\theta \sin^2 \frac{\Delta m^2 L}{4E}. \tag{4.8}$$

The oscillations are the effect of the phase increase which changes the interference pattern. The depth of oscillations is the measure of mixing.

4.3. Evolution equation

In vacuum the mass states are the eigenstates of Hamiltonian. So, their propagation is described by independent equations

$$i d\nu_i/dt = E_i \nu_i \approx (p_i + m_i^2/2p_i)\nu_i,$$

where we have taken ultrarelativistic limit and omitted the spin variables which are irrelevant for these oscillations. In the matrix form for three neutrinos $\nu \equiv (\nu_1, \nu_2, \nu_3)^T$, we can write

$$i \frac{d\nu}{dt} \approx \left(pI + \frac{|M_{diag}|^2}{2E} \right) \nu, \tag{4.9}$$

where $M_{diag}^2 = diag(m_1^2, m_2^2, m_3^2)$. Using the relation $\nu = U_{PMNS}^\dagger \nu_f$ (3.1), we can write the equation for the flavor states:

$$i \frac{d\nu_f}{dt} \approx \frac{|M|^2}{2E} \nu_f, \tag{4.10}$$

where $M^2 = U_{PMNS} |M_{diag}|^2 U_{PMNS}^\dagger$ is the mass matrix squared in the flavor basis. In (4.10) we have omitted the term proportional to the unit matrix which does not produce any phase difference and can be absorbed in the renormalization of the neutrino wave functions. So, the Hamiltonian of the neutrino system in vacuum is

$$H_0 = \frac{|M|^2}{2E}. \tag{4.11}$$

In the 2ν mixing case we have explicitly:

$$H_0 = \frac{\Delta m^2}{4E} \begin{pmatrix} -\cos 2\theta & \sin 2\theta \\ \sin 2\theta & \cos 2\theta \end{pmatrix}. \tag{4.12}$$

Solution of the equation (4.10) with this Hamiltonian leads to the standard oscillation formula (4.8).

4.4. Matter effect

Refraction. In matter, neutrino propagation is affected by interactions. At low energies the *elastic forward scattering* is relevant only (inelastic interactions can be neglected) [5]. It can be described by the potentials V_e, V_a. In usual medium difference of the potentials for ν_e and ν_a is due to the charged current scattering of ν_e on electrons ($\nu_e e \to \nu_e e$) [5]:

$$V = V_e - V_a = \sqrt{2} G_F n_e, \tag{4.13}$$

where G_F is the Fermi coupling constant and n_e is the number density of electrons. The result follows straightforwardly from calculation of the matrix element $V = \langle \Psi | H_{CC} | \Psi \rangle$, where Ψ is the state of medium and neutrino. Equivalently, one can describe the effect of medium in terms of the refraction index: $n_{ref} - 1 = V/p$.

The difference of the potentials leads to an appearance of additional phase difference in the neutrino system: $\phi_{matter} \equiv (V_e - V_a)t$. The difference of potentials (or refraction indexes) determines the *refraction length*:

$$l_0 \equiv \frac{2\pi}{V_e - V_a} = \frac{\sqrt{2}\pi}{G_F n_e}. \tag{4.14}$$

l_0 is the distance over which an additional "matter" phase equals 2π.

In the presence of matter the Hamiltonian of system changes:

$$H_0 \to H = H_0 + V, \tag{4.15}$$

where H_0 is the Hamiltonian in vacuum. Using (4.11) we obtain (for 2ν mixing)

$$H = \frac{|M|^2}{2E} + V, \quad V = diag(V, 0). \tag{4.16}$$

The evolution equation for the flavor states in matter then becomes

$$i\frac{d\nu_f}{dt} = \left[\frac{\Delta m^2}{4E} \left(\begin{array}{cc} -\cos 2\theta & \sin 2\theta \\ \sin 2\theta & \cos 2\theta \end{array} \right) + V \right] \nu_f. \tag{4.17}$$

The eigenstates and the eigenvalues change:

$$\nu_1, \; \nu_2 \;\; \to \;\; \nu_{1m}, \; \nu_{2m}, \tag{4.18}$$

$$\frac{m_1^2}{2E}, \; \frac{m_2^2}{2E} \;\; \to \;\; H_{1m}, \; H_{2m}. \tag{4.19}$$

The mixing in matter is determined with respect to the eigenstates of the Hamiltonian in matter ν_{1m} and ν_{2m}. Similarly to (3.3) the mixing angle in matter, θ_m, gives the relation between the eigenstates in matter and the flavor states:

$$\nu_e = \cos\theta_m \nu_{1m} + \sin\theta_m \nu_{2m}, \quad \nu_a = \cos\theta_m \nu_{2m} - \sin\theta_m \nu_{1m}. \tag{4.20}$$

The angle θ_m in matter is obtained by diagonalization of the Hamiltonian in matter (4.16)

$$\sin^2 2\theta_m = \frac{\sin^2 2\theta}{(\cos 2\theta - 2VE/\Delta m^2)^2 + \sin^2 2\theta}. \tag{4.21}$$

In matter both the eigenstates and the eigenvalues, and consequently, the mixing angle depend on matter density and neutrino energy. It is this dependence that activates new degrees of freedom of the system and leads to qualitatively new effects.

Resonance. Level crossing. According to (4.21), the dependence of the effective mixing parameter in matter, $\sin^2 2\theta_m$, on density, neutrino energy as well as the ratio of the oscillation and refraction lengths:

$$x \equiv \frac{l_\nu}{l_0} = \frac{2EV}{\Delta m^2} \propto E n_e \tag{4.22}$$

has a resonance character. At

$$l_\nu = l_0 \cos 2\theta \qquad \text{(resonance condition)} \tag{4.23}$$

the mixing becomes maximal: $\sin^2 2\theta_m = 1$. For small vacuum mixing the condition (4.23) reads:

Oscillation length $\approx$ *Refraction length.* $\tag{4.24}$

That is, the eigen-frequency which characterizes a system of mixed neutrinos, $1/l_\nu$, coincides with the eigen-frequency of medium, $1/l_0$.

For large vacuum mixing (for solar LMA: $\cos 2\theta = 0.4 - 0.5$) there is a significant deviation from the equality (4.24). Large vacuum mixing corresponds to the case of strongly coupled system for which, as usual, the shift of frequencies occurs.

The resonance condition (4.23) determines the resonance density:

$$n_e^R = \frac{\Delta m^2}{2E} \frac{\cos 2\theta}{\sqrt{2}G_F}. \tag{4.25}$$

The width of resonance on the half of the height (in the density scale) is given by

$$2\Delta n_e^R = 2n_e^R \tan 2\theta. \tag{4.26}$$

Similarly, one can introduce the resonance energy and the width of resonance in the energy scale.

In medium with varying density, the layer where the density changes in the interval

$$n_e^R \pm \Delta n_e^R \tag{4.27}$$

is called the resonance layer.

In resonance, the level splitting (difference of the eigenstates $H_{2m} - H_{1m}$) is minimal [7,8] and therefore the oscillation length being inversely proportional to the level spitting, is maximal.

4.5. Degrees of freedom

An arbitrary neutrino state can be expressed in terms of the instantaneous eigenstates of the Hamiltonian, ν_{1m} and ν_{2m}, as

$$\nu(t) = \cos\theta_a \nu_{1m} + \sin\theta_a \nu_{2m} e^{i\phi}, \tag{4.28}$$

where

• $\theta_a = \theta_a(t)$ determines the admixtures of eigenstates in $\nu(t)$;

• $\phi(t)$ is the phase difference between the two eigenstates (phase of oscillations):

$$\phi(t) = \int_0^t \Delta H dt' + \phi(t)_T, \tag{4.29}$$

here $\Delta H \equiv H_{1m} - H_{2m}$. The integral gives the adiabatic phase and $\phi(t)_T$ can be related to violation of adiabaticity. It may also have a topological contribution (Berry phase) in more complicated systems;

• $\theta_m(n_e(t))$ determines the flavor content of the eigenstates: $\langle \nu_e | \nu_{1m} \rangle = \cos\theta_m$, etc.

Different processes are associated with these three different degrees of freedom.

4.6. Oscillations in matter. Resonance enhancement of oscillations

In medium with constant density the mixing is constant: $\theta_m(E, n) = const$. Therefore

• the flavors of the eigenstates do not change;

• the admixtures of the eigenstates do not change; there is no $\nu_{1m} \leftrightarrow \nu_{2m}$ transitions, ν_{1m} and ν_{2m} are the eigenstates of propagation;
• monotonous increase of the phase difference between the eigenstates occurs: $\Delta\phi_m = (H_{2m} - H_{1m})t$.

This is similar to what happens in vacuum. The only operative degree of freedom is again the phase. Therefore, as in vacuum, the evolution of neutrino has a character of oscillations. However, parameters of oscillations (length, depth) differ from the parameters in vacuum. They are determined by the mixing in matter and by the effective energy splitting in matter:

$$\sin^2 2\theta \rightarrow \sin^2 2\theta_m, \qquad l_\nu \rightarrow l_m = \frac{2\pi}{H_{2m} - H_{1m}}. \qquad (4.30)$$

For a given density of matter the parameters of oscillations depend on the neutrino energy which leads to a characteristic modification of the energy spectra. Suppose a source produces the ν_e- flux $F_0(E)$. The flux crosses a layer of length, L, with a constant density n_e and then detector measures the electron component of the flux at the exit from the layer, $F(E)$. In fig. 2 we show dependence of

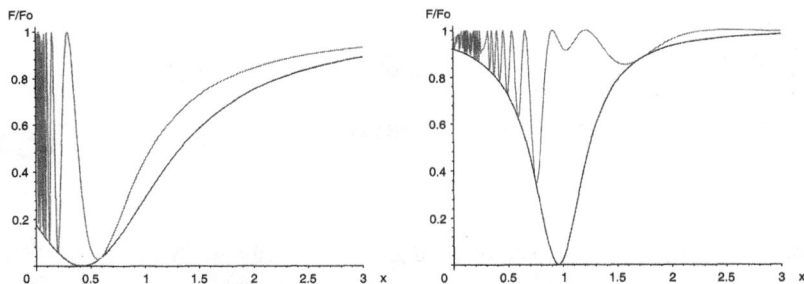

Fig. 2. Resonance enhancement of oscillations in matter with constant density. Shown is a dependence of the ratio of the final and original fluxes, F/F_0, on energy ($x \propto E$) for a thin layer, $L = l_0/\pi$ (left panel) and thick layer $L = 10l_0/\pi$ (right panel). l_0 is the refraction length. The vacuum mixing equals $\sin^2 2\theta = 0.824$.

the ratio $F(E)/F_0(E)$ on energy for thin and thick layers. The oscillatory curve is inscribed in to the resonance curve $(1 - \sin^2 2\theta_m)$. The frequency of the oscillations increases with the length L. At the resonance energy, the oscillations proceed with maximal depths. Oscillations are enhanced in the resonance range:

$$E = E_R \pm \Delta E_R, \qquad \Delta E_R = \tan 2\theta \ E_R = \sin 2\theta \ E_R^0, \qquad (4.31)$$

where $E_R^0 = \Delta m^2 / 2\sqrt{2} G_F n_e$. Notice that for $E \gg E_R$, matter suppresses the oscillation depth; for small mixing the resonance layer is narrow, and the oscillation length in the resonance is large. With increase of the vacuum mixing: $E_R \to 0$ and $\Delta E_R \to E_R^0$.

The oscillations in medium with nearly constant density are realized for neutrinos of different origins crossing the mantle of the Earth.

4.7. MSW: adiabatic conversion

In non-uniform medium, density changes on the way of neutrinos: $n_e = n_e(t)$. Correspondingly, the Hamiltonian of system depends on time, $H = H(t)$, and therefore,

(i) the mixing angle changes during propagation: $\theta_m = \theta_m(n_e(t))$;

(ii) the (instantaneous) eigenstates of the Hamiltonian, ν_{1m} and ν_{2m}, are no more the "eigenstates" of propagation: the transitions $\nu_{1m} \leftrightarrow \nu_{2m}$ occur.

However, if the density changes slowly enough the transitions $\nu_{1m} \leftrightarrow \nu_{2m}$ can be neglected. This is the essence of the adiabatic condition: ν_{1m} and ν_{2m} propagate independently, as in vacuum or uniform medium.

Evolution equation for the eigenstates. Adiabaticity. Let us consider the adiabaticity condition. If external conditions (density) change slowly, the system (mixed neutrinos) has time to adjust this change.

To formulate this condition let us consider the evolution equation for the eigenstate of the Hamiltonian in matter. Inserting $\nu_f = U(\theta_m)\nu_m$ in to equation for the flavor states (4.17) we obtain

$$i \frac{d\nu_m}{dt} = \begin{pmatrix} H_{1m} & -i\dot{\theta} \\ i\dot{\theta} & H_{2m} \end{pmatrix} \nu_m. \tag{4.32}$$

As follows from this equation for the neutrino eigenstates [6,9], $|\dot{\theta}_m|$ determines the energy of transition $\nu_{1m} \leftrightarrow \nu_{2m}$ and $|H_{2m} - H_{1m}|$ gives the energy gap between levels.

If [9]

$$\gamma = \left| \frac{\dot{\theta}_m}{H_{2m} - H_{1m}} \right| \ll 1, \tag{4.33}$$

the off-diagonal terms can be neglected and equations of the eigenstates split. The condition (4.33) means that the transitions $\nu_{1m} \leftrightarrow \nu_{2m}$ can be neglected and the eigenstates propagate independently (the angle θ_a (4.28) is constant).

For small mixing angles the adiabaticity condition is crucial in the resonance layer where (i) the level splitting is small and (ii) the mixing angle changes rapidly. If the vacuum mixing is small, the adiabaticity is critical in the resonance point. It takes the form [6]

$$\Delta r_R > l_R, \tag{4.34}$$

where $l_R = l_\nu / \sin 2\theta$ is the oscillation length in resonance, and $\Delta r_R = n_R / (dn_e/dr)_R \tan 2\theta$ is the spatial width of resonance layer.

MSW-effect. Dynamical features of the adiabatic evolution can be summarized in the following way:
• The flavors of the eigenstates change according to density change. The flavor composition of the eigenstates is determined by $\theta_m(t)$.
• The admixtures of the eigenstates in a propagating neutrino state do not change (adiabaticity: no $\nu_{1m} \leftrightarrow \nu_{2m}$ transitions). The admixtures are given by the mixing in production point, θ_m^0.
• The phase difference increases; the phase velocity is determined by the level splitting (which in turn, changes with density (time)).

Now two degrees of freedom become operative: the relative phase and the flavors of neutrino eigenstates. The MSW effect is driven by the change of flavors of the neutrino eigenstates in matter with varying density. The change of phase produces the oscillation effect on the top of the adiabatic conversion.

Let us derive the adiabatic formula [6,8,10,11]. Suppose in the initial moment the state ν_e is produced in matter with density n_0. Then the neutrino state can be written in terms of the eigenstates in matter as

$$\nu_i = \nu_e = \cos\theta_m^0 \nu_{1m} + \sin\theta_m^0 \nu_{2m}, \tag{4.35}$$

where $\theta_m^0 = \theta_m(n_0)$ is the mixing angle in matter in the production point. Suppose this state propagates adiabatically to the region with zero density (as it happens in the case of the Sun). Then, the adiabatic evolution will consists of transitions $\nu_{1m} \to \nu_1$, $\nu_{2m} \to \nu_2$, and no transition between the eigenstates occurs, so the admixtures are conserved. As a result the final state is

$$\nu_t = \cos\theta_m^0 \nu_1 + \sin\theta_m^0 \nu_2 e^{i\phi}, \tag{4.36}$$

where ϕ is the adiabatic phase. The survival probability is then given by

$$P = |\langle \nu_e | \nu_t \rangle|^2. \tag{4.37}$$

Plugging ν_t (4.36) and ν_e given by (3.3) into this expression and performing averaging over the phase which means that the contributions from ν_1 and ν_2 add incoherently, we obtain

$$P = (\cos\theta \cos\theta_m^0)^2 + (\sin\theta \sin\theta_m^0)^2 = \sin^2\theta + \cos 2\theta \cos^2\theta_m^0. \tag{4.38}$$

This formula gives description of the solar neutrino conversion with accuracy 10^{-7}-corrections due to adiabaticity violation are extremely small [12].

Physical picture of the adiabatic conversion. According to the dynamical conditions, the admixtures of eigenstates are determined by the mixing in neutrino production point. This mixing in turn, depends on the density in the initial point, n_e^0, as compared to the resonance density. Consequently, a picture of the conversion depends on how far from the resonance layer (in the density scale) a neutrino is produced.

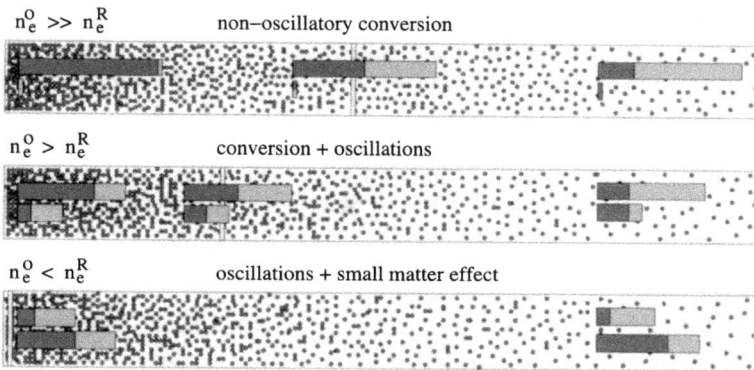

Fig. 3. Adiabatic evolution of neutrino state for three different initial condition (n_e^0). Shown are the neutrino states in different moments of propagation in medium with varying (decreasing) density. The vertical line indicates position of resonance. The initial state is ν_e in all the cases. The sizes of the boxes do not change, whereas the flavors follow the density change.

Three possibilities relevant for solar neutrino conversion are shown in fig. 3. The state produced as ν_e propagates from large density region to zero density. Due to adiabaticity the sizes of boxes which correspond to the neutrino eigenstates do not change.

1) $n_e^0 \gg n_e^R$-production far above the resonance (the upper panel). The initial mixing is strongly suppressed, consequently, the neutrino state, ν_e, consists mainly of one (ν_{2m}) eigenstate, and furthermore, one flavor dominates in a given eigenstate. In the resonance (its position is marked by the yellow line) the mixing is maximal: both flavors are present equally. Since the admixture of the second eigenstate is very small, oscillations (interference effects) are strongly suppressed. So, here we deal with the non-oscillatory flavor transition when the flavor of whole state (which nearly coincides with ν_{2m}) follows the density change. At zero density we have $\nu_{2m} = \nu_2$, and therefore the probability to find

the electron neutrino (survival probability) equals

$$P = |\langle \nu_e | \nu(t) \rangle|^2 \approx |\langle \nu_e | \nu_{2m}(t) \rangle|^2 = |\langle \nu_e | \nu_2 \rangle|^2 \approx \sin^2 \theta. \qquad (4.39)$$

This result corresponds to $\theta_m^0 = \pi/4$ in formula (4.38).

The value of final probability, $\sin^2 \theta$, is the feature of the non-oscillatory transition. Deviation from this value indicates a presence of oscillations.

2) $n_e^0 > n_e^R$ production above the resonance (middle panel). The initial mixing is not suppressed. Although ν_{2m} is the main component, the second eigenstate, ν_{1m}, has appreciable admixture; the flavor mixing in the neutrino eigenstates is significant. So, the interference effect is not suppressed. As a result, here an interplay of the adiabatic conversion and oscillations occurs.

3) $n_e^0 < n_e^R$: production below the resonance (lower panel). There is no crossing of the resonance region. In this case the matter effect gives only corrections to the vacuum oscillation picture.

The resonance density is inversely proportional to the neutrino energy: $n_e^R \propto 1/E$. So, for the same density profile, the condition 1) is realized for high energies, regime 2) for intermediate energies and 3) – for low energies. As we will see all three case are realized for solar neutrinos.

The adiabatic transformations show universality: The averaged probability and the depth of oscillations in a given moment of propagation are determined by the density in a given point and by initial condition (initial density and flavor). They do not depend on density distribution between the initial and final points. In contrast, the phase of oscillations is an integral effect of previous evolution and it depends on a density distribution.

Universal character of the adiabatic conversion can be further generalized in terms of variable [6]

$$n = \frac{n_e^R - n_e}{\Delta n_e^R} \qquad (4.40)$$

which is the distance (in the density scale) from the resonance density in the units of the width of resonance layer. In terms of n the conversion pattern depend only on initial value n_0.

In fig. 4 we show dependences of the average probability and depth of oscillations, that is, $\bar{P}$, $P^{\max}$, and $P^{\min}$, on n. The probability itself is the oscillatory function which is inscribed into the band shown by solid lines. The average probability is shown by the dashed line. The curves are determined by initial value n_0 only, in particular, there is no explicit dependence on the vacuum mixing angle. The resonance is at $n = 0$ and the resonance layer is given by the interval $n = -1 \div 1$. The figure corresponds to $n_0 = -5$, i.e., to production above the

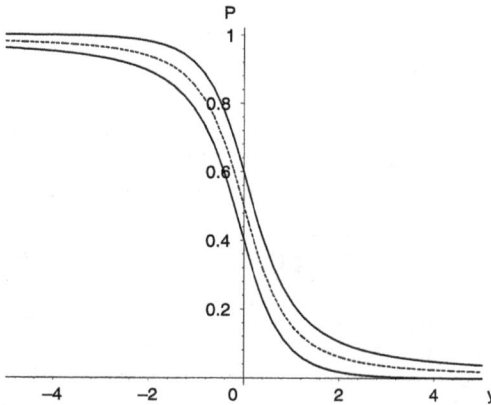

Fig. 4. The dependence of the average probability (dashed line) and the depth of oscillations (P^{max}, P^{min} solid lines) on n for $n_0 = -5$. The resonance layer corresponds to $n = 0$. For $\tan^2 \theta = 0.4$ (large mixing MSW solution) the evolution stops at $n_f = 0.47$.

resonance layer; the oscillation depth is relatively small. With further decrease of n_0, the oscillation band becomes narrower approaching the line of non-oscillatory conversion. For zero final density we have

$$n_f = \frac{1}{\tan 2\theta}. \tag{4.41}$$

So, the vacuum mixing enters final condition. For the best fit LMA point, $n_f = 0.45 - 0.50$, and the evolution should stop at this point. The smaller mixing the larger final n_f and the stronger transition.

4.8. Adiabaticity violation

In the adiabatic regime the probability of transition between the eigenstates is exponentially suppressed $P_{12} \sim exp(-\pi/2\gamma)$ and γ is given in (4.33) [10, 11]. One can consider such a transition as penetration through a barrier of the height $H_{2m} - H_{1m}$ by a system with the kinetic energy $d\theta_m/dt$.

If density changes rapidly, so that the condition (4.33) is not satisfied, the transitions $\nu_{1m} \leftrightarrow \nu_{2m}$ become efficient. Therefore admixtures of the eigenstates in a given propagating state change. In our pictorial representation (fig. 3) the sizes of boxes change. Now all three degrees of freedom of the system become operative.

Typically, adiabaticity breaking leads to weakening of the flavor transition. The non-adiabatic transitions can be realized inside supernovas for the very small 1-3 mixing.

5. Determination of the oscillation parameters

5.1. Solar neutrinos

Analysis includes results from the Homestake experiment [13], from Kamio-kande and SuperKamiokande [14], from radiochemical Gallium experiments SAGE [15], Gallex [16] and GNO [17] and from SNO [18]. The information we have collected can be described in three-dimensional space:

1. Type of events: νe scattering (SK, SNO), CC-events (Cl, Ga, SNO) and NC events (SNO).

2. Energy of events: radiochemical experiments integrate effect over the energy from the threshold to the maximal energy in the spectrum. Also NC events are integrated over energies. The CC events in SNO and νe events at SuperKamiokande give information about the energy spectrum of original neutrinos.

3. Time dependence of rates (searches for time variation of the flux).

Evidence of conversion. There are three types of observations which testify for the neutrino conversion:

1) Deficit of signal which implies the deficit of the electron neutrino flux. It can be described by the ratio $R \equiv N^{obs}/N^{SSM}$, where N^{SSM} is the signal predicted according to the Standard solar model fluxes [19]. The deficit has been found in all (but SNO neutral current) experiments.

2) Energy spectrum distortion – dependence of the suppression factor on energy. Indirect evidence is provided by comparison of the deficits in experiments sensitive to different energy intervals:

$$Low\ energies\ (Ga) \qquad : \quad R = 0.5 - 0.6 \qquad\qquad (5.1)$$

$$High\ energies\ (Cl,\ SK,\ SNO) : \qquad R \approx 0.3. \qquad\qquad (5.2)$$

So the deficit increases with neutrino energy.

3) Smallness of ratio of signals due to charged currents and neutral currents [18]:

$$\frac{CC}{NC} = 0.340 \pm 0.023\ (stat.)\ {}^{+0.029}_{-0.031}(syst.). \qquad\qquad (5.3)$$

The latter is considered as the direct evidence of the flavor conversion since NC events are not affected by this conversion, whereas the number CC events is suppressed.

All this testifies for the LMA MSW solution.

Till now there is no statistically significant observations of other signatures of the conversion, namely,

– distortion of the boron neutrino spectrum: up turn at low energies in SK and SNO; (significant effect should be seen below 5 MeV);

– day-night effect (recall that SK agrees with predictions however significance is about 1σ);

– time variations (semiannual) on the top of annual variations (due to eccentricity of the Earth orbit.

Physics of conversion [20]. Physics can be described in terms of three effects
1) Adiabatic conversion (inside the Sun);
2) Loss of coherence of the neutrino state (on the way to the Earth);
3) Oscillations of the neutrino mass states in the matter of the Earth.

According to LMA, inside the Sun the initially produced electron neutrinos undergo the highly adiabatic conversion: $\nu_e \rightarrow \cos\theta_m^0 \nu_1 + \sin\theta_m^0 \nu_2$, where θ_m^0 is the mixing angle in the production point. On the way from the central parts of the Sun the coherence of neutrino state is lost after several hundreds oscillation lengths [20], and incoherent fluxes of the mass states ν_1 and ν_2 arrive at the surface of the Earth. In the matter of the Earth ν_1 and ν_2 oscillate partially regenerating the ν_e-flux. With regeneration effects included the averaged survival probability can be written as

$$P = \sin^2\theta + \cos^2\theta_{12}^{m0}\cos 2\theta_{12} - \cos 2\theta_{12}^{m0} f_{reg}. \tag{5.4}$$

Here the first term corresponds to the non-oscillatory transition (dominates at the high energies), the second term is the contribution from the averaged oscillations which increases with decrease of energy, and the third term is the regeneration effect f_{reg}. At low energies P reduces to the vacuum oscillation probability with very small matter corrections.

There are three energy ranges with different features of transition:

1. In the high energy part of spectrum, $E > 10$ MeV ($x > 2$), the adiabatic conversion with small oscillation effect occurs. At the exit, the resulting averaged probability is slightly larger than $\sin^2\theta$ expected from the non-oscillatory transition. With decrease of energy the initial density approaches the resonance density, and the depths of oscillations increases.

2. Intermediate energy range $E \sim (2-10)$ MeV ($x = 0.3-2$) the oscillation effect is significant. The interplay of the oscillations and conversion takes place.

For $E \sim 2$ MeV neutrinos are produced in resonance.

3. At low energies: $E < 2$ MeV ($x < 0.3$), the vacuum oscillations with small matter corrections occur.

Inside the Earth. Entering the Earth the state ν_2 (which dominates at high energies) splits in two matter eigenstates:

$$\nu_2 \rightarrow \cos\theta_m' \nu_{2m} + \sin\theta_m' \nu_{1m}. \tag{5.5}$$

It oscillates regenerating partly the ν_e-flux. In the approximation of constant density profile the regeneration factor equals

$$f_{reg} = 0.5 \sin^2 2\theta \, \frac{l_\nu}{l_0}. \tag{5.6}$$

Notice that the oscillations of ν_2 are pure matter effect and for the presently favored value of Δm^2 this effect is small. According to (5.6), $f_{reg} \propto 1/\Delta m^2$ and the expected day-night asymmetry of the charged current signal equals

$$A_{DN} = f_{reg}/P \sim (3-5)\%. \tag{5.7}$$

Apparently the Earth density profile is not constant and it consists of several layers with slow density change and jumps of density on the borders between layers. It happens that for solar neutrinos one can get simple analytical result for oscillation probability for realistic density profile. Indeed, the solar neutrino oscillations occur in the so called low energy regime when

$$\epsilon \equiv \frac{2EV(x)}{\Delta m^2} \ll 1, \tag{5.8}$$

which means that the potential energy is much smaller than the kinetic energy. For the LMA oscillation parameters and the solar neutrinos: $\epsilon(x) = (1-3) \cdot 10^{-2}$. In this case one can use small parameter $\epsilon(x)$ (5.8) to develop the perturbation theory [21]. The following expression for the regeneration factor

$$f_{reg} = \frac{1}{2} \sin^2 2\theta \int_{x_0}^{x_f} dx V(x) \sin \phi_m (x \to x_f). \tag{5.9}$$

Here x_0 and x_f are the initial and final points of propagation correspondingly, and $\phi_m(x \to x_f)$ is the adiabatic phase acquired between a given point of trajectory, x, and final point, x_f. The latter feature has important consequence leading to the attenuation effect – weak sensitivity to the remote structures of the density profile when non-zero energy resolution of detector is taken into account.

Another insight into phenomena can be obtained using the adiabatic perturbation theory which leads to [12]

$$f_{reg} = \frac{2E \sin^2 2\theta}{\Delta m^2} \sin \frac{\phi_0}{2} \sum_{j=0...n-1} \Delta V_j \sin \frac{\phi_j}{2}. \tag{5.10}$$

Here ϕ_0 and ϕ_j are the phases acquired along whole trajectory and on the part of the trajectory inside the borders j. This formula corresponds to symmetric profile with respect to the center of trajectory. Using (5.10) one can easily infer the attenuation effect. The formula reproduces precisely the results of exact numerical calculations.

Determination of the solar oscillation parameters. Knowledge of the energy dependence of the adiabatic conversion allows one to connect the oscillation parameters with observables immediately.

1) Determination of the mixing angle. To explain stronger deficit at higher energies one needs to have $\theta < \pi/4$ or $\sin^2 \theta < 1/2$. Furthermore, using the fact that $P_h > \sin^2 \theta$ and $P_l < 0.5 \sin^2 2\theta$ we find

$$\frac{P_h}{P_l} \geq \frac{\sin^2 \theta}{0.5 \sin^2 2\theta} = \frac{1}{2 \cos^2 \theta}, \tag{5.11}$$

where on the RHS we have taken the asymptotic values of the survival probability. Consequently,

$$\sin^2 \theta \leq 1 - \frac{P_l}{2P_h} \sim 0.1 \pm 0.2. \tag{5.12}$$

The ratio of CC to NC events determines the survival probability:

$$P = \sin^2 \theta + \cos 2\theta \langle \cos^2 \theta_m^0 \rangle = \frac{CC}{NC}. \tag{5.13}$$

For high energies and without Earth matter regeneration effect $P = \sin^2 \theta$. Since no significant distortion of the energy spectrum is seen at SK and SNO the Boron neutrino spectrum should be in the flat part (bottom of the "suppression pit"). In this region the deviation from asymptotic value is weak. For $\Delta m^2 \sim 8 \cdot 10^{-5}$ eV^2 the averaged oscillation effect is about 10%. Therefore

$$\sin^2 \theta_{12} \approx 0.9 \frac{CC}{NC} \approx 0.31. \tag{5.14}$$

2) Determination of Δm^2. Suppression P_l in the Gallium experiment implies that the pp- spectrum is in the vacuum dominated region, whereas stronger suppression of SK and SNO signals (together with an absence of distortion) means that the boron neutrino flux is in the matter dominated region. So the transition region should be $E_{tr} \sim (1 - 4)$ MeV. On the other hand the expression for the middle energy of the transition region equals (it corresponds to neutrino production in resonance)

$$E_{tr} = \frac{\Delta m^2 \cos 2\theta}{2 V_{prod}}, \tag{5.15}$$

A.Yu. Smirnov

where V_{prod} is typical potential in the neutrino production region in the Sun. From (5.15) we obtain

$$\Delta m^2 = \frac{2E_{tr}V_{prod}}{\cos 2\theta}, \tag{5.16}$$

which gives $\Delta m^2 = (3 - 15) \cdot 10^{-5}$ eV^2 in the correct range.

Another way to measure Δm^2 is to study of the high energy effects: according to LMA Δm^2 is restricted from below by increasing day-night asymmetry and from above by absence of the significant up turn of spectrum at low energies.

5.2. KamLAND

KamLAND (Kamioka Large Anti-neutrino detector) is the reactor long baseline experiment [22]. Few relevant details: 1kton liquid scintillator detector situated in the Kamioka laboratory detects the antineutrinos from surrounding atomic reactors (about 53) with the effective distance (150–210) km. The classical reaction of the inverse beta decay, $\bar{\nu}_e p \rightarrow e^+ n$, is used. The data include

(i) the total rate of events;

(ii) the energy spectrum (fig. 5);

(iii) the time dependence of the signal which is due to variations of the reactors power. (Establishing the correlation between the neutrino signal and power of reactors is important check of the whole experiment). In fact, this change also influences the oscillation effect since the effective distance from the reactors changes (e.g., when power of the closest reactor decreases).

In the oscillation analysis the energy threshold $E > 2.6$ eV is established.

The physics process is essentially the vacuum oscillations of $\bar{\nu}_e$. The matter effect, about 1%, is negligible at the present level of accuracy.

The evidences of the oscillations are

1) The deficit of the number of the $\bar{\nu}_e$ events

$$R_\nu = \frac{N_{obs}}{N_{expect}} = \frac{258}{365.2 \pm 23.7} \sim 0.7 \tag{5.17}$$

for $E > 2.6$ MeV.

2) The distortion of the energy spectrum or L/E dependence (when some reactors switch off the effective distance changes). Notice the absence of strong spectrum distortion excludes large part of the parameter space Δm^2.

Oscillation parameters are related to the observables in the following way. The main features of the L/E dependence – maximum at $(L/E)_{max} = 32$ km/MeV (phase $\phi = 2\pi$) and minima at $L/E_m = 16,\ 48$ km/MeV ($\phi = \pi, 3\pi$) fit well

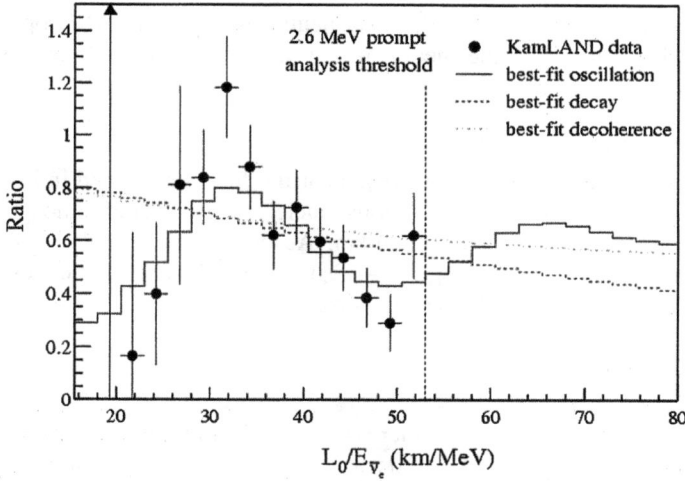

Fig. 5. The L/E distribution of events in the KamLAND experiment; from [22].

the expected oscillation pattern. Taking the first maximum we find

$$\Delta m^2 = \frac{4\pi}{(L/E)_{max}} = 8 \cdot 10^{-5} \text{ eV}^2.$$

The deficit of the signal determines (for a given Δm^2) the value of mixing angle:

$$\sin^2 2\theta_{12} = \frac{1 - R_\nu}{\langle \sin^2 \phi \rangle}, \qquad (5.18)$$

where the averaged over the energy interval oscillatory factor can be evaluated for the KamLAND detector as $\langle \sin^2 \phi \rangle \sim 0.6$. Notice that sensitivity to mixing angle is not high.

Our estimations are a very good agreement with results of statistical analysis of the KamLAND data. Also values of the oscillation parameters are in a very good agreement with those obtained from the solar neutrino analysis. This comparison implies the CPT conservation.

Combined analysis of the solar neutrino data and the KamLAND can be performed in assumption of the CPT conservation. The mixing angle is mainly determined by the solar neutrino data, whereas Δm^2 is fixed by the KamLAND.

Comparison of results from the solar neutrinos and KamLAND open important possibility to check the theory of neutrino oscillation and conversion test CPT, search for new neutrino interactions and new neutrino states.

5.3. Atmospheric neutrinos

Experimental results. The atmospheric neutrino flux is produced in interactions of the high energy cosmic rays (protons, nuclei) with nuclei of atmosphere. The interactions occur at heights (10–20) km. At low energies the flux is formed in the chain of decays: $\pi \rightarrow \mu \nu_\mu$, $\mu \rightarrow e \nu_e \nu_\mu$. So, each chain produces $2\nu_\mu$ and $1\nu_e$, and correspondingly, the ratio of fluxes equals

$$r \equiv \frac{F_\mu}{F_e} \approx 2. \tag{5.19}$$

With increase of energy the ratio increases since the lifetime acquires the Lorentz boost and muons have no time to decay before collisions: they are absorbed or loose the energy. As a consequence, the flux of the electron neutrinos decreases.

In spite of the long term efforts, the predicted atmospheric neutrino fluxes have still large uncertainties (about 20% in overall normalization and about 5% in the so called "tilt" parameter which describes the uncertainty in energy dependence of the flux). The origin of uncertainties is twofold: original flux of the cosmic rays and cross sections of interactions.

The recent analyses include the data from Baksan telescope, SuperKamio-kande [23], MACRO [24], SOUDAN [25]. The data can be presented in the three dimensional space which includes

– type of events detected: e-like events (showers), μ-like events, multi-ring events, NC events (with detection of π^0), τ enriched events.

– energy of events: widely spread classification includes the sub-GeV and multi-GeV events, stopping muons, through-going muons, *etc.*

– zenith angle (upward going, down going, *etc*).

Now MINOS experiment [26] provides with some early information on effects for neutrinos and antineutrinos separately.

The evidence of the atmospheric neutrino oscillations includes:
1) Smallness of the double ratio of numbers of μ-like to e-like events [23]:

$$R_{\mu/e} \equiv \frac{N_\mu^{obs}/N_\mu^{th}}{N_e^{obs}/N_e^{th}}. \tag{5.20}$$

The ratio weakly depends on energy. Apparently in the absence of oscillations (or other non-standard neutrino processes) the double ratio should be 1. The smallness of the ratio testifies for disappearance of the ν_μ flux.

2) Distortion of the zenith angle dependence of the μ-like events (see fig. 6). The up-down asymmetry is defined as

$$A_{up/down} \equiv \frac{N_{up}}{N_{down}}. \tag{5.21}$$

Due to complete up-down symmetric configuration for the production, in the absence of oscillations or other non-standard effect the asymmetry should be absent: $A_{up/down} = 1$.

The zenith angle dependence for different types of events in different ranges of energies is shown in fig. 6 from [23]. The zenith angle of the neutrino trajectory is related to the baseline L as $L = D\cos\theta_z$. So, studying the zenith angle distributions we study essentially the distance dependence of the oscillation probability.

Substantial distortion of the zenith angle distribution is found. The deficit of numbers of events which increases with decrease of $\cos\theta_Z$ and reaches about 1/2 in the upgoing vertical direction for multi-GeV events. The distortion increases with energy. That is, the up-down asymmetry increases with energy:

In contrast to the μ-like, the e-like events distribution does not show any anomaly. Though one can mark some excess (about 15%) of the e-like events in the sub-GeV range.

3) Appearance of the τ-like events [23].

4) L/E dependence shows the first oscillation minimum (fig. 7).

In the first approximation all these data can be consistently described in terms of the $\nu_\mu - \nu_\tau$ vacuum oscillations. Notice that for pure 2ν oscillations of this type no matter effect is expected: the matter potentials of the ν_μ and ν_τ are equal. In the context of three neutrino mixing, for non-zero values of $\sin\theta_{13}$ the matter effect should be taken into account for the $\nu_\mu - \nu_\tau$ channel

Notice that unique description is valid for different types of events and in a very wide range of energies: from 0.1 to more than 100 GeV.

Determination of the atmospheric neutrino oscillation parameters. Let us describe how the oscillation parameters can be immediately related to observables. We will use here the interpretation of the results in terms of 2ν-oscillations $\nu_\mu - \nu_\tau$.

The most clean way to determine parameters is to use the zenith angle distribution of the multi-GeV μ-like events. As follows from fig. 6 for down-going events, $\cos\theta_Z \sim 0.5 - 1$ the oscillation effects are negligible (good agreement with the no-oscillation predictions). For the up-going events $\cos\theta_Z \sim -0.5 - -1$ there is already the averaging oscillation effect. Transition region is for the horizontal events $\cos\theta_Z \sim 0.0 - 0.2$. For these events the baseline $L = 500$ km

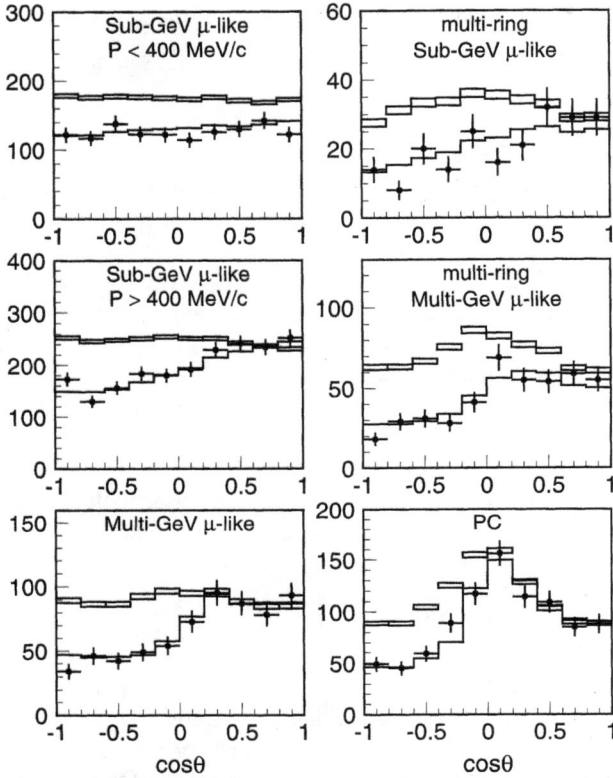

Fig. 6. The zenith angle distribution of the atmospheric μ-like events in different energy ranges; from [23].

should be comparable with the oscillation length $L = l_\nu$, so that

$$\Delta m^2 = \frac{4\pi E_{multi-GeV}}{L_{horizon}}. \tag{5.22}$$

Taking $E = (1 - 2)$ GeV we find $\Delta m^2 = (1 - 4) \cdot 10^{-3}$ eV^2. The uncertainty in the neutrino direction and the fact that distance strongly depends on $\cos\theta_Z$ in the horizontal direction leads to the uncertainty in the determination of the atmospheric Δm^2.

For the upward-going μ-like events the oscillations are averaged (no dependence of the suppression factor on $\cos\theta_Z$), so that $N^{obs}(up)/N^{th}(up) = 1 -$

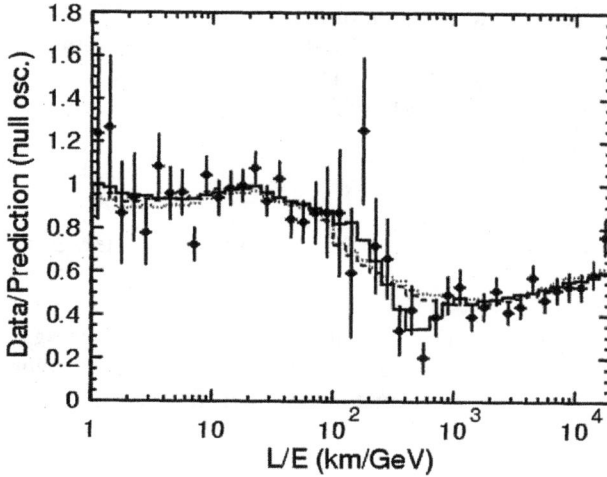

Fig. 7. L/E distribution of the atmospheric μ like events; from [27].

$\sin^2 2\theta$. This allows us to determine the mixing angle:

$$\sin^2 2\theta = 2[1 - N^{obs}(up)/N^{th}(up)].\qquad(5.23)$$

Fron the fig. 6 $N^{obs}(up)/N^{th}(up) \approx 0.5$, and consequently, $\sin^2 2\theta = 1$.

Other independent determinations are possible: in the sub-GeV range the zenith angle dependence is weak because of strong averaging effect: (i) the oscillation length is shorter and therefore the oscillations develop already for the large part of the downgoing events; (ii) the angle between neutrino and detected muon is very large so that directionality is essentially lost. So, taking the deficit of the total rate we obtain

$$\sin^2 2\theta \geq 2[1 - R],\qquad(5.24)$$

where equality corresponds to the developed oscillations for all directions. From fig. 6 we find $R = 0.67$, and therefore $\sin^2 2\theta \geq 0.7$.

To determine mixing angle one can use also the double ratio. As follows from (5.20)

$$\sin^2 2\theta = \frac{1 - R_{\mu/e}}{\langle \sin^2 \phi/2 \rangle_z},\qquad(5.25)$$

where $\langle \sin^2 \phi/2 \rangle_z$ is the averaged over the energy and zenith angle oscillatory factor. For multi-GeV events $\langle \sin^2 \phi/2 \rangle_z = 0.20 - 0.25$ and therefore from (5.25) we obtain $\sin^2 2\theta \sim 1$.

The most precise determination of Δm^2 follows from the L/E - dependence of the events (fig. 7) which is considered as the direct observation of the neutrino oscillations – oscillatory effect [27]. In the first oscillation minimum - dip in the survival probability the phase of oscillations equals $\phi = \pi$. Therefore

$$\Delta m^2 = \frac{2\pi}{(L/E)_{dip}}. \tag{5.26}$$

From fig. 7: $(L/E)_{dip} = 500$ km/GeV. This gives immediately $\Delta m^2 = 2.5 \cdot 10^{-3}$ eV^2.

Results of the 3ν analysis from [28] and [29] with effects of the 1-2 sector are included show one important systematic effect: the shift of mixing from maximal one when the effect of the 1-2 sector is included. Also whole allowed region is shifted. Even larger shift has been found in [29]. Essentially this result is related to the excess of the e-like events in the sub-GeV range.

5.4. K2K

The $\nu_\mu-$ beam with typical energies $E = (0.5 - 3)$ GeV created at KEK was directed to Kamioka and its interations were detected at SuperKamiokande [30]. The baseline (the source-detector distance) is about 250 km. The oscillations of muon neutrinos, $\nu_\mu \rightarrow \nu_\mu$, (as well as $\nu_\mu \rightarrow \nu_e$ - transition probability) have been studied by comparison of the detected number of μ-like events and the energy spectrum with the predicted ones. The predictions have been made by extrapolating of the results from the "front" detector to the SK place. The front detector similar to SK (but of smaller scale) was at about 1 km distance from the source and detected the μ-like events.

The evidence of oscillations was (i) the deficit of the total number of events: 107 events have been observed whereas 151^{+12}_{-10} have been expected. (ii) The spectrum distortion has been found (fig. 8).

The data are interpreted as the non averaged vacuum oscillations $\nu_\mu - \nu_\tau$.

The energy distribution of the detected μ -like events gives an evidence of the first oscillation dip at $E \sim 0.5$ GeV (see fig. 8). This allows to evaluate the value of Δm^2: Using the relation (5.26) with $L/E = 250$km/0.5GeV $= 500$ km/GeV (apparently the same as in the atmospheric neutrino case), we obtain $\Delta m^2 = 2.5 \cdot 10^{-3}$ eV^2 in perfect agreement with the atmospheric neutrino result. (In fact the data stronger exclude other values of Δm^2 than favor the best one.)

The substantial oscillation suppression is present in the low energy part of the spectrum (E < 1 GeV) only. Therefore the deficit of events ~ 0.67 corresponds to large or nearly maximal mixing.

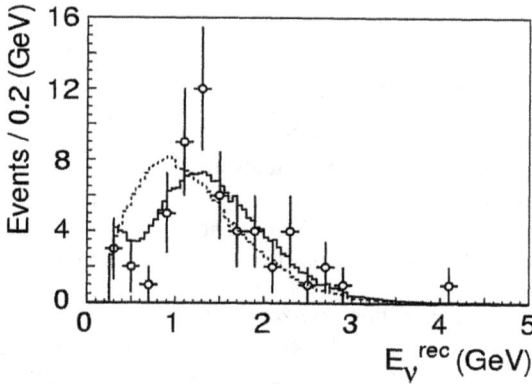

Fig. 8. The energy spectrum of events in the K2K experiment, from [30].

Comment. Simple relations we have presented here allow to understand where sensitivity to different parameters comes from. These relations are embedded in precise statistical analysis. They allow to control the outcome of this analysis, understand uncertainties and give confidence in the results of more sophisticated analysis.

They show robustness of the results and their interpretation.

5.5. 1-3 mixing: effects and bounds

θ_{13} is the last unknown angle in the mixing matrix of active neutrinos;

θ_{13} has important phenomenological consequences: it can produce leading effects for supernova electron (anti) neutrinos and sub-leading effects in the solar and atmospheric neutrinos;

θ_{13} controls the CP-violation in the leptonic sector;

θ_{13} produces the sub-leading effects in the neutrino mass matrix in the flavor basis;

as we will see θ_{13} provides crucial test of mechanism of the lepton mixing enhancement;

non-zero values of θ_{13} can be related to (flavor?) symmetry breaking in the leptonic sector, thus providing tests of this violation.

The direct bounds on 1-3 mixing are obtained in the CHOOZ experiment [31]. This is the experiment with a single reactor and single detector with baseline about 1 km. The effect is vacuum non-average oscillations with survival probability given by the standard oscillation formula

$$P = 1 - \sin^2 2\theta_{13} \sin^2 \phi/2. \tag{5.27}$$

The baseline is comparable with the half oscillation length: For the bf value of Δm^2 from the atmospheric neutrino studies at $E \sim 2$ MeV the oscillation length equals ~ 2 km.

The signature of the oscillations consists of distortion of the energy spectrum described by (5.27). No distortion has been found within the error bars.

In the atmospheric neutrinos non-zero 1-3 mixing will drive oscillations of the electron neutrinos. One of the effects would be $\nu_\mu \leftrightarrow \nu_e$ oscillations in the matter of the Earth. The resonance enhancement of oscillations in neutrino or antineutrino channels should be observable depending on the type of mass hierarchy. That can produce an excess of the e-like events mostly in multi-GeV range where the mixing can be matter enhanced. No substantial effect is found. Notice that in the analysis [29] - the best fit value $\sin \theta_{13}$ is non-zero due to some distortion of the zenith angle dependence. in the multi-GeV range.

In solar neutrinos, the non-zero 1-3 mixing leads essentially to the averaged vacuum oscillations with large frequency and small oscillation depth. The effect is reduced to change of the overall normalization of the flux. The combined analysis of all solar neutrino data gives zero bf value of 1-3 mixing. Whereas the CC/NC measurements at SNO and Gallium results (which depend on the astrophysical uncertainties less) give $\sin^2 \theta_{13} = 0.017 \pm 0.026$.

5.6. Degeneracy of oscillation parameters and global fits

In the previous section we have analyzed various data in the 2ν context. Essentially the 3ν system splits in to two sectors: "solar" sector probed by the solar neutrino data and the "atmospheric" sector probed by the atmospheric neutrino data and K2K. This is justified if 1-3 mixing is zero or small and if in the atmospheric sector studies the effect of 1-2 sector can be neglected. That could happen, e.g., because in the specific experiments the baselines are small or the energies are large, so that the oscillation effects due to 1-2 mixing and 1-2 split have no time to develop.

In the next order in the precision studies when subleading effects, e.g. induced by $\sin \theta_{13}$, become important the split of 3ν problem into two sectors is not possible. At this subleading level the problem of determination of the neutrino parameters becomes much more complicated.

In the table 2 we indicate relevant parameters for different studies.

The same observables depend on several parameters so that the problem of degeneracy of the parameters appears. In such a situation one needs to perform

Table 2

Experiments and relevant oscillation parameters.

Experiments	parameters of leading effects	parameters of sub-leading effects
Solar neutrinos, KamLAND	$\Delta m_{12}^2, \theta_{12}$	θ_{13}
Atmospheric neutrinos	$\Delta m_{23}^2, \theta_{23}$	$\Delta m_{12}^2, \theta_{12}, \theta_{13}, \delta$
K2K	$\Delta m_{23}^2, \theta_{23}$	θ_{13}
CHOOZ	$\Delta m_{23}^2, \theta_{13}$	strongly suppressed
MINOS	$\Delta m_{23}^2, \theta_{23}$	θ_{13}

the global fit of all available data. The advantages are (1) no information is lost; (2) dependence of different observables on the same parameters is taken into account; (3) correlation of parameters and their degeneracy is adequately treated.

There are however some disadvantages. In particular, for some parameters the global fit may not be the most sensitive method, and certain subset of the data can restrict a given parameter much better (*e.g.*, Δm_{23}^2 in atmospheric neutrinos).

In fig. 9 we show the results of the global fit of the oscillation data performed in [29].

Results of global fits of the other groups (see [32]) agree very well. Different types of experiments confirm each other: KamLAND confirms solar neutrino results, K2K – the atmospheric neutrino results *etc.* Furthermore, unique interpretation of whole bulk of the data in terms of vacuum masses and mixings provides with the overall confirmation of the picture So, the determination of the parameters is rather robust, and it is rather non-plausible that future measurements will lead to significant change.

The most probable values of parameters equal

$$\Delta m_{12}^2 = (7.9 - 8.0) \cdot 10^{-5} \text{ eV}^2, \tag{5.28}$$
$$\sin^2 \theta_{12} = 0.310 - 0.315, \tag{5.29}$$
$$\Delta m_{23}^2 = (2.4 - 2.5) \cdot 10^{-3} \text{ eV}^2, \tag{5.30}$$
$$\sin^2 \theta_{23} = 0.44 - 0.50. \tag{5.31}$$

The parameter which describes the deviation of the 23 mixing from maximal equals

$$D_{23} \equiv 0.5 - \sin^2 \theta_{23} = 0.03 - 0.06. \tag{5.32}$$

For 1-3 mixing we have

$$\sin^2 \theta_{23} = 0.00 - 0.01, \qquad 1\sigma = 0.011 - 0.013. \tag{5.33}$$

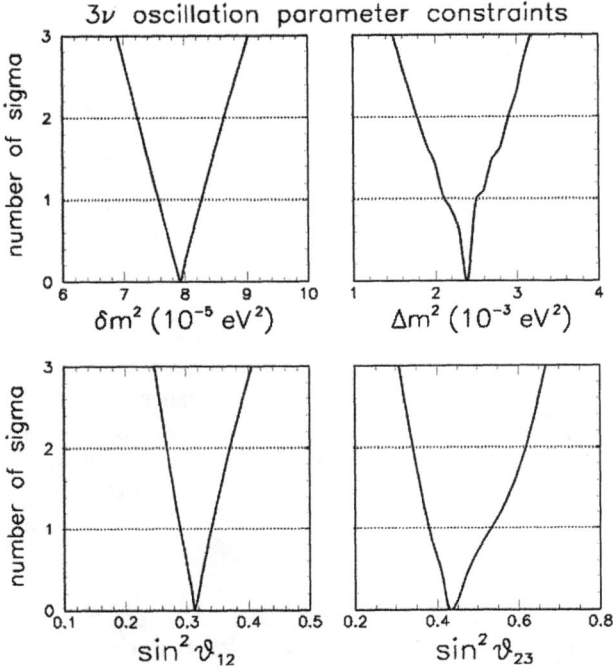

Fig. 9. The results of global 3ν analysis for 1-2 and 2-3 mass splits and mixings; from [29].

The ratio of mass squared differences important for theoretical implications equals

$$r_\Delta \equiv \frac{\Delta m^2_{12}}{\Delta m^2_{23}} = 0.031 - 0.033. \qquad (5.34)$$

6. Neutrino mass and flavor spectrum

6.1. Spectrum

Information obtained from the oscillation experiments allows us to make significant progress in reconstruction of the neutrino mass and flavor spectrum (Fig. 10).

The unknowns are:

(i) admixture of ν_e in ν_3, U_{e3};

(ii) type of mass spectrum: hierarchical, non-hierarchical with certain ordering, degenerate, which is related to the value of the absolute mass scale, m_1;

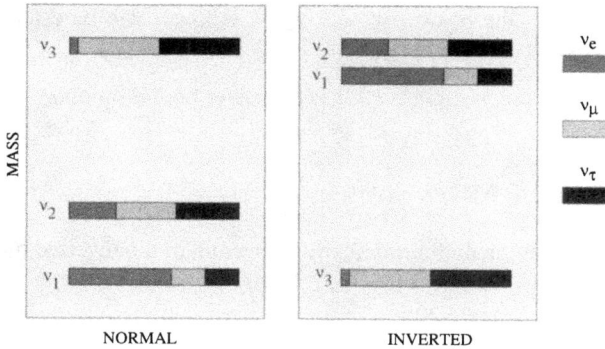

Fig. 10. Neutrino mass and flavor spectra for the normal (left) and inverted (right) mass hierarchies. The distribution of flavors (colored parts of boxes) in the mass eigenstates corresponds to the best-fit values of mixing parameters and $\sin^2 \theta_{13} = 0.05$.

(iii) type of mass hierarchy (ordering): normal, inverted (partially degenerate);
(iv) CP-violation phase δ.

Information described in the previous sections can be summarized in the following way.

1. The observed ratio of the mass squared differences (5.34) implies that there is no strong hierarchy of neutrino masses:

$$\frac{m_2}{m_3} > \sqrt{\frac{\Delta m_{12}^2}{\Delta m_{23}^2}} = 0.18 \pm 0.02. \tag{6.1}$$

For charge leptons the corresponding ratio is 0.06, and even stronger hierarchies are observed in the quark sector.

2. There is the bi-large or large-maximal mixing between the neighboring families (1-2) and (2-3). Still rather significant deviation of the 2-3 mixing from the maximal one is possible.

3. Mixing between remote (1-3) families is weak.

6.2. Absolute scale of neutrino mass

Direct kinematic methods – measurements of the Curie plot of the ^3H decay near the end point – give $m_e < 2.05$ eV (95%), Troitsk after "anomaly" subtraction [33]. And the updated in 2004 result from Mainz experiment [34] $m_e < 2.3$ eV (95%). Future KATRIN experiment [35] aims at one order of magnitude better upper bound: $m_e < 0.2$ eV (90%). The discovery potential is estimated so that the positive result $m_e = 0.35$ eV can be established at 5σ (statistical) level.

From oscillation experiments we get the lower bound on mass of the heaviest neutrino:

$$m_h > \sqrt{\Delta m^2_{atm}} = 0.04 \text{ eV} \quad (95\%). \tag{6.2}$$

In the case of normal mass hierarchy $m_h = m_3$ and in the inverted hierarchy case $m_h = m_1 \approx m_2$.

6.3. Neutrinoless double beta decay

Results. The rate neutrinoless double beta decay is determined by effective Majorana mass of electron neutrino

$$m_{ee} = \left| \sum_k U^2_{ek} m_k e^{i\phi(k)} \right|, \tag{6.3}$$

$\Gamma \propto m^2_{ee}$. Here $\phi(k)$ is the phase of the k eigenvalue.

The best present bound on m_{ee} is given by the Heidelberg-Moscow experiment: $m_{ee} < 0.35 - 0.50$ eV [36]. Part of collaboration claims evidence of a positive signal [37, 38]. Some details follow.

The Heidelberg-Moscow collaboration searched for the mode of the decay

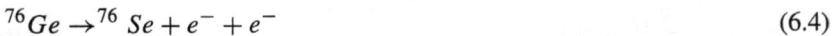

$$^{76}Ge \rightarrow ^{76}Se + e^- + e^- \tag{6.4}$$

with the end point $Q_{ee} = 2039$ keV. The total statistics collected from 5 enriched Ge detectors is 71.7 kg yr. The peak at the end point of spectrum has been found and interpreted in [38] as due to neutrinoless double beta decay.

There is a number of arguments *pro and contra* of such interpretation.

Number of events in the peak (interpreted as $\beta\beta_{0\nu}$ decay gives the half-lifetime

$$T_{1/2} = 1.19 \cdot 10^{25} \text{ y}, \quad 3\sigma \text{ range} : (0.69 - 4.18) \cdot 10^{25} \text{ y}. \tag{6.5}$$

The significance of the peak depends on model of background and quoted by the authors as 4.2σ.

If the exchange of light Majorana neutrino is the dominant mechanism of decay, the measured life time corresponds to the effective mass of the Majorana neutrino:

$$m_{ee} = 0.44 \text{ eV}, \quad 3\sigma \text{ range} : (0.24 - 0.58) \text{ eV}. \tag{6.6}$$

Other groups do not see signal of the $\beta\beta_{0\nu}$ decay though their sensitivity is somehow lower.

Measurements of the neutrinoless double beta decay are of the fundamental importance: apart from checks of the total lepton number conservation and Majorana nature of neutrinos they can provide information about properties of the neutrino mass spectrum: the absolute mass scale and type of mass hierarchy.

To see this we construct the $m_{ee} - m_L$ plot where m_L is the lightest mass eigenstate.

The double beta decay and oscillations. $m_{ee} - m_1$ plot. Introducing the individual contributions of the mass eigenstates $m_{ee}^{(k)}$, $k = 1, 2, 3$, we can write

$$m_{ee} = \sum_k m_{ee}^{(k)} e^{i\phi(k)}. \tag{6.7}$$

Consider first the case of normal mass hierarchy, when $m_3 > m_2 > m_1 = m_L$. The contribution from different eigenstates can be written in terms of the oscillation parameters as

$$m_{ee}^{(1)} = U_{e1}^2 m_L$$
$$m_{ee}^{(2)} = U_{e2}^2 \sqrt{m_L^2 + \Delta m_{21}^2},$$
$$m_{ee}^{(2)} = U_{e3}^2 \sqrt{m_L^2 + \Delta m_{31}^2}. \tag{6.8}$$

The important feature is that

$$U_{e1}^2 > U_{e2}^2 > U_{e3}^2. \tag{6.9}$$

The dependences of $m_{ee}^{(k)}$ on m_L are shown in fig. 11. The change of slope of the curves occurs at $m_L \sim \sqrt{\Delta m_{12}^2}$ and $\sqrt{\Delta m_{13}^2}$. Varying unknown phases ϕ_i we obtain the range of possible values of the effective mass m_{ee} for a given m_L. Apparently maximal value of m_{ee} corresponds to the case when the phase factors are equal. Minimal values correspond to destructive interference of the contributions. Notice in the crossing points of lines which correspond to ν_1 and ν_2 the complete cancellation is possible.

At small m_L (strong mass hierarchy) the contribution $m_{ee}^{(2)}$ dominates:

$$m_{ee} \sim m_{ee}^{(2)} \approx \sin^2 \theta_{12} \sqrt{\Delta m_{12}^2} = (2 - 3) \cdot 10^{-3} \text{eV}. \tag{6.10}$$

Future stronger bounds on $\sin \theta_{13}$ will reduce uncertainty.

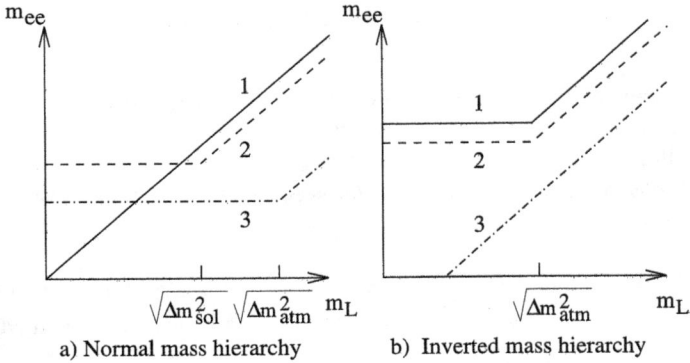

a) Normal mass hierarchy   b) Inverted mass hierarchy

Fig. 11. Contributions from different mass eigenstates to the effective Majorana mass of electron neutrino as functions of the smallest neutrino mass. a) normal mass hierarchy, b). Inverted mass hierarchy.

At $m_L > \sqrt{\Delta m_{12}^2}$ (first two states are degenerate) the first state gives the main contribution. Here the effect of the third state can be neglected and the effective mass is

$$m_{ee} \approx m_L \cos 2\theta_{12} \div m_L. \tag{6.11}$$

In the case of inverted mass hierarchy we have $m_2 > m_1 > m_3 = m_L$. The contributions of different eigenstates equal

$$m_{ee}^{(3)} = U_{e3}^2 m_L,$$
$$m_{ee}^{(1)} = U_{e1}^2 \sqrt{m_L^2 + \Delta m_{31}^2},$$
$$m_{ee}^{(2)} = U_{e2}^2 \sqrt{m_L^2 + \Delta m_{31}^2}. \tag{6.12}$$

Now the first state gives the dominant contribution in whole the range of m_L. There are the following obvious relations:

$$m_{ee}^{(2)}/m_{ee}^{(1)} = U_{e2}^2/U_{e1}^2 = \tan^2 \theta_{12}, \quad m_{ee}^{(3)}/m_{ee}^{(2)} < U_{e3}^2/U_{e1}^2 < 1/7. \tag{6.13}$$

No strong cancellation is possible in this case and the lower bound on the effective mass equals

$$m_{ee} \geq m_L \cos 2\theta_{12} = \cos 2\theta_{12}\sqrt{\Delta m_{12}^2} = 0.015 \text{ eV}. \tag{6.14}$$

The experimental upper bound $m_{ee} < 0.01$ eV will exclude the case of inverted mass hierarchy.

This consideration allows us to understand the $m_{ee} - m_L$ plot fig. 12 from [32] which includes the uncertainties of the oscillation parameters.

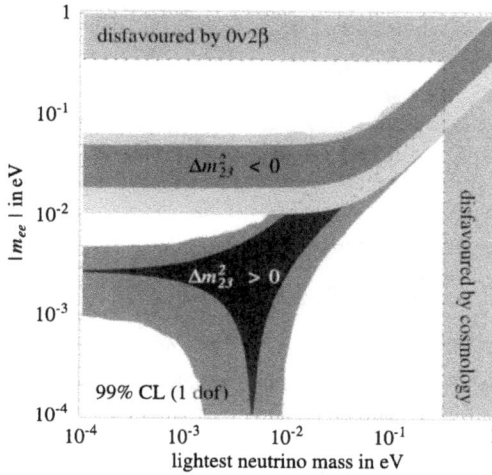

Fig. 12. The 90% CL range for m_{ee} as a function of the lightest neutrino mass for the normal ($\Delta m_{23}^2 > 0$) and inverted ($\Delta m_{23}^2 < 0$) mass hierarchies.[28] The darker regions show how the allowed range for the present best-fit values of the parameters with negligible errors; from [32].

There are several benchmark values of m_{ee}. Apparently if the Heidelberg-Moscow positive result is confirmed, and if it is due the light Majorana neutrino mass, the neutrino mass spectrum should be quasidegenerate.

1) The bound $m_{ee} < 0.05$ eV will exclude the degenerate spectrum;

2) The bound $m_{ee} < 0.01$ eV will exclude the inverted hierarchy.

3) For the plausible scenario with normal mass hierarchy one expects $m_{ee} = 0.003 \pm 0.002$ eV.

4) Strong cancellation, $m_{ee} < 0.001$ eV, is expected for $m_1 = 0.002 - 0.008$ eV.

Determination of m_{ee} faces several problems. In particular, (i) uncertainties of the nuclear matrix elements, (ii) identification of the mechanism of the $\beta\beta_{0\nu}$ decay: other contributions apart from the light Majorana neutrino exchange are possible.

Precise measurements of m_{ee} will allow to check certain test equalities which will allow in principle to get important information on neutrino mass spectrum and CP-violation phases:

1. For the normal mass hierarchy with $m_2 \gg m_1$ and $U_{e3}^2 \ll 0.04$ one expects

$$m_{ee} = \sin^2 \theta_{sol} \sqrt{\Delta m_{sol}^2}. \qquad (6.15)$$

2. The equality

$$m_{ee} = \cos 2\theta_{sol} \sqrt{\Delta m_{atm}^2} \qquad (6.16)$$

will testify for the inverted mass hierarchy and opposite CP-phases of the two heavy states (with strong degeneracy). In the case of the same phases one obtains
$$m_{ee} = \sqrt{\Delta m_{atm}^2}.$$

3. For the degenerate spectrum

$$m_{ee} = m_e \ \ (same\ CP\ phases),$$
$$m_{ee} = \cos 2\theta_{sol} m_e \ \ (opposite\ CP\ phases), \qquad (6.17)$$

where m_e is the absolute scale of masses which can be measured in the β decay experiments.

6.4. Cosmology and neutrino mass

Massive neutrinos influence the Large scale structure formation in the Universe. Let us describe physics of the effect following considerations in [39] and [40,41].

1) Let us define the relative density perturbations as

$$\delta \equiv \frac{\delta \rho}{\rho}. \qquad (6.18)$$

1) If all components in the Universe cluster then the perturbations grow as the scale parameter $a(t)$:

$$\delta \propto a(t). \qquad (6.19)$$

Indeed, using the Einstein equation

$$\rho = 3H^2 \frac{m_{Pl}^2}{8\pi} + \frac{c}{a^2}, \qquad (6.20)$$

we find

$$\delta \rho \approx \frac{\delta c}{a^2}. \qquad (6.21)$$

Here H is the Hubble constant, $c \equiv 3m_{Pl}^2 k/8\pi$ is the curvature and k is the parameter of the Friedman-Robertson-Walker metric which determines geometry of the Universe ($k = 0$ corresponds to the flat Universe).

Since in the matter dominated epoch $\rho \propto a^{-3}$ we obtain from (6.21)

$$\delta = \frac{\delta \rho}{\rho} \propto \delta c \, a. \tag{6.22}$$

2) If only fraction Ω_* of total matter density clusters, the perturbations grow slower:

$$\delta \propto a^p, \quad p \propto \Omega_*^{3/5} < 1. \tag{6.23}$$

Qualitatively effect is clear but obtaining precise powers in this equation is beyond the scope of this lecture.

3) Neutrinos do not cluster on the small enough scales even if they are massive and non-relativistic due to high velocities, v. These scales are determined by the free streaming scale λ_{free}. The latter can be estimated as the distance neutrinos travel while the Universe expands by factor 2:

$$\lambda_{free} \approx v t_2. \tag{6.24}$$

Then on small scales, $\lambda < \lambda_{free}$, neutrino clustering is suppressed (escape velocity is smaller than typical neutrino velocity). On large scale $\lambda > \lambda_{free}$ neutrinos cluster as the cold dark matter and for them $p = 1$. As we will see, this changes the *shape* of spectrum of perturbations in contrast to dark energy.

4) Let us now consider the growth of perturbations in the case when only part of the components in the Universe can cluster. Apparently photons and the dark energy do not cluster. (In fact, effect of photons can be neglected in the first approximation.) When the DE dominates (as it happens now), $\Omega_* \approx 0$ and clustering stops. So, essentially clustering occurs in the period between the epoch characterized by the scale parameter, a_{MD}, when matter starts to dominate, and the epoch $a_{\Lambda D}$ when DE start to dominate.

The perturbation growth factor equals

$$\delta \sim \left(\frac{a_{\Lambda D}}{a_{MD}} \right)^{p(k)}, \tag{6.25}$$

where $k \sim 1/\lambda$ is the wavenumber (inverse of the scale).

In the interval $a_{MD} - a_{\Lambda D}$ only neutrinos can not cluster, so that

$$\Omega_* \approx 1 - f_\nu(k), \tag{6.26}$$

where $f_\nu(k)$ is the energy density in neutrinos for which $1/k > \lambda_{free}$. The growth factor can be written as

$$\left(\frac{a_{\Lambda D}}{a_{MD}} \right)^{(1-f_\nu(k))^{3/5}} \approx \left(\frac{a_{\Lambda D}}{a_{MD}} \right)^{(1-3/5 f_\nu(k))} \sim e^{-4 f_\nu(k)}. \tag{6.27}$$

the last equality takes into account that $a_{\Lambda D}/a_{MD} \approx 4700$.

5) The power spectrum of fluctuation is defined as

$$P(k) = \langle \delta^2(k) \rangle, \tag{6.28}$$

where the averaging should be performed over whole space. Then according to (6.27) we have the suppression of the power spectrum:

$$\frac{P(k, f)}{P(k, 0)} \approx e^{-8 f_\nu(k)}. \tag{6.29}$$

Thus suppression of the power spectrum is determined by the energy density in the non-clustering component for a given wavenumber k. For the non-relativistic neutrinos

$$f_\nu(k) = \sum_i m_i n_i(k). \tag{6.30}$$

For very large k: $k > 1h/Mpc$ (small scales, see fig. 13), all the neutrinos in spectrum satisfy inequality $1/k < \lambda_{free}$ and therefore do not cluster. So we need to take $n_i = 112$ cm^{-3}. If neutrino mass spectrum is degenerate $f_\nu(k) = 3mn$.

With decrease of k only part of neutrino spectrum will satisfy inequality $1/k < \lambda_{free}$ and this part will decrease with k. Consequently, the suppression of the power spectrum decreases with k. For $k < 0.01h/Mpc$ the suppression is practically absent.

The power spectra for zero neutrino mass and $\sum_i m_i = 1$ eV are shown in fig. 13. According to (6.29) for $\sum_i m_i = 1$ eV the power is suppressed by factor 2.

Analysis of the cosmological data which includes CMB, SDSS of galaxies, Lyman alpha forest observations and weak lensing lead to the upper bound [42]

$$m < 0.13, \quad \text{eV}, \quad 95\%. \tag{6.31}$$

Apparently the positive claim of observation of neutrinoless double beta decay is disfavored by the cosmological data.

In future, the weak lensing will allow to perform direct measurements of clustering of all matter and not just luminous one. This will improve the sensitivity down to $\sum_i m_i \sim 0.03$ eV.

6.5. Physics of the long baseline experiments

In what follows we summarize the parameters, physics goals and physics reach of the next generation (already approved) LBL experiments. In each case we give short description of the project, baseline, L, the mean energy of neutrino $\langle E_\nu \rangle$ and the goals. All the estimations are given for the 90% C.L.

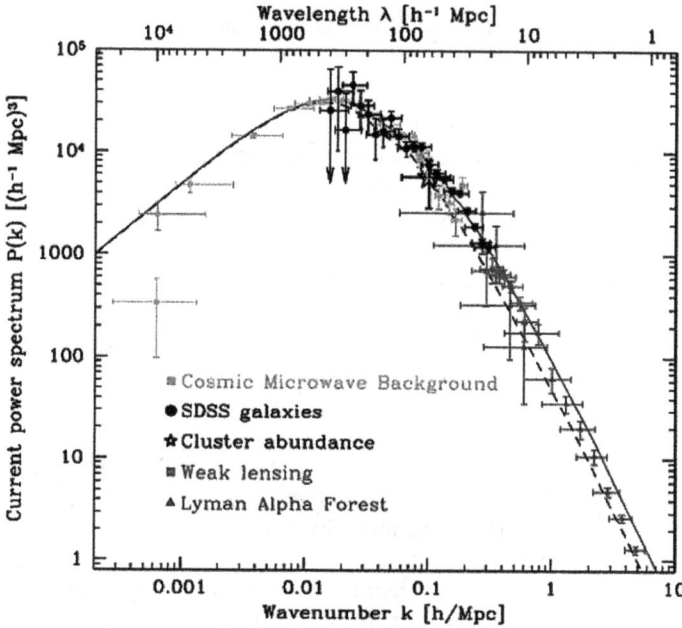

Fig. 13. Spectrum of perturbations: data versus predictions for zero and non-zero neutrino mass; from [41].

1) T2K ("Tokai to Kamioka"): JPARK → SuperKamiokande [43]. This is the accelerator off-axis experiment on searches for $\mu_\mu \to \nu_\mu$ and $\mu_\mu \to \nu_e$ oscillations; parameters of the experiment: $L = 295$ km, $\langle E_\nu \rangle = 0.7$ GeV. The goal is to reach sensitivity to the ν_e appearance which will allow to put the bound $\sin^2 \theta_{13} < 0.005$ (or discover the 1-3 mixing if the angle is larger), to measure 2-3 mass split and mixing with accuracy: $\delta(\Delta m_{23}^2) \sim 0.1$ meV, $\delta(\sin^2 2\theta_{23}) = 0.01$ near the maximal mixing. The latter corresponds to $\delta(\sin^2 \theta_{23}) = D_{23} = 0.05$. If 1-3 mixing is near the present bound the hope is to get some information about the mass hierarchy. The measurements will start in 2009.

2) NOνA: Fermilab → Ash River [44]. This is also the accelerator off-axis experiment on $\nu_\mu \to \nu_\mu$ and $\mu_\mu \to \nu_e$ oscillation searches. Parameters: $L = 810$ km, $\langle E_\nu \rangle = 2.2$ GeV. The goals are the bound on 1-3 mixing $\sin^2 \theta_{13} < 0.006$, precise measurement of Δm_{23}^2 and possibly determination of the mass hierarchy. Start: 2008–2009.

3) Double CHOOZ reactor experiment [45] will search for $\bar{\nu}_e \to \bar{\nu}_e$ oscillation disappearance. Parameters: $L = 1.05$ km, $\langle E_\nu \rangle = 0.004$ GeV, $L/E =$

250 km/GeV; the goal is to reach the bound $\sin^2 \theta_{13} < 0.005 - 0.008$. Start: 2008, results: 2011.

6.6. *Expecting the supernova neutrino burst*

Detection of the Galactic supernova can substantially contribute to determination of the neutrino parameters and reconstruction of the neutrino mass spectrum. This study will contribute to determination of the 1-3 mixing and type of the neutrino mass hierarchy.

In supernovas one expects new elements of the MSW dynamics. Whole 3ν level crossing scheme can be probed and the effects of both resonances (due to Δm_{12}^2 and Δm_{13}^2) should show up. Various effects associated to the 1-3 mixing can be realized, depending on value of θ_{13}. The SN neutrinos are sensitive to $\sin^2 \theta_{13}$ as small as 10^{-5}. Studies of the SN neutrinos will also give an information the type of mass hierarchy [46–49].

The small mixing MSW conversion can be realized due to the 1-3 mixing and the "atmospheric" mass split Δm_{13}^2. The non-oscillatory adiabatic conversion is expected for $\sin^2 \theta_{13} > 10^{-3}$. Adiabaticity violation occurs if the 1-3 mixing is small $\sin^2 \theta_{13} < 10^{-3}$.

Another possible interesting effect is related to shock wave propagation. The shock wave can reach the region of the neutrino conversion, $\rho \sim 10^4$ g/cc, after $t_s = (3-5)$ s from the bounce (beginning of the $\nu-$ burst) [50]. Changing suddenly the density profile and therefore breaking the adiabaticity, the shock wave front influences the conversion in the resonance characterized by Δm_{13}^2 and $\sin^2 \theta_{13}$, if $\sin^2 \theta_{13} > 10^{-4}$.

Monitoring the shock wave with neutrinos can shed some light on the mechanism of explosion [46, 51–53].

6.7. *LSND result and new neutrinos*

Large Scintillator Neutrino Detector collaboration studied interactions of neutrinos from Los Alamos Meson Physics Facility. In particular, neutrinos from the decay chain: $\pi^+ \rightarrow \mu^+ + \nu_e$, $\mu^+ \rightarrow e^+ + \nu_e + \bar{\nu}_\mu$. The excess of the $(e^+ + n)$ events has been observed in the detector which could be due to inverse beta decay: $\bar{\nu}_e + p \rightarrow e^+ + n$ [54]. In turn $\bar{\nu}_e$ could appear due to oscillations $\bar{\nu}_\mu - \bar{\nu}_e$ in the original $\bar{\nu}_\mu$ beam. If confirmed the LSND result may substantially change implications of the discussed results.

Interpretation of the excess in terms of the $\bar{\nu}_\mu - \bar{\nu}_e$ oscillations would correspond to the transition probability

$$P = (2.64 \pm 0.76 \pm 0.45) \cdot 10^{-3}. \tag{6.32}$$

The allowed region is restricted from below by $\Delta m^2 > 0.2$ eV^2.

This result is clearly beyond the "standard 3ν picture. It implies new sector and new symmetries of the theory.

The situation with this ultimate neutrino anomaly [54] is really dramatic: all suggested physical (not related to the LSND methods) solutions are strongly or very strongly disfavored now. At the same time, being confirmed, the oscillation interpretation of the LSND result may change our understanding the neutrino (and in general fermion) masses.

Even very exotic possibilities are disfavored. An analysis performed by the KARMEN collaboration [55] has further disfavored a scenario [56] in which the $\bar{\nu}_e$ appearance is explained by the anomalous muon decay $\mu^+ \rightarrow \bar{\nu}_e \bar{\nu}_i e^+$ ($i = e, \mu, \tau$).

The CPT-violation scheme [57] with different mass spectra of neutrinos and antineutrinos is disfavored by the atmospheric neutrino data [58]. No compatibility of LSND and "all but LSND" data have been found below 3σ [59].

The main problem of the $(3 + 1)$ scheme with $\Delta m^2 \sim 1$ eV^2 is that the predicted LSND signal, which is consistent with the results of other short baseline experiments (BUGEY, CHOOZ, CDHS, CCFR, KARMEN) as well as the atmospheric neutrino data, is too small: the $\bar{\nu}_\mu \rightarrow \bar{\nu}_e$ probability is about 3σ below the LSND measurement.

Introduction of the second sterile neutrino with $\Delta m^2 > 8$ eV^2 may help [60]. It was shown [61] that a new neutrino with $\Delta m^2 \sim 22$ eV^2 and mixings $U_{e5} = 0.06$, $U_{\mu 5} = 0.24$ can enhance the predicted LSND signal by $(60 - 70)\%$. The $(3 + 2)$ scheme has, however, problems with cosmology and astrophysics. The combination of the two described solutions, namely the $3 + 1$ scheme with CPT-violation has been considered [62].

Some recent proposals including the mass varying neutrinos MaVaN [63] and decay of heavy sterile neutrinos [64] also have certain problems.

MiniBooNE [65] is expected to clarify substantially interpretation of the LSND result. MiniBooNE searches for ν_e appearance in the 12 m diameter tank filled in by the 450 t of mineral oil scintillator and covered by 1280 PMT. The flux of muon neutrinos with the average energy $\langle E_\nu \rangle \approx 800$ MeV is formed in π decays (50m decay pipe) which are in turn produced by 8 GeV protons from the Fermilab Booster. The 541 m baseline is about half of the oscillation length for $\Delta m^2 \sim 2$ eV^2. The results of (blind) oscillation analysis will be published in 2006.

Of course, confirmation of the LSND (in terms of oscillations) would be most decisive (though the problem with background should be scrutinized). The negative result still left the situation ambiguous: (in some cases the signal is expected in the antineutrino channel only.)

In fig. 14 the sensitivity limits and discovery potential of MiniBooNE are shown.

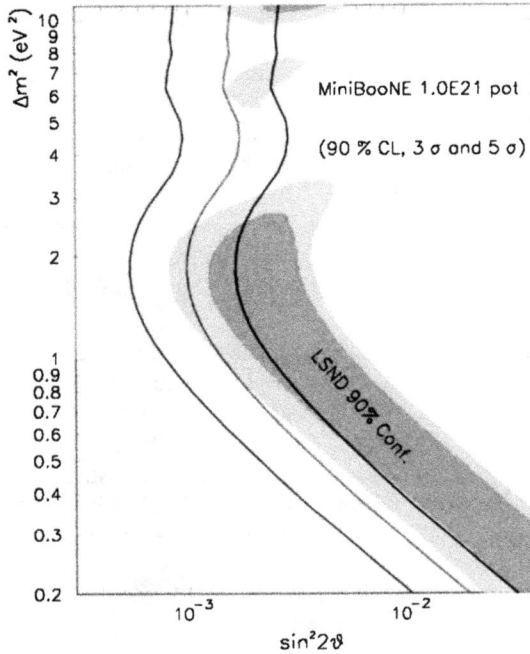

Fig. 14. The region of oscillation parameters selected by LSND result versus sensitivity of the Mini-BooNE experiment; from [65].

7. Toward the underlying physics

7.1. Mass and mixing

There are two salient features related to neutrinos:
- smallness of neutrino mass
- peculiar mixing pattern.

It would be natural to assume that both originate from the same mechanism which is, in fact, related to the neutrality of neutrinos. At the same time the situation can be much more complicated - mass and mixing may not be immediately related For instance the mixing pattern can be determined by some particular symmetries which do not determine masses, smallness of neutrino mass and mixing pattern decouple.

Remarks. Suppose the SM particles are the only light degrees of freedom. Then at low energies (after integrating out the heavy degrees of freedom) one can get the operator: [66]

$$\frac{\lambda_{ij}}{M}(L_i H)^T (L_j H), \quad i, j = e, \mu, \tau, \tag{7.1}$$

where L_i is the lepton doublet, λ_{ij} are the dimensionless couplings and M is the cut-off scale. After EW symmetry breaking it generates the neutrino masses

$$m_{ij} = \frac{\lambda_{ij}\langle H \rangle^2}{M}. \tag{7.2}$$

For $\lambda_{ij} \sim 1$ and $M = M_{Pl}$ we find $m_{ij} \sim 10^{-5}$ eV [67]. Several important conclusions follow immediately from this consideration.

The Planck scale (gravitational) interactions are not enough to generate the observed values of the masses. So, new scales of physics below M_{Pl} should exist.

It has been found that contributions to the neutrino masses of the order $\sim 10^{-5}$ eV are still relevant for phenomenology. Furthermore the sub-dominant structures of the mass matrix can be generated by the Planck scale interactions [68]. So, the neutrino mass matrix can get observable contributions from all possible energy/mass scales from the EW scale (or even lower) to the Planck scale. As a consequence, the structure of the mass matrix can be rather complicated.

Seesaw. The see-saw (type I) mechanism [69, 70] implements the neutrality in full strength (Majorana nature, heavy RH components). Let us introduce the Dirac mass matrix, $m_D = Y v_{EW}$, where Y is the matrix of Yukawa couplings and v_{EM} is the electroweak VEV, and the Majorana mass matrix for the RH neutrinos M. Then in in the basis v, N, ($N = (v_R)^c$) we have the mass matrix

$$\begin{pmatrix} 0 & m_D^T \\ m_D & M \end{pmatrix}. \tag{7.3}$$

For $m_D \ll M$ the diagonalization gives the mass matrix of light neutrinos

$$m = -m_D^T M_R^{-1} m_D \quad (type\ I). \tag{7.4}$$

If the $SU(2)$ triplet, Δ_L, exists which develops a VEV $\langle \Delta_L \rangle$, the left-handed neutrinos can get a direct mass m_L via the interaction $f_\Delta L^T L \Delta_L$. If Δ_L is very heavy, it can develop the induced VEV from interactions with a doublet: $\langle \Delta_L \rangle = v_{EW}^2 / M$. So that

$$m_L = f_\Delta \frac{v_{EW}^2}{M} \quad (type\ II), \tag{7.5}$$

and here we deal with the see-saw of VEV's [71].

In $SO(10)$ with 126_H-plet of Higgses we have $M_R = f v_R$, where f is the Yukawa coupling of the matter 16-plet with 126_H and v_R is the VEV of the $SU(5)$ singlet component of 126_H. Now $f_\Delta = f$, and the general mass term which contains both types of contributions can be written as

$$m = \frac{v_{EM}^2}{v_R}(f\lambda - Y^T f^{-1} Y). \tag{7.6}$$

Here λ is the coupling of 10- and 126-plets of Higgses responsible for the induced VEV of triplet in 126. According to this expression the flavor structure of the two contributions may partially correlate.

GU theories provide with a large mass scale comparable to the scale of RH neutrino masses. Furthermore, one can argue that GUT + see-saw can naturally lead to the large lepton mixing in contrast to the quark mixing. or inversely, one can say that the large lepton mixing testifies for Grand Unification. Indeed, suppose that all quarks and leptons of a given family are in a single multiplet F_i (as 16 of SO(10)). Suppose also that all Yukawa couplings are of the same order thus producing matrices with generically large mixing.

If in the first approximation the Dirac masses are generated by an unique Higgs multiplet, say 10_H of SO(10), the mass matrices of the up and down components of the weak doublets have identical structures, and so, will be diagonalized by the same rotations. As a result, no mixing appears for quarks, and masses of up and down components will be equal to each other.

In contrast to other fermions, the RH neutrinos acquire Majorana masses via the additional Yukawa couplings (with 126_H of SO(10)). If those couplings are also of the generic form, they produce M_R with large mixing which leads then to non-zero lepton mixing. So in the lowest approximation the quark mixing is zero and the lepton mixing can be large. Then the quark mixing appears as correction.

The problem of this scenario is the strong hierarchy of the quark and lepton masses. Indeed, taking the neutrino Dirac masses as $m_D \sim diag(m_u, m_c, m_t)$ in a spirit of GU, we find that for generic M_R the see-saw type I formula (7.4) produces strongly hierarchical mass matrix with small mixings unless M_R has a special structure which compensates the strong hierarchy in m_D.

Other solutions include a substantial difference in the Dirac matrices of quarks and leptons: $m_D(q) \neq m_D(l)$ or a type II see-saw for which there is no relation to m_D. Let us consider the first possibility.

See-Saw enhancement of mixing [72]. Can the same mechanism (see-saw) which explains the smallness of the neutrino mass also explain the large lepton mixing?

The idea is that due to the (approximate) quark-lepton symmetry, the Dirac mass matrices of the quarks and leptons have the same (similar) structure $m_D \sim m_{up}$, $m_l \sim m_{down}$ leading to small mixing in the Dirac sector. The special structure of M_R (which has no analogue in the quark sector) leads to an enhancement of lepton mixing. Two different possibilities have been found [72]:

• strong (nearly quadratic) hierarchy of the RH neutrino masses: $M_{iR} \sim (m_{iup})^2$; and

• strong interfamily connection (pseudo Dirac structures) like

$$M_R \approx \begin{pmatrix} A & 0 & 0 \\ 0 & 0 & B \\ 0 & B & 0 \end{pmatrix}, \quad \text{or} \quad \begin{pmatrix} 0 & A & 0 \\ A & 0 & 0 \\ 0 & 0 & B \end{pmatrix}. \tag{7.7}$$

(Small corrections should be introduced to these matrices.) In the three neutrino context both possibilities can be realized simultaneously, so that the pseudo Dirac structure leads to maximal 2-3 mixing, whereas the strong hierarchy $A \ll B$ enhances the 1-2 mixing.

8. New symmetry of nature?

There are various approaches to perform analysis of the mixing and mass matrices. One can search for particular features of the matrices like equalities, zeros and hierarchies of its elements. Those may testify for certain exact or approximate symmetries. One can try to decompose matrices into the dominant structure and small corrections, identify small parameters, *etc.*

What are results of this "bottom-up" analysis?

The data show that two types of mixing matrices or the corresponding mass matrices can play the role of the dominant structures.

8.1. Bi-maximal mixing [73]

$$U_{bm} \equiv U_{23}^m U_{12}^m = \frac{1}{2} \begin{pmatrix} \sqrt{2} & \sqrt{2} & 0 \\ -1 & 1 & \sqrt{2} \\ 1 & -1 & \sqrt{2} \end{pmatrix}. \tag{8.1}$$

Identification $U_{PMNS} = U_{bm}$ is not possible due to strong (5 - 6) σ deviation of the 1-2 mixing from maximal. However, U_{bm} can play a role of matrix in the lowest order. Correction can originate from the charged lepton sector (mass matrix), so that $U_{PMNS} = U' U_{bm}$ and in analogy with quark mixing $U' \approx U_{12}(\theta_C)$. It generates simultaneously deviation of the 1-2 mixing from maximal and non-zero 1-3 mixing, which are related.

The mass matrix which corresponds to the bi-maximal mixing has the following general form

$$
m_{bm} = \begin{pmatrix} A & B & -B \\ B & 0.5(D+A) & 0.5(D-A) \\ -B & 0.5(D-A) & 0.5(D+A) \end{pmatrix}, \tag{8.2}
$$

where the parameters are related to the mass eigenvalues as

$$
A = \frac{1}{2}(m_1 + m_2), \quad B = \frac{1}{2\sqrt{2}}(m_1 - m_2), \quad D = m_3. \tag{8.3}
$$

The mixing is determined by general structure of the mass matrix (equalities of some elements) and does not depend on their absolute values. Indeed, the permutation symmetry S_2 determines the general structure and therefore the mixing but not the masses which depend on the absolute values of parameters. Masses and mixing are unrelated their explanations decouple.

The matrix (8.2) can be considered as the dominant structure which should be corrected.

More insight can be obtained for particular mass relations or hierarchies.

1) For normal mass hierarchy, $m_1 \ll m_2$:

$$
m_{bm} = \begin{pmatrix} A & A/\sqrt{2} & -A/\sqrt{2} \\ A/\sqrt{2} & 0.5D & 0.5D \\ -A/\sqrt{2} & 0.5D & 0.5D \end{pmatrix}, \quad A \ll D. \tag{8.4}
$$

2) For the inverted mass hierarchy, $|m_1| \approx |m_2|$, $m_3 \approx 0$, there are two substantially different cases depending on the relative CP- violating phases

$$
m_{bm}^{(-)} \approx D \begin{pmatrix} 0 & 1 & 1 \\ 1 & 0 & 0 \\ 1 & 0 & 0 \end{pmatrix}, \quad m_{bm}^{(+)} \approx D \begin{pmatrix} 1 & 0 & 0 \\ 0 & 0.5 & -0.5 \\ 0 & -0.5 & 0.5 \end{pmatrix}. \tag{8.5}
$$

The first matrix corresponds to the opposite phases $m_1 \approx -m_2$, whereas the second one to the same phases $m_1 \approx m_2$ of the first and the second mass eigenstates.

3) For the quasi-degenerate mass spectrum the form of the mass matrix also depends strongly on the CP-violating phases. For $m_1 \approx m_2 \approx -m_3$, we obtain an important example of the "triangle" matrix

$$
m_\triangle \approx A \begin{pmatrix} 1 & 0 & 0 \\ 0 & 0 & 1 \\ 0 & 1 & 0 \end{pmatrix}. \tag{8.6}
$$

And for $m_1 \approx -m_2 \approx -m_3$ we find

$$D \begin{pmatrix} 0 & \frac{1}{\sqrt{2}} & -\frac{1}{\sqrt{2}} \\ \frac{1}{\sqrt{2}} & \frac{1}{2} & \frac{1}{2} \\ -\frac{1}{\sqrt{2}} & \frac{1}{2} & \frac{1}{2} \end{pmatrix}. \tag{8.7}$$

Notice that $m_{ee} = m$ in the first case $m_{ee} = m$, whereas $m_{ee} = 0$ in the second one.

From this consideration some important conclusions can be drown immediately:

1) Form of the matrix depends strongly on the non-oscillation parameters: the absolute mass scale and CP-violating phases.

2) The matrices have reacher symmetry (beside the permutation symmetry marked before). E.g. the matrix (8.6) has S_3 and A_4 symmetry.

The first matrix in (8.5) with the inverted hierarchy obeys also the $L_e - L_\mu - L_\tau$ symmetry [76] (which however does not imply the equality of the 12 and 13 matrix elements).

Notice that all these matrices in the flavor basis lead to the maximal 1-2 mixing and therefore should be corrected. For instance the matrix (8.5) gives $m_1 = m_2 = \sqrt{2}m_0 = \sqrt{\Delta m_{atm}^2}$ and $m_3 = 0$. Corrections can be introduced to generate the solar mass split. These corrections lead also to deviation of the 1-2 mixing from maximal, $1 - \tan^2 \theta_{12} \sim \Delta m_{sol}^2 / \Delta m_{atm}^2$, which is still too small, as well as to small 1-3 mixing: $\theta_{13} \sim \Delta m_{sol}^2 / 2\Delta m_{atm}^2$.

Correct mixing can be obtained if additional contribution from the charge lepton mass matrix is taken into account, so that

$$U_{PMNS} = U' U_{bm}.$$

However in this case the symmetry basis does not coincide with the flavor basis and one should speak on 2-3 permutation symmetry and $L_1 - L_2 - L_3$ symmetry. symmetry.

8.2. Tri-bimaximal mixing [74]

$$U_{tbm} \equiv U_{23}^m U_{12}(\theta_{12}) = \frac{1}{\sqrt{6}} \begin{pmatrix} 2 & \sqrt{2} & 0 \\ -1 & \sqrt{2} & \sqrt{3} \\ 1 & -\sqrt{2} & \sqrt{3} \end{pmatrix}, \tag{8.8}$$

where $\sin^2 \theta_{12} = 1/3$. Here ν_2 is tri-maximally mixed: in the middle column three flavors mix maximally, whereas ν_3 (third column) is bi-maximally mixed.

This matrix is in a good agreement with data, in particular, $\sin^2 \theta_{12}$ is close to the present best fit value 0.31.

The mass matrix which generates the tribimaximal mixing has (in the flavor basis) the following form:

$$m_{tbm} = \begin{pmatrix} A & B & -B \\ B & 0.5(D+A+B) & 0.5(D-A-B) \\ -B & 0.5(D-A-B) & 0.5(D+A+B) \end{pmatrix}, \tag{8.9}$$

and the mass eigenvalues are given by

$$A = \frac{1}{3}(2m_1 + m_2), \quad B = \frac{1}{3}(m_2 - m_1), \quad D = m_3. \tag{8.10}$$

The matrix has also S_2 permutation symmetry $\nu_\mu \leftrightarrow \nu_\tau$ [75].

For normal mass hierarchy, $m_1 \ll m_2$, it reduces (after some rephasing) to

$$m_{tbm} = \begin{pmatrix} A & A & A \\ A & 0.5D+A & -0.5D+A \\ -A & -0.5D+A & 0.5D+A \end{pmatrix}$$

$$= A \begin{pmatrix} 1 & 1 & 1 \\ 1 & 1 & 1 \\ 1 & 1 & 1 \end{pmatrix} + 0.5D \begin{pmatrix} 0 & 0 & 0 \\ 0 & 1 & -1 \\ 0 & -1 & 1 \end{pmatrix}, \tag{8.11}$$

which looks very suggestive.

For the inverted mass hierarchy and the degenerate spectrum (where 1-2 mixing is not determined in the lowest approximation) the matrices coincide with those for the bi-maximal mixing.

8.3. Neutrino mass and horizontal symmetry

Do the results on neutrino masses and mixing indicate certain regularities or symmetry? Can the dominant structures of the mass matrix be explained by a symmetry with the sub-dominant elements appearing as a result of violations of the symmetry? Is the neutrino mass matrix consistent with symmetries suggested for quarks? In this context the following symmetries have been considered.

1) $L_e - L_\mu - L_\tau$ [76]. This symmetry supports, in particular, the structure with an inverted mass hierarchy. However, the rather large element m_{ee} required by the data shows strong violation of this symmetry.

2) Discrete symmetries: A_4 [77], S_3 [79], Z_4 [80], and D_4 [81] see also [82]. They reproduce successfully the dominant structures of the mass matrix the.

Both classes of symmetries 1) and 2) typically treat quarks and leptons differently.

3) $U(1)$ [83]: In the Froggatt-Nielsen context [84] this symmetry can describe mass matrices of both quarks and leptons. The symmetry can explain general structure of the mass matrix – hierarchy of its elements. However, the predictability of this approach is substantially restricted by unknown coefficients (prefactors) of the order 1 (1/2–2) in front of powers of the expansion parameter (usually – Cabibbo angle). The outcome is that the mixing pattern depends substantially on values of these unknown prefactors. Furthermore, the $U(1)$ charges should be considered as discrete free parameters.

4) $SU(2)$, [85] $SO(3)$, [86], and $SU(3)$ [87] require a complicated Higgs sector to break the symmetry. Often models are too restrictive and predictions are on the borders of allowed regions. The problem of Yukawa coupling structure here is reduced to the problem of complicated scalar potential which should produce certain alignment of VEV's.

8.4. Symmetry case

What testifies for the symmetry in the neutrino sector?

(1) Maximal 2-3 mixing

(2) Zero (small) 1-3 mixing;

(3) Particular value 1-2 mixing.

Clearly strong degeneracy of the mass spectrum, if established, will imply symmetry.

It was observed some time ago that the two facts: maximal 2-3 mixing and zero 1-3 mixing can originate both from the same symmetry: invariance of the neutrino mass matrix under $\nu_\mu - \nu_\tau$ permutations in the flavor basis [75]. This permutation symmetry can be a part of larger symmetry which includes also ν_e.

In this connection the flavor symmetry A_4 looks very appealing [77]. It has one triplet representation and three different singlet representations, **1**, **1'**, **1''**, which provides with enough freedom to explain data. Three leptonic doublets form the triplet of A_4: $L_i = (\nu_i, l_i) \sim$ **3**, $i = 1, 2, 3$. Required lepton mixing is generated due to different A_4 transformation properties of the right handed components of charged leptons and neutrinos. It is this difference which eventually leads to mixing. In some models: $l_i^c \sim$ **1**, **1'**, **1''**, whereas $N_i^c \sim$ **3**. In other models *vice versa*: $l_i^c \sim$ **3**, $N_i^c \sim$ **1**, **1'**, **1''**.

Let us consider two examples of the models which illustrate existing achievements and problems.

1) *Model A* [77,78]. The right handed components of charged leptons are three different singlets: $l_i^c \sim$ **1**, **1'**, **1''**. In contrast, the RH components of neutrinos form triplet of A_4: $N_i^c \sim$ **3**. This is one of the most important differences.

Higgs doublets, are invariant under A_4: $H_{1,2} \sim \mathbf{1}$. (This leads to necessity of introduction of new charged leptons; an alternative would be A_4 triplet of the Higgs bosons.) To construct A_4-invariant Yukawa couplings for charged leptons one needs to introduce Higgs EW singlets which transform as triplets of A_4: $\xi_i \sim \mathbf{3}$. Both neutrinos and charged leptons get masses via the see-saw but the chain of couplings for the two are substantially different. For charge leptons one needs to introduce new heavy charged leptons E_i, $E_i^c \sim \mathbf{3}$ and the chain of couplings is

$$l_i - \langle H_1 \rangle - E_i^c - [M_E] - E_i - \langle \xi_i \rangle - l_i^c. \tag{8.12}$$

The mixing is generated in the last step by ξ_i. For neutrinos the mass is formed as

$$\nu_i - \langle H_2 \rangle - N_i^c - [M_M] - N_i^c - \langle H_2 \rangle - \nu_i. \tag{8.13}$$

Extra symmetry is required to forbid unwanted couplings.

The A_4 is broken by VEV of ξ_i which couples to charged leptons and not to neutrinos. This and also the fact that different RH components l_i^c transform according to different singlet representations allows one to reach the goals:

– generate different masses for different charged leptons;
– obtain mixing of charged leptons of specific form which does not depend on mass eigenvalues:

$$U_L = U_{tm} \equiv \begin{pmatrix} 1 & 1 & 1 \\ 1 & \omega & \omega^2 \\ 1 & \omega^2 & \omega \end{pmatrix}, \qquad \omega \equiv e^{-2i\pi/3}. \tag{8.14}$$

Note, in this way we produce mixing matrix which does not depend on masses.

Neutrino mass matrix is diagonal and degenerate. So, in the flavor basis (where the charged leptons are diagonal), the neutrino mass matrix has the form

$$m_0 \propto U_L^T U_L = \begin{pmatrix} 1 & 0 & 0 \\ 0 & 0 & 1 \\ 0 & 1 & 0 \end{pmatrix}. \tag{8.15}$$

This matrix gives maximal 2-3 mixing and zero 1-3 mixing. However corrections should be introduced to generate mass split and 1-2 mixing.

It is interesting that in the same context the mixing matrix of quarks is given by $U_L^\dagger U_L = I$ and the CKM matrix should appear due to corrections. That realizes an idea that strong difference of the quark and lepton mixings appears because in zero order of some approximation the quark mixing is zero whereas the lepton mixing is non-zero and large. The large lepton mixing is related to the Majorana nature of neutrinos. To generate neutrino mass split 1-2 leptonic mixing and quark mixing one need to introduce corrections to the above mass matrices. In [78] the radiative mechanism has been proposed to generate these corrections.

Model B: Getting tribimaximal mixing [88]. It is modification of the model A: specifically – modification of the Majorana mass matrix of the RH neutrinos. The charged lepton sector and mechanism of generation of masses coincide with those of the Model A. Additional Higgs multiplets are introduced: A_4 triplet (and SM singlet) ξ_i' and two singlets $S_{1,2}$ which couple to the RH neutrino only.

The scheme of neutrino mass generation is

$$\nu_i - \langle H_2 \rangle - N_i^c - (\langle \xi_2 \rangle, \langle S_{1,2} \rangle) - N_i^c - \langle H_2 \rangle - \nu_i. \tag{8.16}$$

This produces non-diagonal RH mass matrix which (via seesaw) leads to light neutrino mass matrix with maximal 1-3 rotation: $U_\nu = V_{13}^m$. For this the crucial condition is that only the second component of ξ_2 acquires non-zero VEV.

As a result after rephasing the lepton mixing matrix equals

$$U_{PMNS} = U_{tm} V_{13}^m = U_{tbm}. \tag{8.17}$$

It is interesting that tri/bimaximal mixing equals the product of the trimaximal and maximal 1-3 rotations.

On symmetry approach. The main question here is whether the "neutrino" symmetries are accidental or real, that is, have some physics behind. Models proposed so far are rather complicated with a number of *ad hoc* assumptions. It is difficult to include quarks in these models. Further (Grand) unification looks rather problematic. Asymmetries between neutrinos and leptons are embedded into theory from the beginning. This shows the price one should to pay for realization of the "neutrino" symmetries.

Furthermore, the facts behind the symmetries - maximal 2-3 mixing and relatively small 1-3 mixing are not yet well established. Still significant deviation of 2-3 mixing is possible and 1-3 mixing can be not so small. Structure of the neutrino mass matrix depends substantially on these deviations. So, it may happen that symmetry constructions are simply misleading.

On the other hand if symmetries are not accidental, they have consequences of the fundamental importance as the models constructed show. New structures and particles are predicted, unification path may differ substantially from what we are considering now, *etc.* The symmetries may give some clue for understanding fermion masses in general.

The key question is how to test this? Obviously, we need to search for and measure deviations: of 2-3 mixing from maximal, D_{23}, and 1-3 mixing, $\sin\theta_{13}$, from zero. In the context of specific models the deviations (though small) are expected anyway. The facts we are discussing can originate from the same symmetry and violation of this symmetry will lead then to relations between D_{23} and $\sin\theta_{13}$.

9. Leptons and quarks

There is apparent correspondence between quarks and leptons. Each quark has its own counterpartner in the leptonic sector. Leptons can be treated as the 4th color [89] following the Pati-Salam $SU(4)$ unification symmetry. Unification is possible, so that quarks and leptons form multiplets of the extended gauge group. The most appealing one is SO(10) [90], where all known components of quarks and leptons (including the RH neutrinos) form unique 16-plet. It is difficult to believe that these features are accidental. Though it is not excluded that the quark-lepton connection has some more complicated form, e.g., of the quark - lepton complementarity [91, 92]. We will consider the quark-lepton symmetry and unification later.

9.1. Comparing results

The mixing patterns of leptons and quarks is strongly different: the lepton mixings are large whereas quark mixings are small. The only common feature is that the 1-3 mixing (between the "remote" generations) is small in both cases. Two other angles look complementary in a sense that they sum up to maximal mixing:

$$\theta_{12} + \theta_C \approx \frac{\pi}{4}, \qquad\qquad (9.1)$$

and similar approximate relation is satisfied for the 2-3 mixings. For various reasons it is difficult to expect precise relation but qualitatively one can say that, the 2-3 mixing in the lepton sector is close to maximal because the corresponding quark mixing is very small, the 1-2 mixing deviates from maximal substantially because the 1-2 (Cabibbo) quark mixing is relatively large. It seems that for the third angle we do not expect simple relation and apparently the quark feature $\theta_{13} \sim \theta_{12} \times \theta_{23}$ does not work in the lepton sector.

The ratio of neutrino masses (6.1) can be compared with ratios for charged leptons and quarks (at m_Z scale): $m_\mu/m_\tau = 0.06$, $m_s/m_b = 0.02 - 0.03$, $m_c/m_t = 0.005$. The neutrino hierarchy – see eq. (6.1) (if exists at all - still the degenerate spectrum is not excluded) is the weakest one. This is consistent

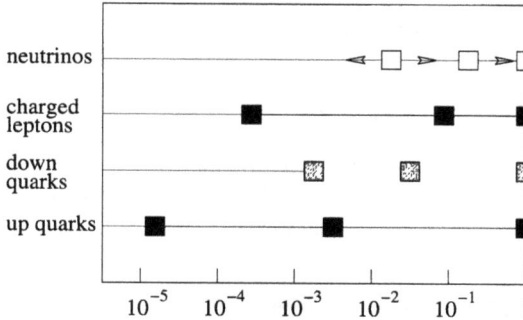

Fig. 15. Mass hierarchies of quarks and leptons. The mass of heaviest fermion of a given type is taken to be 1.

with possible mass-mixing relation: large mixings are associated to weak mass hierarchy.

In fig. 15 we show the mass ratios for three generations. The strongest hierarchy and geometric relation $m_u \times m_t \sim m_c^2$ exist for the upper quarks. Apart from that no simple relations show up.

What is behind this picture? Symmetry, regularities, relation? In the quark sector we can speak about fermion families with weak interfamily connection (mixing) which means strong flavor alignment. In the lepton sector the alignment is weaker.

Furthermore, peculiar situation with fermion masses is that spectra have small number of states (levels) - 3 (in contract to atomic or nuclear levels), and on the other hand there is no simple relations between parameters of spectra. This may indicate that physics behind fermion masses is rather complicated. It looks like the observed pattern is an interplay of some regularities and randomness ("anarchy").

9.2. Quark-lepton universality

The picture described in the previous section is still consistent with the approximate quark-lepton symmetry or universality. However, the symmetry is realized in terms of mass matrices (matrices of the Yukawa couplings) and not in terms of observables - mass ratios and mixing angles.

The key point is that similar mass matrices can lead to substantially different mixing angles and masses (eigenvalues) if the matrices are nearly singular (rank-1) [93, 94]. The singular matrices are "unstable" in a sense that small perturbations can lead to strong variations of mass ratios and mixing angles (the latter - in the context of seesaw.

Let us consider the universal structure for the Yukawa coupling matrices of all quarks and leptons [94]:

$$Y_u \sim Y_d \sim Y_D \sim Y_M \sim Y_L \sim Y_0, \tag{9.2}$$

where Y_D is the Dirac type neutrino Yukawa matrix, Y_M is the Majorana type matrix for the RH neutrinos and Y_0 is the singular matrix. As an important (though may be not the best) example we can take

$$Y_0 = \begin{pmatrix} \lambda^4 & \lambda^3 & \lambda^2 \\ \lambda^3 & \lambda^2 & \lambda \\ \lambda^2 & \lambda & 1 \end{pmatrix}, \quad \lambda \sim 0.2 - 0.3. \tag{9.3}$$

This matrix has only one non-zero eigenvalue and since all matrices have the same structure, the mixing is zero.

Let us introduce perturbations ϵ in the following form

$$Y_{ij}^f = Y_{ij}^0 (1 + \epsilon_{ij}^f), \quad f = u, d, e, \nu, N, \tag{9.4}$$

where Y_{ij}^0 is the element of the original singular matrix. This form can be justified, e.g. in the context of the Froggatt-Nielsen mechanism [84]. (The key element is the form of perturbations (9.4) which distinguishes the ansatz (9.3) from other possible schemes with singular matrices.) It has been shown that small perturbations $\epsilon \le 0.25$ are enough to explain large difference in mass hierarchies and mixings of quarks and leptons [94].

Smallness of neutrino mass is explained by the seesaw mechanism. Furthermore, nearly singular matrix of the RH neutrinos leads to enhancement of the lepton mixing [72] and to flip of sign of mixing angle which comes from diagonalization of the neutrino mass matrix. So, the angles from the charged leptons and neutrinos sum up, whereas in quark sector mixing angles from up and down quark mass matrices subtract.

Keeping this possibility in mind one can consider the following "working" hypothesis:
1) No particular "neutrino" symmetry exists, and in general one expects some deviation of the 2-3 mixing from maximal as well as non-zero 1-3 mixing. Nearly maximal 2 -3 mixing would be accidental in this case.
2) Seesaw mechanism with the scale of RH neutrino masses $M \sim (10^7 - 10^{15})$ GeV explains smallness of neutrino mass. The upper part of this range is close to the GU scale and can be considered as indication of the Grand Unification.
3) The quark-lepton unification or Grand Unification are realized in some form, e.g. $SO(10)$.

4) The quark-lepton symmetry is (weakly) broken and some observable consequences like, $m_b = m_\tau$, exist.

5) Large lepton mixing is a consequence of the seesaw type-I mechanism - the seesaw enhancement of lepton mixing due to special structure of the RH neutrino mass matrix, (or/and of the contribution from the type II seesaw).

6) Flavor (family) symmetry or/and physics of extra dimensions determine this special structure.

Testing scenario. This is the key question which requires essentially the test of existence of the heavy Majorana RH neutrinos. The RH neutrinos can produce renormalization effects above the scale of their masses: between M_R and, say, the GUT scale. In particular, they can renormalize the $m_b - m_\tau$ mass relation [95] which leads to the observable effect in the assumption of $m_b - m_\tau$ unification at the GUT scale. Another possibility is that the renormalization due to RH neutrinos modifies masses and mixing of the light neutrinos, e.g., enhances the mixing [96].

Several indirect) possibilities to test seesaw are known at present:

– Leptogenesis [97];
– Neutrinoless beta decay;
– Rare lepton number violating decays [98].

Radical solution – the "Low scale seesaw" with masses of the RH neutrinos at 1 TeV or 1 keV or even 1 eV have been proposed. One of the motivations is to make things testable.

9.3. Quark-lepton complementarity (QLC)

The complementarity condition (9.1) would require certain modification of the picture described above [91, 92]. The latest determination of the solar mixing angle gives

$$\theta_{12} + \theta_C = 46.7° \pm 2.4° \quad (1\sigma),$$

which is consistent with maximal mixing angle within 1σ. If not accidental, the QLC relation implies that there is some structure in the theory which generates maximal or bi-maximal mixing and it should be non-trivial quark-lepton connection which communicates the quark mixing to the lepton sector. The fact that for the 2-3 mixings the approximate complementarity is also fulfilled hints some more serious reason than just numerical coincidence.

A general scheme is that

$$\text{"lepton mixing} = \text{bi} - \text{maximal mixing} - \text{CKM"}. \tag{9.5}$$

There is a number of non-trivial conditions for the exact QLC relation to be realized.

(i) Order of rotations: apparently U_{12}^m and $U_{12}^{CKM\dagger}$ should be attached

$$U_{PMNS} \equiv U_L^\dagger U_\nu = \ldots U_{23}^m \ldots U_{12}^m U_{12}^{CKM\dagger} \qquad (9.6)$$

(two last rotations can be permuted). Different order leads to corrections to the exact QLC relation;

(ii) Matrix with CP violating phases should not appear between $U_{12}^{CKM\dagger}$ and U_{12}^m, or the corresponding mixing should be small enough.

(iii) Presumably the quark-lepton symmetry which leads to the QLC relation is realized at high mass scales. Therefore the renormalization group effects should be small enough, *etc.*

Let us describe two possible scenarios which differ by origin of the bi-maximal mixing and lead to different predictions [92].

1) QLC1: In the symmetry basis, the bi-maximal mixing is generated by the neutrino mass matrix, presumably due to seesaw. The charged lepton mass matrix produces the CKM mixing as a consequence of the q-l symmetry: $m_l \approx m_d$. In this case the order of matrices (9.6) is not realized (U_{12}^{CKM} should be permuted with U_{23}^m), and consequently the QLC relation is modified:

$$\sin\theta_{12} = \sin(\pi/4 - \theta_C) + 0.5\sin\theta_C(\sqrt{2} - 1). \qquad (9.7)$$

Numerically we find $\tan^2\theta_{12} = 0.495$ which is practically indistinguishable from the tri-bimaximal mixing with $\tan^2\theta_{12} = 0.50$.

2) QLC2: Maximal mixing comes from the charged lepton mass matrix and the CKM mixing originates from the neutrino mass matrix due to the q-l symmetry: $m_D \sim m_u$ (assuming also that in the context of seesaw the RH neutrino mass matrix does not influence mixing). In this case the QLC relation is satisfied precisely: $\sin\theta_{12} = \sin(\pi/4 - \theta_C)$, and the 1-3 mixing is very small.

There are two main issues related to the QLC relation:

(1) origin of the bi-maximal mixing;

(2) mechanism of propagation of the CKM mixing from the quark to the lepton sector. The problem here is large difference of mass ratios in the quark and lepton sectors: $m_e/m_\mu = 0.0047$, $m_d/m_s = 0.04 - 0.06$, as well as difference of masses of muon and s-quark at the GU scale. This means that mixing should weakly depend or be independent on masses.

The mass matrices are different for quarks and leptons and "propagation" of the CKM mixing leads to corrections to the QLC relation of the order $\Delta\theta_{12} \sim \theta_C m_d/m_s \sim 0.5 - 1.0°$ [92].

The Cabibbo mixing can be transmitted to the lepton sector in more complicated way (than via the q-l symmetry). In fact, $\sin\theta_C$ may turn out to be the generic parameter of theory of fermion masses and therefore to appear in various places: mass ratios, mixing angles. The relation: $\sin \theta_C \approx \sqrt{m_\mu / m_\tau}$ is in favor of this possibility. On the other hand, this relation may indicate that the QLC relation is accidental. Indeed, it can be rewritten as a pure leptonic relation $\theta_{12} + \theta_{\mu\tau} = \pi/4$, where $\tan \theta_{\mu\tau} \equiv \sqrt{m_\mu / m_\tau}$. Though this relation may even be more difficult to realize.

So, if not accidental the QLC relation may have two different implications: One includes the quark-lepton symmetry, existence of some additional structure which produces the bi-maximal mixing, and mass matrices with weak correlation of the mixing angles on mass eigenvalues. Alternatively, it may imply certain flavor physics with $\sin \theta_C$ being the "quantum" of this physics.

10. See-saw and GUT's

10.1. Seesaw: variations on the theme

The number of RH neutrinos (or SM singlets involved in generation of neutrino mass) can differ from 3. In fact minimal number of the RH neutrinos needed to generate masses of light neutrinos via type I seesaw is 2. In this case we have the 3×2 see-saw [99]. Such a possibility can be realized in the limit when one of the RH neutrinos is very heavy: $M \sim M_{Pl}$, being, unprotected by, *e.g.*, the $SU(2)_H$ horizontal symmetry. It leads to one massless LH neutrino and smaller number of free parameters.

The number of SM singlets involved in the neutrino mass generation can be larger than 3, moreover additional singlets may not be related to the family structure.

Alternatively, three additional singlets, S, which belong to families, can couple to the RH neutrinos. In the latter case the double see-saw can be realized [100].

In the basis (v, v^c, S) the mass matrix may have the form

$$\begin{pmatrix} 0 & m_D & 0 \\ m_D^T & 0 & M \\ 0 & M^T & M_S \end{pmatrix} \tag{10.1}$$

due to certain symmetries including the lepton number one. It leads to the light neutrino masses:

$$m = -m_D^T (M^{-1})^T M_S M^{-1} m_D. \tag{10.2}$$

Two interesting limits are:

(i) $M_S \ll M$, it allows one to reduce all high mass scales for the same values of the light neutrino masses. In this case in each generation one has heavy pseudoDirac neutrino with mass $\sim M$. Such a possibility has some justification in the string theory.

(ii) $M_S \gg M$ produces the "cascade" seesaw: the mass of RH neutrino also appears as a result of seasaw:

$$M_R = M M_S^{-1} M^T. \tag{10.3}$$

For $M_S = M_{Pl}$, and $M = M_{GU}$ we obtain the required intermediate mass scale for these masses $M_R = M_{GU}^2/M_{Pl} = (10^{12} - 10^{14})$ GeV.

Seesaw type III [101]. Three additional singlets can couple both to the LH and RH neutrinos, so that the mass matrix in the basis (ν, N, S) becomes

$$m_\nu \approx \begin{pmatrix} 0 & m_D & m \\ m_D^T & 0 & M \\ m^T & M^T & 0 \end{pmatrix}, \tag{10.4}$$

where $m \sim v_{EW}$. The lepton number is violated in this system since S couples with both ν and N which have the lepton numbers $+1$ and -1 correspondingly.

The Majorana mass matrix of light neutrinos becomes

$$m_\nu = m_D (M^T)^{-1} m^T + (transponent). \tag{10.5}$$

The new feature of this matrix is that it is linearly proportional to the Dirac mass matrix in contrast to the seesaw type I. Here the cancellation of the mass hierarchies occurs if $M \propto m$. The spectrum of the heavy components consists of three pairs of the preudoDirac neutrinos which can lead to the resonance leptogenesis.

Screening of Dirac structure. The quark-lepton symmetry manifests as certain relation (similarity) between the Dirac mass matrices of quarks and leptons, and it is this feature which creates problem for explanation of strongly different mixings and possible existence of the "neutrino" symmetries. Let us consider an extreme case when in spite of the q-l unification, the Dirac structure in the lepton sector is completely eliminated – "screened" [102].

Consider the double seesaw structure (10.1). Suppose that due to some horizontal symmetry or Grand unification which includes also new singlets S, the two Dirac mass matrices in the double seesaw are proportional to each other:

$$M_D = A^{-1} m_D, \quad A \equiv v_{EW}/V_{GU}. \tag{10.6}$$

Then they cancel each other in (10.2) and for the light neutrinos we obtain

$$m_\nu = A^2 M_S. \tag{10.7}$$

That is, the structure of light neutrino mass matrix is determined by M_S immediately and does not depend on the Dirac mass matrix. In this case the seesaw mechanism provides with the scale of neutrino masses but not the flavor structure of the mass matrix. It can be shown that at least in SUSY version the radiative corrections do not destroy screening [102].

Structure of the light neutrino mass matrix depends now on M_S which can be related to some physics at the Planck scale, and consequently lead to "unusual" neutrino properties. In particular, (i) M_S can be the origin of the "neutrino" symmetry; (ii) the matrix $M_S \propto I$ leads to the quasi-degenerate spectrum; (iii) M_S can be the origin of the bi-maximal or maximal mixing thus leading to the QLC relation if the charged lepton mass matrix generates the CKM rotation.

10.2. GUT's and neutrino mass

GUT's naturally provides us with
- the RH neutrino components,
- large mass scale,
- lepton number violation.

So, it contains all ingredients needed for realization of the seesaw mechanism. What else GUT's can do for neutrinos? Generically GUTs give relations between masses and mixings of the quarks and leptons (see for review [103]). Nature of the relations, however, is model dependent. It is determined by the gauge symmetry, representations of fermions and Higgses and number of various Higgs representations.

The highest predictivity is of course when all fermions are in the same multiplet (like **16**-plet of $SO(10)$) and only one higgs multiplet generates masses (unless some additional principles are introduced on the top of GUT). At this point we can discuss "Minimal $SO(10)$" model with only two Higgs multiplets which generate the fermion masses: $\mathbf{10}_H$ and $\overline{\mathbf{126}}_H$ [104]. With only one $\mathbf{10}_H$ the predictions are the most stringent but contradict observations: all up masses are equal, all down masses are equal, mass hierarchies are the same for all fermions and there is no mixing.

So one needs to introduce other sources of fermion masses and the straightforward step is to add $\overline{\mathbf{126}}_H$. Now predictivity becomes weaker: instead of equalities of masses we get the "sum rules" [105].

Indeed, with 10_H and $\overline{126}_H$ the following mass matrices are generated:

$$M_u = Y_{10}v_{10}^u + Y_{126}v_{126}^u,$$
$$M_d = Y_{10}v_{10}^d + Y_{126}v_{126}^d,$$
$$M_l = Y_{10}v_{10}^d - 3Y_{126}v_{126}^d,$$
$$M_\nu = Y_{126}k, \tag{10.8}$$

where Y_{10} and Y_{126} are the matrices of the Yukawa couplings and v_{10}^u, v_{10}^d and v_{126}^u, v_{126}^d VEV's are the VEV's of 10_H and $\overline{126}_H$ correspondingly. It is assumed in (10.8) that seesaw type II, (due to the EW triplet in $\overline{126}_H$) gives the main contribution to neutrino mass and k denotes the induced VEV of this triplet.

Excluding product of Yukawas and VEV's in the above system of equations we find the sum rule

$$M_\nu \propto M_l - M_d \tag{10.9}$$

– relation between mass matrices of the charged leptons, down quarks and neutrinos. The $b - \tau$ unification (that is $m_b \approx m_\tau$ at the GUT scale) implies $(M_l)_{33} \approx (M_d)_{33}$. Consequently from the relation (10.9) we obtain $(M_\nu)_{33} \approx 0$. In fact, numerically $(M_\nu)_{33} \approx (M_\nu)_{22}$. This leads to large 2-3 leptonic mixing. Notice that with Higgs sector containing both 10 plet and 126 plet, the $b - \tau$ unification is not the consequence of theory but phenomenological input.

Considering the sum rule (10.9) for the matrices of second and third generations one finds [106]

$$\tan 2\theta_{23} = \frac{2\sin\theta_{23}^q}{2\sin^2\theta_{23}^q - (m_b - m_\tau)/m_b}. \tag{10.10}$$

Here θ_{23}^q is the quark mixing m_b is the mass of b-quark and m_τ is the mass of τ-lepton. This relation connect large leptonic 2-3 mixing and an approximate equality, $m_b \approx m_\tau$, at the GUT scale. Indeed, only for $(m_b - m_\tau)/m_b \ll 1$ one gets large θ_{23}. Essentially, the point is that $\overline{126}_H$ should give small contribution to the 33 element of the mass matrix, otherwise masses m_b and m_τ will be different. In fact, numerically one needs to have $(Y_{126})_{33} \sim (Y_{126})_{22}$. In the assumption that $\overline{126}_H$ gives the main contribution to the neutrino mass this implies large neutrino 2-3 mixing.

It seems the minimal SO(10) has problems in explaining all the fermion masses so that further extension is needed. Introduction of 120_H in addition leads to further loss of predictivity.

11. Landscape of models and mechanisms

As we have stressed before, the neutrino mass matrix can obtain substantial contributions from new physics at all possible scales from the EW (or even lower) to the Planck scale and from various mechanisms. We can write the following "superformula" for neutrino masses:

$$m_\nu = \sum m_{seesaw} + m_{triplet} + \sum m_{rad} + m_{SUSY} + m_{Planck} + \dots, \qquad (11.1)$$

where in order the terms correspond to contributions from (1) the seesaw realized at different energy scales, (2) the Higgs triplet, (3) one, two, *etc.* loops effects, (4) SUSY contributions, (5) Planck scale physics, *etc.*

One can imagine two possibilities: (i) The seesaw gives the leading contribution, whereas other mechanisms produce sub-leading effects.

(ii) The seesaw may turn out to be the sub-leading mechanism and play the role of the suppression mechanism for the Dirac neutrino masses.

Let us come to our main question: do neutrino masses imply new physics, at least something new in comparison with other fermion masses? What is special about neutrinos? Can they have simply the Dirac mass as other fermions have?

11.1. Small Dirac mass

Suppression of Dirac masses. If ν_R exist, why the Dirac mass terms are small or absent? There are two possible answers to this question:

1) The Dirac masses are forbidden by symmetry with immediate objection that this is unnatural - why neutrino but not other fermion masses are suppressed?

2) Dirac mass contributions are suppressed by couplings with the heavy degrees of freedom. Here again one can consider two possibilities:

The first is the introduction of large Majorana mass of the RH neutrinos. In this way we come back to the seesaw. So, the seesaw can be the mechanism of suppression of the Dirac mass term but not the main contribution to the neutrino mass;

There is another possibility: let us introduce another (large) Dirac mass terms formed by ν_R and new singlet N. In the basis (ν, ν_R, N) consider the mass matrix

$$m = \begin{pmatrix} 0 & m_D^T & 0 \\ m_D & 0 & M_D \\ 0 & M_D^T & 0 \end{pmatrix}, \qquad (11.2)$$

which leads to one strictly massless neutrino [100]. For $m_D \ll M$ the admixture of the heavy lepton is negligible. We will refer to this possibility as to the *multi-singlet* mechanism of the suppression.

A general context for consideration of neutrino masses can be formulated in the following way. Beyond the Standard Model there are three RH neutrinos, ν_{Rj}, and also a number of other SM singlets, S_i. The Yukawa couplings of these singlets with the active neutrinos,

$$h_{kj}\bar{l}_k \nu_{Rj} H + f_{ik}\bar{l}_k S_i H, \tag{11.3}$$

are small due to symmetry, or their contributions to neutrino masses are suppressed by the seesaw or by "multi-singlet" mechanism.

Small Yukawa couplings. Observed neutrino masses can be reproduced if $h_{ij} \sim 10^{-13}$ in (11.3). For usual Dirac type Yukawa couplings similar to the quark or charged lepton couplings these values look very unnatural and require some explanation.

One can consider the following scenario: the usual Yukawa couplings for the ν_L and ν_R are not small (of the same size as quark and lepton couplings). However the corresponding masses are strongly suppressed by the seesaw of multi-singlet mechanisms.

Neutrino masses which we observe in the oscillation experiments are formed by ν_L and new singlets, S, (see second term in (11.3)) which have no analogy in the quark sector. These singlets may have some particular symmetry properties or/and come from the hidden sector of theory. As a consequence, their couplings, f_{ij}, can be small.

In this case f_{ij} can appear as the effective couplings generated by the high dimensional operators. Indeed, the non-renormalizable operators

$$a_{ij}\bar{l}_i S_j H \frac{S}{M} \tag{11.4}$$

generate small effective Yukawa couplings

$$f_{ij} = a_{ij} \frac{\langle S \rangle}{M}. \tag{11.5}$$

even for $a_{ij} \sim O(1)$, provided that $\langle S \rangle / M \sim 10^{-13}$. (Renormalizable coupling can be suppressed by symmetry). One can consider SUSY or GUT scales for M, if $\langle S \rangle$ is at the electroweak scale or take $m_{3/2}/M_{Pl}$. Another possibility is to take small VEV of S [107]. Hierarchy $\langle S \rangle / M$ can be substantially reduced if the effective coupling appears in higher order non-renormalizable interactions:

$$h_{ij} = a_{ij} \frac{\Pi_{k=1...n}\langle S_k \rangle}{M^n}. \tag{11.6}$$

Clearly, scenario with small Dirac couplings will be excluded if the neutrino-less double beta decay is discovered and it will be shown that the decay is due to light Majorana neutrinos.

11.2. Higgs triplet mechanism

The Majorana neutrino mass can be generated at tree level by coupling with Higgs triplet [108–111] $\Delta \equiv (\Delta^{++}, \Delta^{+}, \Delta^{0})$:

$$g_{\alpha\beta} l_\alpha^T l_\beta \Delta. \tag{11.7}$$

The electroweak precision measurements give $\langle \Delta \rangle / \langle H \rangle < 0.03$. To avoid appearance of the triplet Majoron [109] the coupling $\mu \Delta H H$ with the Higgs doublets should be introduced. If $g_{\alpha\beta} \sim 1$, then $\langle \Delta^0 \rangle \sim 1$ eV.

Various scenarios depend on the triplet mass M_Δ. If $M_\Delta, \mu \gg \langle H \rangle$, the induced VEV appears $\langle \Delta^0 \rangle \sim \langle H \rangle^2 \mu / M_\Delta^2$ [108] and we arrive at the seesaw type-II. If in contrast, $M_\Delta \sim \langle H \rangle$ and $\mu \ll \langle H \rangle$, we find $\langle \Delta^0 \rangle \sim \mu$. The pseudo-Majoron mass $\sim \mu \langle H \rangle^2 / \langle \Delta^0 \rangle$ can be made large enough to avoid the experimental bounds, in particular, from measured Z^0 width [110, 111].

One can consider the effective coupling of neutrinos with triplet which arises from the non-renormalizable interactions:

$$g_{\alpha\beta} \frac{S}{M} l_\alpha^T l_\beta \Delta, \tag{11.8}$$

where the singlet S acquires VEV $\langle S \rangle \ll M$. This allows us to increase the required VEV of Δ. Another possibility appears in models with the triplet and two Higgs doublets [111].

11.3. Radiative mechanisms

Zee mechanism. There is no RH neutrinos, instead new scalar bosons are introduced: the charged singlet of SU(2), η^+, and second Higgs doublet H_2. Their couplings

$$l^T \hat{f} i \sigma_2 l \eta^+ + \sum_{i=1,2} \bar{l} \hat{f}_i e H_i, \tag{11.9}$$

where $\hat{f}_i$ is the matrix of the Yukawa couplings of Higgs H_i (i = 1,2), generate neutrino masses in one loop [112]

$$m_\nu = A[(\hat{f}\hat{m}^2 + \hat{m}^2 \hat{f}^T) - v(\cos \beta)^{-1}(\hat{f}\hat{m}\hat{f}_2 + \hat{f}_2^T \hat{m}\hat{f}^T)]. \tag{11.10}$$

Here

$$A \equiv \frac{\sin 2\theta_Z}{8\pi^2 v \tan \beta} \ln(M_2/M_1),$$

θ_Z is the mixing angle of charged bosons, $\hat{m} = diag(m_e, m_\mu, m_\tau)$, $\tan \beta \equiv v_1/v_2$, $v^2 \equiv v_1^2 + v_2^2$. In the minimal version with only one Higgs doublet coupled to leptons, $\hat{f}_2 = 0$. The neutrino mass matrix has zero diagonal elements and experimentally excluded [113]: Such a matrix can not reconcile two large mixings, one small mixing and hierarchy of Δm^2.

So one needs the non-zero couplings of both Higgs doublets with leptons (non-zero second term in (11.10)). This leads to non-zero diagonal mass terms of the mass matrix and a possibility to describe all experimental results. The model predicts decays $\tau \to \mu\mu\mu$, $\mu \to eee$, $\tau \to \mu\mu e$ due to the Higgs exchange at the level of 2 - 3 orders of magnitude below the present experimental bounds.

Other possibilities include an additional mechanisms which can give contributions to the neutrino mass matrix, in particular, to the diagonal terms: Higgs triplet, scalar singlet, two loop contribution [113]. Additional contributions to the mass matrix can appear if new leptons, in particular, sterile neutrinos, exist.

The model is testable in the precision electroweak measurements, searches for charged Higgses and rare decays. Still the problem exists with explanation of smallness of the couplings: the neutrino data require inverse flavor hierarchy of $f_{\alpha\beta}$ and $f_{\alpha\beta} \sim 10^{-4}$.

Zee-Babu mechanism. There is no RH neutrinos. New scalar bosons, singlets of SU(2) η^+ and k^{++} are introduced with the following couplings

$$l^T \hat{f} l \eta^+ + l_R^T \hat{h} l_R k^{++}. \tag{11.11}$$

Here $\hat{f}$ and $\hat{h}$ are the Yukawa matrices in the flavor basis. The Majorana neutrino masses are generated in two loops [112, 114]:

$$m_\nu \sim 8\mu \hat{f} \hat{m}_l \hat{h} \hat{m}_l \hat{f} I, \tag{11.12}$$

where $\hat{m}_l \equiv diag(m_e, m_\mu, m_\tau)$.

The main features of the model (see [115, 116] for recent discussion) are: one massless neutrino; inverted hierarchy of the couplings in the flavor basis; values of the couplings: $f, h \sim 0.1$. The model is testable: new charged scalar bosons exist at the electroweak scale, the decay rates for $\mu \to e\gamma$, and $\tau \to 3\mu$ are within a reach of the forthcoming experiments.

11.4. Neutrino mass and SUSY

Supersymmetric models provide with
  – new mass scales, *e.g.* the gravitino mass $m_{3/2}$, the μ-term mass, the scale of SUSY violation in the hidden sector of supergravity, *etc.*
  – new neutral fermions which can mix with neutrinos: Higgsinos, gauginos (this implies the R-parity violation);

– new sources of the lepton number violation;
– superpartners of neutrinos – sneutrinos which can develop non-zero VEV'c and contribute to the neutrino mass generation.

In what follows we will describe some representative models which implement these features.

R-parity violation SUSY. The terms of superpotential

$$W = -\mu_\alpha l_\alpha H_u - 0.5\lambda_{\alpha\beta m} + \lambda'_{\alpha n m} l_\alpha Q_n d_m^c + h_{mn} H_u Q_n u_m^c \qquad (11.13)$$

violate the lepton number. Here $\alpha = 0, 1, 2, 3$, $l_0 \equiv H_d$, $m = 1, 2, 3$. No RH neutrinos are introduced.

The bi-linear terms in (11.13) [117] give the dominant tree level contribution to neutrino masses. In the basis where sneutrinos have zero VEV's, $\langle \tilde{\nu} \rangle = 0$, the masses are produced via mixing with Higgsinos by the diagram (fig. 16):

$$m_{ij} = \mu_i \mu_j \frac{\cos^2 \beta}{m_\chi}. \qquad (11.14)$$

In the basis with $\mu_m = 0$, the neutrino masses are generated by the electroweak seesaw: light neutrinos are mixed with wino (neutralino) after sneutrinos get VEV's (fig. 16b): $m_{ij} = A \langle \tilde{\nu}_i \rangle \langle \tilde{\nu}_j \rangle$. Here $A = h_b^2/(16\pi^2 m_{\tilde{W}}^2)$.

Only one neutrino acquires mass at this tree level (in assumption of universality of the soft symmetry breaking terms) [117–121]. Moreover, mixing is determined by the ratios of the mass parameters: μ_i/μ_j.

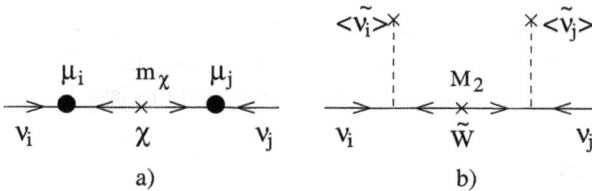

Fig. 16. Diagrams for the neutrino mass generation at tree level in the model with the R-parity violation. a) In the basis $\langle \tilde{\nu} \rangle = 0$. b) In the basis $\mu_m = 0$.

The trilinear RpV couplings in (11.13) and soft symmetry breaking terms (characterized by the mass parameters B_i) generate one loop contributions (fig. 17). As a result, natural neutrino mass hierarchy appears: at tree level one (largest) mass as well as one large mixing are generated; loop contributions generate other small masses and small mixings.

Correct scale of the neutrino masses requires $\mu_m \sim 10^{-4}$ GeV, which in turn, implies further structuring – explanation of the hierarchy $\mu_m \ll m_0$.

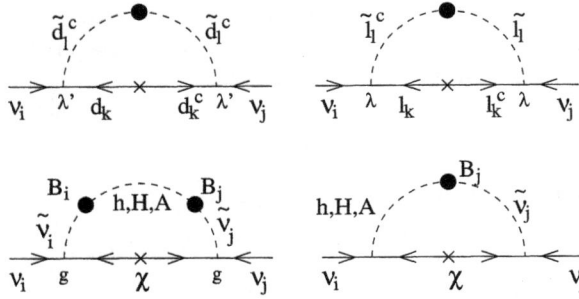

Fig. 17. One loop diagrams for the neutrino mass generation in the model with R-parity violation.

In the original version (universal SUSY breaking at high scale), generically only one large mixing can be obtained. Explanation of neutrino data with two large mixings requires violation of universality of the soft symmetry breaking terms. Both the Higgs-lepton ($\mu - \mu_i$) and flavor universality should be broken [120, 121], that is, B_m should be different at the high scale.

The RpV models have very rich phenomenology: new physics at colliders, relatively fast flavor violating decays, new neutrino interactions *etc.*

SUSY violation and neutrino mass [122–127]. It was observed long time ago [122] that

$$m = \frac{m_{3/2} v_{EW}}{M_{Pl}} \sim 10^{-4} \text{ eV}, \tag{11.15}$$

where $m_{3/2} \sim 1$ TeV is the gravitino mass and $v_{EW} \equiv \langle H \rangle$ is the electroweak VEV. The value (11.15) is close to the scale of observed neutrino masses and certainly is interesting from the phenomenology point of view. It can be generated by the Yukawa interaction

$$\lambda \bar{l} S H, \quad \text{with} \quad \lambda = \frac{m_{3/2}}{M_{Pl}}. \tag{11.16}$$

It can mix active neutrinos with singlets of SM, *e.g.*, form usual mass term, or mix neutrino with modulino [122].

The interaction (11.16) may follow from non-renormalizable term in the superpotential or from the Kähler potential similarly to an appearance of the μ - term in the Giudice-Masiero mechanism:

$$K = \frac{1}{M_{Pl}} P(S, z, z^*) \bar{l} H + h.c., \tag{11.17}$$

where z are the Wilson lines [122, 124].

The mass (11.15) is too small to explain observations. But there are various ways to enhance it. In general, the mass can be written as

$$m = \frac{\alpha \eta m_{3/2} v_{EW}}{M_{Pl}}, \tag{11.18}$$

where η describes the renormalization group effect and α is an additional numerical factor.

1) The gravitino mass can be larger: the value $m_{3/2} \sim 10^2$ TeV brings the mass to the correct range 10^{-2} eV. Such a scale appears, *e.g.*, in the model of "consistent anomaly mediation" [125].

2) One can take M_{GUT} instead of M_{Pl} which leads to $\alpha \sim 10^2$.

3) Large factor α may appear as a consequence of particular mechanism of mass generation. For instance, the terms in the Kähler potential

$$K = \frac{1}{M} P(S, \sigma, \sigma^*) \bar{l} H N + h.c. \tag{11.19}$$

may have the cut-off parameter $M = 10^{17}$ GeV – below the Planck mass. Here σ are the fields of the hidden sector. Then the dominant contribution to the neutrino mass is given by [126] $v_{EW} F_\sigma / M^2$, where $F_\sigma = \sqrt{3} M_{Pl} m_{3/2}$. It leads to a correct range

$$m = \frac{\eta m_{3/2} v_{EW}}{M} \frac{\sqrt{3} M_{Pl}}{M} \sim 0.05 \text{ eV}. \tag{11.20}$$

Another possibility is that small Dirac type Yukawa couplings appear from the superpotential, whereas the Majorana masses follow from the Kähler potential [127]:

$$W = g \frac{X}{M} l N H, \quad K = h \frac{Y^*}{M} N N. \tag{11.21}$$

Here $M = M_{Pl}/\sqrt{8\pi}$, X and Y are the fields of hidden sector, with VEV's $\langle X_A \rangle = m_I$ and $\langle Y_F \rangle = m_I^2$ at the intermediate mass scale $m_I = \sqrt{m_{3/2} M_{Pl}}$.

μ-term mixing. Small Dirac mass of neutrino can be related to (protected by) a small value of the μ-term in the context of SUSY models and not to the EW scale [128] or $m_{3/2}$. Though we do not know the origin of smallness of the μ term mass, we can connect smallness of neutrino mass with this smallness in comparison with *e.g.* GUT scale.) Such a possibility is realized in the SU(5) GUT with R-symmetry [128].

11.5. Extra dimensions and neutrino mass

Theories with extra space dimensions provide qualitatively new mechanism of generation of the small *Dirac* neutrino mass. There are different scenarios, however their common feature can be called the overlap suppression: the overlap of wave functions of the left, $\nu_L(y)$, and right, $\nu_R(y)$ handed components in extra dimensions (coordinate y). The suppression occurs due to different localizations of the $\nu_L(y)$ and $\nu_R(y)$ in the extra space. The effective Yukawa coupling is proportional to the overlap. One can introduce also suppression of overlap of neutrinos with Higgs field. Let us consider realizations of the overlap suppression mechanism in different extra dimensional scenarios.

Large flat extra dimensions: ADD scenario. The setup is the $3D$ spatial brane embedded in $(3+\delta)D$ bulk [129]. Extra dimensions have large radii $R_i \gg 1/M_{Pl}$ which allows one to reduce the fundamental scale of theory down to $M^* \sim 10 - 100$ TeV [129]. The left handed neutrino is localized on the brane, whereas the right handed component (being the singlet of the gauge group) propagates in the bulk (see fig. 18a).

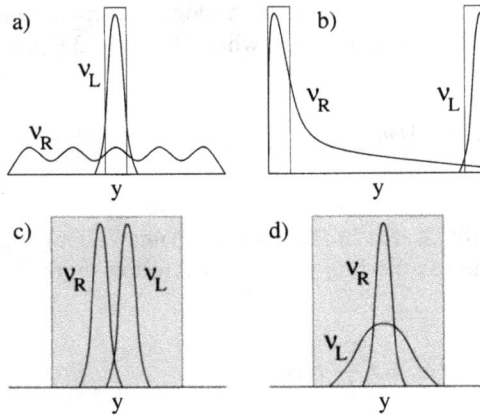

Fig. 18. The overlap mechanism of small Dirac neutrino mass generation in models with extra spatial dimensions. a) Large flat extra dimensions. b) Warped extra dimensions. c–d) Models of "fat" branes.

To clarify the mechanism of suppression let us consider one extra dimension of radius R. (Generalization to several extra dimensions is straightforward.) Since the RH component is not localized, we find from the normalization condition that its wave function has typical value $\nu_R(y) \sim 1/\sqrt{R}$. The effective width of the brane is of the order $d \sim 1/M^*$, so the amplitude of probability to find the

RH neutrino on the brane equals

$$d^{1/2}\nu_R \sim \frac{1}{\sqrt{M^*R}}. \tag{11.22}$$

Since the LH neutrino is localized on the brane, this factor describes the overlap of the wave functions.

Generalization to δ extra dimensions is straightforward: for the overlap factor we get

$$\frac{1}{\sqrt{M^{*\delta}V_\delta}}, \tag{11.23}$$

where V_δ is the volume of extra dimensions. Using the relation between the effective Planck mass and the fundamental mass scale M^*:

$$M_{Pl}^2 = M^{*2+\delta}V_\delta, \tag{11.24}$$

we can rewrite the overlap factor as

$$\frac{M^*}{M_{Pl}}, \tag{11.25}$$

which does not depend on the number of extra dimensions explicitly. If λ is the Yukawa coupling for neutrinos in the $(4+\delta)D$ theory, the effective coupling in 4D will be suppressed by this overlap factor. Consequently,

$$m_D = \lambda v_{EW}\frac{1}{\sqrt{M^{*\delta}V_\delta}} = \lambda v_{EW}\frac{M^*}{M_{Pl}}. \tag{11.26}$$

For $M^* \sim 100$ TeV and $\lambda \sim 1$ we obtain from (11.26) $m_D \sim 10^{-2}$ eV.

Warped extra dimensions: Randall-Sundrum scenario. A set up is one extra dimension compactified on the S^1/Z_2 orbifold, and non-factorizable metric. The coordinate in the extra dimension is parameterized by $r_c\phi$, where r_c is the radius of extra dimension and the angle ϕ changes from 0 to π. Two branes are localized in different points of extra dimension: the "hidden" brane is at $\phi = 0$ and the observable one is at $\phi = \pi$ [130]. The wave function of the RH neutrino $\nu_R(\phi)$ is centered on the hidden brane, whereas the LH one – on the visible brane (see fig. 18 b). Due to warp geometry $\nu_R(\phi)$ exponentially decreases from the hidden to the observable brane. On the observable brane it is given by

$$\nu_R(\pi) \sim \epsilon^{\nu-1/2}, \quad \epsilon = e^{-kr_c\pi} = \frac{v_{EW}}{M_{Pl}}. \tag{11.27}$$

Here M_{Pl} is the Planck scale, $k \sim M_{Pl}$ is the curvature parameter. In (11.27) $\nu \equiv m/k$ and $m \sim M_{Pl}$ is the Dirac mass in 5D. Essentially $\nu_R(\pi)$ gives the overlap factor and the Dirac mass on the visible brane equals

$$m_D = \lambda \nu_R(\pi) \nu_{EW} \sim M \left(\frac{V_{EW}}{M} \right)^{\nu+1/2} . \qquad (11.28)$$

For $\nu = 1.1 - 1.6$ we get the mass in the phenomenologically required range. Notice that expression for the mass is of the seesaw type with, however, arbitrary power of small ratio. Now explanation of smallness of the neutrino mass is reduced to explanation of particular values of the mass m. Small variations of m can produce strong change in the light neutrino masses. Though the mass parameters can be all of the same order their particular values should be fine tuned.

Different realization is when both ν_L and ν_R are on the TeV brane whereas the lepton number is violated on the Planck brane [131].

Fat brane scenarios. The LH and RH neutrino wave functions can be localized differently on the same "fat" brane [132]. There are various possibilities to suppress the overlap:

1) localize ν_L and ν_R in different places of the brane (fig. 18c);

2) arrange parameters in such a way that the RH neutrino is localized in the narrow region of the fat brane, whereas the LH neutrino wave function is distributed in whole the brane (fig.18d) [133] *etc.*

These attempts are less advanced than seesaw – GUT scenario. They provide a context with interesting possibilities for construction of specific models.

12. Conclusion

Last 5–7 years was epoch of great achievements in neutrino physics:

– discovery of neutrino oscillations,

– resolution of the solar neutrino problem and establishing the matter effect,

– measurements of neutrino parameters.

As a result of these discoveries amazing pattern of the lepton mixing has emerged.

Clear program of future phenomenological and experimental studies has been elaborated. At the same time implications and identification of the underlying physics is a big challenge, and it may happen that something important (in principles and context) is still missed. Substantial input from high energy experiments astrophysics and cosmology will be helpfull.

The hope is that neutrinos will uncover something simple and illuminating before we will be lost in the string landscape.

References

[1] R. E. Marshak and R. N. Mohapatra, Phys. Lett. B **91** (1980) 222.

[2] B. Pontecorvo, Zh. Eksp. Theor. Fiz. **33** (1957); *ibidem*, **34** (1958) 247.

[3] Z. Maki, M. Nakagawa and S. Sakata, Prog. Theor. Phys. **28** (1962) 870.

[4] B. Pontecorvo, ZETF **53** (1967) 1771 [Sov. Phys. JETP **26** (1968) 984]; V. N. Gribov and B. Pontecorvo, Phys. Lett. **28B** (1969) 493.

[5] L. Wolfenstein, Phys. Rev. D **17** (1978) 2369; in: *Neutrino 78*, Purdue Univ. C3, 1978, Phys. Rev. D **20** (1979) 2634.

[6] S. P. Mikheyev and A. Yu. Smirnov, Sov. J. Nucl. Phys. **42** (1985) 913; Nuovo Cim. C **9** (1986) 17; S.P. Mikheev and A.Yu. Smirnov, Sov. Phys. JETP **64** (1986) 4.

[7] N. Cabibbo, summary talk given at 10th Int. Workshop on Weak Interactions and Neutrinos, Savonlinna, Finland, June 1985.

[8] H. Bethe, Phys. Rev. Lett. **56** (1986) 1305.

[9] A. Messiah, in: *Proc. of the 6th Moriond Workshop on Massive Neutrinos in Particle Physics and Astrophysics*, eds O. Fackler and J. Tran Thanh Van, Tignes, France, Jan. 1986, p. 373; S. P. Mikheev and A. Yu. Smirnov, Sov. Phys. JETP **65** (1987) 230.

[10] S. J. Parke, Phys. Rev. Lett. **57** (1986) 1275.

[11] W. C. Haxton, Phys. Rev. Lett. **57** (1986) 1271.

[12] P. C. de Holanda, Wei Liao and A. Yu. Smirnov, Nucl. Phys. B **702** (2004) 307.

[13] B. T. Cleveland et al., Astrophys. J. **496** (1998) 505.

[14] J. Hosaka et al. [Super-Kamiokande Collaboration], hep-ex/0508053.

[15] J. N. Abdurashitov et al. [SAGE Collaboration], J. Exp. Theor. Phys. **95** (2002) 181 [Zh. Eksp. Teor. Fiz. **122** (2002) 211].

[16] W. Hampel et al. [GALLEX Collaboration], Phys. Lett. B **447** (1999) 127.

[17] M. Altmann et al. [GNO Collaboration], Phys. Lett. B **616** (2005) 174.

[18] B. Aharmim et al. [SNO Collaboration], Phys. Rev. C **72** (2005) 055502.

[19] J. N. Bahcall and M.H. Pinsonneault, Phys. Rev. Lett. **92** (2004) 121301; J. N. Bahcall, A. M. Serenelli and S. Basu, astro-ph/0511337.

[20] P. C. de Holanda and A.Yu. Smirnov, Astropart. Phys. **21** (2004) 287.

[21] A. N. Ioannisian and A.Yu. Smirnov, Phys. Rev. Lett. **93** (2004) 241801.

[22] T. Araki et al. [KamLAND Collaboration], Phys. Rev. Lett. **94** (2005) 081801.

[23] Y. Ashie et al. [Super-Kamiokande Collaboration], Phys. Rev. D **71** (2005) 112005.

[24] M. Ambrosio et al. [MACRO Collaboration], Eur. Phys. J. C **36** (2004) 323.

[25] M. C. Sanchez et al. [Soudan 2 Collaboration], Phys. Rev. D **68** (2003) 113004.

[26] P. Adamson et al. [MINOS Collaboration], hep-ex/0512036.

[27] Y. Ashie et al. [Super-Kamiokande Collaboration], Phys. Rev. Lett. **93** (2004) 101801

[28] M. C. Gonzalez-Garcia, M. Maltoni and A. Yu. Smirnov, Phys. Rev. D **70** (2004) 093005.

[29] G. L. Fogli et al., hep-ph/0506083.

[30] E. Aliu et al. [K2K Collaboration], Phys. Rev. Lett. **94** (2005) 081802.

[31] M. Apollonio et al., Eur. Phys. J. C **27** (2003) 331.

[32] A. Strumia and F. Vissani, Nucl. Phys. B **726** (2005) 294.

[33] V. M. Lobashev et al., Nucl. Phys. Proc. Suppl. **91** (2001) 280.

[34] C. Kraus et al., Eur. Phys. J. C **40** (2005) 447.

[35] A. Osipowicz et al. [KATRIN Collaboration], hep-ex/0109033.

[36] H.V. Klapdor-Kleingrothaus et al., Eur. Phys. J. A **12**, 147 (2001); A. M. Bakalyarov et al., talk given at the 4th International Conference on Non-accelerator New Physics (NANP 03), Dubna, Russia, 23-28 Jun. 2003 [hep-ex/0309016].

[37] H.V. Klapdor-Kleingrothaus et al., Mod. Phys. Lett. A **16** (2001) 2409.

[38] H.V. Klapdor-Kleingrothaus et al., Phys. Lett. B **586** (2004) 198.

[39] A. D. Dolgov, Phys. Rept. **370** (2002) 333

[40] M. Tegmark, hep-ph/0503257.

[41] M. Tegmark et al, ApJ **606** (2004) 702.

[42] U. Seljak et al., Phys. Rev. D **71** (2005) 103515.

[43] Y. Itow et al., hep-ex/0106019.

[44] D. S. Ayres et al. [NOvA Collaboration], hep-ex/0503053.

[45] F. Ardellier et al., *Letter of intent for double-CHOOZ*, hep-ex/0405032.

[46] A. S. Dighe and A. Yu. Smirnov, Phys. Rev. D **62** (2000) 033007; C. Lunardini and A. Yu. Smirnov, JCAP **0306** (2003) 009.

[47] H. Minakata and H. Nunokawa, Phys. Lett. B **504** (2001) 301.

[48] V. Barger, D. Marfatia and B.P. Wood, Phys. Lett. B **532** (2002) 19.

[49] K. Takahashi, K. Sato, A. Burrows and T. A. Thompson, Phys. Rev. D **68** (2003) 113009.

[50] R.C. Schirato and G. M. Fuller, astro-ph/0205390.

[51] K. Takahashi, K. Sato, H. E. Dalhed and J.R. Wilson, Astropart. Phys. **20** (2003) 189.

[52] G.L. Fogli, E. Lisi, D. Montanino and A. Mirizzi, Phys. Rev. D **68** (2003) 033005.

[53] R. Tomas et al., JCAP **0409** (2004) 015.

[54] A. Aguilar et al. [LSND Collaboration], Phys. Rev. D **64** (2001) 112007.

[55] B. Armbruster et al. [KARMEN Collaboration], Phys. Rev. Lett. **90** (2003) 181804.

[56] K. S. Babu and S. Pakwasa, hep-ph/0204226.

[57] G. Barenboim, L. Borissov and J. Lykken, hep-ph/0212116.

[58] A. Strumia, Phys. Lett. B **539** (2002) 91.

[59] M.C. Gonzalez-Garcia, M. Maltoni and T. Schwetz, Phys. Rev. D **68** (2003) 053007.

[60] O. L. G. Peres and A.Yu. Smirnov, Nucl. Phys. B **599** (2001) 3.

[61] M. Sorel, J. Conrad and M. Shaevitz, Phys. Rev. D **70** (2004) 073004.

[62] V. Barger, D. Marfatia and K. Whisnant, Phys. Lett. B **576** (2003) 303.

[63] D. B. Kaplan, A. E. Nelson and N. Weiner, Phys. Rev. Lett. **93** (2004) 091801.

[64] S. Palomares-Ruiz, S. Pascoli and T. Schwetz, JHEP **0509** (2005) 048.

[65] M. H. Shaevitz [MiniBooNE Collaboration], Nucl. Phys. Proc. Suppl. **137** (2004) 46 [hep-ex/0407027].

[66] S. Weinberg, Phys. Rev. Lett. **43**, 1566 (1979).

[67] R. Barbieri, J. Ellis and M. K. Gaillard, Phys. Lett. B **90** (1980) 249; E. Kh. Akhmedov, Z. G. Berezhiani and G. Senjanović, Phys. Rev. Lett. **69** (1992) 3013.

[68] F. Vissani, M. Narayan and V. Berezinsky, Phys. Lett. B **571** (2003) 209.

[69] P. Minkowski, Phys. Lett. B **67** (1977) 421. For earlier work, see H. Fritzsch, M. Gell-Mann and P. Minkowski, Phys. Lett. B **59** (1975) 256; H. Fritzsch and P. Minkowski, Phys. Lett. B **62** (1976) 72.

[70] M. Gell-Mann, P. Ramond and R. Slansky, in: *Supergravity*, eds P. van Niewenhuizen and D. Z. Freedman (North Holland, Amsterdam 1980); P. Ramond, *Sanibel talk*, retroprinted as hep-ph/9809459; T. Yanagida, in: *Proc. of Workshop on Unified Theory and Baryon number in the Universe*, eds. O. Sawada and A. Sugamoto, KEK, Tsukuba, 1979; S. L. Glashow, in: *Quarks and Leptons*, in: Cargèse lectures, eds M. Lévy, (Plenum, 1980, New York) p. 707; R. N. Mohapatra and G. Senjanović, Phys. Rev. Lett. **44**, (1980) 912.

[71] R. N. Mohapatra and G. Senjanović, Phys. Rev. D **23** (1981) 165; C. Wetterich, Nucl. Phys. B **187** (1981) 343.

[72] A. Yu. Smirnov, Phys. Rev. D **48** (1993) 3264; M. Tanimoto, Phys. Lett. B **345** (1995) 477; T. K. Kuo, G.-H. Wu and S. W. Mansour, Phys. Rev. D **61** (2000) 111301; G. Altarelli, F. Feruglio and I. Masina, Phys. Lett. B **472** (2000) 382; S. Lavignac, I. Masina and C. A. Savoy, Nucl. Phys. B **633** (2002) 139. A. Datta, F. S. Ling and P. Ramond, Nucl. Phys. B **671** (2003) 383; M. Bando et al., Phys. Lett. B **580** (2004) 229.

[73] F. Vissani, hep-ph/9708483; V. D. Barger et al., Phys. Lett. B **437** (1998) 107.

[74] L. Wolfenstein, Phys. Rev. D **18** (1978) 958; P. F. Harrison, D. H. Perkins and W. G. Scott, Phys. Lett. B **458** (1999) 79, Phys. Lett. B **530** (2002) 167.

[75] T. Fukuyama and H. Nishiura, hep-ph/9702253; R. N. Mohapatra and S. Nussinov, Phys. Rev. D **60** (1999) 013002; E. Ma and M. Raidal, Phys. Rev. Lett. **87** (2001) 011802; C. S. Lam, Phys. Lett. B **507** (2001) 214.

[76] S. T. Petcov, Phys. Lett. B **110** (1982) 245; R. Barbieri et al., JHEP **9812** (1998) 017.

[77] E. Ma, Mod. Phys. Lett. A **17** (2002) 2361; E. Ma and G. Rajasekaran, Phys. Rev. D **64** (2001) 113012.

[78] K.S. Babu, E. Ma and J.W.F. Valle, Phys. Lett. B **552** (2003) 207.

[79] J. Kubo et al., Prog. Theor. Phys. **109** (2003) 795.

[80] E. Ma and G. Rajasekaran, hep-ph/0306264.

[81] W. Grimus and L. Lavoura, hep-ph/0305046.

[82] For some recent publications, see: W. Grimus and L. Lavoura, JHEP **0508** (2005) 013; E. Ma, Mod. Phys. Lett. A **20** (2005) 2601.

[83] J. Bijnens and C. Wetterich, Nucl. Phys. B **292** (1987) 443; M. Leurer, Y. Nir and N. Seiberg, Nucl. Phys. B **398** (1993) 319; *ibidem*, **420** (1994) 468; L. E. Ibanez and G.G. Ross, Phys. Lett. B **332** (1994) 100; P. Binetruy and P. Ramond, Phys. Lett. B **350** (1995) 49. For references and recent discussion, see G. Altarelli and F. Feruglio, hep-ph/0306265; P. H. Chankowski, K. Kowalska, S. Lavignac and S. Pokorski, Phys. Rev. D **71** (2005) 055004.

[84] C. D. Froggatt and H. B. Nielsen, Nucl. Phys. B **147** (1979) 277.

[85] R. Kuchimanchi and R. N. Mohapatra, Phys. Rev. D **66** (2002) 051301.

[86] R. Barbieri, L. J. Hall, G. L. Kane and G. G. Ross, hep-ph/9901228.

[87] G. Kribs, hep-ph/0304256; S. F. King, Phys. Lett. B **520** (2001) 243; S.F. King and G.G. Ross, Phys. Lett. B **574** (2003) 239.

[88] K. S. Babu and X. G. He, hep-ph/0507217.

[89] J. C. Pati and A. Salam, Phys. Rev. D **10** (1974) 275.

[90] H. Georgi, in: *Coral Gables 1979 Proceeding, Theory and experiment in high energy physics* (New York, 1975), p. 329; H. Fritzsch and P. Minkowski, Annals Phys. **93** (1975) 193.

[91] A. Yu. Smirnov, hep-ph/0402264; M. Raidal, Phys. Rev. Lett. **93** (2004) 161801.

[92] H. Minakata and A. Yu. Smirnov, Phys. Rev. D **70** (2004) 073009.

[93] E. K. Akhmedov et al., Phys. Lett. B **498** (2001) 237; R. Dermisek, Phys. Rev. D **70** (2004) 033007.

[94] I. Dorsner and A.Yu. Smirnov, Nucl. Phys. B **698** (2004) 386.

[95] F. Vissani and A. Yu. Smirnov, Phys. Lett. B **341** (1994) 173; A. Brignole, H. Murayama and R. Rattazzi, Phys. Lett. B **335** (1994) 345.

[96] M. Lindner, S. Antusch, J. Kersten, M. Lindner and M. Ratz, Phys. Lett. B **544** (2002) 1.

[97] These proceedings.

[98] These proceedings.

[99] P. H. Frampton, S. L. Glashow and T. Yanagida, Phys. Lett. B **548** (2002) 119.

[100] R. N. Mohapatra, Phys. Rev. Lett. **56** (1986) 561; R. N. Mohapatra and J. W. F. Valle, Phys. Rev. D **34** (1986) 1642.

[101] S. M. Barr, Phys. Rev. Lett. **92** (2004) 101601.

[102] M. Lindner, M. A. Schmidt and A. Yu. Smirnov, JHEP **0507** (2005) 048.

[103] For recent review of neutrino masses in GUTs, see M-C. Chen and K. T. Mahanthappa, Int. J. Mod. Phys. A **18** (2003) 5819.

[104] K. S. Babu and R. N. Mohapatra, Phys. Rev. Lett. **70** (1993) 2845.

[105] B. Bajc, G. Senjanovic and F. Vissani, Phys. Rev. Lett. **90** (2003) 051802.

[106] S. Bertolini, M. Frigerio and M. Malinsky, Phys. Rev. D **70** (2004) 095002.

[107] Z. Chacko et al., Phys. Rev. Lett. **94** (2005) 111801.

[108] M. Magg and C. Wetterich, Phys. Lett. **94B** (1980) 61; J. Schechter and J. W. F. Valle, Phys. Rev. D **22** (1980) 2227; R. N. Mohapatra and G. Senjanovic, Phys. Rev. D **23** (1981) 165.

[109] G. B. Gelmini and M. Roncadelli, Phys. Lett. B **99** (1981) 411.

[110] E. Ma and U. Sarkar, Phys. Rev. Lett. **80** (1998) 5716.

[111] E. Ma, Phys. Rev. D **66** (2002) 037301; P. H. Frampton, M. C. Oh and T. Yoshikawa, Phys. Rev. D **66** (2002) 033007.

[112] A. Zee, Phys. Lett. B **93** (1980) 389; *ibidem*, **161** (1985) 141.

[113] For the latest discussion, see e.g. P. H. Frampton, M. C. Oh and T. Yoshikawa, Phys. Rev. D **65** (2002) 073014; T. Kitabayashi and M. Yasue, Int. J. Mod. Phys. A **17** (2002) 2519; X.-G. He, Eur. Phys. J. C **34** (2004) 371.

[114] K. S. Babu, Phys. Lett. B **203** (1988) 132.

[115] K.S. Babu and C. Macesanu, Phys. Rev. D **67** (2003) 073010.

[116] I. Aizawa et al., Phys. Rev. D **70** (2004) 015011.

[117] L. J. Hall and M. Suzuki, Nucl. Phys. B **231** (1984) 419; A. Joshipura and M. Nowakowski, Phys. Rev. D **51** (1995) 2421; A. Yu. Smirnov and F. Vissani, Nucl. Phys. B **460** (1996) 37; R. Hempfling, Nucl. Phys. B **478** (1996) 3.

[118] A. S. Joshipura, R. D. Vaidya and S. K. Vempati, Nucl. Phys. B **639** (2002) 290.

[119] For a review, see R. Barbier et al., Phys. Rept. **420** (2005) 1, and references therein.

[120] Y. Grossman and S. Rakshit, Phys. Rev. D **69** (2004) 093002.

[121] M. A. Diaz et al., Phys. Rev. D **68** (2003) 013009.

[122] K. Benakli and A. Yu. Smirnov, Phys. Rev. Lett. **79** (1997) 4314.

[123] N. Arkani-Hamed, L. J. Hall, H. Murayama, D. R. Smith and N. Weiner, Phys. Rev. D **64** (1997) 115011; hep-ph/0007001.

[124] F. Borzumati and Y. Nomura, Phys. Rev. D **64** (2001) 053005; F. Borzumati et al., hep-ph/0012118.

[125] H. Murayama, hep-ph/0410140.

[126] S. Abel, A. Dedes and K. Tamvakis, Phys. Rev. D **71** (2005) 033003.

[127] J. March-Russell and S. M. West Phys. Lett. B **593** (2004) 181.

[128] R. Kitano, Phys. Lett. B **539** (2002) 102.

[129] N. Arkani-Hamed, S. Dimopoulos, G. R. Dvali and J. March-Russell, Phys. Rev. D **65** (2002) 02432; K. R. Dienes, E. Dudas and T. Ghergetta, Nucl. Phys. B **557** (1999) 25.

[130] Y. Grossman and M. Neubert, Phys. Lett. B **474** (2000) 361.

[131] T. Gherghetta, Phys. Rev. Lett. **92** (2004) 161601.

[132] N. Arkani-Hamed and M. Schmaltz, Phys. Rev. D**61** (2000) 033005.

[133] P.Q. Hung, Phys. Rev. D **67** (2003) 095011.

Course 12

BARYOGENESIS VIA LEPTOGENESIS

Alessandro Strumia

Dipartimento di Fisica dell'Università di Pisa, Italia

D. Kazakov, S. Lavignac and J. Dalibard, eds.
Les Houches, Session LXXXIV, 2005
Particle Physics Beyond the Standard Model
© *2006 Elsevier B.V. All rights reserved*

655

Contents

1. Introduction

The universe contains various relict particles: γ, e, p, ν, ^{4}He, Deuterium, ...,
plus likely some Dark Matter (DM). Their abundances are mostly understood,
with the following main exception:

$$\frac{n_B - n_{\bar{B}}}{n_\gamma} = \frac{n_B}{n_\gamma} = (6.15 \pm 0.25) \cdot 10^{-10}, \tag{1.1}$$

where n_γ and n_B and are the present number densities of photons and baryons
(anti-baryons have negligible density). This is the problem of baryogenesis. Before addressing it, let us briefly summarize the analogous understood issues.

As suggested by inflation, the total energy density equals the critical energy
density, discussed later. Next, almost all relative abundances can be understood
assuming that these particles are thermal relics of a hot Big-Bang phase. The
number densities of electrons and protons are equal, $n_e = n_p$, because nothing violated electric charge. The relative proton/neutron abundancy was fixed
by electroweak processes such as $n\nu_e \leftrightarrow pe$ at $T \sim$ few MeV. Neutrons get
bound in nuclei at $T \sim 0.1$ MeV: the measured nuclear primordial nuclear abun-
dances agree with predictions: $n_{^4\text{He}}/n_p \approx 0.25/4$, $n_\text{D}/n_p \approx 3 \; 10^{-5}/2$, etc. The
neutrino density, predicted to be $n_{\nu_{e,\mu,\tau}} = n_{\bar{\nu}_{e,\mu,\tau}} = 3n_\gamma/22$, is too low to be
experimentally tested: the baryon asymmetry problem is more pressing than the
analogous lepton asymmetry problem because we do not know how to measure
the lepton asymmetry. Finally, the DM abundancy suggested by present data is
obtained if DM particles are weakly-interacting thermal relics with mass

$$m \sim \sqrt{T_\text{now} \cdot M_\text{Pl}} \sim \text{TeV}, \tag{1.2}$$

where $T_\text{now} \approx 3$ K is the present temperature and $M_\text{Pl} \approx 1.2 \; 10^{19}$ GeV is the
Planck mass. The LHC collider might soon produce DM particles and test this
speculation. It is useful to digress and understand eq. (1.2), because the necessary
tools will reappear, in a more complicated context, when discussing leptogenesis.

1.1. The DM abundancy

First, we need to compute the expansion rate $H(t)$ of the universe as function
of its energy density $\rho(t)$. Let us study how a homogeneous $\rho(t)$ made of non-

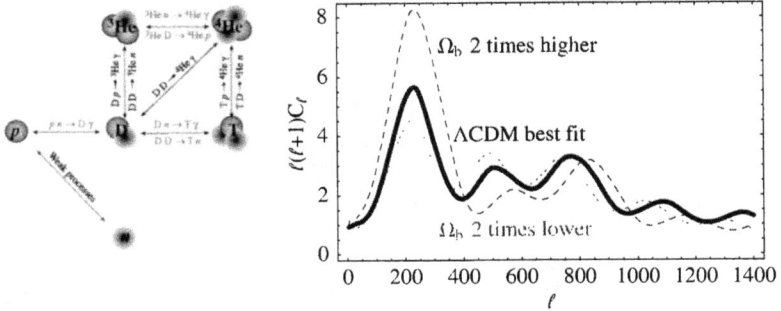

Fig. 1. (a) Main reactions that determine primordial nuclear abundances. (b) How CMB anisotropies depend on the baryon abundancy $\Omega_B = \rho_B/\rho_{cr}$.

relativistic matter evolves according to gravity. A test particle at distance R from us feels the Newton acceleration

$$\ddot{R} = -\frac{GM(R)}{R^2} = -\frac{4\pi G\rho(t)}{3}R, \tag{1.3}$$

where $M(R)$ is the total mass inside a sphere of radius R and $G = 1/M_{\rm Pl}^2$ is the Newton constant. By multiplying both sides of eq. (1.3) times $\dot{R}$ and integrating taking into account that $\rho(t) \propto 1/R^3(t)$ one obtains as usual the 'total energy' constant of motion, here named k:

$$\frac{d}{dt}\left[\frac{1}{2}\dot{R}^2 - \frac{4\pi}{3}G\rho R^2\right] = 0, \quad \text{so that} \quad H^2 \equiv \frac{\dot{R}^2}{R^2} = \frac{8\pi G}{3}\rho - \frac{k}{R^2}. \tag{1.4}$$

Let us discuss the special case $k = 0$. It is obtained when the density ρ equals the 'critical density' $\rho = \rho_{cr} \equiv 3H^2/8\pi G$. $k = 0$ is special because means zero 'total energy' (the negative gravitational potential energy compensates the positive matter energy): a universe with critical density that expands getting big for free could have been theoretically anticipated since 1687. Today we abandoned prejudices for a static universe, and more advanced theories put the above discussion on firmer grounds. In general relativity eq. (1.4) holds for more general sources of energy (relativistic particles, cosmological constant, ...), and the constant k gets the physical meaning of curvature of the universe. The inflation mechanism generates a smooth universe with negligibly small k.

Second, we need to know that a gas of particles in thermal equilibrium at temperature $T \gg m$ has number density $n_{\rm eq} \sim T^3$ and energy density $\rho_{\rm eq} \sim T^4$: one particle with energy $\sim T$ per de-Broglie wavelength $\sim 1/T$. The number density of non relativistic particles ($T \ll m$) is suppressed by a Boltzmann factor, $n_{\rm eq} \sim (mT)^{3/2}e^{-m/T}$, and their energy density is $\rho_{\rm eq} \simeq mn_{\rm eq}$.

We can now understand eq. (1.2), by studying what happens to a DM particle of mass m when the temperature T cools below m.

Annihilations with cross section $\sigma(\text{DM DM} \to$ SM particles) try to maintain thermal equilibrium, $n_{\text{DM}} \propto \exp(-m/T)$. But they fail at $T \lesssim m$, when n_{DM} is so small that the collision rate Γ experienced by a DM particle becomes smaller than the expansion rate H:

$$\Gamma \sim n_{\text{DM}}\sigma \lesssim H \sim T^2/M_{\text{Pl}}.$$

As illustrated in the picture, annihilations become ineffective, leaving the following out-of-equilibrium relic abundancy of DM particles:

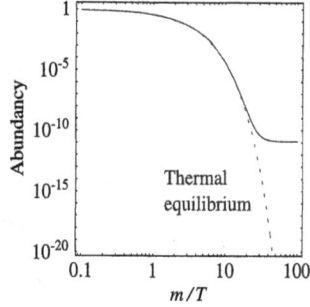

$$\frac{n_{\text{DM}}}{n_\gamma} \sim \frac{m^2/M_{\text{Pl}}\sigma}{m^3} \sim \frac{1}{M_{\text{Pl}}\sigma m} \quad \text{i.e.} \quad \frac{\rho_{\text{DM}}(T)}{\rho_\gamma(T)} \sim \frac{m}{T}\frac{n_{\text{DM}}}{n_\gamma} \sim \frac{1}{M_{\text{Pl}}\sigma T}. \quad (1.5)$$

Inserting the observed DM density, $\rho_{\text{DM}} \sim \rho_\gamma$ at $T \sim T_{\text{now}}$, and a typical cross section $\sigma \sim g^4/m^2$ gives eq. (1.2) for a DM particle with weak coupling $g \sim 1$. A precise computation can be done solving Boltzmann equations for DM.

1.2. The baryon asymmetry

Let us summarize how the value of the baryon asymmetry in eq. (1.1) is measured. The photon density directly follows from the measurement of the CMB temperature and from Bose-Einstein statistics: $n_\gamma \sim T^3$. Counting baryons is more difficult. Direct measurements are not accurate, because only some fraction of baryon formed stars and other luminous objects. Two different indirect probes point to the same baryon density, making the result trustable. Each one of the two probes would require a dedicated lesson:

1. Big-Bang-Nucleosynthesis (BBN) predictions depend on n_B/n_γ. Fig. 1a illustrates the main reactions. The important point is that the presence of many more photons than baryons delays BBN, mainly by enhancing the reaction $pn \leftrightarrow D\gamma$ in the $\leftarrow$ direction, so that D forms not when the temperature equals the Deuterium binding energy $B \approx 2\,\text{MeV}$, but later at $T \approx B/\ln(n_B/n_\gamma) \approx 0.1\,\text{MeV}$, giving more time to free neutrons to decay. A precise computation can be done solving Boltzmann equations for neutrino decoupling and nucleosynthesis.

2. Measurements of CMB anisotropies [1], among many other things, allow us to probe acoustic oscillations of the baryon/photon fluid happened around photon last scattering. A precise computation can be done evolving Boltzmann equations for anisotropies, assuming that they are generated by quantum fluctuations during

inflation: fig. 1b illustrates how the amount of anisotropies with angular scale $\sim 1/\ell$ depends on n_B/n_γ. Acoustic oscillations have been seen also in matter inhomogeneities [1].

2. Baryogenesis

The small baryon asymmetry $n_B/n_\gamma \ll 1$ can be obtained from a hot big-bang as the result of a small excess of baryons over anti-baryons. We would like to understand why, when at $T \lesssim m_p$ matter almost completely annihilated with anti-matter, we survived thanks to the 'almost':

$$n_B - n_{\bar{B}} \propto 1000000001 - 1000000000 = 1.$$

This might be the initial condition at the beginning of the big-bang, but it would be a surprisingly small excess. In inflationary models it is regarded as a surprisingly large excess, since inflation erases initial conditions.

In absence of a baryon asymmetry an equal number of relic baryons and of anti-baryons survive to annihilations at $T \lesssim m_p$. This process is analogous to DM annihilations studied in the previous section, so we can estimate the relic baryon density by inserting $m = m_p$ and a typical $p\bar{p}$ cross section $\sigma \sim 1/m_p^2$ in eq. (1.5), obtaining $n_p/n_\gamma \sim m_p/M_{\text{Pl}} \sim 10^{-19}$. Therefore this is a negligible contribution.

Assuming that the hot-big-bang started with zero baryon asymmetry at some temperature $T \gg m_p$, can the baryon asymmetry can be generated dynamically in the subsequent evolution? Once that one realizes that this is an interesting issue (this was done by Sakharov), the answer is almost obvious: yes, provided that at some stage [3]
1. baryon number B is violated;
2. C and CP are violated (otherwise baryons and antibaryons behave in the same way);
3. the universe was not in thermal equilibrium (we believe that CPT is conserved, so that particles and antiparticles have the same mass, and therefore in thermal equilibrium have the same abundance).
Having discussed in section 1.1 a concrete out-of-equilibrium situation, it should be clear what the general concept means.

2.1. Baryogenesis in the SM?

A large amount of theoretical and experimental work showed that, *within the SM, Sakharov conditions are not fulfilled*. At first sight one might guess that the only problem is 1.; in reality 2. and 3. are problematic.

1. Within the SM B is violated in a non trivial way [3], thanks to quantum anomalies combined with extended $SU(2)_L$ field-configurations: the anomalous B and L symmetry are violated, while $B - L$ is a conserved anomaly-free symmetry. The basic equation is $\partial_\mu J^\mu_B \sim N_{gen} F^a_{\mu\nu} \tilde{F}^a_{\mu\nu}$, and means that there are many vacua that differ by their B, L content, separated by a potential barrier of electroweak height. At temperatures $T \ll 100\,\text{GeV}$ transition between different vacua are negligible because suppressed by a quantum-tunneling factor $e^{-2\pi/\alpha_2}$. If $N_{gen} = 1$ this would imply proton decay with an unobservably slow rate; since there are 3 generations and all of them must be involved, proton decay is kinematically forbidden. This suppression disappears at high temperature, $T \gtrsim 100\,\text{GeV}$, and the space-time density of B, L-violating 'sphaleron' interactions is $\gamma \sim \alpha_2^5 T^4$, faster than the expansion rate of the universe up to temperatures of about $T \sim 10^{12}\,\text{GeV}$ [3].[1]

3. SM baryogenesis is not possible due to the lack of out-of equilibrium conditions. The electroweak phase transition was regarded as a potential out-of equilibrium stage, but experiments now demand a higgs mass $m_h \gtrsim 115\,\text{GeV}$, and SM computations of the Higgs thermal potential show that, for $m_h \gtrsim 70\,\text{GeV}$, the higgs vev shifts smoothly from $\langle h \rangle = 0$ to $\langle h \rangle = v$ when the universe cools down below $T \sim m_h$ [3].

2. In any case, the amount of CP violation provided by the CKM phase would have been too small for generating the observed baryon asymmetry, because it is suppressed by small quark masses. Indeed CP-violation would be absent if the light quarks were massless.

Many extensions of the SM could generate the observed n_B. 'Baryogenesis at the electroweak phase transition' needs new particles coupled to the higgs in order to obtain a out-of-equilibrium phase transition and to provide extra sources of CP violation. This already disfavored possibility will be tested at future accelerators. 'Baryogenesis from decays of GUT particles' seems to conflict with non-observation of magnetic monopoles, at least in simplest models. Furthermore minimal GUT model do not violate $B - L$, so that sphaleron processes would later wash out the eventually generated baryon asymmetry.

The existence of sphalerons suggests *baryogenesis through leptogenesis*: lepton number might be violated by some non SM physics, giving rise to a lepton asymmetry, which is converted into the observed baryon asymmetry by sphalerons.

---

[1] A real understanding of these issues needs advanced quantum field theory. This sector lead to one observable consequence: the η' mass, that is related to the QCD analogous of the $SU(2)_L$ effects we are considering. Therefore there should be no doubt that B, L are violated, and this is almost all what one needs to know to understand leptogenesis quantitatively.

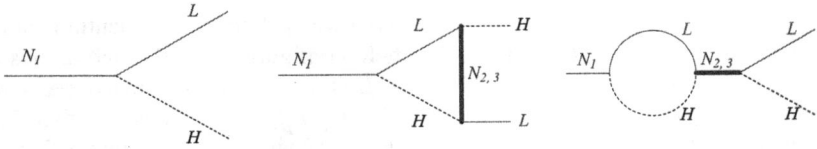

Fig. 2. CP-violating N_1 decay.

This scenario can be realized in many different ways [3]. Majorana neutrino masses violate lepton number and presumably CP, but do not provide enough out-of-equilibrium processes. The minimal successful implementation [4] needs just the minimal amount of new physics which can give the observed small neutrino masses via the see-saw mechanism [2]: heavy right-handed neutrinos N with masses M. 'Baryogenesis via thermal leptogenesis' [4] proceeds at $T \sim M$, when out-of-equilibrium (condition 3) CP-violating (condition 2) decays of heavy right-handed neutrinos generate a lepton asymmetry, converted in baryon asymmetry by SM sphalerons (condition 1).

3. Thermal leptogenesis: the basic physics

We now discuss the basic physics, obtaining estimates for the main results. The SM is extended by adding the heavy right-handed neutrinos suggested by see-saw models. To get the essential points, we consider a simplified model with one lepton doublet L and two right-handed neutrinos, that we name 'N_1' and '$N_{2,3}$', with N_1 lighter than $N_{2,3}$. The relevant Lagrangian is

$$\mathscr{L} = \mathscr{L}_{\mathrm{SM}} + \bar{N}_1 i \not{\partial} N_1 + \lambda_1 N_1 H L + \frac{M_1}{2} N_1^2 +$$
$$+ \bar{N}_{2,3} i \not{\partial} N_{2,3} + \lambda_{2,3} N_{2,3} H L + \frac{M_{2,3}}{2} N_{2,3}^2 + \mathrm{h.c..}$$

By redefining the phases of the N_1, $N_{2,3}$, L fields one can set M_1, $M_{2,3}$, λ_1 real leaving an ineliminabile CP-violating phase in $\lambda_{2,3}$.

3.1. CP-asymmetry

The tree-level decay width of N_1 is $\Gamma_1 = \lambda_1^2 M_1 / 8\pi$. The interference between tree and loop diagrams shown in fig. 2 renders N_1 decays CP-asymmetric:

$$\varepsilon_1 \equiv \frac{\Gamma(N_1 \to LH) - \Gamma(N_1 \to \bar{L}\bar{H})}{\Gamma(N_1 \to LH) + \Gamma(N_1 \to \bar{L}\bar{H})} \sim \frac{1}{4\pi} \frac{M_1}{M_{2,3}} \mathrm{Im}\, \lambda_{2,3}^2.$$

In fact

$$\Gamma(N_1 \to LH) \propto |\lambda_1 + A\lambda_1^*\lambda_{2,3}^2|^2, \qquad \Gamma(N_1 \to \bar{L}\bar{H}) \propto |\lambda_1^* + A\lambda_1\lambda_{2,3}^{2*}|^2,$$

where A is the complex CP-conserving loop factor. In the limit $M_{2,3} \gg M_1$ the sum of the two one loop diagrams reduces to an insertion of the $(LH)^2$ neutrino mass operator mediated by $N_{2,3}$: therefore A is suppressed by one power of $M_{2,3}$. The intermediate states in the loop diagrams in fig. 2 can be on shell; therefore the Cutkosky rule guarantees that A has an imaginary part. Inserting the numerical factor valid in the limit $M_{2,3} \gg M_1$ we can rewrite the CP-asymmetry as

$$\varepsilon_1 \simeq \frac{3}{16\pi} \frac{M_1 \text{Im}\, \tilde{m}_{2,3}}{v^2} = 10^{-6} \frac{\text{Im}\, \tilde{m}_{2,3}}{0.05\,\text{eV}} \frac{M_1}{10^{10}\,\text{GeV}}, \tag{3.1}$$

where $\tilde{m}_{2,3} \equiv \lambda_{2,3}^2 v^2 / M_{2,3}$ is the contribution to the light neutrino mass mediated by the heavy $N_{2,3}$.

The operator argument implies that eq. (3.1) holds in any model where particles much heavier than M_1 mediate a neutrino mass operator with coefficient $\tilde{m}_{2,3}$. In the past it was debated about if only the 'vertex' diagram in fig. 2 or also the 'self-energy' diagram should be included when computing the CP asymmetry: the operator argument makes clear that both diagrams contribute, since in the limit $M_{2,3} \gg M_1$ the two diagrams reduce to the same insertion of the $(LH)^2$ operator.

The final amount of baryon asymmetry can be written as

$$\frac{n_B}{n_\gamma} \approx \frac{\varepsilon_1 \eta}{g_{\text{SM}}}, \tag{3.2}$$

where $g_{\text{SM}} = 118$ is the number of spin-degrees of freedom of SM particles (present in the denominator of eq. (3.2) because only N_1 among the many other particles in the thermal bath generates the asymmetry) and η is an efficiency factor that depends on how much out-of-equilibrium N_1-decays are.

3.2. Efficiency

We now discuss this issue. If $N_1 \to LH$ decays are slow enough, the N_1 abundancy does not decrease according to the Boltzmann equilibrium statistics $n_{N_1} \propto e^{-M_1/T}$ demanded by thermal equilibrium, so that late out-of-equilibrium N_1 decays generate a lepton asymmetry. Slow enough decay means N_1 lifetime longer than the inverse expansion rate. At $T \sim M_1$ one has

$$R \equiv \frac{\Gamma_1}{H(M_1)} \sim \frac{\tilde{m}_1}{\tilde{m}^*}, \quad \text{where} \quad \tilde{m}^* \equiv \frac{256\sqrt{g_{\text{SM}}}v^2}{3M_{\text{Pl}}} = 2.3\,10^{-3}\,\text{eV}$$

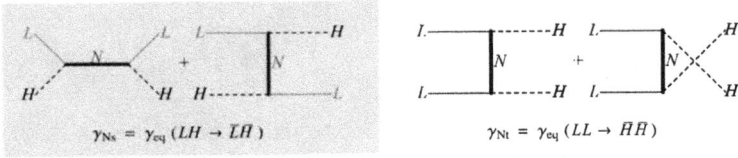

Fig. 3. Wash-out $LH \leftrightarrow \bar{L}\bar{H}$ and $LL \leftrightarrow \bar{H}\bar{H}$ $\Delta L = 2$ scatterings.

is fixed by cosmology. All the dependence on the mass and Yukawa couplings of N_1 is incorporated in $\tilde{m}_1 \equiv \lambda_1^2 v^2 / M_1$, the contribution to the light neutrino mass mediated by N_1. Unfortunately $\tilde{m}_1$ and $\tilde{m}_{2,3}$ are only related to the observed atmospheric and solar mass splittings in a model-dependent way. Unless neutrinos are almost degenerate (and unless there are cancellations) $\tilde{m}_1$ and $\tilde{m}_{2,3}$ are smaller than $m_{\text{atm}} \approx 0.05$ eV.

If $R \ll 1$ (i.e. N_1 decays strongly out-of-equilibrium) then $\eta = 1$.

If instead $R \gg 1$ the lepton asymmetry is only mildly suppressed as $\eta \sim 1/R$. The reason is that N_1 inverse-decays, which tend to maintain thermal equilibrium by regenerating decayed N_1, have rates suppressed by a Boltzmann factor at $T < M_1$: $R(T < M_1) \approx R \cdot e^{-M_1/T}$. The N_1 quanta that decay when $R(T) < 1$, i.e. at $T < M_1/ \ln R$, give rise to unwashed leptonic asymmetry. At this stage the N_1 abundancy is suppressed by the Boltzmann factor $e^{-M_1/T} = 1/R$. In conclusion, the suppression factor is approximately given by

$$\eta \sim \min(1, \tilde{m}^* / \tilde{m}_1) \qquad \text{(if } N_1 \text{ initially have thermal abundancy).}$$

Furthermore, we have to take into account that virtual exchange of $N_{1,2,3}$ gives rise to $\Delta L = 2$ scatterings (see fig. 3) that wash-out the lepton asymmetry. Their thermally-averaged interaction rates are relevant only at $M_1 \gtrsim 10^{14}$ GeV, when $N_{1,2,3}$ have large $\mathcal{O}(1)$ Yukawa couplings. When relevant, these scatterings give a strong exponential suppression of the baryon symmetry, because their rates are not suppressed at $T \lesssim M_1$ by a Boltzmann factor (no massive N_1 needs to be produced).

So far we assumed that right-handed neutrinos have thermal initial abundancy. Let us discuss how the result depends on this assumption. If $\tilde{m}_1 \gg \tilde{m}^*$ (in particular if $\tilde{m}_1 = m_{\text{atm}}$ or m_{sun}) the efficiency does not depend on the assumed initial conditions, because decays and inverse-decays bring the N_1 abundancy close to thermal equilibrium. For $\tilde{m}_1 \lesssim \tilde{m}^*$ the result depends on the unknown initial condition.

• If N_1 have negligible initial abundancy at $T \gg M_1$ and are generated only by the processes previously discussed, their abundancy at $T \sim M_1$ is suppressed by

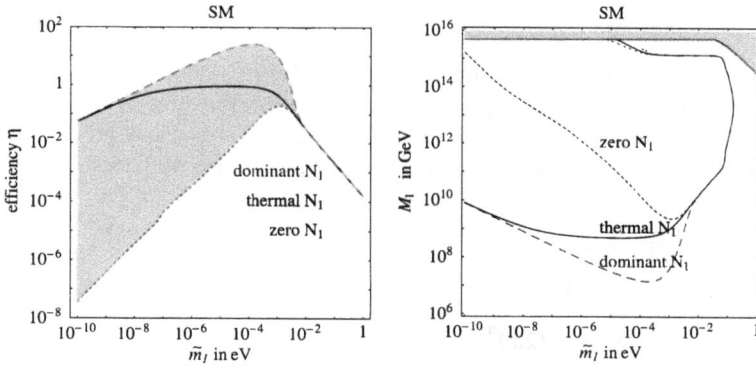

Fig. 4. Fig. 4a: efficiency η as function of $\tilde{m}_1$ for $M_1 = 10^{10}\,\text{GeV}$ and for different assumptions about the initial N_1 abundancy. Fig. 4b: the regions in the $(\tilde{m}_1, M_1)$ plane inside the curves can lead to successful leptogenesis.

$\tilde{m}_1/\tilde{m}^*$. Therefore the efficiency factor is approximatively given by

$$\eta \sim \min(\tilde{m}_1/\tilde{m}^*, \tilde{m}^*/\tilde{m}_1) \qquad \text{(if } N_1 \text{ initially have zero abundancy).}$$

• Finally, in the opposite limit where N_1 initially dominate the energy density of the universe, the suppression factor $1/g_{\text{SM}}$ in eq. (3.2) no longer applies, and the efficiency factor can reach $\eta \sim g_{\text{SM}}$.

$$\eta \sim \min(g_{\text{SM}}, \tilde{m}^*/\tilde{m}_1) \qquad \text{(if } N_1 \text{ initially have dominant abundancy).}$$

These estimates agree with the results of a detailed numerical computation, shown in fig. 4a. Notice that if $\tilde{m}_1 \gg \tilde{m}^*$ the N_1 abundancy gets close enough to thermal equilibrium, such that the lepton asymmetry generated by N_1 decays does not depend on the initial N_1 abundancy. Furthermore, N_1 decays and inverse-decays typically wash-out a possible pre-existing lepton asymmetry.

In the next section we describe how a precise computation can be done. This is not necessary for understanding the final discussion of section 5.

4. Thermal leptogenesis: precise computation

As previously discussed many important computations in cosmology are done using Boltzmann equations. So it is useful to have this tool.

4.1. Boltzmann equations

In absence of interactions the number of particles in a comoving volume V remains constant. Boltzmann equations allow to follow the effect of different interactions. Let us study e.g. how $1 \leftrightarrow 2+3$ decay and inverse-decay processes affect the number n_1 of '1' particles (in the case of leptogenesis we have $N_1 \leftrightarrow LH$):

$$\frac{d}{dt}(n_1 V) = V \int d\vec{p}_1 \int d\vec{p}_2 \int d\vec{p}_3 (2\pi)^4 \delta^4(p_1 - p_2 - p_3) \qquad (4.1)$$
$$\times |A|^2[-f_1(1 \pm f_2)(1 \pm f_3) + (1 \pm f_1)f_2 f_3],$$

where $d\vec{p}_i = d^3 p_i/2E_i(2\pi)^3$ is the relativistic phase space, $|A|^2$ is the squared transition amplitude summed over initial and final spins, and f_i are the energy (and eventually spin, flavour, color, ...) distributions of the various particles.

In line of principle we should study the evolution of all f's in order to obtain the total densities $n = \sum \int f \, d^3p/(2\pi)^3$. In practice elastic scatterings (i.e. interactions that do not change the number of particles) are typically fast enough that they maintain *kinetic equilibrium*, so that the full Boltzmann equations for f are solved by $f(p) = f_{eq}(p)n/n_{eq}$ where $f_{eq} = [e^{E/T} \pm 1]^{-1}$ are the Bose-Einstein and Fermi-Dirac distributions. Each particle species is simply characterized by its total abundance n, that can be varied only by inelastic processes.

The factors $1 \pm f_i$ in eq. (4.1) take into account Pauli-Blocking (for fermions) and stimulated emission (for bosons). Since the average energy is $\langle E \rangle \sim 3T$ within 10% accuracy one can approximate with the Boltzmann distribution $f_{eq} \approx e^{-E/T}$ and set $1 \pm f \approx 1$. This is a significant simplification.

When inelastic processes are sufficiently fast to maintain also *chemical equilibrium*, the total number n_{eq} of particles with mass M at temperature T are

$$n_{eq} = g \int \frac{d^3p \, f_{eq}}{(2\pi\hbar)^3} = \frac{gM^2T}{2\pi^2}K_2\left(\frac{M}{T}\right) = \begin{cases} gT^3/\pi^2 & T \gg M \\ g(MT/2\pi)^{3/2}e^{-M/T} & T \ll M \end{cases}$$

where g is the number of spin, gauge, etc degrees of freedom. A right handed neutrino has $g_N = 2$, a photon has $g_\gamma = 2$, the 8 gluons have $g_{G^a} = 16$, and all SM particles have $g_{SM} = 118$. The factor $\hbar = h/2\pi$ has been explicitly shown to clarify the physical origin of the 2π in the denominator.

The Boltzmann equation for n_1 simplifies to

$$\frac{1}{V}\frac{d}{dt}(n_1 V) = \int d\vec{p}_1 \int d\vec{p}_2 \int d\vec{p}_3 (2\pi)^4 \delta^4(p_1 - p_2 - p_3) \qquad (4.2)$$
$$\times |A|^2\left[-\frac{n_1}{n_1^{eq}}e^{-E_1/T} + \frac{n_2}{n_2^{eq}}\frac{n_3}{n_3^{eq}}e^{-E_2/T}e^{-E_3/T}\right].$$

One can recognize that the integrals over final-state momenta reconstruct the decay rate Γ_1, and that the integral over $d^3 p_1/E_1$ averages it over the thermal distribution of initial state particles; the factor $1/E_1$ corresponds to Lorentz dilatation of their life-time. Therefore the final result is

$$\frac{1}{V}\frac{d}{dt}(n_1 V) = \langle\Gamma_1\rangle n_1^{\text{eq}}\left[\frac{n_1}{n_1^{\text{eq}}} - \frac{n_2}{n_2^{\text{eq}}}\frac{n_3}{n_3^{\text{eq}}}\right], \qquad (4.3)$$

where $\langle\Gamma_1\rangle$ is the thermal average of the Lorentz-dilatated decay width

$$\Gamma_1(E_1) = \frac{1}{2E_1}\int d\vec{p}_2\, d\vec{p}_3 (2\pi)^4\delta^4(p_1 - p_2 - p_3)|A|^2. \qquad (4.4)$$

Analogous results holds for scattering processes.

If $\langle\Gamma_1\rangle \gg H$ the term in square brackets in eq. (4.3) is forced to vanish. This just means that interactions much faster than the expansion rate force chemical equilibrium, giving $n = n_{\text{eq}}$.

In the case of leptogenesis $2, 3 = L, H$ have fast gauge interactions. Therefore we do not have to write and solve Boltzmann equations for L, H, because we already know their solution: L, H are kept in equilibrium. We only need to insert this result in the Boltzmann equation for N_1, that simplifies to

$$\dot{n}_1 + 3Hn_1 = \langle\Gamma_1\rangle(n_1 - n_1^{\text{eq}}), \qquad (4.5)$$

having used $\dot{V}/V = 3H = -\dot{s}/s$.

In computer codes one prefers to avoid very big or very small numbers: it is convenient to reabsorb the $3H$ term (that accounts for the dilution due to the overall expansion of the universe) by normalizing the number density n to the entropy density s. Therefore we study the evolution of $Y = n/s$, as function of $z = m_N/T$ in place of time t ($H\, dt = d\ln R = d\ln z$ since during adiabatic expansion sV is constant, i.e. $V \propto 1/T^3$).

Using $Y(z)$ as variables, the general form of Boltzmann equations is

$$sHz\frac{dY_1}{dz} = \sum \Delta_1 \cdot \gamma_{\text{eq}}(12\cdots \leftrightarrow 34\cdots)\left[\frac{Y_1}{Y_1^{\text{eq}}}\frac{Y_2}{Y_2^{\text{eq}}}\cdots - \frac{Y_3}{Y_3^{\text{eq}}}\frac{Y_4}{Y_4^{\text{eq}}}\cdots\right] \qquad (4.6)$$

where the sum runs over all processes that vary the number of '1' particles by Δ_1 units (e.g. $\Delta_1 = -1$ for $1 \to 23$ decay, $\Delta_1 = -2$ for $11 \to 23$ scatterings, etc.) and γ_{eq} is the specetime density (i.e. the number per unit volume and unit time) in thermal equilibrium of the various processes.

Neglecting CP-violating effects, direct and inverse processes have the same reaction densities. For a scattering and for its inverse process one gets the previous result:

$$\gamma_{eq}(1 \to 23) = \int d\vec{p}_1 \, f_1^{eq} \int d\vec{p}_2 \, d\vec{p}_3 \, (2\pi)^4 \delta^4(p_1 - p_2 - p_3)|A|^2 = \gamma_{eq}(23 \to 1)$$

The thermal average of the decay rate can be analytically computed in terms of Bessel functions:

$$\gamma_{eq}(1 \to 23 \cdots) = \gamma_{eq}(23 \cdots \to 1) = n_1^{eq} \frac{K_1(z)}{K_2(z)} \Gamma(1 \to 23 \cdots) \tag{4.7}$$

For a 2-body scattering process

$$\gamma_{eq}(12 \leftrightarrow 34) = \int d\vec{p}_1 \, d\vec{p}_2 \, f_1^{eq} f_2^{eq} \int d\vec{p}_3 \, d\vec{p}_4 \, (2\pi)^4 \delta^4(p_1 + p_2 - p_3 - p_4)|A|^2$$

When there are n identical particles in the final or initial states one should divide by a $n!$ symmetry factor. One can analytically do almost all integrals, and obtain

$$\gamma_{eq}(12 \to 34 \cdots) = \frac{T}{32\pi^4} \int_{s_{min}}^{\infty} ds \, s^{3/2} \, \lambda(1, M_1^2/s, M_2^2/s)\sigma(s) \, K_1\left(\frac{\sqrt{s}}{T}\right)$$

which is the thermal average of $v \cdot \sigma$, summed over initial and final state spins.

4.2. Boltzmann equations for leptogenesis

We now specialize to leptogenesis. The main process is $N_1 \to HL, \bar{H}\bar{L}$ decay. We denote by γ_D its equilibrium density rate, computed inserting $\Gamma(N_1) = \lambda_1^2 M_1/8\pi$ in eq. (4.7). The Boltzmann equation for the N_1 abundancy is

$$sHz\frac{dY_{N_1}}{dz} = -\gamma_D\left(\frac{Y_{N_1}}{Y_{N_1}^{eq}} - 1\right). \tag{4.8}$$

Fig. 5 shows how Y_{N_1} evolves for different values of $\tilde{m}_1$. As expected if $\tilde{m}_1 \gg 10^{-3}$ eV one gets a result close to thermal equilibrium independently of the initial condition.

In order to get the Boltzmann equation for the lepton asymmetry one needs to take into account the small CP-violating terms. Let us start by including only the $\Delta L = 1$ CP-violating $N_1 \to HL, \bar{H}\bar{L}$ decays. We write the decay rates in terms of the CP-conserving total decay rate γ_D and of the CP-asymmetry $\varepsilon_1 \ll 1$:

$$\begin{aligned}
\gamma_{eq}(N_1 \to LH) &= \gamma_{eq}(\bar{L}\bar{H} \to N_1) = (1 + \varepsilon_1)\gamma_D/2, \\
\gamma_{eq}(N_1 \to \bar{L}\bar{H}) &= \gamma_{eq}(LH \to N_1) = (1 - \varepsilon_1)\gamma_D/2.
\end{aligned} \tag{4.9}$$

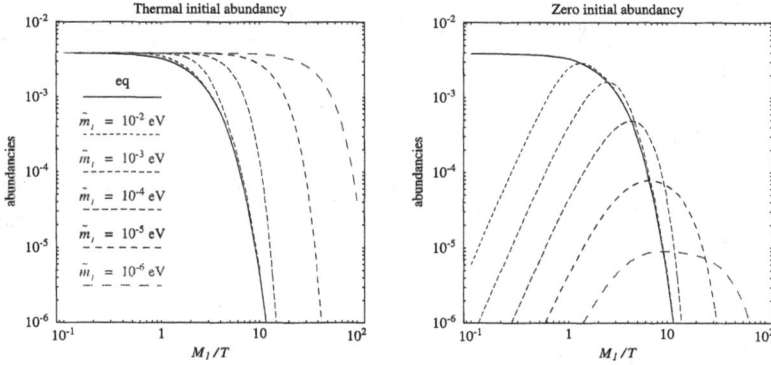

Fig. 5. Evolution of $Y_{N_1}(M_1/T)$ for different values of $\tilde{m}_1$ starting from different initial conditions. The continuous line is the abundancy in thermal equilibrium.

In this approximation the number of leptons L and anti-leptons $\bar{L}$ evolve as

$$zHsY'_L = \frac{\gamma_D}{2}\left[\frac{Y_{N_1}}{Y_{N_1}^{eq}}(1+\varepsilon_1) - \frac{Y_L}{Y_L^{eq}}(1-\varepsilon_1)\right], \tag{4.10}$$

$$zHsY'_{\bar{L}} = \frac{\gamma_D}{2}\left[\frac{Y_{N_1}}{Y_{N_1}^{eq}}(1-\varepsilon_1) - \frac{Y_{\bar{L}}}{Y_{\bar{L}}^{eq}}(1+\varepsilon_1)\right]. \tag{4.11}$$

Here $Y_{N_1}^{eq}$, Y_L^{eq} and $Y_{\bar{L}}^{eq}$ are equilibrium densities each with 2 degrees of freedom. Ignoring $\mathcal{O}(\varepsilon_1^2)$ terms, the lepton asymmetry $\mathcal{L} = Y_L - Y_{\bar{L}}$ evolves as

$$sHz\mathcal{L}' = \varepsilon_1\gamma_D\left(\frac{Y_N}{Y_N^{eq}}+1\right) - \frac{\mathcal{L}}{2Y_L^{eq}}\gamma_D. \tag{4.12}$$

The second term describes how γ_D tends to restore thermal equilibrium, washing out the lepton asymmetry. *The first term makes no sense*: it would generate a lepton asymmetry even in thermal equilibrium, $Y_{N_1} = Y_{N_1}^{eq}$, violating Sakharov conditions. An acceptable Boltzmann equation would contain $Y_N/Y_N^{eq} - 1$, but we made no sign error. Indeed, taking into account only decays and inverse decays, an asymmetry is really generated even in thermal equilibrium, since CPT invariance implies that if N_1 decays preferentially produce L, than inverse decays preferentially destroy $\bar{L}$ i.e. they have the same net effect.

A subtlety: avoiding overcounting

To obtain correct Boltzmann equations one must include all processes that contribute at the chosen order in the couplings. The CP-asymmetry is generated at

$\mathcal{O}(\lambda^4)$: at this order we must include also the $\Delta L = 2$ scatterings of fig. 3; we name their rates as

$$\gamma_{Ns} \equiv \gamma_{eq}(LH \leftrightarrow \bar{L}\bar{H}), \qquad \gamma_{Nt} \equiv \gamma_{eq}(LL \leftrightarrow \bar{H}\bar{H}), \qquad (4.13)$$

and $\gamma_{\Delta L=2} \equiv 2(\gamma_{Ns} + \gamma_{Nt})$. At first sight it is enough to include these scatterings at tree level, obtaining CP-conserving reaction densities $\gamma = \mathcal{O}(\lambda^4)$ that cannot correct our non-sensical CP-violating term. This is basically right, although the true argument is more subtle.

Indeed $LH \leftrightarrow \bar{L}\bar{H}$ can be mediated by on-shell N_1 exchange (see fig. 3a): as usual in these situations (e.g. the Z-peak) resonant enhancement gives $\sigma_{peak} \propto \lambda^0$ in an energy range $\Delta E \propto \lambda^2$, so that $\gamma_{Ns} \propto \sigma_{peak} \cdot \Delta E \propto \lambda^2$. (We will soon obtain the exact result). Nevertheless one can proof that the reaction density remains CP-conserving up to one-loop order: unitarity demands $\sum_j |M(i \rightarrow j)|^2 = \sum_j |M(j \rightarrow i)|^2$, so

$$\sigma(LH \rightarrow LH) + \sigma(LH \rightarrow \bar{L}\bar{H}) = \sigma(LH \rightarrow LH) + \sigma(\bar{L}\bar{H} \rightarrow LH)$$

(at higher order states with more particles allow a negligible CP asymmetry).

The key subtlety is that the $LH \leftrightarrow \bar{L}\bar{H}$ scattering rate mediated by s-channel exchange of N_1 shown in fig. 3a, must be computed by subtracting the CP-violating contribution due to on-shell N_1 exchange, because in the Boltzmann equations this effect is already taken into account by successive decays, $LH \leftrightarrow N_1 \leftrightarrow \bar{L}\bar{H}$. Since the on-shell contribution is

$$\gamma_{Ns}^{on-shell}(LH \rightarrow \bar{L}\bar{H}) = \gamma_{eq}(LH \rightarrow N_1)BR(N_1 \rightarrow \bar{L}\bar{H}),$$

where $BR(N_1 \rightarrow \bar{L}\bar{H}) = (1 - \varepsilon_1)/2$, we obtain

$$\gamma_{eq}^{sub}(LH \rightarrow \bar{L}\bar{H}) = \gamma_{Ns} - (1 - \varepsilon_1)^2 \gamma_D/4, \qquad (4.14)$$

$$\gamma_{eq}^{sub}(\bar{L}\bar{H} \rightarrow LH) = \gamma_{Ns} - (1 + \varepsilon_1)^2 \gamma_D/4. \qquad (4.15)$$

Including subtracted scatterings at leading order in ε_1 gives the final Boltzmann equation [4]

$$zHs\mathcal{L}' = \gamma_D\left[\varepsilon_1\left(\frac{Y_{N_1}}{Y_{N_1}^{eq}} - 1\right) - \frac{\mathcal{L}}{2Y_L^{eq}}\right] - 2\gamma_{\Delta L=2}^{sub}\frac{\mathcal{L}}{Y_L^{eq}}, \qquad (4.16)$$

where $\gamma_{\Delta L=2}^{sub} = \gamma_{\Delta L=2} - \gamma_D/4$.

Equivalently, one can more simply not include the decay contribution γ_D to washout of $\mathcal{L}$ because it is already accounted by resonant $\Delta L = 2$ scatterings. Then one gets

$$zHs\mathcal{L}' = \gamma_D \varepsilon_1 \left(\frac{Y_{N_1}}{Y_{N_1}^{\text{eq}}} - 1 \right) - 2\gamma_{\Delta L=2} \frac{\mathcal{L}}{Y_L^{\text{eq}}}, \tag{4.17}$$

which is equivalent to (4.16). Using eq. (4.17) it is not necessary to compute subtracted rates. (Subtraction is performed incorrectly in works before 2003).

Including sphalerons and Yukawas

Finally, we have to include sphaleronic scatterings, that transmit the asymmetry from left-handed leptons to left-handed quarks, generating a baryon asymmetry. Similarly we have to include SM Yukawa couplings, that transmit the asymmetry to right-handed quarks and leptons.

In theory one should enlarge Boltzmann equations adding all these processes. In practice, depending on the value of M_1, during leptogenesis at $T \sim M_1$ these process often give reaction densities which are either negligibly slower or much faster than the expansion rate: in the first case they can be simply neglected, in the second case they simply enforce thermal equilibrium. One can proceed by converting the Boltzmann equation for $\mathcal{L}$ into a Boltzmann equation for $\mathcal{B} - \mathcal{L}$: since $\mathcal{B} - \mathcal{L}$ is not affected by these redistributor processes we only need to find how processes in thermal equilibrium relate $\mathcal{B} - \mathcal{L}$ to $\mathcal{L}$. Sphalerons and the $\lambda_{t,b,c,\tau}$ Yukawas are fast at $T \lesssim 10^{11 \div 12}$ GeV. At larger temperatures all redistribution processes are negligibly slow and one trivially has $\mathcal{B} - \mathcal{L} = -\mathcal{L}$. At intermediate temperatures flavour issues can be important, and one should consider the evolution of the full flavour 3×3 density matrix [4].

It is interesting to explicitly compute redistribution factors at $T \sim$ TeV when all redistributor processes are fast. Each particle $P = \{L, E, Q, U, D, H\}$ carries an asymmetry A_P. Interactions equilibrate 'chemical potentials' $\mu_P \equiv A_P/g_P$ as

$$\begin{cases} ELH \text{ Yukawa :} & 0 = \mu_E + \mu_L + \mu_H \\ DQH \text{ Yukawa :} & 0 = \mu_D + \mu_Q + \mu_H \\ UQ\bar{H} \text{ Yukawa :} & 0 = \mu_U + \mu_Q - \mu_H \\ QQQL \text{ sphalerons :} & 0 = 3\mu_Q + \mu_L \\ \text{No electric charge :} & 0 = N_{\text{gen}}(\mu_Q - 2\mu_U + \mu_D - \mu_L + \mu_E) - 2N_{\text{Higgs}}\mu_H \end{cases}$$

Solving the system of 5 equations and 6 unknowns, one can express all asymmetries in terms of one of them, conveniently chosen to be $\mathcal{B} - \mathcal{L}$:

$$\mathcal{B} = N_{\text{gen}}(2\mu_Q - \mu_U - \mu_D) = \frac{28}{79}(\mathcal{B} - \mathcal{L}),$$

$$\mathcal{L} = \mathcal{B} - (\mathcal{B} - \mathcal{L}) = -\frac{51}{79}(\mathcal{B} - \mathcal{L}).$$

The efficiency η is precisely defined such that the final baryon asymmetry is

$$\frac{n_B}{s} = \mathcal{B} = -\frac{28}{79}\epsilon\eta Y_{N_1}^{\text{eq}}(T \gg M_1), \quad \text{i.e.} \quad \frac{n_B}{n_\gamma}\bigg|_{\text{today}} = -\frac{\epsilon_1\eta}{103}. \tag{4.18}$$

in agreement with the estimate (3.2).

Various extra processes give corrections of relative order $g^2/\pi^2, \lambda_t^2/\pi^2 \sim$ few %. Some of these corrections have been already computed: scattering involving gauge bosons and/or top quarks. Others have not yet been included: three body N_1-decays, one-loop correction to the $N_1 \to LH$ decay and its CP-asymmetry. Thermal corrections have not been fully included. The fact that γ_D is the only really relevant rate makes a full inclusion of these subleading corrections feasible.

5. Testing leptogenesis?

Unfortunately speculating that neutrino masses and the baryon asymmetry are produced by the see-saw mechanism and by thermal leptogenesis is much easier than testing them.

A direct test seems impossible, because right-handed neutrinos are either too heavy or too weakly coupled to be produced in accelerators.

What about indirect tests? We trust BBN because it predicts the primordial abundances of several nuclei in terms of known particle physics. Leptogenesis explains a single number, n_B/n_γ, in terms of speculative physics at high energies: the see-saw model has 18 free parameters. Neutrino masses only allow to measure 9 combinations of these parameters, and thereby provide a too weak link. The situation might improve if future experiments will confirm certain supersymmetric models, in which quantum corrections imprint neutrino Yukawa couplings $\lambda_{ij} N_i L_j H$ in slepton masses, inducing lepton flavour violating (LFV) processes such as $\mu \to e\gamma$, $\tau \to \mu\gamma$, $\tau \to e\gamma$ with possibly detectable rates. Measuring them, in absence of other sources of LFV, would roughly allow us to measure the 3 off-diagonal entries of $\lambda^\dagger \cdot \lambda$. Detectable LFV rates are obtained if $\lambda \gtrsim 10^{-(1\div2)}$. In any case, reconstructing all see-saw parameters in this way is unrealistic. Maybe future experiments will discover supersymmetry, LFV in charged leptons, and will confirm that neutrino masses violate lepton number and CP, and we will be able to convincingly argue that this can be considered as circumstantial evidence for see-saw and thermal leptogenesis. Archeology is not an exact science.

Another possibility is that we might find a correctly predictive model of flavour. Presently three approaches give some predictions: symmetries, numerology, zerology. Symmetries can be used to enforce relations like $\theta_{23} = \pi/4$, $\tan^2 \theta_{12} = 1/2$, $\theta_{13} = 0$ where θ_{ij} are the neutrino mixing angles. Numerology can suggest relations like $\theta_{12} + \theta_C = \pi/4$. Zerology consists in assuming that flavour matrices have many negligibly small entries; for example one can write see-saw textures with only one CP-violating phase. This scheme does not allow to predict the sign of CP-violation in neutrino oscillations in terms of the sign of the baryon asymmetry, because the sign of the baryon asymmetry also depends on which right-handed neutrino is the lightest one.

5.1. Constraints from leptogenesis

We here discuss testable constraints. Although this topic is tortuous, we avoid over-simplifications, at the price of obtaining a tortuous section.

As discussed in section 3, assuming that N_1 is lighter enough than other sources of neutrino masses that their effects can be fully encoded in the neutrino mass operator, the CP-asymmetry ε_1 is directly connected with it. Under the above hypothesis, one can derive constraints from leptogenesis. To do so, we need to generalize eq. (3.1) taking flavour into account. Denoting flavour matrices with boldface, we define $\tilde{\boldsymbol{m}}_i$ to be the contribution to the neutrino mass matrix mediated by the right-handed neutrino N_i, so that $\boldsymbol{m}_\nu = \tilde{\boldsymbol{m}}_1 + \tilde{\boldsymbol{m}}_2 + \tilde{\boldsymbol{m}}_3$. Then

$$|\varepsilon_1| = \frac{3}{16\pi} \frac{M_1}{v^2} \frac{|\mathrm{Im}\,\mathrm{Tr}\,\tilde{\boldsymbol{m}}_1^\dagger (\tilde{\boldsymbol{m}}_2 + \tilde{\boldsymbol{m}}_3)|}{\tilde{m}_1} \leq \frac{3}{16\pi} \frac{M_1}{v^2} (m_{\nu_3} - m_{\nu_1}), \tag{5.1}$$

where m_{ν_3} (m_{ν_1}) denotes the mass of the heaviest (lightest) neutrino. Rather than rigorously proofing the constraint (5.1) [5], let us understand its origin and limitations in a simpler way.

1) Let us start considering the case of hierarchical neutrinos: $m_{\nu_1} \ll m_{\nu_3}$: the constraint is obtained by substituting $\tilde{\boldsymbol{m}}_2 + \tilde{\boldsymbol{m}}_3 = \boldsymbol{m}_\nu - \tilde{\boldsymbol{m}}_1$, and holds whatever new physics produces $\tilde{\boldsymbol{m}}_{2,3}$. Since $|\varepsilon_1| \propto M_1$, one can derive a lower bound on the mass M_1 of the lightest right-handed neutrino by combining eq. (5.1) with a precise computation of thermal leptogenesis and with the measured baryon asymmetry and neutrino masses:

$$M_1 > \begin{cases} 2.4 \times 10^9 \,\mathrm{GeV} & \text{if } N_1 \text{ has zero} \\ 4.9 \times 10^8 \,\mathrm{GeV} & \text{if } N_1 \text{ has thermal} \quad \text{initial abundancy} \\ 1.7 \times 10^7 \,\mathrm{GeV} & \text{if } N_1 \text{ has dominant} \end{cases} \tag{5.2}$$

and assuming $M_1 \ll M_{2,3}$.

2) The factor $m_{\nu_3} - m_{\nu_1}$ is specific to the see-saw model with 3 right-handed neutrinos. To understand it, let us notice that 3 right-handed neutrinos can produce

the limiting case of degenerate neutrinos $m_{\nu_1} = m_{\nu_2} = m_{\nu_3} = m_\nu$ only in the following way: each N_i gives mass m_ν to one neutrino mass eigenstate. Since they are orthogonal in flavour space, the CP-asymmetry of eq. (5.1) vanishes due to flavour orthogonality: this is the origin of the $m_{\nu_3} - m_{\nu_1}$ suppression factor.

The bound (5.1) implies an upper bound on the mass of quasi-degenerate neutrinos: $m_\nu \lesssim 0.2\,\text{eV}$ [5]. Indeed for large m_ν both the efficiency and the maximal CP-asymmetry become smaller, because heavy neutrinos must be quasi-degenerate, and $\tilde{m}_1 \geq m_{\nu_1}$, $m_{\nu_3} - m_{\nu_1} \simeq \Delta m_{\text{atm}}^2/2m_\nu$. Furthermore, the bound can be improved up to $m_{\nu_3} < 0.15\,\text{eV}$ [5] (3σ confidence level) by computing the upper bound on $|\varepsilon_1|$ for given $\tilde{m}_1$ and maximizing n_B with respect to $\tilde{m}_1$ taking into account how the efficiency of thermal leptogenesis decreases for large $\tilde{m}_1$. The leptogenesis constraint is very close to observed neutrino masses, and stronger than experimental bounds.

However, this leptogenesis constraint holds under the dubious assumption that hierarchical right-handed neutrinos produce quasi-degenerate neutrinos, while good taste suggests that quasi-degenerate neutrinos are more naturally produced by quasi-degenerate right-handed neutrinos. In general the constraint (5.1) evaporates if the particles that mediate $\tilde{m}_{23}$ are so light that their effects cannot be encoded in $\tilde{m}_{23}$: the CP-asymmetries becomes sensitive to the detailed structure of the neutrino mass model. Suppression due to flavor-orthogonality was the most delicate consequence of our initial assumption, and is the first result that disappears when they are relaxed, allowing $M_{2,3}$ to be not much heavier than M_1. Correspondingly, the constraint on quasi-degenerate neutrino masses, that heavily relies on the factor $m_{\nu_3} - m_{\nu_1}$, becomes weaker than experimental constraints.

Furthermore, in the extreme situation where right-handed neutrinos are very degenerate, $M_2 - M_1 \lesssim \Gamma_{1,2}$ a qualitatively new effect appears: CP violation in $N_1 \leftrightarrow N_2$ mixing. This phenomenon is fully analogous to $K^0 \leftrightarrow \bar{K}^0$ mixing, and for $M_2 - M_1 \sim \Gamma_{1,2}$ it allows a maximal CP-asymmetry, $|\varepsilon_1| \sim 1$. This means that with a tiny $M_2 - M_1$ one can have successful thermal leptogenesis even at the weak scale.

The constraint (5.2) is more robust, but we still have to clarify what the assumption $M_{2,3} \gg M_1$ means in practice. Surely one needs $M_2 - M_1 \gg \Gamma_{1,2}$ such that only CP-violation in N_1 decay is relevant. The issue is: is $M_{2,3}/M_1 \sim 10$ (a hierarchy stronger than the one present in left-handed neutrinos) hierarchical enough to guarantee that the constraint holds, up to $1/10^2 = \%$ corrections? The answer is no: an operator analysis allows to understand how the constraint can be completely relaxed. The physics that above M_1 produces the dimension-5 neutrino mass operator $(LH)^2/2$ can also produce a related dimension-7 operator $\Upsilon \equiv (LH)\partial^2(LH)/2$, that contributes to ε_1. Since neutrino masses do not constrain Υ, it can be large enough to over-compensate the suppression due to its higher dimension. We make the argument more explicit, by considering the concrete case of see-saw models, where above M_1 there are two other right-handed neutrinos with masses M_2 and M_3. Including 'dimension-7' terms suppressed by $M_1^2/M_{2,3}^2$ and dropping inessential $\mathcal{O}(1)$ and flavour factors, the CP-asymmetry becomes:

$$\varepsilon_1 \sim \frac{3}{16\pi} \frac{M_1}{v^2} \text{Im}\left[\tilde{m}_2 \left(1 + \frac{M_1^2}{M_2^2} \right) + \tilde{m}_3 \left(1 + \frac{M_1^2}{M_3^2} \right) \right]. \tag{5.3}$$

At leading order ε_1 depends only on $\tilde{m}_2 + \tilde{m}_3$, that, as previously discussed, cannot be large and complex. At higher order ε_1 depends separately on $\tilde{m}_2$ and $\tilde{m}_3$, that can be large and complex provided that their sum stays small. One can build models where this naturally happens, obtaining an ε_1 orders of magnitudes above the DI bound. The enhancement is limited only by perturbativity, $\lambda_{2,3} \lesssim 4\pi$. In supersymmetric models, large Yukawa couplings lead to the testable effects previously discussed.

5.2. Leptogenesis and supersymmetry

Adding supersymmetry affects some $\mathcal{O}(1)$ factors. Eq. (4.18) remains almost unchanged, because adding spartners roughly doubles both the number of particles that produce the baryon asymmetry and the number of particles that share it. Ignoring small supersymmetry-breaking terms, right-handed neutrinos and sneutrinos have equal masses, equal decay rates and equal CP-asymmetries. Both Γ_{N_1} and ε_1 become 2 times larger, because there are new decay channels. As a consequence of more CP-asymmetry compensated by more wash-out, the constraints on right-handed and left-handed neutrino masses discussed in the nonsupersymmetric case remain essentially unchanged.

The leptogenesis constraint (5.2) on M_1 acquires a new important impact: in many supersymmetric models the maximal temperature at which the Big-Bang started (or, more precisely, the 'reheating temperature' T_{RH}) must be less than about $T_{\text{RH}} \lesssim 10^7$ GeV, in potential conflict with the constraint (5.2).

Let us discuss the origin of the supersymmetric constraint on T_{RH}. Gravitinos are the supersymmetric partner of the graviton: they have a mass presumably not much heavier than other sparticles, $m_{\tilde{G}} \lesssim$ TeV, and gravitational couplings to SM particles. Therefore gravitinos decay slowly after BBN (lifetime $\tau \sim M_{\text{Pl}}^2/m_{\tilde{G}}^3 \sim \sec(100\,\text{TeV}/m_{\tilde{G}})^3$): their decay products damage the nuclei generated by BBN. The resulting bound on the gravitino abundancy depends on unknown gravitino branching ratios.[2] The gravitino interaction rate is

---

[2]Gravitinos might be stable if they are the lightest supersymmetric particle. But in this case dangerous effects are produced by gravitational decays of the next-to-lightest SUSY particle into gravitinos.

A. Strumia

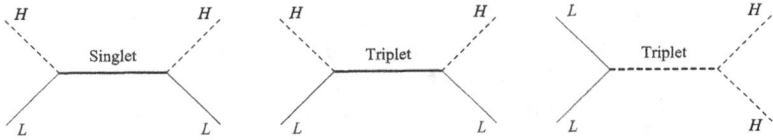

Fig. 6. The Majorana neutrino mass operator $(LH)^2$ can be mediated by tree level exchange of: *I)* a fermion singlet; *II)* a fermion triplet; *III)* a scalar triplet.

$\gamma_{\tilde{G}}(T) \sim T^6/M_{\text{Pl}}^2$, which means that gravitinos have been generated around the reheating temperature T_{RH}, with abundancy $n_{\tilde{G}}/n_\gamma \sim \gamma_{\tilde{G}}/Hn_\gamma \sim T_{\text{RH}}/M_{\text{Pl}}$. Therefore gravitinos suggest an upper bound on T.

This is why in the previous section we carefully discussed specific scenarios that allow low-scale thermal leptogenesis. Supersymmetry suggests a new scenario named 'soft leptogenesis': complex soft terms in the see-saw sector give new contributions to the CP-asymmetry, $\varepsilon_1 \sim \alpha_2 m_{\text{SUSY}}^2/M_1^2$, which can be significant if M_1 is not much larger than the SUSY-breaking scale m_{SUSY} (presumed to be below 1 TeV). At larger M_1 'soft leptogenesis' can still be relevant, but only in a fine-tuned range of parameters.

5.3. Leptogenesis in alternative neutrino-mass models

Generic neutrino masses can be mediated by tree-level exchange of three different kinds of new particles [2]: I) at least three fermion singlets; II) at least three fermion SU(2)$_L$ triplets; III) one scalar SU(2)$_L$ triplet (or of combinations of the above possibilities). Fig. 6 shows the relevant Feynman diagrams.

So far we studied case I). Can leptogenesis be used to distinguish between these possibilities? Leptogenesis can be produced in decays of P, the lightest particle that mediates neutrino masses. The neutrino-mass contribution to its CP-asymmetry is given by expressions similar to eq. (5.1), with $\tilde{m}_1$ generalized to be the contribution to neutrino masses mediated by P, and $\tilde{m}_2 + \tilde{m}_3$ generalized to be the contribution of all heavier sources.

It was expected that the efficiency η can be high enough only if P is a right-handed neutrino: as discussed in section 3.2 it easily decays out-of-equilibrium giving

$$\eta(\text{fermion singlet}) \approx \min\left[X, \frac{H}{\Gamma}\right] \tag{5.4}$$

where Γ is its decay rate, H is the expansion rate at $T \sim M$ and

$$X = \begin{cases} 1 & \text{for thermal} \\ \Gamma/H & \text{for negligible} \quad \text{initial abundancy.} \\ g_{\text{SM}} & \text{for dominant} \end{cases} \tag{5.5}$$

A $SU(2)_L$ triplet (scalar or fermion) has gauge interactions that keep its abundancy close to thermal equilibrium so that the 3rd Sakharov condition cannot be fulfilled. This suppression is present, and a quantitative analysis is needed to see if/when it is strong enough. The Boltzmann equation for the triplet abundancy Y has an extra term γ_A that accounts for annihilations of two triplets into gauge bosons:

$$sHz\frac{dY}{dz} = -\gamma_D\left(\frac{Y}{Y_{eq}} - 1\right) - 2\gamma_A\left(\frac{Y^2}{Y_{eq}^2} - 1\right). \tag{5.6}$$

The term $\gamma_A \sim g^4 T^4$ is dominant only at $T \gtrsim M$, where M is the triplet mass; at lower temperatures it is strongly suppressed by a double Boltzmann factor $(e^{-M/T})^2$, because gauge scattering must produce 2 triplets. The resulting efficiency can be approximated as

$$\eta(\text{fermion triplet}) \approx \min\left[1, \frac{H}{\Gamma}, \frac{M}{10^{12}\,\text{GeV}}\max\left(1, \frac{\Gamma}{H}\right)\right]. \tag{5.7}$$

η is univocally predicted, because at $T \gg M$ gauge interactions thermalize the triplet abundancy. At $T \sim M_1$ gauge scatterings partially annihilate triplets: in section 1.1 we learnt how to estimate how many particles survive to annihilations, and this is the origin of the factor $M/10^{12}\,\text{GeV}$ in η. The last factor takes into account that annihilations are ineffective if triplets decay before annihilating.

As illustrated in fig. 6 a scalar triplet separately decay to leptons, with width $\Gamma_L(T^* \to LL)$, and in Higgses, with width $\Gamma_H(T \to HH)$. Lepton number is effectively violated only when both processes are faster than the expansion of the universe, giving

$$\eta(\text{scalar triplet}) \approx \min\left[1, \frac{H}{\min(\Gamma_L, \Gamma_H)}, \frac{M}{10^{12}\,\text{GeV}}\max\left(1, \frac{\Gamma_L + \Gamma_H}{H}\right)\right]. \tag{5.8}$$

This means that a quasi-maximal efficiency $\eta \sim 1$ is obtained when $T^* \to LL$ is faster than gauge annihilations while $H \to TT$ is slower than the expansion rate. In conclusion, leptogenesis from decays of a $SU(2)_L$ triplet can be sufficiently efficient even if triplets are light enough to be tested at coming accelerators, $M \sim \text{TeV}$.

Acknowledgments

I thank S. Davidson and T. Kashti for useful discussions. Since these are lessons, the bibliography prefers later systematic works to pioneering imperfect works.

References

[1] CMB anisotropies: C. L. Bennett et al. [WMAP collaboration], Astrophys. J. Suppl. **148** (2003) 1 [astro-ph/0302207]. Matter inhomogeneities: D. J. Eisenstein et al. [SDSS collaboration], astro-ph/0501171. The physics necessary to understand these measurements is clearly presented in the book *Modern Cosmology* by S. Dodelson.

[2] Neutrino masses from right-handed neutrinos ('see-saw'): P. Minkowski, Phys. Lett. B **67** (1977) 421. Neutrino masses from scalar triplets: M. Magg and C. Wetterich, Phys. Lett. B **94** (1980) 61. Neutrino masses from fermion triplets: R. Foot, H. Lew, X.-G. He and G.C. Joshi, Z. Phys. C **44** (1989) 441.

[3] **Baryogenesis.** Sakharov conditions: A. D. Sakharov, JETP Lett. **91B** (1967) 24. Anomalies: G. 't Hooft, Phys. Rev. Lett. **37** (1976) 37. G. 't Hooft, Phys. Rev. D **14** (1976) 3432. Sphalerons: N. S. Manton, Phys. Rev. D **28** (1983) 2019. V. Kuzmin, V. A. Rubakov and M. E. Shaposhnikov, Phys. Lett. B **155** (1985) 36. J. Ambjýrn, T. Askgaard, H. Porter and M. E. Shaposhnikov, Nucl. Phys. B **353** (1991) 346. P. Arnold, D. Son and L. G. Yaffe, Phys. Rev. D **55** (1997) 6264 [hep-ph/9609481].

[4] **Thermal leptogenesis.** M. Fukugita and T. Yanagida, Phys. Lett. B **174** (1986) 45. The CP-asymmetry has been computed in E. Roulet, L. Covi and F. Vissani, Phys. Lett. B **424** (1998) 101 [hep-ph/9712468]. See also W. Buchmüller and M. Plümacher, Phys. Lett. B **431** (1998) 354 [hep-ph/9710460 version 2]; M. Flanz and E. A. Paschos, Phys. Rev. D **58** (1998) 11309 [hep-ph/9805427]. First attempts of solving Boltzmann equations: M. A. Luty, Phys. Rev. D **45** (1992) 455; M. Plümacher, Z. Phys. C **74** (1997) 549 [hep-ph/9604229]. Flavour effects and RGE corrections have been included in R. Barbieri, P. Creminelli, N. Tetradis and A. Strumia, Nucl. Phys. B **575** (2000) 61 [hep-ph/9911315]. Thermal corrections, correct subtraction of resonant processes, reheating effects have been included in G. F. Giudice, A. Notari, M. Raidal, A. Riotto and A. Strumia, Nucl. Phys. B **685** (2004) 89 [hep-ph/0310123]. Supersymmetric leptogenesis has been first studied in M. Plümacher, Nucl. Phys. B **530** (1998) 207 [hep-ph/9704231]. 'Resonant leptogenesis' with quasi-degenerate right-handed neutrinos: see e.g. A. Pilaftsis, Phys. Rev. D **56** (1997) 5431 [hep-ph/9707235]. A. Pilaftsis and T. Underwood, Nucl. Phys. B **692** (2004) 303 [hep-ph/0309342]. G. C. Branco et al., hep-ph/0507092. 'Soft leptogenesis': for a systematic treatment see Y. Grossman, T. Kashti, Y. Nir and E. Roulet, JHEP **0411** (2004) 080 [hep-ph/0407063].

[5] **Constraints from leptogenesis.** The maximal CP asymmetry ε_1 of eq. (5.1) was rigorously proofed in S. Davidson and A. Ibarra, Phys. Lett. B **535** (2002) 25 [hep-ph/0202239]. Constraint on neutrino masses: see W. Buchmuller, P. Di Bari and M. Plümacher, Nucl. Phys. B **665** (2003) 445 [hep-ph/0302092] and T. Hambye et al., Nucl. Phys. B **695** (2004) 169 [hep-ph/0312203]. Mildly hierarchical right-handed neutrinos allow a much larger ε_1, as discussed in the previous paper. Leptogenesis from fermion triplets was studied in the previous paper. Leptogenesis from scalar triplets: for a systematic treatment see T. Hambye, M. Raidal and A. Strumia, hep-ph/0510008.

www.ingramcontent.com/pod-product-compliance
Lightning Source LLC
Chambersburg PA
CBHW050452190326
41458CB00005B/1250